Postharvest Technology of Horticultural Crops

Third Edition

Adel A. Kader
Technical Editor

University of California
Agriculture and Natural
Resources
Publication 3311

For information about ordering this publication, contact
University of California
Agriculture and Natural Resources
Communication Services
6701 San Pablo Avenue, 2nd Floor
Oakland, California 94608-1239

Telephone 1-800-994-8849
(510) 642-2431
FAX (510) 643-5470
E-mail: danrcs@ucdavis.edu
Visit the ANR Communication Services website at http://anrcatalog.ucdavis.edu

Publication 3311

This publication has been anonymously peer reviewed for technical accuracy by University of California scientists and other qualified professionals. This review process was managed by the ANR Associate Editor for Pomology, Viticulture, and Subtropical Horticulture.

ISBN 1-879906-51-1

Library of Congress Control No. 2001087610

⊛ Printed in the United States of America on recycled paper.

To simplify information, trade names of products have been used. No endorsement of named or illustrated products is intended, nor is criticism implied of similar products that are not mentioned or illustrated.

7m-rev-1/02-SB/CR/KT

WARNING ON THE USE OF CHEMICALS

Pesticides are poisonous. Always read and carefully follow all precautions and safety recommendations given on the container label. Store all chemicals in the original labeled containers in a locked cabinet or shed, away from food or feeds, and out of the reach of children, unauthorized persons, pets, and livestock.

Confine chemicals to the property being treated. Avoid drift onto neighboring properties, especially gardens containing fruits or vegetables ready to be picked.

Do not place containers containing pesticide in the trash nor pour pesticides down sink or toilet. Either use the pesticide according to the label or take unwanted pesticides to a Household Hazardous Waste Collection site. Contact your county agricultural commissioner for additional information on safe container disposal and for the location of the Hazardous Waste Collection site nearest you.

Dispose of empty containers by following label directions. Never reuse or burn the containers or dispose of them in such a manner that they may contaminate water supplies or natural waterways.

Authors

James E. Adaskaveg

Associate Professor, AES*

Department of Plant Pathology
University of California
Riverside, CA 92521
(909) 787-7577
(909) 787-3880 FAX
jim.adaskaveg@ucr.edu

Mary Lu Arpaia

Subtropical Horticulturist, CE† Department of Botany and Plant Science, UCR

UC Kearney Agricultural Center
9240 S. Riverbend Avenue
Parlier, CA 93648-9757
(559) 646-6521
(559) 646-6593 FAX
arpaia@uckac.edu

Diane M. Barrett

Fruit & Vegetable Products Specialist, CE

Department of Food Science and Technology
University of California
Davis, CA 95616-8598
(530) 752-4800
(530) 752-4759 FAX
dmbarrett@ucdavis.edu

Christine M. Bruhn

Consumer Food Marketing Specialist, CE

Department of Food Science and Technology
University of California
Davis, CA 95616-8598
(530) 752-2774
(530) 752-3975 FAX
cmbruhn@ucdavis.edu

Marita I. Cantwell

Vegetable Specialist, CE

Department of Vegetable Crops
University of California
Davis, CA 95616-8631
(530) 752-7305
(530) 752-4554 FAX
micantwell@ucdavis.edu

Roberta L. Cook

Economist, CE

Department of Agricultural and Resource Economics
University of California
Davis, CA 95616-8512
(530) 752-1531
(530) 752-5614 FAX
cook@primal.ucdavis.edu

Carlos H. Crisosto

Postharvest Physiologist, AES & CE

Department of Pomology, UCD
UC Kearney Agricultural Center
9240 S. Riverbend Avenue
Parlier, CA 93648
(559) 646-6586
(559) 646-6596 FAX
carlos@uckac.edu

Donald C. Edwards

Photographer

Department of Pomology
University of California
Davis, CA 95616-8683
(530) 752-0932
(530) 752-8502 FAX
dcedwards@ucdavis.edu

Helga Förster

Staff Research Associate

Department of Plant Pathology
University of California
Riverside, CA 92521
(909) 787-3880
(909) 787-3880 FAX
helgaf@ucr.edu

Robert J. Fortlage

Staff Research Associate (Retired)

Department of Pomology
University of California
Davis, CA 95616-8683
(530) 752-0508
(530) 752-8502 FAX

James R. Gorny

Technical Director

International Fresh-Cut Produce Association
430 Grande Avenue
Davis, CA 95616
(530) 756-8900
(530) 756-8901 FAX
jgorny@fresh-cuts.org

Linda J. Harris

Food Safety/Microbiology Specialist, CE

Department of Food Science and Technology
University of California
Davis, CA 95616-8598
(530) 754-9485
(530) 752-4759 FAX
ljharris@ucdavis.edu

Adel A. Kader

Professor and Pomologist, AES & CE

Department of Pomology
University of California
Davis, CA 95616-8683
(530) 752-0909
(530) 752-8502 FAX
aakader@ucdavis.edu

Lisa Kitinoja

Consultant

Extension Systems International
73 Antelope Street
Woodland, CA 95695
(530) 668-6407
kitinoja@ix.netcom.com

Elizabeth J. Mitcham

Postharvest Physiologist, AES & CE

Department of Pomology
University of California
Davis, CA 95616-8683
(530) 752-7512
(530) 752-8502 FAX
ejmitcham@ucdavis.edu

F. Gordon Mitchell
Emeritus Pomologist, CE & AES

> Department of Pomology
> University of California
> Davis, CA 95616-8683
> (530) 752-0508
> (530) 752-8502 FAX

Jeffrey P. Mitchell
Extension Vegetable Specialist, CE

> Department of Vegetable Crops, UCD
> UC Kearney Agricultural Center
> 9240 S. Riverbend Avenue
> Parlier, CA 93648
> (559) 646-6565
> (559) 646-6593 FAX
> mitchell@uckac.edu

Michael S. Reid
Professor and Postharvest Physiologist, AES & CE

> Department of Environmental Horticulture
> University of California
> Davis, CA 95616-8587
> (530) 754-6751
> (530) 754-6753 FAX
> msreid@ucdavis.edu

Noel F. Sommer
Emeritus Lecturer and Postharvest Pathologist, AES

> Department of Pomology
> University of California
> Davis, CA 95616-8683
> (530) 752-0508
> (530) 752-8502 FAX

Trevor V. Suslow
Postharvest Specialist, CE

> Department of Vegetable Crops
> University of California
> Davis, CA 95616-8631
> (530) 754-8313
> (530) 752-4554 FAX
> tvsuslow@ucdavis.edu

James F. Thompson
Agricultural Engineer, CE

> Department of Biological and Agricultural Engineering
> University of California
> Davis, CA 95616-5294
> (530) 752-6167
> (530) 752-2640 FAX
> jfthompson@ucdavis.edu

Devon Zagory
Consultant

> Davis Fresh Technologies
> P.O. Box 72711
> Davis, CA 95617
> (530) 756-2720
> (530) 756-4174 FAX
> dzagory@davisfreshtech.com

*AES = Agricultural Experiment Station
†CE = Cooperative Extension

Contents

Preface

I estimate that about one-third of fresh produce harvested worldwide is lost at various points in the distribution system between production and consumption sites. While it may be impossible and uneconomical to completely eliminate these losses, it is possible and desirable to reduce them by 50%. Minimizing postharvest losses of food that has already been produced is more sustainable and environmentally sound than increasing production areas to compensate for these losses. The strategies for attaining these goals include selecting genotypes with good quality when harvested at optimal maturity and with longer postharvest life, using an integrated crop management system that maximizes yield without sacrificing quality, and using optimal postharvest handling procedures to maintain the quality and safety of the produce.

Based on our improved understanding of postharvest biology, great advances have been made in postharvest technology to maintain the quality and safety of fresh horticultural crops and their products during distribution. The most useful technological innovations in production, harvesting, and postharvest handling systems have resulted from interdisciplinary research and development approaches; this will likely continue to be the case in the future. Concurrently, we need to expand our efforts in extending the available information about optimal production, harvesting, and postharvest handling procedures to producers, handlers, marketers, and consumers, worldwide. I hope that this book will continue to serve as a tool for achieving these goals.

Postharvest Technology of Horticultural Crops is the outcome of a syllabus that was developed for a short course initiated in 1979 and offered annually since then in two modes: as a regular University of California at Davis course (Plant Biology 196) for advanced undergraduate and graduate students interested in postharvest biology and technology of horticultural crops, and as a short course organized through University Extension for participants who are not current UCD students. The latter group usually includes research and extension workers, consultants, quality management personnel, and other people concerned with postharvest handling of fresh horticultural perishables.

The first and second editions (published in 1985 and 1992, respectively) of this book have been well received and widely distributed and used throughout the world. In this third edition, all chapters have been updated, many were expanded, and five new chapters added. Although the emphasis is on current postharvest technology procedures for fresh fruits, vegetables, and ornamentals in California, all the principles discussed are applicable to postharvest handling of fresh horticultural crops worldwide.

Thirty-eight chapters are included, of which 21 chapters present various aspects of postharvest technology of horticultural commodities and 12 chapters briefly cover postharvest handling systems for certain commodities or commodity groups. It was not possible to include every horticultural crop in these 12 chapters and keep the book to a reasonable length, but a summary table of optimal storage conditions for most fruits and vegetables is included in the appendix. The remaining 5 chapters deal with sources of information, an overview of the fresh produce industry, consumer issues, processing methods, and extension efforts. We are continuously working on improving all aspects of this book, and we welcome comments and suggestions for incorporation into future editions.

I dedicate this edition to the memory of our colleague and co-author Robert F. Kasmire, who contributed to the revisions of several chapters before his death on December 22, 1998. On behalf of the authors, I wish to thank all those individuals who assisted us. I especially want to acknowledge the tireless efforts of Sean Versoza (Department of Pomology, UC Davis) in word processing, and organizing production of this book. Thanks are also due to the University of California Agriculture and Natural Resources Communication Services staff members who participated in the production of this book.

ADEL A. KADER
TECHNICAL EDITOR

Numerous sources of information related to postharvest biology and technology of horticultural crops can be used to supplement and expand the information in this book.

A STARTING POINT

The website of the University of California Postharvest Technology Research and Information Center (http://postharvest.ucdavis.edu) is an excellent starting point for accessing information and linking to a large number of relevant Internet sites.

Another very useful website (http://www.fao.org/inpho/) for postharvest information worldwide was established by the United Nations Food and Agriculture Organization in 1998. It can be searched by topic, by information providers, and according to type of information. It is linked to most of the postharvest-related websites available worldwide.

Textbooks and general references related to postharvest biology and technology of horticultural crops are listed at the end of this chapter. Furthermore, each subsequent chapter includes a list of references for further reading. The following reference list (available on our website) provides a good starting point for developing background information: A. A. Kader, L. L. Morris, and M. Cantwell, *Postharvest Handling and Physiology of Horticultural Crops—A List of Selected References*, 18th ed., Univ. Calif. Postharv. Hort. Ser. 2 (Davis: Univ. Calif., Davis, Dept. of Pomology, 2001).

ADDITIONAL LIBRARY RESOURCES

To review the published literature about a specific topic or a given commodity, use a computerized information service (if one is available) or consult one or more of the following abstracting journals (in electronic or printed formats):

Biological Abstracts
Bulletin of the International Institute of Refrigeration
Chemical Abstracts
Food Science and Technology Abstracts
Horticultural Abstracts
Postharvest News and Information (initiated in 1990)
Review of Plant Pathology

Review journals, including the following, periodically contain review articles on topics related to postharvest biology and technology:

Advances in Food Research
Annual Review of Phytopathology
Annual Review of Plant Physiology and Molecular Biology
Critical Reviews in Food Science and Nutrition
Horticultural Reviews
Trends in Food Science and Technology

To follow current publications, search (either in printed or electronic version) the weekly publication *Current Contents: Agriculture, Biology & Environmental Sciences*, published by Institute for Scientific Information, 3501 Market St., Philadelphia, PA 19104 (www.isinet.com), which includes tables of contents from many periodicals, including most of those listed below.

Sources of Information Related to Postharvest Biology and Technology

Adel A. Kader

Until 1991, no single scientific journal specialized in postharvest biology and technology of horticultural crops. Research reports are currently published in a wide range of scientific journals (many are available in both electronic and printed formats), including the following:

Acta Horticulturae
Agricultural Engineering
American Potato Journal
Dairy Food and Environmental Sanitation
Food Quality and Preference
Food Technology
Fruits d'Outre Mer
HortScience
HortTechnology
International Journal of Food Science and Technology
International Journal of Refrigeration
Journal of Agricultural and Food Chemistry
Journal of Economic Entomology
Journal of Food Biochemistry
Journal of Food Protection
Journal of Food Quality
Journal of Food Science
Journal of Horticultural Science and Biotechnology
Journal of Textural Studies
Journal of the American Society for Horticultural Science
Journal of the Japanese Society for Horticultural Science
Journal of the Science of Food and Agriculture
Phytochemistry
Phytopathology
Plant Disease
Plant Physiology
Postharvest Biology and Technology (starting 1991)
Proceedings of the Florida State Society for Horticultural Science
Proceedings of the Tropical Region of the American Society for Horticultural Science (Interamerican Society for Tropical Horticulture)
Scientia Horticulturae
Transactions of the American Society of Agricultural Engineers
Tropical Agriculture
Tropical Science

Semitechnical and popular periodicals include:

Americafruit, Asiafruit, Eurofruit (www.fruitnet.com)
American Fruit Grower
American Vegetable Grower
California Agriculture (http://danr.ucop.edu/calag/)
Citrus and Vegetables Magazine
Florida Grower and Rancher
Florists Review (www.floristsreview.com)
Fresh Cut (www.freshcut.com)
Global Produce (www.globalproduce.com)
Produce Business
The Good Fruit Grower
The Packer (a weekly newspaper) (www.thepacker.com)
Western Grower and Shipper

Newsletters and other periodicals published by Cooperative Extension specialists in postharvest biology and technology of horticultural crops at various locations include:

The Central Valley Postharvest Newsletter (available from Carlos Crisosto, Kearney Agricultural Center, 9240 S. Riverbend Ave., Parlier, CA 93648; E-mail: carlos@uckac.edu)
Packinghouse Newsletter (available from Mark Ritenour, Indian River REC, 2199 South Rock Road, Ft. Pierce, FL 34945-3138; E-mail: mrit@gnv.ifas.ufl.edu)
Perishables Handling Quarterly (available from Postharvest Technology RIC, Department of Pomology, University of California, One Shields Ave., Davis, CA 95616; E-mail: postharvest@ucdavis.edu)
Tree Fruit Postharvest Journal (available from Gene Kupferman, Tree Fruit Research and Extension Center, 1100 N. Western Ave., Wenatchee, WA 98801-1299; E-mail: kupfer@wsu.edu)

Other sources include publications of the U.S. Department of Agriculture; agricultural experiment stations and cooperative extensions in California, Florida, New York, Michigan, Oregon, Washington, and other states; International Institute of Refrigeration (Paris, France; Internet site: http://www.iifiir.org); Natural Resources Institute (Chatham, Kent, England; Internet site: http://www.nri.org); Postharvest Institute for Perishables (University of Idaho, Moscow, ID 83843); and other organizations.

Industry organizations that offer publications include United Fresh Fruit and Vegetable Association (UFFVA) (722 North Washington, Alexandria, VA 22314; Internet site: www.uffva.org), Produce Marketing

Association (PMA) (P. O. Box 6036, Newark, DE 19714-6036; Internet site: http://www.pma.com), and International Fresh-Cut Produce Association (1600 Duke St., Suite 400, Alexandria, VA 22314; Internet site: http://www.fresh-cuts.org).

VISUAL AIDS

To find out about audiovisual programs (slide sets, videotapes, and so on) that deal with various aspects of postharvest technology of horticultural crops, consult the publications catalog of the Division of Agriculture and Natural Resources, University of California. This information is also available on our website (http://postharvest.ucdavis.edu)

REFERENCES

Barkai-Golan, R. 2001. Postharvest diseases of fruits and vegetables: development and control. amsterdam: Elsevier Science. 432pp.

Burton, W. G. 1982. Postharvest physiology of food crops. London and New York: Longman. 339 pp.

Dennis, C. 1983. Postharvest pathology of fruits and vegetables. London: Academic Press. 264 pp.

Eskin, N. A. M., ed. 1989. Quality and preservation of vegetables. Boca Raton, FL: CRC Press. 313 pp.

———. 1991. Quality and preservation of fruits. Boca Raton, FL: CRC Press. 176 pp.

Friend, J., and M. J. C. Rhodes, eds. 1981. Recent advances in the biochemistry of fruit and vegetables. New York: Academic Press. 278 pp.

Hardenburg, R. E., A. E. Watada, and C. Y. Wang. 1986. The commercial storage of fruits, vegetables, and florist and nursery stocks. USDA Handb. 66. 130 pp. (A new edition will be published in both electronic and printed formats in late 2001).

Hulme, A. C., ed. 1970–71. The biochemistry of fruits and their products. 2 vols. New York: Academic Press. 1,408 pp.

Kays, S. J. 1991. Postharvest physiology and handling of perishable plant products. New York: Van Nostrand Reinhold. 532 pp.

Knee, M., ed. 2001. Fruit quality and its biological basis. Sheffield: Sheffield Academic Press. 320pp.

Mitra, S., ed. 1997. Postharvest physiology and storage of tropical and subtropical fruits. Wallingford, UK: CAB International. 423 pp.

Nagy, S., and P. E. Shaw, eds. 1980. Tropical and subtropical fruits: Composition, properties, and uses. Westport, CT: AVI. 570 pp.

Nagy, S., P. E. Shaw, and W. F. Wardowski, eds. 1990. Fruits of tropical and subtropical origin: Composition, properties and uses. Lake Alfred, FL: Florida Science Source. 391 pp.

Nowak, J., and R. M. Rudnicki. 1990. Postharvest handling and storage of cut flowers, florist greens, and potted plants. Portland, OR: Timber Press. 210 pp.

O'Brien, M., B. F. Cargill, and R. B. Fridley. 1983. Principles and practices for harvesting and handling of fruits and nuts. Westport, CT: AVI. 636 pp.

Pantastico, E. B., ed. 1975. Postharvest physiology, handling and utilization of tropical and subtropical fruits and vegetables. Westport, CT: AVI. 560 pp.

Peleg, K. 1985. Produce handling, packaging and distribution. Westport, CT: AVI. 625 pp.

Ryall, A. L., and W. J. Lipton. 1979. Handling, transportation and storage of fruits and vegetables. Vol. 1, Vegetables and melons. 2nd ed. Westport, CT: AVI. 588 pp.

Ryall, A. L., and W. T. Pentzer. 1982. Handling, transportation and storage of fruits and vegetables. Vol. 2, Fruits and tree nuts. Westport, CT: AVI. 610 pp.

Salunkhe, D. K., and B. B. Desai. 1984a. Postharvest biotechnology of fruits. 2 vols. Boca Raton, FL: CRC Press. 352 pp.

———. 1984b. Postharvest biotechnology of vegetables. 2 vols. Boca Raton, FL: CRC Press. 520 pp.

Salunkhe, D. K. and S. S. Kadam, eds. 1995. Handbook of fruit science and technology: Production, composition, storage, and processing. New York: Marcel Dekker. 611 pp.

Salunkhe, D. K., N. R. Bhat, and B. B. Desai. 1990. Postharvest biotechnology of flowers and ornamental plants. New York: Springer-Verlag. 192 pp.

Salunkhe, D. K., H. R. Bolin, and N. R. Reddy. 1991. Storage, processing, and nutritional quality of fruits and vegetables. 2nd ed. 2 vols. Boca Raton, FL: CRC Press. 520 pp.

Seymour, G. B., J. E. Taylor, and G. A. Tucker, eds. 1993. Biochemistry of fruit ripening. London: Chapman and Hall. 454 pp.

Shewfelt, R. L., and S. E. Prussia, eds. 1993. Postharvest handling: A systems approach. San Diego, CA: Academic Press. 358 pp.

Smith, D. S., J. N. Cash, A. Nip, and Y. H. Hui, eds. 1997. Processing vegetables science and technology. Lancaster, PA: Technomic. 448 pp.

Snowden, A. L. 1990–92. A color atlas of postharvest diseases and disorders of fruits and vegetables. 2 vols. Boca Raton, FL: CRC Press. 718 pp.

Somogyi, L. P., D. M. Barrett, H. Ramaswany, and Y. H. Hui, eds. 1996. Processing fruits science and technology. 2 vols. Lancaster, PA: Technomic. 1,038 pp.

Thompson, A. K. 1996. Postharvest technology of fruits and vegetables. Oxford: Blackwell Science. 410 pp.

Weichman, J., ed. 1987. Postharvest physiology of vegetables. New York: Marcel Dekker. 616 pp.

Wiley, R. C., ed. 1994. Minimally processed refrigerated fruits and vegetables. New York: Chapman and Hall. 368 pp.

Wills, R., B. McGlasson, D. Graham, and D. Joyce. 1998. Postharvest: An introduction to the physiology and handling of fruit, vegetables, and ornamentals. Wallingford: CAB International. 262 pp.

The U.S. Fresh Produce Industry: An Industry in Transition

Roberta L. Cook

As the U.S. fresh fruit and vegetable marketing system enters the twenty-first century, there is increasing focus within the industry on adding value and decreasing costs by streamlining distribution and understanding customer needs. This dynamic system has evolved toward predominantly direct sales from shippers to final buyers, both foodservice and retail, with foodservice channels absorbing a growing share of total volume. Product form and packaging are changing as more firms introduce value-added products such as fresh-cut produce that are designed to respond to the growing demand for convenience in food preparation and consumption. Fresh produce continues to be a critical element in the competitive strategy of retailers, and its year-round availability is now a necessity for both foodservice and retail buyers.

OVERVIEW OF KEY TRENDS AND PRODUCE INDUSTRY FUNDAMENTALS

INTERNATIONAL TRADE

The challenge to supply seasonal, perishable products year-round has favored imports and increased horizontal and vertical integration among shippers regionally, nationally, and internationally. Generally speaking, no country produces all of the fresh fruits and vegetables it demands in every week of the year, creating the opportunity for trade. Other countries are responding to the U.S. market's growing demand for imports, aggressively developing their horticultural industries consistent with the implementation of broader export-led economic growth and diversification strategies. Simultaneously, the United States is investing in the long-term development of new export markets, in response to slowing consumer demand at home and the growth in year-round demand for produce in other countries. Indeed, the United States is the dominant player in the international trade of horticultural commodities, ranked number one as both importer and exporter, accounting for about 18% of the $44 billion in world horticultural trade.

Seasonality in the production and consumption of perishable commodities, combined with natural climatic production advantages, are the driving forces behind horticultural trade. Trade is often contraseasonal, such as the shipment of Southern Hemisphere grapes, stone fruits, and avocados from Chile to the United States and Europe in order to meet consumer demand during the Northern Hemisphere's winter, when domestic supplies are low. Similarly, the United States imports grapes from Mexico in the spring and exports them to Mexico in the fall. Differences in natural climatic and growing conditions between countries can provide competitive advantages that lead to trade in complementary products, such as U.S. apples and stone fruits shipped to Costa Rica and Costa Rican bananas and pineapples exported to the United States. Contraseasonal and complementary trade is generally rather uncontentious, as long as imports do not overlap to a significant extent with domestic shipping seasons. Most contentious is trade caused by differing levels of relative competitiveness between producers of the same or similar products during the same season. For example, trade disputes between the winter tomato industries in Florida and West Mexico have escalated as Mexican tomato exports increased due to the improved competitiveness of Mexican extended-shelf-life tomatoes relative to Florida mature green tomatoes.

As trade liberalization progresses, more disputes will arise; as greater market access is achieved through tariff reduction and the tariffication of nontariff trade barriers, more trade disputes will center around sanitary and phytosanitary (SPS) concerns. However, under World Trade Organization (WTO) and North American Free Trade Agreement (NAFTA) rules, SPS measures must be scientifically justified, and formal dispute settlement mechanisms exist to test their validity. Clearly, there is no turning back, and ongoing trade liberalization and improved transportation services, along with improved temperature management and modified atmosphere technology, will facilitate even greater world trade in fruits and vegetables well into the twenty-first century.

Nevertheless, world trading rules on SPS issues are evolving in response to challenges over issues such as the use of genetically modified organisms (GMOs) in food production. For example, a new biosafety protocol was adopted on January 29, 2000, in Montreal, Canada, by more than 130 countries. The provisions of the protocol include the establishment of a biosafety clearinghouse to help countries assess risks from bioengineered organisms, and a requirement for exporters to seek consent from importers before shipping living GMOs for intentional release into the environment (such as seeds for transplanting). The procedure does not apply to GMO commodities destined for consumption or contained use, or commodities in transit.

DIFFERENTIATED PRODUCTS VERSUS COMMODITY ORIENTATIONS

Another key trend over the last decade was the attempt made by many suppliers to differentiate fresh produce. Despite these efforts, the difficulty of controlling the quality and volume of perishable items, both intra- and interseasonally, has limited the evolution of true consumer franchises for specific brands of consistently different products. Nature may change the appearance and eating quality of the same variety of the same produce item at any time in the production process. The riskiness of branding as well as selling private labels is further increased by the opportunity for improper temperature management throughout the distribution system. While firms may strive to perfect control over the ripening process and product quali-

ty throughout distribution, the reality is that if temperature abuse occurs, the image is tarnished of the firm whose name is on the product. For these reasons, the dynamics of fresh produce markets are still largely commodity-like, with relatively low levels of consumer advertising and most firms acting as price-takers.

The major exception to this is the fresh-cut produce sector, which includes value-added items such as bagged salads, washed baby carrots, and fresh-cut melons. Many fresh-cut products are regularly available in consistent quantities and qualities. Hence, fresh-cut produce is frequently marketed more like manufactured food products, often branded and backed by higher promotion budgets with shippers having a greater ability to influence price.

RISK AND INDUSTRY DYNAMICS

The notoriously high level of risk observed in the fresh produce sector arises from the combination of product perishability and weather variability. Weather factors can always undo the best-laid plans by unexpectedly shifting short-run supply or demand. Perishability limits storability and the ability of firms to adjust to short-run disequilibria in supply and demand, other than through price.

Understanding this fundamental characteristic of the fresh produce industry helps explain the common grower-shipper practice of selling below total costs. Since shippers fiercely compete to retain buyer loyalty and buyers rank consistency of supply highly as a supplier attribute, shippers never want to risk shorting customers. Therefore, they tend to err on the side of excess plantings in order to be assured of meeting the firm-level demand for their products, even if weather, disease, or management factors should decrease their yields and production. In the aggregate, this creates a tendency for excess supply, as defined by market-clearing prices below grower-shipper F.O.B. break-even levels. Since most fresh fruits and vegetables can't be stored until supply declines relative to demand and prices improve, shippers facing long market conditions and are compelled to "sell it or smell it." This industry maxim captures the dynamics behind supplier behavior in the fresh produce sector, explaining why firms frequently sell at prices barely covering variable costs.

On the other hand, exogenous supply shocks caused by random weather events can significantly reduce total supply overnight. Given the relatively inelastic nature of the demand for fresh produce, equally rapid and dramatic increases in prices can occur. Weather conditions can also unexpectedly shift short-run demand. For example, when severe and extended storms in the Northeast keep people housebound and impede the ability of trucks to reach the largest market in the United States, demand and prices may decline significantly for winter produce shippers based in Florida, California, and Mexico.

The price volatility common to fresh produce markets has contributed to a heavy reliance on spot market (daily) sales, as opposed to forward contracting between shippers and buyers. However, this is changing as food markets become more consolidated, with fresh produce increasingly expected to fit within the paradigm of procurement practices for nonperishables in the food industry as a whole. A recent national study (Calvin and Cook et al. 2001) surveying shippers of five fruit and vegetable commodities found that daily sales had declined from 72% of the sample's total dollar volume in 1994 to 58% in 1999, replaced by advance pricing for advertising (lid prices) and contracts.

The high level of risk and price volatility at the produce shipper or supplier level does not encourage dominance by publicly traded companies concerned with quarterly profit reports to shareholders. Despite the entrance of multinational food processors into the production and shipping levels of the fresh produce industry during the 1980s, a sizable portion of fresh produce sales at the first-handler level still remains in the hands of relatively specialized, frequently family-controlled grower-shippers.

Furthermore, as the overall U.S. food market matures, competition for the consumer's food dollar is increasing, continually challenging fresh produce to compete with other more highly advertised food products. Fortunately, both the positive health messages associated with fruits and vegetables and their availability in more convenient forms have continued to stimulate per capita consumption of fresh produce. Still, fresh produce per capita consumption grew at an average annual rate of only 1% per year from 1989 to 1999 (U.S. Department of Agricul-

ture, Economic Research Service [ERS] 2000a, 2000b). In mature (slow growth) markets there is less room for marginal players. At all levels of the vertical food system, the playing field has been "leveled upward" in terms of the quality and service demanded.

Additionally, the maturation of the food industry has led to the entrance of new competitors playing by new rules, encouraging mergers among existing firms as companies attempt to thwart the new competitive pressures. Chief among these new competitors are mass merchandisers introducing supply chain management, a procurement model designed to streamline the distribution system by eliminating non-value-adding transaction costs. Wal-Mart leads this trend with its penchant for contracting with suppliers and its growing use of the co-vendor-managed automatic inventory replenishment model. Investment of European supermarket chains in the U.S. market has likely reinforced this trend, as many European chains are further along in the implementation of supply chain management than conventional U.S. retailers. Today, 4 of the top 12 chains operating in the United States have European ownership, and the fourth largest chain is Ahold, a Dutch firm invested in the U.S. foodservice and online food shopping industries as well. Indeed, retailers have also been faced with the challenge of positioning themselves in a marketplace that offers nascent business to consumer food marketing choices (online food shopping) and emerging e-commerce procurement options, both of whose impacts and roles are as yet uncertain.

These new competitive pressures and others have contributed to retail, wholesale, and foodservice consolidation and an increase in upstream buying power. Greater buying power has caused an increase in the level and types of fees and services being requested from suppliers and is leading to more closely coordinated relationships between buyers and sellers. Shippers must adapt by adopting information technology and developing the systems and services capable of serving the needs of fewer, larger buyers. Shipper consolidation is a part of this process, whether through ownership or strategic alliances, although to date consolidation at the shipper level varies greatly by crop.

The current trend toward fewer, larger buyers and suppliers offers the industry the opportunity to rethink standard operating practices and to adopt new coordination mechanisms designed to improve vertical coordination, including contracts with preferred suppliers, category management, and loyalty marketing, all components of efficient consumer response (ECR). A unified set of coordination mechanisms, ECR has been in use for some time in the dry grocery sector, with some of its elements now being applied to fresh produce. For example, category management has been used for an average of only 3 years in the produce department but is being rapidly adopted, with 65% of supermarkets reporting that they are implementing it, albeit to varying degrees (Progressive Grocer 2000). This chapter provides a snapshot of the evolving fresh produce industry at the outset of the new millennium, in transition to a more globalized, databased, technology- and information-intensive system.

FRUIT AND VEGETABLE DEMAND

GENERAL TRENDS

The immediate post–World War II era in the United States was characterized by accelerating population growth, rising affluence, and a relatively homogeneous population. Under these conditions, mass-marketing strategies for food became the norm, and emphasis was put on products that could be marketed nationwide and in large volumes. Much less variety was available than today in terms of the number, form, and quality of food products, and exporting was not a priority since there was a large and growing market right at home.

Since the 1970s, demographic and lifestyle trends have segmented the U.S. market, causing a marked increase in the diversity of consumers and the products they demand. Targeted marketing began to replace mass marketing in the 1980s and 90s, and even more finely tuned segmentation strategies can be expected in the future as information technology assists marketers. For example, the rapid expansion in the use of supermarket customer cards (which generate electronic records of individual consumer purchases) now enables retailers to micromarket in order to increase customer loyalty in the saturated, intensely competitive U.S. retail food market.

Fresh produce has participated much less in the movement toward targeted marketing given its reliance on product sales in bulk form (without UPC bar codes), making produce subject to a paucity of electronically available sales data by product and consumer type. However, this has recently changed with the advent of standardized product-look-up (PLU) codes for fresh produce. Supermarkets report that they have been using standardized PLU codes for an average of 5 years, with a 92% adoption rate (Progressive Grocer 2000). The use of standardized PLUs allows for instantaneous data collection at checkout and facilitates better information management, such as by benchmarking relative product and category performance both within and between stores, markets, and chains. At least three commercial providers now sell electronic produce retail sales data, permitting general access to formerly unavailable data. The availability of timely data should contribute to improved performance by highlighting poor sales results and documenting effective merchandising and pricing strategies. The combination of store-level data with consumer data generated from customer cards should contribute to more targeted produce marketing in the future, by both retailers and suppliers, although supplier access to customer data may be more limited. The potential for more strategic, consumer-specific fresh produce marketing is arguably largely untapped.

CONSUMPTION

Two key lifestyle trends continue to affect food consumption: the ongoing entrance and advancement of women in the work force, increasing the demand for foods of high and predictable quality that offer convenience and variety; and the growth in public knowledge about how diet and health are linked and the importance of maintaining physical fitness throughout life. These trends have influenced the mix and form of foods consumed in the United States.

Per capita consumption trends

In part as a response to health concerns, per capita consumption of fruits and vegetables, in both fresh and processed form, increased 17% from 1976 to 1999, reaching 331 kg (730 lb) in 1999, as shown in table 2.1 (ERS 2000a, c). There was a general shift in product form toward the fresh and "natural." Many marketers incorporated "lite" or "natural" on their labels, along with stronger health claims such as reduction of heart disease or prevention of cancer. These claims benefited fresh fruits and vegetables proportionally more than processed ones, with 56% of total fruit and vegetable consumption in processed form in 1999, compared to 60% in 1976.

Vegetable consumption, in both fresh and processed form, grew much more rapidly from 1976 to 1999 than did fruit consumption. Vegetable per capita consumption increased 24% to 202 kg (445 lb), an average annual

rate of 0.9%. Per capita fruit consumption grew by only 8%, to 129 kg (284 lb), an average annual rate of 0.33%. Consumption of fresh fruits and vegetables grew more rapidly than that of processed fruits and vegetables. Fresh vegetable per capita consumption grew at an annual average rate of 1.3% over this period, compared to 0.6% for processed vegetables. Fresh fruit consumption increased at an average annual rate of 0.9%, compared to only 0.03% for processed fruits.

Still, processed fruit consumption far outweighs fresh, at 83 kg (183 lb) per capita in 1999, compared to 46 kg (101 lb) for fresh

Table 2.1. U.S. per capita fruit and vegetable consumption (kg), 1976–1999, and growth rates

Item	1976	1986	1989	1990	1991	1992	1993	1994	1995	1996	1997	1998	1999	Growth 1989–1999	Avg. growth per year 1989–1999	Growth 1976–1999	Avg. growth per year 1976–1999
VEGETABLES																	
Vegetables, excluding potatoes																	
Fresh	52.0	59.4	66.9	65.2	63.2	66.6	68.6	72.1	70.1	73.3	75.8	74.3	78.0	16.6%	1.5%	49.9%	1.8%
Processed	54.1	54.0	56.2	60.5	62.0	60.7	61.9	61.2	61.0	60.3	59.3	60.4	59.7	6.3%	0.6%	10.3%	0.4%
Subtotal	106.1	113.4	123.1	125.8	125.1	127.2	130.5	133.3	131.1	133.6	135.0	134.6	137.7	11.9%	1.1%	29.7%	1.1%
Potatoes																	
Fresh	22.4	22.1	22.7	21.2	22.9	22.0	22.9	22.8	22.6	23.0	22.0	21.7	21.9	−3.4%	−0.3%	−2.2%	−0.1%
Processed	34.4	35.0	34.9	35.1	38.1	37.2	39.6	40.0	40.3	43.8	42.1	42.2	42.5	21.6%	2.0%	23.5%	0.9%
Subtotal	56.8	57.1	57.6	56.3	61.0	59.2	62.5	62.7	63.0	66.8	64.1	63.9	64.4	11.7%	1.1%	13.3%	0.5%
All vegetables																	
Total fresh	74.4	81.5	89.6	86.5	86.0	88.6	91.5	94.9	92.8	96.3	97.8	95.9	99.9	11.5%	1.1%	34.2%	1.3%
Total processed	88.5	89.0	91.1	95.6	100.1	97.9	101.5	101.1	101.3	104.1	101.4	102.6	102.1	12.1%	1.2%	15.4%	0.6%
Grand total	162.9	170.5	180.7	182.1	186.2	186.5	193.0	196.0	194.1	200.4	199.2	198.5	202.0	11.8%	1.1%	24.0%	0.9%
FRUITS																	
Citrus																	
Fresh	12.9	11.0	10.7	9.7	8.6	11.0	11.8	11.3	10.9	11.3	12.2	12.3	9.4	−12.0%	−1.3%	−27.3%	−1.4%
Processed	46.4	43.2	40.0	39.5	39.0	33.7	40.8	39.8	42.8	42.6	43.1	44.7	39.5	−1.2%	−0.1%	−15.0%	−0.7%
Subtotal	59.4	54.2	50.7	49.2	47.7	44.8	52.5	51.2	53.8	54.0	55.4	57.0	48.9	−3.5%	−0.4%	−17.6%	−0.8%
Noncitrus																	
Fresh	24.7	31.5	32.8	32.1	32.0	33.8	33.1	34.3	33.5	33.5	34.6	34.6	36.9	12.7%	1.2%	49.3%	1.8%
Processed	35.5	42.6	42.7	42.7	41.4	44.2	43.7	42.3	41.5	42.0	45.3	42.0	43.1	1.0%	0.1%	21.4%	0.8%
Subtotal	60.3	74.0	75.5	74.8	73.4	78.0	76.8	76.6	75.0	75.5	79.9	76.5	80.1	6.1%	0.6%	32.8%	1.2%
All fruits																	
Total fresh	37.7	42.5	43.5	41.8	40.7	44.8	44.9	45.6	44.4	44.8	46.8	46.9	46.4	6.6%	0.6%	23.0%	0.9%
Total processed	82.0	85.8	82.6	82.2	80.4	77.9	84.4	82.1	84.4	84.6	88.4	86.7	82.6	−0.1%	0.0%	0.8%	0.0%
Grand total	119.6	128.2	126.1	124.0	121.0	122.8	129.3	127.8	128.8	129.4	135.3	133.5	128.9	2.2%	0.2%	7.8%	0.3%
FRUITS AND VEGETABLES																	
Fresh fruits and vegetables	112.1	124.0	133.1	128.2	126.7	133.4	136.4	140.5	137.2	141.1	144.6	142.8	146.3	9.9%	1.0%	30.5%	1.2%
Processed fruits and vegetables	170.5	174.7	173.7	177.8	180.5	175.8	185.9	183.2	185.7	188.7	189.8	189.3	184.7	6.3%	0.6%	8.4%	0.4%
All fruits and vegetables	282.6	298.7	306.8	306.1	307.2	309.2	322.3	323.8	322.9	329.8	334.5	332.1	331.0	7.9%	0.8%	17.1%	0.7%

Source: USDA ERS 2000a, 2000c. *Note*: Data may not sum to 100 due to rounding.

fruits. This is largely due to the importance of processed citrus consumption in the American diet, which totaled 39.5 kg (87 lb) in 1999. While processed citrus consumption is important, it declined by 15% over the period in question. Fresh citrus consumption declined even more, by 27% from 1976 to 1999, to 9 kg (20 lb) per capita in 1999. Growth in fresh fruit consumption came entirely from the noncitrus category, which grew by 49% to 37 kg (82 lb) in 1999.

Although per capita consumption of processed vegetables is still slightly larger than that of fresh vegetables, the gap has almost been eliminated, with processed vegetable consumption totaling 102 kg (225 lb) per capita in 1999, compared to 100 kg (220 lb) for fresh.

Total (as opposed to per capita) fresh produce consumption in the United States amounted to 40.2 billion kg (88.6 billion lb) in 1999, with fresh vegetable consumption outweighing that of fruit (ERS 2000a and 2000c). Fresh vegetable and melon consumption totaled 26.8 billion kg (59 billion lb), including 6 billion kg (13.2 billion lb) of fresh potato consumption. Consumption of fresh fruits was approximately half that of fresh vegetables, at 13.4 billion kg (29.5 billion lb), 3.9 billion kg (8.6 billion lb) of which was bananas.

Broccoli, carrots, peppers, onions, tomatoes, melons, bananas, grapes, strawberries, and kiwifruit led fresh produce consumption gains over the past 20 years. Several factors contributed to consumption growth for these commodities, including improved varieties and greater variety selection (grapes, tomatoes, melons, peppers); introduction of convenient fresh-cut forms (washed, peeled carrots); the development of year-round availability (broccoli, strawberries), in some cases through imports (grapes, melons); new uses through foodservice channels (tomatoes, broccoli, onions); and new consumer awareness of the nutritional benefits of the item (bananas, broccoli, carrots).

Lettuce is conspicuous in its absence from this list of key produce gainers. Despite the introduction of fresh-cut bagged salads, iceberg lettuce consumption declined over the last decade. Although romaine and specialty lettuce consumption grew dramatically by about 138%, the growth was from a very small base of 1.6 kg (3.5 lb) per capita in 1989 to 3.8 kg (8.4 lb) in 2000. Until recently, this growth merely cannibalized iceberg lettuce without contributing to expansion of the total lettuce category. However, in 2000 total lettuce consumption reached 15.1 kg (33.2 lb), finally surpassing the 1989 peak of 14.7 kg (32.3 lb) per capita. This indicates that adding value and convenience may indeed be enhancing demand.

The demand for specialty and ethnic fresh fruits and vegetables is growing, albeit from a very small base. While per capita consumption of specialty and ethnic fresh produce is likely underreported, for those specialty noncitrus fruits monitored by USDA, per capita consumption totaled 2 kg (4.4 lb) in 1999 compared to 0.86 kg (1.9 lb) in 1976 (ERS 2000a), and consumption of nontraditional fresh vegetables increased from 1.4 kg to 2.7 kg (3.1 to 5.9 lb) over the same period (ERS 2000c).

Influence of demographics on fresh produce consumption

In 1998, the average household size was 2.6 people, with an average of $48,100 in income and $4,810 spent on food (The Food Institute 2000a). Fresh produce consumption has been favorably affected by numerous demographic trends, including declining household size, rising income levels, the consumption habits of baby boomers, and the growth in the numbers of Hispanic American and Asian American consumers.

Today, single-person households are one of the largest groups, representing 26% of the 102.5 million total households in 1998, next to married couples with children at 27% of total households. Husbands and wives without children account for 26% of households, while single parents are 6% of the total; other households (such as people living as roommates) account for 15% (The Food Institute 2000a). In 1998, the expenditure on fresh produce for single-person households was $164 per year, double that of per capita expenditures for households with five or more people and well above the per capita average of $118 for all households (The Food Institute 2000a). If household units continue to decline in size, fresh produce consumption should be further stimulated, given the generally greater discretionary income of smaller households to spend on high-value foods, as well as their lesser ability to exploit economies of scale in purchasing.

While the share of national aggregate personal income received by each quintile of consumers was relatively stable between 1995 and 1998 (U.S. Bureau of the Census 1995–1998), the distribution of households by income level and food spending changed. The economic expansion of the 1990s increased the relative share of higher-income consumers in both the number of households and total food spending, while low-income households declined as a share of both. The following discussion highlights the current importance of higher-income consumers in food and produce spending.

In 1998, households earning over $50,000 per year represented 30% of U.S. households and accounted for an impressive 46% of total food spending, up from 25% and 35%, respectively, in 1995 (The Food Institute 2000a). In contrast, households earning under $15,000 represented 25% of the total number of households yet accounted for only 14% of food spending, down from 28% and 20%, respectively, in 1995.

Higher income levels have also stimulated fresh produce consumption. In 1998 the average household spent $294 to $305 per year on fresh produce in retail food outlets, roughly evenly divided between fruits and vegetables, out of total annual grocery store food expenditures of $2,780 (The Food Institute 2000a). Consumption of both fruits and vegetables is positively correlated with income. In 1998, households earning $70,000 and over spent an average of $475 on fresh produce compared to $194 for households earning under $15,000 per year. Middle-income consumers in the $30,000 to $39,999 income bracket spent close to the overall average of $315 on fresh produce. Incidentally, although fresh fruit and vegetable dollar expenditures are relatively similar, in physical volume, vegetable consumption is about double that of fruit, as noted earlier. Hence, fruits have a substantially higher average price per unit than vegetables.

Clearly, today there are more households with the ability to pay for high-quality food and value-added product forms, including items in the produce department. U.S. and Canadian consumers experience the lowest share of food expenditures relative to disposable personal income in the world, at 11 and

10%, respectively, in 1998 (ERS 1999). Despite the ability to pay for high-quality produce, were the economy to enter a recession, consumer expenditures on produce and willingness-to-pay for convenience should decline, just as they did in the recession of the early 1990s.

Households headed by consumers 55 years and older represent 32% of the total and account for the same percentage of fresh fruit and vegetable expenditures (The Food Institute 2000a). On the other hand, households headed by consumers 34 to 55 years old (the broadly defined "baby boomer" cohort) represent about 41% of the total while contributing nearly 48% of fresh produce spending. Conversely, households headed by people under age 35 amount to 26% of the total but proportionately contribute only 20% of fresh fruit and vegetable spending. The future of fresh produce consumption should remain strong if baby boomers continue to consume at above-average rates as they age. The lower consumption rates of younger consumers emphasize the importance of educating people about the benefits of fresh produce consumption, starting from youth. Such programs are currently underway by the 5-A-Day for Better Health Foundation, among others.

The changing ethnic makeup of the U.S. population is also favorable to fresh produce consumption, since Hispanic and Asian Americans consume fruits and vegetables at higher rates than African Americans and whites. In 1998, white households on average consumed $292 of fresh produce per year, compared to $408 for Hispanic Americans and $217 for African Americans (statistics are unavailable for Asian Americans). Over the last twenty years Hispanic and Asian Americans have consistently increased their share of the U.S. population, with 31 million Hispanics representing 11% of the 273 million U.S. residents in 2000, compared to 7% in 1980 (U.S. Bureau of the Census 1980, 1995–1998). In contrast, the share of African Americans was flat at 12% over the same period; Asians grew from 1 to 4% of the population (U.S. Bureau of the Census 1980).

The highest average household expenditures on fresh produce are now in the West, in part given to the higher concentration of Hispanic and Asian Americans there. The South still lags the nation in produce expenditures, with the Northeast ranked second in importance, followed by the Midwest. The long-term movement of the population to the West and Southwest is likely to continue to benefit fresh produce consumption as regional migration exposes consumers to different eating patterns.

KEY TRENDS IN MARKETING STRATEGIES

The moderate rate of growth in aggregate per capita consumption of fruits and vegetables is not surprising given the maturity of the U.S. food market. In high-income countries with slow rates of population growth (under 1% annually), total food consumption tends to be relatively stable since people are already well fed. Hence, food marketers compete for "share of stomach" (percentage of the total food consumed by a person), and although consumption of certain items may grow, it is generally at the expense of others. Because firms operating in highly competitive, saturated markets can't rely on population growth to expand sales quickly, they focus on three broad marketing strategies: new product introductions, market share growth, and development of new markets, including export markets and foodservice.

NEW PRODUCT INTRODUCTIONS

The development of new food products occurred at a record rate after 1980 (when only about 1,000 new products were introduced), peaking in 1995 at 16,863 new products. Since the average supermarket carries around 20,000 products, competition for shelf space is increasingly keen. The negotiating power of food retailers has grown as the battle for their limited shelf space by food marketing firms has intensified, resulting in costly slotting fees (fees paid by suppliers to secure shelf space). To date, these fees have been confined mainly to the grocery section of the store; within the produce department they are only used for fresh-cut, branded products, where they may reach up to $2 million to acquire the business of a large multiregional chain (Calvin and Cook et al. 2001).

Since 1995, new product introductions declined to 9,664 products in 1999 (The Food Institute 2000c), highlighting the shift by many food marketing firms to more targeted strategies. As consumer segments and their food product needs are better understood, suppliers tend to introduce somewhat fewer but better targeted products. The high cost of new product launchings, in part due to slotting fees, coupled with a high new product failure rate, are compelling reasons for marketers to become more focused.

Fruits and vegetables were part of the growth trend in new product introductions, as well as part of the recent slowdown in this trend as firms became more market research–based in their new product offerings: 254 new products were introduced in this category in 1999 compared with 251 in 1997, and down from a 1996 peak of 552. While the proliferation of fresh-cut items, such as bagged salads, shredded broccoli, microwave-ready fresh vegetables, and washed baby carrots, fueled new product growth, many specialty items were also a factor.

In addition to conventional broccoli, today consumers may select from broccoli romanesco, purple broccoli, and broccoflower. Similarly, the watermelon category has been differentiated to include seedless, icebox, and yellow watermelons, and the tomato category has been expanded to over 15 offerings, including various colors of round, pear-shaped, and round cherry tomatoes, and heirloom tomatoes of various types. Many more tropical and subtropical fruits and vegetables are available today as well, including passion fruit, cherimoya, carambola, mamey, jicama, tomatillos, cactus leaves, and specialty squashes such as chayote. Many more nontropical ethnic varieties of produce are also commonly marketed now, including numerous types of Italian and Japanese eggplants and squashes, and many specialty leafy greens and cabbages including arugula, mizuna, several varieties of radicchio, mache, savoy cabbage, bok choy, and baby bok choy, and many Italian and Asian specialty mushrooms.

Product diversity

After more than a decade of a high level of both new product introductions and failures, the average number of items handled in a U.S. fresh produce department is up dramatically. Estimates of the average number of products handled in a U.S. fresh produce department vary from 345 in 1998 (Supermarket Business 1999) to 431 in 1999

(McLaughlin et al. 1999). This compares to 173 in 1987 (Litwak 1998) and 312 in 1994 (McLaughlin et al. 1999). The U.S. consumer probably now enjoys the greatest level of product diversity available anywhere in the world. Yet six commodity groups still make up 41% of total sales, the same as in the 1980s. However, the product mix and rankings of the top six have changed slightly from (in descending order) bananas, apples, citrus, potatoes, lettuce, and tomatoes, to bananas, lettuce, apples, tomatoes, potatoes, and grapes in 1999 (Progressive Grocer 2000).

Fresh-cut produce of all types has grown from virtually a zero base at the outset of the 1990s to account for an estimated 15% of the average retailer's sales in 1999 (McLaughlin et al. 1999). Specialty produce represented 2.6 to 3.7% of retail produce department sales in 1998, with independents selling at the high end of the range and chains at the low (Produce Marketing Association 2000). This likely highlights the strategy of independents who offer a more diverse product mix as a means for competing with chains.

The fact that the rapid pace of new product introductions has not stimulated even a remotely proportional increase in overall fresh produce consumption clearly indicates that fresh produce marketers are no different than other food marketers competing for a relatively fixed share of stomach.

Competition for retail shelf space is fierce, causing a "densing-up" phenomenon, with retailers introducing multideck cases to accommodate more products in the same linear space. The shelf space battle continues to prompt shippers and grower commodity groups, such as marketing commissions and orders (mandated-marketing programs), to commit more resources to in-store merchandising programs targeted to the needs of individual retailers. Marketers will increasingly seek to be "shelf captains," targeting those retail accounts and specific stores with the consumer demographics and pyschographics suited to their product. They will then attempt to influence shelf-space decisions and achieve a dominant position in that product category.

MARKET SHARE GROWTH OR MERGER MANIA

Firms in the U.S. food marketing sector view a large market share, including, if possible, the position of market leader, as a key requisite to success. Since the 1980s, pursuit of market share has led to a dramatic consolidation in the U.S. food chain at all levels, from the farm through food retailing. Rather than competing to capture market share from rival firms, U.S. food marketers have often pursued share growth through mergers and acquisition of rivals. Beginning in 1997, merger and acquisition activity in the food sector rebounded strongly from a decade-long slump after intense activity throughout most of the 1980s. Mergers and acquisitions reached a historical high of 813 mergers in 1998, then retreated to 753 in 1999 (The Food Institute 2000b) with 630 estimated for 2000 (The Food Institute 2000e).

Despite the recent slowing in merger rates, the absolute level of mergers is still quite high, and these mergers have had important implications for the structure of competition in the U.S. food sector. In 1999, the four largest food retailers' share of grocery store sales was 27%, up from 18% in 1987; the 8 largest retailers' share was 38%, up from 27%; and the 20 largest retailers' share was 52%, up from 39% (Calvin and Cook et al. 2001).

NEW MARKET DEVELOPMENT

Exports represent an important growth market for U.S. produce marketers. Although the importance of the export market varies widely by commodity, in general, exports were traditionally a small share of the market for perishable fruits and vegetables, owing in large part to trade barriers and the difficulty and expense of long-distance shipping. Trade liberalization negotiated under the recent Uruguay Round of the GATT and implemented under the new World Trade Organization, as well as through regional trade agreements such as NAFTA, has expanded market access and provided strengthened mechanisms for combating nontariff trade barriers such as scientifically unfounded phytosanitary restrictions. Advances in postharvest technology such as the development of container-level modified atmosphere technologies have also facilitated exporting to distant markets. Total U.S. horticultural exports, including fresh and processed fruits, vegetables, and nuts, were $10.5 billion in fiscal year 2000, up from $2.7 billion in 1985 (U.S. Department of Agriculture, Foreign

Agricultural Service [FAS] 2000). Processed horticultural crop exports far outweigh fresh, and within the fresh category, fresh fruit exports exceed those of fresh vegetables. Fresh fruit exports were $2 billion in 2000, compared to $1.3 billion worth of fresh vegetable exports.

More than 20% of the production of numerous fruits and vegetables is now exported, with the highest export propensity in the fresh table grape subsector, which on average has sent over 45% of total production abroad in recent years (FAS 2000). In export dollar value, the most important fresh fruit and vegetable exports are table grapes, apples, oranges, grapefruit, and lettuce. Fruits continue to play a greater role than do vegetables in our export trade. The top five U.S. horticultural export markets are Canada, Japan, Mexico, the United Kingdom, and the Netherlands.

The foodservice market is also becoming more important to fresh produce marketers, and more shippers are selling directly to foodservice distributors rather than through intermediaries (Calvin and Cook et al. 2001). In 1999 the consumer dollar was almost evenly divided between retail and foodservice expenditures, with the latter accounting for 48% of the total. However, the amount of value added to the product in foodservice channels is much greater than in retail since the food is prepared and served to consumers. Hence, foodservice's large share of total food expenditures substantially overrepresents the share of physical product volume sold through this channel.

For example, since consumers eat out on average 2.5 times per week and there are 1,095 potential meal occasions in a year, this implies that only 11% of total meals are eaten outside of the home. On the other hand, more and more consumers are purchasing meals to eat at home, and 1997 was the first year that restaurants sold more meals for takeout than for on-premises consumption. Although foodservice growth rates have been declining, down from 11% per year in 1978 to 5% in 1999, foodservice continues to grow at a more rapid rate than the retail food industry. In 2000, 76% of consumers ate out at least once per week, with 40% eating out two or three times per week (Food Marketing Institute 1999–2000). This highlights the importance of shippers further developing this important and growing market.

There continues to be potential for the addition of produce to menus and for the substitution of fresh fruits and vegetables for processed products. For example, when a large pizza chain substitutes fresh for processed mushrooms, the new volume may measurably increase fresh mushroom demand. In addition, the convenience store sector is still an untapped market, offering a potential new distribution channel for convenience-oriented fresh produce.

FRUIT AND VEGETABLE INDUSTRY PROFILE

PRODUCTION LEVEL

Total production of fresh-market vegetables, excluding potatoes, reached 20.4 billion kg (50 billion lb) in 1999 (ERS 2000b), up from 11 billion kg (24.3 billion lb) in 1976 (ERS 1977). Potato production was 18.5 billion kg (40.8 billion lb) in 1999 for fresh market and processing uses only, excluding seed, animal feed and other uses, about one-third of which was destined for the fresh market (U.S. Department of Agriculture, National Agricultural Statistics Service [NASS] 2000b). Production of fruit, for both processing and fresh markets, totaled 29.2 billion kg (65.9 billion lb) in 1999, compared to 24.4 billion kg (53.8 billion lb) in 1976 (ERS 1997, 2000a). In 1999, fruit was grown on approximately 1.3 million ha (3.2 million acres) (ERS 2000a) while vegetables, excluding potatoes, were produced on 1.4 million ha (3.5 million acres), 768,825 ha (1.89 million acres) of which were destined for the fresh market (ERS 2000b). Potato area harvested totaled 539,676 ha (1.33 million acres) in 1999, for both fresh and processed uses.

In 1999 the farm-level utilized value of production was $2.6 billion for citrus fruits, $8.3 billion for noncitrus fruits (ERS 2000a), and $15.2 billion for vegetables (ERS 2000c), making fruit and vegetable production destined for both the fresh market and processing a $26.1 billion industry. In 1999 the farm gate value of the twenty-five major fresh-market vegetables and melons, excluding potatoes, totaled $7.5 billion (ERS 2000b), and fresh-market potato production was estimated by ERS at $933 million. Fresh noncitrus fruit farm gate value was $5.6 billion, and fresh citrus production was valued at $1.2 billion (ERS estimate). Total fresh-market fruit and vegetable farm gate production was valued at $16.5 billion in 1999, including minor vegetables (ERS estimate).

RETAIL AND FOODSERVICE LEVELS

Total 2000 fresh produce sales through all channels are estimated to be $75.8 billion. Sales of fresh produce in grocery stores are estimated to have reached $40.6 billion in 2000 (fig. 2.1). The estimated 2000 value of

Figure 2.1

U.S. fresh fruit and vegetable value chain, 2000, estimated $75.8 billion.

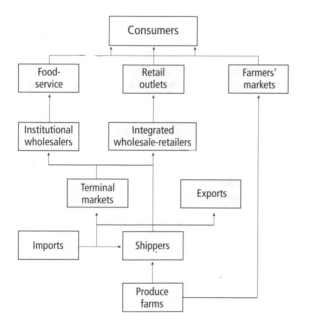

produce sold through foodservice channels was $34.1 billion, and an estimated additional $1.1 billion of fresh produce was sold directly from farmers to consumers via farmers' markets, "u-pick" operations, and roadside stands. The farmgate value of fresh produce was $16.6 billion in 2000, with exports valued at $3.2 billion and imports valued at $5 billion.

The entire U.S. food system totaled $843.2 billion in final sales in 2000, including $449.5 billion of food sold through retail channels and $393.7 sold through foodservice channels (ERS estimate). Produce is estimated to represent 9% of the total. The producer share of the final value of fresh produce sold through all channels was 24.5% in 1999, compared to 20% for food products as a whole.

LOCATION OF PRODUCTION

In 1999, California, the largest producer of horticultural commodities in the United States, contributed 54% each, respectively, to the nation's production of major fresh vegetables and fresh fruits, by value (NASS 2000c). California is the nation's exclusive supplier of clingstone peaches, dates, figs, kiwifruit, olives, pomegranates, prunes, and raisins. California's share of U.S. production exceeds 70% for each of the following fruits and vegetables: lettuce, processing tomatoes, broccoli, cauliflower, carrots, celery, strawberries, grapes, nectarines, plums, apricots, avocados, lemons, and honeydew melons. California's dominant position in the horticultural industry is explained by climatic, technological, and infrastructure advantages, as well as the market-and consumer-driven orientation of its agribusiness managers.

Florida, the second-largest producer of horticultural crops, produced 14% of U.S. fresh vegetables in 1999 (NASS 2000c) and 8% of national fresh fruit production, by value (ERS estimate). In the fresh fruit industry Florida's role is more important in citrus, where it contributed 22% of national value in 1999 (led by grapefruit), than in noncitrus, where it contributed 3% of the production value. While Florida accounts for only 18% of the total U.S. fruit industry in fresh and processed form (ERS estimate), it dominates in citrus. In 1999 Florida produced 76% of U.S. citrus, with oranges and grapefruit as Florida's leading citrus crops (NASS 2000a). Although Florida is the primary U.S. producer of oranges, with 9,512 metric tons (10,482 tons) of production in 1999/00, most of this went to processing, making California, at 2,280 metric tons (2,513 tons), the dominant producer for the fresh market (ERS 2000a). Florida leads in the production of several vegetables, including fresh-market tomatoes, snap beans, watermelons, and cucumbers, and accounts for more than half of the nation's production of fresh escarole, endive, and eggplant.

The remainder of U.S. fruit and vegetable production is dispersed among other states, primarily Arizona, Texas, Georgia, Washington, Wisconsin, Oregon, Minnesota, Michigan, New York, Idaho, and Hawaii.

IMPORTS

Imports of fresh fruits and vegetables into the United States have expanded rapidly since 1980, when they totaled 3.7 billion kg (8.2 billion lb) and accounted for a 15.4% share of total consumption. In 1996 imports were 7.6 billion kg (16.8 billion lb), amounting to an import market share of 21% of the total 37 billion kg (81.5 billion lb) of fresh fruits and vegetables consumed in the United States (Lucier, Pollack, and Perez, 1997). Imports have continued to grow in absolute volume, but their share of consumption has increased only slightly. In 1999 fresh fruit and vegetable imports totaled 8.9 billion kg (19.6 billion lb), representing 22% of the 40.2 billion kg (88.6 billion lb) of U.S. fresh produce consumption (ERS estimate). Bananas, a crop that essentially does not compete with domestic production, represented a sizable 43% of imports at 3.9 billion kg (8.6 billion lb), and a 41.5% share of fresh fruit consumption. When bananas are excluded, fresh produce imports represented a much smaller 14% market share of U.S. produce consumption. Hence, despite the rapid growth in imports, the vast majority of fresh produce consumed in the United States is still domestically produced.

The principal foreign suppliers are Mexico, South America, Canada, and the Caribbean Basin Initiative (CBI) countries, but suppliers vary significantly between the vegetable and fruit categories. For example, while Mexico dominates U.S. fresh vegetable imports, totaling $2.1 billion in 1998, it is not a principal supplier of fresh fruit (FAS 1999). In 1998, Mexico accounted for 68% of fresh vegetable import value, including melons and potatoes, compared to only 23% of the $2.7 billion in fresh fruit imports. Mangoes lead Mexico's contribution to fresh fruit imports. For bananas, the principal suppliers are Ecuador, Costa Rica, and Guatemala, accounting for 30, 30, and 15% respectively of the $1.1 billion in total banana imports. Chile supplied 16% of fresh fruit imports in 1998 and is the leading foreign supplier of table grapes, deciduous fruit, and kiwifruit. Chilean produce ships during the off-season, when U.S. supply of these crops is low to nonexistent.

SUPERMARKET PRODUCE DEPARTMENT PROFILE

Since 1990, the quality of fresh produce has been one of the principal factors influencing where consumers shop for food, with 88% of consumers ranking it as important in 2000, making it the number-two factor in importance after a clean, neat store (Food Marketing Institute 1990–2000). Many retailers have repositioned their store formats and image around the produce department, and produce is a critical element in their competitive strategy. In 1999, 91% of supermarkets placed the produce department prominently in the front of the store (Progressive Grocer 2000). Fresh produce sales now eclipse meat sales, traditionally the most important department in the supermarket. High awareness of the health benefits of produce and improved produce quality and merchandising should continue to reinforce the produce department's role in attracting customers to the store.

PROFITABILITY AND SIZE

Data on the share of total store sales of the average produce department vary, in 1999 reaching an estimated 12.8% of total store sales, occupying 12.9% of store space and generating 20.9% of store profits according to *Fresh Track 1999* (McLaughlin et al. 1999). The 1997 U.S. Census of Retail Trade reported that supermarket and supercenter produce departments on average contributed 9.5% of store sales and 17.2% of store profits, and accounted for 12.7% of total store space (Kaufman et al. 2000). The proportionately greater contribution to profits than to sales and space is due to both the high turnover and high average gross margin (33.2%) of the produce department compared to the storewide average margin of 26% (Kaufman et al. 2000). Indeed, some estimates of produce department gross margins are as high as 44.1% (Supermarket Business 1999).

According to *Produce Merchandising*, average weekly fresh produce sales were $27,780 in the third quarter of 2000. The average size of a produce department was 327 m² (3,516 ft²) in 1999, up from 237 m² (2,548 ft²) and 10.3% of total store space in 1994 (McLaughlin et al. 1999).

CHANGING STORE FORMATS

The produce department's larger average share of total store space represents an expanded share of increasingly larger stores. The superstore format now represents about 26% of all food retailer sales, compared to 12% in 1980, with an average size of 4,608 m² (51,200 ft²). Conventional supermarkets (2,322 m² [25,800 ft²] on average) have dramatically lost share, down from 55% of all food retailer sales in 1980 to 19% in 2000 (The Food Institute 2001). The superstore share of retail food sales is projected to remain stable, compared to a 14% share for conventional supermarkets by 2005. The larger size of superstores permits greater product offerings, including specialty food and service departments such as delicatessens, seafood departments, and bakeries, as well as nonfood departments. Other types of store formats have also gained share, including combination food and drug stores and super warehouses. Food-drug combinations are projected to account for 15% of food retail sales in 2005, up from 2% in 1980 (The Food Institute 2001). Given the greater product offerings in most store formats today, fresh produce's growing share of total store sales is especially impressive.

A nontraditional type of retail outlet also evolved rapidly during the 1990s. Supercenters (not to be confused with superstores) are a type of mass-merchandise outlet, combining a full-line supermarket with a full-line discount department store. Supercenters range up to 18,000 m² (194,000 ft²) in size. The maturation of the U.S. discount market has induced mass merchandisers, such as Wal-Mart, K-Mart, and Target, to diversify into food marketing via the supercenter format, emerging as a major new force in food retailing. Wal-Mart has also recently entered the conventional food retailing industry with a 3,600-m² (38,700-ft²) Neighborhood Market format.

The supercenter format is led by Wal-Mart, with estimated 1999 grocery-equivalent supercenter sales of $15.7 billion and total supercenter sales of $39.1 billion, 56% of the total national supercenter industry sales of $69.8 billion (The Food Institute 1999c). Wal-Mart's development of the supercenter concept has propelled it into the number five ranking among U.S. food retailers in 1999, compared to $45.3 billion for Kroger, the nation's number-one retailer. Its rapid rate of new store openings and the large size of the outlets means that Wal-Mart and other mass merchandisers are having a far bigger competitive effect on the food retailing industry than store numbers would imply. It is estimated that in 2000, 9.3% of national food store sales were generated by the grocery-equivalent sales of only 1,300 supercenters, compared to 31,500 supermarkets. This share may reach 16% by 2000 given the higher growth rate of the segment relative to the mature conventional retailing industry. Total U.S. supercenter sales are forecast to reach $112 billion in 2002, including $57 billion of food sales (The Food Institute 1999a). Despite the large nonfood sales of supercenters, fresh produce still represents 10% of the grocery-equivalent sales, just under the average level for supermarkets, again highlighting the importance of this new format to the produce industry (The Food Institute 1999c).

Another type of mass merchandiser is the membership club store that small businesses, individuals, or groups pay a fee to join. The club store focus is on high-volume sales of large-sized packs at relatively low margins in a warehouse format with minimal customer service. National club store sales, which were recently eclipsed by supercenter sales, totaled $60.7 billion in 1999, divided between Costco (49.3% of market share), Sam's Club (also owned by Wal-Mart, with a 43% market share), and BJ's, with a 6.7% market share (The Food Institute 1999b).

While only 4.3% of club store sales are fresh produce (The Food Institute 1999a), this is still equivalent to an impressive $2.6 billion worth of produce sales in 1999. Selling to membership clubs often requires providing special packs, which can imply additional risk for shippers given the difficulty in shifting these packs to other channels if sales don't materialize. Club stores generally don't have distribution centers, preferring just-in-time inventory systems. Hence, if store movement is slower than expected shippers generally absorb the slack as club stores are unable to hold excess inventory.

The focus of all mass merchandisers on streamlining the supply chain and eliminating non-value-adding costs will continue to exert competitive pressure on conventional retailers over the next decade, with super-center growth outpacing that of club stores.

MARKETING CHANNELS AND PROCUREMENT PRACTICES

The principal marketing channels in the U.S. fresh fruit and vegetable marketing system are shown in figure 2.1. The three primary sales outlets to consumers are retail food stores; foodservice establishments, hotels, restaurants, and institutions (schools, the military, hospitals, nursing homes, shelters, and prisons); and direct farmer-to-consumer sales via "u-pick" operations, farmers' markets, and roadside stands. Although the majority of produce still moves through retail channels, foodservice may now account for 45% of total volume, and direct sales may account for 1.5%.

Produce sold in retail or foodservice outlets may be procured directly from shippers or via intermediaries such as wholesalers operating in terminal (wholesale) markets or in independent warehouses in local communities. According to the PMA *Fresh Track 1999* national survey of retailers, 43% of produce procured by the retailer respondents came directly from grower-shippers located in the production area (McLaughlin et al. 1999). Since the 1950s, terminal markets have steadily declined in importance; today there are only 22 major terminal markets, and the volume of produce sales they handle is an estimated 30% of the national total. Product formerly moving through terminal markets now goes directly from shippers to final buyers or via nonterminal market wholesalers, leaving these markets to handle the residual fresh-market production that cannot be marketed directly to retail or foodservice buyers. Terminal markets do play a dominant role in handling imported produce, especially markets located at or near ports.

Exceptions to the dominant marketing channel (directly from the production region to the final buyer) do exist, most notably for fresh-market tomatoes. The tomato ripening process and the long distances from production source to market make it difficult to achieve uniform color upon arrival. Consequently, repackers or wholesalers, both on and off terminal markets, handle a sizable portion of tomato shipments. These handlers repack to meet the color, size, and pack style requirements of specific retail and foodservice buyers.

The decline in terminal market share is largely a result of the increased buying power of integrated wholesale-retail buying entities. Integrated wholesale-retailers are self-distributing, operating large-volume centralized buying operations, making it more efficient for them to buy directly from the source, thereby avoiding intermediary margins and handling costs. Also, buyers are able to communicate directly with suppliers concerning important issues such as desired product quality characteristics and timing of production and delivery, without the information being diffused and possibly distorted by middlemen. For fresh products, production-source-to-buyer shipments have the additional advantage of not breaking the cold chain, better preserving product quality. Therefore, integrated wholesale-retail buyers use terminal markets primarily to balance short orders and to procure small-volume exotic or specialty items.

INTEGRATED WHOLESALE-RETAILERS

Integrated wholesale-retailers include the centralized buying operations of corporate chains (11 or more stores) and affiliated groups comprised of voluntary chains and retail cooperatives. Voluntary chains consist of sponsoring wholesalers who supply independent retailers (retailers operating fewer than 11 stores) or small chains, as well as their own stores. Retail cooperatives are essentially member-owned wholesalers, since they consist of groups of retailers who vertically integrate, jointly owning a central buying and warehousing facility. These vertical coordination strategies give affiliated groups the benefits of joint buying, advertising, and merchandising programs, enabling them to compete with corporate chains despite the smaller size of individual members. Well-known examples of affiliated groups are SUPERVALU, Certified Grocers of California, and Independent Grocers Association (I.G.A.) in the United States; Spar, operating throughout Europe; and Lecler in France. Independent affiliated retailers are common in rural areas underserved by corporate chains.

Chains have steadily grown in importance since their origins in the early 1900s, when unaffiliated independent wholesalers and retailers were the norm. In 1999, chains accounted for 80% of supermarket sales compared to 74% in 1994, 62% in 1974, and 58% in 1954 (Progressive Grocer Annual Report 1954, 1974, 1994). In each of these years the remainder was accounted for by sales through mainly affiliated independents, with unaffiliated groups currently representing fewer than 3% of all U.S. grocery sales.

Retailers in affiliated groups and corporate chains may differ in their procurement practices. Because a corporate chain owns all of its stores, it controls the products it handles and essentially exercises forced distribution. With standardized store formats, chains have more consistent quality needs than do affiliated groups who serve a wide diversity of independent retail members. Chains also typically have less ordering flexibility than affiliated groups, who can make more rapid store-level adjustments to accommodate sudden shipping-point changes in product availability and quality (McLaughlin 1983).

As the U.S. market has matured, mergers and acquisitions in the food industry have increased. As the industry has undergone consolidation and larger operators have acquired smaller firms, the number of integrated wholesale-retailer centralized buying operations has declined and sales per firm have increased. It is estimated that fewer than 250 centralized buying operations supply 127,000 food stores, including 20,300 chain supermarkets, 11,200 independent supermarkets (supermarkets are grocery stores with over $2 million in annual sales), 37,200 other grocery stores (food stores with less than $2 million in annual sales), 57,500 convenience stores, and 800 membership wholesale club stores (Progressive Grocer Annual Report 2000). As noted, estimates do vary, and *The Food Institute Report* (The Food Institute 1999d) indicates that there were 914 membership warehouse clubs in 1999. Some of the smaller grocery stores include greengrocers; the 1997 *Census of Retail Trade* identified 3,179 stores (compared to 2,971 stores in 1992) that specialized completely in fruits and vegetables, with $2.1 billion in sales. Supermarkets account for the bulk of food store sales, estimated by ERS at 70% in 1999.

The United States traditionally has not had any truly national supermarket chains; chains have tended to be regional in focus probably due to the nation's large geographic size. While this continues to be the case, it is rapidly changing, with a few chains now approaching national scope. Five chains have over 1,000 stores each, and Kroger, the market leader, surpasses 2,300 stores. The numerous recent retail mergers and the emergence of the supercenter concept are increasingly concentrating buying power in the hands of a few very large players, influencing the way firms deal with produce shippers.

Retailers often cite the potential for lowering procurement, marketing, and distribution costs as motivations for mergers and acquisitions. By purchasing more volume directly from larger shippers, retailers hope to gain greater efficiency in procurement by eliminating intermediaries and lowering the per-unit cost of goods. Large retailers also desire large volumes of consistent product to provide uniformity across all their stores, which may be more easily supplied by larger shippers. In return for consistent supply, retailers may offer shippers preferential procurement agreements such as partnering, contracts, or other strategic alliances that can be mutually beneficial. Large retailers can also achieve marketing efficiencies such as lower costs for advertising.

Although the economic effects of the recent mergers on fresh produce have not yet been determined, many suppliers fear that competition will erode. To date, many recently merged chains are still in the process of integrating their buying systems, and some still buy produce on a division basis (with divisions defined along the lines of the incorporated chains), lessening the effect of consolidation. However, this is changing, with corporate buying growing in importance at most chains and field buying declining somewhat (Calvin and Cook et al. 2001). Grower-shippers can expect consolidated food retailers to gradually reduce the number of buying offices and combine orders into larger volumes. If e-commerce platforms take hold, the procurement practices of integrated wholesale-retailers may become even more centralized.

Supply chain management practices such as continuous or automatic inventory replenishment are becoming more common. Under this system, shippers have access to retail sales data and are responsible for providing the correct amount of produce to each distribution center served on a just-in-time basis, potentially reducing the size and cost of retail distribution centers. This system also allows retailers to streamline and downsize their produce buying offices. However, to date, mainly mass merchandisers rather than conventional grocery retail chains have implemented automatic inventory replenishment systems in fresh produce.

Clearly, the magnitude of produce that must be procured by large retailers today points to the need for closer coordination with preferred suppliers. There are now ten integrated wholesale-retailers, each with over 1,000 stores and selling a total of over $1 billion in fresh produce annually. For each of the two largest supermarket chains, fresh produce sales are estimated to exceed $4 billion. This makes consistent, predictable supply imperative, highlighting the need for retailers to work with suppliers as partners rather than adversaries. Shippers who are not equipped to sell to these very large buyers must focus their efforts on the remaining more fragmented portion of the food system, both retail and other outlets.

WHOLESALERS AND BROKERS

While it is difficult to determine the total number of produce wholesalers, brokers, and distributors, the number may reach 6,000 (McLaughlin et al. 1997). Brokers are noteworthy players in fresh produce distribution, and their role has grown in importance since World War II. Brokers help negotiate sales on behalf of buyers or sellers for a percentage sales commission or a flat fee per unit. They do not physically handle or take title of the merchandise; thus, their fees are substantially lower than those charged by commission merchants. The use of brokers varies greatly by type of buyer and commodity, but buyers or sellers at any level of the distribution system may use brokers. As buyers procure broader product lines of both domestic and imported produce, many brokers have become global in their sourcing abilities and are increasingly oriented to meet specialized buyer needs. In 1999 it was estimated that brokers were involved in the trading of $8.9 billion worth of produce. However, the more consolidated marketing system poses new challenges to brokers, as larger buyers and sellers deal directly with each other and e-commerce procurement options evolve.

Today, terminal market and other whole-salers focus on independent retailers and foodservice accounts. Primary market handlers (receivers, merchant wholesalers, and commission merchants) procure more than half of their product from the shipping point. Receivers and merchant wholesalers buy and resell products, and commission merchants operate on a consignment basis. Secondary market handlers (jobbers and purveyors) procure more than half of their product from other wholesalers, principally primary handlers. They serve small-volume accounts such as greengrocers and restaurants, which require frequent deliveries of small lots. Purveyors focus almost exclusively on foodservice accounts.

While terminal markets in the Midwest and East are primarily destination markets, those located near the production regions on the West Coast and in Florida ship significant volumes to terminal markets and other wholesalers in the destination markets. Wholesalers in all regions have expanded customer services to include such functions as ripening, sizing, repacking, consumer packaging, and suggested advertising for retail accounts.

While food retailers have been consolidating, so have other produce buyers, such as broad-line wholesalers who sell to retail buyers. Grocery-oriented wholesalers undertook 32 mergers and acquisitions in 1999 and have undertaken a cumulative total of 105 since 1997. Foodservice wholesalers completed 31 mergers and acquisitions in 1999. Still, foodservice wholesalers remain relatively fragmented. In 1998, the 4 largest foodservice wholesalers accounted for 21% of the $147 billion in total foodservice wholesale industry sales, followed by the top 8 and top 20 firms with shares of 25 and 27% respectively (Calvin and Cook et al. 2001). Ongoing consolidation in the general-line, produce (specialized), and foodservice wholesaling industries will continue to contribute to a more consolidated marketplace, even though consolidation at the wholesale level still lags behind retail.

GROWERS, SHIPPERS, AND NEW ENTRANTS

The number of U.S. farms of all types has been steadily declining for many years, including fruit and vegetable farms. In 1997 there were a total of 53,641 farms producing vegetables for both the fresh and processed markets, compared to 61,924 in 1992; 85,973 farms were producing fruits, berries, and nuts in 1997, down from 89,417 in 1992 (U.S. Bureau of the Census 1992a, 1997a). Despite the decline, farm production of most commodities remains atomized in the sense that producer volumes, although often large in absolute terms, are small relative to the size of the market. However, increasingly, there are larger farmers concentrated in key production areas such as California and Florida that account for a growing share of the total farm value of fruits and vegetables. For example, according to the 1997 *Census of Agriculture* roughly 2,500 vegetable growers in California accounted for almost half of the total value of vegetable production captured in the census. Just over 14,000 California growers contributed 60% of the national value of fruit, berry, and nut production in 1997.

Furthermore, in key production regions such as California and Florida, a few large growers are forward-integrated into the marketing of their own production and the production of other growers—hence their designation as "grower-shippers." These grower-shippers control production, packing, and cooling facilities, and also arrange for both domestic and export sale, transportation, and promotion. Sales at the shipper level are quite concentrated relative to the grower level, but the shipper structure for many crops is still quite fragmented relative to structure at the buying end of the marketing system (although this varies substantially by crop). For example, in 1999 there were approximately 149 California table grape shippers, with none estimated to account for over 6% of total industry sales. In contrast, there were only 25 California tomato shippers, with the top 4 shippers handling 43% of industry sales (Calvin and Cook et al. 2001).

In general, consolidation at the buying end of the food marketing system is driving consolidation at the shipping level as suppliers structure their operations to attain operating scales consistent with the needs of the fewer, larger buyers. Retailers and foodservice users continue to demand more services, including year-round availability of a wide line of consistent quality fruits and vegetables, ripening and other special handling and packaging, assistance in category management, product stickering with PLU codes, and information on product attributes, recipes, and merchandising.

Many grower-shippers have become multiregional, and some have become multicommodity in order to maintain a year-round presence in the marketplace. This enables them to extend shipping seasons and sell products produced in several locations via one centralized marketing organization. For example, lettuce–leafy green and cole crop shippers headquartered in Salinas, California also commonly ship out of the San Joaquin Valley, Imperial Valley, and southwestern Arizona. Also, to achieve year-round volume, key Florida tomato shippers produce in several Florida locations during the winter and produce in the East Coast and California during the summer and fall.

As year-round shipping has become common for many produce firms, so has international off-season sourcing, in particular for California shippers. Shippers usually obtain offshore produce either through joint ventures with foreign producers, as exclusive or preferential importer-marketers, or on a contract basis. Many California grape, stone fruit, and kiwifruit shippers maintain a consistent market presence with buyers by acting as importer-marketers for contraseasonal production of the same produce from Chile. Research done by Alston et al. (1996) indicates that year-round sourcing actually increased demand for California table grapes, most likely because the year-round availability reinforces consumer buying habits.

The rapid growth in multilocation firms has also contributed to the integration of the Mexico-California-Arizona vegetable industries (Cook 1990). Because most vegetable crops are not perennials, the location of production can shift readily, based on the relative costs of production and marketing and the growing season. Despite this flexibility to source elsewhere and the greater market access resulting from NAFTA, most California-Arizona firms still produce the bulk of their fresh vegetables domestically. This is due to infrastructure, technology, and efficiency advantages relative to producing in Mexico. Seasonal climatic considerations generally remain the primary reason for sourcing in Mexico, rather than cost competitiveness; early-season table grapes, asparagus, radishes, and green onions are exceptions.

New entrants to the produce industry have challenged independent shippers and grower cooperatives. Multinational food processors entered the fresh produce market during the 1980s as consumption of canned produce declined. These firms began applying their branded marketing strategies to produce, contracting with producers here and in foreign nations to ensure a year-round market presence for their brands. Several acquired produce wholesalers and shippers to broaden their base of commodities and distribution channels. Three notable multinational players stand out in the produce industry: Dole, Del Monte, and Chiquita. Still, many multinationals have failed in the risky fresh produce arena, so far precluding a transformation of the fresh sector of the fruit and vegetable industry to the dominant multinational structure now observed in the processed sector.

FOODSERVICE

The growth in fresh produce items handled on foodservice menus has affected distribution channels. In the 1980s, many fast-food outlets added salad bars, and in the 1990s they added fresh-cut salads as well as other menu items that include produce. In 1987, McDonald's already reportedly used 2% of the total U.S. lettuce crop and 1% of the fresh tomato crop. Simultaneously, upscale "white tablecloth" restaurants were expanding the demand for premium quality and exotic produce.

Prior to the 1980s, broadline institutional wholesalers (foodservice distributors) supplying the foodservice industry with dry and packaged groceries did not handle produce; foodservice users procured their produce largely through wholesalers specializing in produce. The rising volume of produce handled by foodservice establishments presented an opportunity for broadline institutional wholesalers, and virtually all the leading distributors formed entire divisions to procure and merchandise produce. Sysco is North America's largest foodservice distributor, with fiscal year 2000 sales of $19.3 billion, 104 distribution centers, servicing over 356,000 commercial and noncommercial foodservice establishments. Promotional programs to further stimulate fresh produce sales through this type of entity obviously hold great potential.

These changes in produce buying have enabled a growing portion of foodservice produce to be procured directly from the shipping point. Some foodservice distributors invested in shipping-point firms or formed joint buying groups in production regions (e.g., Markon and Pro-Act). This has increased their negotiating power and enabled them to exert better control over product quality, packaging, and consistency of supply, no longer allowing retailers to freely dictate standards in the produce trade.

Many shippers have introduced special foodservice packs (smaller than retail packs). The development of the fresh-cut sector during the 1980s was aimed almost entirely at the foodservice market, where convenience of preparation was already recognized as an asset. Products such as cored or chopped lettuce, peeled garlic, and broccoli florets were designed to cut waste and labor at the operator level. Limited availability of labor and high worker turnover are major problems for foodservice operators. Furthermore, liability costs for restaurants and institutions are rising due to employee accidents from handling knives, and the cost of kitchen space is increasing in urban areas. Consequently, foodservice demand for fresh-cut produce is expected to continue to expand.

The efforts of shippers to meet the special needs of foodservice users are complicated by the fragmented nature of the foodservice industry. In 1997 there were 385,400 restaurant establishments, with the top 100 chains accounting for half of sales, smaller chains and independents contributing 45.5%, and the second 100 chains accounting for the remainder (The Food Institute 1999a). There were a total of 201,520 quickservice ("fast food") restaurants and 183,880 full service restaurants, with the latter accounting for 51% of total restaurant sales on 48% of total units. Noncommercial units totaled 111,000 in 1999, with educational establishments the leading contributor to noncommercial sales, accounting for 38%, followed by extended care facilities at 9% (ERS estimate). The commercial sector accounts for 81% of the dollar volume of food and drink sold through foodservice channels, with the noncommercial sector accounting for the remainder (ERS estimate). The noncommercial sector accounts for a higher percentage of purchases than sales since commercial sales include higher profit margins. Some estimates indicate that the noncommercial sector accounts for about two-thirds of food-service purchases.

CHANGING PROCUREMENT PRACTICES AND BARGAINING POWER

Increasingly, buyers are contracting with grower-shippers for high-volume perishable items in order to stabilize prices, qualities, and volumes. While contracts have been common in the foodservice sector, they are new to retail. In 1999, 49% of retailers surveyed in *Fresh Track 1999* reported that they used contracts for 11 to 25% of their purchases, while 16% of retailers reported that over 25% of their purchases were under contract (McLaughlin et al. 1999). Both of these rates are up from 8.5% and 2.1%, respectively, in 1994. The heavier users of contracts are the very largest firms (those with over $1.5 billion in annual sales), with 35% of this group purchasing over 25% of volume under contract, up from 0% as recently as 1994.

The introduction of contracting is likely to have structural implications at the grower-shipper level since shippers need to have sufficient scale to offer large, consistent, year-round volumes to meet buyer contracting requisites. To finance production in numerous production regions and manage complicated distribution logistics, shippers must meet "the test of capital," an especially formidable challenge for family-controlled firms (Wilson, Thompson, and Cook 1997).

The evolution of the produce industry has improved efficiency by cutting marketing costs and enhanced the communication of consumer demand back to growers. However, the consolidation of purchasing in the hands of a few large buyers raises concerns about oligopsony exploitation of producers. As noted earlier, perishable crops, which must be harvested, sold, and marketed within a very short time, tend to give growers relatively little bargaining power in dealings with buyers. Sexton and Zhang (1996) analyzed this issue in the California lettuce industry and found that buyers were able to reduce growers' profit to essentially zero.

In a recent study most shippers and retailers reported that the incidence and magnitude of fees (such as volume discounts and rebates) and services (such as third-party food safety certification and special packaging requests) associated with transactions had increased

over the last 5 years (Calvin and Cook et al. 2001). Data were collected from commodity shippers on actual fees paid to the top five retailer and mass merchandiser accounts. They were usually around 1 to 2% of sales for most commodities. Bagged salad firms reported that fees ranged from 1 to 8% for all retail accounts. Fees paid to all retailer and mass merchandiser accounts averaged $5,200 and $8,700 per million dollars of sales for the interviewed grape and orange shippers, respectively, compared with $10,100 for the grapefruit shippers and only $1,300 for California tomato shippers. Services per million dollars of sales were less than fees for all the commodity samples, averaging from $1,200 for grapes to $4,400 for grapefruit. However, many firms did not keep close track of the cost of fees and, in particular, services. Hence, these data likely underestimate total costs.

Nevertheless, this research indicates that the increasing fees and services requested by retailers of shippers are potentially sufficient to make the difference between profit and loss, given the thin margins typically prevailing at the shipper level. This is especially true for commodity (as opposed to value-added) shippers, who act as price takers and are less able to pass costs along to customers.

VALUE-ADDED PRODUCTS, BRANDING, POSTHARVEST HANDLING, AND SPECIALTY PRODUCE

CONVENIENCE-ORIENTED FRESH-CUT PRODUCE

Prior to the early to mid-1990s, the vast majority of fresh-cut produce was sold in foodservice channels. Then, growing consumer demand for healthful and convenient food began to merge with advances in postharvest technology and handling that improved the quality, presentation, and shelf life of fresh-cut produce at the retail outlet. Bagged salads, broccoli and cauliflower florets, sliced mushrooms, cored pineapples, fresh-cut melons, stir-fry vegetable mixes, packaged baby carrots, carrot and celery sticks, and precut vegetables with cheese sauces in microwaveable trays are all examples of the attempt to add value to produce without losing its fresh, natural image.

According to *Fresh Trends 1998* (The Packer 1998), 84% of consumers had purchased precut vegetables at least once in the prior 6 months, with 76% purchasing bagged salads and 42% purchasing fruit (The Packer 2000). In 1999, 93% of consumers reported having purchased either precut fresh produce in a bag or whole items in a bag or other container in the prior 6 months (ERS 1999).

The exact size of the rapidly growing U.S. fresh-cut industry is unknown. However, sales of precut vegetables at retail were $1.4 billion in 1997, according to Information Resources, Inc. (IRI) scanner data, and are now estimated to reach $2 billion, while Nielsen reported that bagged salads topped $1.73 billion in supermarket sales in 2000. Although retail fresh-cut annual growth rates have been slowing dramatically relative to 3 years ago, they are still impressive in the context of a mature food market. For example, the retail bagged salad category grew at an average annual rate of 61.5% from 1993 to 1996 in value, compared to 12% between 1998 and 1999 (IRI data). While fresh-cut fruit lags fresh-cut vegetables and bagged salads due to greater postharvest technology challenges, fresh-cut fruit sales via supermarkets were estimated to represent 3.1% of total retail produce sales in 1999 (Progressive Grocer 2000), or approximately $1.26 billion. Industry experts estimate that sales of all types of fresh-cut produce through foodservice channels are at least equal to sales via retail channels, although foodservice is now growing at a lower rate, estimated at 3 to 5% per year. The total estimated size of the U.S. fresh-cut industry in 1999 was $9 to $12 billion (IRI data).

The fresh-cut vegetable and salad industry has consolidated in response to the slowing growth rates, with many local and regional players being acquired by larger firms and marginal players being squeezed out of the business entirely. In 1999 three California-based firms controlled 86% of total bagged salad sales through mainstream supermarkets (IRI). The number of competitors outside the top five firms selling bagged salads to retailers shrank from 58 to 48 between 1994 and 1999 (Calvin and Cook et al. 2001). Some of the remaining processors who were unable to compete with the market leaders have shifted production away from branded products to private label (store label) or foodservice. Private label grew from 2.4% of national supermarket bagged salad sales in 1994 to 10% in 2000 (IRI data). Private label sales enable processors to utilize plant capacity without incurring the marketing costs associated with supporting brands, including slotting fees paid to retailers to secure shelf space (Calvin and Cook et al. 2001).

In some instances, major California processors have developed joint ventures with regional processors to expand distribution of their brands into new geographic markets. Throughout the 1990s a major industry debate existed over whether it was preferable to process at the shipping point, where product freshness is at its maximum level, or at the destination, where product reworking can occur. Net shipping costs are also lower if processing is done in the production point, since the finished product rather than the raw product is shipped. Both require optimal temperature management throughout the distribution system to maximize marketable yield.

It now appears evident that regional processing plants will play an important role due to their proximity to market and the demand for just-in-time deliveries. Local processors will continue to have a niche in supplying the more perishable fresh-cut products, such as chopped tomatoes and diced fruit, where proximity to market is a strategic advantage. Fruits are often still processed at the store level, and this practice is likely to decline as food safety regulations become more stringent, especially if HACCP programs were to be required of retailers. This would likely strengthen demand for the services of local and regional processors.

In short, better film technology and store-level temperature management have helped the fresh-cut industry to overcome its initial growing pains. Although it is clear that an increasing number of consumers and foodservice users are willing to pay for convenience-oriented produce, further consolidation among processors is expected as the industry matures.

CONCEPT OF PRODUCE BRANDS
As vertical and horizontal integration increase in the fresh produce industry, investment in value-added products is stimulating new marketing and distribution strategies. For example, produce marketers are working with biotechnology firms to develop convenience-oriented products with unique

flavor attributes. Nonbiotech firms are exploring the link between improved, proprietary varieties and branding. However, while several firms have launched proprietary varieties, few have succeeded in either consistent sourcing of these products (both quality and volumes), or in protecting the integrity of their branded varieties from the intrusions of competitors with "me-too" products. Still, 19% of retail fresh produce sales were branded in 1997, up substantially from 7% in 1987 (Kaufman et al. 2000). Also, while produce is still sold predominantly in bulk form, packaged sales (including branded products) are growing, accounting for 41% of produce department sales in 1997 (Kaufman et al. 2000).

In general, successful produce brands have been limited because of the need for year-round availability; a consistent, high-quality supply; a differentiated product; and proper handling throughout the cold chain. Indeed, despite the intensified efforts to market branded produce during the 1980s, *Fresh Trends 1990* reported that most consumers still viewed branded produce as about the same quality as nonbranded produce (The Packer 1990). Furthermore, branding ranked last among numerous factors that influence produce purchases (see table 2.2).

The *Fresh Trends 2000* survey indicates that little had changed after several more years of branded marketing by more firms of more products. Only 28% of respondents stated that branded produce had better appearance, and only 15% ranked it as having better overall value than nonbranded produce. Just over two-thirds of respondents felt that the appearance and overall value of branded produce was about the same as nonbranded (The Packer 2000).

As recently as the *Fresh Trends 1998* survey, 60% of respondents said they do not seek out branded produce when they shop (The Packer 1998). Of those seeking a brand, consumers are much more likely to seek out branded fruit than vegetables. In 2000, bananas continued to lead the way among consumers seeking out a fruit brand, mentioned by 22%, followed by oranges at 5% and pineapples at 4%. To date, most brand recognition is for fruit items with few value-added attributes, with the exception of pineapple, which is sold in both bulk and fresh-cut form. On the vegetable side, the principal brand recognition was for lettuce and bagged salads, with 3% each of consumers expressing a preference for either a lettuce or bagged salad brand. This low recognition for bagged salads is despite rapid growth in branded bagged salad sales, highlighting the difficulty of obtaining true consumer franchises for fresh produce, even when sold in value-added fresh-cut form.

According to *Fresh Trends 2000*, the specific brands with the greatest recognition were Dole, with 21% of consumers saying they sought this label; Chiquita, at 12%; Sunkist, at 6%; and Del Monte, at 4%.

The many obstacles to developing widely recognized consumer brands for fresh produce has meant that fresh produce has been undermerchandised and underpromoted relative to packaged food products. This gap has traditionally been somewhat mitigated through generic promotion programs paid for by commodity growers and shippers via mandated marketing programs. However, this type of program has recently become more controversial, raising questions about the future role of generic promotion, in particular in light of the continued growth in attempts at branding and product differentiation on the part of shippers in many commodity subsectors. Recent research at UC Davis documented the net benefits to producers who pay for generic advertising, based on results from numerous studies of a variety of commodities (Crespi 2000). However, it also showed that although not necessarily the case, it is possible for an increase in generic advertising to differentially affect the profits of competing firms, such that a firm selling higher-quality goods would prefer less generic advertising than a firm selling lower-quality goods. Another challenge to the future of generic promotion is the increasing cost, as retailers now tend to require more funds than in the past in exchange for participating in joint promotion programs.

While generic promotion programs have played an important role in promoting unbranded produce, successful brands have the potential to stimulate even greater consumption of produce. Brands tend to bring additional promotional dollars, injecting a positive advertising and merchandising jolt to their categories, independent of their success in achieving consumer recognition of a specific brand.

For example, banana consumption appears to have benefited from the presence of several strong brands, with U.S. per capita consumption growing from 8.8 kg (19.4 lb) in 1976 to 14.2 kg (31.3 lb) in 1999. Brands may be even more effective when combined with other value-added features, such as washed baby carrots, carrots in snack packs with dips, or carrots in resealable packaging. The introduction of new presentations of branded carrots may have played a positive role in stimulating carrot demand, with fresh per capita consumption up from 3.5 kg (7.7 lb) in 1991 to 5 kg (11.3 lb) in 2000.

Whether brands can profitably develop a product category is another matter. The interaction between produce brands and slotting fees varies by product form and may potentially influence the profitability of branded strategies. Branding of produce commodities as opposed to fresh-cut produce does not appear to have led to significant use of slotting fees. In contrast, branded fresh-cut produce is treated more like a manufactured food product (where slotting fees are common) because it is consistently available year-round with standardized quality and requires dedicated shelf space. This similarity has likely contributed to the growth of slotting fees in the fresh-cut category. Market share battles between fresh-cut processors have also likely contributed to slotting fees, as the industry structure is quite concentrated for those processors who focus on retail sales, and rivalry between firms is high. Slotting fees are sometimes offered as a tool for capturing market share from competitors. Yet the inability of processors to achieve strong consumer brand recognition arguably means that they are still subject in part to commodity market dynamics. This may make them vulnerable to greater marketing costs without reaping all of the rewards normally associated with branded food marketing.

POSTHARVEST HANDLING

The importance of proper postharvest handling and temperature management in stimulating sales is highlighted by the results of two *Fresh Trends* surveys, the first conducted in 1990 and the second a decade later in 2000 (published in 2001) (table 2.2). In both years the top three factors ranked by consumers as most influencing their buying decisions were taste or flavor, ripeness, and appearance. Consumers clearly base buying decisions on what looks good and appears likely to taste good. Least influential are the geographic origin of the product and whether it was organically grown. Indeed, these are ranked as even less important today than in the past, with price and nutritional value also becoming less important to consumers. Declining consumer concern about nutrition has been documented in many surveys pertaining to food products in general, likely because today many consumers feel that they are making healthier food choices, making this issue of less concern. Appearance became less important to consumers in 2000 than in 1990 but more important than ripeness, moving up to the number two ranking.

The importance of in-store merchandising, appearance, and freshness is underscored by the extent of impulse purchases of produce. Only 32% of consumers report shopping with a written or mental list of the produce items they plan to buy, while 39% decide in the produce department and 29% just know the general category of produce they plan to buy. Shelf positioning and merchandising are critical given the high level of product diversity, and optimal postharvest handling is equally critical to maximizing sales potential.

The rise in product diversity has also greatly increased the volume of mixed (consolidated) load shipments from production regions. Mixed loads, due to temperature and ethylene incompatibilities, create notable postharvest handling challenges. There is an ongoing need for training and education in appropriate handling methods, and there is also a growing market demand for innovative handling technologies, such as pallet-level modified atmospheres.

SPECIALTY PRODUCE

During the last 15 years, a niche market has rapidly developed for unusual or exotic produce. Larger ethnic populations and the growth in their cultural expression have augmented the demand for product diversity as these consumers seek out traditional foods. Furthermore, a broader portion of the population is consuming foods once considered ethnic or regional. About 75% of ethnic food sales are estimated to be destined for

Table 2.2. Factors indicated by consumers as influencing produce purchases, 1990 and 2000

Factor	Rating of extremely or very important (%)	
	1990	2000
Taste or flavor	96	87
Ripeness	96	70
Appearance or condition	94	83
Nutritional value	65	57
Price	63	47
In-season	38	41
Growing region, state, or country of origin	17	14
Organically grown	17	12
Brand name	9	NA*

Source: The Packer 1990, 2001.

Note: *NA = not available

mainstream consumers or ethnic consumers outside the original audience (Produce Marketing Association 2000). Ethnic food sales are expected to reach $383 million in 2001, up from $272 million in 1996, and about 15% of the growth in food sales over the next 10 years is forecast to come from ethnic foods (Produce Marketing Association 2000). Italian food is the most frequently consumed ethnic food in the United States, followed by Mexican.

Another expanding segment of specialty produce is varieties of traditional items grown primarily for their eating characteristics (superior taste) rather than for yield or shipping attributes. Common examples are Blenheim apricots, special varieties of vine-ripened tomatoes, tree-ripened peaches (including white-fleshed cultivars), donut peaches, Pink Lady and other specialty apples, and super-sweet white and yellow sweet corns. Many of these, available years ago, are marketed as "heirloom" varieties. "Boutique growers," farmers who target restaurant chefs and upscale consumers that are willing to pay a premium, produce them. Indeed, specialty products are generally introduced to the American palate first through upscale and ethnic restaurants and farmers' markets, and then through exotic produce sections in supermarkets. Most successful items eventually are included in conventional produce displays.

Information on specialty produce volume is available primarily for vegetables and herbs rather than specialty fruit. Shipments of specialty fresh vegetables reached 1,247.4

million kg (2,750 million lb) in 2000, up from 258.9 million kg (570.8 million lb) in 1984 (ERS 2001). Despite the rapid growth, specialty vegetables still represent only a 6% share of U.S. vegetable shipments, which totaled 21,708.6 million kg (47,858.5 million lb) in 2000. Included in the specialty shipments are less-than-exotic items, such as romaine lettuce, representing 34% of total shipments. Next in importance are tropical vegetables, followed by other specialty lettuces, and then chile peppers. Forty-one percent of 1999 fresh specialty vegetable shipments were imported (ERS 2000c), compared with 14% of all fresh vegetable consumption (ERS 2000d). California is the largest producer of specialties, in 1999 harvesting 158,257 ha (391,054 acres) of specialty and minor vegetables for both the fresh and processed markets, yielding total production of 2,582 million kg (5,692 million lb) valued at $1.9 billion (ERS 2001).

Another type of specialty produce experiencing rapid growth is organically grown fruits and vegetables. In 1999 organic fresh produce was estimated by Progressive Grocer (2000) to account for 1.5% of supermarket produce department sales, approximately equivalent to $609 billion, while *Fresh Track 1999* estimated organic fresh fruit and vegetable sales at 1.7% of produce department sales. Hence, despite the rapid growth of the organic fresh produce industry, it is still a niche market and therefore easily saturated. Still, the quality and availability of organic produce is improving, which should continue to stimulate distribution. Organic products that tend to be most successful are those that are not significantly more expensive than conventional produce, have similar appearance, and are consistently available, such as organic bagged salads and carrots.

Of the shoppers surveyed in the *Fresh Trends 2000* study, 35 and 82%, respectively, said that they bought organically grown fruits and vegetables in the prior 6 months, and satisfaction with product quality and value was high. On the other hand, in 2000 only 12% of consumers ranked as important whether produce was organically grown, down from 17% in 1990 (table 2.2), helping to explain the small organic sales relative to total produce sales. Still, today there appears to be a consumer segment more loyal to organics, with a higher purchase frequency than a decade ago.

The Hartman Group (2000) estimates that 18% of U.S. consumers are strongly interested in buying organic produce, meaning that they are interested enough to be willing to pay a price premium, accept lower quality, or seek it out in less convenient outlets. There is a larger segment of consumers, 28%, that is generally interested in organic produce but tends to purchase only sporadically when organic produce is conveniently available at a price and quality similar to conventional. The remaining 54% of the population is either ambivalent or uninterested. To grow the organic industry into more than a niche market it will be necessary to motivate the generally interested consumers into becoming more frequent users.

The total 2000 U.S. organic foods industry was estimated at $7.8 billion, up from $1 billion in 1990 (Organic Trade Assn.), equivalent to 1.4% of retail food sales that year. The Henry A. Wallace Institute estimated the larger natural foods market, of which organic foods are a part, at $11 billion in 2000. Organic foods are still sold predominantly through retail channels, with foodservice sales generally limited to a segment of very exclusive restaurants. In California, which is the leading producer of organic fresh fruits and vegetables, the 1998 farm gate value of organically grown vegetables was $86.1 million, while fruits and nuts contributed $48.1 million (Klonsky et al. 2001), with the combined sales equivalent to about 0.5% of California's agricultural output.

As of February 20, 2001, the first-ever National Organic Program (NOP) went into effect, with full implementation of the rules expected at the end of 18 months. The NOP will standardize production, handling, labeling, certification, and other requirements for organic foods and is expected to stimulate the development of the organic market by increasing consumer and trade confidence in organic foods.

CONCLUSIONS

Per capita consumption of fresh produce expanded over the last 25 years, even as the U.S. food market matured. Still, fresh produce firms face numerous challenges as they attempt to stimulate greater fresh produce consumption, given the array of food alternatives available to increasingly time-pressed consumers. However, the emergence of the fresh-cut industry, the still-rising consumer awareness of the health benefits of fresh produce, and continued improvements in postharvest handling and transportation technologies should further improve the distribution system for highly perishable fruit and vegetable commodities, potentially stimulating demand. Certainly, demand for even better performance will increase as product diversity grows, postharvest fungicides become less available, and world trade expands. Successful produce marketing firms will become more market-driven, identifying and meeting the specific needs of each market segment for quality, packaging, product form, merchandising, and information. The most proactive firms will go even further, becoming account driven and acting as partners helping to meet the needs of individual accounts. This is part of a supply chain management approach, emphasizing faster delivery, more accurate temperature management, improved packaging technologies, and creative merchandising, all based on better demand information.

REFERENCES

Alston, J. M., J. A. Chalfant, J. E. Christian, E. Meng, and N. E. Piggott. 1996. The California Table Grape Commission's promotion program: An evaluation. Davis: Univ. Calif. Davis Dept. of Agricultural and Resource Econ. 121 pp.

Calvin, L., and R. Cook (coordinators), with M. Denbaly, C. Dimitri, L. Glaser, C. Handy, M. Jekanowski, P. Kaufman, B. Krissoff, G. Thompson, and S. Thornsbury. 2001. U.S. fresh fruit and vegetable marketing: Emerging trade practices, trends and tssues. U.S. Department of Agriculture, Economic Research Service, Agricultural Economic Report No. 795. 52 pp.

Cook, R. L. 1990. Evolving vegetable trading relationships. J. Food Distrib. Res. 21(1): 31–46.

Crespi, J. 2000. Generic commodity promotion and product differentiation. Unpub. PhD diss., Univ. Calif. Davis Dept. Agricultural and Resource Econ. 128 pp.

The Food Institute. 1999a. The food industry review. Fair Lawn, NJ: The Food Institute.

———. 1999b. The food institute report. Fair Lawn, NJ: The Food Institute. May 1.

———. 1999c. The food institute report. Fair Lawn, NJ: The Food Institute. May 24.

———. 1999d. The food institute report. Fair Lawn, NJ: The Food Institute. June 7.

———. 2000a. Demographics of consumer food spending 2000. Fair Lawn, NJ: The Food Institute. 62 pp.

———. 2000b. The food institute report. Fair Lawn, NJ: The Food Institute. January.

———. 2000c. The food institute report. Fair Lawn, NJ: The Food Institute. February 7.

———. 2000d. The food institute report. Fair Lawn, NJ: The Food Institute. May 1.

———. 2000e. The food institute report. Fair Lawn, NJ: The Food Institute. December 25.

———. 2001. The food institute report. Fair Lawn, NJ: The Food Institute. July 30.

Food Marketing Institute. 1999–2000. Trends: Consumer attitudes and the supermarket. Washington, D.C.: Food Marketing Institute.

Hartman Group. 2000. Organic lifestyle shopper study: Understanding key factors of brand success for organic foods and beverages. August. Available via Internet at www.Hartman-group.com

Kaufman, P., C. Handy, E. McLaughlin, K. Park, G. Green. 2000. Understanding the dynamics of produce markets: Consumption and consolidation grow. U.S. Department of Agriculture, Economic Research Service, Agricultural Information Bulletin 758. 17 pp.

Klonsky, K., R. Kosloff, L. Torte, B. Shouse. 2001. Statistical review of California's organic agriculture, 1995–1998. Davis: Univ. Calif. Ag. Issues Ctr.

Litwak, D. 1998. Is bigger better? Supermarket Business 53(10) (October).

Lucier, G., S. Pollack, and A. Perez. 1997. Import penetration in the U.S. fruit and vegetable industry. U.S. Department of Agriculture, Economic Research Service, Vegetables and Specialties Situation and Outlook Report VGS-273. 53 pp.

Mayer, S. D. 1988. U.S. foodservice industry: Responsive and growing. In Marketing U.S. agriculture: 1988 yearbook of agriculture. Washington, D.C.: U.S. Department of Agriculture. 86–90.

McLaughlin, E. W. 1983. Buying and selling practices in the fresh fruit and vegetable industry: Implications for vertical coordination. Unpub. PhD diss., Michigan State University, Department of Agricultural Economics. 484 pp.

McLaughlin, E. W., K. Park, and D. Perosio. 1997. Fresh track 1997: Marketing and performance benchmarks for the fresh produce industry. Newark, DE: Produce Marketing Association (PMA). 125 pp.

McLaughlin, E. W., K. Park, D. Perosio, and G. Green. 1999. Fresh track 1999: New dynamics of produce buying and selling. Newark, DE: Produce Marketing Association and Food Industry Management. 67 pp.

Organic Trade Association. 2001. Consumer facts and market information. Available via Internet at www.ota.com

Produce Marketing Association. 2000. Fresh specialty produce trends, 2000. Available via Internet at http://www.pma.com

Produce Merchandising. 2001. Benchmark, quarterly sales review. January.

Progressive Grocer. 2000. 2000 Produce annual report. 79(10) (October).

Progressive Grocer Annual Report. 1954. Annual report of the grocery industry. Supplement to Progressive Grocer. April.

———. 1974. Annual report of the grocery industry. Supplement to Progressive Grocer. April.

———. 1994. Annual report of the grocery industry. Supplement to Progressive Grocer. April.

———. 2000. Annual report of the grocery industry. Supplement to Progressive Grocer. April.

Sexton, R. J., and M. Zhang. 1996. A model of price determination for fresh produce with application to California iceberg lettuce. Amer. J. Agric. Econ. 78:924–934.

Supermarket Business. 1999. 12th annual produce operations review. Supermarket Business 54(10) (October).

The Packer. 1990. Fresh trends '90: A profile of the fresh produce consumer. Reports 1–4.

———. 1998. Fresh trends '98: A profile of the fresh produce consumer. Lincolnshire, IL: Vance Publishing. 89 pp.

———. 1999. Fresh trends 1999: Detailed demographic tabulations for purchase preferences and influences. Lincolnshire, IL: 178 pp.

———. 2000. Fresh trends 2000: Detailed demographic tabulations for purchase preferences and influences. 287 pp.

———. 2001. Fresh trends 2001: A profile of the fresh produce consumer. 70 pp.

U.S. Bureau of the Census. 1980. Population statistics. Washington, D.C.: Government Printing Office.

———. 1992a. Census of agriculture. Washington, D.C.: U.S. Government Printing Office.

———. 1992b. Census of retail trade. Washington, D.C.: Government Printing Office.

———. 1995–1998. Population Statistics. Washington, D.C.: Government Printing Office.

———. 1997a. Census of agriculture. Washington, D.C.: Government Printing Office.

———. 1997b. Census of retail trade. Washington, D.C.: Government Printing Office.

U.S. Department of Agriculture, Economics Research Service (ERS). 1977. Vegetables and specialties situation and outlook report. Washington, D.C.: Government Printing Office. October.

———. 1999. Annual Spotlight on the U.S. Food System. Food Review 22(3): 42 (September-December).

———. 2000a. Fruit and tree nuts situation and outlook yearbook. Washington, D.C.: Government Printing Office. October.

———. 2000b. Vegetables and specialties situation and outlook report. Washington, D.C.: Government Printing Office. April.

———. 2000c. Vegetables and specialties situation and outlook yearbook. Washington, D.C.: Government Printing Office. July.

———. 2000d. Vegetables and specialties situation and outlook report. Washington, D.C.: Government Printing Office. November.

———. 2001. Vegetables and specialties situation and outlook yearbook. Washington, D.C.: Government Printing Office. July.

U.S. Department of Agriculture, Foreign Agricultural Service (FAS). 1999. U.S. fruit and vegetable imports, calendar year 1998. Horticultural and Tropical Products Division Report (March, posted in April). Available via Internet at http://www.fas.usda.gov/htp

———. 2000. U.S. horticultural exports increase slightly in fiscal year 2000. Washington, D.C.: Government Printing Office. December.

U.S. Department of Agriculture, National Agricultural Statistics Service (NASS). 2000a. Citrus fruits, 2000 summary. Washington, D.C.: Government Printing Office. September.

———. 2000b. Potatoes, 1999 summary. Washington, D.C.: Government Printing Office. September.

———. 2000c. Vegetables, 2000 summary. Washington, D.C.: Government Printing Office. January.

Wilson, P., G. Thompson, and R. Cook. 1997. Mother nature, business strategy, and fresh produce. Choices, First Quarter, 18–25.

3

Consumer Issues in Quality and Safety

Christine M. Bruhn

Quality produce is very important to consumers and retailers, and it is a key factor consumers use in evaluating a supermarket. High-quality produce is second only to a clean, neat store as top factors in selecting a supermarket. High-quality fruits and vegetables are rated very important by all income and geographic groups, but they are especially valued among the highest income households, where 97% rated it very important in 2000. From 1992 to 2000, 99% of consumers rated high-quality produce as very or somewhat important in supermarket selection (Abt Associates 1997; Research International 2000).

COMPONENTS OF QUALITY

Although eating quality is a combination of characteristics, attributes, and properties that lead to enjoyment, consumers say that appearance and freshness are most important in initial purchase. They select products that are the appropriate color, size, and shape, with the proper firmness. Expectation of nutritional value and health-enhancing properties are also of importance. The Food Marketing Institute's annual survey of 1,000 households consistently indicates that good taste is the most important factor influencing purchase, followed by nutritional value, safety, and price. Consumers also expect food in the supermarket to be safe.

Appropriate color, shape, and size are important quality criteria. Although color varies by produce and variety, red blush is preferred in some products, like peaches and nectarines. A characteristic odor is desirable, as it indicates ripeness and reflects eating quality. Generally, larger-sized products are priced at a premium; however, some people prefer medium or smaller sizes depending on intended use. Scars, scratches, and other marks lower quality rating, but some consumers will purchase lower grades if the price is sufficiently low and other factors indicate good eating quality.

Attitude studies indicate that consumers tend to prefer locally grown produce, both due to perceptions of higher quality and to support the local economy. Many people don't know what produce is grown locally, however.

At this time, branding does not appear to be a major factor related to consumer perceptions of quality. Almost 90% of consumers believe branded and nonbranded items are about the same in nutritional value, and about 80% consider them comparable in storage life and taste. In regards to safety, 75% of consumers consider branded and nonbranded items comparable (The Packer 2000).

When asked how produce can be promoted to encourage purchase, consumers suggest that quality be the focus, with spoiled produce kept out of the display, products held at proper temperature, and tasting offered so consumers can verify quality. Consumers name several products they would eat more often if quality were higher (table 3.1). Price, however, is the most frequently cited reason for not purchasing a favorite item.

NUTRITION

Produce is viewed as a healthy food choice. In each of the last 10 years, 70% or more consumers responding to the Food Marketing Institute's annual survey indicate that they have increased produce consumption to obtain a healthier diet. Consumers view fruits and

Table 3.1. Percentage of consumers who would buy selected produce if the produce were of higher quality

Fruits (%)	
Peaches	25
Apples	24
Bananas	22
Oranges	21
Grapes	17
Strawberries	15
Vegetables (%)	
Tomatoes	20
Lettuce	17
Broccoli	15
Green beans	13
Cauliflower	10

Source: The Packer 1996.

vegetables as good sources of vitamins, minerals, and fiber; helpful in calorie control; and a possible cancer preventative.

As a result of the U.S. Nutrition Labeling and Education Act of 1990 (NLEA), most supermarkets post nutrition facts on the 20 top-selling fruits and 20 top-selling vegetables. In 1995, 78% of consumers in a national survey indicated that they were aware of in-store nutritional information. Thirty-six percent of women and 19% of men responded that this information was easy to understand, but only 17% indicated that they bought a particular fruit or vegetable *because of* nutrition information (The Packer 1995).

Although consumers consistently cite flavor as the primary force that guides food selection, perception of health does have an influence. In 1997 (Foerster et al. 1998), Californians said they were eating more fruits and vegetables because they were
- "trying to eat healthier," 30%
- "liking the taste," 30%
- "lowering disease risk," 6%
- "weight reduction," 5%
- "availability," 4%

Primary barriers for consumption included:
- "hard to get at work," 60%
- "hard to get at restaurants," 50%
- "not knowing how to fix fruits and vegetables," 33%
- "not in the habit," 29%
- "cost," 29%

NLEA also permits health claims that relate consumption of produce items with specific nutrient profiles and possible prevention of cancer or heart disease. Although several produce items meet the nutrient requirements, at this writing, few claims are posted in the supermarket.

5-A-DAY

The food pyramid, developed by the U.S. Department of Agriculture and released to the public in 1992, provides general guidelines for selecting a healthful diet. Fruits and vegetables are featured as the second-largest segment of the pyramid, with three to five servings of vegetables and two to four servings of fruit recommended daily.

The Produce for Better Health Foundation was established in 1991 to lead a national generic promotion program for fresh produce. The program's goal is to reduce the incidence of cancer and chronic disease by increasing produce consumption from 2.5 servings a day to 5. The 5-A-Day message was presented on national television through a series of public service announcements. Retailers present the message through logos and ads. Restaurants can also participate in the program through promotions, advertising copy posters, brochures, and table tents. Awareness of the 5-A-Day Program peaked in 1998, with 86% of consumers hearing of the program, then fell to 79% in 1999 and 77% in 2000. About 50% of consumers correctly respond that the recommended number of servings of fresh produce is five or more (The Packer 1998, 1999, 2000).

An American Dietetic Association survey indicates that even though 92% of consumers recognize fruits and vegetables as healthful, consumers are not consuming the recommended number of servings (American Dietetic Association 1997). A USDA report (Cleveland et al. 1997) based on 1994 dietary intake noted that Americans eat $3\frac{1}{3}$ servings of vegetables daily. Intake of dark-green leafy vegetables highlighted in the pyramid was low, making up only 3% of vegetable servings, while potatoes made up 33% of servings. More than half, 59%, of consumers failed to meet this recommendation based on their caloric intake. Intake of fruit was even lower, with an average of $1\frac{2}{3}$ servings daily. Only 24% of Americans met the fruit recommendations, and almost half, 48%, did not consume even one serving of fruit daily. The produce industry has tremendous growth potential if consumers follow dietary recommendations.

SAFETY

Most consumers are confident in the safety of the food supply, yet their perception varies over time, likely depending on news coverage. The Food Market Institute annual survey indicated that in 2000, 74% of consumers were completely or mostly confident that food in the supermarket is safe. Confidence peaked at 84% in 1996 and was at 72% in 1992 (Abt Associates 1997; Research International USA 2000).

Concern about pesticide residue is the most frequently volunteered food safety concern associated with produce, followed by mishandling and cleanliness. In the past, consumers have not realized that microbiological hazards could be associated with fresh produce. Modifications by biotechnology and treatment by food irradiation are newer technologies that generate concern among some consumers. When potential food safety problems were specifically identified in a 1997 survey, 66% considered pesticide residues a serious hazard, compared to 82% for bacterial contamination, 33% for food irradiation, 21% for food additives, and 15% for biotechnology (table 3.2).

PESTICIDE RESIDUES

Concern about pesticide residue was highest in 1989 at the time of the controversy over use of the growth regulator Alar on apples. Over time, confidence in the safety of produce and belief in the health-enhancing

value of produce increased due to concerted educational efforts by the produce industry and health professionals. Some supermarkets advertise the use of a certification system to verify that produce meets legal pesticide residue minimums or contains no residues detectable by test sensitivity. Many supermarkets also offer organic produce.

ORGANIC FOOD

The organic market grew from $178 million in 1980 to $2.3 billion in 1994 and 6.7 billion in 2000 (Sloan 1999). In 1995 the organic industry represented 0.5% of U.S. farmers and included over 5,000 certified and up to 6,000 uncertified producers. In 2000, organic produce was estimated to represent approximately 2% of retail produce sales and is projected to increase by 10 to 12% yearly. This growth has been facilitated by improved distribution channels and entree into upscale supermarkets and restaurants.

Consumers cite several reasons for selecting organic produce. About a quarter of consumers said they purchased organic produce because the produce looked good, 17% thought organic produce looked fresher, 16% indicated that they purchased organic produce out of curiosity, 15% were motivated by taste, and 12% purchased because of pesticide concerns. Price is the greatest barrier to purchasing organic produce. In 2000, about half of consumers who purchased organic items indicate that they are extremely or very likely to buy organic items again (The Packer 2000).

Consumers with children under 18 years of age are less likely to buy organic produce than those without young children. Consumers in the West are more likely to buy organic produce than consumers in other regions, and households with incomes of $75,000 or more are most likely to purchase organic produce (The Packer 2000).

Many consumers perceive organic production to be a pesticide-free production method. The Organic Foods Production Act of 1990 directed the U.S. Department of Agriculture to implement federal rules covering this method of farming. The advisory body, the National Organic Standards Board, developed a set of recommendations to be used as a basis of USDA regulation. The National Organic Standards Board clearly indicates that "organic" is a not a pesticide-free claim, but rather a system of managing crops and live-

Table 3.2. Consumer perception of potential health risks, 1997

Question: "I am going to read a list of food items that may or may not constitute a health risk. For each one, please tell me if you believe it presents a serious health risk, somewhat of a health risk, a slight health risk, or no health risk at all?"

	Percentage*				
Health risk	Serious	Somewhat	Slight	No hazard	Not sure
Contaminated by bacteria	82	13	5	1	1
Residues such as pesticides	66	24	8	2	2
Food handling in supermarket	45	36	15	3	1
Irradiation	33	23	13	8	24
Additives and preservatives	21	50	19	7	2
Food produced by biotechnology	15	31	16	10	28

Source: Abt Associates 1997.

*Note:** May not add up to 100% due to rounding.

stock that emphasizes natural feeds, medications, pest control methods, and soil inputs. Care must be taken to accurately position the organic approach of production.

MICROBIOLOGICAL SAFETY

Consumers volunteer that the greatest threat to food safety is microbiological (table 3.3). Outbreaks of foodborne illness have occurred in which fresh produce was identified as the source of the pathogen. Cantaloupes were the source of salmonella, hepatitis A contaminated frozen strawberries, and lettuce, sprouts, fresh apple juice, and fresh basil were implicated in E. coli 0157:H7 outbreaks. Consumers responded by avoiding the implicated product. In 1998, about 60% of consumers indicated that they were more concerned about bacterial contamination of fresh produce than in the previous year (The Packer 1998). While consumers believe produce grown in the United States is safer than imported produce, the U.S. Economic Research Service notes that outbreaks occur from domestic as well as imported produce. Care must be taken in production and processing to avoid or destroy potential pathogens.

WAXING

Waxing is a latent concern for many consumers. A national survey (The Packer 1995) indicates that only 35% of women and 43% of men say that they definitely would eat fresh produce knowing that it is coated with an approved food-grade wax. Consumer attitudes center around the preference for natural, unmodified produce, doubt as to the long-term safety of ingesting wax, and the perception that the wax doesn't taste good.

This is a paradox for the industry, since waxing extends shelf life and improves the flavor, freshness, and appearance of produce. The FDA mandatory wax labeling laws, effective in August 1994, require all shippers, packers, retailers, convenience stores, roadside stands, and other outlets to label for wax coatings.

BIOTECHNOLOGY

Biotechnology or genetic modification can be used to create new varieties that have better quality or that can be grown with reduced herbicide or pesticide use. Although few consumers consider themselves knowledgeable about biotechnology, awareness has increased due to news coverage. In May 2000, 43% of consumers knew that there were foods produced through biotechnology in the supermarket, with vegetables mentioned by 45% of these consumers as an example of a modified product (Wirthlin Group 2000). Most consumers have a positive view of biotechnology, with 59% believing the technology will benefit them or their family in the next five years.

Acceptance of the concept of gene transfer is increased when tied to a specific benefits that consumers value. Generally, consumers are most accepting of benefits applied to human medicine, followed by environmental stewardship. In May 2000, 69% of U.S. consumers indicated they would purchase a food modified by biotechnology to reduce pesticide use, and 54% said they would buy a product modified for improved flavor (table 3.4). Approximately 10% fewer people expressed interest in buying foods modified by biotechnology in 2000 compared to the beginning of 1999. This decline could be related to concerns about the environmental impact on nontarget beneficial insects (e.g., butterflies) widely presented in the media. Media coverage of subsequent studies disputing harm as been limited.

Consumers opposed to applications of biotechnology fear that modifications could disrupt nature, leading to unforeseen consequences. Some were concerned that pesticide use would increase or that manufacturers would make inappropriate profits. Others stated that they felt that it is inappropriate for humans to modify nature. Those actively lobbying against genetic engineering (GE) describe the technology as revolutionary, in

Table 3.3. Consumer perception of serious threats to food safety, 1997

Question: "What if anything do you feel are the greatest threats to the safety of the food you eat?"

Food safety concern	Percentage								
	1989	1990	1991	1992	1993	1994	1995	1996	1997
Spoilage or germs	36	29	27	36	46	41	52	49	69
Pesticide residues	16	19	20	18	13	14	15	17	10
Chemicals	11	16	15	13	8	12	11	10	6
Tampering	20	14	8	6	7	4	4	4	1
Preservatives	7	8	7	6	6	7	6	5	>1
Irradiation	1	1	1	1	0	0	<0.5	<0.5	>1
N=	772	1,005	1,004	1,000	1,006	1,008	1,011	1,007	1,011

Source: Abt Associates 1997.

Table 3.4. Consumer likelihood to buy foods modified by biotechnology

Questions:

A. "All things being equal, how likely would you be to buy a variety of produce, like tomatoes or potatoes, if it had been modified by biotechnology to taste better or fresher? Would you be very likely, somewhat likely, not too likely, or not at all likely to buy these items?"

B. "All things being equal, how likely would you be to buy a variety of produce, like tomatoes or potatoes, if it had been modified by biotechnology to be protected from insect damage and require fewer pesticide applications? Would you be very likely, somewhat likely, not too likely, or not at all likely to buy these items?"

Survey date	A. To taste better (%)	B. To reduce pesticides (%)
1997	55	77
1999 (Feb)	62	77
1999 (Oct)	51	67
2000 (May)	54	69

Source: Wirthlin Group 2000.

the early experimental stages of development, and likely to endanger human health and the environment.

Widespread planting of corn modified with Bt approved for animal but not human food (StarLink) led to a recall of numerous corn-based products in the fall of 2000. A consumer survey in October 2000 found that although 53% of consumers had read about the recall, no one volunteered that they had avoided foods with genetically modified ingredients (Grabowski 2000). Consumers remained positive toward applications they considered important, with 67% saying they would purchase produce modified to reduce pesticide use and 66% indicating they would purchase produce modified to contain more vitamins and nutrients. Only 5% of consumers surveyed indicated they had taken actions because of concerns about genetically modified foods.

Acceptance of biotechnology-modified grains and oil crops in the European and Asian markets is controversial at this time, and a comprehensive labeling program has been mandated. Puree labeled as originating from biotechnology-modified tomatoes was well received in the United Kingdom, but the product is no longer in the market due to limited production. Many countries have initiated or are considering mandatory labeling. Labels must be carefully designed to be truthful and not misleading. Consumer research indicates terms like "genetically modified organism" are misunderstood and carry a negative connotation. To increase consumer acceptance the reason for modification should be indicated.

FOOD IRRADIATION

In recent years, interest in food irradiation has centered around destruction of food-borne pathogens in meat and poultry. This food safety application has value for the sprout industry, since irradiation can destroy pathogens even if they are located under the seed coat. The FDA approved irradiation of seeds in October 2000. Food irradiation is also an effective quarantine treatment for some insects, prevents the sprouting of tubers like potatoes and onions, and extends the shelf life of some produce items.

Interest in purchasing foods irradiated for increased safety varies, apparently due to media coverage of the consequences of food-borne disease and the safety of irradiated food. The Food Marketing Institute's annual survey indicates that those expressing concern for this process have decreased from 40 to 42% in the 1980s to 33% in 1997. Over 80% of consumers expressed an interest in purchasing irradiated meat and poultry after the FDA approved the process at the end of 1997. Similarly, over 77% indicated that it was very important to irradiate to kill disease-causing bacteria, 64% to control insect infestation, and 60% to reduce the need for pesticides. Only 40% indicated it was very important to irradiate to obtain longer shelf life for perishables.

Over the years, various surveys have indicated that about 60 to 70% of consumers would purchase irradiated produce, with the percentage increasing to 80% or more after completing an educational program and 90% or more after sampling irradiated produce.

Irradiated tropical and other fruits have sold well in a limited number of markets since 1995. In 1995 tropical fruit from Hawaii was sold at several Midwest markets in collaboration with a study to determine quarantine treatment. Papaya, atemoya, rambutan, lychee, starfruit, banana, Chinese taro, oranges, and other fruits were shipped to an Isomedix gamma ray facility near Chicago for irradiation between 0.25kGy and 1.0kGy. The fruits were well received by consumers, but one retailer withdrew due to threats from an activist organization. Beginning in November 1996, a small quantity of irradiated fruit was also made available in specialty markets in San Francisco and Los Angeles. In subsequent years about 91,000 kg (200,000 lb) per year of fruits from Hawaii were shipped to the mainland, irradiated, and distributed to select markets. In the summer of 2000, an X-ray facility was opened on the big island of Hawaii to process fruit on the island and ship it directly to the mainland.

CONVENIENCE

Women's increased participation in the paid work force has led to a significant change in American lifestyles. In 1950 only 30% of working-age women were employed. In 1994, 59% of adult women and 77% of adult men participated in the labor force (U.S. Department of Labor 1995). Increased demand on personal time has led consumers to seek convenience in meal preparation. Consumers are purchasing an increasing amount of partially or fully prepared items from the supermarket, and they are increasing the number of meals purchased from fast-food or regular restaurants. In 1970 about 33% of the food dollar was spent on food away from home, while in 1998, this amount increased to 47%. People say they eat out primarily because they don't have time to cook, but 18% simply don't want to cook and 5% acknowledge that they don't know how.

In the fresh produce area, the demand for convenience has led to growth in prewashed greens for salads, peeled carrots, and other cut and washed vegetables, some of which may be packaged with sauce or accompaniments. In addition to convenience, consumers perceive fresh-cut produce to be a good value because of price, less wastage, and high quality. In 1998 carrots led the list of frequently purchased convenient vegetables at 69%, followed by broccoli at 44%, cauliflower at 28%, and other vegetables at 33%. Iceberg lettuce or salad mix leads the salad category with 60% of consumers indicating they have purchased this item in the last six months, followed by specialty salad at 33%, coleslaw mix at 25%, romaine lettuce at 24%, and washed spinach at 22%. Of consumers who purchase fresh-cut fruits, 27% have purchased melon, 17% have purchased pineapple, and 19% have purchased other fruits (The Packer 1998).

BUYING DECISIONS

Women most often make produce-buying decisions. In 1996, *The Packer* national survey found that 14% of fresh produce purchasers were men, most of whom are unmarried. Of the women purchasers, more than 8 of 10 are buying for households of at least 2 people. Potential consumer and eating occasions influenced produce choice. Bananas, apples, grapes, and tree fruits led the list of favorite produce children request (table 3.5). Preferred snacks include bananas, apples, and seedless grapes. Bananas and strawberries are preferred fruit for breakfast.

Table 3.5. Produce items requested by children

Produce	Response (%)
Apples	58
Bananas	58
Grapes (seedless)	44
Oranges	31
Plums	25
Peaches	23
Carrots	21
Lettuce	18
Strawberries	17
Tomatoes	15
Broccoli	13
Cucumbers	13
Nectarines	13
Sweet corn	13

Source: The Packer 1995.

REFERENCES

Abt Associates. 1997. Trends in the United States: Consumer attitude and the supermarket, 1997. Washington, D.C.: Food Marketing Institute. 91 pp.

American Dietetic Association Nutrition Trends Survey. 1997. Executive Summary. Chicago: American Dietetic Association.

Bruhn, C. M. 1995. Consumer attitudes and market response to irradiated food. J. Food Protection 58(2): 175–181.

Clauson, A. 1999. Share of food spending for eating out reaches 47%. Food Review 22(3): 20–22.

Cleveland, L. E., A. J. Cook, J. W. Wilson, J. E. Friday, J. W. Ho, and P. S. Chahil. 1997 Pyramid servings data results from USDA's 1994 Continuing Survey of Food Intakes by Individuals. Riverdale, MD: USDA Agricultural Research Service. 23 pp. Available via Internet at http://www.barc.usda.gov/bhnrc/foodsurvey/dhks95.html

Foerster, S. B., J. Gregson, S. Wu, and M. Hudes. 1998. California dietary practices survey; focus on fruits and vegetables trends among adults, 1989–1997. Sacramento: California Department of Health Services Public Health Institute. 41 pp.

Food Chemical News. 1996. Organic standards back-grounder. Food labeling and nutrition news special report. Boca Raton, FL: CRC Press. 21 pp.

Food Marketing Institute. 1998. Consumers' views on food irradiation. Washington, D.C. 46 pp.

Grabowski, G. 2000. FMA survey shows Americans learning more about biotechnology; food consumption patterns unchanged. Press release, Oct. 12. Available via Internet at http://www.gmabrands.com

Hoban, T. J. 1998. International acceptance of agricultural biotechnology. Agricultural Biotechnology and Environmental Quality Gene Escape and Pest Resistance. NABC Report 10, pp. 59–73. Ithaca, NY: National Agricultural Biotechnology Council.

The Packer. 1995. Fresh trends: A profile of fresh produce consumers. Lincolnshire, IL: Vance Publishing. 104 pp.

———. 1996. Fresh Trends: A profile of fresh produce consumers. Lincolnshire, IL: Vance Publishing. 96 pp.

———. 1998. Fresh Trends: A profile of fresh produce consumers. Lincolnshire, IL: Vance Publishing. 88 pp.

———. 1999. Fresh Trends: A profile of fresh produce consumers. Lincolnshire, IL: Vance Publishing. 88 pp.

———. 2000. Fresh Trends: A profile of fresh produce consumers. Lincolnshire, IL: Vance Publishing. 72 pp.

Packwood Research Corporation. 1994. Shopping for health, eating in America: Perception and reality. Washington, D.C., and Emmaus, PA: Food Marketing Institute and Prevention Magazine. 40 pp.

Putnam, J. J. 1994. American eating habits changing: Part 2: Grains, vegetables, fruits, and sugars. Food Review 17(2): 36–47.

Research International USA. 2000. Trends in the United States: Consumer attitudes and the supermarket. Washington, D.C.: Food Marketing Institute. 90 pp.

Sloan, A. E. 1999. Top ten trends to watch and wrok on for the millennium. Food Technol. 53(8): 40–60.

U.S. Department of Labor, Bureau of Labor Statistics. 1995. Employment and Earnings 42(1) (January).

Wirthlin Group. 2000. Consumer survey on biotechnology. International Food Information Council (IFIC). Available via Internet at www:/ificinfo.health.org/foodbiotech/survey.htm

4

Postharvest Biology and Technology: An Overview

Adel A. Kader

Losses in quantity and quality affect horticultural crops between harvest and consumption. The magnitude of postharvest losses in fresh fruits and vegetables is an estimated 5 to 25% in developed countries and 20 to 50% in developing countries, depending upon the commodity, cultivar, and handling conditions. To reduce these losses, producers and handlers must first understand the biological and environmental factors involved in deterioration, and second, use postharvest techniques that delay senescence and maintain the best possible quality. This chapter briefly discusses the first item and introduces the second, which is covered in detail in subsequent chapters.

Fresh fruits, vegetables, and ornamentals are living tissues that are subject to continuous change after harvest. While some changes are desirable, most—from the consumer's standpoint—are not. Postharvest changes in fresh produce cannot be stopped, but they can be slowed within certain limits. Senescence is the final stage in the development of plant organs, during which a series of irreversible events leads to breakdown and death of the plant cells.

Fresh horticultural crops are diverse in morphological structure (roots, stems, leaves, flowers, fruits, and so on), in composition, and in general physiology. Thus, commodity requirements and recommendations for maximum postharvest life vary among the commodities. All fresh horticultural crops are high in water content and are subject to desiccation (wilting, shriveling) and to mechanical injury. They are also susceptible to attack by bacteria and fungi, with pathological breakdown the result.

BIOLOGICAL FACTORS INVOLVED IN DETERIORATION

RESPIRATION

Respiration is the process by which stored organic materials (carbohydrates, proteins, fats) are broken down into simple end products with a release of energy. Oxygen (O_2) is used in this process, and carbon dioxide (CO_2) is produced. The loss of stored food reserves in the commodity during respiration means the hastening of senescence as the reserves that provide energy to maintain the commodity's living status are exhausted; reduced food value (energy value) for the consumer; loss of flavor quality, especially sweetness; and loss of salable dry weight, which is especially important for commodities destined for dehydration. The energy released as heat, known as vital heat, affects postharvest technology considerations, such as estimations of refrigeration and ventilation requirements.

The rate of deterioration (perishability) of harvested commodities is generally proportional to the respiration rate. Horticultural commodities are classified according to their respiration rates in table 4.1. Based on their respiration and ethylene (C_2H_4) production patterns during maturation and ripening, fruits are either climacteric or nonclimacteric (table 4.2). Climacteric fruits show a large increase in CO_2 and C_2H_4 production rates coincident with ripening, while nonclimacteric fruits show no change in their generally low CO_2 and C_2H_4 production rates during ripening.

ETHYLENE PRODUCTION

Ethylene (C_2H_4), the simplest of the organic compounds affecting the physiological processes of plants, is a natural product of plant

Table 4.1. Horticultural commodities classified according to respiration rates

Class	Range at 5°C (41°F) (mg CO_2/kg-hr)*	Commodities
Very low	<5	Dates, dried fruits and vegetables, nuts
Low	5–10	Apple, beet, celery, citrus fruits, cranberry, garlic, grape, honeydew melon, kiwifruit, onion, papaya, persimmon, pineapple, pomegranate, potato (mature), pumpkin, sweet potato, watermelon, winter squash
Moderate	10–20	Apricot, banana, blueberry, cabbage, cantaloupe, carrot (topped), celeriac, cherry, cucumber, fig, gooseberry, lettuce (head), mango, nectarine, olive, peach, pear, plum, potato (immature), radish (topped), summer squash, tomato
High	20–40	Avocado, blackberry, carrot (with tops), cauliflower, leek, lettuce (leaf), lima bean, radish (with tops), raspberry, strawberry
Very high	40–60	Artichoke, bean sprouts, broccoli, Brussels sprouts, cherimoya, cut flowers, endive, green onions, kale, okra, passion fruit, snap bean, watercress
Extremely high	>60	Asparagus, mushroom, parsley, peas, spinach, sweet corn

Note: *Vital heat (Btu/ton/24 hrs) = mg CO_2/kg-hr × 220.
Vital heat (kcal/1,000 kg/24 hrs) = mg CO_2/kg-hr × 61.2.

metabolism and is produced by all tissues of higher plants and by some microorganisms. As a plant hormone, C_2H_4 regulates many aspects of growth, development, and senescence and is physiologically active in trace amounts (less than 0.1 ppm). It also plays a major role in the abscission of plant organs.

The amino acid methionine is converted to *S*-adenosylmethionine (SAM), which is the precursor of 1-aminocyclopropane-1-carboxylic acid (ACC), the immediate precursor of C_2H_4. ACC synthase, which converts SAM to ACC, is the main site of control of ethylene biosynthesis. The conversion of ACC into ethylene is mediated by ACC oxidase. The synthesis and activities of ACC synthase and ACC oxidase are influenced by genetic factors and environmental conditions, including temperature and concentrations of oxygen and carbon dioxide.

Horticultural commodities are classified according to their C_2H_4 production rates in table 4.3. There is no consistent relationship between the C_2H_4 production capacity of a given commodity and its perishability; however, exposure of most commodities to C_2H_4 accelerates their senescence.

Generally, C_2H_4 production rates increase with maturity at harvest and with physical injuries, disease incidence, increased temperatures up to 30°C (86°F), and water stress. On the other hand, C_2H_4 production rates by fresh horticultural crops are reduced by storage at low temperature, by reduced O_2 levels (less than 8%), and elevated CO_2 levels (more than 2%) around the commodity.

COMPOSITIONAL CHANGES

Many changes in pigments take place during development and maturation of the commodity on the plant; some may continue after harvest and can be desirable or undesirable:

- Loss of chlorophyll (green color) is desirable in fruits but not in vegetables.

- Development of carotenoids (yellow and orange colors) is desirable in fruits such as apricots, peaches, and citrus. Red color development in tomatoes and pink grapefruit is due to a specific carotenoid (lycopene); beta-carotene is provitamin A and thus is important in nutritional quality.

- Development of anthocyanins (red and blue colors) is desirable in fruits such as apples (red cultivars), cherries, strawberries, cane berries, and red-flesh oranges. These water-soluble pigments are much less stable than carotenoids.

- Changes in anthocyanins and other phenolic compounds may result in tissue browning, which is undesirable for appearance quality. On the other hand, these constituents contribute to the total antioxidant capacity of the commodity, which is beneficial to human health.

Changes in carbohydrates include starch-to-sugar conversion (undesirable in potatoes, desirable in apple, banana, and other fruits); sugar-to-starch conversion (undesirable in peas and sweet corn; desirable in potatoes); and conversion of starch and sugars to CO_2 and water through respiration. Breakdown of pectins and other polysaccharides results in softening of fruits and a consequent increase in susceptibility to mechanical injuries. Increased lignin content is responsible for toughening of asparagus spears and root vegetables.

Changes in organic acids, proteins, amino acids, and lipids can influence flavor quality of the commodity. Loss in vitamin content, especially ascorbic acid (vitamin C) is detrimental to nutritional quality. Production of flavor volatiles associated with ripening of fruits is very important to their eating quality.

GROWTH AND DEVELOPMENT

Sprouting of potatoes, onions, garlic, and root crops greatly reduces their food value and accelerates deterioration. Rooting of

Table 4.2. Fruits classified according to respiratory behavior during ripening

Climacteric fruits		Nonclimacteric fruits	
Apple	Muskmelon	Blackberry	Lychee
Apricot	Nectarine	Cacao	Okra
Avocado	Papaya	Carambola	Olive
Banana	Passion fruit	Cashew apple	Orange
Biriba	Peach	Cherry	Pea
Blueberry	Pear	Cranberry	Pepper
Breadfruit	Persimmon	Cucumber	Pineapple
Cherimoya	Plantain	Date	Pomegranate
Durian	Plum	Eggplant	Prickly pear
Feijoa	Quince	Grape	Raspberry
Fig	Rambutan	Grapefruit	Strawberry
Guava	Sapodilla	Jujube	Summer squash
Jackfruit	Sapote	Lemon	Tamarillo
Kiwifruit	Soursop	Lime	Tangerine and mandarin
Mango	Sweetsop	Longan	
Mangosteen	Tomato	Loquat	Watermelon

Table 4.3. Classification of horticultural commodities according to ethylene (C_2H_4) production rates

Class	Range at 20°C (68°F) ($\mu L\ C_2H_4$/kg-hr)	Commodities
Very low	Less than 0.1	Artichoke, asparagus, cauliflower, cherry, citrus fruits, grape, jujube, strawberry, pomegranate, leafy vegetables, root vegetables, potato, most cut flowers
Low	0.1–1.0	Blackberry, blueberry, casaba melon, cranberry, cucumber, eggplant, okra, olive, pepper (sweet and chili), persimmon, pineapple, pumpkin, raspberry, tamarillo, watermelon
Moderate	1.0–10.0	Banana, fig, guava, honeydew melon, lychee, mango, plantain, tomato
High	10.0–100.0	Apple, apricot, avocado, cantaloupe, feijoa, kiwifruit (ripe), nectarine, papaya, peach, pear, plum
Very high	More than 100.0	Cherimoya, mammee apple, passion fruit, sapote

onions and root crops is also undesirable. Asparagus spears continue to grow after harvest; elongation and curvature (if the spears are held horizontally) are accompanied by increased toughness and decreased palatability. Similar geotropic responses occur in cut gladiolus and snapdragon flowers stored horizontally. Seed germination inside fruits such as tomatoes, peppers, and lemons is an undesirable change.

TRANSPIRATION OR WATER LOSS

Water loss is a main cause of deterioration because it results not only in direct quantitative losses (loss of salable weight), but also in losses in appearance (wilting and shriveling), textural quality (softening, flaccidity, limpness, loss of crispness and juiciness), and nutritional quality.

The commodity's dermal system (outer protective coverings) governs the regulation of water loss. It includes the cuticle, epidermal cells, stomata, lenticles, and trichomes (hairs). The cuticle is composed of surface waxes, cutin embedded in wax, and a layer of mixtures of cutin, wax, and carbohydrate polymers. The thickness, structure, and chemical composition of the cuticle vary greatly among commodities and among developmental stages of a given commodity.

The transpiration rate (evaporation of water from the plant tissues) is influenced by internal, or commodity, factors (morphological and anatomical characteristics, surface-to-volume ratio, surface injuries, and maturity stage) and by external, or environmental, factors (temperature, relative humidity [RH], air movement, and atmospheric pressure). Transpiration is a physical process that can be controlled by applying treatments to the commodity (e.g., waxes and other surface coatings or wrapping with plastic films) or by manipulating the environment (e.g., maintaining high RH and controlling air circulation).

PHYSIOLOGICAL BREAKDOWN

Exposure of the commodity to undesirable temperatures can result in physiological disorders:

- Freezing injury results when commodities are held below their freezing temperatures. The disruption caused by freezing usually results in immediate collapse of the tissues and total loss of the commodity.

- Chilling injury occurs in some commodities (mainly those of tropical and subtropical origin) held at temperatures above their freezing point and below 5° to 15°C (41° to 59°F), depending on the commodity. Chilling injury symptoms become more noticeable upon transfer to higher (nonchilling) temperatures. The most common symptoms are surface and internal discoloration (browning), pitting, watersoaked areas, uneven ripening or failure to ripen, off-flavor development, and accelerated incidence of surface molds and decay (especially the incidence of organisms not usually found growing on healthy tissue).

- Heat injury is induced by exposure to direct sunlight or excessively high temperatures. Its symptoms include bleaching, surface burning or scalding, uneven ripening, excessive softening, and desiccation.

Certain types of physiological disorders originate from preharvest nutritional imbalances. For example, blossom end rot of tomatoes and bitter pit of apples result from calcium deficiency. Increasing calcium content by preharvest or postharvest treatments can reduce the susceptibility to physiological disorders. Calcium content also influences the textural quality and senescence rate of fruits and vegetables; increased calcium content has been associated with improved firmness retention, reduced CO_2 and C_2H_4 production rates, and decreased decay incidence.

Very low O_2(<1%) and high CO_2 (>20%) atmospheres can cause physiological breakdown of most fresh horticultural commodities, and C_2H_4 can induce physiological disorders in certain commodities. The interactions among O_2, CO_2, and C_2H_4 concentrations, temperature, and duration of storage influence the incidence and severity of physiological disorders related to atmospheric composition.

PHYSICAL DAMAGE

Various types of physical damage (surface injuries, impact bruising, vibration bruising, and so on) are major contributors to deterioration. Browning of damaged tissues results from membrane disruption, which exposes phenolic compounds to the polyphenol oxidase enzyme. Mechanical injuries not only are unsightly but also accelerate water loss, provide sites for fungal infection, and stimulate CO_2 and C_2H_4 production by the commodity.

PATHOLOGICAL BREAKDOWN

One of the most common and obvious symptoms of deterioration results from the activity of bacteria and fungi. Attack by most organisms follows physical injury or physiological breakdown of the commodity. In a few cases, pathogens can infect apparently healthy tissues and become the primary cause of deterioration. In general, fruits and vegetables exhibit considerable resistance to potential pathogens during most of their postharvest life. The onset of ripening in fruits, and senescence in all commodities, renders them susceptible to infection by pathogens. Stresses such as mechanical injuries, chilling, and sunscald lower the resistance to pathogens.

ENVIRONMENTAL FACTORS INFLUENCING DETERIORATION

Temperature. Temperature is the environmental factor that most influences the deterioration rate of harvested commodities. For each increase of 10°C (18°F) above optimum, the rate of deterioration increases by two- to threefold (table 4.4). Exposure to undesirable temperatures results in many physiological disorders, as mentioned above. Temperature also influences the effect of C_2H_4, reduced O_2, and elevated CO_2. The spore germination and growth rate of pathogens are greatly influenced by temperature; for instance, cooling commodities below 5°C (41°F) immediately after harvest can greatly reduce the incidence of Rhizopus rot. Temperature effects on postharvest responses of chilling-sensitive and nonchilling-sensitive horticultural crops are compared in table 4.5.

Relative humidity. The rate of water loss from fruits and vegetables depends on the vapor pressure deficit between the commodity and the surrounding ambient air, which is influenced by temperature and RH. At a given temperature and rate of air movement, the rate of water loss from the commodity depends on the RH. At a given RH, water loss increases with the increase in temperature.

Atmospheric composition. Reduction of O_2 and elevation of CO_2, whether intentional

Table 4.4. Effect of temperature on deterioration rate of a non-chilling-sensitive commodity

Temperature (°F)	(°C)	Assumed Q_{10}*	Relative velocity of deterioration	Relative shelf life	Loss per day (%)
32	0	—	1.0	100	1
50	10	3.0	3.0	33	3
68	20	2.5	7.5	13	8
86	30	2.0	15.0	7	14
104	40	1.5	22.5	4	25

Note: $*Q_{10} = \dfrac{\text{Rate of deterioration at temperature (T)} + 10°C}{\text{Rate of deterioration at T}}$

Table 4.5. Fruits and vegetables classified according to sensitivity to chilling injury

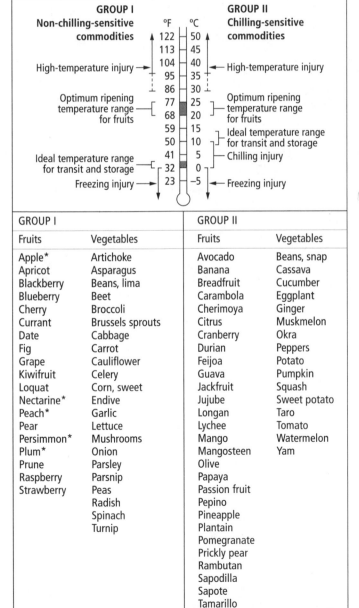

GROUP I		GROUP II	
Fruits	Vegetables	Fruits	Vegetables
Apple*	Artichoke	Avocado	Beans, snap
Apricot	Asparagus	Banana	Cassava
Blackberry	Beans, lima	Breadfruit	Cucumber
Blueberry	Beet	Carambola	Eggplant
Cherry	Broccoli	Cherimoya	Ginger
Currant	Brussels sprouts	Citrus	Muskmelon
Date	Cabbage	Cranberry	Okra
Fig	Carrot	Durian	Peppers
Grape	Cauliflower	Feijoa	Potato
Kiwifruit	Celery	Guava	Pumpkin
Loquat	Corn, sweet	Jackfruit	Squash
Nectarine*	Endive	Jujube	Sweet potato
Peach*	Garlic	Longan	Taro
Pear	Lettuce	Lychee	Tomato
Persimmon*	Mushrooms	Mango	Watermelon
Plum*	Onion	Mangosteen	Yam
Prune	Parsley	Olive	
Raspberry	Parsnip	Papaya	
Strawberry	Peas	Passion fruit	
	Radish	Pepino	
	Spinach	Pineapple	
	Turnip	Plantain	
		Pomegranate	
		Prickly pear	
		Rambutan	
		Sapodilla	
		Sapote	
		Tamarillo	

Note: *Some cultivars are chilling sensitive.

(modified or controlled atmosphere storage) or unintentional (restricted ventilation within a shipping container or transport vehicle), can either delay or accelerate the deterioration of fresh horticultural crops. The magnitude of these effects depends on the commodity, cultivar, physiological age, O_2 and CO_2 levels, temperature, and duration of holding.

Ethylene. Because the effects of C_2H_4 on harvested horticultural commodities can be desirable or undesirable, C_2H_4 is of major concern to all produce handlers. Ethylene can be used to promote faster and more uniform ripening of fruits picked at the mature-green stage. On the other hand, exposure to C_2H_4 can be detrimental to the quality of most nonfruit vegetables and ornamentals.

Light. Exposure of potatoes to light should be avoided because it results in greening due to formation of chlorophyll and solanine (toxic to humans). Light-induced greening of Belgian endive is also undesirable.

Other factors. Various kinds of chemicals (e.g., fungicides, growth regulators) may be applied to the commodity to affect one or more of the biological deterioration factors.

POSTHARVEST TECHNOLOGY PROCEDURES

TEMPERATURE MANAGEMENT PROCEDURES

Temperature management is the most effective tool for extending the shelf life of fresh horticultural commodities. It begins with the rapid removal of field heat by using one of the following cooling methods: hydrocooling, in-package icing, top-icing, evaporative cooling, room cooling, forced-air cooling, serpentine forced-air cooling, vacuum cooling, or hydro-vacuum cooling.

Cold storage facilities should be well-engineered and adequately equipped. They should have good construction and insulation, including a complete vapor barrier on the warm side of the insulation; strong floors; adequate and well-positioned doors for loading and unloading; effective distribution of refrigerated air; sensitive and properly located controls; enough refrigerated coil surface to minimize the difference between the coil and air temperatures; and adequate capacity for expected needs. Commodities should be stacked in the cold room with air spaces between pallets and room walls to ensure good air circulation. Storage rooms should not be loaded beyond their limit for proper cooling. In monitoring temperatures, commodity temperature rather than air temperature should be measured.

Transit vehicles must be cooled before loading the commodity. Delays between cooling after harvest and loading into transit vehicles should be avoided. Proper temperature maintenance should be ensured throughout the handling system.

CONTROL OF RELATIVE HUMIDITY

Relative humidity can influence water loss, decay development, incidence of some physiological disorders, and uniformity of fruit ripening. Condensation of moisture on the commodity (sweating) over long periods of time is probably more important in enhancing decay than is the RH of ambient air. Proper relative humidity is 85 to 95% for fruits and 90 to 98% for vegetables except dry onions and pumpkins (70 to 75%). Some root vegetables can best be held at 95 to 100% RH.

Relative humidity can be controlled by one or more of the following procedures:
- adding moisture (water mist or spray, steam) to air by humidifiers
- regulating air movement and ventilation in relation to the produce load in the cold storage room
- maintaining the refrigeration coils within about 1°C (2°F) of the air temperature
- providing moisture barriers that insulate storage room and transit vehicle walls; adding polyethylene liners in containers and plastic films for packaging
- wetting floors in storage rooms
- adding crushed ice in shipping containers or in retail displays for commodities that are not injured by the practice
- sprinkling produce with water during retail marketing (use on leafy vegetables, cool-season root vegetables, and immature fruit vegetables such as snap beans, peas, sweet corn, summer squash

SUPPLEMENT TEMPERATURE MANAGEMENT

Many technological procedures are used commercially as supplements to temperature management. None of these procedures, alone or in their various combinations, can substitute for maintenance of optimal temperature and RH, but they can help extend the shelf life of harvested produce beyond what is possible using refrigeration alone (table 4.6).

Treatments applied to commodities include
- curing of certain root, bulb, and tuber vegetables
- cleaning followed by removal of excess surface moisture
- sorting to eliminate defects
- waxing and other surface coatings, including film wrapping
- heat treatments (hot water or air, vapor heat)
- treatment with postharvest fungicides
- sprout inhibitors
- special chemical treatments (scald inhibitors, calcium, growth regulators, anti-ethylene chemicals for ornamentals)
- fumigation for insect control
- ethylene treatment (de-greening, ripening)

Treatments to manipulate the environment include
- packaging
- control of air movement and circulation
- control of air exchange or ventilation
- exclusion or removal of C_2H_4
- controlled or modified atmospheres (CA or MA)
- sanitation

RECENT TRENDS IN PERISHABLES HANDLING

SELECTION OF CULTIVARS

For many commodities, producers are using cultivars with superior quality and/or long postharvest life, such as "super-sweet" sweet corn, long-shelf-life tomatoes, and sweeter melons. Plant geneticists in public and private institutions are using molecular biology methods along with plant breeding procedures to produce new genotypes that taste better, maintain firmness better, are more disease resistant, have less browning potential, and have other desirable characteristics.

PACKING AND PACKAGING

The produce industry is increasingly using plastic containers that can be reused and recycled in order to reduce waste disposal problems. For example, standard-sized (48 by 40 in., about 120 by 100 cm) stacking (returnable) pallets are becoming more widely used. There is continued increase in use of modified atmosphere and controlled atmosphere packaging (MAP and CAP) systems at the pallet, shipping container (fiberboard box liner), and consumer package levels. Also, the use of absorbers of C_2H_4, CO_2, O_2, and/or water vapor as part of MAP and CAP is increasing.

COOLING AND STORAGE

The current trend is towards increased precision in temperature and relative humidity (RH) management to provide the optimal environment for fresh fruits and vegetables during cooling and storage. Precision temperature management (PTM) tools are becoming more common in cooling and storage facilities. Forced-air cooling continues to be the predominant cooling method for horticultural perishables. Operators can ensure that all produce shipments leave the

Table 4.6. Fresh horticultural crops classified according to relative perishability and potential storage life in air at near-optimal temperature and RH

Relative perishability	Potential storage life (weeks)	Commodities
Very high	<2	Apricot, blackberry, blueberry, cherry, fig, raspberry, strawberry; asparagus, bean sprouts, broccoli, cauliflower, cantaloupe, green onion, leaf lettuce, mushroom, pea, spinach, sweet corn, tomato (ripe); most cut flowers and foliage; fresh-cut (minimally processed) fruits and vegetables
High	2–4	Avocado, banana, grape (without SO_2 treatment), guava, loquat, mandarin, mango, melons (honeydew, crenshaw, Persian), nectarine, papaya, peach, pepino, plum; artichoke, green beans, Brussels sprouts, cabbage, celery, eggplant, head lettuce, okra, pepper, summer squash, tomato (partially ripe)
Moderate	4–8	Apple and pear (some cultivars), grape (SO_4-treated), orange, grapefruit, lime, kiwifruit, persimmon, pomegranate, pummelo; table beet, carrot, radish, potato (immature)
Low	8–16	Apple and pear (some cultivars), lemon, potato (mature), dry onion, garlic, pumpkin, winter squash, sweet potato, taro, yam; bulbs and other propagules of ornamental plants
Very low	>16	Tree nuts, dried fruits and vegetables

cooling facility within 0.5°C (about 1°F) of the optimal storage temperature. Periodic ventilation of storage facilities is effective in maintaining C_2H_4 concentrations below 1 ppm, which permits mixing of temperature-compatible, ethylene-producing, and ethylene-sensitive commodities.

POSTHARVEST INTEGRATED PEST MANAGEMENT (IPM)

Controlled atmosphere (CA) conditions delay senescence, including fruit ripening, and consequently reduce the susceptibility of fruits to pathogens. On the other hand, CA conditions unfavorable to a given commodity can induce physiological breakdown and render it more susceptible to pathogens. Calcium treatments have been shown to reduce decay incidence and severity; wound healing following physical injury has been observed in some fruits and has reduced their susceptibility to decay. Biological control agents are being used alone or in combination with reduced concentrations of postharvest fungicides, heat treatments, and/or fungistatic CA for control of postharvest diseases.

Chemical fumigants, especially methyl bromide, are still the primary method used for insect control in harvested fruits when such treatment is required by quarantine authorities in importing countries. Many studies are under way to develop alternative methods of insect control that are effective, not phytotoxic to the fruits, and present no health hazard to the consumer. These alternatives include cold treatments, hot water or air treatments, ionizing radiation (0.15–0.30 kilogray) and exposure to reduced (less than 0.5%) O_2 and/or elevated CO_2 (40–60%) atmospheres. This is a high-priority research and development area because of the possible loss of methyl bromide as an option for insect control.

USE OF CONTROLLED AND MODIFIED ATMOSPHERES

The use of CA during transport and/or storage of fresh fruits and vegetables (marketed intact or lightly processed) continues to expand because of improvements in nitrogen-generation equipment and in instruments for monitoring and maintaining desired concentrations of oxygen and carbon dioxide. Controlled atmosphere is a useful supplement to the proper maintenance of optimal temperature and RH during transport and storage of many fresh fruits and vegetables. It allows use of marine transport instead of air transport of some commodities.

Several refinements in CA storage have been made in recent years to improve quality maintenance. These include creating nitrogen by separation from compressed air using molecular sieve beds or membrane systems; low O_2 (1.0–1.5%) storage; low ethylene CA storage; rapid CA (rapid establishment of the optimal levels of O_2 and CO_2); and programmed (or sequential) CA storage (e.g., storage in 1% O_2 for 2 to 6 weeks followed by storage in 2 to 3% O_2 for the remainder of the storage period). Other developments, which may expand use of MA during transport and distribution, include using edible coatings or polymeric films with appropriate gas permeabilities to create a desired MA around and within the commodity. Modified atmosphere packaging is widely used in marketing fresh-cut fruits and vegetables.

Successful application of atmospheric modification depends on the commodity, cultivar, maturity stage at harvest, and a positive return on investment (benefit-cost ratio). Commercial use of CA storage is greatest worldwide on apples and pears; less on kiwifruits, avocados, persimmons, pomegranates, nuts, and dried fruits and vegetables. Atmospheric modification during long-distance transport is used on apples, asparagus, avocados, bananas, broccoli, cane berries, cherries, figs, kiwifruits, mangoes, melons, nectarines, peaches, pears, plums, and strawberries. Continued technological developments in the future to provide CA during transport and storage at a reasonable cost are essential to greater CA applications on fresh fruits and vegetables.

TRANSPORTATION

Improvements are continually being made in attaining and maintaining the optimal environmental conditions (temperature, RH, and concentrations of O_2, CO_2, and C_2H_4) in transport vehicles. Produce is commonly cooled before loading and is loaded with an air space between the palletized produce and the walls of the transport vehicles to improve temperature maintenance. In some cases, vehicle and produce temperature data are transmitted by satellite to a control center,

allowing all shipments to be continuously monitored. Some new trucks have air ride suspension, which can eliminate transport vibration damage. As the industry realizes the value of air ride, its popularity will increase.

HANDLING AT WHOLESALE AND RETAIL

Wholesale and retail markets have been increasingly using automated ripening, in which the gas composition of the ripening atmosphere, the room temperature, and fruit color are continuously monitored and modulated to meet desired ripening characteristics. Improved ripening systems will lead to greater use of ripening technology to deliver products that are ripened to the ideal eating stage. Better-refrigerated display units, with improved temperature and RH monitoring and control systems, are being used in retail markets, especially for fresh-cut fruit and vegetable products. Many retail and food service operators are using Hazard Analysis Critical Control Points (HACCP) Programs to assure consumers that food products are safe.

FOOD SAFETY ASSURANCE

During the past few years, food safety became and continues to be the number-one concern of the fresh produce industry. U.S. trade organizations such as the International Fresh-Cut Produce Association (IFPA), Produce Marketing Association (PMA), United Fresh Fruit and Vegetable Association (UFFVA), and Western Growers Association (WGA) have taken an active role in developing voluntary food safety guidelines for producers and handlers of fresh fruits and vegetables. The U.S. Food and Drug Administration (FDA) published in October 1998 the *Guide to Minimize Microbial Food Safety Hazards for Fresh Fruits and Vegetables*. This guide should be used by all handlers of fresh produce to develop the most appropriate agricultural and management practices for their operations.

The FDA guide is based on the following basic principles and practices associated with minimizing microbial food safety hazards from the field through distribution of fresh fruits and vegetables.

Principle 1. Prevention of microbial contamination of fresh produce is favored over reliance on corrective actions once contamination has occurred.

Principle 2. To minimize microbial food safety hazards in fresh produce, growers, packers, or shippers should use good agricultural and management practices in those areas over which they have control.

Principle 3. Fresh produce can become microbiologically contaminated at any point along the farm-to-table food chain. The major source of microbial contamination of fresh produce is associated with human or animal feces.

Principle 4. Whenever water comes in contact with produce, the quality of the water dictates the potential for contamination. Minimize the potential of microbial contamination from water used with fresh fruits and vegetables.

Principle 5. Practices using animal manure or municipal biosolid wastes should be managed closely to minimize the potential for microbial contamination of fresh produce.

Principle 6. Worker hygiene and sanitation practices during production, harvesting, sorting, packing, and transport play a critical role in minimizing the potential for microbial contamination of fresh produce.

Principle 7. Follow all applicable local, state, and federal laws and regulations or corresponding or similar laws, regulations, or standards for operators outside the United States, for agricultural practices.

REFERENCES

Brady, C. J. 1987. Fruit ripening. Annu. Rev. Plant Physiol. 38:155–178.

Cappellini, R. A., and M. J. Ceponis. 1984. Postharvest losses in fresh fruits and vegetables. In H. E. Moline, ed., Postharvest pathology of fruits and vegetables: Postharvest losses in perishable crops. Oakland: Univ. Calif. Bull. 1914. 24–30.

Giovannoni, J. 2001. Molecular biology of fruit maturation and ripening. Annu. Rev. Plant Physiol. Plant Mol. Biol. 52:725–749.

Grierson, D. 1987. Senescence in fruits. HortScience 22:859–862.

Harvey, J. M. 1978. Reduction of losses in fresh market fruits and vegetables. Annu. Rev. Phytopathol. 16:321–341.

International Institute of Refrigeration. 2000. Recommendations for chilled storage of perishable produce. Paris: International Institute of Refrigeration. 219 pp.

Kader, A. A. 1983. Postharvest quality maintenance of fruits and vegetables in developing countries. In M. Lieberman, ed., Postharvest physiology and crop preservation. New York: Plenum. 520–536.

Kantor, L. S., K. Lipton, A. Manchester, and V. Oliveira. 1997. Estimating and addressing America's food losses. Food Review 20:3–11.

Kitinoja, L., and A. A. Kader. 1995. Small-scale postharvest handling practices: A manual for horticultural crops. 3rd ed. Davis: Univ. Calif. Postharv. Hort. Ser. 8. 231 pp.

Kitinoja, L., and J. R. Gorny. 1999. Postharvest technology for small-scale produce marketers: Economic opportunities, quality and food safety. Davis: Univ. Calif. Postharv. Hort. Ser. 21.

Lidster, P. D., P. D. Hilderbrand, L. S. Bérard, and S. W. Porritt. 1988. Commercial storage of fruits and vegetables. Can. Dept. Agric. Publ. 1532. 88 pp.

Lipton, W. J. 1987. Senescence in leafy vegetables. HortScience 22:854–859.

Mayak, S. 1987. Senescence in cut flowers. HortScience 22:863–865.

National Academy of Sciences. 1978. Postharvest food losses in developing countries (Science and Technology for International Development). Washington, D.C.: Natl. Acad. Sci. 202 pp.

Rhodes, M. J. C. 1980a. The maturation and ripening of fruits. In K. V. Thimann, ed., Senescence in plants. Boca Raton, FL: CRC Press. 157–205.

———. 1980b. The physiological basis for the conservation of food crops. Prog. Food Nutr. Sci. 4(3–4): 11–20.

Romani, R. J. 1987. Senescence and homeostasis in postharvest research. HortScience 22:865–868.

Shewfelt, R. L. 1986. Postharvest treatment for extending the shelf-life of fruits and vegetables. Food Technol. 40(5): 70–89.

Tindall, H. D., and F. J. Proctor. 1980. Loss prevention of horticultural crops in the tropics. Prog. Food Nutr. Sci. 4(3–4): 25–40.

United Nations Food and Agriculture Organization (FAO). 1981. Food loss prevention in perishable crops. FAO Agric. Serv. Bull. 43. 72 pp.

Wang, C. Y., ed. 1990. Chilling injury of horticultural crops. Boca Raton, FL: CRC Press. 313 pp.

5

Preharvest Factors Affecting Fruit and Vegetable Quality

Carlos H. Crisosto and

Jeffrey P. Mitchell

Although fruit or vegetable quality can only be maintained, not improved, after harvest, little research has been conducted on the influence of preharvest factors on postharvest quality of fruits other than citrus and pome fruits. Because vegetables are typically produced during short growing seasons with intensive inputs, the role of preharvest factors on vegetable quality and potential postharvest life has been studied in more detail. In general, preharvest factors that can be managed should be aimed to optimize their impact on postharvest quality.

Preharvest factors often interact in complex ways that depend on specific cultivar characteristics and growth or development stage sensitivities. The tremendous diversity of fruits and vegetables that are produced commercially and the general lack of research relating preharvest factors to postharvest quality precludes generalizations about preharvest influences that uniformly apply to all fruits and vegetables. Maximum postharvest quality for any cultivar can be achieved only by understanding and managing the various roles that preharvest factors play in postharvest quality.

CULTIVAR AND ROOTSTOCK GENOTYPE

Cultivar and rootstock genotype have an important role in determining the taste quality, yield, nutrient composition, and postharvest life of fruits and vegetables. The incidence and severity of decay, insect damage, and physiological disorders can be reduced by choosing the correct genotype for given environmental conditions. Breeding programs are constantly creating new cultivars and rootstocks with improved quality and better adaptability to various environmental and crop pest conditions.

Some experts consider the most important cultivar characteristic for fruits and vegetables to be disease resistance, including resistance to diseases that diminish postharvest quality. Control of some postharvest diseases may include breeding for resistance to the vector (e.g., aphid, nematode, leafhopper, or mite), rather than just for the pathogen.

Nutritional quality may also vary greatly according to cultivar. In potatoes, Granola and Russet Norkotah had higher antioxidant activity than Yukon Gold and Viking (antioxidant index = 88–89 versus 65–68). L-ascorbic acid levels in different pepper types also vary considerably. For example, in jalapeño peppers, the highest ascorbic acid levels were in Jaloro (131 mg·100g^{-1}) and the lowest were in Mitla (49 mg·100g^{-1}). Wide variation in beta-carotene content of several cultivars of sweet potato has similarly been reported; Georgia Jet, suggested for processing, contained low concentrations of beta-carotene (6.9 mg·100g^{-1}). There is a need to identify and develop cultivars that are suitable for processing and high in antioxidant vitamin content.

Genetic engineering can be a successful tool in altering the quality and yield of certain vegetables, but its commercial application will depend largely on consumer acceptance and food safety issues. Future advances will depend on successful team efforts between plant breeders, plant pathologists, molecular geneticists, and consumer education programs.

MINERAL NUTRITION

Nutritional status is an important factor in quality at harvest and postharvest life of various fruits and vegetables. Deficiencies, excesses, or imbalances of various nutrients are known to result in disorders that can limit the storage life of many fruits and vegetables. Fertilizer application rates vary widely among growers and generally depend upon soil type, cropping history, and soil test results, which help indicate nitrogen (N), phosphorous (P), and potassium (K) requirements. To date, fertilization recommendations for fruits and vegetables have been established primarily for productivity goals, not as diagnostics for good flavor quality and optimal postharvest life.

The nutrient with the single greatest effect on fruit quality is nitrogen. Research performed over the last 12 years at the Kearney Agricultural Center in Parlier, CA, has established that peaches and nectarines grown under California conditions should be kept between 2.6 and 3.0% leaf nitrogen for best fruit quality (see Crisosto et al. 1997; Crisosto et al. 1995; Daane et al. 1995). Response of peach and nectarine trees to nitrogen fertilization is dramatic. High nitrogen levels stimulate vigorous vegetative growth, causing shading and death of lower fruiting wood. Although high-nitrogen trees may look healthy and lush, excess nitrogen does not increase fruit size, production, or soluble solids content (SSC). Furthermore, excessive nitrogen delays stone fruit maturity, induces poor red color development, and inhibits ground color change from green to yellow. However, nitrogen deficiency leads to small fruit with poor flavor and unproductive trees. The postharvest fruit water loss from fruit from the highest nitrogen rate tested (3.6% leaf N) was greater than that from the lowest rate (2.6% leaf N).

The relationship between fruit nitrogen concentration and fruit susceptibility to decay caused by brown rot (*Monilinia fructicola* [Wint.] Honey) has been extensively studied on stored nectarines (see Daane et al. 1995). Wounded and brown-rot-inoculated fruit from Fantasia and Flavortop nectarine trees having more than 2.6% leaf nitrogen were more susceptible to brown rot than fruit from trees with 2.6% or less leaf nitrogen. Anatomical observations and cuticle density measurements on the fruit indicated differences in cuticle thickness among Fantasia fruit from the low, middle, and high nitrogen treatments. But this can only partially explain the differences in fruit susceptibility to this disease.

In vegetable crops, excessive nitrogen levels induce delayed maturity and increase several disorders that diminish postharvest quality. Disorders such as gray wall or internal browning in tomato, hollow stem of broccoli, lower soluble solids concentration in potato, fruit spot in peppers, and growth cracks and hollow heart in broccoli and cauliflower have been associated with high nitrogen. High nitrogen has also been associated with increased weight loss during storage of sweet potatoes and soft rot in tomatoes.

Excessive soil nitrogen can negatively impact vegetable quality in several ways. High nitrogen can result in composition changes such as reduced ascorbic acid (vitamin C) content, lower sugar content, lower acidity, and altered ratios of essential amino acids. In leafy green vegetables grown under low light, it can result in the accumulation of nitrates in plant tissues to unhealthy levels. High nitrogen fertilization can lead to reduced volatile production and changes in the characteristic flavor of celery. In table beets, high nitrogen can lead to increased glutamine levels that result in off flavors in the processed beet puree.

Although calcium (Ca) is classified as a secondary nutrient, it is involved in numerous biochemical and morphological processes in plants and has been implicated in many disorders of considerable economic importance to the production and postharvest quality of fruits and vegetables. Bitter pit in apple, corkspot in pear, blackheart in celery, blossom end rot in tomato, cavity spot and cracking in carrot, and tipburn of lettuce are calcium deficiency disorders that reduce the quality and marketability of these commodities. Certain calcium deficiency disorders, such as bitter pit in apples and blossom end rot in tomatoes, may be lessened through proper irrigation, fertilizer management, and supplemental fertilization. However, for tipburn of lettuce, a physiological disorder caused by the lack of mobility of calcium in the heads during warm weather and rapid growing conditions, there is currently no preharvest control practice.

There is mounting evidence that soil cation balance directly impacts the postharvest quality of several vegetables. Recent research (see Hartz et al. 1998) has demonstrated that the incidence of yellow eye and white core, two color defects in tomato, are correlated with soil cation balance. Increasing levels of soil potassium (expressed as ppm extractable or as a percentage of base exchange) decrease color disorders, while higher soil magnesium levels increase them. Applications of gypsum and potassium amendments may be helpful in reducing the incidence of these color defects, but they may not be economically practical in soils that tightly fix potassium, causing potassium to be available at such low levels that the treatment may result in little benefit for tomatoes grown in those locations.

FOLIAR NUTRIENT SPRAYS

Calcium is often considered to be the most important mineral element in determining fruit quality, especially in apples and pears, where it has been demonstrated to reduce metabolic disorders, maintain firmness and reduce decay. In apples, to significantly affect fruit firmness or decrease fruit decay caused by postharvest wound pathogens, it is necessary to raise the level of flesh calcium (dry weight basis) to 800 to 1,000 $\mu g \cdot g^{-1}$. Concentrations significantly higher than 1,000 $\mu g \cdot g^{-1}$ can result in surface injury to the fruit. Bitter pit, however, can be alleviated with a flesh tissue concentration of only about 250 $\mu g \cdot g^{-1}$. While various spray programs may not be able to raise the tissue calcium level high enough to affect firmness or decay by pathogens through wounds, these programs may be able to increase the calcium concentration enough to prevent bitter pit. Thus, the fruit-flesh calcium concentration that is necessary to reduce diseases and physiological disorders is usually difficult to obtain through normal fertilizer regimes.

Several studies have investigated the effects of the direct application of calcium salts to fruits (see Fallahi et al. 1997; Ferguson et al. 1999; Sams 1999). Calcium chloride sprays are widely used to reduce bitter pit and cork symptoms in apples and pears, respectively. In some cases, it has been reported to improve fruit firmness and reduce the incidence and severity of physiological disorders during and after storage.

Little research has been done on the effect of foliar calcium sprays on stone fruit quality. The limited published research suggests that these sprays have little effect on stone fruit quality (see Crisosto et al. 1997). Work done in California screening several

commercial calcium foliar sprays on peaches and nectarines (applied every 14 days, starting 2 weeks after full bloom and continuing until 1 week before harvest) showed there was no effect on the fruit quality of mid- or late-season cultivars. These foliar sprays did not affect the soluble solids content, firmness, decay incidence, or fruit flesh calcium concentration. Fruit flesh calcium concentration measured at harvest varied among cultivars from 200 to 300 $\mu g \cdot g^{-1}$ (dry weight basis). A lack of decay control was also reported on Jerseyland peaches grown in Pennsylvania and treated with 10 weekly preharvest calcium sprays of $CaCl_2$ at 0, 34, 67, or 101 $kg \cdot ha^{-1}$. Even fruit treated at a rate of 101 $kg \cdot ha^{-1}$ that had 70% more flesh calcium (490 versus 287 $\mu g \cdot g^{-1}$, dry weight basis) than untreated fruit showed no reduction in decay severity. Recent research suggests that these sprays on peaches and nectarines should be treated with caution because their heavy metal (Fe, Al, Cu, etc.) content may contribute to peach and nectarine skin discoloration (inking) (see Crisosto et al. 1997).

Postharvest vacuum infiltration of 1, 2 and 4% $CaCl_2$ solutions into mature peaches (increasing flesh calcium concentration from 287 to 1088 $\mu g \cdot g^{-1}$, dry weight basis) maintained higher fruit flesh firmness during cold storage but did not show a reduction in decay incidence. However, these potential benefits were negated by skin injury and sanitation problems.

IRRIGATION

Despite the important role of water in fruit growth and development, few studies have been done on the influence of the amount and the timing of water applications on fruit and vegetable quality at harvest and during postharvest.

In peaches growing under San Joaquin Valley conditions, the irrigation regimes of 100%, 50% and 150% evapotranspiration (ET) applied 4 weeks before harvest affected O'Henry peach size and soluble solids content (SSC), but did not affect internal breakdown incidence or severity. In general, fruit from the 50% ET treatment were small in size but had high SSC.

An increase in fruit defects such as deep suture and double-fruit formation has been reported for early-season Regina peaches as a consequence of imposing a postharvest water stress (50% ET) in mid and late summer during the previous season. These defects reduce the final packout. A similar regulated water stress regime applied to early-season plums did not affect the number of double and deep-sutured fruit of the Red Beaut, Ambra, and Durado cultivars.

In Bartlett pear, size and SSC were closely related to the level of water stress experienced by the tree. Increased tree water stress was associated with increases in fruit SSC, firmness, and yellow fruit color and in decreased fruit size and vegetative growth. There were no evident effects on postharvest disorders such as softening, internal breakdown, scald, or decay.

Water management as a direct determinant of postharvest quality has also been investigated for a number of vegetables produced in semiarid irrigated regions such as California and Israel. Except for a few studies, however, which have comprehensively tested a broad range of water management practices and conditions and their impacts on postharvest quality, it is often difficult to generalize about the effects of water management from the site-specific irrigation regimes that have been reported.

There is considerable evidence that water stress at the end of the season, which may be achieved by irrigation cutoff or deficit irrigation relative to evapotranspirative demand for generally more than 20 days prior to harvest, may markedly improve SSC in tomatoes. Irrigation cutoffs may also facilitate harvests and minimize soil compaction from mechanical harvest operations. Late-season irrigations with saline water have also been shown to increase tomato SSC. Although a higher SSC may result in premiums paid to producers, because of the link between applied water and yield, irrigation practices typically aim at the best overall economic balance between productivity and quality.

Melon postharvest quality is also quite sensitive to water management. Overirrigation can result not only in low SSC in melons but also unsightly ground spots and fruit rots (and measles in honeydews). Rapid growth resulting from irrigations following extended periods of soil water deficits may result in growth cracks in carrots, potatoes, tomatoes, and several other vegetable crops. Uneven irrigation management may also increase the incidence of "spindle"- or "dumb-bell"-shaped potatoes, depending on the growth stage during which soil water was limited.

Postharvest losses due to storage diseases such as neck rot, black rot, basal rot, and bacterial rot of onions can be influenced by irrigation management. Selecting the proper irrigation system relative to the crop stage of growth, reducing the number of irrigations applied, and assuring that onions cure adequately prior to harvest can help prevent storage losses.

Management of water frequently poses a dilemma between yield and postharvest quality. A deficiency or excess of water may influence postharvest quality of berry crops. Extreme water stress reduces yield and quality; mild water stress reduces crop yield but may improve some quality attributes in the fruit; and no water stress increases yield but may reduce postharvest quality. In strawberries, reduction of water stress by natural rainfall or irrigation during maturation and ripening decreases firmness and sugar content and provides more favorable conditions for mechanical fruit injury and rot. If strawberry plants are overirrigated, especially at harvest, the fruit is softer and more susceptible to bruising and decay.

CANOPY MANIPULATIONS

Crop Load. In most fruit, fruitlet thinning increases fruit size while also reducing total yield; a balance between yield and fruit size must be achieved. Generally, maximum profit does not occur at maximum marketable yield since larger fruit bring a higher market price. For example, leaving too many fruit on a tree reduces fruit size and SSC in the early-ripening May Glo nectarine and the late-ripening O'Henry peach. Crop load on O'Henry peach trees affected the incidence of internal breakdown measured after 1, 2, and 3 weeks at 5°C (41°F). Despite a large amount of mealy fruit in all lots, the overall incidence of mealiness and flesh browning in fruit from the high crop load was low, intermediate in fruit from the commercial crop load, and highest in fruit from the low crop load.

Also, it is well known that the fruit count–leaf count ratio (F:L) influences high bush blueberry fruit quality more than mineral nutrition. A high F:L results in later ripening, lower SSC, and smaller berries. During the harvest season, as berries are picked, SSC increases when the F:L drops to between 1:1 and 2:1. In general, berry crops have better postharvest fruit quality when the plant microclimate is improved by having an open canopy and maximum air circulation.

Fruit Canopy Position. Large differences in SSC, acidity, and fruit size were detected between fruit obtained from the outside versus inside canopy positions of open-vase-trained peach, nectarine, and plum trees. Peaches grown under a high light environment (outside canopy) have a longer storage and market life than peaches grown under a low light environment (inside canopy). The use of more efficient training systems that allow sunlight penetration into the center and lower canopy areas is recommended to reduce the number of shaded fruit.

Numerous studies have shown that improving light penetration into the canopy improves fruit composition of grapes, including increased SSC, aroma, anthocyanins, and total soluble phenols; but it reduced titratable acidity and potassium content (see Prange and DeEll 1997). In kiwifruit, shading reduces fruit count rather than individual fruit weight, delays harvest maturity, decreases SSC, and accelerates the rate of fruit softening during storage.

In grape vines, open canopies and optimal air circulation can be attained by the proper combination of plant spacing, vegetative thinning, and training. Vine vigor can be controlled by stem training and by avoiding high levels of nitrogen. This improves light penetration to leaves, ensuring that they continue to produce photosynthate and do not prematurely senesce and become pathogen hosts. An open canopy lowers the humidity around the plant, reduces wetting periods, and improves spray penetration, which reduces disease and insect problems and improves foliar nutrient application. An open canopy also enables the pickers to harvest more rapidly, decreasing the likelihood of overripe fruit.

Leaf Removal. Summer pruning and leaf pulling around the fruit increases fruit light exposure and, when performed properly, can increase fruit color without affecting fruit size and SSC. Excessive leaf pulling or leaf pulling done too close to harvest, however, can reduce both fruit size and SSC in peaches and nectarines.

Girdling. Girdling (a commercial practice in which the phloem of the tree or vine is removed) 4 to 6 weeks before harvest can increase peach and nectarine fruit size and advance and synchronize maturity. In some cases, girdling increases fruit SSC but also increases fruit acidity and phenolics so that the taste resulting from the additional sugars may be masked. Girdling can also cause the pits of peach and nectarine fruits to split, especially if it is done too early during pit hardening. Fruit with split-pits soften more quickly than intact fruit. Split-pits, as a consequence of girdling, have not been observed in Black Amber, Santa Rosa, Friar, or Royal Diamond plum cultivars; however, rapid fruit softening and severe tree weakening have been noted.

In grapes, the balance between vegetative and fruit growth can be altered by girdling vines that have excessive vigor and a history of poor berry and bunch size. In these cases, girdling may improve bunch shape and berry size.

CROP ROTATIONS

Crop rotation may be an effective management practice for minimizing postharvest losses by reducing decay inoculum in a production field. Because soilborne fungi, bacteria, and nematodes can build up to damaging levels with repeated cropping of a single vegetable crop, rotations out of certain vegetables are commonly recommended in intensive vegetable production regions. Four-year rotations with noncucurbit crops are routinely recommended for cucurbit disease management, as are 4-year rotations for garlic to decrease postharvest disease incidence.

There is also evidence that the use of plastic mulches can increase postharvest losses from decay in vegetables such as tomatoes. The impacts of cover crop–derived mulches on postharvest quality have not been well evaluated for vegetable crops. A number of destructive postharvest diseases of vegetables can be spread from infested fields to clean fields in soil and crop debris carried by workers and on equipment. Sanitation efforts such as working clean fields before entering infested fields and washing equipment and clothes to remove soil and debris when leaving infested fields can help reduce contamination and postharvest losses in crops grown in clean fields.

REFERENCES

Arpaia, M. L. 1994. Preharvest factors influencing postharvest quality of tropical and subtropical fruit. HortScience 29:982–985.

Crisosto, C. H., F. G. Mitchell, and R. S. Johnson. 1995. Factors in fresh market stone fruit quality. Postharv. News and Info. 6:17N–21N.

Crisosto, C. H., R. S. Johnson, T. DeJong, and K. R. Day. 1997. Orchard factors affecting postharvest stone fruit quality. HortScience 32:820–823.

Crisosto, C. H., R. S. Johnson, J. G. Luza, and G. M. Crisosto. 1994. Irrigation regimes affect fruit soluble solids content and the rate of water loss of 'O'Henry' peaches. HortScience 29:1169–1171.

Daane, K. M., R. S. Johnson, T. J. Michailides, C. H. Crisosto, J. W. Dlott, H. T. Ramirez, G. T. Yokota, and D. P. Morgan. 1995. Excess nitrogen raises nectarine susceptibility to disease and insects. Calif. Agric. 49(4): 13–17.

Fallahi, E., W. S. Conway, K. D. Hickey, and Carl E. Sams. 1997. The role of calcium and nitrogen in postharvest quality and disease resistance of apples. HortScience 32:831–835.

Ferguson, I., R. Volz, and A. Woolf. 1999. Preharvest factors affecting physiological disorders of fruit. Postharv. Biol. Technol. 15:255–262.

Hartz, T. K., K. S. Mayberry, and J. Valencia. 1996. Cantaloupe production in California. Oakland: Univ. Calif. Div. Ag. and Nat. Res. Publ. 7218. 3 pp. Available via Internet at http://anrcatalog.ucdavis.edu

Hartz, T. K., C. Giannini, G. Miyao, J. Valencia, M. Cahn, R. Mullen, and K. Brittan. 1998. Soil cation balance affects tomato fruit color disorders. HortScience 33:445–446.

Jackson, L., K. Mayberry, F. Laemmlen, S. Koike, K. Schulbach, and W. Chaney. 1996. Iceberg lettuce production in California. Oakland: Univ. Calif. Div. Ag. and Nat. Res. Publ. 7215. 4 pp. Available via Internet at http://anrcatalog.ucdavis.edu

Johnson, R. S., D. F. Handley, and T. DeJong. 1992. Long-term response of early maturing peach trees to postharvest water deficit. J. Amer. Soc. Hort. Sci. 69:1035–1041.

Kader, A. A. 1988. Influence of preharvest and postharvest environment on nutritional composition of fruits and vegetables. In B. Quebedeaux and F. A. Bliss, eds., Horticulture and human health-contributions of fruits and vegetables. Englewood Cliffs, NJ: Prentice-Hall. 18–22.

Kays, S. J. 1999. Preharvest factors affecting appearance. Postharv. Biol. Technol. 15:233–247.

Mattheis, J. P., and J. K. Fellman. 1999. Preharvest factors influencing flavor of fresh fruits and vegetables. Postharv. Biol. Technol. 15:227–232.

Mayberry, K. S., T. K. Hartz, and J. Valencia. 1996. Mixed melon production in California. Oakland: Univ. Calif. Div. Ag. and Nat. Res. Publ. 7209. 3 pp. Available via Internet at http://anrcatalog.ucdavis.edu

Prange, R., and J. R. DeEll. 1997. Preharvest factors affecting quality of berry crops. HortScience 32:824–830.

Prashar, C. R. K., R. Pearl, and R. M. Hagan. 1976. Review on water and crop quality. Scientia Hort. 5:193–205.

Sams, C. E. 1999. Preharvest factors affecting postharvest texture. Postharv. Biol. Technol. 15:249–254.

Weston, L. A., and M. M. Barth. 1997. Preharvest factors affecting postharvest quality of vegetables. HortScience 32:812–816.

6

Maturation and Maturity Indices

Michael S. Reid

The first step in the postharvest life of the product is the moment of harvest. For most fresh produce, harvest is manual, so the picker is responsible for deciding whether the produce has reached the correct maturity for harvest. The maturity of harvested perishable commodities has an important bearing on their storage life and quality and may affect the way they are handled, transported, and marketed. An understanding of the meaning and measurement of maturity is therefore central to postharvest technology. The meaning of the term *mature*, the importance of maturity determination, and some examples of approaches to determining and applying a satisfactory index of maturity, are discussed in this chapter.

DEFINITION OF MATURITY

To most people *mature* and *ripe* mean the same thing when describing fruit. For example, *mature* is defined in Webster's dictionary as: "mature (fr. L *maturus* ripe):1: Based on slow, careful consideration; 2a (1): having completed natural growth and development: RIPE (2): having undergone maturation, b: having attained a final or desired state; 3a: of or relating to a condition of full development." In postharvest physiology we consider *mature* and *ripe* to be distinct terms for different stages of fruit development (fig. 6.1). *Mature* is best defined by 2a (1) above as "having completed natural growth and development"; for fruits, it is defined in the U.S. Grade standards as "that stage which will ensure proper completion of the ripening process." This latter definition lacks precision in that it fails to define "proper completion of the ripening process." Most postharvest technologists consider that the definition should be "that stage at which a commodity has reached a sufficient stage of development that after harvesting and postharvest handling (including ripening, where required), its quality will be at least the minimum acceptable to the ultimate consumer."

Horticultural maturity is the stage of development at which a plant or plant part possesses the prerequisites for use by consumers for a particular purpose. A given commodity may be horticulturally mature at any stage of development (see fig. 6.1). For example, sprouts or seedlings are horticulturally mature in the early stage of development, whereas most vegetative tissues, flowers, fruits, and underground storage organs become horticulturally mature in the midstage, and seeds and nuts in the late stage, of development. For some commodities, horticultural maturity is reached at more than one stage of development, depending on the desired use of the product. In zucchini squash, for example, the mature product can be the fully open flower, the young fruit, or the fully developed fruit.

A qualitative difference in the relationship between maturity and edibility distinguishes many fruits from vegetables. In many fruits, such as mature (but green) bananas, the eating quality at maturity will be far less than optimal. The fruit becomes edible only after proper ripening has taken place. In contrast, in most vegetables, optimal maturity coincides with optimal eating quality.

INDICES OF MATURITY

The definition of maturity as the stage of development giving minimum acceptable quality to the ultimate consumer implies a measurable point in the commodity's development, and it also implies the

Figure 6.1

Horticultural maturity in relation to developmental stages of the plant. (Watada et al. 1984)

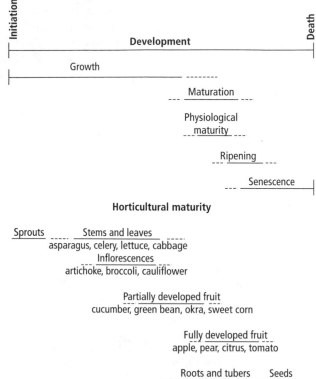

need for techniques to measure maturity. The maturity index for a commodity is a measurement or measurements that can be used to determine whether a particular example of the commodity is mature. These indices are important to the trade in fresh fruits and vegetables for several reasons.

Trade regulations. Regulations published by grower groups, marketing orders, or legally appointed authorities (such as the state departments of agriculture and the USDA) frequently include a statement of the minimum (and sometimes maximum) maturity acceptable for a given commodity. Objective maturity standards are available for relatively few commodities, and most regulations rely on subjective judgments related to the broad definitions quoted above.

Marketing strategy. In most markets the laws of supply and demand create price incentives for the earliest (or sometimes the latest) shipments of particular commodities.

This encourages growers and shippers to expedite or delay harvesting their crop to take advantage of premium prices. The minimum maturity statements in the grade standards exist to prevent the sale of immature or overmature product and the consequent loss of consumer confidence. Objective maturity indices enable growers to know whether their commodity can be harvested when the market is buoyant.

Efficient use of labor resources. With many crops the need for labor and equipment for harvesting and handling is seasonal. In order to plan operations efficiently, growers need to predict the likely starting and finishing dates for the harvest of each commodity. Objective maturity indices are vital for accurate prediction of harvest dates.

CHARACTERISTICS OF A MATURITY INDEX

Maturity measures made by producers, handlers, and quality control personnel must be simple, readily performed in the field or orchard, and require relatively inexpensive equipment. The index should preferably be objective (a measurement) rather than subjective (an evaluation). The index must consistently relate to the quality and postharvest life of the commodity for all growers, districts, and years. If possible, the index should be nondestructive.

The search for an objective determination of maturity has occupied the attention of many horticulturists working with a wide range of commodities for many years. The number of satisfactory indices that have been suggested is nevertheless rather small, and for most commodities the search for a satisfactory maturity index continues.

Two rather different problems will be addressed here. The first problem is how to measure maturity at harvest or at a subsequent inspection point. The second and more complex problem is how to predict the time at which a commodity will mature. For both problems, similar techniques may be appropriate, but the ways in which they are applied differ.

DEVELOPING A MATURITY INDEX

Many features of fruits and vegetables have been used in attempting to provide adequate estimates of maturity. Examples of those that have been proposed, or that are presently in use, are shown in table 6.1. The wide range of methods that have been devised to measure these features are summarized in table 6.2.

The strategy for developing a maturity index is

- To determine changes in the commodity throughout its development.

- To look for a feature (size, color, solidity, etc.) whose changes correlate well with the stages of the commodity's development.

- To use storage trials and organoleptic assays (taste panels) to determine the value (or level) of the maturity index that defines minimum acceptable maturity.

Table 6.1. Maturity indices for selected fruits and vegetables

Index	Examples
Elapsed days from full bloom to harvest	Apples, pears
Mean heat units during development	Peas, apples, sweet corn
Development of abscission layer	Some melons, apples, feijoas
Surface morphology and structure	Cuticle formation on grapes, tomatoes Netting of some melons Gloss of some fruits (development of wax)
Size	All fruits and many vegetables
Specific gravity	Cherries, watermelons, potatoes
Shape	Angularity of banana fingers Full cheeks of mangoes Compactness of broccoli and cauliflower
Solidity	Lettuce, cabbage, Brussels sprouts
Textural properties:	
Firmness	Apples, pears, stone fruits
Tenderness	Peas
External color	All fruits and most vegetables
Internal color and structure	Formation of jellylike material in tomato fruits
	Flesh color of some fruits
Compositional factors:	
Starch content	Apples, pears
Sugar content	Apples, pears, stone fruits, grapes
Acid content, sugar/acid ratio	Pomegranates, citrus, papaya, melons, kiwifruit
Juice content	Citrus fruits
Oil content	Avocados
Astringency (tannin content)	Persimmons, dates
Internal ethylene concentration	Apples, pears

- When the relationship between changes in the maturity index quantity and the quality and storage life of the commodity has been determined, an index value can be assigned for the minimal acceptable maturity.

- To test the index over several years and in several growing locations to ensure that it consistently reflects the quality of the harvested product.

FEATURES USED AS MATURITY INDICES
Chronological features
For certain crops (fast-rotation vegetables, such as radish, and perennial tree crops growing in short summer environments), maturity can be defined chronologically, for example, as days from planting or as days from flowering. Chronological indices are seldom perfect, but they do permit a degree of planning, and they are widely used. For some crops, the chronological method is refined by calculating heat units accumulated during the growing period, which modulates the chronological index according to the weather pattern during the growing season.

Physical features
A wide range of physical features are used to assess the maturity of various commodities.

Size, shape, and surface characteristics. Changes in the size, shape, or surface characteristics of fruits and vegetables are commonly used as maturity indices. For example, vegetables in particular are harvested when they have reached a marketable size and before they become too large. Maturity in bananas is determined by measuring the diameter of the fingers; changes in the surface gloss or feel (waxiness) are used as a practical tool in harvesting of some melons such as honeydew (see chapter 33, table 33.4).

Abscission. In many fruits, during the later stages of maturation and the start of ripening, a special band of cells, the abscission zone, develops on the stalk (pedicel) that attaches the fruit to the plant. The abscission zone permits the fruit to separate from the plant. Measuring the development of this zone (degree of separation) is possibly the oldest of all maturity indices. Abscission force (the force required to pull the fruit from the tree) is not generally used as a formal maturity index, but the development of the abscission zone, or "slip," in the netted muskmelons (see chapter 33, fig. 33.9) is used to determine their maturity.

Color. The color change that accompanies maturation in many fruits is widely used as a maturity index. Objective measurement of color requires expensive equipment (fig 6.2), and although the human eye is unable to give a good evaluation of a single color, it is extremely sensitive to differences between colors. Color comparison techniques are therefore commonly used to assess fruit maturity (fig 6.3). Color swatches may be used to determine external or internal color.

Accurate devices employing state-of-the-art electronics and optics now permit objective color measurements. As the price of such devices has fallen, they have replaced comparison techniques in many cases. For example, digital color examination is now used in the sorting of mechanically harvested processing tomatoes.

Texture. Maturation of fruits is often accompanied by softening; overmature vegetables frequently become fibrous or tough. These textural properties can be used to determine maturity. They are measured with

Table 6.2. Methods of maturity determination

Index	Method of determination	Subjective	Objective	Destructive	Non-destructive
Elapsed days from full bloom	Computation		×		×
Mean heat units	Computation from weather data		×		×
Development of abscission layer	Visual or force of separation	×	×		×
Surface structure	Visual	×			×
Size	Various measuring devices, weight		×		×
Specific gravity	Density gradient solutions, flotation techniques, vol/wt		×		×
Shape	Dimensions, ratio charts	×	×		×
Solidity	Feel, bulk density, gamma rays, X-rays	×	×		×
Textural properties:					
Firmness	Firmness testers, deformation		×	×	
Tenderness	Tenderometer		×	×	
Toughness	Texturometer, fibrometer (also: chemical methods for determination of polysaccharides)		×	×	
Color, external	Light reflectance		×		×
	Visual color charts	×			×
Color, internal	Light transmittance, delayed light emission		×		×
	Visual examination	×		×	
Compositional factors:					
Dry matter	Sampling, drying		×	×	
Starch content	KI test, other chemical tests		×	×	
Sugar content	Hand refractometer, chemical tests		×	×	
Acid content	Titration, chemical tests		×	×	
Juice content	Extraction		×	×	
Oil content	Extraction, chemical tests		×	×	
Tannin content	Ferric chloride test		×	×	
Internal ethylene	Gas chromatography		×	×	×

instruments that measure the force required to push a probe of known diameter through the flesh of the fruit or vegetable (fig 6.4). The solidity of lettuce, cabbage, and Brussels sprouts is an important quality and maturity characteristic. In the case of lettuce, gamma-ray equipment has been devised to measure head firmness, but the technique has not been adopted commercially.

Chemical changes. The maturation of fruits and vegetables is often accompanied by profound changes in their chemical composition. Many of these changes have been used in studies of maturation, but relatively few have provided satisfactory maturity indices because they usually require destructive sampling and complex chemical analysis. Chemical changes that are used for

Figure 6.4

Using the UC firmness tester to measure the flesh firmness of apples.

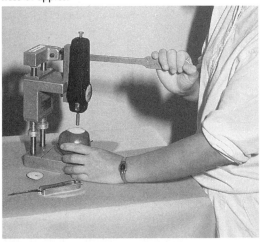

Figure 6.2

Colorimeter used to measure surface color of apples.

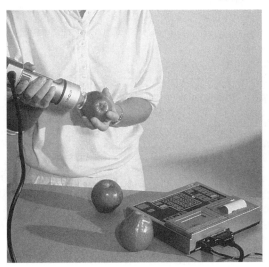

Figure 6.5

Measuring soluble solids content with a refractometer.

Figure 6.3

Color matching used for maturity grading.

Figure 6.6

Treating cut apples with an iodine solution reveals the disappearance of starch (which stains dark) as apples mature.

maturity estimation include the change in total soluble solids, measured using a refractometer (fig. 6.5); changes in the distribution of starch in the flesh of the commodity, measured using a starch-iodine reaction (fig. 6.6); acidity, determined by titration; and the sugar to acid ratio, which is used as the legal maturity index for citrus.

The unsatisfactory nature of chemical tests for maturity is exemplified by the old oil content measurement for avocados, which has been replaced by the determination of percent dry weight because of the time-consuming and complex nature of oil determination.

French scientists have developed an interesting approach to objective maturity (and quality) determination of harvested melons; they remove a slender cylinder of flesh from each melon and rapidly determine its sugar content by measuring the refractive index of the juice. The outer portion of the cylinder is replaced, and the melon is accepted or rejected based on the sugar reading.

New opportunities in chemical analysis are exemplified by the development of near-infrared technologies for examining the composition of fruits and vegetables, and rapid sensor technology for determining volatile profiles in harvested products. The former is able to measure sugars in fruits nondestructively, and the latter is sufficiently rapid to enable determination of melon maturity in the field. Researchers have found, for example, that the sugar content of peaches can be accurately determined using near-infrared absorption profiles. As melons mature, their production of aroma volatiles increases dramatically. An instrument has been developed that enables this increased production to be an indicator of harvest readiness.

Physiological changes. The maturation of commodities is associated with changes in their physiology, as measured by changing patterns of respiration and ethylene production. The problem with using these characteristics in assessing maturity is the variability in absolute rates of ethylene production and respiration among similar individuals of the same commodity. The techniques are also complex and expensive to implement on a commercial scale. Nevertheless, the rate of ethylene production of a sample of apples is used by some producers to establish the maturity of the apples, and it is particularly

used to identify those that will be suited to long-term controlled atmosphere storage.

PREDICTING MATURITY

Predicting when a commodity will mature is more complex than assessing its maturity at or after harvest. The basic requirement for prediction is a measurement whose change during the commodity's development can be modeled mathematically to reveal a pattern or patterns of change. Once the pattern of change is established for the measurement, measurements made early in the season can be compared with the pattern in order to predict the date at which the commodity should reach minimum acceptable maturity. The way in which this strategy has been applied can best be illustrated by the following examples.

APPLES
Although the literature on the prediction of maturity in apples is voluminous, no truly satisfactory method has yet been proposed. The use of climatic data to predict the date of harvest by a modification of the "days from full bloom" index noted in table 6.1, even when adapted by using "days from the 'T' stage" has provided only general predictions of the harvest date.

In an attempt to provide a more satisfactory prediction of the maturation date, researchers have examined a number of changes that occur during fruit development. Measurement of respiration, ethylene production, sugar content, starch content, and the changing firmness of fruit each failed to meet some of the criteria for a satisfactory maturity index, and they proved to be too variable to permit prediction of maturation date. The "starch pattern," an old method of determining apple maturity, refined by assigning scores to a range of patterns, has proved to be a good index. Changes in the mean starch index score during the period prior to harvest are readily analyzed as a linear regression, and the date of minimal acceptable maturity and can be predicted several weeks in advance (fig. 6.7).

AVOCADOS
The State of California for many years promulgated a minimum oil content as the maturity standard for avocados. This index

has been unsatisfactory, since it is difficult to apply and because some avocados that have more than the minimum oil requirement may be lacking in flavor quality. However, raising the minimum oil content might eliminate from the market particular avocado varieties whose flavor quality is adequate at a low oil content. Using taste panel evalua-tions to determine quality, researchers have shown that the patterns of dry weight accu-mulation, or the growth of avocado fruit, can be used not only to determine when mini-mum acceptable maturity has been achieved but also to predict the date when it will be achieved (fig 6.8). Taste panel scores increased as oil content increased, and oil

Figure 6.7

Change in starch index values for maturing Granny Smith apples.

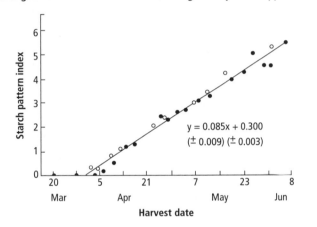

Figure 6.9

Relationship between dry weight and oil content of avocado. (Lee et al. 1983)

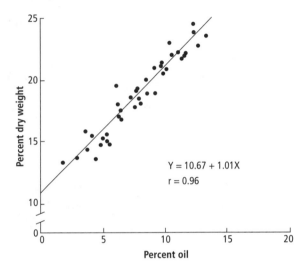

Figure 6.8

Changing acceptability and oil content of maturing avocados. (Lee and Young 1983)

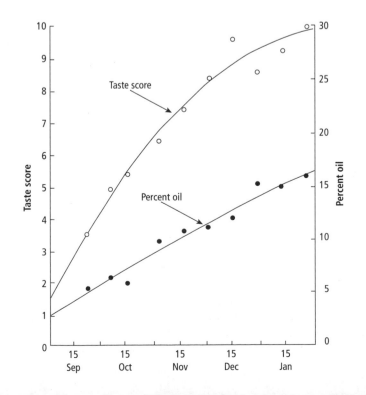

Figure 6.10

Changes in selected chemical and physical parameters of kiwifruit during growth and development.

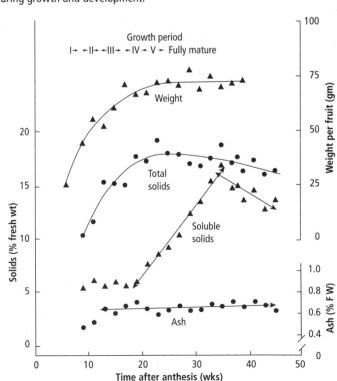

content was found to be closely correlated with the percent dry weight (fig 6.9). Consequently, the California minimum maturity index was changed from oil content to percent dry weight.

KIWIFRUIT

As a prelude to developing a maturity index for kiwifruit, researchers measured changes in a wide range of chemical and physical characteristics during growth and development (fig. 6.10). This information was compared to storage life and taste panel results to decide on possible methods for determining, and if possible predicting, the time of minimal acceptable maturity for this crop. It seemed possible that soluble solids content and fruit firmness might provide a suitable maturity index, but in repeating the experiments over several seasons, changes in the firmness of the fruit were found to be highly variable and unrelated to fruit quality (fig. 6.11). In New Zealand, a minimum maturity index of 6.25% soluble solids has now been used for many years. The change in soluble solids in the 6 weeks prior to the normal harvest date can be used, with regression analysis, to predict the date of harvest for different orchards, seasons, and growing districts (fig. 6.12).

REFERENCES

Arthey, V. D. 1975. Quality of horticultural products. New York: Wiley. 228 pp.

Eskin, N. A. M., ed. 1989. Quality and preservation of vegetables. Boca Raton, FL: CRC Press. 313 pp.

———. 1991. Quality and preservation of fruits. Boca Raton, FL: CRC Press. 212 pp.

Hulme, A. C., ed. 1971. The biochemistry of fruits and their products. Vol. 2. New York: Academic Press. 788 pp.

Kader, A. A. 1999. Fruit maturity, ripening, and quality relationships. Acta Hort. 485:203–208.

Lee, S. K., and R. E. Young. 1983. Growth measurement as an indication of avocado maturity. J. Am. Soc. Hort. Sci. 108:395–397.

Lee, S. K., R. E. Young, P. M. Schiffman, and C. W. Coggins Jr. 1983. Maturity studies of avocado fruit based on picking dates and dry weight. J. Am. Soc. Hort. Sci. 108:390–394.

Pattee, H. E., ed. 1985. Evaluation of quality of fruits and vegetables. Westport, CT: AVI. 410 pp.

Ryall, A. L., and W. J. Lipton. 1979. Handling, transportation and storage of fruits and vegetables. Vol. 1, Vegetables and melons. 2nd ed. Westport, CT: AVI. 588 pp.

Ryall, A. L., and W. T. Pentzer. 1982. Handling, transportation and storage of fruits and vegetables. Vol. 2, Fruits and tree nuts. 2nd ed. Westport, CT: AVI. 610 pp.

Seymour, G. G., J. E. Taylor, and G. A. Tucker, eds. 1993. Biochemistry of fruit ripening. London: Chapman and Hall. 454 pp.

Watada, A. E., R. C. Herner, A. A. Kader, R. J. Romani, and G. L. Staby. 1984. Terminology for the description of developmental stages of horticultural crops. HortScience 19:20–21.

Figure 6.11

Changes in fruit firmness during maturation of kiwifruit.

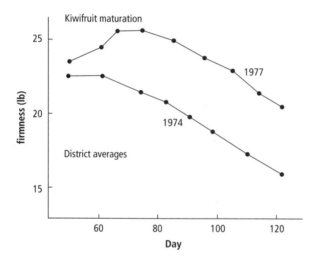

Figure 6.12

Changes in soluble solids content during the 6 weeks before harvest of kiwifruit can be used to predict harvest date.

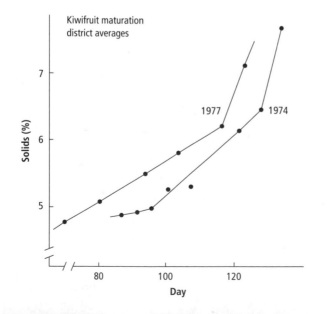

Harvesting Systems

James F. Thompson

The goals of harvesting are to gather a commodity from the field at the proper level of maturity with a minimum of damage and loss, as rapidly as possible, and at a minimum cost. Today, as in the past, these goals are best achieved through hand-harvesting in most fruit, vegetable, and flower crops.

HAND-HARVESTING

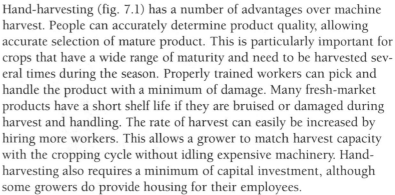

Hand-harvesting (fig. 7.1) has a number of advantages over machine harvest. People can accurately determine product quality, allowing accurate selection of mature product. This is particularly important for crops that have a wide range of maturity and need to be harvested several times during the season. Properly trained workers can pick and handle the product with a minimum of damage. Many fresh-market products have a short shelf life if they are bruised or damaged during harvest and handling. The rate of harvest can easily be increased by hiring more workers. This allows a grower to match harvest capacity with the cropping cycle without idling expensive machinery. Hand-harvesting also requires a minimum of capital investment, although some growers do provide housing for their employees.

The main problems with hand-harvesting center around labor management. Labor supply is a problem for growers who cannot offer a long employment season. Labor strikes during the harvest period can be costly. In recent years, costs associated with complying with U.S. and state government labor regulations have increased significantly. In spite of these problems, quality is so important to marketing fresh-market commodities successfully that hand-harvesting remains the dominant method of harvest of most fruits and vegetables and for all cut flowers.

Effective use of hand-labor requires careful management. New employees must be trained to harvest the product at the required quality and at an acceptable rate of productivity. Employees must know what level of performance is expected of them and must be encouraged and trained to reach that level. Well-managed employees enjoy their jobs more and can be more productive than those who are poorly managed. Benefits such as paid vacations and health insurance help ensure the return of already-trained employees.

Maintaining product sanitation requires that workers be provided with regularly cleaned, supplied, and emptied toilet facilities. Workers also need a source of potable water and washing supplies and must be trained in proper hygiene procedures. Complete documentation of worker hygiene procedures is needed in case of a sanitation problem with a batch of product.

Machines are used to aid hand-harvest in some commodities. Belt conveyors are used in vegetable crops such as lettuce and melons to move them to a central loading or in-field handling station. Scoops with rods protruding from the end are used by workers to comb through some berry crops. Platforms or moveable worker positioners have been used in place of ladders in crops such as dates, papayas, and bananas. Lights are used in California for night harvest of some crops, when temperatures are cool and when worker effectiveness and product quality are at their best. Numerous other mechanical aids have been tried, but few increase productivity enough to warrant their expense.

Figure 7.1

Hand-harvesting, shown here for strawberries, is the primary method for harvesting fresh-market horticultural commodities in the United States.

MECHANICAL HARVESTING

Mechanical harvest is currently used for fresh-market crops that are roots, tubers, or rhizomes and for nut crops. Vegetables that are grown below ground (radishes, potatoes, garlic, carrots, beets, and others) are always harvested only once (once-over), and the soil can be used to cushion the product from machine-caused mechanical injury. Tree nuts and peanuts are protected by a shell and easily withstand mechanical handling. A number of products destined for processing such as tomatoes, wine grapes, beans, peas, prunes, clingstone peaches, olives, and some leafy green vegetables are machine harvested because harvest damage does not significantly affect the quality of processed product. This is often because the product is processed quickly after harvest. These crops have also been amenable to new production techniques and breeding that allow the crop to be better suited to mechanical harvest.

A main advantage of mechanical harvest equipment is that machines can often harvest at high rates. Tree nut harvesters, for example, attach a shaking mechanism to the tree and remove most of the nuts in a few seconds. The nuts are either caught on a fabric-covered frame or picked up from the ground by other machines. This allows an orchard to be harvested very quickly compared to hand-shaking with poles.

Machine harvest also reduces management problems associated with workers. In some cases the need for workers is reduced to the point where full-time employees can do the bulk of the work. This eliminates problems associated with locating and training a temporary work force during harvest. Machines usually improve working conditions because the strenuous jobs of product lifting and transporting are mechanized. Workers are needed mainly for operating the equipment and perhaps for sorting. This may allow a greater employment pool because physically weaker people will be able to function well in the mechanized job.

Effective use of mechanical harvesters requires operation by dependable, well-trained people. Improper operation results in costly damage to expensive machinery and can quickly cause great crop damage. Both regular and emergency maintenance must be performed. The commodity must be grown to accept mechanical harvest. For example, trees must be pruned for strength and to minimize fruit damage caused by fruit falling through the tree canopy. Maximum and uniform stand establishment is necessary for vegetable crops. Cropping patterns must also be set up to use the expensive equipment as long as possible to pay for the high capital investment. This can severely limit the production choices of some farmers.

MECHANICAL HARVEST PROBLEMS

Machines are rarely capable of selective harvest. This often means that mechanical harvest will not be feasible until the crop or production techniques can be modified to allow once-over harvest. Harvest machinery often causes excessive product damage to perennial crops (e.g., bark damage from a tree shaker). Also, it is often quite expensive, and there can be concern that the machines may become obsolete before they are paid for. Handling capacity may not be able to cope with the high rate of harvest. For example, in many vegetables, box filling and product preparation such as trimming and plastic film wrapping may be much more time-consuming than picking. Automating picking alone may not significantly reduce the amount of labor required.

There may be unintended social impacts to lower labor requirements. For example, eliminating harvest jobs during one part of the year may reduce employment in the general area enough so that farmworker families leave to find a location with greater seasonal employment.

The prospects for increased mechanized harvest in the future are uncertain. Equipment has already been developed for the easiest crops to mechanically harvest, and much of this machinery was successfully tested many years ago. Since the 1960s there have been very few additional crops that have been successfully machine harvested, in spite of considerable research in the intervening years. Even robotics have not been effective in successfully adding new crops to those that can be machine harvested. High labor costs or lack of labor availability were thought to be factors that would force mechanization; but it may be that they cause crop production to move to countries that do not have these constraints.

REFERENCES

American Society of Agricultural Engineers (ASAE). 1983. Status of harvest mechanization of horticultural crops. St. Joseph, MI: ASAE. 78 pp.

Grierson, W., and W. C. Wilson. 1983. Influence of mechanical harvesting on citrus quality: Cannery vs. fresh fruit crops. HortScience 18:407–409.

Kader, A. A. 1983. Influence of harvesting methods on quality of deciduous tree fruits. HortScience 18:409–411.

Kasmire, R. F. 1983. Influence of mechanical harvesting on quality of nonfruit vegetables. HortScience 18:421–423.

Morris, J. R. 1983. Influence of mechanical harvesting on quality of small fruits and grapes. HortScience 18:412–417.

_____ 1990. Fruit and vegetable harvest mechanization. Food Technol. 44(2): 97–101.

O'Brien, M., B. F. Cargill, and R. B. Fridley. 1983. Principles and practices for harvesting and handling of fruits and nuts. Westport, CT: AVI. 636 pp.

Peterson, D. L. 1992. Harvest mechanization for deciduous tree fruits and brambles. HortTechnology 2:85–88.

Rosenberg, H. R., V. J. Horwitz, D. L. Egan. 1995. Labor management laws in California agriculture. Oakland: Univ. Calif. Div. Ag. and Nat. Res. Leaflet 21404.

Sarig, Y. 1993. Robotics of fruit harvesting—A state-of-the-art review. J. Agr. Eng. Res. 54:265–280.

Studer, H. E. 1983. Influence of mechanical harvesting on the quality of fruit vegetables. HortScience 18:417–421.

8

Preparation for Fresh Market

James F. Thompson,

Elizabeth J. Mitcham, and

F. Gordon Mitchell

The ability to deliver high-quality fresh fruits and vegetables to market requires attention to details beginning with cultural practices in the field and continuing until the produce is consumed. Poor cultural practices during production, such as improper pruning, thinning, fertilization, and disease control, can reduce the quality of the harvested produce. Quality loss is also caused by rough handling during and after harvest, as illustrated by cumulative data on impact bruising to Bartlett pears at various locations in a pear handling operation in California (table 8.1). Although this is an extreme example of what might be experienced commercially, it shows the effects of repeated mechanical damage to the product during postharvest handling. Protection is vital, in both production and postharvest handling, to avoid immediate causes of deterioration and to slow deterioration that may occur later in the distribution channels.

FIELD PACKING

Soft fruits such as berries and many vegetables are packed directly in the field. This eliminates product handling, can reduce the time between picking and cooling, and eliminates the expense of having a packing facility. The disadvantages are that quality control is more difficult to achieve in the field than in a packinghouse; mechanical grading, sorting, or trimming cannot be done; postharvest chemicals cannot be applied; and workers must work in a less comfortable environment.

Field-packing can be as simple as a picker with a box or as complex as a specially designed machine that carries supplies and has individual packing stations. For example, strawberry pickers place the fruit into baskets that are already loaded into a master container; the container is often carried on a small wheeled dolly (fig. 7.1). Strawberries that do not meet fresh-market grade standards are collected in a can and are used for processed product. As a tray is filled, the picker takes it to a truck, where a supervisor checks for quality and tallies the picker's tray count for payment (fig. 8.1). Table grapes are packed at the end of a vine row on a wheeled cart under a sun shade (fig. 8.2). The cart holds field boxes of picked fruit and often has places for first- and second-quality packed boxes. The packer grades, trims, and bags the product; packed boxes are placed on the ground to be picked up and palletized by another crew. Many leafy vegetables are packed on mobile packing lines (fig. 8.3). Pickers select mature product and place it next to a packer on the mobile unit. The packer trims and sometimes wraps the product and places it in a box. Boxes are assembled on the machine, and packed boxes are palletized. When a load of pallets is ready, a transfer truck backs up to the slowly moving packing line and the pallets are transferred to the truck, which takes them to a cooler. The time from picking to cooling can be less than 1 hour.

HARVESTING FOR PACKING IN A CENTRAL FACILITY

FIELD CONTAINERS

Most fresh-market products are currently harvested by hand into buckets or bags, which are then emptied into field bins for transport to packing or storage operations. Some vegetables (such as tomatoes and melons) are loaded into a fiber-reinforced plastic gondola attached to a highway trailer. Metal or plastic picking buckets are typically used for

Table 8.1. Cumulative levels of impact bruising on Bartlett pears during postharvest handling

Location	Bruised fruits (%)
Tree	0
Picker bag	14
Field bin	26
After dumping	38
After sizing	82

Figure 8.1

Almost all strawberries produced in California are field-packed. The picker places harvested berries directly in a consumer package. Corrugated fiberboard trays are palletized at the edge of the field, and pallets are taken directly to a cooler.

Figure 8.2

California table grapes are usually packed on a wheeled cart at the end of the vine row. Workers on a flatbed truck periodically pick up and palletize packed boxes.

Figure 8.3

Many green vegetables are field-packed on a packing line that moves through the field as the crop is picked. The crew of a transfer truck unloads pallets for transport to the cooling facility as the packing line continues to move through the field.

the softer fruit (such as cherries), and bottom-dump picking bags are used for fruit with less potential for compression bruising (such as citrus). Certain delicate fruits are transferred from buckets to 9-kg (20-lb)-capacity field boxes (some sweet cherries), or are picked directly into field lugs (table grapes), or are picked into buckets and packed directly into packages from the buckets (some stone fruits).

In California, most field bins are standardized at 119 cm (47 in) by 119 cm (47 in) outside length and width and 61 cm (24 in) inside depth. Some kiwifruit and sweet cherries are handled in bins 30 cm (12 in) deep to avoid compression bruising. Wooden bins are made of 19-mm (¾-in) plywood that is smooth or coated on the inside. The bins are vented; to avoid cutting the fruit, ventilation slots are normally cut with the inside edge tapered. Bin surfaces should be clean and smooth. Frequent washing, water dumping, or hydrocooling can cause the surface of plywood bins to become rough, increasing fruit abrasion and cutting problems. Coatings

(paint or varnish-type) are available to reduce this problem. Using separate plastic liners in wooden bins, especially bubble liners, effectively reduces abrasion injury. However, special care is needed to maintain side venting, and the liner must be removed when submersion water dumping is practiced.

Plastic bins are also commonly used. They have many advantages: they are lighter, cause less product abrasion, have significantly lower maintenance costs, have increased ventilation, do not absorb product moisture during storage, and are easily cleaned. However, they cost more than wooden bins. As with wooden bins, plastic bins need periodic washing to remove soil from interior walls. (Ventilation requirements for cooling are discussed in chapter 11.)

Careful field supervision is critical in protecting fruits and vegetables from injury. Physical injuries can result from dropping

Table 8.2. Effect of drop height on incidence and severity of impact bruising on Bartlett pears

Drop height		Fruit bruised (%)	Bruise severity (score)*
(in)	(cm)		
0	0	0	0
4	10	40	0.6
6	15	44	0.6
9	23	56	1.0
12	30	78	1.2
16	41	100	2.3

Note: *Scored on 0–5 scale: 0 = no damage, 5 = unmarketable.

produce into picking buckets or bags, overfilling picking containers, striking the containers (especially soft-sided bags) against limbs and ladders, leaning against picking bags while picking, transferring product into field bins or lugs without care, and overfilling field containers. Even short drops can cause substantial impact bruising of commodities, such as Bartlett pears (table 8.2).

TRANSPORT TO THE PACKINGHOUSE

Mechanical damage to products can occur during field transport. Impact bruises occur when bins or boxes are dropped or bounced. Compression bruises result when overfilled field containers are stacked. Bins should be filled level or slightly underfilled to avoid compressing the products in lower layers. Abrasion or vibration bruises can occur when perishables move or vibrate against rough surfaces or against other products during transport.

Reducing physical injuries to products during transportation can be achieved by

- minimizing the distance that forklifts move bins in the field

- reducing rough handling or dropping of bins during loading

- grading farm roads to eliminate ruts, potholes, and bumps

- routing trucks to avoid rough roads or driving slower on rough roads

- using suspension systems on all vehicles (air suspension systems greatly reduce or eliminate transport vibration damage)

- reducing tire air pressure on transport vehicles

- installing plastic liners on the sidewalls of wooden bins (bottom liners are not needed; plastic bins should be cleaned to remove dirt, which may abrade the product)

- using plywood bin covers for long-distance transport (covers are made of 10-mm (3/8-in) plywood cut to fit inside the bin, faced with a double layer of bubble liner that is 13-mm (1/2-in) thick, and held against the fruit by short rubber straps)

TEMPERATURE PROTECTION

Reduce the time between harvest and transport. Many products, such as strawberries, lose measurable amounts of quality during just 1 hour of delay between harvest and cooling. Frequent transfer of products to the cooler or packinghouse minimizes the time for heating and deterioration. If the harvest rate is slow, use several small trucks to ensure frequent trips to the packinghouse rather than keeping a larger truck waiting until it is full to make a trip.

While in the field, shade product after harvest. Product can warm up considerably above the air temperature (fig. 8.4), and dark-colored products can warm even faster than light-colored ones. In orchards and vineyards, product can be placed in the shade of the trees or vines. If natural shade is not available, use portable shading to reduce exposure to the sun. Portable shading may be more effective than natural shading, because it will not shift as much as the sun moves in the sky. Placing empty packages or lugs over the top of stacks of packages provides some protection. Shading products keeps them from warming above the ambient air temperature. However, even a mild breeze causes harvested product in the shade to quickly warm to near the ambient air temperature. During periods of high field temperatures, product should be harvested early in the day to reduce product warming. Early harvest also increases product turgidity (firmness) and size and reduces the amount and cost of cooling needed. Because of the high turgidity early in the day, some products, such as citrus, are purposely harvested later in the day to avoid injury to the skin.

Rapid transport to the cooler is important. Where transport time is extended because of distance or delays, covering the load can help by reducing product exposure to the sun and reducing airflow through and over

Figure 8.4

Effect of sun exposure and position in box on field warming of sweet cherries.

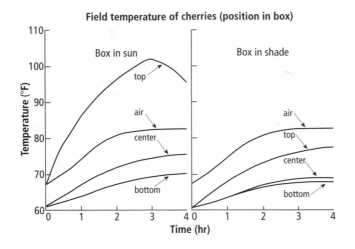

the load. Excessive airflow through the load during transport can quickly warm the product to the ambient air temperature and increase moisture loss through evaporation. If transport of cool, early-harvested product is delayed until ambient temperatures are high, warming during transport can be significant, speeding product deterioration and increasing the cost of subsequent cooling.

Tarpaulins used to cover loads during transport should be a light color (white or silver are best) and should be kept clean to maintain good heat reflection and sanitation. Tarpaulins should be supported by a frame to maintain an air space over the load. The only exception is when insulated tarpaulins are used to delay heating of the product. If the tarpaulin extends down over all sides of the load, it blocks airflow through the load and limits product warming when ambient air temperatures are high. Wetting fabric tarpaulins can further reduce warming by providing an evaporative cooling surface. Take care to ensure that any load cover does not completely seal the load and confine heat within the load. Under difficult transport conditions, place wetted pads, ice packs, or crushed ice over the product in field bins before transport.

PREPARATION FOR PACKING

DELIVERY TO THE PACKINGHOUSE

The typical operations in a packinghouse are illustrated in figure 8.5. Product received from the field is usually weighed, and lot and ranch identification data are recorded. Sometimes a quality inspection occurs at this point, although more commonly the grower is paid on the basis of the grade and quantity

Figure 8.5

Schematic of the typical unit operations in a mechanized packinghouse.

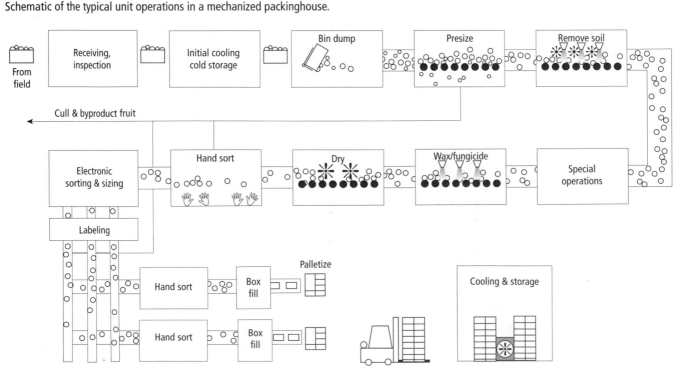

of the final packed product. In this case, lot identification is maintained throughout the packing process.

INITIAL COOLING AND STORAGE

In some operations, the product may be cooled before packing. This is common for products that are stored to extend the marketing period, such as apples, pears, kiwifruit, citrus, potatoes, cabbage, sweet potatoes, garlic, dry onions, winter squashes, and carrots. For other products, cold storage allows them to be held for a short period before packing without appreciable loss of quality, and allows the packing line to operate at an even pace. It also allows the packing operation to begin at a regular time the next day without waiting for that day's first harvested product to arrive. If the storage time will be short, the products are cooled to an intermediate temperature before packing, and cooling is finished after packing. Many products are more susceptible to mechanical damage when they are cold. Intermediate temperatures slow the loss of quality during short-term storage, reduce mechanical damage susceptibility due to low temperatures, and reduce energy costs associated with cooling cull fruit.

Some products are cured after harvest and before storage or marketing. For example, onions and garlic are cured to dry the necks and outer scales; new-crop potatoes and sweet potatoes are cured to develop wound periderms over cut, broken, or scuffed surfaces. Curing helps heal injuries inflicted during harvest, reduces water loss, and prevents entry of decay-causing organisms during storage. Curing may be done in the field in hot, dry areas (garlic, onions), in curing rooms (sweet potatoes), or during transport (new-crop potatoes).

BIN DUMPING

Most products are dumped directly onto a conveyor belt. This was once done by hand-dumping of field boxes, which is still occasionally used. In large operations, box dumping is mechanized to provide uniform product flow and to reduce product damage by careless handling. Commodities that are easily injured do not tolerate dumping.

Most fruits and vegetables are now handled in pallet bins and dumped by either dry bin dumps or water dumps. In dry bin dumps (fig. 8.6) the bin is covered with a padded lid then slowly tilted, and the product is delivered through a controlled opening in the lid. Electronic control of the delivery belt allows control of product flow to the sorting line. Properly designed dry bin dumps provide a uniform flow of product with minimum injury to the product.

Several types of water dumps are used. Products may be dumped from the bin directly into water. The water reduces product-to-product impact damage if care is taken to not dump too much product into the dump tank. Flotation dumps submerge the bins and the product floats free (fig. 8.7). The most common flotation dumps used for tree fruits submerge the entire bin as it travels along a conveyor. Pumps circulate the dump-tank water to move the floating product to an elevator, where it is rinsed and transferred to the sorting line. For flotation of commodities that are heavier than water, a salt (often sodium sulfate) must be added to the dump water to increase its density.

Figure 8.6

Inversion-type dry bin dumps are widely used in fruit packing lines. The padded lid is clamped over the bin before inversion; the gate in the lid is then opened to allow fruit to flow out during dumping.

Figure 0.7

Flotation dumps are used especially for apples and pears. Filled bins are plunged underwater to allow fruit to float. The empty bin is then raised out of the tank.

Some vegetables are transported to the packing house in bulk highway trailers (gondolas) holding about 9 to 11 metric tons (10 to 12 tons) of product. These gondolas often have sidewall doors near the bottom of the trailer. Round products, such as melons, will roll out of the door; in some cases the trailer is tilted slightly to make unloading easier. Products that do not easily roll, such as carrots, tomatoes, and white potatoes, may be emptied from the trailer using a flume system. Large volumes of water are manually directed into the trailer to wash the product out through the sidewall door.

Sanitation is important in water dumps because dump water quickly accumulates a high concentration of fungal spores and bacteria, which can infect fresh wounds from harvesting and handling. Dump tanks should be designed for rapid draining and filling and for easy cleaning. Chlorine at a concentration of 50 to 200 ppm and a water pH of 6.5 to 7.5 is often used to control decay-causing organisms, but not all products tolerate chlorine exposure. Chlorine can be applied as liquid sodium hypochlorite or gaseous chlorine. When sodium hypochlorite is used, water systems must be changed frequently to prevent sodium accumulation and potential product damage. Chlorine reacts with organic matter in the water and is degraded over time. Automatic chlorination systems maintain chlorine levels continuously at a desired concentration, while periodic additions of chlorine generally result in lower-than-desired concentra-

tions prior to the addition of chlorine. Water ozonization is used as an alternative to chlorination in some packinghouses.

PRESIZING

Presizers, usually located immediately after the dump, are designed to eliminate all product below a minimum size. This reduces the amount of product flowing over the balance of the packing line and increases overall equipment capacity. Presizers often eliminate product at only a single size designation and are usually fairly simple designs. A commonly used unit has a series of equally spaced rollers that allow product with a minimum dimension less than the roller spacing to fall through and be conveyed to processing outlets or the cull bin. The rollers move forward, conveying the adequately sized fruit to the next operation. In some operations, some hand-sorting is also done at this point to remove obviously unpackable product. Decayed product may also be removed to reduce contamination of the rest of the facility.

CLEANING AND WASHING

Some products, particularly root and tuber vegetables grown close to the soil, may need cleaning to remove soil and other contaminants. Detergent washes are sometimes used with soft brushes or sponges followed by clear water rinsing. Many peaches receive wet brushing to remove the trichomes (fuzz). Oranges are sometimes washed with a high-pressure spray to remove scale insects and surface mold.

SPECIAL OPERATIONS

A wide range of special operations may be needed to prepare the products for final sorting. For example, unwanted leaves, stems, and roots are removed from some vegetables; asparagus spears are trimmed to length; and clusters of cherries are separated into single fruits using a system of saws.

DISEASE CONTROL

Some postharvest disease control treatments may be applied during packing. Heat treatment, especially hot water treatment, has been studied for many products. It is widely used for papayas and oranges, typically before or at the start of packing. Fungicide applications, if used, are commonly applied while the fruit is spread on the conveyor belts or rollers, often immediately after washing. Fungicides are often incorporated into fruit waxes to aid in achieving a uniform surface application. All chemical applications must be made in strict conformity to government regulations.

WAXING

Some fruits and fruit-type vegetables are waxed (fig. 8.8) to reduce water loss, replace natural waxes removed during washing, cover injuries such as those caused by peach de-fuzzing, act as carriers for fungicides, or improve the product's cosmetic appearance.

Figure 8.8

Some fruits and vegetables are washed and waxed during preparation for market. Here, a cold-water emulsion wax, often containing fungicides, is applied to fruit after washing.

Figure 8.9

Hand-sorting of fruit is used mainly for removal of blemished product and grading. This sorting table reduces the amount of reaching workers must do. The dark background allows easier viewing of blemishes on dark red fruit.

Waxes must be approved "food grade" materials. The two most common types are mineral-oil-based waxes and natural waxes such as carnauba. Both can be formulated as water-based emulsions or with a solvent. Solvent-based formulations dry quickly, but their use may be restricted by air pollution regulations. Water emulsions are commonly used in California; they require hot air drying after application. Studies indicate that waxes reduce the rate of water loss by more than about one-third, but they also reduce product gas exchange and may interfere with normal aerobic respiration.

DRYING

Waxes formulated in a water emulsion must be dried after application to achieve proper wax properties. Convection dryers are most commonly used. Air is heated with hot water or a gas flame using a heat exchanger. Most of the drying air is recirculated to reduce energy use and to control its humidity. A wax emulsion on oranges can be dried with an exposure of 2.5 minutes to air at 49°C (120°F).

SORTING LINE
Hand-sorting

Hand-sorting can be used to segregate product by color, size, and grade. Hand-sorting demands good equipment design (fig. 8.9), and the design should have adequate space for sorting personnel. The number of sorters depends on the amount of product per hour; amount of diverted product (subgrade, alternate grade, and so on); the number of decisions or separations required by color, shape, or defect; and the size of the product being sorted. Small fruits and vegetables require more sorting decisions per package than large product. The speed of the sorting belt should be variable to adjust for differences in product quality. For accurate sorting, conveying equipment should allow sorters to see the entire surface of the product. Systems that both turn and rotate each piece provide the greatest surface visibility. These systems must be carefully adjusted to avoid fruit injury.

Taking steps to reduce worker fatigue increases sorting accuracy and consistency. Workers are less subject to fatigue if they have adjustable platforms so they can work at a comfortable height in relation to the conveyor (fig. 8.10). Foot rails and stools allow people to change position during the day. Sorting and removal belts and chutes should be designed to eliminate unnecessary reaching and stretching. A comfortable work area allows people to work with their upper arms positioned nearly vertical and their forearms nearly horizontal. Worker fatigue can also be reduced by use of effective noise control, proper lighting, and air temperature control.

Light levels of 500 to 1,000 lux at the sorting surface are usually adequate; older workers may need twice as much light as younger ones. Workers should not have surfaces in their field of view that vary in luminance (the level of reflected light) by more than 3 to 1. For example, dark-colored product should not move past the sorters on a bright white belt. Commonly available fluorescent lamps usually have adequate color rendition for most sorting tasks. Lights should be placed to reduce glare to workers' eyes. Shields on the sides of the light fixtures can guide light to the sorting table.

Workers must have clearly specified responsibilities. They should be responsible for only a limited number of sorting decisions for each product and should observe product only on a specific area, zone, or lane of the sorting belt. Periodically rotating worker positions on the sorting line can reduce monotony and fatigue. Workers should be familiar with defects and with segregation categories and limits. Posting visual aids helps in worker training. Supervisors should be familiar with worker performance limits and must be able to identify "under sorting" and "over sorting."

Machine sizing and sorting

Many kinds of mechanical sizers are used. All segregate by weight, or dimension (figs. 8.11 and 8.12). Traditional dimension sizers measure the product by two-, three-, or four-point contact. Electronic sizers capture several video images of each piece and calculate volume based on these images. Weight sizers weigh each piece of product as it moves over a load cell. Signals from the size sensor are sent to a computer, where the product size decision is made. The computer then sends a signal to release the product at the preselected cross-conveyor position as the product moves away from the sensor. If labeling machines are used, the computer also signals which pieces should be labeled. The density of the product must be known for accurate weight sizing.

Video imaging systems can also be used to detect misshapen product and product color. Companies are experimenting with using similar systems for determining sugar content and the presence of mechanical damage. These improvements will reduce the need for hand-sorting, although a limited amount of hand-sorting may be needed before box filling.

Electronic sizers operate at speeds of about 5 to 10 pieces per second. It is fairly common for a large-scale facility to have ten or more lanes operating simultaneously. Electronic sizers are easily adjusted to accommodate changes in product size distribution. The operator can increase the box-filling capacity for a particular size by typing a few instructions into the control computer, directing product of that size to drop at an additional box filling position. Since the machines size each fruit individually, they are also able to keep a good accounting of the amount and grade of product packed.

A properly selected sizer must have adequate capacity and accuracy, and it must not injure the product. The practical commercial capacity of a sizer is about two-thirds of the theoretical rated capacity. Because sizer capacity is related to product size, determine the capacity requirement based on the smallest average product size anticipated. Sizers

Figure 8.10

Schematic of a well-designed work area in packing operations. The details are similar for hand-sorting and hand-packing operations.

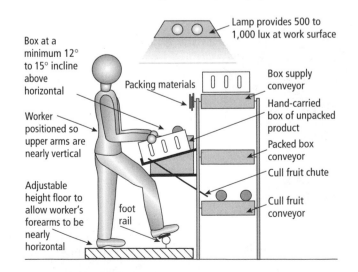

Box at a minimum 12° to 15° incline above horizontal

Lamp provides 500 to 1,000 lux at work surface

Packing materials

Box supply conveyor

Hand-carried box of unpacked product

Worker positioned so upper arms are nearly vertical

Packed box conveyor

Cull fruit chute

Adjustable height floor to allow worker's forearms to be nearly horizontal

foot rail

Cull fruit conveyor

Figure 8.11

Weight sizers segregate products by mass. Pieces are distributed into individual cups that carry them past a weighing section. An operator-controlled electronic system drops each product onto a cross-flow belt (selected based on the product's weight) for transport to a packing station.

Figure 8.12

Electronic sizers are becoming common. Video images of the product are analyzed to estimate volume, which is closely correlated with size. These machines are also capable of measuring product color and detecting a limited number of defects.

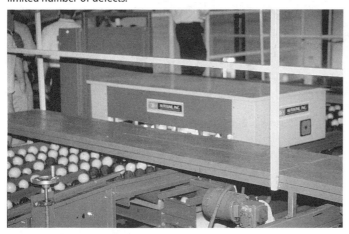

must segregate product with the required accuracy to meet uniformity, marketing, and legal requirements. They must meet these performance requirements without injuring the product, including all varieties and maturities that are commercially packed. Sizers work efficiently only if the product delivery system is properly designed and adjusted to deliver a uniform flow of properly separated product to each sizing lane.

PACKING THE PRODUCT

Although most fruits and vegetables are packed in corrugated fiberboard boxes, the

use of reusable plastic boxes is increasing. Some products are machine-filled into boxes to a predetermined weight; this is called volume filling. Tight-fill packing is a modification of volume filling in which the box is slightly overfilled, and then vibration-settled. A pad is placed on top of the product and the box is tightly sealed. Product may also be hand-packed into plastic film bags or placed into trays. Automated bagging machines are commercially available; these are particularly valuable because most are designed to place a fairly precise product weight in the bag. The bags or trays are then placed in a master container. A few products, such as carrots and potatoes, may be mechanically placed in consumer-sized bags, and then these bags are placed into larger bags.

HAND-PACKING OPERATIONS

Product is hand-packed in operations that cannot afford the expense of packing machinery or those that pack product requiring extra care in handling. These packinghouses may be quite large, employing several hundred packers (fig. 8.13). Hand-packing may be used to create a pack with a special product presentation. The goal is often to pack a fixed count in trays or boxes and always to immobilize the product within the package (fig. 8.14). In these facilities, the packer may also sort, grade, and trim the product before placing it in a box. (See figure 8.10 for design details of a hand-pack operation with limited mechanical equipment.) Workers need adequate lighting, and their workspace should be organized to reduce physical stress by minimizing reaching. Since the height of conveyors cannot be easily changed, platforms should be provided so that workers of various heights are positioned correctly.

Machinery is used in hand-packing mainly for materials handling, not for sorting, grading, or box filling. Product is either hand-carried to the packer, or a return-flow belt is used in which the packer selects product of a desired size from the belt. Adequate delivery of product and packaging materials to packers and removal of filled packages are important to packing efficiency.

Hand-packing may also be used in mechanized operations. In these operations, product arrives at the packer trimmed, sized, and graded, and the packer is responsible only

Figure 8.13

Top view of a hand-pack operation. In this facility, packers are responsible for sorting, grading, trimming, and packing the product.

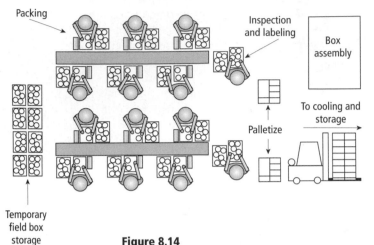

Figure 8.14

The plastic tray pack used here for peaches illustrates one type of hand-packing. Fruit are selected and placed into cups in trays to provide uniform lateral tightness within the package. After packing, a pad is placed over the top layer and the package flaps are folded and glued closed.

for placing the product in a box. Product is supplied from a variety of tubs, bins, or chutes. A modification for nonwrapped packs is the rapid-pack system, in which the work area is designed so that the worker faces the product for easy two-hand access. Tray filling can also be semiautomated: The tray passes on a belt just below the product delivery chute. By carefully regulating the speed of the tray belt, the operator can cause product to fill most tray cups. Other workers orient the product, fill empty cups, remove extra product, and place trays into boxes.

Hand-packs require fairly precise sizing of product, at least within single layers of the package; they also require packing to lateral tightness. Oversized product in tray packs

may prevent top pads or trays from contacting surrounding smaller fruit, which may then be subject to transport vibration damage. Similarly, undersized product may negate lateral tightness, allowing surrounding product to turn and be injured. Packaging materials such as trays, cups, wraps, liners, or pads help isolate or immobilize product. These materials are often as important as pattern packing in preventing product damage, although they often add significantly to packaging costs.

MECHANICAL PACKS

Mechanical packing systems (volume-fill or tight-fill) deliver carefully sorted and sized product, along with empty packages, to automatic fillers. After filling, the packages pass through inspection, marking, and closing operations and may receive special top padding, vibration settling, or lid fastening (tight-fill). A mechanical place-packing system is available for citrus packing. It accumulates a single layer of sized fruit into the desired packing pattern and uses rubber suction cups to transfer each layer of fruit into the package. Alternating layers have different fruit patterns to achieve product immobilization.

Mechanical packing systems normally handle large volumes of product at high speed, and they afford no opportunity for further grading at the filling chute. Arriving fruit must be adequately sorted and sized before box filling to meet the desired grade. Because peak sizes vary with the product, variety, and growing conditions, the delivery system to the fillers must be easily adjustable. The use of electronic sizers greatly increases the flexibility in adjusting for changes in product size.

Mechanical packers should be adjusted to properly fill the volume of the package. Most box fillers are designed to use weight as an estimate of volume (fig. 8.15). Some fill to within a small amount of the desired weight, and the final adjustment is done by hand as the packages pass over scales. Other fillers are designed to adjust the fill weight to the nearest fruit, so that only check-weighing is necessary. Lack of accuracy in fill weight adjustment is a major cause of poor performance of volume-fill packs.

A major problem with mechanical fillers is the height from which products drop into

Figure 8.15

Automatic volume-filling is widely used for fruit packing. Here, the empty package is tilted up to the filling belt and returned to horizontal position during filling with presized fruit. The scale is adjusted to the desired fill weight.

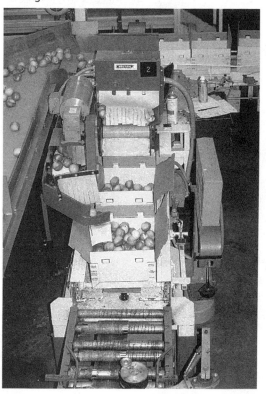

Figure 8.16

Tight-fill packing unit. The machine vibrates fruit into place, seals package flaps, and fastens a lid to the package.

the package. This can be minimized by properly designed equipment. Modifications such as tilting packages during initial filling to reduce drop heights are common. Padding filler chutes and using decelerator curtains may reduce impact force. Cushioning pads in the bottom of the package can further reduce impact injury to product during filling.

Tight-fill packing (fig. 8.16) is a special modification of volume-filling designed to ensure product immobilization during transport. It involves volume-filling product into the package to an exact weight, settling with a few seconds of carefully controlled vibration to eliminate voids, top padding with a special pad that nests around the top fruit, and tightly fastening the lid. To facilitate uniform settling, containers should be three to four times as deep as individual fruits are wide. Box width and length are important only in that they may effect box strength and therefore box integrity. If all of these steps are properly performed and the package is properly designed for the packing system,

the products will be held tightly in place without compression bruising. There are specific requirements for package and pad design, fill density, and vibration characteristics that must be met for successful tight-fill packing. This method is described in more detail in the UC publication *Tight-Fill Fruit Packing* (Mitchell et al. 1968).

After filling, the boxes or bags are then loaded onto pallets and the pallet loads are secured with bands or wrapping. Palletizing may be done manually or by mechanical palletizers. Palletizing increases handling speed and reduces handling damage from loading and unloading individual boxes, and it also helps to maintain load integrity during transport.

RE-SORTING AND QUALITY ASSURANCE

Some mechanized operations have limited product re-sorting before packing. This may be needed in volume-fill packing systems, where large volumes of fruit may be delivered, taxing the system's sorting capacity. Because the fruit have been previously sized and graded, final sorting may be more properly considered a quality assurance procedure to ensure that fruit quality standards have been met.

Product leaving the packing stations is usually inspected before final padding and lidding. This final quality control procedure assures that product and packing procedures meet grade and quality requirements. In some larger operations, third-party grading is used, with the inspector sampling to assure compliance with legal grades and standards. Here also, packages are labeled with information on variety, size, grade, and so on.

EQUIPMENT DESIGN TO PREVENT PRODUCT DAMAGE

The packing line must be designed to minimize product injury. Product can be damaged by impacts if it drops too far onto a hard surface, gains too much speed as it rolls down a ramp and then strikes a solid object or another piece of product, or is accelerated too quickly by a solid object such as a sizer cup. Minimize drop heights as much as possible. This is particularly important if the product hits a round surface like a roller. If drop heights cannot be lowered enough to prevent damage, hard surfaces should be padded with a closed-cell foam 6 mm (0.25 in) thick with a density of 89 to 240 kg/m³ (5.5 to 15 lb/ft³). Impact force is also lessened if product falls onto an unsupported section of a conveyor belt or if product falls through two slowly turning counterrotating brushes (fig. 8.17). Brushes wear out and must be replaced on a regular basis. Keep product speeds on ramps low by installing retarding flaps, curtains, or blankets.

Easily damaged products, such as cherries and apples, are often moved through the packing line in water flumes rather than with conveyor belts. This allows the fruit to fall into water and change direction without transfer impacts. Water falling more than 20 cm (8 in) onto a product can damage the most fragile products, such as cherries or leafy vegetables. Also, product may be subject to compression damage or shearing if it accumulates where flow of product slows, such as at sorting operations, when changing directions sharply, or during box filling. Regular cleaning of the facility to eliminate accumulated dirt also helps reduce product injury.

COOLING AND STORAGE

After packing, pallets of product are usually cooled and then placed temporarily in a cold storage room. Other special operations after packing may include application of ethylene to stimulate ripening of products such as tomatoes, kiwifruit, and pears. For more details on cooling, storage, and ripening, see chapters 11, 12, and 16, respectively.

REFERENCES

Gaffney, J. J., compiler. 1976. Quality detection in food. St. Joseph, MI: ASAE Publ. 1-76. 241 pp.

Gentry, J. P., F. G. Mitchell, and N. F. Sommer. 1965. Engineering and quality aspects of deciduous fruit packed by volume filling and hand placing methods. Trans. ASAE 8:584–589.

Guyer, D. E., N. L. Shulte, E. J. Timm, and G. K. Brown. 1991. Minimizing apple bruising in the packing line. Mich. St. Univ. Ext. Bull. E-2290.

Figure 8.17

Methods for reducing bruising in packing line transfers.

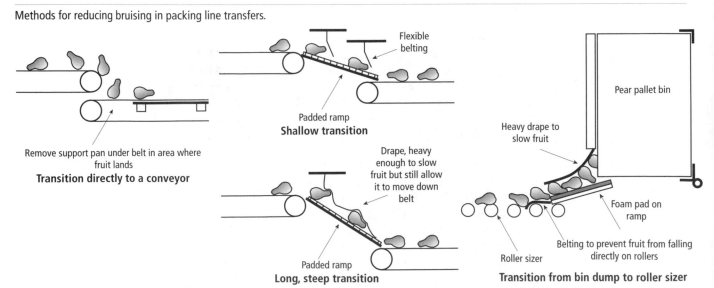

Hardenburg, R. E. 1967. Wax and related coatings for horticultural products; a bibliography. USDA ARS 51-15.

LaRue, J. H., and F. G. Mitchell. 1964. Bulk handling of shipping fruits. Calif. Agric. 18(6): 6–7.

Mitchell, F. G., J. H. LaRue, J. P. Gentry, and M. H. Gerdts. 1963. Packing nectarines to reduce shrivel. Calif. Agric. 17(5): 10–11.

Mitchell, F. G., N. F. Sommer, J. P. Gentry, R. Guillou, and G. Mayer. 1968. Tight-fill fruit packing. Univ. Calif. Agric. Exp. Sta. Ext. Ser. Circ. 548. 24 pp.

O'Brien, M., B. F. Cargill, and R. B. Fridley, eds. 1983. Principles and practices for harvesting and handling fruits and nuts. Westport, CT: AVI. 636 pp.

Ryall, A. L., and W. J. Lipton. 1979. Handling, transportation, and storage of fruits and vegetables. Vol. 1, Vegetables and melons. 2nd ed. Westport, CT: AVI. 43–117.

Sanders, M. S., and E. J. McCormack. 1993. Human factors in engineering and design. New York: McGraw-Hill.

Smith, R. J. 1963. The rapid pack method of packing fruit. Univ. Calif. Agric. Exp. Sta. Circ. 521. 20 pp.

Thompson, J. F., J. A. Grant, E. M. Kupferman, and J. Knuston. 1997. reducing cherry damage in postharvest operations. HortTechnology 7: 134–138.

Wardowski, W. F., S. Nagy, and W. Grierson. 1986. Fresh citrus fruits. Westport, CT: AVI.

9

Waste Management and Cull Utilization

James F. Thompson

About 10 to 15% of the fruit and vegetables delivered to California packinghouses are rejected as culls. They are culled because of scars, split pits, deformities, mechanical injuries, sunburn or sunscald, mold or insect damage, immaturity, overripeness, softness, or small size. These defects result in a huge quantity of material that must be utilized or disposed of.

PRINCIPLES FOR REDUCING WASTE

Take steps to reduce waste whenever possible (table 9.1):

Improve housekeeping. For example, some operations allow product to spill onto the floor. This wastes good-quality product, produces solid waste, and unnecessarily uses water. Conveyers and transition points should be designed to prevent product spill.

Change processing methods. In some situations there may be several alternative methods for accomplishing the same process. Consider waste production in the selection process. Poorly designed machinery may damage product by allowing it to drop onto hard surfaces or be sheared as it moves past a protrusion. Some damaged product is removed as waste in packing. Other damage is not noticed until later in the handling chain, and product may become unsalable at wholesale or retail marketing. Select packing and handling methods that produce less waste or produce waste that is easier to manage. For example, hydrocooling produces spent cooling water that must be disposed of and requires water-resistant packaging (such as waxed fiberboard boxes) that may not be easily recyclable. In contrast, forced-air cooling requires no cooling medium disposal and allows the use of regular fiberboard, which is recyclable.

Minimize the use of chemicals that contaminate waste and require special handling or disposal. Lye processing produces waste with a low pH that may need to be neutralized. Waste brine can be especially difficult to dispose of.

Minimize wastewater flow. This is usually done by reusing and recycling water. Sometimes wash water can reused if solids are screened or settled out and disinfectants are kept at effective levels. As a general rule, the lowest-quality water should be used for initial cleaning of the product, and best-quality water should be used for the last step in processing. Minimizing water use may not decrease the total dry weight of solid waste but it does concentrate it. This may allow the use of smaller, less-expensive treatment systems and perhaps encourage economical recovery of product because solid waste is more concentrated.

Segregate wastes. Wastes that require special processing or disposal should be kept separate from other wastes. Some wastes can be sold for byproducts and should be separated from wastes that must be disposed of.

Recycle, reuse, or reclaim. Although waste products usually have lower purity than they did when they entered the process, their purity may be adequate for some downstream uses.

DISPOSITION OF CULLED PRODUCT

The most obvious way to reduce culls is to reduce the number of culls that reach the packinghouse. The producer should use the best cultural practices to produce a well-sized, unblemished commodity. It

Table 9.1. Examples of wastes produced by postharvest handling processes and possible methods of improved waste management

Process	Waste product	Remediation process
Harvest	Substandard product, leaves, stems	Separate from product and leave in field Use manual harvest which has more potential for product separation
In-field processing	Trimmed material, substandard product	Leave in field
Transport	None	
Unloading	Spent flume water	Use dry unloading Treat and reuse flume water Treat and dispose of waste water
Conveying	Spent flume water, spilled product, spilled water	Collect spilled product separately and add to culls Redesign to prevent spillage
Washing, rinsing	Wastewater	Reuse wash water and use high-quality rinse water
Sizing, sorting, grading	Undersized product, culls	Use culls for by-products or cattle feed
Peeling, cutting	Peels, lye water, pits, seeds, stems	Separate solids and waste water Compost solids Use steam or mechanical peeling
Packaging, unitizing	Spilled product	Collect and add to culls
Cooling	Spent hydrocooling water	Use spent hydrocooling water for washing or flume water
Storage	Out-of-date product	Donate to food bank if it is still fit for human consumption
Distribution	Bruised, decayed product	Work with suppliers to minimize amounts
Marketing	Out-of-date, bruised, decayed product	Donate to food bank if it is still usable Work with suppliers to minimize amounts

should be harvested and handled carefully to minimize injury, and harvesting personnel should be encouraged to discard poor-quality fruits and vegetables in the field.

Culls can also be reduced by lowering quality standards; however, this often results in the poorer-quality fruits and vegetables being culled by the retail distributor or the consumer. The added costs of handling and shipping poor-quality fruits and vegetables can increase the cost of good-quality produce. A packer shipping high-quality fruits and vegetables has a competitive advantage over ones that ship poor-quality product.

However, even careful attention in the field does not eliminate the need for management of culls at the packinghouse. The cost

liability to the packinghouse operator can be minimized if culls are sold as a by-product rather than returned to the field or sent to a landfill.

In California, the largest use of fruit that is not suitable for the fresh market is as processed fruit product or as a by-product source. Fruit that is too ripe for long-distance shipment can be sent to local markets. In many instances undersized product can be given to food banks. Some culls can be processed into juice or a canned or frozen product. A large quantity of culled apples, pears, oranges, and papayas are used for juice extraction. Culled avocados are made into guacamole, and undersized artichokes are marinated and canned. Some culls can be turned into dried fruit for human consumption. However, good-quality dried fruit is made only from good-quality fresh fruit. Only undersized or slightly overripe fruit should be considered for drying. Citrus culls are used as a source for flavorings (lemon and orange oil), pectin, and juice. Other culls can be used as a source of natural food colors.

CATTLE FEED

Culled fruit is palatable and is a good source of energy for animals, but it is low in protein and has other characteristics that make it different from other feed sources. For example, stone fruits (peaches, plums, nectarines) contain 85% water, 9% digestible dry matter, 4% pits, and 2% indigestible dry matter. The high water content diminishes the real value as feed because it makes culls expensive to transport, requires large trough volumes, and allows the feed to spoil quickly. If fed in large proportions, culled fruit causes almost continuous urination, and consequently the animals require a high amount of salt. The only potential advantage to the high water content is that animals in a remote, dry location will not need extra water hauled to them.

Low protein levels in culled fruit limit the quantity that can be fed. Where rapid weight gain is important, in feed lots, for example, only about 20% of the ration can be composed of culled fruit. As a maintenance ration, up to 80% of the feed can be culls.

Stone fruit pits rarely cause internal injuries or choking. Cattle spit out some pits while eating, and many of the remaining pits are regurgitated with the cud and spit out. In fact, the main problem with pits is disposing of them, as they tend to fill feed troughs.

Culled fruit is typically bought for $2 to $5 per ton. In terms of feed value, this is equivalent to buying barley for $20 to $50 per ton. However, the costs of handling and transporting culls must be added to this cost. Also, some cost must be added to account for the uncertain effects of using a feed that has not been thoroughly tested for nutrient levels and trace chemicals.

Culled potatoes are good source of feed for animals. Like stone fruits, they are high in water content (about 77%), high in energy value, and low in protein. Beef steers can be fed up to 50% potato waste in finishing rations and still have acceptable weight gain. However, the steers must be carefully adapted to a potato ration, and the ration should not be changed rapidly.

To a limited extent cattle are also fed culled cantaloupes and other muskmelons. Cool-season vegetable culls have also been used as feed. These culls all have the same general limitations already discussed.

ALCOHOL PRODUCTION

Most fruit and some culled vegetables (especially roots and tubers) can be used for alcohol production. Alcohol for human consumption has a much higher value than alcohol for motor fuel. Some culled pears, kiwifruits, and apples are used for fruit wine production in California, and some of the apple wine produced is converted to cider vinegar.

The use of culls for fuel alcohol production is limited mainly by the low sugar content of most fruits and vegetables. The 8 to 12% sugar content of most culled fruits results in an alcohol yield of about 42 l/metric ton (10 gal/ton) of fruit. Potatoes have one of the best yields of alcohol for culls at 83 to 104 l/metric ton (20 to 25 gal/ton), but this is still low compared to better feedstocks such as corn, which yield 375 l/metric ton (90 gal/ton). The low yield makes it uneconomical to haul culls any significant distance. If production of fuel alcohol from culls is to be economical, it must be done near the packinghouse. Low sugar content also results in 4 to 5% alcohol "wines," which require considerable energy per gallon of alcohol to process and distill.

Waste left after distillation of alcohol pre-sents a waste management problem. It has very little protein, so it is not suitable as an animal feed, but it has a high pollution potential as measured by biological O_2 demand (BOD). Stillage waste may be usable as a feedstock for methane generation. The effluent from a methane generator is low in pollution potential.

In general, there are several key constraints to any method of cull utilization. First, transportation is expensive, as 80 to 90% of the weight of culls is water, which usually has little value to the user of the culls. Second, culls are produced in large quantities during a short time period. A year-round operation must be able to store the culls, usually by drying, which adds a cost to the product, or by using other products during the off-season. Operations that run only during the season cannot make large capital investments in equipment that will be used a few months a year. Finally, there must be a market for the by-product. For example, in California, dried fruit has a specialized and limited market that cannot absorb extra amounts of product made from culls.

CULL DISPOSAL

Unfortunately, the limits to the use of culls often result in large portions of them being discarded. Improper disposal can cause sanitary and pollution problems.

Flies and odor problems can be prevented by ensuring rapid drying. Fly maggots hatch into adults within 7 to 10 days, and odor problems can develop before flies appear. The culls should be crushed and spread no more than one or two layers deep; sometimes this is done on orchard roads or fallow fields. Culls can be disked into the soil, although this tends to cover the fruit with soil and slows drying; also, insects or diseases that may have caused the fruit to be culled in the first place may infect a future crop. Disposal sites should be as far away from neighbors as possible. Flies can travel up to 8 km (5 mi) from the place where they hatch.

Culls should not be dumped near streambeds. Fruit dump sites can attract the dumping of many other kinds of refuse. If culls are deposited away from the point of production, use municipal solid waste disposal sites if available.

REFERENCES

Barnes, D., C. F. Forster, and S. E. Hurdey. 1984. Surveys in industrial wastewater treatment. Vol. 1. Food and allied industries. London: Pitman Publishing. 376 pp.

Higgins, T. E. 1995. Pollution prevention handbook. Boca Raton, FL: CRC Press. 556 pp.

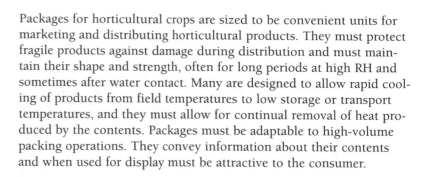

Packages for Horticultural Crops

James F. Thompson and

F. Gordon Mitchell

Packages for horticultural crops are sized to be convenient units for marketing and distributing horticultural products. They must protect fragile products against damage during distribution and must maintain their shape and strength, often for long periods at high RH and sometimes after water contact. Many are designed to allow rapid cooling of products from field temperatures to low storage or transport temperatures, and they must allow for continual removal of heat produced by the contents. Packages must be adaptable to high-volume packing operations. They convey information about their contents and when used for display must be attractive to the consumer.

TODAY'S HORTICULTURAL PACKAGES

Many materials, sizes, and shapes are used in packages for horticultural products. In the United States alone, more than 500 different packages are used for produce. Past efforts at standardization have had limited success, but this situation is beginning to change in response to pressures from produce buyers. Major changes have been in response to economic considerations, the use of less-expensive materials, and the need to adapt to new packing and handling procedures. In the United States, most perishables are packed in corrugated fiberboard boxes, with a limited use of plastic and wood boxes. Hand-packing is used mainly for field-packed product, and most packing-houses employ mechanical packing. Most products are unitized and shipped on pallets.

PRODUCT REQUIREMENTS

Developing successful packages for horticultural products emphasizes the different requirements of various products. Although these requirements vary widely with the commodity, marketing program, packing method, and so on, there are many generalities that apply to most commodities.

PROTECTION FROM INJURIES

Physical injuries to the product must be avoided wherever possible during handling and distribution. Some of the more obvious open wounds (e.g., cuts or punctures) often occur before packaging and can be eliminated by good supervision and sorting. Certain bruises, however, may accumulate throughout all stages of handling, including packaging and distribution.

Impact bruises (fig. 10.1). Impact bruising results from dropping the product onto a hard surface. Because impact injury may not be immediately visible on the surface, careful quality control is needed to protect against it. Dropping the product into the package is a common cause of impact injury during packing. Installing decelerator strips at filling chutes and designing fillers to raise empty boxes to reduce drop heights during volume-filling reduce the incidence and severity of impact bruising. Packaged products can also receive impact bruising from drops during manual or mechanical handling with chutes or conveyors, or during transport as a vehicle runs over a curb or pothole. Unit handling reduces the number of times an individual package is handled and thus the number of impacts. Rough handling by machinery can also cause impact bruising. Corrugated fiberboard can absorb some shock, but it does so by permanently compressing; it will

Figure 10.1

Impact bruise on Anjou pears. Bruising extends into the flesh and may or may not be visible on the surface.

Figure 10.2

Compression bruise on Golden Delicious apples. Bruise damage occurs on surface and extends into flesh of fruit.

not protect against repeated impacts. Perishable commodities usually have too little value to warrant the use of shock-absorbing packaging materials. Careful management of package handling is the main method of reducing impact damage to perishables.

Compression bruises. Compression bruising (fig. 10.2) can result from overfilling boxes or allowing too great a product depth. Soft commodities, such as bush berries and grapes, require shallow packing depths to avoid damage. Box dimensions and fruit volume must be carefully matched to avoid overfilling a container (fig. 10.3). Probably the greatest cause of compression bruising is intentional overpacking to create an over-

weight pack. Overfilled corrugated fiberboard boxes deform and are weak. If a box fails it allows the product to support the weight of packages stacked above it, causing compression damage. Buyers may receive more product in an overpacked box, but the product may be damaged and therefore rejected.

Compression bruising also occurs if a box is not strong enough to support boxes stacked on top of it. Avoid stacking boxes beyond their design limit. It is not economically feasible to design corrugated fiberboard boxes to withstand stacking three or four pallets high during storage. Even two-high stacking should be a temporary situation, although shippers should recognize that temporary stacking often occurs during distribution, and they should allow for this in package design. For multipallet stacking and for storage, pallet racks or corner supports are less expensive than extra-strength packages.

Vibration or abrasion damage. Some products are damaged when they move within the box during transit (fig. 10.4). For example, table grapes drop from the bunch (shatter), and many other products are bruised by excessive vibration. Bruising is usually restricted to the product surface, and because it becomes easily visible it greatly reduces salability. On soft commodities, deep flesh bruising can result. This damage is caused only when product is exposed to acceleration greater than one times the acceleration of gravity (1 g). Below this level, product does not move with respect to the box or neighboring product. In highway vehicles, this level of acceleration is experienced only by the top one or two layers of boxes loaded directly over steel spring suspensions. If all product in a load is bruised, damage probably occurred before transport.

Vibration injury can be prevented by shipping perishables in trailers equipped with air-ride suspension. This equipment is becoming more common and should be specified for products that are sensitive to vibration bruising.

Packing the product so that it is immobilized in the box also prevents vibration damage. Tests have shown that packing Bartlett pears and Thompson seedless grapes in loosely-filled, consumer-sized polyethylene bags prevents vibration damage. In hand-packing, immobilization begins by achieving lateral tightness within the package. Supplemental

Figure 10.3

Relation between net fruit weight of table grapes packed in a 1,210-in³ (19,828-cm³) box and the amount of fruit bruising on top and bottom of fruit mass.

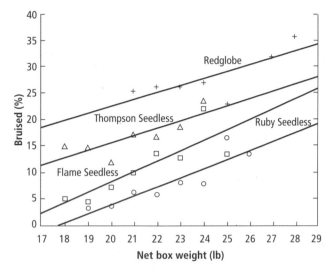

Figure 10.4

Vibration bruise on Bartlett pears. Bruise damage appears on fruit surface and usually does not extend into flesh.

materials, such as wraps, trays, cups, shims, liners, and pads, may be needed to ensure that product is immobilized. A special procedure called tight-fill packing (see chapter 8) is designed specifically to immobilize volume-filled product. To be effective it requires a minimum box depth of three to four times product diameter, precise fill weights, boxes that have a consistent interior volume and do not deform in handling and transport, and padding that fills in any free volume that may develop as product loses weight in storage.

TEMPERATURE MANAGEMENT

Packages must accommodate the temperature management requirements of the product. Good temperature management depends on good contact between the product in the package and the external environment. For some products, providing for airflow past package surfaces may be sufficient. Ventilation is needed for boxes that will be placed in a forced-air cooler. Within limits, increasing the ventilation area speeds heat exchange. For example, 1-lb lidded strawberry containers forced-air-cool faster with 7% sidewall venting compared to lesser amounts of venting, but the cooling time is not reduced with venting greater than 13%. For corrugated boxes, 5% venting of side or end panels allows rapid cooling without overly weakening the package. A few large vents perform better than many small vents. Vertical slots kept at least 5 cm (2 in) from the package edges perform well. The optimal performance of these vents during highway transport requires a loading pattern that lets cold air reach the vent openings. Hydrocooling and transport in marine containers or break-bulk ships requires venting on box tops and bottoms for vertical flow of water or air.

Certain fruits require ripening before retail marketing. They must be uniformly warmed to ripening temperature, and they often need treatment with ethylene gas. A package that is properly vented for cooling performs well for both warming and gassing.

Package vents should be unobstructed by internal packaging materials. Liners, wraps, trays, shims, or pads should minimize the blockage of airflow or waterflow through the box. If these restrictive materials are essential, then the handling operation must allow for longer cooling times.

Some packages are designed to restrict heat flow into the box. Packages used for air transport may be designed without ventilation (and sometimes with insulation) to delay product warming during transit without refrigeration. Flower boxes are designed so that vents can be sealed after forced-air cooling. A sealed box retains heat from product respiration. Package icing for water-tolerant products and boxed or packaged gel ice for others may be useful in preventing product warming in a sealed box, but only if the commodity is packaged while it is cold.

PROTECTION FROM WATER LOSS

Many horticultural products suffer wilting, shriveling, or drying as a result of water loss during handling and marketing. Water loss occurs because of a water vapor pressure difference between the product, which is normally near saturation (100% RH), and the surrounding drier environment. Water from the product is absorbed by wood or paper packaging materials and condensed on storage or transport refrigeration coils. Plastic foam and solid plastic boxes do not absorb water, but allow moisture loss through vents. During storage it is desirable to hold most products at a high RH to minimize water loss. Packaging may be designed to provide a partial barrier to movement of water vapor from the product.

Many package moisture barriers are available. Plastic (polyethylene) liners, usually with small perforations to allow some gas exchange, maintain an essentially saturated atmosphere within the package but can cause surface cracking in some commodities. Plastic film curtains, which may be open at the package ends and folded around the product sides and top, provide a partial moisture barrier and are successful for some fruits. The inside surface of corrugated boxes can be coated with various materials that act as a moisture barrier. The most widely used material is a polyethylene-wax emulsion.

Moisture barriers inside a package must not impede essential airflow through the vents. Unperforated liners are particularly troublesome because they completely block all vents. Perforated liners can allow enough airflow for satisfactory forced-air cooling while still retarding moisture loss in storage. Tests of polyethylene curtains used with tray-packed peaches show that increased fan capacity during cooling and storage can compensate for air blockage. Products can be effectively vacuum-cooled if the wrap is slightly perforated.

Fiberboard packages with surface coatings can be vented normally, so that the moisture barrier does not affect air movement. Under most conditions, package venting causes only a limited increase in water loss from the product. For long-term storage, water loss may be reduced in these packages by reducing airflow in cold storage.

FACILITATING SPECIAL TREATMENTS

Certain commodities have special treatments that must be considered in packaging selection and design. Examples are sulfur dioxide fumigation of grapes for decay control and methyl bromide fumigation of export commodities for insect control. These treatments require well-vented packages through which the fumigant can readily flow. Venting sufficient for rapid cooling is more than adequate for fumigation. Some grapes are packed with pads containing sodium metabisulfite that slowly release sulfur dioxide. This system requires a plastic liner with restricted ventilation.

A few fruit and fruit-type vegetables are ripened at temperatures of 15° to 20°C (59° to 68°F) using ethylene gas. Packages for these products need venting for uniform warming and ethylene treatment. Some commodities are damaged by ethylene and must be protected from it. In-package ethylene scrubbing products, which are in limited use, perform best if package ventilation is restricted. Room scrubbers that circulate air in the storage room through an ethylene scrubbing unit require good package venting to be effective.

Other special packaging requirements may exist for certain commodities. For example, gladiolus and asparagus must be packed upright to avoid curvature caused by geotropism. Asparagus is packed with a moist pad to reduce dehydration and to slow toughening. It must also be packed with some headspace above the spear tips to allow for growth and elongation.

COMPATIBILITY WITH HANDLING SYSTEMS

Most boxes are hand-lifted at some point in the marketing chain, so package weight must be limited. A few commodities are packaged in units designed only for mechanical lifts. For example, watermelons, bagged apples, and head lettuce for processing are shipped in pallet bins.

Packaging may need special design features to make it compatible with packing equipment and handling procedures. For example, top flaps that extend upward may interfere with hand-packing. The package must be sized to facilitate unitization and mixed-load handling. Generally this means that the package must be sized to fit a pallet that is 1,200 by 1,000 mm (48 by 40 in). Possible problems with weather and contamination may require that the package be water-resistant or have a well-covered top. Packages may be constructed of various materials to meet special requirements. Wood packages (fig. 10.5), long standard in the horticultural industry, are still used for long-term storage or high-moisture conditions. Plastic or plastic foam packages also perform well in high-humidity storage and are finding increased use.

Some products, especially root, bulb, and tuber crops, are sometimes handled in large plastic or fabric bags without other packaging. Bags are an inexpensive package and can be packed quickly. However, they provide little protection from mechanical damage and are often not well suited to forced-air or hydrocooling.

PACKING FACILITIES

In designing a package, its compatibility with conveyors and other packing equipment must be considered. Modifying equipment to accommodate a new package can be expensive. In volume-fill operations, package top flaps can be important in containing the product during filling. This advantage may justify the cost of redesigning conveyor systems to accommodate the height of the flaps.

A package used for field packing must be compatible with field conditions. Some fiberboard box designs can be shipped to the field flat and hand-assembled without stapling or gluing. Expanded polystyrene boxes and thermoformed consumer baskets are lightweight when empty, and precautions may be needed to stabilize them during strong winds. If the packages may be exposed to rain or heavy dew, they need to be moisture-resistant or protected from moisture exposure with tarps or stored in covered trailers. Packages placed on the ground usually collect soil that may contaminate product in neighboring packages. For example, strawberry trays should not be placed on the ground because their open-top design allows fruit on lower trays to be exposed to soil on the bottom of upper trays. Closed-topped containers allow soil-to-box contact and prevent product contamination.

When a new package is introduced, inventory problems must be considered. New packages should simplify operations, so early elimination of existing packages should be scheduled. Careful planning can minimize the inventory problem during package changeover.

UNITIZED HANDLING

Most horticultural packages in the industrially developed world must be designed for secure palletization, with packages either stacked in-register or cross-stacked. Stacking is difficult if boxes are overfilled or have bulging sidewalls, so attention to product mass and volume requirements must be considered when designing a new package. The package must withstand expected stresses in the stacking column. If packages are cross-stacked, their vents for vertical or horizontal airflow must align to allow air circulation through all packages.

Package dimensions must be compatible with pallet dimensions. Corrugated fiberboard boxes are seriously weakened if they are not completely supported under their load-bearing corner sections. U.S. receivers are asking for boxes that fit the Grocery Manufacturer's of America standard 48- by 40-in. pallet. This is causing packers to switch to boxes with the horizontal dimensions listed in table 10.1. The metric 1,200 by 1,000 mm pallet is 47.2 by 39.4 in., slightly smaller than the U.S. standard. An 800- × 1,200-mm pallet (31.5 by 47.2 in) is also a standard in Europe. Although standard pallet dimensions are desired, the paramount need is for the package to be compatible with the pallet size likely to be used for that commodity. State or national regulations may restrict the types of packaging that can be used for a particular product.

Table 10.1. Horizontal box dimensions for standard-sized pallets

Boxes per layer	Nominal outside box dimensions for 48″ × 40″ pallet* length × width (in)	Nominal outside box dimensions for 1,000 × 1,200 mm pallet length × width (mm)
4	24 × 12	500 × 300
5	24 × 16	600 × 400
6	20 × 16	500 × 400
8	20 × 12	500 × 300
9	19 × 13.3	400 × 333

Note: * Fiberboard boxes often bulge after packing; outside dimensions may need to be as much as 10 mm (0.4 in) shorter to allow boxes to fit completely on the pallet.

Figure 10.5

Wood lug used for shipping grapes. These packages often have wood ends, a wood/paper veneer side, and bottom wrap.

Figure 10.6

Corrugated "Bliss type" lug used for shipping many fruits. Package shown is equipped with plastic trays often used for hand or hand-assisted packs of fruit. End-stacking tabs are used to stabilize loaded pallets.

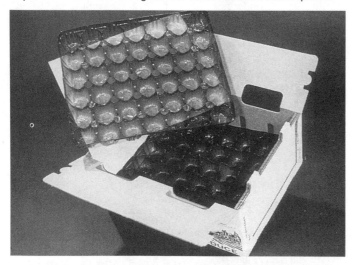

Check the regulations of the locations of origin and destination for restrictions on package dimensions or design.

Machine-applied mesh pallet wrap is often used to stabilize pallet loads. It can be applied quickly and allows airflow through boxes. Some boxes are designed to help stabilize the palletized load. An example is the use of stacking tabs (fig. 10.6) or wire ties used with berry trays. Packages with a covered top can be glue-bonded between layers on the pallet with special breakaway palletizing

glues. Normally, at least one horizontal plastic or steel pallet strap is required to assure stability of a glue-bonded pallet unit.

Although pallets are usually made of wood, pallets made of plastic are increasingly being used. The least-expensive pallets are made of soft woods and are not designed for reuse. Pallet disposal is a major cost to receivers. Hardwood pallets can be reused, but they are significantly more expensive. Companies that specialize in managing pallet reuse are becoming more popular in some areas. They provide a reusable pallet to a shipper at a fixed cost and arrange for pallet repair and reuse after each trip. Reused pallets require regular maintenance, particularly to reinstall fasteners that work their way out of the wood. Reused pallets must be guaranteed to not have toxic residues from nonfood items carried on previous trips.

PACKAGE STANDARDIZATION

A large number of package types, shapes, and dimensions are in use today (see fig. 10.5, 10.6, 10.7, 10.8, 10.9), and most are not designed to be compatible with each other in loading. Unnecessary and often serious product injury occurs because incompatibly sized packages are loaded together for retail distribution. Most products would fit into packages of just a few standard horizontal dimensions (see table 10.1).

ADAPTABILITY TO HANDLING REQUIREMENTS

Packaging must perform efficiently in product marketing and distribution, and in inventory and handling of packaging materials.

MOISTURE

Most horticultural packaging must tolerate exposure to high RH. Storage facilities often have at least 85 to 90% RH, and water released from the product creates near 100% RH within the package. Moisture from the air may condense on cold surfaces of packages when they are removed from refrigerated transport vehicles or storage rooms. Boxes must be able to withstand direct water exposure when commodities, especially certain vegetables, are hydrocooled, when loose ice is placed into the package (package-icing), or when top ice is placed over a load during transport.

Figure 10.7

Full telescope corrugated package used to ship apples. Similar packages are used for many other horticultural commodities. Apples are layer-packed into trays shown.

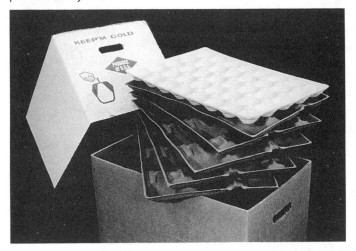

Figure 10.8

Triple-wall corrugated pallet bin used for shipping melons.

Figure 10.9

One-pound, lidded strawberry baskets in a fiberboard tray.

Plastic and wood are generally well suited to high humidity and direct water contact. Fiberboard can be made water resistant by treating it with proprietary blends of wax, polyethylenes, resins, and other plastics, generally called waxes.

HIGH TEMPERATURE

Packages for field use must be able to avoid deterioration, and care must be taken during their use to prevent deterioration. Wax coatings deteriorate at high temperatures, discoloring the packages and destroying much of their effectiveness. High ambient temperatures can damage carry-over inventories of corrugated packages by causing delamination of the corrugated board and creating potential problems during the next season.

Boxes may also need to be able to protect product from heat gain or freezing. Light-colored fiberboard reflects more radiant heat than natural-colored kraft board. Flower boxes are sometimes insulated with sprayed-on urethane foam to protect flowers from heat or cold when they are transported or temporarily stored at ambient temperature conditions.

HANDLING

After packing, a box may be handled a number of times, and each time the box and its contents are subject to damage. Manual handling during palletizing or restacking is particularly damaging. Workers usually drop heavy boxes into position in order to prevent back strain. Boxes should not weigh more than 20 kg (44 lb). Machine unitization and handling is often more gentle than manual operations.

Bagged product is particularly subject to damage in handling. Minimize damage by placing a fiberboard sheet between the pallet and the first layer of bags, using bag materials with a slide angle of at least 30°, stabilizing pallet loads with vented pallet wrap, using square-end bag designs, and using bag material that is strong enough so that the bag can be picked up by a single corner without tearing.

Packages need to be clearly labeled so that warehouse personnel can locate product. In some cases, there are government regulations specifying labeling requirements. Labels are usually needed on box ends and sides because boxes rarely have a single orientation in handling.

Assembling packages from various orders often causes many box styles to be mixed on a pallet. This invariably causes damage, as boxes are not properly supported below and may even be stacked sideways to fit on the pallet.

INSPECTION

The package should facilitate easy inspection of the contents. Telescope-style corrugated packages are easily inspected by removing the lid. Snap-on lids and inspection ports serve the same function. Whatever the design, the package must securely re-close and protect the product during the balance of the distribution period.

RETAIL DISPLAY

Some packaging systems are designed for use in retail displays. Most notable examples are berries that are packed in small baskets, and apples, oranges, carrots that are bagged in consumer-sized units. Advantages of consumer packages are that they can be designed to modify the atmosphere around the product and extend shelf life, improve sanitation by reducing opportunity for human contact, increase RH around the product and reduce moisture loss, and protect the product from mechanical injury. Their disadvantages are that it may be difficult to sort out the occasional poor-quality product, high RH may foster growth of decay, and bags or consumer containers may slow cooling. Packages to be used in retail display must fit the needs of the retailer, so package appearance becomes much more important than for other types of packaging.

PACKAGE DISPOSAL AND REUSE

In recent years, package disposal has become increasingly expensive. With incineration restricted or eliminated in most metropolitan areas, disposal choices include reuse, recycling, or landfill dumping. Pallet reuse is becoming more common in North America and in Europe. Nonwaxed, corrugated fiberboard packages are commonly recycled. Recycling of the material in plastic foam packages and the sale of ground foam for insulation has been attempted only on a limited basis. The use of returnable plastic containers is increasing.

ECONOMIC CONSIDERATIONS

In considering the cost of any new package, all of the costs of adapting it into the marketing system must be considered. These costs include packaging material, labor, modifications in packing and handling operations, and potential changes in product condition.

Packaging costs include the costs of

- package components
- transportation
- package make-up, labor, and materials
- internal packaging materials (if needed), including liners, shims, pads, trays, and wraps
- storage of package components

Packing costs include the costs of

- adapting to mechanized package distribution
- new or modified equipment and facilities at packing operation
- possible reduction in packing labor efficiency
- increased number of packing steps required

Palletizing and handling costs include the costs of

- changes in pallet stacking efficiency
- labor, materials, and equipment for unitizing pallets
- compatibility with various pallet materials and substitutes

Marketing costs include the costs of

- decreased load density in storage and in transport vehicles
- special labor or equipment for handling
- suitability of package as a display unit
- disposal of packages

Product value costs include the costs of

- possible increase in product deterioration
- discounts in sale price due to package failures
- possible negative impact on brand reputation related to package performance

The switch to reusable plastic boxes in place of fiberboard containers in the United States is a good example of a cost and performance comparison between two package types. Currently, reusable plastic containers are common in Europe and are being tried on a limited basis in the United States. They have significant advantages for the retailer: they have no disposal cost and are often used to display product, reducing costs of managing produce displays; they are also quite strong and do not collapse in high-humidity environments as fiberboard does occasionally. They weigh more than fiberboard and increase transportation costs that are usually paid by the retailer. They are usually well-vented and allow rapid cooling (although the large vent areas may allow excessive product moisture loss in later storage and handling). A potential advantage over fiberboard is that plastic does not absorb water and may slow product loss. On the other hand, growers and packers see a number of disadvantages to reusable plastic containers. They often cost more than an equivalent fiberboard package. They are usually dark-colored and often do not have lids, which can allow excessive product heating in field-packing operations. They often require redesign of box-handling equipment in packing houses. Other kinds of packaging changes are often similar to this situation, in which the costs and benefits of change are not evenly distributed. The switch to reusable boxes in the United States will ultimately be driven by retailers wishing to reduce their costs.

TRANSIT TESTING

Testing is an important aspect of any package development program. Objective laboratory testing is the first step in evaluating new packages and packing procedures; it can reduce costly and time-consuming trial shipments and allow more precise evaluation of a large number of variables. After laboratory testing, a promising package or treatment can be compared with the standard in a trial shipment. The use of laboratory tests involves a minimum expenditure of time and provides confidence that major problems will not occur in subsequent usage.

BASIS FOR TEST PROCEDURES

To be effective, laboratory test procedures must meet certain conditions:

- The types of abuses to products that occur in actual transport and handling must be duplicated.
- Laboratory treatment should equal severe transport or handling conditions.
- Laboratory procedure should allow rapid testing of a large number of variables.
- Tests should emphasize the effects of transport and handling on the product and on packages and packaging materials.

TEST PROCEDURES

Tests are done to determine package and product response to: vibration, vertical drops, horizontal impacts if the package may be shipped by rail, and compression. A number of standard laboratory testing procedures can evaluate packages. Most procedures are developed by the American Society for Testing and Materials (ASTM) and a few by the International Safe Transit Association (ISTA).

Normally the best components of all variables tested in the laboratory are combined in the new package or packing method. This is first compared with the industry standard in a laboratory test. If the results are promising, it is then tested in a well-replicated trial shipment in comparison with the industry standard.

CORRUGATED FIBERBOARD FACTS

Corrugated fiberboard was first used commercially in 1903, and by the end of World War II it had become the dominant type of shipping container used in the United States. Figure 10.10 shows its basic construction and some of the standard terminology describing it. Corrugated fiberboard is manufactured with four flute types (table 10.2). Perishables are usually packed in boxes made with a B-flute medium with a basis weight of 26 lb per 1,000 ft^2 (125 g/m^2). Liners are usually of equal weight, and the weight is selected based on the minimum weight needed to allow the product to be marketed without damage caused by box failure (table 10.3). Board strength is often

Figure 10.10

Basic construction and standard terminology describing a corrugated fiberboard box.

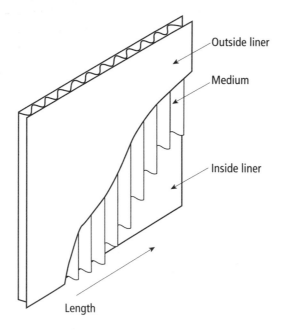

Outside liner

Medium

Inside liner

Length

Table 10.2. Common corrugations used in corrugated fiberboard

Flute	Medium height* (in)	Medium height (mm)	Flutes/ft	Flutes/m
A	³⁄₁₆	4.7	33–39	108–128
B	⅛	2.4	47–53	154–174
C	⁵⁄₃₂	3.6	39–45	128–148
E	¹⁄₁₆	1.6	90–98	295–321

Note: *Does not include liners

Table 10.3. Commonly used liner board grades

North American grade basis weight (lb)	European grade grammage (g)
26	125
33	150
38	—
42	200
—	225
—	250
69	300

Note: — = Not applicable.

described by the Mullen burst test value. For example, a board made of 26-lb liners would have a Mullen value 125 lb, and a board made of 42-lb liners would have a Mullen value of 200 lb.

Fiberboard box strength is also affected by design, use, and environmental conditions including storage time, humidity, corrugation orientation, location and size of vent holes, and stacking arrangement. Fiberboard loses strength over time when it is supporting weight. This is called fatigue. For example, a box supporting weight for 10 days will have only 65% of its original, laboratory-determined strength; after 100 days its strength is only 55% of the original strength.

Fiberboard absorbs moisture based on the RH around it. Laboratory testing is based on material in equilibrium with air at 50% RH and 23°C (73°F). At 90% RH a box will have only 40% of its original stacking strength. Absorbed moisture also causes fiberboard to expand slightly, increasing box dimensions and sometimes causing warping.

Proprietary blends of wax, low-molecular-weight polyethylenes, resins, and plastics can be added to fiberboard during manufacture to slow the rate of moisture gain in high humidity environment. Waxing can be accomplished during board formation by

adding up to 12 to 18% wax (carton weight basis) to the liner or medium, making the fiberboard resistant to high humidity. Waxing can also be done after the board is formed by passing it through a curtain of molten wax. This treatment makes the board resistant to direct water contact. Greater amounts of wax (45% to 60% of board weight) can be added by dipping the finished box in molten wax. This process is used for boxes that will be hydrocooled or package-iced. Waxing adds strength to fiberboard and slows the rate of moisture uptake, allowing a box to maintain strength for a longer period. Waxed corrugated boxes are considerably more expensive than packages suited only for dry conditions. Heavy waxing turns kraft paper dark brown. Waxed boxes are recyclable only in special processes and cannot be mixed with regular corrugated for recycling.

Corrugations are usually oriented vertically in the critical load-bearing regions in a box. Horizontal orientation reduces fiberboard strength to 80% of vertical for A-flute and to 90% of vertical for C-flute boards. Horizontally oriented B-flute is 120% stronger than vertically oriented B-flute.

Most of the strength in a corrugated box is near the corners. Vents in vertical walls should be located away from corners and oriented vertically to minimize strength loss. Box corners should be supported by pallet boards and stacked boxes should align. A sidewall overhang of 2.5 cm (1 in) on a pallet reduces stacking strength by 14 to 34%.

REFERENCES

American Society for Testing and Materials (ASTM). 1996. Annual book of ASTM standards. Philadelphia: ASTM.

Anon. 1976. Fibre box handbook. Chicago: Fibre Box Assn.

Ben Yehoshua, S. 1985. Individual seal-packaging of fruit and vegetables in plastic film, a new postharvest technique. HortScience 20:32–38.

Boustead, P. J., and J. H. New. 1986. Packaging of fruit and vegetables: A study of models for the manufacture of corrugated fiberboard boxes in developing countries. London: Tropical Development and Research Institute Publ. G199. 44 pp.

Gentry, J. P., F. G. Mitchell, and N. F. Sommer. 1965. Engineering and quality aspects of deciduous fruit packed by volume filling and hand placing methods. Trans. ASAE 8:584–589.

Guillou, R., N. F. Sommer, and F. G. Mitchell. 1962. Simulated transit testing for produce containers. TAPPI 45(1): 176–179A.

Hanlon, J. F. 1995. Handbook of package engineering. 2nd ed. Lancaster, PA: Technomic.

Hardenburg, R. E. 1966. Packaging and protection. In Protecting our food supply. USDA Yearb. 1966. 102–117.

Hochart, B. 1972. Wood as a packaging material in developing countries. United Nations Publ. E.72.II.B.12. 111 pp.

International Safe Transit Association. 1992. Preshipment test procedures. Chicago: ISTA.

Mitchell, F. G., N. F. Sommer, J. P. Gentry, R. Guillou, and G. Mayer. 1968. Tight-fill fruit packing. Univ. Calif. Agric. Exp. Sta. Circ. 548. 24 pp.

O'Brien, M., and R. Guillou. 1969. An in-transit vibration simulator for fruit handling studies. Trans. ASAE 12:94–97.

O'Brien, M., J. E. Gentry, and R. C. Gibson. 1965. Vibrating characteristics of fruits as related to in-transit injury. Trans. ASAE 8:241–243.

Selke, S. E. 1990. Packaging and the environment: Alternatives, trends and solutions. Lancaster, PA: Technomic. 178 pp.

Slaughter, D. C., R. T. Hinsch, and J. F. Thompson. 1993. Assessment of vibration injury to Bartlett pears. Tran. ASAE 36:1043–1047.

Smith, R. J. 1963. The rapid pack method of packing fruit. Univ. Calif. Agric. Exp. Sta. Circ. 521. 20 pp.

Sommer, N. F., and D. A. Luvisi. 1960. Choosing the right package for fresh fruit. Pack. Eng. 5:37–43.

Soronka, W. 1995. Fundamentals of packaging technology. Herndon, VA: Inst. of Packaging Professionals.

Cooling Horticultural Commodities

James F. Thompson, F. Gordon Mitchell,

and Robert F. Kasmire

Controlling product temperature and reducing the amount of time that product is at less-than-optimal temperatures are the most important methods of slowing quality loss in perishables (see chapter 4). Postharvest temperature management begins with planning the harvest and field handling. Some products are so sensitive to temperature abuse that they should not be harvested when temperatures are too warm. For example, table grapes show signs of stem shrivel at about 2% weight loss, and if stem quality is to be maintained at consumer level, the fruit should not be subject to more than about 0.5% weight loss between harvest and the beginning of cooling. Although grapes can be held for more than 8 hours at 20°C (68°F) before cooling, at 30°C (86°F) cooling should begin within 1.5 hours after picking (fig. 11.1). A few growers harvest at night to prevent exposure to excessive heat after harvest.

Other methods of protecting product from temperature-caused damage are to

- Make frequent trips between the field and the cooler to minimize temporary field storage.

- Pack in light-colored containers.

- Cover containers with lids if left in direct sun.

- Use a shaded area for temporary field storage. Remember, shade cast by a tree moves with the sun during the day.

- For short trips, use covered trucks to transport product to cooler. Long trips need refrigerated trucks.

- Begin cooling as soon as possible after product arrives at the cooling facility.

Some commodities can withstand a fairly long time between harvest and cooling. For example, apples placed in controlled atmosphere storage often do not reach optimal storage temperature until several days after harvest; exported California oranges may not reach best storage temperature until they have been at sea for several days. Products that do not require fast cooling generally have slow respiration rates, low moisture loss (transpiration) rates, and are often grown in climates with mild temperatures.

The first part of this chapter describes the variety of cooling systems available for horticultural commodities and the issues that need to be understood in their use. The second part describes a systematic approach for selecting a cooling system for a particular operation.

COOLING METHODS

Initial cooling of horticultural products to near their optimal storage temperature can be done with several cooling methods, including room cooling, forced-air cooling, hydrocooling, package icing, and vacuum cooling. Mechanical refrigeration in ships or refrigerated marine containers may be used for cooling a few commodities during transport. A few cooling methods (e.g., room cooling, forced-air cooling, and hydrocooling) are used with a wide range of commodities. Some commodities can be cooled by several methods, but most commodities respond best to one or two cooling methods.

Most users are concerned with the time to "complete cooling," which usually means the time to reach a desired temperature before

Figure 11.1

Effect of grape temperature at harvest on the time needed for the berries to lose 0.5% of their harvested weight.

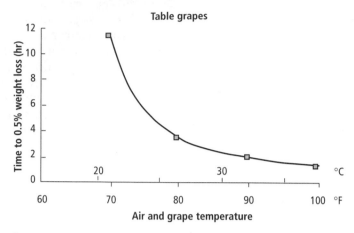

Figure 11.2

Typical cooling curve for perishable products. Cooling times are typical for large fruit, like peaches, exposed to moderate amounts of airflow.

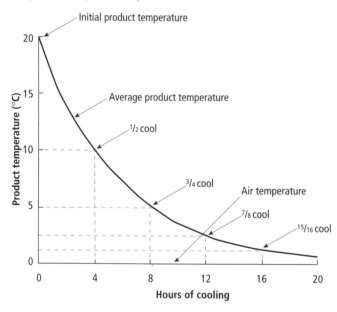

transfer to storage or transport. Yet cooling times are usually reported as "half-cooling" or "seven-eighths-cooling" times. Half-cooling time is the time to cool the product halfway from its initial temperature to the temperature of the cooling medium. Seven-eighths cooling time is three times longer than half cooling and is the time needed for the product temperature to drop by seven-eighths of the difference between the initial product temperature and the temperature of the cooling medium. Both of these cooling times are constant values for a given package type in a given cooling system and are not affected by varying initial product temperatures or varying cooling medium temperature.

As horticultural products cool, their rate of temperature drop slows as cooling progresses. For example, for peaches with an initial pulp temperature of 20°C (68°F), half-cooling them in a forced-air cooler at 0°C (32°F) (i.e., cooling them to 10°C [50°F]) takes 4 hours. It takes an additional 4 hours to cool them to 5°C (41°F) and 4 more hours to reach seven-eighths cooling (about 2.5°C [36.5°F]) (fig. 11.2). Seven-eighths-cooling, or three half-cooling times (in this example, 12 hours), is often used as a reference cooling time.

Both initial product temperature and coolant temperature influence cooling time of a product. The peaches in the example mentioned above might be harvested in the morning, but by late afternoon on a hot California day they could have pulp temperatures near 40°C (104°F), in which case, with 0°C (32°F) cooling air, one additional 4-hour half-cooling period (16 hours total time) would be required to reach the same 2.5°C (36.5°F) pulp temperature as fruit harvested in the morning. If the cooling air temperature was 1.2°C (about 34°F), cooling peaches with initial temperature of 20°C would also require 4 half-cooling periods, or 16 hours.

In room and forced-air coolers, the product closest to the cold air cools noticeably faster than the product farthest from the cold air. Coolers should be managed so that the warmest product reaches acceptably low temperatures before the cooling process is halted.

ROOM COOLING

This widely used cooling method involves placing field or shipping containers of produce into a cold room. Its most common use is for products with a relatively long storage life that are stored in the same room in which they are cooled. Examples include cut flowers before packing, potatoes, sweet potatoes, citrus fruits, apples, and pears.

In room cooling, cold air from the evaporator coils sweeps past the produce containers and slowly cools the product (fig. 11.3). The main advantage of room cooling is that produce can be cooled and stored in the same room without the need for transfer. Its disadvantages are that it is too slow for most commodities; it may initially require empty floor

Figure 11.3

Diagrammatic view of air path during room cooling of produce in bins. Air circulating through the room passes over surfaces and through forklift openings in returning to the cooling coils. In this system the air takes the path of least resistance in moving past the product. Cooling from the surface to the center of bins is largely by conduction.

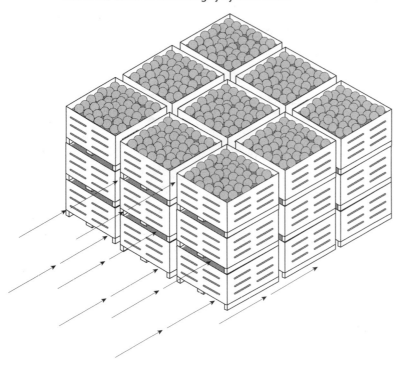

Cooling bays

For both cooling and storage, a single large room is divided into bays by installing partitions partway into the room from each side (fig. 11.4). Air supply channels direct the air into the back of each bay. When a single bay is filled with warm product, supply ducts are opened to direct a large volume of cold air behind the product. Air return occurs down the center forklift aisle. When cooling is completed, the air supply is reduced in that one bay to create the desired storage conditions. If each cooling bay has a separate cold air supply, cold product in one bay is not warmed by warm product in other bays.

FORCED-AIR COOLING

Forced-air cooling is adaptable to a wider range of commodities than any other cooling method. It is much faster than room cooling because it causes cold air to move through, rather than around, containers. This allows cold air to be in direct contact with warm product. With proper design, fast, uniform cooling can be achieved through stacks of pallet bins or unitized pallet loads of containers. Water loss varies with the moisture loss characteristics of individual products and can range from virtually none to 1 to 2% of initial weight. Forced-air is the most widely adaptable and fastest cooling method for small-scale operations.

The speed of forced-air cooling is controlled by the volume of cold air passing over the product. Maximum feasible cooling requires about 0.001 to 0.002 $m^3 \cdot sec^{-1} \cdot kg^{-1}$ of product (1 to 2 cfm/lb). Rates greater than this only slightly reduce cooling time, but as the air volume increases, the static pressure required greatly increases, raising the energy consumption of the fan. Some products can withstand slower cooling and use air volumes of 0.00025 to 0.0005 $m^3 \cdot sec^{-1} \cdot kg^{-1}$ of product (0.25 to 0.5 cfm/lb). Static pressure needed to produce the desired airflow is very dependent on container vent design and the use of interior packaging materials. A complete description of this issue is included in the UC ANR publication *Commercial Cooling of Fruits, Vegetables, and Flowers* (Thompson et al. 1998).

Various airflow designs can be used, depending on specific needs. Converting existing cooling facilities to forced-air cooling is often simple and inexpensive if enough

area between stacked containers for air channels to speed cooling and subsequent rehandling after cooling is finished; and for some products, it can result in excessive water loss compared with faster cooling systems. Room cooling requires days for packed product to reach desired temperature, but it can be faster for unpacked products with good exposure to the cold air. For example, bunched flowers in buckets can cool in 15 minutes, but the same flowers packed in boxes and loaded on a pallet take days to cool.

For best results, containers should be stacked so that the moving cold air can contact all container surfaces. Total fan airflow should be at least 0.3 m³/min per ton of product storage capacity (100 cfm/ton) for adequate heat removal. (For several commonly used airflow systems for room cooling, see fig. 12.3 in chapter 12.) Well-vented containers with vent alignment between containers greatly speed room cooling by allowing air movement through containers. After cooling is complete, airflow can be reduced to 20 to 40% of that needed for initial cooling.

Figure 11.4

Top view of a cold room divided into cooling bays.

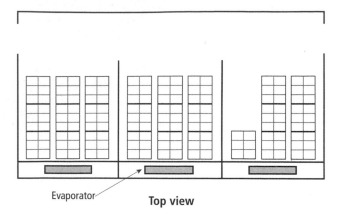

Evaporator   **Top view**

Figure 11.5

Diagrammatic view of a forced-air cooling tunnel. Either bins or palletized containers can be placed to form a tunnel from which air is exhausted. The negative pressure then causes cold air from the room to pass through ventilation slots to directly contact the warm product.

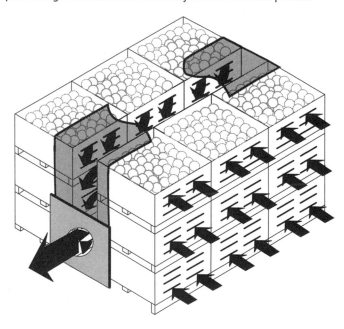

refrigeration capacity is available. A well-designed forced-air cooler is a separate room from cold storage rooms.

Tunnel-type forced-air cooling

In tunnel-type forced-air cooling, the most-used forced-air cooling system, a row of palletized containers or bins is placed on either side of an exhaust fan, leaving an aisle between the rows. The aisle and the open end are then covered to create an air plenum tunnel (fig. 11.5). The exhaust fan creates

Figure 11.6

Forced-air cooling tunnel is in operation, cooling packaged produce on unitized pallets. Air-circulating fan circulates air through fruit and over cooling coils. Canvas plenum cover is designed to fit varying cooling loads.

low air pressure within the tunnel. Cold air from the room moves through the openings in or between containers toward the low-pressure zone, sweeping heat away from the product. The exhaust fan is usually a permanent unit that also circulates air over the refrigeration coils and returns it to the cold room (fig. 11.6). The exhaust fan can also be a portable unit that is placed to direct the warm exhaust air toward the air return of the cold room or refrigeration evaporators.

Cold wall

This forced-air cooling system uses a permanent air plenum equipped with exhaust fans (fig. 11.7). The air plenum is often located at one end or side of a cold room, with the exhaust fans designed to move air over the refrigeration coils. Because openings are located along the room side of the plenum, against which stacks or pallet loads of containers can be placed (fig. 11.8), this method is not often used for products in bins. Various damper designs can be used to ensure that airflow is blocked except when a pallet is in place. Each pallet starts cooling as soon as it is in place, so there is no need to await deliveries to complete a tunnel. Shelves may be built so that several layers of pallets can be cooled. Different packages and even partial pallets

Figure 11.7

Cross section of a cold-wall type forced-air cooler.

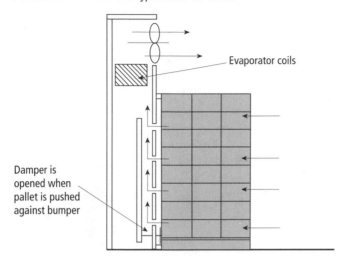

Damper is opened when pallet is pushed against bumper

Evaporator coils

Figure 11.8

Cold-wall type forced-air cooler for use with stacks of flower containers. Open end-vents allow air to be pulled through the containers for rapid cooling but allow closing during shipment.

can be accommodated by proper design of the damper system. This is a benefit in operations that handle a range of commodities. Each pallet must be independently monitored for temperature and promptly moved from the cooler as soon as it is cold in order to avoid desiccation from continued rapid airflow over the product.

Serpentine cooling

The serpentine system is used for forced-air cooling of produce in bins. Bins must have bottom ventilation slots. Ventilation on the sidewalls of the bins does not aid cooling, although it probably does not hinder it either. The system requires modification of the cold-wall design to allow the forklift openings between bins to be used as air supply and return plenums. The cold air moves vertically through the product in each bin, causing a slight pressure difference between plenums (fig. 11.9). Bins may be stacked up to six rows deep against the cold wall, depending on the cooling speed desired and the available airflow. The airflow capacity of the small forklift opening plenums usually limits airflow to less than 0.0005 $m^3 \cdot sec^{-1} \cdot kg^{-1}$ of product (0.5 cfm/lb), and cooling is usually slower than tunnel coolers. To achieve the desired airflow pattern, openings are placed in the cold wall to match alternate forklift openings, starting one bin up from the floor. On the room side of the bins, these same openings are then blocked (fig. 11.10). Air flows into an open slot between bins and passes up or down through one bin of prod-

uct to reach the return plenum. This system requires no space between rows of bins, and bins can be stacked in even numbers as high as the cold store or forklifts can reach.

Forced-air evaporative cooling

In forced-air evaporative cooling, the air is cooled with an evaporative cooler instead of with mechanical refrigeration. If designed and operated correctly, an evaporative cooler produces air a few degrees above the outside wet bulb temperature, and the cooler air is above 90% RH. In most areas of California, product temperatures of 16° to 21°C (60° to 70°F) can be achieved. A typical forced-air evaporative cooler is shown in figure 11.11. This cooling method may be adequate for some products that are best held at moderate temperatures such as tomatoes or for products that are marketed quickly after harvest. In most cases, growers can build their own forced-air evaporative coolers. They are much more energy-efficient than mechanical refrigeration, and if properly designed they provide high-humidity cooling air.

Container venting

Effective container venting is essential for forced-air cooling to work efficiently. Cold air must be able to pass through all parts of a container. For this to happen, container vents must remain unblocked after stacking. If containers are palletized in-register (aligned with one another), container side or end vents will suffice, provided they are properly located in relation to trays, pads,

Figure 11.9

Pattern of airflow in a serpentine forced-air cooling system. This system is specific for cooling fruit in field bins. By blocking alternate forklift openings on cold-wall and room sides, with fans operating, air is forced to pass vertically through bins to cool fruit.

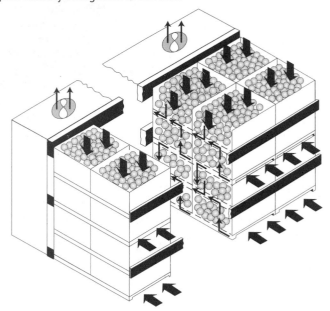

Figure 11.10

Serpentine forced-air cooler in operation. Plastic straps are placed over every other forklift opening from bottom to top. These close off the openings in the room side of the cooler. Air entering the open channels then must move up and down through the product to return to the cold wall. Note that bins can be tightly stacked in rows since no center airflow plenum is needed.

and other packaging. If cross-stacking is used, matching side and end vents is essential. For the 400 by 300 mm (or 16 by 12 in) container cross-stacked on the 1,200 by 1,000 mm (or 48 by 40 in) pallet, vertical vent slots on 100-mm (4-in) centers around the container perimeter should be considered, because they remain matched when cross-stacked (fig. 11.12).

Too little venting restricts airflow; too much venting weakens the container. A reasonable compromise appears to be about 5 to 6% side or end wall venting. A few large vents are more effective than many small vents for speeding the cooling rate. Locating vents midway from top to bottom is adequate unless trays or other packing materials isolate some of the product. Vertical slots at least 12 mm (½ in) wide are better than round vents. Vent design should minimize the effect of product blocking vents. Any type of unvented bag, liner, or vertical divider inside the package may block vents and reduce airflow. If a solid liner is used, slow but acceptable cooling times can be obtained if the box is designed so that cold air can flow over and under each box, and boxes are fairly shallow.

Flower boxes often have closable vents. This allows closing after cooling so that

flowers shipped on unrefrigerated transport can maintain temperatures longer than if shipped in a box with open vents.

Cooling in transport

Marine containers and break-bulk ships have enough airflow and refrigeration capacity to achieve slow forced-air cooling. In-transit cooling is used for products produced in areas without cooling facilities. But it is usually better to cool the product before it is loaded in the ship or container, since most transport modes do not have the extra refrigeration capacity needed for fast cooling.

Refrigerated ships and most marine containers have a bottom-delivery air supply system. Air travels from the refrigeration system to a floor plenum. It then flows upward through the boxes and returns to the refrigeration system in the space above the product. To work well, boxes must have vent holes on the top and bottom, and the vent holes must align if boxes are cross-stacked. The floor must be covered to prevent air from bypassing the load.

Refrigerated highway trailers with a top-delivery air supply system do not have enough airflow to allow transport cooling.

HYDROCOOLING

Cold water is an effective method for quickly cooling a wide range of fruits and vegetables in containers or in bulk (figs. 11.13 and 11.14). Typical seven-eighths cooling times are 10 minutes for small-diameter products like cherries and up to 1 hour for large products such as melons. Hydrocoolers can use either an immersion or a shower system to bring products in contact with the cold water. Hydrocooling avoids water loss and may even add water to a slightly wilted commodity, as is often done with leafy green vegetables. Hydrocoolers can be portable, extending the cooling season. Containers used in hydrocooling must be water-tolerant.

In a typical shower-type hydrocooler, cold water is pumped to an overhead, perforated distribution pan. The water showers over the commodity, which may be in bins or boxes, or loose on a conveyer belt. The water leaving the product may be filtered to remove debris, then passed over refrigeration coils (or ice), where it is recooled. The refrigeration coils are located under or beside the conveyer or above the shower pan. Some commodities, such as leafy vegetables and cherries, are sensitive to water-beating damage. For products like these, the distribution pan should be no more than 20 cm (8 in) above the exposed product. These products can also be cooled in an immersion hydrocooler.

Efficient cooling depends upon adequate water flow over the product surface. For product in bins or boxes, water flows of 13.6 to 17.0 l·sec^{-1}·m^{-2} (20 to 25 gal/min/ft^2) of surface area are generally used. Bin hydrocoolers are often designed to accommodate two-high stacking of bins on the conveyors. Bulk product in shallow layers on a conveyor belt requires 4.75 to 6.80 l·sec^{-1}·m^{-2} (7 to 10 gal/min/ft^2). Water is usually cooled by mechanical refrigeration, but ice may be used if it can be broken into fist-sized pieces and added fast enough to produce adequate cooling. In some areas, clean well water is cold enough to do initial cooling or even complete cooling. Hydrocoolers should be drained and cleaned at least daily or be equipped with special filters to clean the water. Low concentrations (100 to 150 ppm)

Figure 11.11

Cutaway view of an evaporative forced-air cooler. Air is cooled by passing through the wet pad before it passes through packages and around the product.

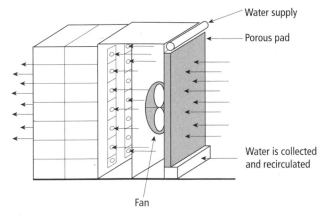

Figure 11.12

Recommended box vent design to allow good airfow or water flow while maintaining package strength.

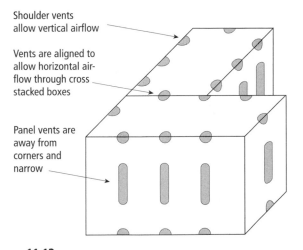

Figure 11.13

Side view of a batch-type hydrocooler for pallet bins.

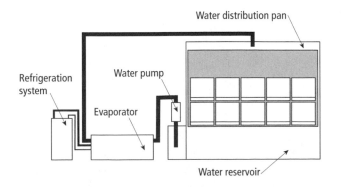

Figure 11.14

Conveyor-type bin hydrocooler in operation. Ice water is pumped into the top pan, where it runs down through the product in the bin. Dwell time in the cooler is controlled by conveyor speed.

of active chlorine are usually used to disinfect the water and minimize the spread of postharvest decay of products.

Hydrocooling has some potential limitations. The product and any packages and packing materials must be tolerant of wetting, and they must also be tolerant of chlorine (apricots sometimes show chlorine damage) or other chemicals that are used to sanitize the hydrocooling water. Shower-pan holes must be cleaned regularly to avoid plugging, which causes uneven water flow over the product. Arriving warm produce may have to remain at ambient temperatures for some time when the hydrocooler is operating at peak capacity. Cooled product must be moved quickly to a cold room or else rapid rewarming occurs. Hydrocooling operations can also require rehandling of the pallet bins before packing or storage.

Hydrocooling can be energy-efficient provided that the hydrocooler is operated continuously at maximum capacity and is inside a cold room or an insulated enclosure.

Shower-type hydrocoolers (conveyor or batch units) are the most commonly used hydrocooling systems, but immersion hydrocoolers are sometimes used. Fresh-cut vegetables are commonly cooled as they are conveyed in a water flume. In this case, the product, normally in bulk, is in direct contact with the cold water as it moves through a long tank of cold water. This method is best suited for products that do not float. Because slow cooling would result if the product simply moved with the water, immersion hydrocoolers convey product against the direction of water flow and often have a system for agitating the water. Conveyors must be designed for positive movement of the product through and out of the water.

PACKAGE-ICING

Some commodities are cooled by filling packed containers with crushed or flaked ice. Initially, the direct contact between product and ice causes fast cooling. However, as the ice in contact with the product melts, the cooling rate slows considerably. The constant supply of meltwater keeps a high RH around the product. Liquid ice, a slurry of ice and water, distributes ice throughout the box, achieving better contact with the product (figs. 11.15 and 11.16). Ice can be produced during off-peak hours when electricity is cheapest and stored for daytime use.

Package-icing requires expensive, water-tolerant packages. The packages should be fairly tight but should have enough holes to drain meltwater. In small operations the ice is hand-raked or shoveled into containers. Large operations use liquid-ice machines to automatically ice pallet loads of packed cartons. The process, which requires only a few minutes, is used for cooling some field-packed vegetables, particularly broccoli. The iced packages should be placed into a cold room after filling to minimize ice melt.

The product must be tolerant of prolonged exposure to wet conditions at 0°C (32°F). Some low-density products have excess space in which to load ice within the package, and ice not melted during cooling can remain in the package even after transport. This excess ice can keep the product cold if the cold chain is broken. However, this is an inefficient use of ice, and the weight of the ice can add significantly to the freight load, sometimes limiting the amount of product hauled. An ice weight equal to 20 to 30% of the product weight is needed for initial cooling, but liquid icing often adds an ice weight equal to the product weight. Also, during transport of mixed loads, water from melting ice can damage neighboring boxes that are not water-tolerant, and vehicle insulation can become wet. Ice and meltwater can be a safety hazard at wholesale distribution.

Figure 11.15

Pallet liquid-icing machine in operation. The high-volume flow of the ice-water mix is pumped into chamber and flows through container vents to deposit the ice throughout the package.

Figure 11.16

A package of liquid-iced broccoli is opened to show the penetration of ice throughout the package.

Cut flowers are often forced-air cooled, but if the box is handled in an unrefrigerated environment, bags of ice may be added to the box to prevent heating. This system allows a measured amount of ice to be added, and the melt water is contained, preventing damage to the product and fiberboard boxes.

VACUUM COOLING

Vacuum cooling takes place by evaporating water from the product at very low atmospheric pressure. Products that easily release water may cool in 20 to 30 minutes. Vegetables that have a high surface-to-mass ratio and that release water rapidly, such as leafy green vegetables (especially iceberg lettuce), are best suited to this method. It is also sometimes used to cool celery, some sweet corn, green beans, carrots, and bell peppers. It is used with carrots and peppers primarily to dry the surface and stems, respectively, and to inhibit postharvest decay. Even boxes of film-wrapped products can cool quickly provided the film allows easy movement of water vapor.

Moisture loss and consequent cooling is achieved by pumping air out of a large steel chamber containing the product (fig. 11.17). Reducing the pressure of the atmosphere around the product lowers the boiling temperature of its water, and as the pressure falls, the water boils, quickly removing heat from the product. Water vapor is removed by condensing it on refrigerated coils located between the ports and the vacuum pump. Vacuum cooling causes about 1% product weight loss (mostly water) for each 6°C (11°F) of cooling. This amount of weight loss can be objectionable for green onions, celery, and some leaf lettuces. Some coolers are equipped with a water spray system that adds water to the surface of the product during the cooling process. Like hydrocooling water, this water must be disinfected if it is recirculated. Water can also be sprayed on the product before it enters the cooler. The rapid release of the vacuum at the end of the process can force surface water into some vegetables, giving them a water-soaked appearance.

A typical vacuum tube, sometimes called a retort, holds 800 boxes of iceberg lettuce (20 pallets). Some small vacuum coolers hold only a single pallet. Most vacuum cooling equipment is portable and is used in two or more production areas each year, allowing the high capital cost of vacuum coolers to be amortized over a longer operating season. Most coolers used today have mechanical refrigeration and rotary vacuum pumps.

COOLING BEFORE PACKING

Cooling problems with products in unitized pallets, or liner-packed products, can be avoided by cooling the products before packing. However, this increases cooling costs if the products are cooled before culling and sorting operations. If 20% cullage occurs after cooling, the cooling cost increases 25%. If 50% is culled (for example, diverting pears to a processor), the cooling cost per ton of packed product

Figure 11.17

Vacuum cooler being loaded. Batches of product are filled into the chamber, which is then closed and the vacuum is drawn. This unit uses a patented process that introduces water during the cooling cycle to reduce water evaporation from the product.

doubles. Another disadvantage to cooling before packing is that cold fruit is more subject to mechanical damage in packing than warm fruit. For cantaloupes and cherries, these problems are avoided by removing most culls before hydrocooling.

Some rewarming occurs when produce is packed after cooling. A mild breeze can rewarm unpacked products to near ambient temperatures within 30 minutes. Some packers minimize this by only partially cooling the product before packing, followed by complete cooling after packing. One packer solves the problem another way: Fruit arriving from the field is forced-air cooled in bins; the bin dump is located in the forced-air bin cooling room. The cooled fruit moves from the cold room to a nearby packing area, where it is sorted, sized, and volume-filled into containers within 3 or 4 minutes. Packed containers are conveyed into a cold room for palletizing within 6 or 7 minutes of leaving the bin cooler. In this system, product rewarming is minimal; the product must be finish cooled.

SELECTING A COOLING METHOD

The physiological or physical characteristics of a product may limit the suitable cooling methods. For example, strawberries, which cannot tolerate free moisture because of disease and injury problems, cannot be cooled by hydrocooling or package-icing, and because they require fast cooling after harvest,

room cooling is not suitable. Vacuum cooling is fast but causes noticeable moisture loss in berries. Thus, forced-air cooling is the only effective cooling method for strawberries. Other commodities, such as some deciduous fruits and many vegetables, are suited to several cooling methods. Table 11.1 lists the cooling methods commonly used for various types of fruits, vegetables, and flowers.

If a cooling facility is used for several types of commodities, it may or may not be possible to use the same method for all products. Table 11.1 shows that vacuum cooling, package icing, and room cooling are used for only a few products; hydrocooling is suited to a much wider variety; and forced-air cooling is adaptable to most products and is therefore ideal for operations where a wide variety of products must be cooled. This is why forced-air and room cooling are most often recommended for small-scale operations, which typically handle many commodities and may change the products they handle as the market changes from year to year. In some cases the product mix may require that more than one cooling system be used.

PRODUCT TEMPERATURE REQUIREMENTS

A facility that must handle products with very different optimal storage temperatures usually needs separate cooling facilities. Keeping chilling-sensitive commodities below their critical threshold temperature too long will cause damage.

If product temperature requirements are not very different, careful cooler management may allow a common cooler to be used. For example, summer squash can be forced-air cooled in a 0°C (32°F) room if it is removed from the cooler at 7°C (45°F) flesh temperature. It should then be stored at 7°C (45°F). Many chilling-sensitive commodities can be safely kept for short periods below their chilling threshold temperature.

COSTS OF OPERATING COOLERS

Capital costs vary significantly among different types of coolers. Liquid ice coolers are the most expensive to purchase, followed by vacuum coolers (including units equipped with water spray capability), forced-air coolers, and hydrocoolers. Figure 11.18 shows the capital cost, expressed in

Table 11.1. Cooling methods suggested for horticultural commodities

| Commodity | Size of operation | | Remarks |
	Large	Small	
Tree fruits			
Citrus	R, FA	R	
Stone fruits	FA, HC	FA	Apricots cannot be HC
Pome fruits	FA, R, HC	R	
Subtropical	FA, HC, R	FA	
Tropical	FA, R	FA	
Berries	FA	FA	
Kiwifruit	FA	FA	
Grapes	FA	FA	Require rapid cooling facilities adaptable to SO$_2$ fumigation
Leafy vegetables			
Cabbage	VC, FA	FA	
Iceberg lettuce	VC	FA	
Kale, collards	VC, R, WVC	FA	
Leaf lettuces, spinach, endive, escarole, Chinese cabbage, bok choy, romaine	VC, FA, WVC, HC	FA	
Root vegetables			
With tops	HC, PI, FA	HC, FA	Carrots can be VC
Topped	HC, PI	HC, PI, FA	
Irish potatoes	R w/evap coolers,		With evap coolers, facilities should be adapted to curing
Sweet potatoes	HC	R	
Stem and flower vegetables			
Artichokes	HC, PI	FA, PI	
Asparagus	HC	HC	
Broccoli, Brussels sprouts	HC, FA, PI	FA, PI	
Cauliflower	FA, VC	FA	
Celery, rhubarb	HC, WVC, VC	HC, FA	
Green onions, leeks	PI, HC, WVC	PI	
Mushrooms	FA, VC	FA	
Pod vegetables			
Beans	HC, FA	FA	
Peas	FA, PI, VC	FA, PI	
Bulb vegetables			
Dry onions	R	R, FA	Should be adapted to curing
Garlic	R		
Fruit-type vegetables			
Cucumbers, eggplant	R, FA, FA-EC	FA, FA-EC	Fruit-type vegetables are chilling-sensitive but at varying temperatures
Melons			
cantaloupes	HC, FA, PI	FA, FA-EC	
honeydew, casaba, crenshaw	FA, R	FA, FA-EC	
watermelons	FA, HC	FA, R	
Peppers	R, FA, FA-EC, VC	FA, FA-EC	
Summer squashes, okra	R, FA, FA-EC	FA, FA-EC	
Sweet corn	HC, VC, PI	HC, FA, PI	
Tomatillos	R, FA, FA-EC	FA, FA-EC	
Tomatoes	R, FA, FA-EC		
Winter squashes	R	R	

Table 11.1. Cont.

| Commodity | Size of operation | | Remarks |
	Large	Small	
Fresh herbs			
Not packaged	HC, FA	FA, R	Can be easily damaged by water beating in HC
Packaged	FA	FA, R	
Cactus			
Leaves (nopalitos)	R	FA	
Fruit (tunas or prickly pears)	R	FA	
Ornamentals			
Cut flowers	FA, R	FA	When packaged, only use FA
Potted plants	R	R	

KEY:
FA = Forced-air cooling
FA-EC = Forced-air evaporative cooling
HC = Hydrocooling
PI = Package icing

R = Room cooling
VC = Vacuum cooling
WVC = Water spray vacuum cooling

cost per daily cooling capacity, of four types of coolers, based on 1998 data. The wide cost range for liquid icing reflects the variation in the amount of ice that is put in the carton. If just enough ice for product cooling is used, much less refrigerating capacity is needed and capital cost is lower. However, many broccoli shippers add extra ice to handle refrigerating needs in transport, and they also add an extra 4.5 kg (10 lb) of ice so that the box arrives at the market with unmelted ice.

The capital cost per unit cooled can be minimized by using the equipment as much as possible. Vacuum-cooling equipment is very compact and is often portable. In California, vacuum coolers are moved as harvest locations change during the year. It is common in the western United States for portable vacuum coolers to be used more than 10 months per year. Forced-air cooling facilities can be used for short-term storage of product during the harvest season and for long-term storage of product after the season ends.

Energy costs
The energy cost of cooling varies greatly among coolers (fig. 11.19). Energy use is expressed in terms of an energy coefficient (EC), defined as

$$EC = \frac{\text{cooling work done (expressed in kilowatt-hours)}}{\text{electricity purchased (kWh)}}$$

High EC numbers indicate an energy-efficient operation. The range of EC for each type of cooler reflects differences in design and operation procedures between coolers of the same type.

Actual energy costs for operating a cooler can be calculated using the formula below (assuming a value for EC). Energy costs can be less than 5% of total costs in efficient cooling systems.

In English units:

$$\text{Electricity cost} = \frac{W \times TD \times R \times Cp}{3{,}413 \times EC}$$

where
W = weight cooled (lb)
TD = temperature reduction in product (°F)
R = electricity rate ($/kWh)
EC = energy coefficient
Cp = 1 Btu/lb-°F
3,413 Btu/kWh

In SI (metric) units:

$$\text{Electricity cost} = \frac{W \times TD \times R \times Cp}{3.6 \times EC}$$

where
W = weight cooled (kg)
TD = temperature reduction in product (°C)
R = electricity rate ($/kWh)
EC = energy coefficient
Cp = 4,184 J/kg-°C
3.6 J/kWh

Figure 11.18

Capital cost of commonly used cooling systems (in 1998 dollars).

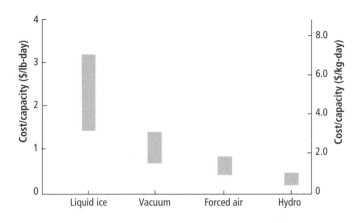

Figure 11.19

Energy use of commonly used cooling systems.

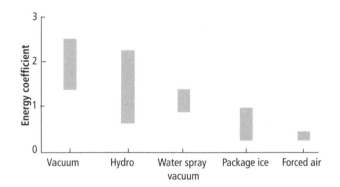

Labor and other equipment costs must be included in calculating total operating costs. Although no specific data are available for these costs, they can vary significantly. For example, a hydrocooler built into a packing line requires very little labor and other equipment, but stand-alone coolers used in field packing operations require operators and lift trucks for moving product in and out of the cooler.

If a cooling method requires that the product be packaged in a special carton, the extra cost of the carton should be included in a comparison of cooler types. For example, package icing, hydrocooling, and water-spray vacuum cooling need water-resistant packaging. This can increase the cost of an individual box by 25 cents to $1 depending on the design, size, and quantity of boxes purchased.

Other considerations

Marketing tradition may dictate the choice of a cooling method. For example, some markets require that broccoli boxes arrive with ice in them. Shippers selling in this market must select a package-ice cooling system.

Existing facilities may determine the type of cooler to be used. An existing cold-storage room can often be used for forced-air cooling of small amounts of product by installing a small portable fan. Larger amounts of product usually require installation of more refrigeration capacity and a permanent air handling system.

A short harvesting season in a particular location may cause an operator to consider using portable cooling equipment. The cooler can be moved to a grower's other production areas or leased to shippers in other areas, eliminating the cost of buying separate permanent coolers for each location. Some portable coolers can be leased or jointly owned by shippers and cooler manufacturers, eliminating or reducing the need for capital expenditure.

Some growers contract with commercial companies to do their cooling. This requires no direct capital investment and no operating or management costs. But the grower loses some control over the product, often cannot control when product is cooled, and loses the chance to make a profit from the cooling operation. Cooling cooperatives can give a grower some of the advantages of owning a cooler while reducing individual investment costs.

ESTIMATING REFRIGERATION CAPACITY

After deciding which cooling method or methods to use, the operator must estimate the amount of refrigeration capacity needed. This will help determine how large a cooler is needed. In general, 2.3 kW of refrigeration requires about 1 kW of compressor capacity, or 1 ton requires a little less than 2 horsepower of compressor capacity. Coolers requiring less than about 40 kW (11.4 tons) of refrigeration capacity can often be built by the grower.

The refrigeration capacity needed for large systems must be determined by a refrigeration engineer. The engineer will consider a number of factors, such as

- amount of product cooled

- temperature of incoming product

- rate at which the product is received at the cooler

- required speed of cooling

- variety of products cooled and their unique cooling requirements

- building design and how it affects heat gain to the refrigerated volume

- heat input from lights, fan motors, fork-lifts, people, etc.

An estimate of the amount of refrigeration capacity needed for small-scale facilities does not require detailed calculations. Figure 11.20 can be used to estimate the refrigeration capacity needed for a cooler handling up to 450 kg/hr (1,000 lb/hr). For example, if a product is cooled from 23.9°C (75°F) to 1.7°C (35°F) (a temperature drop of about 22°C [40°F]), and the cooler must handle a maximum of 400 kg/hr (900 lb/hr), then 16 kW (4.5 tons) of refrigeration capacity is necessary. Estimates from this figure are based on reasonably fast cooling; slow cooling, as is achieved by room cooling, requires slightly less refrigeration capacity. The figure is also based on the assumption that heat input to the cooler from sources other than the product are less than 25% of the total.

Figure 11.20

Approximate mechanical refrigeration requirements for small scale coolers based on maximum hourly product input and product temperature drop.

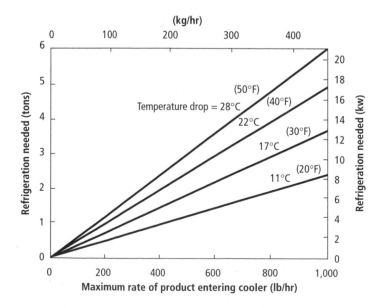

Some small-scale cooling operations purchase ice for cooling. Figure 11.21 can be used to estimate the daily amount of ice needed to operate a small cooler. For example, if 2,000 kg/day (4,400 lb/day) of product are cooled by about 22°C (40°F), a little more than 1,000 kg (1.1 tons) of ice would be melted. The figure is based on 50% of the ice being used for product cooling and the rest of the cooling potential lost to outside heat gain. This efficiency level is common for uninsulated hydrocoolers. For more details on designing and operating coolers, consult the UC ANR publication *Commercial Cooling of Fruits, Vegetables, and Flowers* (Thompson et al. 1998).

EFFECTIVE COOLER MANAGEMENT

Proper management of a cooler involves effective product cooling at minimum cost. Records of cooler operation are vital to enable a manager to evaluate the cooler. Good records should include

- a sampling of incoming and outgoing product temperature for each lot and the type of product cooled

- the temperature of the cooling medium during each cooling cycle

- the length of cooling cycles

- the quantity of product cooled in each cycle

- operating conditions of refrigeration system, such as suction and head pressures

- monthly energy use

Knowledge of incoming product temperature is helpful in estimating the cooling time required; outgoing product temperature is essential for determining the quality of the cooling process. The average product temperature should be within acceptable tolerances, and, just as important, the warmest product temperatures should be within acceptable tolerances. A good operator checks outgoing product temperatures in various parts of the load to determine where the warmest product tends to be, and then controls the operation to get product in this area below required temperature. For example, in tunnel-type forced-air coolers, the warmest product is usually next to the return air tunnel in the pallet farthest from the fan. Hydrocoolers tend to cool product fairly uniformly,

Figure 11.21

Approximate amount of ice needed to operate small-scale coolers based on amount of product cooled per day and product temperature drop.

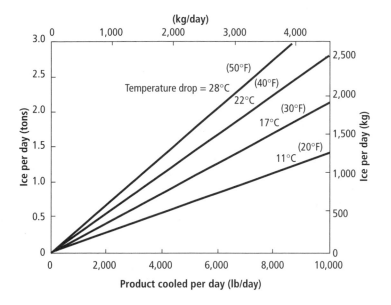

levels and other features of the refrigeration system should be checked daily.

COLD ROOMS

Cooled product will quickly warm up unless it is loaded directly into refrigerated transport vehicles or placed in cold rooms. Rewarming wastes the benefits of cooling; and cooled products left in a warm environment are also subject to condensation, which may lead to disease. To help solve these problems, a cold room should be associated with the cooler. In some cases, the cooler may be a part of the cold room, as with forced-air coolers, but this is not recommended. Small cold rooms can be commercially constructed, purchased in prefabricated form and erected by growers, constructed by growers, or purchased as used refrigerated transport vehicles (rail cars, trailers, or marine containers). The cost of the cold room should be added to the total capital cost of a cooling facility.

SUMMARY

Effective cooling and temperature management requires a complete understanding of product and market requirements, and of the cooling methods available.

- Rapid thorough cooling and good product temperature management are essential for successful produce marketing.

- Cooling is part of the total system of handling perishables. Effects on cooling rate must be considered whenever a change is made in packaging or handling.

- Requirements for cooling and cold storage differ, and they should be considered as two separate operations.

- Four cooling methods and variations are available to achieve rapid cooling. Select a cooling method or methods that fit the needs of your customers and the range of commodities you handle.

- Fast cooling can often be achieved through minor modifications of existing cooling facilities. Design requirements should be determined by a qualified refrigeration engineer after evaluating the complete refrigeration system. The increased costs involved in achieving faster cooling may be relatively small when the total cost of the cooling system is considered.

and the warmest product is in areas with restricted water flow, perhaps caused by misaligned box vents. Vacuum coolers tend to cool very uniformly. The performance of liquid icing systems is determined by the uniformity of ice added to each box.

Other factors are useful in determining the long-term performance of a cooler. For example, if cooling times begin to increase and the temperature of the cooling medium does not change, then there is a good chance that flow of the cooling medium through the product is being restricted (assuming that the type of product and its incoming temperature remain constant). If the temperature of the cooling medium shows a trend of increasing during the cooling cycle, there may be problems in the refrigeration system, or there may be too much product in the cooler. Changes in operating conditions of the refrigeration system can give clues to possible problems and their solutions.

Regular maintenance is important for all types of coolers. In vacuum coolers, door seals must be checked regularly and pressure gauges must be recalibrated about once per year. Daily cleaning is vital for proper hydrocooler operation. Trash screens, the water distribution pan, and the water reservoir must be cleaned each day and chlorine levels must be checked several times a day. Fluid

- Cooling time can often be reduced by attention to details of air or water management, package design, packing material, and pallet stacking patterns.

- Keep careful records of cooling performance. Good cooler management requires systematic measurement and recording of product temperatures.

REFERENCES

Hardenburg, R. E., A. E. Watada, and C. Y. Wang. 1986. The commercial storage of fruits, vegetables, and florist and nursery stocks. USDA Handb. 66. 130 pp.

Isenberg, F. M. R., R. F. Kasmire, and J. E. Parson. 1982. Vacuum cooling vegetables. Cornell Univ. Coop. Ext. Bull. 186. 10 pp.

Jeffrey, J. J. 1977. Engineering principles related to the design of systems for air cooling of fruits and vegetables in shipping containers. Proc. 29th Intl. Conf. on Handling Perishable Agricultural Commodities. East Lansing: Mich. State Univ. 151–164.

Rij, R. E., J. F. Thompson, and D. S. Farnham. 1979. Handling, precooling, and temperature management of cut flower crops for truck transportation. Oakland: Univ. Calif. Coop. Ext. Leaflet 21058.

Sargent, S. A., M. T. Talbot, and J. K. Brecht. 1989. Evaluating precooling methods for vegetable packinghouse operations. Proc. Fla. State Hort. Soc. 101:175–182.

Thompson, J. F., and R. F. Kasmire. 1981. An evaporative cooler for vegetable crops. Calif. Agric. 35(3–4): 20–21.

Thompson, J. F., F. G. Mitchell, T. R. Rumsey, R. F. Kasmire, and C. H. Crisosto. 1998. Commercial cooling of fruits, vegetables and flowers. Oakland: Univ. Calif. Div. of Ag. and Nat. Res. Publ. 21567. 61 pp.

Watkins, J. B., and S. Ledger. 1990. Forced-air cooling. 2nd ed. Brisbane, Australia: Queensland Dept. of Primary Industries. 64 pp.

12

Storage Systems

James F. Thompson

Orderly marketing of perishable commodities often requires some storage to balance day-to-day fluctuations between product harvest and sales; for a few products, long-term storage is used to extend marketing beyond the end of harvest season. The goals of storage are to

- slow biological activity of the product by maintaining the lowest temperature that will not cause freezing or chilling injury and by controlling atmospheric composition

- slow the growth and spread of microorganisms by maintaining low temperatures and minimizing surface moisture on the product

- reduce product moisture loss and the resulting wilting and shrivel by reducing the difference between product and air temperatures and maintaining high humidity in the storage room

- reduce product susceptibility to damage from ethylene gas

With some commodities, the storage facility may also be used to apply special treatments. For example, potatoes and sweet potatoes are held for a few days at high temperature and high RH to cure wounds sustained during harvest; table grapes are fumigated with sulfur dioxide to minimize Botrytis decay damage; and pears and peaches may be warmed and exposed to ethylene to ripen more quickly and uniformly. This chapter describes the equipment and techniques commonly used to control temperature, RH, and atmospheric composition in a storage facility (see fig. 12.1).

STORAGE CONSIDERATIONS

TEMPERATURE

The temperature in a storage facility normally should be kept within about ±1°C (2°F) of the desired temperature for the commodities being stored. For storage very close to the freezing point, a narrower range may be needed. Temperatures below the optimal range for a given commodity can cause freezing or chilling injury; temperatures above it can shorten storage life. In addition, wide temperature fluctuations can result in water condensing on stored products and more rapid water loss from them. Recommended temperatures and humidities for long-term storage of horticultural products are listed in appendix A. In many storage facilities, particularly at wholesale and retail marketing, many different products are held in a common room. Figure 12.2 groups commodities into four temperature and humidity groups for short-term storage.

Maintaining storage temperatures within the prescribed range depends on several important design factors. The refrigeration system must be sized to handle the maximum expected heat load. Undersized systems allow the air temperature to rise during peak heat load conditions, but an oversized system is unnecessarily expensive. The system should also be designed so that air leaving the refrigeration coils is close to the desired temperature in the room. This prevents large temperature fluctuations as the refrigeration system cycles on and off. Large refrigeration coils installed with suction pressure controls allow a small temperature difference to be maintained between the air leaving them and the air in the room while still having adequate refrigeration capacity. A small temperature difference also increases RH in the room and may reduce frost buildup on the coils. A large free space above the stored product allows air from the evaporators to mix with

Figure 12.1

Interior view of a table grape cold storage.

room air before coming in contact with stored product.

Temperature variation is minimized with adequate air circulation. Most storages are designed to provide an airflow of 0.052 m³ per second and per metric ton (100 cfm/ton) of product, based on the maximum amount of product that can be stored in the room. This is needed to cool the product to storage temperature and may be needed if the product has a high respiration rate. Because this high airflow rate can cause excessive weight loss from products and fans are a significant source of heat, the system should be designed to reduce airflow to 0.0208 to 0.0104 m³ per second and per metric ton (20 to 40 cfm/ton) after the product has reached storage temperature. Systems that control motor speed, such as variable frequency controllers for AC motors, are often used to control fan speed. Operate fans at the lowest possible speed that prevents unacceptably warm product in the storage. The warmest product will tend to be near the top of the room next to a warm wall or roof, and farthest from the evaporator fans.

Low circulation rates require that the system be designed to move air uniformly past

all of the stored product. Figure 12.3 shows some systems used to uniformly distribute air in large storage rooms. When air flows past the sides of bins or pallet loads, product must be stacked to form air channels 10 to 15 cm (4 to 6 in) wide past the sides of each unit. These channels should be formed parallel to direction of air movement. There should also be space between the product and the walls to allow refrigerated air to absorb the heat conducted from the outside. Curbs installed on the floor next to walls help assure that the product will be loaded with an air space between it and the walls. When forklift openings are used for air distribution, they must align along the entire length of the air path. The serpentine airflow system is the only one that forces air to flow through the product. This speeds initial cooling but may not be necessary for uniform product temperatures in long-term storage.

Because air takes the path of least resistance, partially filled rooms often have poor air distribution. Large rooms can be divided into sections using uninsulated walls parallel to the direction of airflow. This allows product in one bay to be removed with little effect on airflow in neighboring bays.

Smaller rooms with packaged evaporator coils use an airflow pattern similar to the ceiling plenum design except that the plenum is not needed because high-capacity evaporator fans can discharge air 15 m (50 ft). The wall plenum is formed by stacking product 20 to 25 cm (8 to 10 in) away from the wall under the evaporator. Rooms cooled with roof-mounted packaged evaporative coolers can use ceiling-mounted paddle fans to distribute air downward past product bins.

In some long-term controlled atmosphere storage, airflow is minimized by cycling evaporator fans off for as much as 85 to 90% of the time. In the winter in temperate climates, most refrigeration demand comes from heat released by evaporator fans and motors. Reducing fan operation time reduces heat input, which in turn reduces refrigeration operation. Tests conducted in the northwestern United States showed that fan cycling reduced electricity use by 65%. It may also increase humidity in the storage and reduce product moisture loss.

Thermostat sensors are usually placed 1.5 m (5 ft) above the floor (for ease of checking) in representative locations in the room. They

Figure 12.2. Compatible fresh fruits and vegetables during 10-day storage

Group 1A: 0°–2°C (32°–36°F) and 90–98% RH

VEGETABLES

alfalfa sprouts	Brussels sprouts*	daikon*	leek*	scorzonera
amaranth*✓	cabbage*✓	endive,* chicory	lettuce*✓	shallot*
anise*	carrot*	escarole*	mint*	snow pea*
artichoke*	cauliflower*	fennel*	mushroom	spinach*✓
asparagus*	celeriac	garlic	mustard greens*	sweet pea*
beans: fava, lima	celery*	green onion*	parsley*	Swiss chard*
bean sprouts	chard*	herbs* (not basil)	parsnip	turnip
beet	Chinese cabbage*✓	horseradish	radicchio*	turnip greens*
Belgian endive*	Chinese turnip	Jerusalem artichoke	radish	waterchestnut
bok choy*✓	collard*	kailon✓	rutabaga	watercress*
broccoflower*	corn: sweet, baby	kale*✓	rhubarb	
broccoli*	cut vegetables	kohlrabi	salsify	

Group 1B: 0°–2°C (32°–36°F) and 85–95% RH

FRUITS AND MELONS

apple	cantaloupe	elderberry	lychee	prune
apricot	cashew apple	fig	nectarine	quince
avocado, ripe	cherry	gooseberry	peach	raspberry
Barbados cherry	coconut	grape	pear: Asian, European	strawberry
blackberry	currant	kiwifruit*	persimmon*	
blueberry	cut fruits	loganberry	plum	
boysenberry	date	longan	plumcot	
caimito	dewberry	loquat	pomegranate	

Group 2: 7°–10°C (45°–50°F) and 85–95% RH

VEGETABLES

basil*	okra*
beans: snap, green, wax	pepper: bell, chili
cactus leaves (nopales)*	squash: summer (soft-rind)*
calabaza	tomatillo
chayote*	winged bean
cowpea (Southern pea)	
cucumber*	
eggplant*	
kiwano (horned melon)	
long bean	
malanga*	

FRUITS AND MELONS

avocado, unripe	guava	pineapple
babaco	Juan Canary melon	pummelo
cactus pear, tuna	kumquat	sugar apple
calamondin	lemon*	tamarillo
carambola	lime*	tamarind
cranberry	limequat	tangelo
custard apple	mandarin	tangerine
durian	olive	ugli fruit
feijoa	orange	watermelon
granadilla	passion fruit	
grapefruit*	pepino	

Group 3: 13°–18°C (55°–65°F) and 85–95% RH

VEGETABLES

bitter melon	squash: winter (hard rind)*
boniato*	
cassava	sweet potato*
dry onion	taro (dasheen)
ginger	tomato: ripe, partially ripe, and mature green
jicama	yam
potato	
pumpkin	

FRUITS AND MELONS

atemoya	jaboticaba	rambutan
banana	jackfruit	sapodilla
breadfruit	mamey sapote	sapote
canistel	mango	soursop
casaba melon	mangosteen	
cherimoya	papaya	
crenshaw melon	Persian melon	
honeydew melon	plantain	

Notes:

Ethylene level should be kept below 1 ppm in storage areas.

* Products sensitive to ethylene damage.

Figure 12.3

Airflow systems used in cold storages. The top two designs cause air to flow through planned gaps between lanes of pallets or bins. The bottom designs work with tightly stacked bins and force air through forklift openings.

Airflow through spaces between bins or pallet loads

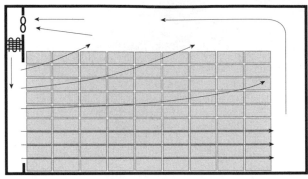

Vertical slots in supply plenum

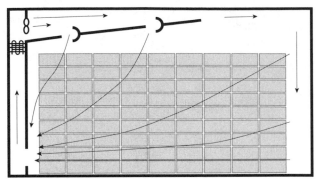

Ceiling plenum with turning vanes

Airflow through forklift openings in bins

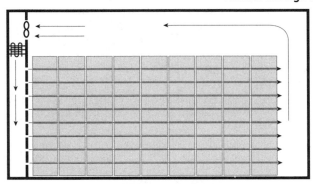

Horizontal slots in supply plenum

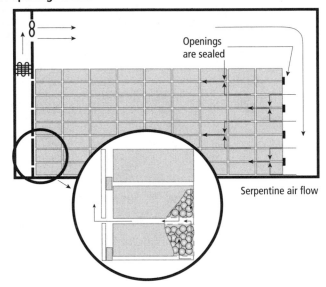

Openings are sealed

Serpentine air flow

should not be placed near sources of heat such as doors or walls with an exterior surface. Nor should they be placed in a cold area such as near the air discharge of the refrigeration unit. A calibrated thermometer should be used to periodically check the thermostat. Errors of only a few degrees can affect product quality.

HUMIDITY

For most perishable commodities, the RH in a long-term storage facility should be kept at 90 to 95%. Humidities below this range result in unacceptable moisture loss. Humidities very close to 100% may cause excessive growth of microorganisms and surface cracking on some fruits, although it is unusual for a storage facility to have relative humidities

that are too high. Partially dried products such as garlic, ginger, and dry onion are held at 65 to 75% RH. Dried fruits and nuts should be held at 55 to 65% RH.

Fiberboard containers are weakened by prolonged exposure to the high humidity recommended for most horticultural commodities. Typical fiberboard exposed to 95% RH has half as much strength as the same material exposed to 50% RH. The storage humidity for perishables can be lowered below recommended levels if the product is packaged in plastic bags or box liners to prevent moisture loss.

Refrigeration equipment must be specially designed to maintain high RH. In systems not designed for horticultural commodities, the evaporator coils (which produce the cold air) operate at a temperature about 6°C (11°F) lower than the desired air temperature in the room. This causes an excessive amount of moisture to condense on the coils and can result in 70 to 80% RH in the storage room. Coils with a large surface area and refrigeration controls to maintain highest possible coil temperatures achieve the same refrigeration capacity as smaller coils but can operate at a higher temperature, thus reducing the amount of moisture removed from the air. The coils should be large enough to operate 3°C (5°F) colder than the room air temperature.

Mechanical humidifiers or fog spray nozzles are sometimes used to add moisture to the storage room and reduce the drying effect of the evaporator coils. However, this added moisture results in the need for more frequent defrosting of coils. Humidifiers are needed in conditions where product is stored at a temperature warmer than the outside environment, such as cold-winter locations where heat must be added to prevent damage from low temperatures.

Some refrigeration systems use a wet coil heat exchanger to maintain humidity. In this system, water is cooled to 0°C (32°F) or a higher temperature if higher room temperatures are desired. The water is sprayed down through a coil, and the storage area air is cooled and humidified to nearly 100% RH as it moves upward through the coil. However, as the air moves through the storage area it picks up heat, and the rise in temperature reduces RH. This system is usually limited to air temperatures above 0.2°C (32.4°F) and does not work well for commodities that are held close to or below 0°C, without the use of compounds that lower the freezing point of water (such as caustic soda or ethylene glycol).

Low humidities for partially dried roots and bulbs and dried fruits and nuts are usually obtained by installing a special evaporator coil. In the first part of the coil the air is cooled considerably below the desired room temperature to dry the air. Then, in a reheat section, the air is heated to the set point air temperature. Heat energy is supplied by an electrical resistance heater or hot refrigerant gas from the compressor discharge piped to a the reheat section of the evaporator coil.

REFRIGERATION

The capacity of a refrigeration system is based on the sum of the heat inputs to a storage area, including heat conducted through walls, floor, and ceiling; field and respiration heat from the product; heat from air infiltration; and heat from personnel and equipment such as lights, fans, and forklifts. Details of heat load calculations are listed in ASHRAE handbooks and in *Industrial Refrigeration Handbook* (Stoecker 1998).

Refrigeration equipment for storage facilities is generally not designed to remove much field heat from the product, since a large capacity would be required; a separate cooling facility is used for this purpose.

MECHANICAL REFRIGERATION

Most storage facilities use a refrigeration system to control storage temperature. This system employs the fact that a liquid absorbs heat as it changes to a gas. The simplest method of this is to allow a controlled release of liquid nitrogen or liquid carbon dioxide in the storage area. As these liquids boil, they cause a cooling effect in the storage area. However, this method requires a constant outside supply of refrigerant; it is used only to a limited extent with highway vans and rail cars. The more common mechanical refrigeration systems use a refrigerant such as ammonia or a variety of halocarbon fluids (sometimes referred to by the trade name "freon"), whose vapor can be easily recaptured by a compressor and heat exchanger.

Figure 12.4 shows the components of a typical vapor recompression (or mechanical) refrigeration system. The refrigerant fluid passes through the expansion valve, where the pressure drops and the liquid evaporates at temperatures low enough to be effective in removing heat from the storage area. Heat from the material to be cooled is transferred to the room air, which is then forced past the evaporator (a cooling coil located in the room), usually a finned tube heat exchanger, which transfers the heat from the air to the refrigerant, causing it to evaporate. After fully changing to a gas, it is repressurized by the compressor and then passed through a condenser, where it is cooled to a liquid. The condenser is located outside the storage area and releases heat. Liquid is stored in the receiver and is metered out as needed for cooling.

Figure 12.4

Schematic of a typical vapor recompression, or mechanical, refrigeration system.

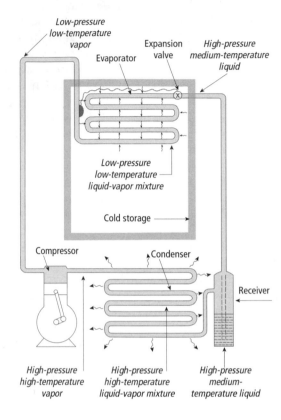

EXPANSION VALVES

Small mechanical refrigeration systems are controlled primarily by an expansion valve, which regulates the pressure of the refrigerant in the evaporator. Low pressures cause the liquid refrigerant to evaporate at low temperatures. The valve also controls the flow of refrigerant, which affects the amount of refrigeration capacity available. Capillary tubes and thermostatic expansion valves are the two most common types of expansion valves.

The capillary tube is used with very small refrigeration equipment (less than 1 HP). It is a tube 0.6 to 6 m (2 to 20 ft) long with a very small inside diameter of 0.6 to 2.3 mm (0.025 to 0.090 in). The resistance of the liquid flowing through the tube creates the needed pressure drop between the low-pressure and high-pressure sides of the system and regulates the flow of refrigerant. A capillary is inexpensive and has no moving parts to maintain, but it cannot be adjusted, is subject to clogging, and requires a relatively constant weight of refrigerant in the system.

A thermostatic expansion valve regulates

the flow of refrigerant to maintain a constant temperature difference between the evaporator inlet (or evaporating temperature) and the coil outlet, maintaining a constant superheat. It allows the low-side pressure to vary, so that when high refrigeration loads are required, the temperature of the evaporator coil increases. This type of expansion valve is not well suited to obtaining high RH needed in long-term storage.

Large refrigeration systems may use a flooded coil, an evaporator coil that is designed to always have liquid refrigerant in it. A flooded coil has a greater heat transfer efficiency than a nonflooded coil of equal size. Refrigerant flow is controlled primarily with a float control that ensures a constant level of refrigerant in the coil. The float control may operate in parallel with a thermostatic expansion valve.

Other controls, such as suction pressure regulators, may be used in conjunction with float controls. These are especially useful in maintaining the highest possible evaporator coil temperature in order to maintain high humidity in the storage room.

EVAPORATORS

Modern cold storages usually use finned tube evaporators. Air from the storage is forced past the tubes by fans, which are a part of a complete evaporator unit. Evaporators operating below 0°C (32°F) build up frost that must be removed to maintain good heat transfer efficiency. Defrosting may be done by periodically flooding the coils with water, by electric heaters, by directing hot refrigerant gas to the evaporators, or by continuously defrosting with a brine or glycol solution.

COMPRESSORS

The most common types of refrigeration compressors are reciprocating (piston) and rotary screw (fig. 12.5). Reciprocating compressors come in a wide range of sizes and can be set up to operate efficiently at varying refrigerant flow rates. Flow rates are varied by shutting off pairs of cylinders in a unit, which may have 6 to 12 cylinders. The main disadvantage of reciprocating compressors is their fairly high maintenance costs. Rotary screw compressors have low maintenance costs but are not available in sizes smaller than about 23 kW (30 HP).

Figure 12.5

Common types of refrigeration compressors.

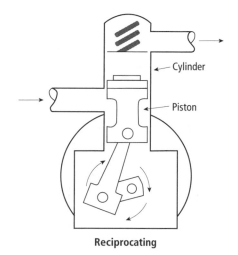

Reciprocating

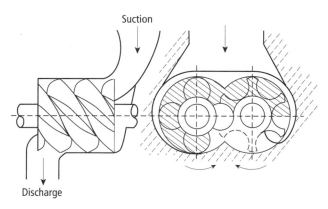

Rotary screw

Maintaining the highest possible suction pressures reduces compressor energy use. Use large evaporator coils and a control system that increases suction pressure as demand on the refrigeration system is reduced. Use a compressor system that operates efficiently over the required range of refrigerant flows. Screw compressors operate efficiently only at flow rates near their maximum capacity. Use several in parallel, shutting down those that are not needed, or consider using reciprocating compressors for peak loads. Reciprocating compressors operate efficiently over a large range of refrigerant flows.

CONDENSERS

Condensers are categorized as air-cooled or water-cooled. Small systems usually use an air-cooled unit. Many home refrigerators, for instance, have a coiled tube in the back that allows a natural draft of air to flow past. Larger systems use a fan to provide airflow past the condenser. Large condensers are more likely to be water cooled. Water is a better heat conductor than air, allowing water-cooled condensers to be smaller than forced-air units of equal capacity. However, water-cooled units may require large quantities of water, which can be expensive to obtain and dispose of. Evaporative condensers reduce water consumption by recycling the heated condenser water; they require close attention to water quality to maintain efficiency and to prevent damage to the heat exchanger.

Energy use is minimized by selecting a condenser that cools the refrigerant fluid to as low a temperature as possible. For example, a facility maintaining 0°C (32°F) and a condensing temperature of 52°C (125°F) requires 50% more power than one that operates at a condensing temperature of 35°C (95°F). In warm areas, well-water-cooled or evaporatively cooled condensers should be selected over air-cooled units.

REFRIGERANTS

The choice of which refrigerant to use in a vapor recompression system is based on the following factors:

- Cost of refrigerant. Halocarbon refrigerants are more expensive than ammonia. Environmental regulations restrict the availability of some halocarbon refrigerants.

- Compatibility. Ammonia cannot be used with metals that contain copper; halocarbon refrigerants cannot be used with alloys containing more than 2% magnesium and may damage some elastomeric materials.

- Toxicity. Ammonia at very low concentrations can injure perishable commodities. It is toxic to humans, and government regulations may require ammonia systems to have equipment to contain accidental releases and a plan for protecting personnel and neighbors from a release.

- Flammability. Ammonia is very flammable. Most commonly used halocarbon refrigerants are not flammable or have a low flammability.

CONTROL SYSTEMS

A large refrigeration system requires a good control system and equipment for displaying the system's operating condition. At a minimum, panel lights should be installed to indicate the operating status of fans and compressors and the fluid levels in surge and receiver tanks. Controls should be set up to allow manual operation of motors.

Microcomputers and programmable controllers allow even more precise control of large refrigeration systems. They are especially valuable in reducing electricity use during peak rate periods. Defrost cycles can be programmed to take place at night, and unnecessary fans and compressor motors can be turned off during peak rate periods.

ABSORPTION REFRIGERATION

Absorption refrigeration, used in a few cold storage operations, differs from mechanical refrigeration in that the vapor is recovered primarily through use of heat rather than mechanical power. It is less energy-efficient than mechanical refrigeration and usually is used only where an inexpensive source of heat is available. Processing facilities with excess low-pressure steam are well suited to absorption refrigeration.

SECONDARY REFRIGERANT

Some storages are cooled with a secondary coolant. A brine solution (sodium chloride or calcium chloride) or glycol (propylene glycol or ethylene glycol) is cooled by a mechanical refrigeration system and pumped to heat exchangers in the cold storage. These systems are a little less energy-efficient than conventional systems, and brine solutions are corrosive. But they dramatically reduce the quantity of first-stage refrigerant needed and confine it to the engine room. This a great asset in dealing with the flammability and safety issues of ammonia. Secondary refrigerant piping does not need to withstand the pressure of primary refrigerants, and plastic piping can sometimes be used. The temperature of the heat exchanger can be precisely controlled with a mixing valve. Brines and glycol solutions are corrosive and must be used with corrosion inhibitors, and they should never come in contact with zinc. Sodium chloride and propylene glycol are food-grade materials.

ALTERNATIVE REFRIGERATION SOURCES

In many developing countries, where mechanical refrigeration is prohibitively expensive to install, maintain, and operate, a number of other techniques can be used to produce refrigeration. In some cases, these techniques can provide cooling levels that approach recommended storage conditions. In others, they are a compromise between proper storage conditions and costs for equipment, capital, and operations.

EVAPORATIVE COOLING

Evaporative cooling techniques are very energy-efficient and economical (fig. 12.6). A well-designed evaporative cooler produces air with a RH greater than 90%. Its main limitation is that it cools air only to the wet-bulb temperature of the outside air. During the harvest season in the United States, wet-bulb temperatures vary from 10° to 25°C (50° to 77°F) depending on location, time of day, and weather conditions. This temperature range is acceptable for some chilling-sensitive commodities.

The water for evaporative cooling systems comes from domestic sources. It is also practical to cool by evaporating water directly from the commodity. Snap beans have been cooled in transit by erecting an air scoop above the cab of the truck that forces outside air through a bulk load of beans. This system prevents heat buildup and keeps the beans at or below the outside air temperature. It is not advisable to use this system for any great length of time in order to avoid excessive water loss.

Figure 12.6

Evaporatively cooled sweet potato storage. Unit evaporative coolers on the roof continuously supply cooled air to the storage. Room air is exhausted from vents along the bottom of the side walls.

NIGHTTIME COOLING

In some parts of the world, significant differences between night and day temperatures allow nighttime ventilation to be a means of refrigeration. In dry Mediterranean or desert climates, the difference between daily maximum and minimum temperatures can be as great as 22°C (40°F) during the summer. Nighttime cooling is commonly used for unrefrigerated storage of potatoes, onions, sweet potatoes, hard-rind squashes, and pumpkins. As a rule, night ventilation effectively maintains a given product temperature when the outside air temperature is below the product temperature for 5 to 7 hours per day.

Low nighttime temperatures can be used to reduce field heat simply by harvesting produce during early morning hours. Some growers in California use artificial lighting to allow nighttime harvest.

It is theoretically possible to produce air temperatures below nighttime minimums by radiating heat to a clear nighttime sky. A clear night sky is very cold, and a good radiating surface such as a black metal roof can cool below the air temperature. Simulations have indicated that this method could cool air about 40°C (70°F) below night air temperatures. This concept is rarely used.

WELL WATER

In some areas, well water can be an effective source of refrigeration. The temperature of the ground greater than about 2 m (6 ft) below the surface is equal to the average annual air temperature. Well water is often very near this temperature.

NATURALLY FORMED ICE

Before the development of mechanical refrigeration, refrigeration was provided by natural ice harvested from shallow ponds during the winter. The ice was stored in straw and hauled to cities as needed during spring and summer. Energy costs today make it unfeasible to transport ice any significant distance. However, cooling facilities in appropriate climates can store ice nearby for summer use. In some cases, it may be feasible to transport perishable commodities to the ice for storage. This would be especially practical where the ice is located between the sites of production and consumption.

HIGH-ALTITUDE COOLING

High altitude can also be a source of cold. As a rule of thumb, air temperatures decrease by 10°C with every 1 km (5°F per 1,000 ft) increase in altitude. It is not possible to bring this air down to ground level because the air heats naturally by compression as it drops in altitude. However, in some cases it may be possible to store commodities at high altitudes in mountainous areas. For example, in California most perishable commodities are grown in the valley floors near sea level. However, much of the production is shipped east across the Sierra Nevada over passes about 1,800 m (6,000 ft) high. Air temperature has the potential of being 18°C (32°F) cooler at these elevations, and it may reduce energy costs to store perishables there rather than on the valley floor.

UNDERGROUND STORAGE

Cellars, abandoned mines, and other underground spaces have been used for centuries to store fruits and vegetables. At certain depths, the underground temperature is near the average annual air temperature. Underground spaces work well for storing produce that has already been cooled but not for removing field heat. Soil is a poor conductor of heat; once an underground storage area warms up, it does not cool down rapidly. This can be overcome by installing a network of buried pipes around the storage. Cooled air is pumped from the pipes to the storage area, drawing cooling capacity from a greater soil volume.

THE STORAGE BUILDING

The storage must be sized to handle peak amounts of product. The floor area can be calculated by dividing the volume of the produce by the maximum product storage height and adding floor area for aisles, forklift maneuvering, and staging areas. The maximum storage height can be increased by using shelves or racks and forklifts with suitable masts. Multistory structures are generally not used for storage because of the difficulty and expense of moving the product between levels.

The storage building should ideally have a floor perimeter in the shape of a square. A rectangular configuration has more wall area per square foot of floor area, resulting in higher construction costs and higher heat loss than a square configuration. Entrances, exits, and storage areas should be arranged so that products generally move in one direction through the facility, especially if the storage facility is used in conjunction with a cooler to remove field heat.

SITE SELECTION

Good utility service must be available for the facility. Extending roads and energy utilities to a facility can be very expensive. Three-phase electrical power is needed to operate refrigeration equipment motors. In some areas a backup power supply may be advisable. There should be enough water to supply the evaporative condensers, personnel needs, and the needs of a packinghouse, if it is a part of the project. Consider the availability of fire protection services, gas supply, and sewer utilities. The area should have good drainage and room for future expansion. There should be enough space around the facility for smooth movement of large highway trucks.

BUILDING LAYOUT

Room layouts with an interior corridor offer better operating conditions for cold storages and better control of controlled atmosphere storages than designs that allow room access only through exterior doors. The interior corridor, seen in layouts 1, 2, and 3 in figure 12.7, allows easy access to piping and controls. Doors and equipment are shielded from the elements, and product observation is easier. Layout 1 is common in small operations where storage and packing are done in one building. Layout 2 allows better prod-

uct flow than layout 3, but layout 3 has less area devoted to corridor. Layout 4 is the least expensive of the designs because none of the cold storage building is used as a permanent corridor.

Refrigerated facilities can be constructed from a wide variety of materials. Floors and foundations are usually concrete. A vapor barrier is installed to prevent moisture from moving through the floor, and rigid insulation is sometimes placed above the barrier and below the concrete. Walls can be made of concrete blocks, tilt-up concrete, insulated metal panels, or wood frame construction. In large facilities in the United States, metal and tilt-up concrete construction is now being used more commonly than wood frame or concrete blocks.

Walls are insulated with fiberglass batts, rigid urethane foam boards, or sprayed-on foam. Batt and board insulation must be protected with a vapor barrier on the warm side. Properly applied sprayed-on foam can be moisture-proof. Exposed foam insulation must be coated with a fire retardant. Some storages combine insulation types; for example, the interior can be sprayed with foam to form a vapor barrier, then covered with batt or board insulation for appearance and fire protection. If modified atmosphere techniques are used in the storage facility, the vapor barrier may also serve as a gas barrier, and special precautions must be taken to ensure a gastight seal.

Total wall insulation level, measured in heat resistance units, is often in the range of 3.5 to 7.0 m²·K/W (R20 to R40). Ceilings can be insulated with rigid board or foam materials, or they can be suspended below

Figure 12.7

Typical layouts of cold storage facilities.

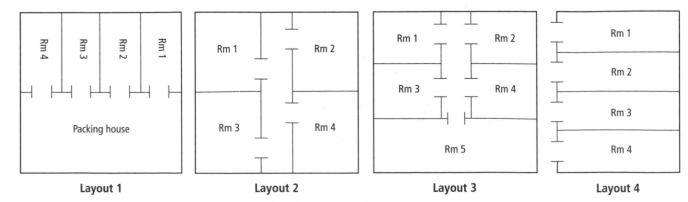

Layout 1   Layout 2   Layout 3   Layout 4

the roof and insulated with loose fill or batts. Ceiling insulation of at least 10.6 m²·K/W (R60) is common in new construction. In general, it is advisable to build with more insulation than utility costs may presently warrant, because energy costs are difficult to predict and it is much less expensive to install insulation during construction than it is to add insulation after construction is completed.

Sunlight falling on walls or roofs can dramatically increase the effective outside air temperature, increasing heat flow into a storage facility. (The effective outside air temperature is the normal air temperature plus a factor to account for the sun shining on a surface.) Table 12.1 shows the relationship between wall or roof orientation, color, and effective outside air temperature. A dark, flat roof can be 38°C (69°F) warmer than the outside air temperature. Painting a south-facing wall a light color can reduce the effective wall temperature by 9°C (16°F) compared with a dark wall. The walls and roof of a cold storage facility should be painted a light color or shaded from the direct sun.

Warm outside air leaking into the cold storage room increases energy use. Use plastic flap doors to reduce infiltration during loading and unloading. Seal around openings for pipes and electrical conduits. Loading docks for outbound product should be fitted with flexible bumpers that provide an airtight seal between the dock and a truck.

Use high-efficiency lighting sources, such as metal halide lamps, to reduce heat in the cold storage. Turn off lights when not needed.

Table 12.1. Effective outside air temperature when direct sunlight strikes various building surfaces, Fresno, CA on July 21 at 40°N latitude

		Effective air temperature	
Surface	Color	°C	°F
Actual air	—	34	93
Horizontal roof	Light	51	124
Horizontal roof	Dark	72	162
South wall	Light	42	108
South wall	Dark	51	124
West wall	Light	51	124
West wall	Dark	71	160

Source: Adapted from ASHRAE 1993.

SMALL-SCALE REFRIGERATED STORAGE

COLD ROOMS
Cold rooms for small-scale operations can be purchased from commercial suppliers, self-built, or made from used refrigerated equipment such as rail cars, marine containers, or highway vans. The choice of which system to use is based on cost and availability of equipment in the area and the amount of time available to invest in cold room installation.

Rail cars
Refrigerated rail cars in the United States are very sturdy and well insulated. Their refrigeration is powered by an electric motor that is powered in turn by a diesel generator. The generator set can be removed from the car and salvaged, and the refrigeration can be connected to the operator's electric utility. The ceiling in the cars is 9 ft, 4 in. (2.8 m), which limits the height that produce can be stacked. The most significant problem and greatest cost of using rail cars for cold storage is getting them from the railroad to the site.

Highway vans
The one unique advantage of using highway vans as cold storage is that they are portable if the wheels are left on. The refrigeration system is powered with a diesel engine. This can be a benefit if utility electricity is not available at the site. In some areas, it may be less expensive to operate the refrigeration unit if it is converted to operate with an electric motor, but a considerable conversion cost must be added to the project. Highway vans are built as light as possible to maximize the load weight they can carry; this often means that used vans are in fairly poor condition. Their insulation, which is limited to begin with, may be deteriorated, and they may have poorly sealed doors that permit excess air leakage. Also, old vans often have fairly small fans that may not provide adequate air circulation.

Marine containers
Used marine containers are available in lengths of 20, 24, and 40 ft (6.1, 7.3, and 12.2 m). Their built-in refrigeration units are powered with 220- or 440-volt three-phase electricity, and they can be plugged directly into utility power. They are usually well

built and have deep T-beam floors and sufficient fan capacity to provide good air circulation; in fact, air circulation is good enough to allow adequate room cooling.

A disadvantage of all transport vehicles is that their refrigeration systems are usually not designed to produce high RH. When product dries due to low humidity, it causes weight loss and poor quality. This is particularly a problem if the cold room is to be used for long-term storage. About the only way to reduce the drying is to keep the floor and walls of the cold room wet, but this causes increased corrosion, reduced equipment life, and increased need for defrosting.

Refrigerated transport vehicles rarely have enough refrigeration capacity to cool produce rapidly. If rapid cooling is needed, extra capacity must be added. Moreover, transportation vehicles are too narrow for the frequent product movement needed in a cooling facility. A separate, self-constructed room is much more convenient for cooling operations.

SELF-CONSTRUCTED COLD ROOMS

For many producers, a self-built cold room is the least expensive option. (See the UC ANR publication *Small-Scale Cold Rooms for Perishable Commodities* [Thompson and Spinoglio 1996] for more details.) Self-built cold rooms usually have a concrete or wood frame floor. The walls and roof are wood frame construction insulated with fiberglass batts. Care must be taken to install a tight vapor barrier on the warm side of the insulation. Refrigeration is provided by a small mechanical refrigeration system. If the room is to be kept above 10°C (50°F), it may be possible to use a room air conditioner. These cost about half as much as a packaged refrigeration system.

CONTROLLED ATMOSPHERE STORAGE

Controlled atmosphere (CA) storage uses oxygen and carbon dioxide concentrations of about 1 to 5% for each gas in most applications. Normal room air has an O_2 concentration of about 21% and CO_2 levels near 0.03%. Low O_2 and high CO_2 levels slow ripening processes, stop the development of some storage disorders such as scald in apples, and slow the growth of decay organisms. All of these effects increase storage life

of fresh produce compared with refrigerated air storage. More details about the potential benefits and hazards of CA storage are presented in chapter 14.

SIMPLE CA SYSTEM

Controlled atmosphere storage has all the design requirements of conventional refrigerated storage plus gas-tight rooms, equipment to create the desired gas concentrations, and equipment to measure and control atmospheric composition. The simplest system for obtaining gas-tight storage uses a plastic tent inside a conventional refrigerated storage room. The tent is made of 3- to 5-mil polyethylene sheeting supported by a wood framework. The sheeting is sealed to the concrete floor by pressing the plastic into a narrow trough and forcing tubing into the trough to keep the plastic in place. A better gas barrier at the floor can be obtained by laying a sheet of plastic on the floor and covering it with wood panels. A seal is obtained by joining the tent to the floor sheet. A fan inside the tent provides air circulation. The O_2 level is reduced initially by allowing fruit respiration to consume O_2 or by using CA generators. Oxygen is kept above the minimum by allowing a controlled amount of outside air to enter the tent. The CO_2 level is maintained by placing bags of fresh hydrated lime (calcium hydroxide) in the tent to absorb excess CO_2.

PERMANENT CA FACILITIES

Permanent facilities require that the storage plant be designed specifically for CA storage. Usually a CA facility costs about 5% more to build than a conventional refrigerated storage facility. The extra cost is in building the storage rooms to be airtight and, in some cases, designing smaller individual rooms than those needed in conventional refrigerated storage.

Room size

Individual rooms should be sized to allow them to be filled in a short time. Many CA rooms for storing apples are built to hold 1 week's fruit harvest. If an operator wishes to use rapid CA, the room should be small enough to be filled in 3 days or sooner. Many facilities have several room sizes to allow for variations in incoming fruit volume, fruit varieties, and marketing strategy.

Today's CA storages are single-story

designs, with individual rooms often tall enough to allow fruit to be stacked 10 bins high, including enough height between the bins and the ceiling to allow air from the evaporator coils to mix with the room air and travel easily to the far end of the room.

Walls and ceilings

Three main types of interior wall and ceiling construction are used in recently built CA storages: foamed-in-place urethane over concrete or steel walls, plywood-covered stud walls with fiberglass insulation, and insulated, metal-covered panels.

Urethane foam serves as a gas seal, insulation, and vapor barrier. But it is expensive, primarily because the urethane must be covered with a fire barrier. Foamed-in-place urethane can be applied directly to any type of masonry, wood, or primed metal. It should not be applied over rigid board insulation as the foam can distort the board, and board materials that are not tightly attached cause the gas seal to fail over time. Figure 12.8 shows a cross-section of a typical CA storage sealing system with urethane form insulation.

Installing a plywood cover over fiberglass insulation is a common method of insulating and sealing CA storage. It is usually the least expensive of the three gas barrier systems. The plywood sheets are sealed with a butyl rubber compound applied between the sheet and the wood framing. Sheets are separated with a gap of 3 mm (⅛ in) to allow for expansion. The gap is filled with butyl rubber, and the joint is covered with fabric and an elastomeric sealer. If regular plywood is used, the whole board must be covered with the sealer. High-density plywood does not need to be treated for gas tightness. In most CA storage, a vapor barrier is installed on the outside (warm side) of the insulation. An outside vapor barrier is not recommended with the plywood system, because there is already a gas-tight seal on the inside of the insulation. A vapor barrier on the outside will trap moisture that might get into the insulation, ruining its insulating value.

Insulated panels are usually a sandwich design in which 8 to 15 cm (3 to 6 in) of rigid foam insulation is covered on both sides with painted metal or fiberglass sheets. Panels are 1.2 to 2 m (4 to 6 ft) wide and usually extend from floor to ceiling. The

Figure 12.8

Gas-tight seal using urethane foam insulation.

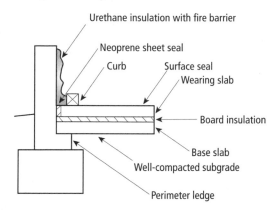

panels are installed inside a building shell and are held together with mechanical fasteners. Joints are sealed with polyvinylacetate copolymer or latex emulsion sealers. Large gaps are backed with a nonwoven fabric.

Floors

Floors can be made gastight using layered or single-slab designs. An insulated floor is usually a sandwich design, with board insulation placed between two layers of concrete (see fig. 12.8). With this design, a gas seal of two layers of hot-mopped asphalt roofing felt is applied to the subfloor. A single-slab floor is used if only perimeter insulation is used. In this case the floor is sealed by applying special materials such as chlorinated rubber compounds to the top surface. In both designs, the wall-to-floor seal is the area that is most likely to fail. Prevent the floor from moving with respect to the wall by carefully backfilling and thoroughly compacting the subgrade. The floor can be tied to the walls with rebar, or the floor can set on a ledge that is 10 cm (4 in) wide built into the foundation wall.

Doors

Many different door designs can be used in CA facilities. In all of them the door is constructed of a solid frame that can be clamped tightly against a gasketed door frame without warping. The frame can be covered with well-sealed plywood or metal. Some of the more expensive doors have aluminum sheets welded to an aluminum frame. The bottom of the door is usually sealed with caulking compound after the door is closed. Most doors are 2.4 to 3 m (8 to 10 ft) wide and tall enough to allow a lift truck with two bins to pass through.

Each room should also have an access door that is 60 × 75 cm (24 × 30 in) to allow for entry for checking fruit and making repairs without opening the main door. Many storages also have a clear acrylic window near the top of a wall to allow the fruit to be inspected without entering the room. The window is usually a concave shape, allowing all areas of the room to be seen.

PRESSURE RELIEF

A pressure difference between the cold room and the outside can develop because of changes in weather or room temperature. This difference can damage the gas seal if it is not relieved. A water trap (fig. 12.9) is usually used to allow pressures to equalize. The trap is often filled with ethylene glycol to prevent the water from evaporating. A spring-loaded or weight-loaded check valve can be used, but they are much more expensive than a water trap. Provide 10 cm² of vent opening for each 40 m³ of room volume (1 in² per 1,000 ft³).

Small changes in pressure can be relieved by using breather bags. These have the advantage of capturing the gas mixture in the room and allowing it to reenter the room at a later time. Bags should have 0.35 to 0.4 m³ of capacity per 100 m³ of room volume (3.5 to 4 ft³/1,000 ft³), although some are designed with several times this capacity. If a CA room is so tightly sealed that air must be regularly bled into the room to maintain O_2 level, breather bags are not necessary.

PRESSURE TEST

The overall gas-tightness of a CA room can be tested by pressurizing it and measuring the rate of pressure drop. After all doors and openings are sealed, a small fan is used to increase the static pressure in the room to 25 mm (1 in) of water column. The fan is then turned off and the room is sealed, and the tester determines how long it takes for the pressure to drop by one-half. If it takes 20 minutes or longer, the room is considered tight enough for a CA system that uses hydrated lime to maintain CO_2 levels. At least 30 minutes are required for rooms that use carbon or water scrubbers to control CO_2. Some researchers believe that rooms should be even more gas-tight than these standards, recommending 45 minutes to 2 hours for the pressure to drop by one-half. Because gas seals deteriorate, new rooms should be much tighter than prevailing standards.

If the room does not meet the desired standard, check for leaks. Leaks are most common around doors, at the wall and floor junctions, near poorly sealed penetrations, and at unsealed electrical boxes. Test for leaks by putting the room under a slight vacuum and then listen for air leaking or spray suspect areas with soapy water and watch for bubbles. Smoke sticks or bee smokers can also be used to detect streams of outside air entering the room.

ATMOSPHERIC MODIFICATION

The least expensive but slowest method of modifying the storage room atmosphere is to let the product do it through natural respiration. Fruits and vegetables use O_2 and release CO_2. The product filling a sealed room eventually lowers the O_2 to the level needed for CA storage; if O_2 drops too low, outside air can be added to restore it to the desired range. However, respiration causes CO_2 levels to rise well above required levels. Bags of hydrated lime can be used to absorb excess CO_2. Lime requirements are 1 to 3 kg per 100 kg (1–3 lb per 100 lb) of product, depending on the product being stored, storage time, surface area, and quality of the lime. Bags can be placed either in single layers in the CA room or in an adjacent lime room that is connected to the CA room with a fan and ductwork. Rooms should be sized to hold about

Figure 12.9

Schematic of water trap pressure relief system.

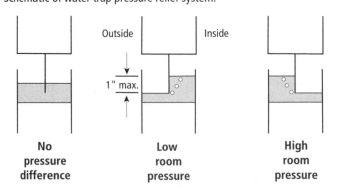

| No pressure difference | Low room pressure | High room pressure |

15 kg of lime per metric ton of product (30 lb/ton). Lime is very effective in producing the low levels of CO_2 that are increasingly used in apple storage. Carbon dioxide levels can also be controlled with activated carbon adsorption systems, also known as molecular sieves. Molecular sieves tend to use more energy than activated carbon systems.

Relying on product respiration to remove O_2 is fairly slow, and the product storage life can be increased if the O_2 is removed faster. Some operations purge the CA room using nitrogen purchased in liquid form or produced on site. One type of nitrogen generator uses ammonia in a combustion process to consume O_2 and produce nitrogen and water. Two other systems use a molecular sieve process (pressure swing adsorption, or PSA system) or a semipermeable membrane to remove O_2. Machines that remove O_2 by combustion of natural gas or propane all produce CO_2, which must be removed by another process. Incomplete combustion in these machines has caused explosions in CA rooms and carbon monoxide poisoning of workers.

REFRIGERATION EQUIPMENT

Refrigeration equipment for CA facilities is the same as for any other cold storage operation. Most storages are designed to maintain 0°C (32°F) at 95% RH. Air circulation during storage should be in the range of 0.0208 to 0.0104 m³/sec/metric ton (20 to 40 cfm/ton) of maximum storage capacity. Up to 0.052 m³/sec/metric ton (100 cfm/ton) is required for product cooling after initial loading.

MONITORING EQUIPMENT

Oxygen and carbon dioxide levels must be monitored daily or more frequently to ensure that they are within prescribed limits. Traditionally, operators have used an Orsat gas analyzer, a wet chemistry system that is fairly time-consuming to use. Automatic equipment is now widely used. Automatic equipment is more accurate than the Orsat system, provides a log of the data, and can be connected to a controller to automatically maintain proper gas concentrations.

Temperature should also be monitored regularly. A minimum of two calibrated dial thermometers should be installed in each room. One should be near eye level with the dial located on the outside of the room. The other should be above the commodity and should

be readable through an observation window near the ceiling. Electronic thermometers allow easier observation of room temperature, and the data can easily be printed out for a permanent record of operating conditions. Most new storages use four probes (or more) per room. Some probes are placed into bins to monitor commodity temperatures.

SAFETY CONSIDERATIONS

The atmosphere in CA rooms will not support human life, and people have died of asphyxia while working in CA rooms without breathing apparatus. A danger sign should be posted on the door. The access hatch in the door should be large enough to accommodate a person equipped with breathing equipment. At least two people with breathing equipment should work together at all times, one inside the room and one outside the room watching the first person.

REFERENCES

ASHRAE (American Society of Heating, Refrigeration, and Air-Conditioning Engineers). 1993. ASHRAE handbook fundamentals. Atlanta: ASHRAE. See especially chapter 20.

Bartsch, J. A., and G. D. Blanpied. 1984. Refrigeration and controlled atmosphere storage for horticultural crops. Cornell Univ. Coop. Ext. Bull. NRAES-22. 42 pp.

Davis, D. C. 1980. Moisture control and storage systems for vegetable crops. In C. W. Hall, ed., Drying and storage of agricultural crops. Westport, CT: AVI. 310–359.

Dellino, C. V. J. ed. 1990. Cold and chilled storage technology. New York: Van Nostrand Reinhold.

Dewey, D. H. 1983. Controlled atmosphere storage of fruits and vegetables. In S. Thorne, ed., Developments in food preservation. London: Applied Science. 1–24.

Fokens, F. H., and H. F. Th. Meffert. 1967. Apparatus for measuring leakage from C.A. storage rooms and other insulated construction. Paper presented at XIIth Int. Congress of Refrigeration, Madrid.

Hallowell, E. R. 1980. Cold and freezer storage manual. Westport, CT: AVI. 356 pp.

Hardenburg, R. E., A. E. Watada, and C. Y. Wang. 1986. The commercial storage of fruits, vegetables, and florist and nursery stocks. USDA Handb. 66. 130 pp.

Hunter, D. L. 1982. C.A. Storage structure. In D. G. Richardson and M. Meheriuk, eds., Controlled atmospheres for storage and transport of perishable agricultural commodities. Beaverton, OR: Timber Press. 13–19.

International Institute of Refrigeration. 1993. Cold store guide. Paris: Internat. Inst. Refrig. 205 pp.

Leyte, J. C., and C. F. Forney. 1999. Controlled atmosphere tents for storing fresh commodities in conventional refrigerated rooms. HortTechnology. 9:672–675.

Stoecker, W. F. 1998. Industrial refrigeration handbook. New York: McGraw-Hill.

Thompson, J. F., and M. Spinogolio. 1996. Small-scale cold rooms for perishable commodities. Oakland: Univ. Calif. Div. Agric. Nat. Res. Leaflet 21449. 8 pp.

Waelti, H., and J. A. Bartsch. 1990. Controlled atmosphere storage facilities. In M. Calderon and R. Barkai-Golan, eds., Food preservation by modified atmospheres. Boca Raton, FL: CRC Press. 373–389.

13

Psychrometrics and Perishable Commodities

James F. Thompson

Psychrometrics is the measurement of the heat and water vapor properties of air. Commonly used psychrometric variables are temperature, RH, dewpoint temperature, and wet-bulb temperature. While these may be familiar, they are often not well understood.

PSYCHROMETRIC CHART

The psychrometric chart describes the relationships between these variables. Figures 13.1 and 13.2 are psychrometric charts in English and metric units, respectively, which will help to illustrate the meaning of various terms.

Temperature, sometimes called dry-bulb temperature after the unwetted thermometer in a psychrometer, is the horizontal axis of the chart. The vertical axis is the moisture content of the air, called the humidity ratio (sometimes called the mixing ratio or absolute humidity). The units of the humidity ratio are mass of water vapor per mass of dry air. Under typical California conditions, the humidity ratio of outside air varies between 0.004 and 0.015 kg/kg. Even though water vapor represents only 0.4 to 1.5% of the weight of the air, this small amount of water vapor plays a very significant role in the postharvest life of perishable commodities.

The maximum amount of water vapor that air can hold at a specific temperature is given by the leftmost, upward-curved line in the psychrometric chart. Notice that air holds more water vapor at increasing temperatures. As a rule of thumb, the maximum amount of water that the air can hold doubles for every 11°C (20°F) increase in temperature. This line is also called the 100% RH line. A corresponding 50% RH line is approximated by the points that represent the humidity ratio when the air contains one-half of its maximum water content. The other RH lines are formed in a similar manner.

Notice that RH without some other psychrometric variable does not determine a specific air condition on the chart and is not very meaningful. For instance, 80% RH at 0°C (32°F) is a much different air condition than 80% RH at 20°C (68°F).

If a mass of air is cooled without changing its moisture content, it loses capacity to hold moisture. If cooled enough, it becomes saturated (has 100% RH) and if cooled further, begins to lose water in the form of dew or frost. The temperature at which condensation begins to form is called the dewpoint temperature if it is above 0°C (32°F) or the frost point temperature if it is below 0°C (32°F).

Another commonly used psychrometric variable is wet-bulb temperature. On the chart this is represented by lines that slope diagonally upward from right to left. These lines represent the temperature and water vapor conditions of a thermometer covered with water-soaked gauze. In practice, wet-bulb lines are used to determine the exact point on the psychrometric chart that represents the air conditions in a given location as measured by a psychrometer. The intersection of the diagonal wet-bulb temperature line (equal to the temperature of a wet-bulb thermometer) and the vertical dry-bulb temperature line defines the temperature and humidity conditions of air.

Water vapor pressure is not usually shown on psychrometric charts, but it is an important concept in handling perishables. It is directly proportional to humidity ratio. The following formula is used to calculate vapor pressure:

$$Vp = \frac{w \times Pa}{e}$$

where
Vp = vapor pressure (same units as Pa)
w = humidity ratio
Pa = atmospheric pressure
e = 0.622 = ratio of the molecular weight of water divided by molecular weight of air

Psychrometric charts and calculators are based on a specific atmospheric pressure, usually a typical sea-level condition. Precise calculations of psychrometric variables require an adjustment for barometric pres-sures that are different from those listed on a standard chart. Consult the ASHRAE Handbook listed in the references for more information on this. Most field measurements do not require adjustment for pressure.

EFFECT OF PSYCHROMETRIC VARIABLES ON PERISHABLE COMMODITIES

TEMPERATURE
Air temperature is the most important variable because it tends to control the flesh temperature of perishable commodities. All perishables have an optimal range of storage

Figure 13.1

A psychrometric chart in English units for sea level elevation.

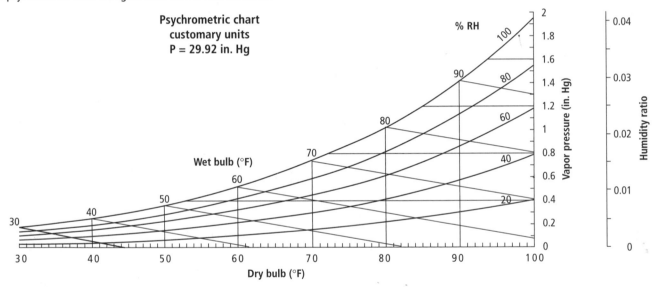

Figure 13.2

A psychrometric chart in SI (metric) units for sea level elevation.

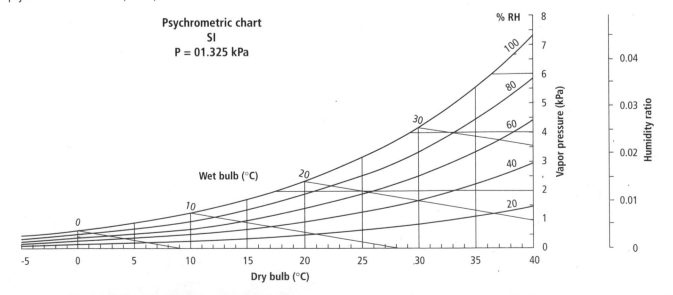

temperatures. Above the optimum, they respire at unacceptably high rates and are more susceptible to damage from ethylene and disease. In fact, horticultural commodities respire at rates that double, triple, or even quadruple for every 10°C (18°F) increase in temperature. Temperatures below the optimum result in freezing or chilling damage. Accurate control of temperature is vital in maintaining maximum shelf life.

VAPOR PRESSURE

The rate of moisture loss from a perishable is primarily controlled by the difference in water vapor pressure between the air in the intercellular spaces of plant material and the air surrounding it. The air in fresh plant material is nearly saturated or, in other words, is close to 100% RH. The water vapor pressure of this air is therefore determined solely by the temperature of the plant material. From the psychrometric chart it is apparent that low temperatures result in low internal water vapor pressures and high temperatures cause high internal water vapor pressure.

Consider several examples of how the drying of perishables is influenced by water vapor pressure differences. An apple cooled to 0°C (32°F) and placed in a refrigerated room with saturated air at 0°C (32°F) would not lose moisture because the water vapor pressures of the air in the apple and of the surrounding air are the same. Table 13.1 lists water vapor pressures for the example cases. However, if the apple were at 20°C (68°F) because it was not cooled before being placed in the coldroom refrigerator, the air in the apple would have a high water vapor pressure compared to the refrigerated air, causing the apple to dry. If the apple were cooled to 0°C (32°F) but the refrigerated air was at 70% RH, drying would occur because the refrigerated air is at a lower water vapor pressure than the nearly saturated air in the

apple. However, the rate of moisture loss is much greater when the apple is not cooled than when the apple is at storage temperature but the storage room air is not saturated. The difference in water vapor pressure between the air in the apple and the storage air is over 9 times more when the apple is not cooled than when it is cooled and put in unsaturated storage air.

Drying is reduced by decreasing the difference in water vapor pressure between the air in the perishable commodity and the air surrounding it. Both the temperature of the commodity and the humidity ratio in the surrounding air must be controlled.

OTHER FACTORS

Relative humidity. Relative humidity (RH) is a commonly used term for describing the humidity of the air, but it is not particularly meaningful without knowing the dry-bulb temperature of the air. Together these two variables allow the water vapor pressure to be determined, which is a better index of the potential for desiccation.

Dewpoint temperature. Condensation of liquid water on perishables can contribute to disease problems. If a commodity is cooled to a temperature below the dewpoint temperature of the outside air and brought out of the cold room, condensation forms. Condensation can also occur in storage if the air temperature fluctuates too greatly.

MEASURING PSYCHROMETRIC VARIABLES

All psychrometric properties of air are determined by measuring two psychrometric variables (three, if barometric pressure is considered). For example, if wet- and dry-bulb temperatures are measured, then RH, vapor pressure, dewpoint, and so on can be determined with the aid of a psychrometric chart. Many variables can be measured to determine the psychrometric state of air, but dry-bulb temperature, wet-bulb temperature, dewpoint temperature, and RH are most commonly measured.

DRY-BULB TEMPERATURE

Dry-bulb temperature can be simply and inexpensively measured by a mercury-in-glass thermometer. The thermometer should be marked in divisions of at most 0.2°C or

Table 13.1. Water vapor pressure of various storage air conditions and product temperatures

Variables	Water vapor pressure (kPa)
Room air at: 0°C (32°F), 100% RH	0.61
0°C (32°F), 70% RH	0.43
Fresh product* at: 0°C (32°F)	0.61
20°C (68°F)	2.34

Note: *Assumes air in product is saturated.

0.5°F if it is used in conjunction with a wet-bulb thermometer for determining cold storage air conditions. The thermometer should be shielded from radiant heat sources such as motors, lights, external walls, and people. This can be done by placing the thermometer where it cannot "see" the warm object or by protecting it with a radiant heat shield assembly.

Hand-held thermistor, resistance bulb, or thermocouple thermometers can also be used. They are more expensive than a mercury-in-glass thermometer but are not necessarily more accurate. Some of these instruments may have a sharp probe, allowing them to be used for measuring product pulp temperature. Mercury-in-glass thermometers should not be used near food products because mercury is toxic. Inexpensive alcohol-in-glass or bimetallic dial thermometers can be used if their calibration has been checked against a calibrated thermometer. In field situations, an ice-water mixture is an easy way to check calibration at 0°C (32°F).

WET-BULB TEMPERATURE

Using a wet-bulb thermometer in conjunction with a dry-bulb thermometer is a common method for determining the state point of the air on the psychrometric chart. The wet-bulb thermometer is basically an ordinary glass thermometer (although electronic temperature sensing elements can also be used) with a wetted cotton wick secured around the mercury bulb. Air is forced over the wick, causing it to cool to the wet-bulb temperature. The wet- and dry-bulb temperatures together determine the state point on the psychrometric chart, allowing all other variables to be determined.

An accurate wet-bulb temperature reading depends on sensitivity and accuracy of the thermometer, maintaining an adequate air speed past the wick, shielding the thermometer from radiation, using distilled or deionized water to wet the wick, and using a cotton wick.

The thermometer sensitivity required to determine an accurate humidity varies according to the temperature range of the air. More sensitivity is needed at low temperatures than at high ones. For example, at 65°C a 0.5°C error in the wet-bulb temperature reading results in a 2.6% error in RH determination, but at 0°C that same error

results in a 10.5% error in RH. In most cases, absolute calibration of the wet- and dry-bulb thermometers is not as important as ensuring that they read the same at a given temperature. For example, if both thermometers read 0.5°C low, this will result in less than a 1.3% error in RH at dry-bulb temperatures between 65°C and 0°C (at a 5°C difference between dry- and wet-bulb temperatures). Before wetting the wick of the wet-bulb thermometer, operate both thermometers long enough to determine if there is any difference between their readings. If there is a difference, assume that one is correct and adjust the reading of the other accordingly when determining RH.

The rate of evaporation from the wick is a function of the air speed past it. A minimum air speed of about 3 m/sec (500 ft/min) is required for accurate readings. An air speed much below this will result in an erroneously high wet-bulb reading. Wet-bulb devices that do not provide a guaranteed airflow cannot be relied on to give an accurate reading.

As with the dry-bulb thermometer, sources of radiant heat such as motors and lights can affect the wet-bulb thermometer. The reading must be taken in an area protected from these sources of radiation, or thermometers must be shielded from radiant energy.

A buildup of salts from impure water or contaminants in the air affects the rate of water evaporation from the wick and results in erroneous data. Distilled or deionized water should be used to moisten the cotton wick, and the wick should be replaced if there is any sign of contamination. The wick material should not have been treated with chemicals such as sizing compounds that affect the water evaporation rate.

Special care must be taken when using a wet-bulb thermometer when the wet-bulb temperature is near freezing. Most humidity tables and calculators are based on a frozen wick at wet-bulb temperatures below 0°C (32°F). At temperatures below 0°C, touch the wick with a piece of clean ice or another cold object to induce freezing, because distilled water can be cooled below 0°C without freezing. The psychrometric chart or calculator must use frost-bulb, not wet-bulb, temperatures below 0°C to be accurate with this method.

Under most conditions wet-bulb temperature data are not reliable when the RH is

SAMPLE PSYCHROMETRIC CALCULATIONS

1. A wet-bulb thermometer reads 18°C and a dry-bulb thermometer reads 25°C. What is the RH?
Solution: On figure 13.3 the diagonal 18°C wet-bulb (wb) line and the vertical 25°C dry-bulb (db) line intersect at point A. Point A falls on the 50% RH line.

2. What is the dewpoint temperature of the air in problem 1?
Solution: If the air represented by point A is cooled without changing its moisture content, it will follow a horizontal line until it reaches 14°C. At that temperature, it has 100% RH, and any further cooling will cause water to condense out of the air (form dew). The dewpoint (dp) temperature is 14°C.

3. What is the humidity ratio and water vapor pressure of the air in problem 1?
Solution: Find the humidity ratio and water vapor pressure of the air represented by point A by reading horizontally across to the vertical axis of the psychrometric chart. The humidity ratio (w) is 0.01 kg/kg and the vapor pressure is 1.6 kPa.

4. If the air in problem 1 is passed through a 100%-efficient evaporative cooler, what will be its temperature after it leaves the cooler?
Solution: Evaporative cooling (and spray humidification) follows the diagonal wet-bulb lines. As air passes through the cooler, it will move from point A along the 18°C wet-bulb line until it reaches 100% RH. At this humidity, it is saturated, will not accept any more water vapor, and will stop cooling. It will leave the cooler at a temperature of 18°C.

5. When air represented by point A (db = 25°C, wb = 18°C) enters a storage room with a temperature of 0°C and a RH of 95%, will it add moisture to the storage room or dry it out?
Solution: The air has a dewpoint temperature of 14°C (see problem 2). When this air has cooled to just less than 14°C, it will begin to lose water and will continue to lose water until it reaches the storage room temperature. If fact, each kilogram of air will lose about 0.006 kg of water as it cools. The air will add moisture to the storage room.

6. If air leaves a wet-coil evaporator at 0°C and 100% RH and heats 2°C before it reaches a stored product, what is the RH of the air that the product is exposed to?
Solution: Sensible heating processes follow horizontal lines on the psychrometric chart. Air leaves the coil at point B and moves horizontally to the right on the chart until it reaches 2°C. At that point, the RH will be 83%.

Figure 13.3

Example of how to use a psychrometric chart.

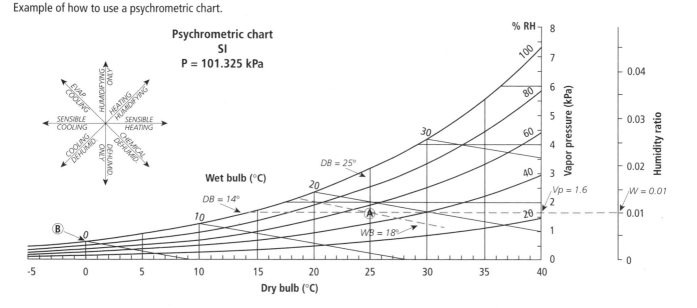

below 20% or the wet-bulb temperature is above 100°C (212°F). At low humidities, the wet-bulb temperature is much lower than the dry-bulb temperature, and it is difficult for the wet-bulb thermometer to be cooled completely because of heat transferred by the glass or metal stem. Water boils above 100°C (212°F), so temperatures above that cannot be measured with a wet-bulb thermometer.

In general, properly designed and operated wet- and dry-bulb psychrometers can operate with an accuracy of less than 2% of the actual RH. Improper operation greatly increases the error.

RELATIVE HUMIDITY

Direct RH measurement usually uses an electric sensing element or a mechanical system. Electric hygrometers are based on substances whose electrical properties change as a function of their moisture content. As the humidity of the air around the sensor increases, its moisture increases, proportionally affecting the sensor's electrical properties. These devices are more expensive than wet- and dry-bulb psychrometers, but their accuracy is not as severely affected by incorrect operation. An accuracy of less than 2% of the actual humidity is often obtainable. Sensors lose their calibration if allowed to become contaminated, and some lose calibration if water condenses on them. Most sensors have a limited life. Mechanical hygrometers usually employ human hairs as a sensing element. Hair changes in length in proportion to the humidity of the air. The response to changes in RH is slow and is not dependable at very high relative humidities. These devices are acceptable as an indicator of a general range of humidity but are not suitable for accurate measurements.

DEW POINT INDICATORS

Two types of dewpoint sensors are commonly used today: a saturated salt system and a condensation dewpoint method. The saturated salt system operates at dewpoints between −12° and 37°C (10° to 100°F) with an error of less than ± 1°C (2°F). The system costs less than the condensation system, is not significantly affected by contaminating ions, and has a response time of about 4 minutes. The condensation-type sensor is very accurate over a wide range of dewpoint temperatures (less than ± 0.5°C from −73° to 100°C or less than ± 1°F from −100° to 212°F). A condensation dewpoint hygrometer can be expensive.

There are a variety of other methods for measuring psychrometric variables. Some are extremely accurate and have characteristics suitable to particular sampling conditions. Most, however, are not commercially available and are used primarily as laboratory instruments.

REFERENCES

ASHRAE (American Society of Heating, Refrigeration, and Air-Conditioning Engineers). 1997. ASHRAE handbook, fundamentals. Atlanta: ASHRAE.

Gaffney, J. J. 1978. Humidity: Basic principles and measurement techniques. HortScience 13:551–555.

Wexler, A. 1965. Humidity and moisture, measurement and control in science and industry. New York: Reinhold.

Wexler, A., and W. G. Brombacher. 1951. Methods of measuring humidity and testing hygrometers. National Bureau of Standards Circ. 512.

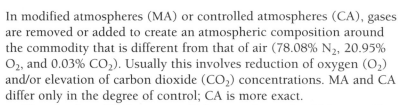

14

Modified Atmospheres during Transport and Storage

Adel A. Kader

In modified atmospheres (MA) or controlled atmospheres (CA), gases are removed or added to create an atmospheric composition around the commodity that is different from that of air (78.08% N_2, 20.95% O_2, and 0.03% CO_2). Usually this involves reduction of oxygen (O_2) and/or elevation of carbon dioxide (CO_2) concentrations. MA and CA differ only in the degree of control; CA is more exact.

The use of modified or controlled atmospheres should be considered as a supplement to proper temperature and relative humidity management (fig. 14.1). The potential for benefit or hazard from using MA depends on the commodity, cultivar or variety, physiological age, atmospheric composition, and temperature and duration of storage. This helps explain the wide variability in results among published reports for MA or CA used on a given commodity.

Continued efforts to develop MA or CA technology have permitted its increased use during transport, temporary storage, or long-term storage of horticultural commodities destined for fresh market or processing. A prestorage treatment with elevated CO_2 can also be used for some fruits. Elevated (above 21%) O_2 in combination with elevated CO_2 concentrations may be applied, in a few cases, during transport or storage or both. Carbon monoxide (CO) is used to a very limited extent as an added component to MA for slowing down brown discoloration and controlling decay.

EFFECTS OF CONTROLLED ATMOSPHERES

POTENTIAL BENEFITS

Used properly, MA or CA can supplement proper temperature management and can result in one or more of the following benefits, which translate into reduced quantitative and qualitative losses during postharvest handling and storage of some horticultural commodities:

- Retardation of senescence (ripening) occurs, along with associated biochemical and physiological changes, i.e., slowed respiration and ethylene production rates, softening, and compositional changes.

- Reduction of fruit sensitivity to ethylene action occurs at O_2 levels below about 8% or CO_2 levels above 1% or their combinations.

- Alleviation of certain physiological disorders can occur, such as chilling injury of various commodities, russet spotting in lettuce, and some storage disorders including scald of apples.

- Modified atmospheres can directly or indirectly affect postharvest pathogens and, consequently, decay incidence and severity. For example, elevated CO_2 levels (10 to 15%) significantly inhibit development of Botrytis rot on strawberries, cherries, and other fruits.

- Atmospheric modification (less than 1% O_2 or 40 to 60% CO_2) can be a useful tool for insect control in some commodities.

POTENTIAL HARMFUL EFFECTS

In most cases, the difference between beneficial and harmful MA combinations is relatively small. Also, MA combinations that are necessary to control decay or insects, for example, cannot always be tolerated by the commodity and may result in faster deterioration. Potential hazards of MA to the commodity include:

- Initiation or aggravation of certain physiological disorders can occur, such as blackheart in potatoes, brown stain on lettuce, and

Figure 14.1

Relative postharvest life of a fresh commodity stored in air or in its optimal modified atmosphere (MA) at room temperature (20° to 25°C [68° to 77°F]), or at its optimal temperature (near 0°C [32°F] for nonchilling-sensitive commodities or 5° to 14°C [41° to 57°F] for chilling-sensitive commodities).

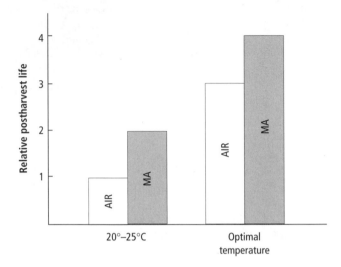

brown heart in apples and pears.

- Irregular ripening of fruits, such as banana, mango, pear, and tomato, can result from exposure to O_2 levels below 2% or CO_2 levels above 5% for more than 2 to 4 weeks.

- Off-flavors and off-odors at very low O_2 or very high CO_2 concentrations may develop as a result of anaerobic respiration and fermentative metabolism.

- Susceptibility to decay may increase when the commodity is physiologically injured by too-low O_2 or too-high CO_2 concentrations.

- Sprouting and retardation of periderm development are stimulated in some root and tuber vegetables, such as potatoes.

CA AND MA REQUIREMENTS AND RECOMMENDATIONS

During the past 50 years, the use of CA and MA has increased steadily, contributing significantly to extending the postharvest life and maintaining the quality of several fruits and vegetables. This trend is expected to continue as technological advances are made in attaining and maintaining CA and MA during transport, storage, and marketing of fresh produce. Several refinements in CA

storage include low-O_2 (1.0 to 1.5%) storage, low-ethylene CA storage, rapid CA (rapid establishment of the optimal levels of O_2 and CO_2), and programmed or sequential CA storage (e.g., storage in 1% O_2 for 2 to 6 weeks followed by storage in 2 to 3% O_2 for the remainder of the storage period). Other developments that will likely continue to expand use of MA during transport and distribution include improved technologies of establishing, monitoring, and maintaining CA and the use of edible coatings or polymeric films to create a desired MA within the commodity.

Fresh fruits and vegetables vary greatly in their relative tolerance to low O_2 concentration (table 14.1) and elevated CO_2 concentrations (table 14.2). These are the levels beyond which physiological damage would be expected. These limits of tolerance can be different at temperatures above or below recommended temperatures for each commodity. Also, a given commodity may tolerate brief exposures to higher levels of CO_2 or lower levels of O_2 than those indicated. The limit of tolerance to low O_2 would be higher as storage temperature or duration increases because O_2 requirements for aerobic respiration of the tissue increase with higher temperatures. Depending on the commodity, damage associated with CO_2 may either increase or decrease with an increase in temperature. CO_2 production increases with temperature, but its solubility decreases; thus, CO_2 in the tissue can be increased or decreased by an increase in temperature. Further, the physiological effect of CO_2 could be temperature-dependent. Tolerance limits to elevated CO_2 decrease with a reduction in O_2 level, and similarly, the tolerance limits to reduced O_2 increase with the increase in CO_2 level.

Current MA and CA recommendations are summarized in table 14.3 (fruits) and table 14.4 (vegetables). Also included is an estimate of the extent of current commercial use. There is no doubt that some of these MA combinations will change as more research is completed. The possibility of adding CO for some commodities may change the potential benefit of MA and CA. Hypobaric or low-pressure systems may also provide new opportunities for making CA a more useful treatment for some commodities.

Table 14.1. Fruits and vegetables classified according to their tolerance to low O_2 concentrations

Minimum O_2 concentration tolerated (%)	Commodities
0.5	Tree nuts, dried fruits, and vegetables
1.0	Some cultivars of apples and pears, broccoli, mushrooms, garlic, onion, most cut or sliced (minimally processed) fruits and vegetables
2.0	Most cultivars of apples and pears, kiwifruit, apricot, cherry, nectarine, peach, plum, strawberry, papaya, pineapple, olive, cantaloupe, sweet corn, green bean, celery, lettuce, cabbage, cauliflower, Brussels sprouts
3.0	Avocado, persimmon, tomato, pepper, cucumber, artichoke
5.0	Citrus fruits, green pea, asparagus, potato, sweet potato

Table 14.2. Fruits and vegetables classified according to their tolerance to elevated CO_2 concentrations

Maximum CO_2 concentration tolerated (%)	Commodities
2	Apple (Golden Delicious), Asian pear, European pear, apricot, grape, olive, tomato, pepper (sweet), lettuce, endive, Chinese cabbage, celery, artichoke, sweet potato
5	Apple (most cultivars), peach, nectarine, plum, orange, avocado, banana, mango, papaya, kiwifruit, cranberry, pea, pepper (chili), eggplant, cauliflower, cabbage, Brussels sprouts, radish, carrot
10	Grapefruit, lemon, lime, persimmon, pineapple, cucumber, summer squash, snap bean, okra, asparagus, broccoli, parsley, leek, green onion, dry onion, garlic, potato
15	Strawberry, raspberry, blackberry, blueberry, cherry, fig, cantaloupe, sweet corn, mushroom, spinach, kale, Swiss chard

Current CA use for long-term storage of fresh fruits and vegetables is summarized in table 14.5. Its use on nuts and dried commodities (for insect control and quality maintenance, including prevention of rancidity) is increasing, since it provides an excellent substitute for chemical fumigants (such as methyl bromide) used for insect control. Also, the use of CA on commodities listed in table 14.5 other than apples and pears is expected to increase as international market demands for year-round availability of various commodities continue to expand and as the technology becomes more cost-effective (positive benefit-cost ratio).

CA and MA use for short-term storage and transport of fresh horticultural crops (table 14.6) will continue to increase supported by technological developments in transport containers, MA packaging, and edible coatings. CO at 5 to 10%, added to O_2 levels below 5%, is an effective fungistat that can be used for decay control on commodities that do not tolerate 15 to 20% CO_2. However, CO is very toxic to humans, and special precautions must be taken.

The major limitation to long-term storage of many cut flowers is pathological breakdown due to infection with *Botrytis cinerea* (gray mold). Atmospheres containing sufficient CO_2 to reduce fungal attack cause severe bronzing of the foliage of some cultivars. The use of CO as a fungistat is limited by its ethylene-mimicking effects. It is not yet possible to identify the best MA or CA combination for each species of ornamentals because of insufficient or inconclusive data.

CA or MA conditions, including MA packaging (MAP), can replace certain postharvest chemicals used for control of some physiological disorders, such as scald on apples. Proper use of CA can also eliminate the need for using preharvest applications of growth regulators to delay ripening of fruits such as apples. Furthermore, some postharvest fungicides and insecticides can be reduced or eliminated where CA or MA provides adequate control of postharvest pathogens or insects.

CA or MA may facilitate picking and marketing of more-mature (better flavor) fruits by slowing their postharvest deterioration to permit transport and distribution. Another potential use for CA or MA is in maintaining the quality and safety of minimally processed (fresh-cut) fruits and vegetables, which are increasingly being marketed as value-added convenience products.

The residual effects of CA and MA on fresh commodities after transfer to air (during marketing) may include reduction of respiration and ethylene production rates, maintenance of color and firmness, and delayed decay. Generally, the lower the concentration of O_2 and the higher the concentration of CO_2 (within the tolerance limits of the commodity), and the longer the exposure to CA or MA conditions, the more prominent are the residual effects.

Table 14.3. Summary of recommended CA or MA conditions during transport and/or storage of selected fruits

Commodity	Temperature range*(°C)	CA† %O$_2$	CA† %CO$_2$	Commercial use as of June 2001
Apple	0–5	1–2	0–3	About 60% of production is stored in CA
Apricot	0–5	2–3	2–3	
Avocado	5–13	2–5	3–10	Used during marine transport
Banana	12–16	2–5	2–5	Used during marine transport
Blackberry	0–5	5–10	15–20	Used within pallet covers during transport
Blueberry	0–5	2–5	12–20	Limited use during transport
Cherimoya and atemoya	8–15	3–5	5–10	
Cherry, sweet	0–5	3–10	10–15	Used within pallet covers or marine containers during transport
Cranberry	2–5	1–2	0–5	
Durian	12–20	3–5	5–15	
Fig	0–5	5–10	15–20	Limited use during transport
Grape	0–5	2–5	1–3	Incompatible with SO$_2$
		or 5–10	10–15	Can be used instead of SO$_2$ for decay control up to 4 weeks
Grapefruit	10–15	3–10	5–10	
Kiwifruit	0–5	1–2	3–5	Expanding use during transport and storage; C$_2$H$_4$ must be maintained below 20 ppb
Lemon	10–15	5–10	0–10	
Lime	10–15	5–10	0–10	
Lychee (litchi)	5–12	3–5	3–5	
Mango	10–15	3–7	5–8	Increasing use during marine transport
Nectarine	0–5	1–2	3–5	Limited use during marine transport
		or 4–6	15–17	Used to reduce chilling injury (internal breakdown) of some cultivars
Nuts and dried fruits	0–10	0–1	0–100	Used in packaging to delay rancidity and control insects
Olive	5–10	2–3	0–1	Limited use to extend processing season
Orange	5–10	5–10	0–5	
Papaya	10–15	2–5	5–8	
Peach, clingstone	0–5	1–2	3–5	Limited use to extend canning season
Peach, freestone	0–5	1–2	3–5	Limited use during marine transport
		or 4–6	15–17	Used to reduce incidence and severity of internal breakdown (chilling injury) of some cultivars
Pear, Asian	0–5	2–4	0–3	Limited use for long-term storage of some cultivars
Pear, European	0–5	1–3	0–3	About 25% of production is stored in CA
Persimmon	0–5	3–5	5–8	Limited use of MA packaging
Pineapple	8–13	2–5	5–10	Waxing is used to create MA and reduce endogenous brown spot
Plum	0–5	1–2	0–5	Limited use for long-term storage of some cultivars
Pomegranate	5–10	3–5	5–10	
Rambutan	8–15	3–5	7–12	
Raspberry	0–5	5–10	15–20	Used within pallet covers during transport
Strawberry	0–5	5–10	15–20	Used within pallet covers during transport
Sweetsop (custard apple)	12–20	3–5	5–10	

Source: Kader 1997; and Kader 2001 (in CD-ROM, Postharvest Horticulture Series 22, University of California, Davis).

Notes:

*Usual or recommended range; a relative humidity of 90–95% is recommended.

† Specific CA combination depends on cultivar, temperature, and duration of storage. These recommendations are for transport or storage beyond 2 weeks. Exposure to lower O$_2$ and or higher CO$_2$ concentrations for shorter durations may be used for control of some physiological disorders, pathogens, and insects.

Table 14.4. Summary of recommended CA or MA conditions during transport and/or storage of selected vegetables

| Vegetable | Temperature (°C)* | | Atmosphere† | | |
	Optimum	range	%O_2	%CO_2	Application
Artichokes	0	0–5	2–3	2–3	Moderate
Asparagus	2	1–5	Air	10–14	High
Beans, green snap	8	5–10	2–3	4–7	Slight
processing	8	5–10	8–10	20–30	Moderate
Broccoli	0	0–5	1–2	5–10	High
Brussels sprouts	0	0–5	1–2	5–7	Slight
Cabbage	0	0–5	2–3	3–6	High
Cantaloupes	3	2–7	3–5	10–20	Moderate
Cauliflower	0	0–5	2–3	3–4	Slight
Celeriac	0	0–5	2–4	2–3	Slight
Celery	0	0–5	1–4	3–5	Slight
Chinese cabbage	0	0–5	1–2	0–5	Slight
Cucumbers, fresh	12	8–12	1–4	0	Slight
pickling	4	1–4	3–5	3–5	Slight
Herbs‡	1	0–5	5–10	4–6	Moderate
Leeks	0	0–5	1–2	2–5	Slight
Lettuce (crisphead)	0	0–5	1–3	0	Moderate
cut or shredded	0	0–5	1–5	5–20	High
Lettuce (leaf)	0	0–5	1–3	0	Moderate
Mushrooms	0	0–5	3–21	5–15	Moderate
Okra	10	7–12	Air	4–10	Slight
Onions (bulb)	0	0–5	1–2	0–10	Slight
Onions (bunching)	0	0–5	2–3	0–5	Slight
Parsley	0	0–5	8–10	8–10	Slight
Pepper (bell)	8	5–12	2–5	2–5	Slight
Pepper (chili)	8	5–12	3–5	0–5	Slight
processing	5	5–10	3–5	10–20	Moderate
Radish (topped)	0	0–5	1–2	2–3	Slight
Spinach	0	0–5	7–10	5–10	Slight
Sugar peas	0	0–10	2–3	2–3	Slight
Sweet corn	0	0–5	2–4	5–10	Slight
Tomatoes (green)	12	12–20	3–5	2–3	Slight
ripe	10	10–15	3–5	3–5	Moderate
Witloof chicory	0	0–5	3–4	4–5	Slight

Source: Saltveit 1997; and Saltveit 2001 (in CD-ROM, Postharvest Horticulture Series 22, University of California, Davis).

Notes:

*Optimum and range of usual or recommended temperatures. A relative humidity of 90–95% is usually recommended.

†Specific CA recommendations depend on cultivar, temperature, and duration of storage.

‡Herbs: chervil, chives, coriander, dill, sorrel and watercress.

CARBON MONOXIDE AS A SUPPLEMENT

Beginning in 1970, CO at 2 to 3% has been used as a supplement to MA during transit of lettuce to inhibit discoloration. Some additional benefits of CO are now known.

BENEFITS

CO (1 to 5%) added to reduced (2 to 5%) O_2 atmospheres inhibits discoloration of lettuce butts and mechanically damaged tissue. Similar effects have been observed on other commodities, including lightly processed (cut, sliced, etc.) fruits and vegetables. This inhibition of discoloration is lost when the commodity is moved from MA to air during destination marketing.

CO (5 to 10%) added to MA has been found to inhibit growth of several important postharvest pathogens and to prevent decay development on several fruits and vegetables. The fungistatic effects of CO are maximized at O_2 levels below 5%.

Although CO alone was not found to be an effective fumigant for insect control in harvested lettuce, its possible use with other CA combinations merits further study.

POSSIBLE HAZARDS

CO may aggravate certain physiological disorders. For example, in a situation where CO_2 accumulates above 2% during transit of lettuce, it increases the severity of brown stain (a CO_2–induced disorder).

CO mimics ethylene (C_2H_4) effects such as enhancing ripening and inducing certain physiological disorders. However, when CO is used in combination with reduced O_2 or elevated CO_2, such effects are minimized to insignificance except for commodities that are extremely C_2H_4-sensitive, such as kiwifruit.

Because of its extreme toxicity to humans and flammability at concentrations between 12.5 and 74.2% in air, strict safety measures should be followed when CO is used.

PRESTORAGE TREATMENTS WITH ELEVATED CARBON DIOXIDE

Tests conducted at several experiment stations indicated that treating apples for 2 weeks or pears for 2 to 4 weeks with 12% CO_2 at 0° to 5°C (32° to 41°F) before CA storage delayed fruit softening. However, this treatment resulted in varying amounts of internal and external CO_2 injury, depending on variety, season, and production area. Its commercial application is currently limited to some Golden Delicious apples in the U.S. Northwest.

Table 14.5. Summary of CA use for long-term storage of fresh fruits and vegetables

Storage duration (months)	Commodities
More than 12	Almond, Brazil nut, cashew, filbert, macadamia, pecan, pistachio, walnut, dried fruits and vegetables
6–12	Some cultivars of apples and European pears
3–6	Cabbage, Chinese cabbage, kiwifruit, some cultivars of Asian pears
1–3	Avocado, banana, cherry, cranberry, grape, mango, olive; some peach, nectarine, and plum cultivars; persimmon, pomegranate

Table 14.6. Summary of CA or MA use for short-term storage and transport of fresh horticultural crops

Primary benefit of CA or MA	Commodities
Delay of ripening and avoiding chilling temperatures	Avocado, banana, mango, melons, nectarine, papaya, peach, plum, tomato (picked mature-green or partially ripe)
Control of decay	Blackberry, blueberry, cherry, fig, grape, raspberry, strawberry
Delay of senescence and undesirable compositional changes (including tissue brown discoloration)	Asparagus, broccoli, lettuce, sweet corn, fresh herbs, minimally processed (fresh-cut) fruits and vegetables

Elevated CO_2 treatments have also been shown to alleviate chilling injury symptoms on some subtropical and tropical fruits, but this treatment is not recommended yet for commercial application.

SUPERATMOSPHERIC OXYGEN ATMOSPHERES

Oxygen concentrations greater than 21% may influence postharvest physiology and the quality maintenance of fresh horticultural perishables either directly (via the action of free radicals) or indirectly (via altered CO_2 and C_2H_4 production rates or actions). Sensitivity to O_2 toxicity varies among species and developmental stages. Ripening of mature-green, climacteric fruits may be slightly enhanced by exposure to 30 to 80% O_2, but levels above 80% may retard their ripening and induce O_2 toxicity disorders on some fruits. Elevated O_2 atmospheres enhance some of the effects of ethylene on fresh horticultural commodities, including ripening, senescence, and ethylene-induced physiological disorders (such as russet spotting on lettuce and bitterness in carrots). While superatmospheric O_2 concentrations

influence the growth of some bacteria and fungi, they are much more effective if combined with elevated (15 to 20%) CO_2 as a fungistatic treatment.

ETHYLENE REMOVAL IN MODIFIED ATMOSPHERE STORAGE

Most researchers have assumed that removing ethylene from MA storage rooms is not necessary because its effects on fruit ripening at 0° to 5°C (32° to 41°F) and under MA conditions are negligible. However, the presence of ethylene at concentrations likely to occur in MA and CA rooms can enhance fruit softening during long-term storage. Thus, ethylene removal is recommended in long-term CA storage of apples and pears. It is particularly important for storage of avocado, kiwifruit, Fuyu persimmon, carnations, and other commodities that are extremely sensitive to ethylene. Further research is needed to evaluate the effects of ethylene and nonethylenic volatiles on other commodities under MA conditions. Also, there is a need for more effective and more economical methods for removing ethylene and other volatiles from MA storage rooms.

ATMOSPHERIC MODIFICATION

ATMOSPHERE GENERATORS

Oxygen control. The oxygen level can be controlled by recirculating air from the CA room into the generator and back into the room, or by a purge system in which fresh air has its oxygen reduced in the generator, then fed into the room. Catalytic burners or converters, used in the past, have been replaced by purging with nitrogen obtained from liquid nitrogen or from separators that separate the nitrogen by circulating air through molecular sieve beds or membrane systems. See chapter 12 for more information.

Carbon dioxide control. Addition of CO_2 is usually from pressurized gas cylinders. Dry ice is sometimes used as a source of CO_2 during transport. CO_2 is reduced by scrubbing methods that use sodium hydroxide, water, activated charcoal, hydrated lime (Ca[OH]$_2$), or a molecular sieve. The most common scrubber uses brine, which is pumped over the evaporator coil, where it absorbs CO_2. A lime box is also commonly used adjacent to the CA room with a circulating system to pass the room atmosphere through it. The box is usually sized to hold about 12 kg (26.4 lb) lime per ton of fruit, and the spent lime is replaced with fresh lime. The amount of lime needed to absorb CO_2 may be placed inside the CA room. Several types of activated carbon scrubbers are currently used.

Carbon monoxide addition. CO can be added from pressurized gas cylinders, blending it with nitrogen to avoid exceeding 10% CO.

Ethylene removal. A few methods can be used to remove ethylene from cold storage facilities. The use of ethylene absorbers, such as potassium permanganate alone or in combination with activated and brominated charcoal, can be effective in CA storage facilities provided that the air is circulated through these materials, which must be replaced when spent. Catalytic burners can be used for ethylene removal from air storage facilities. Ventilation (one air exchange per hour) to reduce C_2H_4 concentration cannot be used in CA storage. Also, the use of ozone to oxidize C_2H_4 requires O_2 levels above those available in CA storage.

HYPOBARIC OR LOW-PRESSURE SYSTEM

Reducing the total pressure (under partial vacuum conditions) results in reducing the partial pressures of individual gases in air. This can be an effective method for reducing O_2 tension and for accelerating the escape of C_2H_4 and other volatiles. This low-pressure system (LPS) has the advantages over other methods of atmospheric modification of more exact control of O_2 concentrations that permit the use of lower O_2 tensions than is possible with CA and removal of ethylene and other volatiles.

However, LPS has limitations when CO_2 or CO addition is important for a given commodity. Transit vehicles with LPS were tested to a limited extent on a commercial scale for transporting some animal and plant products. Stationary storage structures with LPS were being developed by Grumman Dormavac before it terminated its efforts related to hypobaric storage.

COMMODITY-GENERATED MODIFIED ATMOSPHERE

In some cases the commodity itself, through respiration, is used to reduce O_2 and increase CO_2 with restricted air-exchange conditions and barriers, as shown in figure

Figure 14.2

Barriers that can be used to establish a modified atmosphere. B_1: Natural epidermis, skin, peel, or rind; wax coating, film wrap. B_2: Package-wood, paperboard, plastic (may include additional liner in package). B_3: Storage room wall or vehicle wall (may be sealed against gas exchange). Additional barriers may include consumer packages inside the master package and pallet covers over several packages.

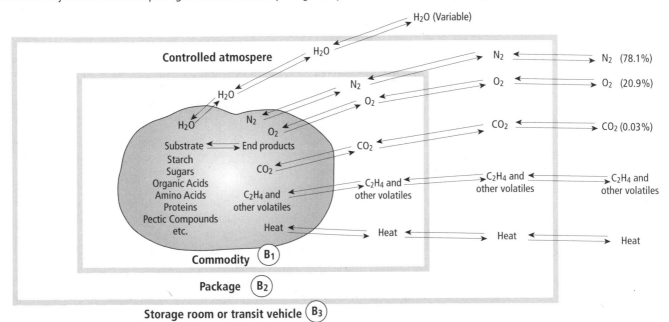

14.2. If elevated CO_2 is not desirable, scrubbers are used. Restricted air exchange may be achieved by

- using airtight cold storage rooms
- packaging in film wraps or bags
- using polyethylene liners in shipping containers
- using pallet shrouds (plastic covers)
- manipulating shipping container vents
- applying waxes and other surface coatings
- using plastic covers with diffusion windows (polymeric membranes)

Atmospheric modification by these methods is usually slow, and much of the benefit of MA may be lost.

ATMOSPHERES DURING TRANSIT

Modified atmospheres in rail cars, trucks, and marine containers. Gastight transit vehicles, essential to the maintenance of MA during transit, are a limiting factor to expanded use of MA. The Tectrol system (TransFresh Corp.) used in rail cars and marine containers is based on reduced O_2 achieved by N_2 flushing, CO_2 and/or CO added using gas blending manifolds, CO_2 removal by placing bags of fresh hydrated lime in the transit vehicle, and breather bags to compensate for barometric pressure fluctuations.

Some systems carry a tank of liquid N_2 along with the marine container or use a membrane N_2-separator unit. In these systems, the container is equipped with an O_2 sensor for controlling N_2 release or for introducing fresh air as needed to maintain the desired O_2 concentration. Scrubbers for CO_2 or C_2H_4 may be used in some marine containers. There are also efforts to use small nitrogen separators in marine containers.

Modified atmosphere in a pallet cover. Polyethylene pallet covers (or shrouds) are used to cover all shipping containers on a pallet and are sealed by various means (tape, heat seal, etc.) onto a plastic sheet placed on the wooden or plastic pallet base (fig. 14.3). The desired gas mixture is introduced within the pallet cover. This method is in common use on strawberries and cane berries, cherries, figs, and other commodities. It can facilitate mixing of commodities that require different MA conditions during transit at the same temperature. Potential problems are

Figure 14.3

A pallet of strawberries and raspberries covered with 5-mil polyethylene pallet cover and ready for introduction of air enriched with 15 to 20% CO_2 as a fungistat.

primarily related to loss of the seal due to tearing of the pallet cover or an imperfect seal at the base.

Modified atmospheres in individual shipping containers. Examples of commercial use of commodity-generated MA during transit include polyethylene liners in cherry boxes, and polyethylene bags for bananas destined for distant markets (Banavac system) and for cut lettuce and other vegetables. Lightly processed (shredded or chopped) lettuce may be packaged into 5-mil plastic bags, and then a partial vacuum is established and a gas mixture (30 to 50% O_2 plus 4 to 6% CO, or air plus 10 to 15% CO_2) is introduced into the bag, which is then sealed.

MODIFIED ATMOSPHERE PACKAGING (MAP)

Modified atmospheres can be created either passively by the commodity or intentionally, as described below.

Commodity-generated or passive MA. If commodity and film permeability characteristics are properly matched, an appropriate atmosphere can passively evolve within a sealed package through consumption of O_2 and production of CO_2 by respiration. The gas permeability of the selected film must allow O_2 to enter the package at a rate offset by the consumption of O_2 by the commodity.

Similarly, CO_2 must be vented from the package to offset the production of CO_2 by the commodity. Furthermore, this atmosphere must be established rapidly and without creating anoxic conditions or injuriously high levels of CO_2.

Active modified atmosphere. Because of the limited ability to regulate a passively established atmosphere, it is likely that atmospheres within MAP will be actively established and adjusted. This can be done by creating a slight vacuum and replacing the package atmosphere with the desired gas mixture. This mixture can be further adjusted through the use of absorbing or adsorbing substances in the package to scavenge O_2, CO_2, or C_2H_4.

Although active modification implies some additional costs, its main advantage is that it ensures the rapid establishment of the desired atmosphere. In addition, ethylene absorbers can help delay the climacteric rise in respiration for some fruits. Carbon dioxide absorbers can prevent the buildup of CO_2 to injurious levels, which can occur for some commodities during passive modification of the package atmosphere.

Many plastic films are available for packaging, but relatively few have been used to wrap fresh produce, and fewer have gas permeabilities that make them suitable to use for MAP. Because the O_2 content in a MA package is typically reduced from an ambient 21% to 2 to 5%, there is a danger that CO_2 will increase from ambient 0.03% to 16 to 19% in the package. This is because there is normally a one-to-one correspondence between O_2 consumed and CO_2 produced. Because such high levels of CO_2 would be injurious to most fruits and vegetables, an ideal film must let more CO_2 exit than it lets O_2 enter. The CO_2 permeability should be about 3 to 5 times the oxygen permeability, depending on the desired atmosphere. Several polymers used in film formulation meet this criterion (table 14.7). Low-density polyethylene and polyvinyl chloride are the main films used in packaging fruits and vegetables. Polystyrene has been used, but Saran and polyester have such low gas permeabilities that they would be suitable only for commodities with very low respiration rates.

Major advances in polymeric film technologies have been made during the past few years, and more improvements are expected in the future in manufacturing, consumer, and cost issues related to film packaging. Some of the new film packages have addressed the need for changes in permeability with temperature, increased gas diffusion by using microperforations, machinability, printability, sealing integrity and ability to reseal the bag, clarity, and anti-fog characteristics. For some applications gas diffusion in and out of the package is controlled primarily through a membrane with product-specific gas transmission rate, such as the FreshHold and the Intellipac membranes. The latter is a temperature-responsive membrane developed by Landec Corporation. Absorbent sachets can be used to absorb off-odors, ethylene, or other gases as needed for specific products. Bio-sensors to detect ethanol or ethyl acetate (as indicators of fermentative metabolism) are being developed for potential use in produce packages along with a mechanism to open a hole to allow O_2 into the package.

MONITORING ATMOSPHERIC COMPOSITION

Accurate monitoring of O_2 and CO_2 concentrations is essential to successful CA or MA storage. It is required for certification of CA storage in some states and by some insurance companies. The various methods of gas sampling and analysis are discussed in chapter 15.

REFERENCES

Baldwin, E. A. 1994. Edible coatings for fresh fruits and vegetables: Past, present, and future. In J. M. Krochta et al., eds., Edible coatings and films to improve food quality. Lancaster, PA: Technomic. 25–64.

Table 14.7. Permeabilities of films available for packaging fresh produce

Film type	Permeabilities (cc/m²/mil/day at 1 atm)		$CO_2{:}O_2$ Ratio
	CO_2	O_2	
Polyester	180–390	52–130	3.0–3.5
Polyethylene, low density	7,700–77,000	3,900–13,000	2.0–5.9
Polypropylene	7,700–21,000	1,300–6,400	3.3–5.9
Polystrene	10,000–26,000	2,600–7,700	3.4–3.8
Polyvinyl chloride	4,263–8,138	620–2,248	3.6–6.9
Saran	52–150	8–26	5.8–6.5

Beaudry, R. M. 1999. Effect of O_2 and CO_2 partial pressure on selected phenomena affecting fruit and vegetable quality. Postharv. Biol. Technol. 15:293–303.

———. 2000. Responses of horticultural commodities to low oxygen: Limits to the expanded use of modified atmosphere packaging. HortTechnology 10:491–500.

Brecht, P. E. 1980. Use of controlled atmospheres to retard deterioration of produce. Food Technol. 34(3): 45–50.

Brody, A. L., ed. 1989. Controlled/modified atmosphere/vacuum packaging of foods. Trumbull, CT: Food and Nutrition Press. 179 pp.

Calderon, M., and R. Barkai-Golan, eds. 1990. Food preservation by modified atmospheres. Boca Raton, FL: CRC Press. 402 pp.

Dalrymple, D. G. 1967. The development of controlled atmosphere storage of fruits. USDA Div. Mktg. Utiliz. Sci. 56 pp.

El-Goorani, M. A., and N. F. Sommer. 1981. Effects of modified atmospheres on postharvest pathogens of fruits and vegetables. Hortic. Rev. 3:412–461.

Gorny, J. R., ed. 1997. CA '97 Proceedings, Vol. 5: Fresh-cut fruits and vegetables and MAP. Davis: Univ. Calif. Postharv. Hort. Ser.19. 168 pp.

Gorris, L. G. M., and H. W. Pepplenbos. 1992. Modified atmosphere and vacuum packaging to extend the shelf life of respiring food products. HortTechnology 2:203–209.

Isenberg, F. M. R. 1979. Controlled atmosphere storage of vegetables. Hortic. Rev. 1:337–394.

Jamison, W. 1980. Use of hypobaric conditions for refrigerated storage of meats, fruits, and vegetables. Food Technol. 34(3): 64–71.

Kader, A. A. 1986. Biochemical and physiological basis for effects of controlled and modified atmospheres on fruits and vegetables. Food Technol. 40(5): 99–100, 102–104.

Kader, A. A., ed. 1997. CA '97 Proceedings, Vol. 3: Fruits, other than apples and pears. Davis: Univ. Calif. Postharv. Hort. Ser. 17. 263 pp.

Kader, A. A., and S. Ben-Yehoshua. 2000. Effects of superatmospheric oxygen levels on postharvest physiology and quality of fresh fruits and vegetables. Postharv. Biol. Technol. 20:1–13.

Kader, A. A., and C. B. Watkins. 2000. Modified atmosphere packaging - Toward 2000 and beyond. HortTechnology 10:483–486.

Kader, A. A., D. Zagory, and E. L. Kerbel. 1989. Modified atmosphere packaging of fruits and vegetables. Crit. Rev. Food Sci. Nutr. 28:1–30.

Lange, D. L. 2000. New film technologies for horticultural products. HortTechnology 10:487–490.

Lougheed, E. C. 1987. Interactions of oxygen, carbon dioxide, temperature, and ethylene that may induce injuries in vegetables. HortScience 22:791–794.

Lougheed, E. C., D. P. Murr, and L. Berard. 1978. Low pressure storage for horticultural crops. HortScience 13:21–27.

Mattheis, J., and J. K. Fellman. 2000. Impacts of modified atmosphere packaging and controlled atmospheres on aroma, flavor, and quality of horticultural commodities. HortTechnology 10:507–510.

Mitcham, E. J., ed. 1997. CA '97 Proceedings, Vol. 2: Apples and pears. Davis: Univ. Calif. Postharv. Hort. Ser. 16. 308 pp.

Rooney, M. L., ed. 1995. Active food packaging. London: Chapman and Hall. 272 pp.

Saltveit, M. E., ed. 1997. CA '97 Proceedings, Vol. 4: Vegetables and ornamentals. Davis: Univ. Calif. Postharv. Hort. Ser. 18. 168pp.

Smith, J. P., Y. Abe, and J. Hoshino. 1995. Modified atmosphere packaging - Present and future uses of gas absorbents and generators. In J. M. Farber and K. L. Dodds, eds., Principles of modified atmospheres and sous vide product packaging. Lancaster, PA: Technomic. 287–323.

Smith, S., J. Geeson, and J. Stow. 1987. Production of modified atmospheres in deciduous fruits by the use of films and coatings. HortScience 22:772–776.

Smock, R. M. 1979. Controlled atmosphere storage of fruits. Hortic. Rev. 1:301–336.

Solomos, T. 1987. Principles of gas exchange in bulky plant tissues. HortScience 22:766–771.

Thompson, A. K. 1998. Controlled atmosphere storage of fruits and vegetables. Walllingford, UK: CAB International. 288 pp.

Thompson, J. F., and E. J. Mitcham, eds. 1997. CA '97 Proceedings, Vol. 1: CA technology and disinfestation studies. Davis: Univ. Calif. Postharv. Hort. Ser. 15. 159 pp.

Vineault, C., V. G. S. Raghavan, and R. Prange. 1994. Techniques for controlled atmosphere storage of fruits and vegetables. Tech. Bull. 1993-18E. Kentville, NS: Agriculture Canada. 15 pp.

Watkins, C. B. 2000. Responses of horticultural commodities to high carbon dioxide as related to modified atmosphere packaging. HortTechnology 10:501–506.

Weichmann, J. 1986. The effect of controlled atmosphere storage on the sensory and nutritional quality of fruits and vegetables. Hortic. Rev. 8:101–127.

Zagory, D. 1995. Principles and practices of modified atmosphere packaging of horticultural commodities. In J. M. Farber and K. L. Dodds, eds., Principles of modified-atmosphere and sous vide product packaging. Lancaster, PA: Technomic. 175–206.

Methods of Gas Mixing, Sampling, and Analysis

Adel A. Kader

PRINCIPLES OF MIXING GASES

Postharvest research and technology is usually concerned with monitoring atmospheric composition and with mixing two or more of the following gases: air, nitrogen, oxygen, carbon dioxide, and ethylene. Procedures for gas mixing are based on mass, volume, or pressure relationships. The various methods of gas mixing, sampling, and analysis rely on the following laws and definitions for gases:

Avogadro's Law:

One mole of any compound contains 6.0228×10^{23} molecules. This quantity of a gas occupies 22.414 l at standard temperature ($0°C = 273°K$) and pressure (760 mm Hg).

Boyle's Law:

$$V = K(1/P); \quad P_1V_1 = P_2V_2$$

where
V = volume (in liters)
K = proportionality constant
P = pressure (in atmospheres)

Charles' Law:

$$PV = KT; \text{ thus: } P_1V_1/T_1 = P_2V_2/T_2$$

where
P = pressure (in atmospheres)
V = volume (in liters)
T = temperature (°K)
K = constant

Ideal gas law:

If K is proportional to the number of moles of gas (n), then: $PV = nRT$

where
R = molar gas constant
P, V, and T = same as above

Density:

$$\text{Density} = \text{mass} \div \text{volume} = P \times M \div R \times T$$

where
P = pressure
M = molecular weight
R = molar gas constant
T = temperature

Graham's Law of Diffusion:

The rate of diffusion of a gas is inversely proportional to the square root of its density.

Dalton's Law of Partial Pressures:

The total pressure of a mixture of gases is the sum of the partial pressures of the component gases.

Concentration (C_a), in percent (ml/100 ml) or in ppm (μL/L):

$$C_a\ (\%) = \frac{100 \times V_a}{V_a + V_b + \ldots + V_n} = \frac{100 \times P_a}{P_a + P_b + \ldots + P_n}$$

where
V_a to V_n = volume of components
P_a to P_n = partial pressures of components

$$C_a\ (\text{ppm}) = \frac{10^6 \times V_a}{V_D + V_a} = \frac{10^6 \times P_a}{P_D + P_a}$$

where
V_D = volume of diluent gas
P_D = partial pressure of diluent gas

Fick's First Law of Diffusion:

Gas diffusion (or rate of transfer from a region of high concentration to a region of lower concentration) is represented by

$$\text{Flux} = -D \times A \times 1 \div T \times (C_i - C_o)$$

where
Flux = the rate of transfer
D = the diffusion coefficient (the negative sign indicates the substance is moving in the direction of decreasing concentration)
A = the area of the barrier to diffusion
T = the thickness of the barrier to diffusion
C_i = the initial, or inside concentration
C_o = the later, or outside concentration

The diffusivity D of most gases is related inversely to the square root of their molecular weight, the pressure, and the absolute temperature. (For a discussion of units, see Banks et al. 1995).

Henry's Law:

The mass of any gas that will dissolve in a given volume of liquid is directly propor-tional to the pressure of the gas. The various components of a gas mixture behave inde-pendently of each other.

GAS MIXING TECHNIQUES

STATIC SYSTEM
Gravimetric procedure (mixing by weight)
This method is independent of temperature, pressure, and compressibility. It involves weighing components into a gas cylinder.

Mixing by volume
Evacuate cylinder to 0.1 mm Hg, flush with diluent gas, evacuate again, inject compo-nent gas using a gastight syringe, and allow diluent gas to pressurize cylinder to the desired pressure.

Mixing by pressure
Because the partial pressure of each compo-nent equals its mole fraction (MF) times total pressure (P_t) of the mixture, a mix of 10% A and 90% B at a total cylinder pressure of 2,000 psia can be prepared as follows:

$$P_A = MF_A \times P_t = 0.10 \times 2,000 = 200\ \text{psia}$$

Add 200 psia of A, then 1,800 psia of B.

Homogenizing the gas mixture
Homogeneity depends on the densities and the relative amounts of the components. Homogenize gas mixtures by rolling cylin-ders or by thermal convection; temperatures above 50°C (122°F) should be avoided. Once the mixture is homogeneous, it remains so and does not separate except in the case of liquefied gases. Liquefied components may partially condense in the cylinder if subject-ed to low temperatures.

Calibration
Gas mixtures should be calibrated (analyzed) using chemical and gravimetric techniques (for some primary standards) or other gas analysis methods mentioned later in this chapter.

Accuracy, purity, and tolerances
Commercially available gas mixtures vary in their accuracy (table 15.1). Even the purest gases and gas mixtures may contain impuri-ties. This is of more concern in research work than in postharvest practice.

Table 15.1. Accuracy of commercial gas mixtures

Designation	Accuracy limits
Primary standards	Within 0.02% absolute or 1% of the component, whichever is smaller
Certified mixtures	Within 2 to 5% of component
Unanalyzed (commercial grade)	Same as certified but without certificate of analysis

Storage and handling of compressed gas cylinders

- Gas cylinders should be tested by hydrostatic pressure for their suitability for use with compressed gases.

- For cylinder filling, the pressure limit is 2,000 psi at 21°C (70°F).

- Cylinder contents and whether the cylinder is full or empty should always be identified clearly.

- Cylinders must be well secured and are best stored at 21°C (70°F).

- Proper transportation procedures should be followed.

- Proper valves and regulators relative to the standardized outlets for various families of gases should be used to prevent interchange of regulator equipment between gases that are not compatible.

- Safety procedures required for toxic and flammable gases (e.g., CO at 12.5 to 75% and C_2H_4 at 3 to 30%) must be adhered to in handling gas cylinders.

DYNAMIC SYSTEM

In the dynamic system, gases are mixed (continuous flow mixing) as needed by volume at constant pressure and temperature using flow-control devices such as capillary tubing and needle valves.

GAS SAMPLING

SAMPLING AND SAMPLE CONTAINERS

Several types of containers may be used for sampling gases, including

- syringes of various volumes (the most commonly used ones are between 1 and 10 ml)

- plastic film gas-impermeable bags with sealable gas inlet and septum for withdrawing subsamples for analysis

- glass containers of various capacities with gas inlet and sampling port

- vacuum containers, which are evacuated 150- to 250-ml cans with septum

- vacuotainers, which are evacuated 20-ml test tubes commonly used for blood sampling

Important points to consider:

- Make sure sample containers are gastight and clean before use to minimize errors.

- When vacuum containers are used, the vacuum should be determined for each container before use, and appropriate correction factors should be applied to the analysis data.

- Samples should be representative of the atmosphere to be analyzed.

GAS ANALYSIS METHODS

METHODS FOR IMMEDIATE AND ON-THE-SPOT ANALYSIS

In commercial situations, the following instruments can be used to measure concentrations of various gases:

- volumetric gas analyzers for O_2 and CO_2 (Orsat, Fyrite, and so on)

- Kitagawa gas sampler and detector tubes for C_2H_4, CO_2, CO, SO_2, and other gases

- portable gas analyzers (O_2, CO_2, CO, C_2H_4, SO_2, NH_3, and other gases)

LABORATORY GAS ANALYSIS INSTRUMENTS

Using gas analysis instruments (table 15.2) is much more accurate than the methods mentioned for on-the-spot analysis. The instruments can be used to monitor atmospheric composition in controlled atmosphere storage facilities, ripening rooms, and SO_2 fumigation chambers.

METHODS FOR MEASURING RESPIRATION RATES

- To determine O_2 consumed: For tissue slices or organelles, use the Warburg method or O_2 electrode; for intact plant organs, use laboratory gas analysis methods for O_2 mentioned above.

Table 15.2. Gas analysis instruments

Gas	Instruments
O_2	Oxygen analyzers (paramagnetic, polarographic, electrochemical) Gas chromatography (thermal conductivity detector)
CO_2	Infrared CO_2 analyzer Gas chromatography (thermal conductivity detector)
CO	Gas chromatography (thermal conductivity detector)
C_2H_4	Gas chromatography (flame ionization detector or photoionization detector)
SO_2	Infrared SO_2 analyzer

- To determine CO_2 produced: Colorimetric method (Claypool and Keefer 1942; Pratt and Mendoza 1979) and methods mentioned above for laboratory gas analysis of CO_2.

- Results are usually expressed as ml O_2 (or CO_2) per kg-hr and are calculated as follows:

$$\frac{\Delta O_2\% \text{ or } CO_2\%}{100} \times \frac{\text{flow rate (ml/hr)}}{\text{sample weight (kg)}}$$

- To convert ml CO_2 to mg CO_2, multiply by appropriate factor for temperature used:

°C	(°F)	mg/ml CO_2
0	(32)	1.98
10	(50)	1.90
20	(68)	1.84
30	(86)	1.78

- Conversion factors for calculating heat production:

mg CO_2/kg-hr × 61.2 = kcal/metric ton-day
mg CO_2/kg-hr × 220 = Btu/ton-day

REFERENCES

Banks, N. H., D. J. Cleland, A. C. Cameron, R. M. Beaudry, and A. A. Kader. 1995. Proposal for a rationalized system of units for postharvest research in gas exchange. HortScience 30:1129–1131.

Barmore, C. R., and T. A. Wheaton. 1978. Diluting and dispensing unit for maintaining trace amount of ethylene in a continuous flow system. HortScience 13:169–171.

Bower, J. H., J. J. Jobling, B. D. Patterson, and D. J. Ryan. 1998. A method for measuring the respiration rate and respiratory quotient of detached plant tissue. Postharv. Biol. Technol. 13:263–270.

Claypool, L. L., and R. M. Keefer. 1942. A colorimetric method for CO_2 determination in respiration studies. Proc. Am. Soc. Hortic. Sci. 40:177–186.

Leshuk, J. A., and M. E. Saltveit Jr. 1990. A simple system for the rapid determination of the anaerobic compensation point of plant tissue. HortScience 25:480–482.

Nelson, G. O. 1972. Controlled test atmospheres—Principles and techniques. Ann Arbor, MI: Ann Arbor Sci. Publ. 247 pp.

Peterson, S. J., W. J. Lipton, and M. Uota. 1989. Methods for premixing gases in pressurized cylinders for use in controlled atmosphere experiments. HortScience 24:328–331.

Pratt, H. K., and D. B. Mendoza Jr. 1979. Colorimetric determination of carbon dioxide for respiration studies. HortScience 14:175–176.

Pratt, H. K., M. Workman, F. W. Martin, and J. M. Lyons. 1960. Simple method for continuous treatment of plant material with metered traces of ethylene or other gases. Plant Physiol. 35:609–611.

Saltveit, M. E. 1978. Simple apparatus for diluting and dispensing trace concentrations of ethylene in air. HortScience 13:249–251.

———. 1982. Procedures for extracting and analyzing internal gas samples from plant tissues by gas chromatography. HortScience 17:878–881.

Saltveit, M. E., Jr., and T. Strike. 1989. A rapid method for accurately measuring oxygen concentrations in milliliter gas samples. HortScience 24:145–147.

Watada, A. E., and D. R. Massie. 1981. A compact automatic system for measuring CO_2 and C_2H_4 evolution by harvested horticultural crops. HortScience 16:39–41.

Young, R. E., and J. B. Biale. 1962. Carbon dioxide effects on fruit respiration: 1. Measurement of oxygen uptake in continuous gas flow. Plant Physiol. 37:409–415.

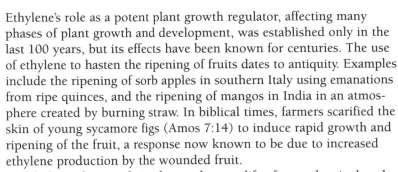

16

Ethylene in Postharvest Technology

Michael S. Reid

Ethylene's role as a potent plant growth regulator, affecting many phases of plant growth and development, was established only in the last 100 years, but its effects have been known for centuries. The use of ethylene to hasten the ripening of fruits dates to antiquity. Examples include the ripening of sorb apples in southern Italy using emanations from ripe quinces, and the ripening of mangos in India in an atmosphere created by burning straw. In biblical times, farmers scarified the skin of young sycamore figs (Amos 7:14) to induce rapid growth and ripening of the fruit, a response now known to be due to increased ethylene production by the wounded fruit.

Ethylene plays a role in the postharvest life of many horticultural crops—often deleterious, speeding senescence and reducing shelf-life, and sometimes beneficial, improving the quality of the product by promoting faster, more uniform ripening before retail distribution. This chapter is concerned with the properties of this gas and with ways to harness its beneficial effects and avoid its deleterious effects during postharvest handling of perishable commodities.

PROPERTIES OF ETHYLENE

The remarkable effects of ethylene on plants were first noted when flammable gas used for lighting and heating was piped through the streets of Europe. This gas contained added ethylene to ensure that lamps burned with a yellow flame, increasing illumination. It was soon noticed that plants growing in the vicinity of leaky pipes showed various abnormalities in growth and development, including premature leaf fall and death of flowers. A Russian graduate student, Neljubow, showed that the cause of these bizarre effects was ethylene. With this finding began the widespread research into the effects of ethylene on plant growth and development that continues today.

ETHYLENE IN PLANT GROWTH AND DEVELOPMENT
Over the years since Neljubow's study, researchers have shown that many phases of plant growth and development are affected by ethylene. It
- stimulates germination of some dormant seeds
- changes the direction of seedling growth to bypass obstacles in the soil
- stimulates growth of special aerating roots in waterlogged soil
- causes abscission of leaves in plants under drought stress
- may stimulate flowering
- is often the trigger for fruit ripening and abscission

In early studies, the effects of ethylene were thought to be an interesting example of growth regulation by a synthetic chemical. In the 1930s, however, it was discovered that ethylene is produced by plants, and it was suggested that the responses to ethylene are part of normal growth and development. Ethylene is now considered a plant hormone, an important part of the mechanisms controlling plant growth and development.

ETHYLENE BIOSYNTHESIS
The unraveling of the biochemical pathway of ethylene biosynthesis in plants (fig. 16.1) has been one of the most interesting biochemical stories of recent years. Researchers in Europe and North America competed

Figure 16.1

Pathway of ethylene biosynthesis.

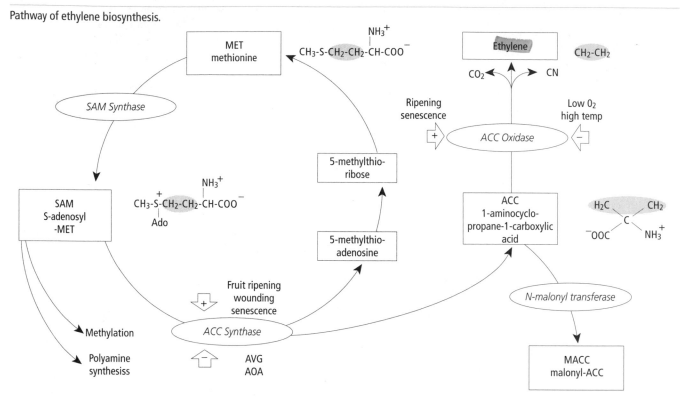

to find each step in the pathway. Lieberman, at the Beltsville laboratories of the USDA, showed that the amino acid methionine was the precursor for ethylene production in apples, and this compound was quickly identified as the starting point for ethylene biosynthesis. Researchers at UC Davis identified SAM (*S*-adenosyl-methionine) as another key compound in the pathway, and then, almost simultaneously, Lürssen in West Germany and Adams and Yang at UC Davis discovered that SAM was converted to an unusual cyclic amino acid, ACC (1-aminocyclopropane-1-carboxylic acid), which is now thought to be the immediate precursor for ethylene. The interesting biochemistry in this pathway had some practical implications. The enzyme that controls the rate at which the pathway operates, ACC synthase, requires pyridoxal phosphate as a cofactor. Inhibitors of enzymes that require pyridoxal phosphate, such as AVG (aminoethoxyvinyl glycine) and AOA (aminooxyacetic acid) can be used to inhibit ethylene production. Cobalt ion and low O_2, which inhibit the final step in the pathway, ACC oxidase, can also reduce ethylene production.

Current research is investigating the way in which ethylene induces such a range of effects. The favored model is that ethylene

binds to a protein, called a binding site (fig. 16.2), thus stimulating release of a so-called second message instructing the DNA (the molecules carrying all the information in the plant cell) to form mRNA (messenger RNA) molecules specific for the effects of ethylene. These molecules are "translated" into proteins by polyribosomes, and the proteins so formed are the enzymes that cause the actual ethylene response. The identification of the gene ETR-1 that encodes the binding site has provided horticulturists with a powerful tool to modulate the action of ethylene in plants. Plants transformed with *etr-1*, a mutated form of this gene, are no longer responsive to ethylene.

OTHER PROPERTIES

Physical. Ethylene is the first member of the unsaturated or olefin series of hydrocarbons. Its properties are summarized in table 16.1.

Toxicological. Ethylene is a gas with a characteristic suffocating, sweetish odor. It is both an anaesthetic and asphyxiant. High vapor concentrations can cause rapid loss of consciousness and perhaps death by asphyxiation. Removal to fresh air usually results in prompt recovery if the person is still breathing. When the gas is handled in liquefied form,

Figure 16.2

Mechanism of ethylene action.

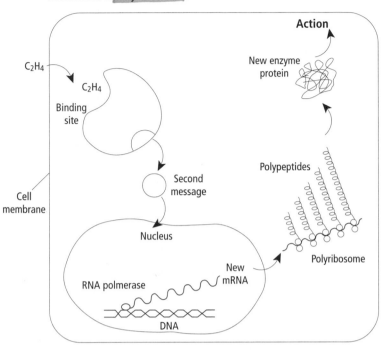

Table 16.1. Physical properties of ethylene gas

Appearance	Colorless, hydrocarbon gas with a faint, sweetish odor that is easily detected in parts per million concentrations
Molecular weight	28.05
Boiling point	
at 760 mm Hg	−103.7°C (−154.6°F)
at 300 mm Hg	−118°C (−180°F)
at 10 mm Hg	−153°C (−243°F)
bp/p at 750 to 770 mm Hg	0.022°C per mm Hg
Freezing point at saturation pressure (triple point)	−169.2°C (−272.6°F)
Surface tension at −103.7°C (−154.6°F)	16.4 dynes/cm
Flammable limits in air*	
lower	3.1% by vol
upper	32% by vol

Note: *All compositions between the upper and lower limits are flammable and can be explosive.

skin and eye burns can result from contact with the liquid. Cases in which liquid ethylene contacts the eye must be treated by a physician.

FDA status. The use of ethylene gas to promote ripening of fruits and vegetables is sanctioned under FDA Regulation 120,1016. Ethylene is exempted from the requirement of a residue tolerance when used as a plant regulator either before or after harvest.

Explosive. Mixtures of ethylene gas and air are potentially explosive when the concentration of ethylene rises above 3.1% by volume. This concentration is at least 30,000 times the concentration required to initiate ripening of most fruits and vegetables. Above 32% by volume, ethylene-air mixtures are not explosive.

MEASUREMENT OF ETHYLENE

A remarkable feature of the effects of ethylene on plants is the minute concentration required. Fruit ripening, for example, typically occurs at the maximum rate at levels of 1 part of ethylene in 1 million parts of air (1 ppm). Effects on opening of roses can be seen at ethylene concentrations as low as 10 parts per billion parts of air (10 ppb). The development of the gas chromatograph and the flame ionization detector, which made it possible to rapidly measure ethylene at such low concentrations, was crucial to our present understanding of the role of ethylene. New techniques, such as laser-acoustic devices, offer even more sensitivity. Unfortunately, sensitive measurement of ethylene is still expensive; satisfactory chromatographs cost from $7,000 to $20,000. For the postharvest technologist, gas-sampling tubes giving a colorimetric reaction can be read reasonably easily down to 1 ppm. These are satisfactory for occasional use by ripening-room operators but cannot monitor the low levels of ethylene that may be of concern in storage rooms and marketing outlets.

POSTHARVEST USES

The wide range of approved uses for ethephon, an ethylene-releasing chemical, in agriculture (table 16.2) indicates the utility of this growth regulator. Fruit ripening is by far the largest application of ethylene gas in postharvest technology, but other responses are also in use for some crops and will be described briefly first.

FLOWER AND SPROUT INDUCTION
The stimulation of flowering of pineapples by ethylene treatment is critical to that industry. Less well known is the spectacular response of some flowering bulbs to ethylene. Japanese bulb growers discovered that iris bulbs from

fields that had been burned at the end of the season to control leaf diseases flowered earlier and more prolifically than controls. It was found that smoke did the same for bulbs that had been harvested, and smoking of bulbs is still practiced in Japan. The primary active ingredient in the smoke is ethylene, and it has now been shown that ethylene treatment of the propagules of a number of flowering crops stimulates flowering. Perhaps the most remarkable example is narcissus. Medium-

sized (6-cm [2.4-in] circumference) narcissus bulbs normally do not flower; treatment with ethylene for a few hours just after lifting induces almost 100% flowering (fig. 16.3).

Now being used commercially, this treatment is applied either using ethephon or fumigating with ethylene gas in rooms similar to those used for ripening fruits. Ethylene is also used as a treatment to enhance the sprouting of seed potatoes—the gas breaks the dormancy of the buds, but prolonged treatment inhibits their extension growth.

Table 16.2. Approved uses for ethephon in U.S. agriculture

Use	Approved crops and states (no parentheses = all states)
Postharvest fruit ripening	Bananas, tomatoes (FL)
Preharvest fruit ripening	Peppers, tomatoes
Fruit removal	Apples, carob, crabapples, olive
Defoliation	Apple, buckthorn, cotton, roses
Fruit loosening	Apples, blackberries (WA, OR), cantaloupes, cherries (CA, AZ, TX), tangerines
Maturity or color development	Apples, cranberries (MA, NJ, WI), figs (CA), filberts (OR), grapes, peppers, pineapple, tomatoes
De-greening (preharvest)	Tangerines
De-greening (postharvest)	Lemons
Dehiscence	Walnuts
Leaf curing	Tobacco
Flower induction	Pineapple and other bromeliads
Sex expression	Cucumber, squash
Flower bud development	Apple
Plant height control	Barley, daffodils, hyacinth, wheat
Stimulate lateral branching	Azaleas, geraniums

Source: Adapted from Kays and Beaudry 1987.

Figure 16.3

Effect of ethylene treatment of bulbs on flowering of narcissus.

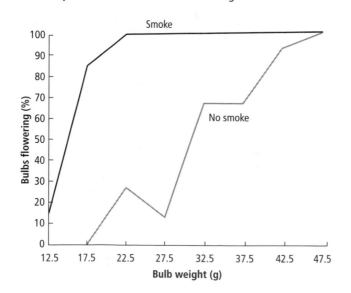

SHUCK LOOSENING AND FRUIT RELEASE

Although physiologists knew for years that ethylene induced abscission of leaves, flowers, and fruit from many plants, it was not until ethylene-releasing compounds such as ethephon and silaid (fig. 16.4) became available that these responses could be tested in horticulture. Ethylene-releasing chemicals are now approved for a variety of uses of interest to postharvest technologists. Preharvest application to walnut and pecan trees induces shuck loosening and improves harvest efficiency. Similarly, these chemicals are used to loosen the abscission zone on the stalk of fruits that are mechanically harvested, notably sour cherries, improving harvest yields.

CHLOROPHYLL DESTRUCTION

In many plant tissues, ethylene treatment results in rapid loss of chlorophyll, the green color in leaves and unripe fruit. This response was previously used to blanch celery before green celery was generally accepted in the United States, and it is still used to accelerate the curing of tobacco. Another important example is the de-greening of citrus, where the orange color is revealed as the chlorophyll is destroyed during ethylene treatment.

FRUIT RIPENING

Physiology. The concentrations of ethylene required for the ripening of various commodities vary (table 16.3) but in most cases are in the range of 0.1 to 1 ppm. The time of exposure to initiate full ripening may vary, but for climacteric fruits exposures of 12 hours or more are usually sufficient. Full ripening may take several days after the ethylene treatment.

Technical considerations. The effectiveness of ethylene in achieving faster and more uniform ripening depends on the type of fruit being treated, its maturity, the temperature and RH of the ripening room, ethylene concentration, and duration of exposure to ethylene. In general, optimal ripening conditions for fruits are

- Temperature: 18° to 25°C (65° to 77°F)
- RH: 90 to 95%
- Ethylene concentration: 10 to 100 ppm

Figure 16.4

Molecules of ethephon, silaid, and alsol, three ethylene-releasing chemicals.

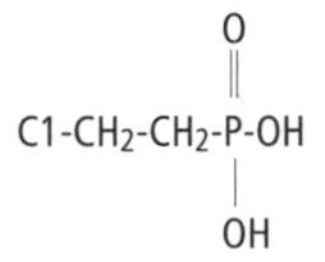
O
C1-CH2-CH2-P-OH
OH

(2-chloroethyl) phosphonic acid

Ethrel

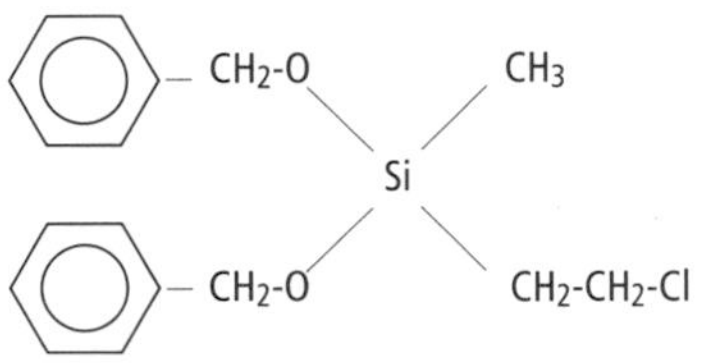
CH2-O
CH3
Si
CH2-O
CH2-CH2-Cl

(2-chloroethyl) methylbis (phenylmethoxy) silane

Silaid

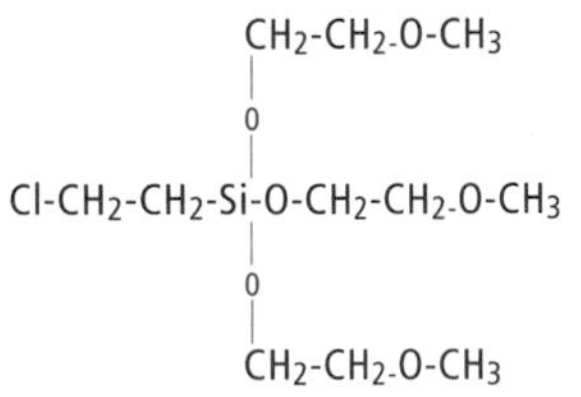
CH2-CH2-O-CH3
O
Cl-CH2-CH2-Si-O-CH2-CH2-O-CH3
O
CH2-CH2-O-CH3

(2-chloroethyl) tris (2-methoxyethoxy) silane

Alsol

Table 16.3. Threshold for ethylene action in various fruits

Fruit	Threshold concentration (ppm)
Avocado (var. Choquette)	0.1
Banana (var. Gros Michel)	0.1–1.0
(var. Lacatan)	0.5
(var. Silk fig)	0.2–0.25
Cantaloupe (var. P.M.R. No. 45)	0.1–1.0
Honeydew melon	0.3–1.0
Lemon (var. Fort Meyers)	0.1
Mango (var. Kent)	0.04–0.4
Orange (var. Valencia)	0.1
Tomato (var. VC-243-20)	0.5

- Duration of treatment: 24 to 72 hours, depending on fruit type and maturity stage
- Air circulation: Sufficient to maintain uniform temperatures within the ripening room
- Ventilation: adequate air exchanges to prevent accumulation of CO_2, which reduces the effectiveness of C_2H_4

For more commodity-specific information, see table 21.1 in chapter 21.

Amount of gas needed. The recommended treatment concentration is 10 to 100 ppm (1 m^3 of C_2H_4 in 10,000 m^3 of room space, or 1 ft^3 of C_2H_4 in 10,000 ft^3 of room space). Lower concentrations are used in well-sealed rooms that will maintain the ethylene concentration, or in rooms where the trickle system (see below) is being used. Higher concentrations are used in leaky rooms to compensate for the fall in concentration during treatment. Concentrations higher than 100 ppm do not speed up the ripening process. Adding too much ethylene may create an explosive air-gas mixture.

Temperature. Control of temperature is critical for optimal ripening with ethylene. The desired ripening temperatures are from 18° to 25°C (65° to 77°F). At lower temperatures ripening is slowed; at temperatures over 25°C (77°F), bacterial growth and rotting may be accelerated, and above 30°C (86°F) ripening may be inhibited. Fruit that have been cool-stored must be warmed to 20°C (68°F) to ensure that ripening proceeds rapidly. As ripening starts, the burst of respiration that accompanies it (the climacteric) generates a burst of heat. Adequate refrigeration equipment controlled with proper thermostats (fig. 16.5) is essential to ensure that this heat does not increase pulp temperatures to the point where ripening is inhibited (above 30° to 35°C, or 86° to 95°F, depending on the commodity). Some ripening-room operators "pigeon-hole" stack banana boxes to ensure that heat generated during ripening is carried away efficiently. This is especially important if boxes of bananas are packed with a polyethylene liner, which restricts airflow and heat removal from the carton.

Modern ripening facilities use the principles of forced-air cooling for maintaining temperature control in the ripening room, eliminating the hand labor required to unload pallets, construct pigeon-hole stacks,

Figure 16.5

Thermostat for temperature control in a banana ripening room.

and reload pallets when ripening is complete. Air is forced or drawn through the fruit at a sufficient rate to ensure control of the ripening temperature.

Safety precautions. Because of the explosion hazard of ethylene mixed with air at concentrations from 3 to 30%, the rules listed below must be followed stringently to prevent buildup of these concentrations and to prevent ignition if they should form.

- Do not permit open flames, spark-producing devices, fire, or smoking in or near a room containing ethylene gas or near the cylinder.

- Use an approved meter for accurately measuring the gas when discharging ethylene from the cylinder.

- Ground all piping to eliminate the danger of electrostatic discharge.

- Store ethylene cylinders in accordance with all instructions and standards of the National Board of Fire Underwriters.

- All electrical equipment, including lights, fan motors, and switches, should comply with the National Electric codes for Class 1, Group D equipment and installation.

- Instruments that detect the concentration of ethylene in air can be set to sound an alarm if the concentration approaches explosive levels.

TREATMENT SYSTEMS

Handlers can equip existing rooms for use as ripening rooms, or they can install specially built chambers that have automatic control of temperature, humidity, and ventilation. It is not essential that the rooms be hermetically sealed, but they should be as tight as practicable to prevent leakage. If pure ethylene is being used to provide the ethylene treatment, it is essential that the rooms adhere to the safety standards above.

Rooms should be heated with hot water or steam pipe systems or with indirect gas or electric heaters that have been examined and listed by Underwriters Laboratories (U.L.), never heated with an open flame. Because of the rapid increase in respiratory heat production following ethylene treatment, ripening rooms should be equipped with refrigeration systems adequate to hold the temperature in the desired range. Room temperature should be continuously monitored using a distant-reading thermometer.

Several methods, varying in sophistication, are used to provide the proper ethylene concentration in the ripening room. The trickle system has become much more common than the "shot" system.

The "shot" system

In the shot system, measured quantities of ethylene are introduced into the room at regular intervals. The shots may be applied by weight (rarely used in the United States today) or by flow, using a gauge that registers the discharge of ethylene in cubic feet per minute. The required ethylene application is made by adjusting the regulator to give an appropriate flow rate, then timing the delivery of the gas. Any piping leading into the ripening room should be grounded to prevent possible electrostatic ignition of the explosive concentrations of ethylene that are always present near the orifice when ethylene is being introduced.

The gas needed for a room is calculated, using the following formulas, where:

C = the amount of ethylene required, usually between 10 and 100 ppm

V = the room volume in thousands of cubic feet

F = the flow rate of gas measured from a flowmeter in cubic feet per minute (cfm)

T = the time in minutes for which the gas is allowed to flow

If the ethylene is administered by weight, the number of pounds of ethylene needed is given by the formula:

$$C \times V/13{,}000{,}000$$

Example: How many pounds of ethylene are required to provide 100 ppm in a room 20 ft high, 100 ft long, and 50 ft wide?

C = 100
Room volume (V) = 20 × 100 × 50 = 100,000 ft³

From the equation, the number of pounds of ethylene required is:

$$(100 \times 100{,}000)/13{,}000{,}000 = 0.77 \text{ lb}$$
$$\text{of ethylene}$$

It is difficult to measure a change of ¾ lb in an 80-lb gas cylinder, so one can see why the weighing method is now considered old-fashioned.

Using the more usual system of metering the flow of ethylene into the room, the required time (in minutes) for which the gas should flow is given by the formula:

$$(C \times V)/(F \times 1{,}000{,}000)$$

For the same room (V = 100,000), a desired ethylene concentration of 100 ppm, and an ethylene flow rate of 20 cfm, the gas should flow for:

$$(100 \times 100{,}000)/(20 \times 1{,}000{,}000) = 0.5$$
$$\text{minute}$$

This is easy to measure with a stopwatch. If the flowmeter is calibrated in milliliters per minute (ml/min), the formula is a little different. The time, in this case, is given by:

$$(C \times V)/(36 \times F)$$

For a flowmeter with a flow rate of 5,000 ml/min, the time to get the same concentration (100 ppm) in the same room (V = 100,000) is:

$$(100 \times 100{,}000)/(36 \times 5{,}000) = 55 \text{ minutes}$$

The long time taken is why many ripening rooms have several flowmeters in parallel. In this situation, add the flows to get the total flow, and use that as *F* in the equations above.

In the "shot" system, the room containing the product being ripened is sealed. Respiration of the fruit produces CO_2 that accumulates in the room and may inhibit the ripening process. It is customary to apply a shot of ethylene twice each day. The room should be well ventilated before each new application, particularly if it is well sealed, by opening the doors for about half an hour. In large ripening rooms, a ventilating fan should be provided. Where the ripening rooms are near rooms used for storage or handling of ethylene-sensitive commodities (for example, in a wholesale distribution center), the rooms should be ventilated to the exterior to prevent contamination.

The trickle, or flow-through, system

The ethylene is introduced into the room continuously, rather than intermittently. As the flow of ethylene is very small, it has to be regulated carefully. This is usually done by reducing the pressure using a two-stage regulator and passing the gas into the room through a metering valve and flowmeter (fig. 16.6).

To prevent a buildup of CO_2, fresh air is drawn into the ripening room at a rate sufficient to ensure a change of air every 6 hours. The air is vented through a small exhaust port to the rear of the room. The fan size in cfm (cubic feet per minute) is calculated by:

$$\text{Volume of room (cubic feet)}/360$$

The ethylene flow rate (in cfm) needed to maintain 100 ppm in the room is calculated by:

$$\text{Ventilation fan delivery (cfm)} \times 0.0001$$

In ml/min, the flow rate is:

$$\text{Ventilation fan delivery (cfm)} \times 2.8$$

A convenient way of monitoring gas being supplied in a trickle system is a simple "sight glass" in which the ethylene bubbles through a water trap on its way to the ripening room (fig. 16.7). As in the shot system, correct temperature maintenance and adequate air circulation are essential for good ripening.

Figure 16.6

Pressure regulators and flowmeters used for controlling and monitoring flow of ethylene into ripening rooms.

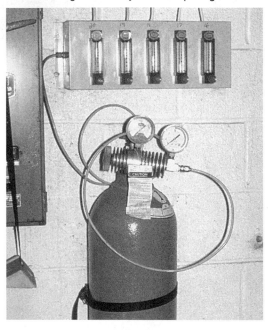

SOURCES OF ETHYLENE

Ethylene gas is a relatively inexpensive industrial chemical, but it is often more convenient or safer to provide ethylene by means other than the gas bottles assumed in the above discussion of treatment systems. Regardless of the source of ethylene, the treatment conditions outlined above are still important to a good result the ripening process.

Explosion-proof ethylene mixtures. The danger of explosions from oversupply of ethylene to a ripening room can be eliminated by using mixtures of ethylene with inert gases. The proportion of the inert gas should be such that at high concentrations of ethylene not enough oxygen remains in the ripening space to provide an explosive mixture. For example, one commercial formulation, Ripegas, contains 6% C_2H_4 in CO_2 by weight. When using these mixtures, the calculation of volumes of gas required must be modified to reflect the composition of the mixture. In the case of Ripegas, for example, the weights, volumes, or flow rates calculated using the formulas above would be increased by a factor of 100/6, or 17, to give the required concentration.

Ethylene generators. Ethylene generators, in which a liquid produces ethylene when

Figure 16.7

A simple sight-glass can be used to monitor flow of ethylene into the ripening room.

heated in the presence of a catalyst, are now widely used for supplying ethylene in ripening rooms. The liquid, a proprietary product, comprises ethanol and agents that assist in catalyzing its dehydration:

$$C_2H_5OH - H_2O \longrightarrow C_2H_4$$

The generator combines a heated platinized asbestos catalyst with a system for attaching a bottle of the generator liquid. The liquid comes in 1-pint and 1-quart bottles and is used up by the generator at the rate of 1 pint every 8 hours. The generator delivers about 14 l (0.5 ft³) of ethylene gas per hour, adjustable on newer models. These figures can be used to determine the number of generators to be used in a given size of room, knowing the air leakage or ventilation rate (gas exchanges per hour). For example, in a 5,000-ft³ ripening room where there is one air exchange per hour, a 1-quart bottle will generate 100 ppm for 16 hours.

Ethephon. Ethephon (2-chloroethane phosphonic acid) is strongly acidic in water solution. When in solutions above a pH of about 5, the ethephon molecule spontaneously hydrolyses, liberating ethylene. Ethephon is commercially available (Ethrel, Florel, Cepa) and is registered for preharvest use on a variety of crops for controlling developmental processes or inducing ripening. As a material for enhancing postharvest ripening, it has the disadvantage that it has to be applied to the fruit in a water solution as a spray or as a dip, an extra step in handling with attendant dangers of microbial infection. In contrast to ethylene treatment, however, no special facilities are required to ripen fruit with ethephon, provided that the ambient temperatures are within the range required to ripen the commodity. Ethephon is approved for postharvest use on only a few commodities (see table 16.2).

On a small scale, commodities can be treated using the shot method with ethylene liberated from ethephon. Place the calculated amount of ethephon (approximately 200 ml [7 fluid oz] of active ingredient to release 27 l [1 ft³] of ethylene gas) in a stainless steel bowl, then, just before closing the room, add enough caustic soda pellets (approximately 85 g [3 oz] or for each 200 ml [7 fl oz] of a.i. of ethephon) to completely neutralize the ethephon. CAUTION! CAUSTIC SODA AND ETHEPHON ARE CORROSIVE. WEAR SAFETY GLASSES AND RUBBER GLOVES.

Calcium carbide. Calcium carbide (Ca_2C), a grayish solid, is readily produced by heating calcium oxide with charcoal under reducing conditions. When hydrolyzed, calcium carbide produces acetylene, containing trace amounts of ethylene that are sufficient to be used in fruit ripening. Simple generators that are used to provide acetylene for lamps can be used in partially vented spaces to ripen or de-green fruits under conditions where ethylene is not available. In some instances Ca_2C wrapped in newspaper can be used as the generator. Packages of Ca_2C are placed with the fruit under a sealed cover (like a polyethylene tarpaulin). Water vapor from the fruit releases sufficient ethylene from Ca_2C to cause ripening.

Use of fruits. Traditionally, ripening has often been stimulated by enclosing unripe fruit with other fruits that are already ripe. This technique forms the basis for a cheap and simple method of fruit ripening. Table 4.3 (in chapter 4) shows the range of ethylene production known for fruits and vegetables. Ripe fruits with high ethylene production can be used in very small scale commercial operations or at home to ripen or de-green other fruits in much the same way as any other ethylene generating system. The desiccating leaves of some tropical plants have also been found to produce large quantities of ethylene and are used in some countries as a source of ethylene for fruit ripening.

UNDESIRABLE EFFECTS OF ETHYLENE

Given the wide range of physiological effects of ethylene in plants and the common occurrence of ethylene and other gases with ethylene-like effects (table 16.4) as air pollutants, it is not surprising that ethylene-mediated growth responses result in quality reduction in a range of commodities.

Accelerated senescence. In green tissues, ethylene commonly stimulates senescence, as indicated by loss of chlorophyll, loss of protein, and susceptibility to desiccation and decay. Ethylene pollution can result in yellowing of leafy vegetables (spinach), fresh herbs (parsley; see fig. 16.8), and other green vegetables (broccoli). The senescence of some flowers is stimulated by ethylene at very low concentrations (fig. 16.9). These effects occur in flowers where increased ethylene production is part of natural senescence (e.g., carnations, sweet peas) and in some where it is not (roses, brodiaeas).

Table 16.4. Comparative effectiveness of ethylene and related analogues in peas stem-section assay

Compound	Relative activity (moles/unit)
Ethylene	1
Propylene	130
Vinyl chloride	2,370
Carbon monoxide	2,900
Acetylene	12,500
1-Butene	140,000

Source: Burg and Burg 1966.

Figure 16.8

Parsley yellows rapidly at room temperature when exposed to low concentrations of ethylene.

Figure 16.9

Carnations senesce rapidly when exposed to low concentrations of ethylene.

Figure 16.10

Cucumbers yellow prematurely when exposed to ethylene.

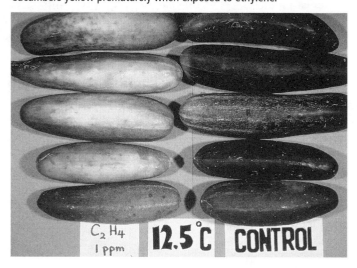

Accelerated ripening. Although acceleration of ripening is a beneficial use of ethylene, it can also be undesirable, as in the premature yellowing of cucumbers in the presence of ethylene (fig. 16.10). Most climacteric fruits respond to ethylene in the atmosphere, even at storage temperatures, so ethylene in the storage area reduces storage life. For example, the firmness of kiwifruit in storage is dramatically reduced if the ethylene concentration in the cold storage room is more than 20 ppb.

Induction of leaf disorders. In many plants, exposure to ethylene results in dark-ening or death of portions of their leaves. This response is commonly seen in foliage plants and is of major economic consequence in lettuce, where ethylene causes the disorder known as russet spotting (fig. 16.11). In lettuce, the browning results from collapse and death of areas of cells following increased synthesis of phenolic compounds in response to ethylene.

Isocoumarin formation. In carrots, ethylene exposure causes the biosynthesis of isocoumarins, which make the carrots bitter. Recently, it has been shown that ethylene concentrations as low as 0.5 ppm cause significantly increased bitterness of carrots within 2 weeks, even when they are stored at 2.5°C (36.5°F) (fig. 16.12).

Sprouting. The ethylene-stimulated sprouting which is useful in propagules is, of course, undesirable in commodities intended for consumption. Sprouting of potatoes, for example, is unsightly and increases water loss, leading to early shriveling.

Abscission of leaves, flowers, and fruits. Ethylene-induced abscission is most often a problem in ornamental plants, where low concentrations can cause complete loss of flowers or leaves. As an example, *Schlumbergera* spp., the Christmas cactus, is sold when the first flowers are open, but it often arrives at the market with all the flowers in the bottom of the box due to exposure to ethylene during transportation.

Toughening of asparagus. Ethylene stimulates the lignification of xylem and fiber elements in the growing asparagus spear, leading to undesirable toughness and reducing the portion of the spear that is edible.

Induction of physiological disorders. Ethylene sometimes induces or hastens the appearance of physiological disorders of stored commodities. Rapid ripening of apples with low calcium contents induces high levels of the bitter pit storage disorder. Similarly, high ethylene levels in the storage chamber reduce the effectiveness of controlled atmospheres in maintaining quality of apples. While useful in inducing flowering in bulbs and other propagules, ethylene damages these propagules after the flowers have started to develop. Ethylene pollution during marketing of tulip bulbs, for example, results in failure of the flowers to develop, a condition called "blasting."

Figure 16.11

Dark brown spotting of the midribs, or "russet spotting," caused by exposure of lettuce to ethylene.

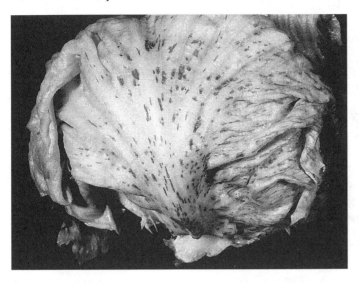

Figure 16.12

Exposing carrots to ethylene turns them bitter due to the synthesis of isocoumarin, an intensely bitter compound. (Lafuente et al. 1989)

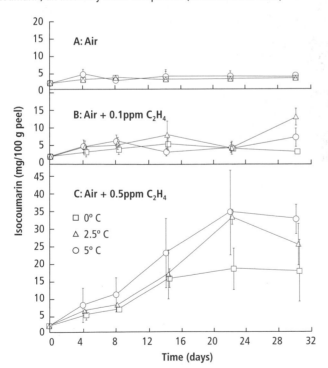

SOURCES OF ETHYLENE IN THE ENVIRONMENT

Ethylene is produced whenever organic materials are stressed, oxidized, or combusted. There are many sources of ethylene pollution during postharvest handling of perish-

ables, but the most important are internal combustion engines, ripening rooms, and ripening fruits. Other sources are aircraft exhaust, fluorescent ballasts, decomposing produce and sometimes fungi growing on it, cigarette smoke, rubber materials exposed to heat or UV light, and virus-infected plants. The sources are not always obvious: supermarkets in Texas traced the problems they were having with their flowers to ethylene contamination from propane-powered floor polishers that were only used at night.

OVERCOMING ETHYLENE'S UNDESIRABLE EFFECTS

A number of techniques have been developed to protect sensitive commodities from the effects of ethylene. Selection of the appropriate method depends on the commodity and the handling techniques used in its marketing.

Removing ethylene from the atmosphere around the commodity is the preferred method of preventing deterioration of ethylene-sensitive produce. Of the available methods, the simplest and cheapest are in many cases the most effective.

ELIMINATING SOURCES OF ETHYLENE

In a great majority of cases, high levels of ethylene in storage and handling areas can be avoided by removing the sources of ethylene. In particular, commodities sensitive to ethylene should be handled using electric forklifts. Vehicles powered by internal combustion should be isolated from handling and storage areas, and engines should never be left idling in an enclosed space during loading and unloading operations. Where these techniques are not feasible, it is possible to fit combustion engine exhausts with catalytic converters, which will reduce C_2H_4 emissions by 90%. Rigorous attention to sanitation will remove ripening fruits, overripe and rotting produce that can be a source of ethylene.

Ventilation

Where the air outside storage and handling areas is not polluted, simple ventilation of these areas can reduce ethylene concentrations. An exchange rate of one air change per hour can readily be provided by installing an intake fan and a passive exhaust. The cost of using such a system (ignoring the small initial capital investment and assuming a power cost of 7 cents per kilowatt-hour) can be determined using the equation:

$$\text{Cost/year (dollars)} = \begin{array}{l} 0.001 \times \text{cooler volume (ft}^3\text{)} \times \\ [\text{outside temperature} - \text{cooler} \\ \text{temperature (°F)}] \end{array}$$

Chemical removal

Ethylene can be removed by a number of chemical processes; the most important are described below.

Potassium permanganate. Commercial materials such as Purafil use the ability of potassium permanganate ($KMnO_4$) to oxidize ethylene to CO_2 and H_2O. The requirements for such materials are a high surface area coated with the permanganate and ready permeability to gases. Many porous materials have been used to manufacture permanganate absorbers, including vermiculite, pumice, and brick. The type of material used may depend on the purpose for which the absorber is required. For removing ethylene from room air, the absorber should be spread out in shallow trays, or air should be drawn through the absorber system. Attempts to develop liquid scrubbers using $KMnO_4$ have been unsuccessful.

Ultraviolet lamps. Australian researchers have developed an effective method for ethylene removal using ultraviolet lamps. In commercial equipment now available using this method, air from the storage room is drawn past the lamps. Ultraviolet lamps produce ozone, which was thought to be the active ethylene removing agent. It now appears that the ethylene is oxidized by a much more reactive intermediate in the formation of ozone. A recent development is the use of titanium dioxide (TiO_2)-coated glass beads illuminated with UV light to remove ethylene, presumably by a similar mechanism. Whatever the mechanism, the ozone produced by the lamps is very toxic to fresh produce and must be removed. Fortunately this is easily achieved by using a rusty steel wool filter at the exit of the unit.

Activated or brominated charcoal. Charcoal air purifiers, especially if brominated, can absorb ethylene from air. These systems are largely confined to use in the laboratory, as potassium permanganate absorbers are cheaper and more widely available.

Catalytic oxidizers. If ethylene and oxygen are combined at high temperature in the presence of a catalyst such as platinized asbestos, ethylene will be oxidized. Ethylene scrubbers using this effect are now available commercially (fig. 16.13). They overcome the difficulty of heating the incoming air by clever use of two beds of ceramic beads as heat sinks and regular reversal of gas flow through the beds. These scrubbers are very efficient, reducing the ethylene concentration in the air to 1% of the input concentration. Because of the small air volume they process, they are most suited to small spaces or long-term CA storage systems.

Bacterial systems. Approximately 30,000 metric tons (33,000 tons) of ethylene is liberated into the atmosphere each day from internal combustion engines, but the concentration of ethylene in air remains very low (nearly undetectable in fresh rural air). This implies that something removes ethylene from the atmosphere. Bacteria that use ethylene as a biochemical substrate have been isolated from soils. It seems possible that a scrubber may be developed in which the bacteria grow at the expense of ethylene in the storage atmosphere.

Hypobaric storage. Removal of endogenous ethylene was the first benefit ascribed to hypobaric (low-pressure) storage: levels of ethylene inside fruit were greatly reduced, and the longer storage life obtained could be reduced by adding ethylene to the atmosphere. There appear to be few commodities where the benefits of reducing the tissue ethylene content warrant the use of this cumbersome and expensive apparatus. Many of

Figure 16.13

This Swingtherm unit removes ethylene from air using a catalytic oxidizer and is equipped with heat exchangers to reduce energy consumption. (Reprinted with permission from Blanpied et al. 1985)

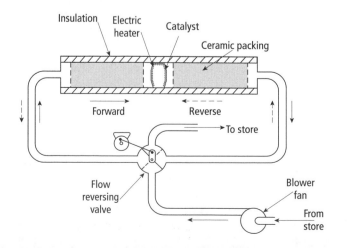

Figure 16.14

Bananas shipped in sealed polyethylene bags with KMnO$_4$-type ethylene absorbers remain green during transportation.

Figure 16.15

Treatment of carnation flowers with silver thiosulfate protects them against ethylene-mediated senescence.

the benefits of hypobaric storage may also be due to reduction in the partial pressure of oxygen that automatically accompanies reduction in the atmospheric pressure.

INHIBITING THE EFFECTS OF ETHYLENE

Sometimes it is not possible to ensure low concentrations of ethylene in the air, as, for example, in supermarkets. For these settings there are several techniques available to inhibit the effects of ethylene.

Controlled atmospheres. Low concentrations of O$_2$ and high concentrations of CO$_2$ in the storage atmosphere reduce the rates of respiration, ethylene production, and other metabolic processes. CO$_2$-enriched atmospheres may also inhibit the action of ethylene on tissues sensitive to it. For example,

bananas packed in polyethylene-lined boxes can be transported at 15° to 25°C (59° to 77°F) in the presence of a potassium permanganate absorber. Even without the absorber, bananas transported in this way arrive in much better condition than control fruit (fig. 16.14) because of the effect of the accumulated CO$_2$ produced by the fruit on preventing the action of ethylene.

Specific anti-ethylene compounds. For nearly 20 years, the ornamentals industry has inhibited the undesirable effects of ethylene with a complex of silver and thiosulfate (STS), which has a very low stability constant and therefore moves readily from the vase solution to the head of cut flowers. Flowers pulsed with this material last two to three times as long as control flowers (fig. 16.15). Potted flowering plants do not lose their flowers during transportation if they are first sprayed with STS (fig. 16.16).

Recently, an ethylene analog, 1-methylcyclopropene (1-MCP, EthylBloc) was shown to be a very effective inhibitor of ethylene action in ornamentals, fruits, and vegetables. The material has been patented and registered for use on ornamentals where it is just as effective as STS in preventing the effects of exogenous ethylene and extending the life of products whose senescence is triggered by endogenous ethylene production. It is likely that 1-MCP will also be registered for use on fruits and vegetables, where it can have an equally beneficial effect.

Inhibition of ethylene biosynthesis. Ethylene may reduce quality even when it is not present as a pollutant if the tissue itself produces ethylene. This may occur, for example, when carnations senesce early or when fruits ripen too fast. Inhibitors of ethylene biosynthesis, such as AVG and AOA, have been used in laboratory experiments to extend flower vase life and fruit storage life. These inhibitors do not prevent the action of ethylene that is present as an environmental pollutant. A flower "preservative," Florish, formulated with an AVG analogue, proved to be only partially beneficial, since it was unable to overcome the effects of exogenous ethylene.

MOLECULAR MANIPULATION OF ETHYLENE RESPONSES

The identification of the genes that encode ACC synthase (the key enzyme in the biosynthesis of ethylene) and ETR-1 (the

Figure 16.16

Pretreatment of potted Christmas cactus plants with STS prevents loss of florets during shipping.

ethylene binding site) have provided biotechnologists with the tools to modify the biosynthesis of ethylene in plants. Endless Summer tomatoes are standard tomatoes in which cosuppression technology has been used to eliminate the plant's own synthesis of ethylene. Fruits of this cultivar have a long shelf life and will ripen rapidly only when exposed to ethylene. Flowers on petunia and carnation plants transformed with *etr-1*, a mutant form of the ethylene binding site, have extended life because they no longer respond to endogenous or exogenous ethylene. Public and private research laboratories are working on applying these technologies to a wide range of crops. In the next few years we may see the usefulness of ethylene extended by molecular modification of its biosynthesis and action in fruits, vegetables, and flowers.

REFERENCES

Abeles, F. B., P. W. Morgan, and M. E. Saltveit Jr. 1992. Ethylene in plant biology. Second edition. San Diego: Academic Press. 414 pp.

Blanpied, G. D., ed. 1985. [Symposium on] Ethylene in postharvest biology and technology of horticultural crops. HortScience 20:39–60.

Blanpied, G. D., J. A. Bartson, and J. R. Turk. 1985. A commercial development programme for low ethylene controlled-atmosphere storage of apples. In J. A. Roberts and G. A. Tucker, eds., Ethylene and plant development. London: Butterworths. 393–404.

Bleeker, A. B. 1999. Ethylene perception and signaling: an evolutionary perspective. Trends in Plant Sci. 4:269–274.

Burg, S. P., and E. A. Burg. 1966. Fruit storage at subatmospheric pressure. Science 153:314–315.

Kays, S. J., and R. M. Beaudry. 1987. Techniques for inducing ethylene effects. Acta Hort. 201:77–116.

Kende, H. 1993. Ethylene biosynthesis. Annu. Rev. Plant Physiol. and Plant Mol. Biol. 44:283–308.

Lafuente, M. T., M. Cantwell, S. F. Yang, and V. Rubatzky. 1989. Isocoumarin content of carrots as influenced by ethylene concentration, storage temperature and stress conditions. Acta Hort. 258:523–534.

Lelievre, J. M., A. Latche, B. Jones, and M. Bouzayen. 1997. Ethylene and fruit ripening. Physiol. Plant. 101:727–739.

Liu, F. 1970. Storage of bananas in polyethylene bags with an ethylene absorbent. HortScience 5:25–27.

Mckeon, T. A., J. C. Fernandez-Maculet, and S. F. Yang. 1995. Biosynthesis and metabolism of ethylene. In P. J. Davies, ed., Plant hormones. 2nd ed. Dordrecht: Kluwer. 118–139.

Pratt, H. K., and J. D. Goeschl. 1969. Physiological roles of ethylene in plants. Annu. Rev. Plant Physiol. 20:541–585.

Reid, M. S. 1995. Ethylene in plant growth, development, and senescence. In P. J. Davies, ed., Plant hormones. 2nd ed. Dordrecht: Kluwer. 486–508.

Saltveit, M. E. 1999. Effect of ethylene on fresh fruits and vegetables. Postharv. Biol. Technol. 15:279–292.

Scott, K. J., and R. B. H. Wills. 1973. Atmospheric pollutants destroyed in an ultraviolet scrubber. Lab. Pract. 22:103–106.

Sherman, M., and D. D. Gull. 1981. A flow-through system for introducing ethylene in tomato ripening rooms. Univ. Florida Veg. Crops. Fact Sheet VC-30. 4 pp.

Sisler, E. C., M. Serek, E. Dupille, and R. Goren. 1999. Inhibition of ethylene responses by 1-methylcyclopropene and 3-methylcyclopropene. Plant Growth Reg. 27:105–112.

Staby, G. L., and J. F. Thompson. 1978. An alternative method to reduce ethylene levels in coolers. Flor. Rev. 163:31, 71.

Union Carbide Corp. 1970. Ethylene for coloring matured fruits, melons, and vegetables. New York: Union Carbide Corp. Product Information. 11 pp.

Watada, A. E. 1986. Effects of ethylene on the quality of fruits and vegetables. Food Technol. 40(5): 82–85.

Wills, R. B. H., V. V. V. Ku, D. Shonet, and G. H. Kim. 1999. Importance of low ethylene levels to delay senescence of non-climacteric fruit and vegetables. Austral. J. Exp. Agr. 39:221–224.

Yang, S. F. 1987. Regulation of biosynthesis and action of ethylene. Acta Hort. 201:53–59.

17

Principles of Postharvest Pathology and Management of Decays of Edible Horticultural Crops

James E. Adaskaveg, Helga Förster,

and Noel F. Sommer

The development of modern agriculture has emphasized high levels of production and distribution of agricultural products to local, national, and international markets. Successful marketing depends on delivering high-quality produce to consumers. With increased emphasis on crop uniformity, bulk handling, tight-fill and cavity-tray packaging, shipment to more distant markets, and longer storage, it has become increasingly important to minimize losses caused by bacteria, filamentous fungi, and yeasts.

Although improvements in growing and postharvest handling of crops may provide more efficient means of food production and distribution at lower costs, they often increase the potential for crop injury and subsequent losses from postharvest decays. With longer storage times and long-distance transportation, the potential for primary decays may increase. Decays that were once secondary or minor may become major limiting factors, or new problems from previously unknown pathogens may develop. Thus, any change in production, handling, or processing of commodities may result in an increased risk for decay that must be carefully evaluated and managed. These changes, however, also promote increased research in improved postharvest disease control that have permitted the development of successful long-distance marketing and shipping.

Any postharvest decay management program needs to begin with preharvest practices that promote a healthy crop, reduce conducive environments for pathogen infection and disease development, and minimize the amount of the pathogen that may infect or contaminate the crop before harvest. Changes in preharvest production practices that have generally increased the inoculum levels of postharvest decay organisms and the potential for postharvest decays to occur include increases in production acreages, diversity of crops, new cultural practices (e.g., high-density plantings, high-angle or overhead irrigation), longer harvest seasons, and planting of new susceptible cultivars. Thus, with increased production areas of a crop, sound preharvest production strategies may be compromised for economic reasons, or localized disease problems may not be noticed in large orchards. With increased crop diversity due to the introduction of new commodities, new or unknown disease problems may be introduced. In contrast, other preharvest practices such as the use of resistant cultivars, irrigation practices that minimize wetness duration, balanced nitrogen fertilization, canopy management (pruning), insect and weed control, and the use of fungicides may reduce the amount of fruit decay before and after harvest and reduce inoculum levels of targeted pathogens (Adaskaveg and Förster 2000).

Preharvest disease management in the field is an important component of the integrated pest management (IPM) strategy to control pathogens. Similarly, postharvest decay control practices should also be considered part of an IPM program that begins at crop planting or the beginning of each growing season, continues through the season, and extends on through postharvest handling, including packaging, storage, transportation, and marketing until the produce reaches the consumer. Postharvest handling practices should focus on maintaining a healthy physiology of the produce and on minimizing losses from decay. Fruits and vegetables that have an active metabolism show considerable resistance to microbial infection and decay, whereas stressed or senescent fruits are prone to disease. In addition, activity of decay microorganisms depends on the presence of conducive environmental conditions. Any environment that slows microbial activity and maintains fruit quality will reduce the amount of decay. Physical methods that maintain the vitality of the crop include temperature management and modification of atmospheres using reduced oxygen (O_2) and elevated carbon dioxide (CO_2). Because fungi cause the majority of postharvest problems, the remainder of this chapter will focus mainly on decays caused by fungi.

Postharvest fungicide treatments effectively reduce decay of temperate, subtropical, and tropical fruits (see Eckert and Ogawa 1985, 1988). Properly applied treatments prevent or impede the development of decay-causing organisms, are generally economical, and are important means of controlling deterioration of fresh and processed food crops. Currently, chemicals are an integral part of disease management programs, and in some instances biological controls have been integrated into existing programs. With higher demands for fresh produce by consumers, the demand for organic or pesticide-free produce has also increased. Furthermore, with heightened awareness regarding worker and consumer safety, restrictions are being placed on the use of many older postharvest chemicals. Thus, development of new treatments for food crops to minimize decay losses has increasingly emphasized safer products, minimal residues, and if possible, methods that do not use pesticides (e.g., temperature management, modified atmospheres, etc.) for specialty markets. Still, the use of pesticides has resulted in a reasonably safe supply of horticultural commodities with minimal health risks from pesticides and decay-causing microorganisms, as well as their by-products (mycotoxins) that are known to be carcinogenic (Phillips 1984).

THE DISEASE TRIANGLE

As for preharvest diseases, postharvest diseases can also be described as the interaction of the disease triangle: host, pathogen, and environment. A change in any of these components in the disease triangle will influence the final amount of disease and will affect the control measures to be taken. The use of postharvest physical (e.g., temperature management, modified atmospheres, etc.) and chemical treatments (e.g., sanitizers, fungi-

cides, and biological control agents) prevents or inhibits the development of decay. Handling and treatment strategies must be based on a sound knowledge of the host-pathogen interrelationships for the periods before and at the time of treatment. The critical points that must be considered are

- the types of pathogens involved
- the location of the pathogen and whether it is contaminating the surface or is established within the host tissue
- time and location of pathogen infection
- the best time for treatment to prevent a secondary disease cycle during storage and transport
- the maturity of the host with reference to susceptibility to specific pathogens
- the environmental conditions during storage, transportation, and marketing
- storage, distribution, and marketing strategies that accelerate delivery of commodities

Specific treatments are selected based on these parameters. In addition, negative effects of any postharvest treatment on fruit quality should also be considered. For example, improper storage temperatures or modified atmospheres may lead to off-flavors or fruit discoloration.

The interactions of the disease triangle that affect decay management by changing host physiology or the physical and chemical environment will be the central focus of this chapter. Principles of disease suppression by

Figure 17.1

Fruit respiration patterns: climacteric (solid line) respiration of peaches and nonclimacteric (broken line) respiration of sweet cherries.

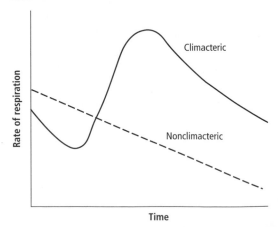

handling practices, modified environments in storage (e.g., temperature management, controlled atmospheres, use of ethylene, etc.), and the influence of the host on decay (e.g., maturity at harvest, climacteric and nonclimacteric fruits, etc.) will be discussed. In addition, chemical and biological treatments available for controlling postharvest decay pathogens, proper usage of these materials and strategies to minimize resistance development within target populations, and the regulatory status of postharvest chemicals and biological control treatments will be reviewed.

POSTHARVEST HOST PHYSIOLOGY

CLIMACTERIC AND NONCLIMACTERIC FRUITS

Respiration is a process by which the captured energy of light stored in organic compounds by photosynthesis is released by oxidation. The respiratory process results in usable chemical energy and heat (vital heat or heat of respiration). While a fruit is still attached to the plant, carbohydrates that are oxidized to CO_2 and water are replaced by photosynthates from green leaves or from stored reserves in the plant. Once separated from the tree, the fruit relies on its carbohydrate reserves. The respiratory process must continue to produce energy for cellular functions or the fruit tissues will die. Postharvest environments are designed to reduce the rate of respiration to the minimum required to maintain vital processes. The stored reserves are thereby conserved and the postharvest life of the fruit is extended to a maximum.

Fruits harvested and placed in respirometers at about 20°C (68°F) exhibit one of two very different respiratory patterns, as shown by CO_2 production (fig. 17.1). Sweet cherries show a gradual decrease in the respiration rate, as indicated by the dotted line, as they ripen and senesce. When harvested before fully ripe, the cherries darken and become soft during the ripening process. Acids may decrease, resulting in a sweeter taste, but large increases in sugar do not occur because of the absence of large reserves of starch at harvest. In addition to sweet cherries, grapes, citrus, strawberries, and pineapples exhibit this nonclimacteric respiratory pattern (for a complete listing, see chapter 4).

Climacteric respiration is exhibited by most deciduous tree fruit species (e.g., apples, pears, apricots, peaches, nectarines, plums), many tropical and subtropical fruits (e.g., bananas, guavas, avocados, mangos), and some fruit vegetables (e.g., tomatoes) (see fig. 17.1, solid line). During the climacteric rise, fruits soften, and yellow colors intensify through loss of chlorophyll and increase of carotenoid pigments. Anthocyanins (red, blue, and purple colors) may be produced at this time. Production of ethylene increases, as does that of other volatiles, including those associated with fruit aromas. The peak of the respiratory curve approximates the time that fruits are considered ripe for consumption. After the climax, respiration gradually decreases as the fruits senesce. In climacteric fruits, harvesting before the start of the climacteric rise in respiration is the best way to attain maximum postharvest life. Studies have shown that the climacteric rise can be initiated prematurely by exposing fruit to ethylene. If some of the fruit have started to ripen, the ethylene they produce may trigger the respiratory rise among the remaining fruit. Similarly, rotting or badly bruised fruit may release sufficient ethylene to trigger ripening.

It is important to maintain fruits in a vigorous condition by lowering the respiration rate to the minimum that still permits normal cellular function. With climacteric fruits it is essential not only to reduce the respiratory rate but, if possible, to minimize and delay the climacteric rise and associated ripening processes. This is best achieved by low temperature (fig. 17.2) or by a combination of low temperature and modified atmospheres during storage.

FRUIT MATURITY AND SUSCEPTIBILITY

Commonly, a high degree of resistance to decay is maintained until the fruit approaches maturity. Resistance is reduced noticeably as the fruit begins to ripen. Not only does the fruit become susceptible to its most common pathogens when it ripens, but it succumbs to attack by pathogens it was resistant to formerly. For example, as stone fruits approach maturity, they are increasingly susceptible to the main diseases of the crop, including brown rot caused by *Monilinia fructicola* and gray mold caused by *Botrytis cinerea*. As these fruits completely ripen or become senescent, however, other fungi such

Figure 17.2

Effect of temperature on suppression and delay of the climacteric rise in rate of respiration. (Redrawn from data of Fidler and North 1967)

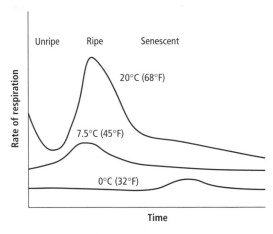

as *Rhizopus* or *Penicillium* species commonly cause postharvest decays. In climacteric fruits, the climacteric rise in respiration roughly coincides with a striking reduction in the fruit's resistance to many pathogens.

Cellular senescence is induced in fruit injuries. Damaged cells have increased respiration, and ethylene production is initiated or stimulated, leading to cellular senescence or death and increased susceptibility to fungal colonization. Wound healing of host injuries is discussed in the next section.

MECHANISMS OF HOST RESISTANCE TO FUNGAL ATTACK

To establish an infection, the pathogen must overcome the natural host defenses of the nonwounded fruit. Essentially, there are preinfectional barriers that prevent or inhibit pathogen penetration and postinfectional responses to disease development (see Adaskaveg 1992). These barriers or responses of host resistance can be structural or biochemical in nature. Preinfectional structural barriers of fruits and other detached plant organs that resist fungal attack include the cuticle and the epidermis, as well as the architecture of stomata (openings in the plant epidermis that allow gas exchange) and the depth and number of trichomes (plant hairs). These structures may resist penetration of the pathogen directly or contribute to modifying the microenvironment on the fruit surface, rendering the fruit either more or less susceptible to infection. Additionally, naturally occurring preformed chemicals (e.g., phenolics, cyanogenic glycosides) or their precursors are present within some fruits that potentially inhibit microbial growth. Often, total polyphenols (i.e., tannins) or specific highly fungitoxic compounds in fruit decrease as fruit mature, ripen, and become susceptible to disease.

As indicated above, postinfectional responses also include structural and biochemical responses. Postinfectional structural barriers in fruits include the development of a resistant corky layer of cells by renewed meristematic activity of certain cells (e.g., wound periderm) and the formation of cell wall appositions (papillae) at the site of fungal penetration. During preharvest development, most fruits have the ability to recover from injuries by wound-healing processes. A fruit injured from biotic or abiotic agents, such as rubbing against other fruit, is usually able to heal the wound by the formation of a barrier of cells that become lignified and suberized. Thus, the fruit effectively regains protection against entrance of microorganisms. In some fruits, individual cells may also form barriers to attempted penetration. For example, in papaya individual epidermal cells respond to infection hyphae from fungal appressoria (swellings on fungal hyphae from which infections are initiated) by producing callose deposits that encase the penetrating hyphae (see Stanghellini and Aragaki 1966).

In postinfectional biochemical responses, fungitoxic compounds not previously present are formed as a consequence of fungal attack. These compounds, called phytoalexins, are mainly polyphenols, isoflavonoids, or terpenoids. The plant's biochemical defense is thus not merely passive but includes the ability to respond with chemical defenses against the invading pathogen. For example, apple cultivars resistant to *Cylindrocarpon mali* (*Nectria galligena*), an important storage disease in Europe, produce two to three times the amount of benzoic acid in response to infection compared to susceptible cultivars (Noble and Drysdale 1983).

Pathogens that depend on wounds in the fruit epicarp to establish an infection may be inhibited if wounds heal prior to potential colonization by fungal pathogens. When fruits are removed from cold storage, wounds often are no longer highly prone to fungal invasion. Some of this resistance is simply due to drying of the wounded area. But injured cells also respond by biochemical mechanisms. Cellular contents are mixed and exposed in the wound area. Enzymes, such as polyphenol oxidases, that are compartmentalized when the cell is alive, are mixed with the polyphenols in the cell sap. Browning in the wound results from enzymatic oxidation of phenolic compounds and lignification. Living cells near the injury become very active metabolically even though they do not show signs of major injury. Repair is set in motion by these stressed cells. Polyphenol synthesis may lead to the accumulation of greater quantities of phenolic compounds than those already present. New compounds, often similar to those that accumulate after infection, may appear in the wound area. These may also be polyphenols. Compounds produced as a result of wounding include some that are highly toxic to fungi. Germinating spores deposited in such "protected" wounds are either inhibited or killed.

PATHOGEN IDENTIFICATION AND BIOLOGY

The accurate identification of the causal pathogen is essential before appropriate treatments can be selected for decay management. Toxicity of chemical and biological treatments may vary between genera or species or even within species of pathogens. Postharvest pathogens are bacteria, yeasts, or more commonly, fungi. Fungi are eukaryotic (have nuclei), heterotrophic (do not produce organic compounds for their growth), single- or multicelled filamentous organisms that are mostly aerobic. Pathogen identification may be difficult for those not familiar with postharvest pathology because fungi growing on stored products or on fruit in the field do not look the same as when grown axenically in culture. Fungal structures that allow easy identification may not be present.

Some fungi can be identified macroscopically, but others must be examined microscopically in pure culture. For example, *Rhizopus stolonifer, R. arrhizus,* and *R. circinans* are very similar morphologically but respond differently to certain fungicides and to temperature (Ogawa et al. 1963b). *Gilbertella persicaria* and *Mucor piriformis* are macroscopically similar and may be found on the same fruit, but they differ microscopically and have different temperature requirements

for mycelial growth (Butler et al. 1960; Smith 1962). *Monilinia* and *Botrytis* species also look similar on decaying fruit but can be easily distinguished in culture. Thus, fungi have been traditionally identified based on their cultural characteristics and microscopic morphology. Bacteria and yeasts must be isolated in pure culture before physiological tests required for identification can be conducted. More recently, molecular methods based on the identification of specific proteins or DNA sequences are being developed (Förster and Adaskaveg 2000; Schots et al. 1994). In the future, these methods will greatly facilitate accurate pathogen diagnosis.

The number of key pathogens and postharvest decays for individual commodities, however, is generally limited; these pathogens and the symptoms they are causing are well described in books and management guides. Thus, their identification often is possible by experienced pest control advisors. Once the pathogen is identified, reasonable approaches for control of key pathogens may be found in management guides. Still, a thorough understanding of their biology and life history is required for determining the appropriate management strategies. Thus, type of infecting propagule, time and location of infection, potential for secondary spread through sporulation or nesting, and presence of survival structures are important characteristics that must be considered. Species of *Colletotrichum, Botrytis,* and *Monilinia* are known to cause quiescent infections. These are visible or nonvisible infections that are established when environmental or host physiological conditions are conducive for penetration but not for active growth or visible decay by the pathogen (Adaskaveg et al. 2000). Thus, for some crops, produce is already infected before or at harvest. Under conducive conditions (e.g. with fruit ripening or with favorable temperatures) these infections may be activated and cause rapid fruit decay. Thus, postharvest losses can be severe even when environmental conditions at harvest time are unfavorable for the infection of ripe fruit.

Postharvest decay originating from quiescent infections emphasizes the importance of an integrated disease management program, as indicated in the introductory paragraph of this chapter. With the pathogen inside the produce, contact treatments will not be successful and other treatments will be required. A description of decay-causing pathogens for different commodities is presented in chapter 18.

ENVIRONMENTAL EFFECTS ON POSTHARVEST FUNGAL PATHOGENS

Fungal growth usually begins with spore germination. In a wet, nutrient-providing environment, spores of many species swell, germ tubes develop after a few hours, and hyphal elongation proceeds. The time it takes to germinate and develop a tiny colony is called the lag phase (fig. 17.3). Growth soon achieves a rapid steady state called the log phase, which continues until growth is slowed, usually from nutrient depletion, at which point the stationary phase of the curve is entered. When growth is plotted against time the shape of the growth curve will be sigmoid, or S-shaped (see fig. 17.3). On fruit, the lag and log phases are usually longer or delayed compared to growth on culture medium because the spore must not only germinate but also must overcome host resistance mechanisms. Depending on the fungal species, the lag phase may last from a few hours to several days at optimal temperatures, to weeks or months at temperatures near the minimum for fungal growth.

Temperature is one of the cardinal factors affecting growth of fungi. Most fungi are mesophiles that grow optimally within a temperature range of 15° to 40°C (59° to 104°F). Some fungi (thermophiles) respond optimally at higher temperatures. The maximum temperatures for growth are about 35° to 50°C (95° to 122°F), but some species can grow at higher temperatures. A few fungi are psychrophilic, with optimums for growth of 0° to 17°C (32° to 62°F) but most are cold-tolerant and can survive exposure to cold temperatures with little growth. Most postharvest pathogens generally grow best at 20° to 25°C (68° to 77°F). Effects of temperature on growth or decay can be illustrated by plotting growth or lesion diameter over a temperature range. The generalized effect of temperatures on growth of *M. fructicola* as measured by lesion diameter of decay is shown in figure 17.4.

Based on the temperature minimum for growth, postharvest decay fungi can be divided into those that have a temperature minimum for growth of about 0°C (32°F) or above and those that can grow at lower temperatures. Non-chilling-sensitive horticultural crops can generally best be stored at the lowest temperature safe from freezing. At −1° to 0°C (30° to 32°F), only a few fungi can be expected to pose difficulties. By far the most notable of these is *B. cinerea*, particularly if the storage period extends for more than 3 to 4 weeks. *Penicillium expansum*, the cause of blue mold disease of fruits of deciduous trees, may also be of concern. Other fungi causing significant rot at 0°C (32°F) include *Alternaria alternata* and *Cladosporium herbarum*. *Monilinia fructicola* grows so slowly at 0°C (32°F) that visible brown rot of stone fruits can be seen only after excessive storage periods. Thus, fungi with a temperature minimum for growth of −5° to −2°C (23° to 28°F) cannot be inhibited by refrigeration without freezing the fruit. Nevertheless, low temperatures are crucial to the suppression of these fungi, because although they are active, their growth rate is only a fraction of that found at higher temperatures. Figure 17.5 shows the extent of rot development in peaches after inoculation with spores of *M. fructicola* and storage at selected temperatures.

Thermal death points of fungi are different for different species of fungi, as well as for different growth stages such as mycelium, spores, or survival structures. Most fungi are sensitive to high temperatures. Thermal death points for spores of many fungi range from 40° to 60°C (104° to 140°F) for 10-minute

Figure 17.3

Sigmoid curve of fungal growth.

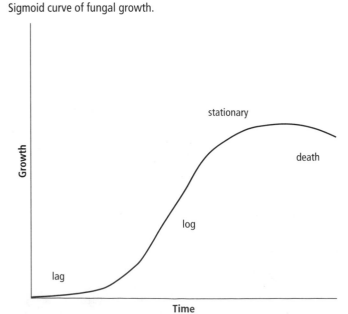

Figure 17.4

Effects of temperature on growth of *Monilinia fructicola* in peach fruits. (Redrawn from Brooks and Cooley 1928)

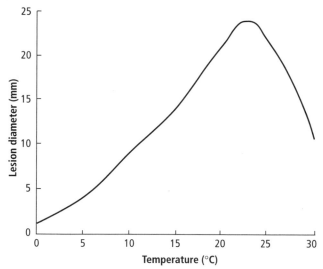

Figure 17.5

Development of brown rot caused by *Monilinia fructicola* in peach fruits at constant temperature. (Redrawn from Brooks and Cooley 1928)

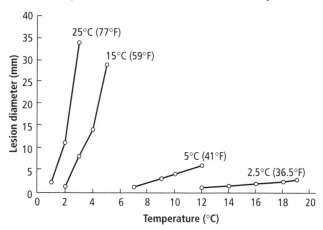

Figure 17.6

Survival of *Rhizopus stolonifer* spores after exposure to 0°C (32°F) after various periods of incubation at 25°C (77°F). (Redrawn from Matsumoto and Sommer 1967)

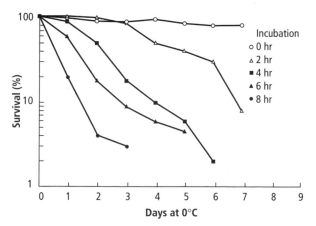

exposures. The effects of extreme temperatures, however, depend on moisture content, metabolic activity, and age of the fungal propagule or somatic tissue. Some postharvest pathogens that can only grow above 5°C (41°F), such as *Rhizopus stolonifer* and *Aspergillus niger*, have developmental stages that are sensitive to lower temperatures. Although nongerminated spores of *R. stolonifer* may not be adversely affected by low temperatures, most spores that have started to germinate are killed within several days at 0°C (32°F) (fig. 17.6). This cold sensitivity, along with wound healing, is believed to be responsible for the general absence of Rhizopus rot after peaches are removed from cold storage.

Most fungi are obligate aerobes that can tolerate low oxygen concentrations. O_2 and CO_2 are required for normal respiration and growth. Although fungi are suppressed by elevated (10–20%) CO_2 levels, many fungi grow poorly in its complete absence because the gas is required in a number of physiological pathways. Oxygen requirements of fungi can differ between species, as well as between growth stages of a single species. For example, mycelial growth and spore germination increase with increasing O_2 concentrations from 0 to 4% or 0 to 1%, respectively, for *B. cinerea* and *R. stolonifer* (Wells and Uota 1970). Thus, mycelial growth generally requires higher levels (4×) of oxygen than spore germination. Suppression of fungi by a 2% O_2 atmosphere is modest, often no more than about 15% below the rate of growth in air (21% O_2), as shown in fig. 17.7 for *B. cinerea* and *M. fructicola* (Sommer et al. 1981). Significant growth reductions result if the O_2 level is lowered to 1%, but this is generally considered too low to be tolerated by most fresh commodities.

MANAGEMENT OF POSTHARVEST DECAYS BY PHYSICAL METHODS AND ENVIRONMENTAL MODIFICATIONS

CROP OR COMMODITY HANDLING

Basic cultural practices such as harvest date (crop maturity) and methods of harvesting have been established for most commodities. A generalized flow chart of postharvest handling and treatment of fruit commodities is

Figure 17.7

Suppression of stone fruit postharvest pathogens by oxygen-modified atmospheres. (Redrawn from Sommer et al. 1981)

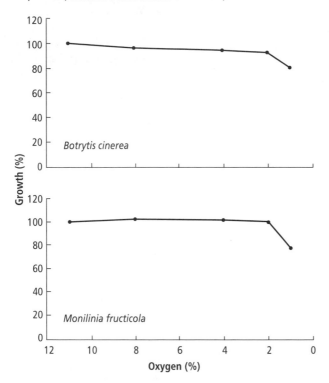

shown in figure 17.8. The most important goals of all harvest and postharvest handling practices are the prevention of injuries and the delay of crop senescence. Most postharvest pathogens enter fruits only through wounds. Thus, handling practices can directly affect the potential for decay to develop by allowing injuries to occur. Specifically, mechanically harvested crops provide more injured sites for infection than hand-harvested crops (Ogawa et al. 1963a). Decay control with chemicals is difficult and often impossible if crops are already infected at harvest, unless a systemic fungicide is used. In addition, all efforts should be taken to avoid bruising and wounding during transport to the packinghouse and subsequent handling. Because bulk bins often increase the potential for bruise damage, smaller containers are often used for bruise-sensitive crops. During the sorting process, it is important to remove all injured and decayed fruit that might develop into foci for secondary infections. Proper packaging to prevent bruise damage is especially important for fruit destined for distant markets (Ogawa et al. 1972). Consequently, handling procedures that minimize injuries enhance the effectiveness of other

postharvest treatments.

To delay senescence, it is important to cool the commodity as quickly as possible to the lowest temperature that does not cause injury. Low temperatures delay senescence in climacteric and nonclimacteric commodities by slowing host metabolism (this is discussed in detail in chapter 13). Modified atmospheres are also important for slowing respiration and other physiological processes that delay senescence (this is discussed briefly elsewhere in this chapter or in more detail in chapter 14). Additionally, treatments with growth regulators or other chemicals that interfere with the production of ethylene can slow senescence of host tissue and reduce postharvest decays (see the section "Plant Growth Regulators" in this chapter).

MODIFICATIONS OF THE PHYSICAL ENVIRONMENT

The environment in which the crop is held after harvest affects both the crop and the pathogen. Usually, quick removal of field heat by hydrocooling or forced-air cooling reduces the rate of ripening, as well as the growth of microorganisms (Mitchell et al. 1972; Thompson 1992). Temperature reductions that do not injure the crop are critical in the management of most pathogens, including those that can grow at low temperatures, such as *M. piriformis* and *B. cinerea*, because the rate of growth of these fungi is greatly decreased. When chilling or freezing injuries occur in cold storage, however, fruits may be predisposed to infection by microorganisms. Other methods for modification of the physical environment include heat treatments. Usually these are short-term treatments to inactivate the pathogen or pest without damaging the commodity. Modification of the environment by chemical or biological treatments that are inhibitory or antagonistic to the decay-causing organism is also an important strategy for postharvest decay management. Various aspects of this strategy are discussed in later sections.

Cold temperature storage

Temperature management is so critical to postharvest disease control that all other control methods have been described as supplements. Without minimizing the importance of other control measures, it can be said that temperature management is central

Figure 17.8

Generalized flow chart of postharvest handling and treatments of temperate fruit crops.

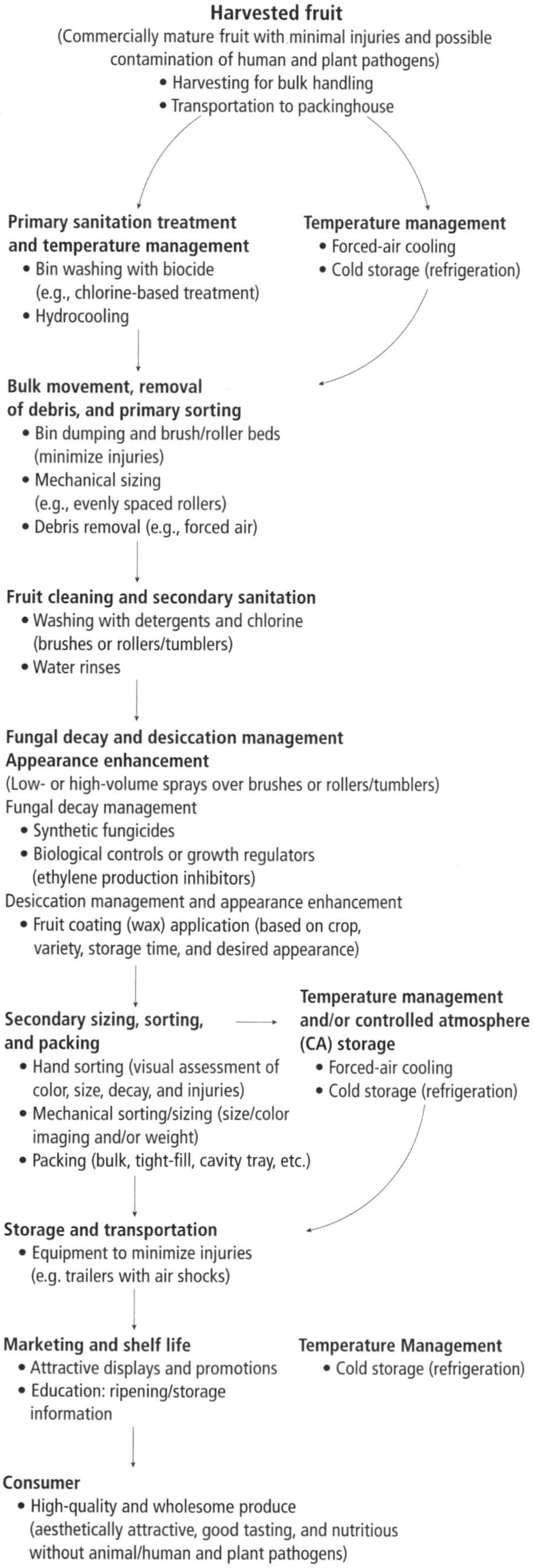

Harvested fruit
(Commercially mature fruit with minimal injuries and possible contamination of human and plant pathogens)
• Harvesting for bulk handling
• Transportation to packinghouse

Primary sanitation treatment and temperature management
• Bin washing with biocide (e.g., chlorine-based treatment)
• Hydrocooling

Temperature management
• Forced-air cooling
• Cold storage (refrigeration)

Bulk movement, removal of debris, and primary sorting
• Bin dumping and brush/roller beds (minimize injuries)
• Mechanical sizing (e.g., evenly spaced rollers)
• Debris removal (e.g., forced air)

Fruit cleaning and secondary sanitation
• Washing with detergents and chlorine (brushes or rollers/tumblers)
• Water rinses

Fungal decay and desiccation management
Appearance enhancement
(Low- or high-volume sprays over brushes or rollers/tumblers)
Fungal decay management
• Synthetic fungicides
• Biological controls or growth regulators (ethylene production inhibitors)
Desiccation management and appearance enhancement
• Fruit coating (wax) application (based on crop, variety, storage time, and desired appearance)

Secondary sizing, sorting, and packing
• Hand sorting (visual assessment of color, size, decay, and injuries)
• Mechanical sorting/sizing (size/color imaging and/or weight)
• Packing (bulk, tight-fill, cavity tray, etc.)

Temperature management and/or controlled atmosphere (CA) storage
• Forced-air cooling
• Cold storage (refrigeration)

Storage and transportation
• Equipment to minimize injuries (e.g. trailers with air shocks)

Marketing and shelf life
• Attractive displays and promotions
• Education: ripening/storage information

Temperature Management
• Cold storage (refrigeration)

Consumer
• High-quality and wholesome produce (aesthetically attractive, good tasting, and nutritious without animal/human and plant pathogens)

to all modern postharvest handling systems. Low temperatures slow fungal development and maximize the potential postharvest life of the commodity. The ideal objectives of refrigeration for disease control are to lower the temperature below the minimum temperature for growth of the fungus and to a point at which infection development will not be completed before the fruit is consumed, or, in the case of cold-sensitive fungi, to kill the spores while they are germinating. These ideals are often unattainable because the pathogens often tolerate lower temperatures better than their hosts. Pathogen development is merely delayed by the low temperatures that are best for maintaining the fruit in good physiological condition.

To obtain the full advantage of refrigeration, it is essential to handle fruits without delay. Field heat should be removed as soon as possible, and the fruit should be cooled to the lowest temperature tolerated by the commodity. The importance of immediate cold temperature storage for fruit is illustrated in figure 17.9. Data show the amount of brown rot that developed in peaches following delays in fruit cooling. After inoculation with spores of *M. fructicola*, the fruit were either placed immediately at 0°C (32°F), or were first stored at 20°C (68°F) for 24 or 36 hours and then placed at 0°C. After removal from cold storage and 3 days of incubation at 15°C (59°F), visible disease lesions had not developed on fruit that was immediately placed at 0°C (32°F), while decay was easily recognized in the other fruit. Data taken after 6 days show that the effects of delayed cooling extended into the normal marketing period.

A high percentage of spore-contaminated wounds may not develop into lesions if cooling is sufficiently prompt. Spore germination is extremely slow and may fail near the minimum temperature for growth of the fungus. Processes involved in establishing the infection are also only marginally functional near growth-stopping temperatures. Low temperatures, while the fungus is still in its early lag phase of growth, may consequently result in fewer fungal lesions and delay their development. Even if subsequent transport is at about 5°C (41°F), there are advantages to first cooling to 0°C because suppression of fungi is more likely to be nearly complete. Decay lesions caused by cold-sensitive fungi may be permanently halted. Furthermore, to

Figure 17.9

Cooling delays and subsequent decay development. Peaches were stored at 0°C (32°F) immediately or after delays of 24 or 36 hours. Data indicate decay development at 3 days (A) and 6 days (B) after removal from cold storage.

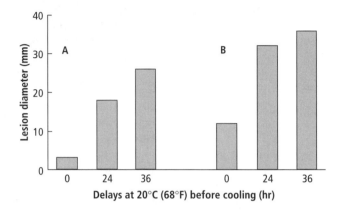

cool to the lowest safe temperature is advantageous for maximizing the life of the fruit.

Heat and radiation treatments

In the past few years, there has been an increasing interest in postharvest heat treatments for management of fruit decays (Barkai-Golan and Phillips 1991; Lurie 1998). This interest is in an effort to find alternatives to postharvest chemical treatments or to enhance the effectiveness of postharvest fungicide treatments. Most commodities tolerate exposure to water temperatures of 50° to 60°C (122° to 140°F) for longer periods than most fungi. Still, hot water treatments have serious limitations and can potentially damage or discolor fruit, shorten storage or shelf life, and increase susceptibility to pathogens. Additionally, the treatment does not provide any residual activity against recontamination of the commodity, and it retards subsequent cooling of the fruit.

Hot water treatments may be noneconomical considering the emphasis and effort that are placed on removal of heat from commodities in cold storage. Hot water dips can be used alone or in combination with chemicals to reduce anthracnose of mangoes (Coates et al. 1993) and decays of stone fruit crops (Smith 1962; Wells and Harvey 1970). For mangoes, treatments of 51° to 55°C (124° to 131°F) for 5 to 30 minutes have been recommended. Additionally, gamma irradiation treatments have excellent penetration properties and kill pathogens by disrupting and energizing chemicals to form reactive molecules that interfere with biochemical processes of the pathogen. However, treatment of fresh fruits and vegetables with gamma radiation often has an undesirable effect on crop texture (Maxie et al. 1971).

Modified or controlled atmospheres

Atmospheres around fruit during storage may be changed by adjusting the concentrations of oxygen, carbon dioxide, carbon monoxide, ethylene, and other gases such as ozone. Additionally, hypobaric atmospheres have also been used to extend storage life and consequently to suppress decay. If a close control of these gases is maintained, the synthetic atmosphere is commonly called a controlled atmosphere (CA). Modified atmosphere (MA) is a term that may designate any synthetic atmosphere, but it often is used if there is little or no possibility of making adjustments in gas composition during storage or transportation. The purpose of these atmospheres is usually to extend the fruit's postharvest life by suppressing the rate of respiration. Another objective is to suppress diseases.

The effects of MA on postharvest diseases can be either indirect or direct. For example, ethylene has an indirect effect on decay by affecting the ripening of the commodity. Ethylene concentrations can be reduced in storage atmospheres by using "scrubbers" (ethylene eliminators), ozone treatments, or by slowing fruit metabolism in controlled atmospheres or temperatures. Maintenance of the fruit in good physiological condition may result in a fruit with considerable disease resistance. O_2, CO_2, and CO can have direct effects on postharvest diseases. Thus, because the fungal pathogen requires O_2, as does the fruit, lowering the O_2 or raising the CO_2 or CO concentrations can slow the growth of a fungal pathogen. Only modified atmospheres that have a direct effect but are not lethal to the pathogen will be discussed in the following sections. Fumigation treatments with fungitoxic gases such as sulfur dioxide or acetic acid (Sholberg et al. 1996) are discussed under "Management of Postharvest Decays with Chemicals," below.

Oxygen reduction, carbon dioxide elevation, and hypobaric atmospheres. Oxygen is required for normal respiration of both the fruit and the fungal pathogens. The beneficial effects of low O_2 on fruit become evident as O_2 in the atmosphere is decreased to 5% or below; benefits increase at lower O_2 levels. In CA storages the level of O_2 is commonly maintained at about 2 to 3%. This level is considered the lowest that can be maintained prudently with the equipment usually available in storage. Anaerobic or fermentative respiration is the consequence of an excessively low O_2 level. The fruit first develops off-flavors as substances, particularly alcohols and acetaldehyde, accumulate in the tissues. Eventually tissues are irreparably damaged and fruit death results. As mentioned above, 2% O_2 concentrations are only modestly effective in suppressing fungal growth (see fig. 17.7).

Air commonly contains about 0.03% CO_2. Elevation of CO_2 above about 5% noticeably suppresses fruit respiration. If the concentration of CO_2 is excessive, however, off-flavors develop and fruit injury results. The relationship of CO_2 concentration to fruit injury is related to time and temperature. Fruit tolerate very high levels of CO_2 (more than 20%) for several days at temperatures between 3° and 5°C (38° and 41°F), but few tolerate those elevated concentrations if storage or transportation in the modified atmosphere is extended for several weeks. The addition of 10 to 15% CO_2 at a temperature of 5°C (41°F) commonly affects both host and pathogens in a manner roughly comparable to a temperature of 0°C (32°F) in air. CO_2 added to air is widely used in transport of Bing cherries, primarily to suppress *B. cinerea* (gray mold) and *M. fructicola* (brown rot), and it is used with strawberries to suppress *B. cinerea*.

The effects of low O_2 and high CO_2 atmospheres are considered to be additive. Commonly used atmospheres with about 2 to 4% O_2 and 5 to 7% CO_2 suppress respiration and delay ripening of fruit, which could not safely be achieved with modification of single gases in the atmosphere. Modification of O_2 alone would likely require 1% O_2 or less to achieve similar effects. CO_2 in air might require 15 to 20% or more to equal the combined effect.

Storage and transport under low pressure or hypobaric conditions has stirred considerable interest in the past. At atmosphere pressures of 0.1 and 0.05, available O_2 is reduced from about 21% of air to about 2% and 1% O_2, respectively. An added benefit to the low O_2 concentration is a very effective removal of

ethylene that is produced by commodities. From the standpoint of postharvest decays, very few critical data are available, but a comparison of results with *B. cinerea* at 0.1 and 0.05 atmosphere suggests that the suppressive effect is similar to CA at 1 or 2% O_2. Currently, the technique is not in commercial practice because of the high costs of equipment.

Controlled atmosphere with carbon monoxide. CO functions physiologically as an enzyme inhibitor and a competitor of O_2. CO (10%) added to air results in only a modest reduction in growth of postharvest pathogens. In general, the suppressive effects of CO increase with lowered O_2, increased CO_2, and cool temperatures (e.g., 5.5°C [42°F]). The modest suppression of fungi by low O_2 and elevated CO_2 concentrations commonly used suggests the desirability of including a fungistatic gas such as CO. Although CO was considered suitable in that role, it is currently not used for postharvest storage of fruits because of the high risks associated with the gas and worker safety. Other gases used as fumigants are discussed in the section "Management of Postharvest Decays with Chemicals," below.

Humidity. Water vapor is a gas that constitutes an important part of the atmospheric environment of harvested perishable commodities. Its amount in the atmosphere as a percentage of saturation (relative humidity, or RH) varies widely with temperature changes. Although the RH of storage facilities is generally never at saturation, free water can occur on the surface of commodities. Liquid water forms if at any time during the normal temperature cycling within a refrigerated storage room the temperature of the commodity surface falls below the dewpoint temperature of the atmosphere. With pathogens such as *M. fructicola*, saturated atmospheres or water on the fruit surface favors spore germination and direct penetration of commodities. With jacketed storage or packages with moisture barriers of plastic film, high RH may be a factor in promoting disease if temperatures are favorable.

Cold commodities removed from refrigeration to ambient temperatures condense moisture on their surface from surrounding warm, humid air. This condensation continues until the fruit has warmed to a temperature above the dewpoint temperature. The duration of wetness depends on the fruit-air temperature difference, the exposure time of the commodity to the air, air movement, and the size or bulk of the commodity. Although moisture from condensation at the fruit surface has often been a concern among handlers, it is likely that the warm-up period of many commodities may be too short to be an important factor in development of decay. The horticultural necessity for maintaining high humidity in commodity environments is primarily to minimize loss of moisture, which results in shrivel and loss of turgidity in tissues. With peaches, for example, a loss of 3 to 4% of the original weight usually results in noticeable shrivel.

MANAGEMENT OF POSTHARVEST DECAYS WITH CHEMICALS

Presently, chemicals are essential for protecting food quality and for preventing crop losses from decay (see Ogawa and Manji 1984). Fresh perishables need protection from the time of harvest until the time of consumption or processing. Without chemicals much of our produce would never reach the consumer. The use of chemicals is a method of modifying the environment of the fruit surface either by removing inoculum as a sanitation practice or by inhibiting the growth and

Table 17.1. Chemicals no longer registered for postharvest use

Chemical, with chemical class	Year introduced*	Crop	Decay/organisms	Methods of application
Benomyl (Benlate) Benzimidazole	1967	Bananas	Crown rots, surface molds, *Colletotrichum* spp.	Dip or spray
		Citrus	*Penicillium* spp.; stem-end rot	Dip or spray
		Pears, apples	*Botrytis, Penicillium* spp.	Dip or spray
		Pineapples	*Thielaviopsis* spp.	Dip or spray
		Apricots, cherries, nectarines, peaches plums, prunes	Fruit rot pathogens (*Monilinia, Botrytis,* and *Penicillium* spp.)	Dip or spray
		Mushrooms	*Trichoderma* spp.	Dip or spray
Biphenyl (Diphenyl) Phenol derivative	1944	Citrus	*Penicillium* spp.; stem-end rot	Pads, wraps, or liners
Captan (N-[(trichloromethyl)thio]-4-cyclohexene-1,2-dicarboximide) Pthalimide	1949	Peaches and nectarines	Storage rot pathogens	Dip or spray
		Citrus	Storage rot pathogens; *Botrytis* and *Rhizopus* spp.	Dip or spray
		Grapes (raisins)	Mold control on fruit on drying trays in fields	Dust
		Pineapples, mangoes	Storage rot pathogens	Dip or wash
		Cantaloupes, cucumbers, potatoes	Storage rot pathogens	Dip or spray
		Onions	Storage rot pathogens	Dip or spray
		Potatoes	Storage rot pathogens	Dip
Dehydroacetic acid, Sodium salt (sodium dehydroacetate) (DHAS) Organic acid	1950s	Strawberries	Postharvest decays (*Aspergillus, Penicillium, Botrytis,* and *Rhizopus* spp.)	Dip
		Squash prepared for packaging crate	Postharvest decays	Dip
Sec-butylamine (2-aminobutane) (Tutane) Aliphatic amine	1962	Citrus and equipment	*Penicillium* spp.; stem-end rot (*Diplodia, Phomopsis* spp.)	Dip, drench, or spray

Table 17.1. Cont.

Chemical, with chemical class	Year introduced*	Crop	Decay/organisms	Methods of application
Iprodione (Rovral) Dicarboximide	1976	Apricots, cherries, nectarines, peaches plums, prunes	Fruit rot pathogens (*Monilinia, Botrytis,* and *Penicillium* spp.)	Dip or spray
Methylene chloride	1957	Citrus	Penicillium decays and decay during de-greening	Fumigate
Phenylphenols: O-phenylphenol (OPP), Sodium o-phenylphenate (tetrahydrate) (SOPP) Phenol derivatives	1936	Bananas	Crown rot, stem rot (*Fusarium, Colletotrichum,* and *Thielaviopsis* spp.)	Brush
		Apples	*Penicillium, Botrytis* spp., and other fungi	Spray, dip
		Nectarines, peaches, plums	*Monilinia* and *Botrytis* spp.; other fungi	Spray, dip
		Carrots, cucumbers, bell peppers, tomatoes	Fungi	Spray, dip
Sodium dimethyldithio-carbamate (Nabam) Dithiocarbamate	1943	Melons	Decay organisms	Spray
Thiophanate-methyl Benzimidazole	1969	Peaches, nectarines, plums, cherries, apricots	*Monilinia* spp.	Dip and spray
Triforine (50WP) Imidazole	1969	Apricots, nectarines, and peaches	*Monilinia* spp.	Dip, drench, or spray
		Cherries	*Monilinia* spp.	Dip or hydrocooler
Zinc ion-maneb complex	1943	Capri figs	Endosepsis (*Fusarium* spp. and other fungi)	Dip
Ziram (zinc dimethyldi-thiocarbamate)	1930	Bananas	Crown rot, stem rot: *Fusarium, Colletotrichum,* and *Thielaviopsis* spp.	Ziram paste with SOPP and sulfur

Source: Adapted from Hopkins 1996; Ogawa and Manji 1984; Spencer 1981.

Note:

*Tolerances may still exist within the U.S. EPA for each fungicide; however, registrants currently do not support postharvest use of their materials in the United States. See U.S. Environmental Protection Agency 1999, 2000.

reproduction of plant pathogenic organisms by prevention (e.g., protection), suppression, or therapy of infections. Suppression is the inhibition of development, whereas therapy is the eradication of established infections. Some treatments have more than one mode of activity and thus fit into more than one category, such as a sanitation and protective treatment. In the following paragraphs, chemicals will be discussed in the section that best describes their usage. Chemical treatments of harvested produce must be used properly for greatest effectiveness and safety. The current status of chemicals available for postharvest use and methods used to obtain effective control of postharvest decay

will be discussed. Some of the early postharvest treatments that do not meet modern safety standards are no longer registered, and these are summarized in table 17.1. Existing registrations and new materials planned for postharvest use on food crops are included in table 17.2.

SANITATION

Sanitation practices include treatments to reduce populations of microorganisms on equipment, on the fruit, and in the wash water used to clean the fruit. Water washes alone will remove nutrients from produce surfaces that allow microorganisms to grow, and it also removes inoculum of postharvest

Table 17.2. Chemicals registered as postharvest treatments to prevent crop decays caused by filamentous fungi, bacteria, or yeasts

Chemical/class	Year introduced	Crop	Decay/organisms	Methods of application	Residue tolerance (ppm)*
Azoxystrobin (Abound) Strobilurin	1996	Pears (?) and citrus	*Penicillium* spp.	Spray	Anticipated (3–5)
Calcium hypochlorite Inorganic Halogen	1798	Potatoes	Bacteria	Wash and rinse	Exempt limits: 25 available Cl in solution (without specific crop label)
		Storage bins, packinghouses	Molds and yeasts	Brush, spray, and drench	Exempt
Captan (N-[(trichloromethyl)thio]-4-cyclohexene-1,2_dicarboximide) Pthalimide	1949	Apples, pears	*Botrytis*, *Rhizopus*, and *Colletotrichum* spp.	Dip or spray	25
		Sweet cherries	Storage rot pathogens	Dip or spray	100
2,6-Dichloro-4-nitroaniline (DCNA, Botran, Allisan) Dicarboximide	1959	Apricots, sweet cherries	*Monilinia*, *Botrytis*, and *Rhizopus* spp.	Spray	20
		Peaches and nectarines	*Monilinia*, *Botrytis*, and *Rhizopus* spp.	Dip, spray, and wrap	20
		Plums, prunes	*Monilinia*, *Botrytis*, and *Rhizopus* spp	Spray-brush and dip	15
		Carrots	*Sclerotinia* spp.	Dip	10
		Sweet potatoes	*Rhizopus stolonifer*	Dip or spray	10
Fenhexamid (Elevate) Hydroxyanilide	1998	Apples, pears	*Botrytis cinerea*	Dip or spray	Anticipated (5-15)
		Kiwifruit	*Botrytis cinerea*	Dip or spray	Anticipated (5-15)
		Apricots, nectarines, peaches, plums, sweet cherries	*Monilinia* and *Botrytis* spp.	Dip or spray	Anticipated (5-15)
Fludioxonil (Scholar) Phenylpyrrol	1990	Apples, pears	*Botrytis*, *Penicillium*, and *Rhizopus* spp.	Dip or spray	Anticipated (5)
		Citrus	*Penicillium* spp.	Spray	Anticipated (5)
		Kiwifruit	*Botrytis cinerea*	Dip or spray	Anticipated (5)
		Apricots, nectarines, peaches, plums, sweet cherries	Fruit rot pathogens (*Monilinia*, *Botrytis*, *Rhizopus*, *Mucor*, and *Gilbertella* spp.)	Spray	Emergency tolerance (5)
		Pomegranates	*Botrytis*	Dip	Emergency tolerance (5)
Formaldehyde Organic compound	1888	Equipment (potatoes)	Bacteria, fungi	Wet	0 (not for food use)
		Storage areas, equipment (citrus)	Bacteria, fungi	Fumigate	0 (not for food use)
Imazalil Imidazole	1974	Citrus	*Penicillium* spp.	Spray	10
Ethylene oxide	1996	Copra, seasonings, spices, walnuts	Molds	Fumigation	50
Propylene oxide	1996	Cocoa, gums, spices, nuts (except peanuts)	Molds	Fumigation	300
Ozone Ozone gas	—	Fresh fruit and vegetables	Bacteria and fungi	Fumigation or ozonated water wash	Exempt

Table 17.2. Cont.

Chemical/class	Year introduced	Crop	Decay/organisms	Methods of application	Residue tolerance (ppm)*
Phenylphenols: O-phenylphenol (OPP), Sodium o-phenylphenate (tetrahydrate) (SOPP) Phenol derivative	1936	Pears	Bacteria and fungi	Dips, sprays, and floods	25
		Citrus	*Geotrichum, Penicillium* spp.; stem-end rots (*Trichoderma* and *Phytophthora* spp.)	Wash, sprays, dips, and foams	10
		Cucumbers, peppers	Fungi	Spray	10
Potassium sorbate (Sorbic acid, potassium salt) Organic salt		Prunes (dried)	Fungi	Dip	GRAS
Sodium borate (Borax) (Sodium tetraborate) Inorganic salt	1938	Grapefruit, oranges	*Penicillium* spp.	Dip; rinse with fresh tap water	Exempt
		Lemons	*Penicillium* spp.	Drench or spray	Exempt
Sodium hypochlorite Inorganic (Halogen)	1798	Fresh fruit and vegetables	Bacteria and fungi	Dip; rinse with fresh tap water	GRAS
		Cannery belts	Fungi	Spray; rinse with fresh tap water	GRAS
Sodium or potassium bisulfite† Inorganic	—	Grapes (fresh)	Fungi	Sawdust mixture, pads	10
Sulfur Inorganic	1800 BC	Bananas	Crown rot fungi	Paste	GRAS
Sulfur dioxide (Sulfurous anhydride)† Inorganic	1928	Grapes (fresh)	Fungi (*Botrytis cinerea*)	Fumigation	10
Tebuconazole (Elite), triazole	1986	Sweet cherry	Brown rot, Rhizopus rot	Spray	California (4) (Section 24C)
2-(4-thiazolyl) benzimidazole (Thiabendazole or TBZ - Mertect 340) Benzimidazole	1968	Bananas	Crown rot	Dip after dehandling and delatexing	3 (0.4 in pulp)
		Citrus	*Penicillium* spp. stem-end rot	Drench or spray	10 (35 in pulp)
		Papayas	*Colletotrichum* spp.	Dip or spray	5
		Pome fruit (apples and pears)	*Penicillium* spp. Bull eye rot, *Botrytis cinerea*, cluster rot, and nest rot	Dip, flood, or spray	10
		Mushrooms	*Dactylium, Mycogone, Trichoderma,* and *Verticillium* spp.	Dip or spray	40
		Carrots	*Botrytis* and *Sclerotinia* spp.	Dip	10
		Cantaloupes	*Fusarium* spp.	Dip	15
		Potatoes	*Fusarium* spp.	Dip	10
Thiram Dithiocarbamate	1931	Bananas	Crown rot, stem-end rot, surface molds (*Fusarium, Colletotrichum,* and *Thielaviopsis* spp.)	Spray, brush, and paste	7
1,1,1-trichloroethane (methyl chloroform)	1930s	Citrus	Penicillium decays during degreening	Fumigant	Exempt

Source: Adapted from Hopkins 1996; Ogawa and Manji 1984; Spencer 1981.

Notes:

* Residue tolerances from U.S. Environmental Protection Agency 1999, 2000. "Exempt" is an EPA classification for exemption from tolerance; "GRAS" is an FDA classification meaning "generally regarded as safe."

† Sulphur dioxide and bisulfites were once included in the GRAS list of chemicals for which no registration is required. Because of heavy usage and because some people are allergic to sulfites, a tolerance of 10 ppm sulfite has been established.

pathogens. Without the use of sanitation treatments, however, the potential for reinoculation of produce is high. Sanitizing treatments include treatments that are used for fruit and equipment and treatments that are used only for equipment and storage facilities. Sanitizers that reduce inoculum levels of decay organisms from fruit surfaces include treatments added to water dumps and spray or dip washes. These treatments inactivate spores brought into solution from fruit or soil and prevent the secondary spread of inoculum in water. Examples of sanitizing washes include halogenated compounds (e.g., hypochlorous acid from chlorine gas or sodium hypochlorite and chlorine dioxide) and ozonated water. Sanitation treatments for equipment and storage facilities include quaternary ammonium washes and fumigations with gases toxic to fungal pathogens such as formaldehyde and ethylene oxide.

Sanitation washes

Chlorination. Aqueous solutions obtained from sodium hypochlorite ($NaOCl$), calcium hypochlorite ($Ca[OCl]_2$), or chlorine gas (Cl_2) produce the microbial biocide hypochlorous acid ($HOCl$). They have been used extensively in postharvest handling of fruit. Sanitation using hypochlorous acid is one of the most effective, inexpensive, nonresidual ways to reduce microbial contamination from wash water, noninjured fruit surfaces, and equipment (White 1992). Hypochlorous acid rapidly and nonspecifically oxidizes carbonaceous materials in aqueous solutions, resulting in fungicidal and bactericidal activity. Solutions of hypochlorous acid, however, are relatively ineffective in reducing decay if the inoculum is inside the wounds of the fruit. The compound is rapidly reduced in fruit injuries before it can inactivate the pathogen. Other oxidizing compounds such as chlorine dioxide, ozonated water, or ozone gas have similar advantages and disadvantages in decay management. Three factors control availability and activity of hypochlorous acid: pH, temperature, and the presence of contaminating organic and inorganic materials. As sodium (or calcium) hypochlorite is added to water, hypochlorous acid and sodium hydroxide are produced:

$$NaOCl + H_2O <> NaOH + HOCl$$

The sodium hydroxide dissociates and OH^- is neutralized with HCO_3^- that naturally occurs in water. Based on pH, hypochlorous acid is in equilibrium with its ions, as shown by the equation

$$HOCl <> H^+ + OCl^-$$

Thus, the pH of the solution determines the proportion of "active" chlorine, as opposed to the "inactive" hypochlorite ion (OCl^-). $HOCl$ and OCl^- together represent the amount of "free" chlorine. Hypochlorous acid, or active chlorine, is an effective disinfectant compound, whereas OCl^- is a poor disinfectant. The concentration of active chlorine determines the oxidizing potential of the solution and its disinfecting power. At higher pH, the amount of active chlorine is dramatically reduced. A low pH (e.g., pH 3–6), however, will result in the volatilization of chlorine as chloramines. At a very low pH (e.g., pH 2), chlorine gas (Cl_2) will be formed that is highly toxic or lethal. Ideally, pH levels between 6.5 and 7.5 should be maintained, because in this range equilibrium exists between hypochlorous acid and the hypochlorite ion in solution. Under packinghouse conditions, a pH range of 7 to 9 is commonly maintained and acceptable considering that longer exposure times or higher concentrations of free chlorine are required than under ideal conditions. Additionally, a higher pH reduces the formation of volatile chloramines, which are eye irritants to packinghouse workers.

Temperature also affects hypochlorous acid concentration and contact times: shorter exposures are required with warmer temperatures but greater volatilization of hypochlorous acid occurs. Commodities may be washed at near 0°C (32°F) (hydrocoolers used in fresh market commodities), 5°C (41°F) (dump tanks), or at ambient temperatures. The effect of pH and temperature on the amount of available chlorine is shown in table 17.3. Furthermore, organic and inorganic materials suspended in water, including flotation salts that are used in the pome fruit industry, may interfere with the oxidation of microbial inoculum. At high levels of these materials, longer contact times or higher concentrations of hypochlorous acid are required for disinfestation. Nitrogenous compounds (e.g., amines, ammonia, and amino acids) in the wash water decrease the amount of active chlorine and may result in the formation of undesirable "combined" chlorine. The amount of chlorine in wash tanks can be described as:

total chlorine = free chlorine (active chlorine + inactive chlorine) + combined chlorine

Combined chlorine includes the formation of chloramines (RNHCl):

$$HOCl + RNH_2 <> RNHCl + H_2O$$

Microbes are not very sensitive to combined chlorine; however they react instantly or within seconds to active chlorine in clean water. High temperatures (>30°C, or 86°F) and organic matter in the chlorinated wash water result in the greatest loss of active chlorine. In recirculating sanitation washes, such as stone fruit hydrocoolers and tomato dump tanks, organic load and combined chlorine are important factors in determining active chlorine concentrations. When chlorination is done using a nonrecirculating wash (e.g., spray bar) on prewashed fruit, the formation of combined chlorine is negligible.

Additional factors affecting the activity of chlorine solutions include the type of microorganism and contact time. In general, higher concentrations of active chlorine result in inactivation of microbial populations with shorter contact times. Microorganisms that form thick-walled survival structures or spores, however, may require longer contact times for disinfestation. Although minute

concentrations of 1 ppm HOCl (1 ppm of active chlorine) are required to kill decay-causing organisms in clean water in the laboratory, higher concentrations of 25 ppm HOCl from generic NaOCl or 50 to 200 ppm from labeled NaOCl are commonly used for most commodities to offset changes in the concentration of active chlorine resulting from the amount of crop treated and organic matter accumulated in the wash water during treatment. Specific crop labels determine the concentration registered for use on a given commodity.

Because the amount of a commodity treated and other factors (e.g., organic matter, pH, temperature) influence the amount of active chlorine in wash waters, the concentration of hypochlorous acid needs to be monitored periodically. Colorimetric test kits are commonly used and are the most accurate method for routine evaluations. Oxidation-reduction potential (ORP or REDOX) is measured by electrical conductivity across a pair of electrodes and is an automatic method commonly used to determine chlorine concentration. Postharvest use of ORP, however, can result in inaccurate measurements due to other salts dissolved in the water from soil particles (inorganic and organic) washed off of the commodity. Thus, ORP is often used as a guide to detect sudden changes in conductivity that may influence active chlorine concentration.

Chlorine dioxide is another biocide that has been evaluated as a postharvest sanitation treatment (Spotts and Peters 1980; Roberts and Reymond 1994). This compound is as effective or more effective than free chlorine. Because it does not mix with water, it does not react with ammonia to form chloramines or with organic compounds to form trihalomethanes (THM) such as chloroform (White 1992). Furthermore, chlorine dioxide is relatively unaffected by pH in a range of 6 to 10. Unfortunately, there are several disadvantages to chlorine dioxide (White 1992), including:

- Because the compound needs to be generated on-site, it is generally more expensive than chlorine.

- Some generators produce free chlorine (in addition to chlorine dioxide) that may also react with organic material to form THM.

- Although chlorine dioxide does not produce THM, the compound produces its

Table 17.3. Effect of pH and temperature on concentration of active chlorine (HOCl)

| pH | Percent active chlorine | | |
	0°C (32°F)	20°C (68°F)	30°C (86°F)
4	—	—	100
5	99.8	99.7	99.6
6	98.5	97.7	96.9
7	87.0	79.3	75.9
8	40.2	27.3	23.9
9	6.3	3.7	3.1
10	0.7	0.4	0.3

Source: White 1992.

own set of breakdown products (e.g., chlorite, chlorate) that may pose a direct threat to human health.

- Simple assays for routine evaluations of concentration are not available.

- Chlorine dioxide is toxic to humans, and it commonly forms noxious odors.

Thus, chlorine dioxide must be used in closed systems or in well-ventilated areas away from packinghouse workers. Because of these concerns, recommended rates are less than 1 ppm; however, rates of 3 to 5 ppm are required for optimal biocidal activity at less than 1-minute exposure times (Roberts and Reymond 1994). Probably the best usage of chlorine dioxide is in foams for washing and disinfecting equipment.

Other sanitizing washes. Ozone is another disinfecting agent recently utilized in postharvest water sanitation systems. This compound is one of the strongest oxidizing agents commonly available. The compound is unstable at ambient temperatures and pressures, with a half-life of about 15 minutes, and it decomposes to O_2 at temperatures greater than 35°C (95°F). Similar to chlorine dioxide, ozone must be generated on-site. Spotts and Cervantes (1992) calculated LD_{95} values (lethal dose to kill 95% of spores) for a 5-minute exposure of spores of *Mucor piriformis* or conidia of *Botrytis cinerea* as 0.69 and 0.99 ppm, respectively. Ozone was ineffective for disinfesting wounds of treated commodities (similar to chlorine and chlorine dioxide). Ozone is generally not affected by pH within a range of 6 to 8, but its decomposition increases with high pH, especially above pH 8. Disinfestation, however, may still occur at a high pH because the biocidal activity of the compound is relatively rapid (White 1992). Wash water must also be thoroughly mixed and filtered for optimal ozone disinfestation. Ozone does not directly form THM, although it may form them indirectly if halogens are present in the wash water. Furthermore, ozone does form a number of nonhalogenated byproducts. The most serious drawbacks of ozone use include:

- Ozone is lethal to humans with continuous exposure at high concentrations (>4 ppm); thus, detectors are required for automatic shutdown of the ozone generator.

- Ozone is highly corrosive to common materials and often requires stainless steel containers.

- Ozonated water must be filtered to remove organic and other particulate materials for effective disinfestation.

Cationic detergents such as quaternary ammonium compounds, isopropyl alcohol, live steam, or hot water are currently used for equipment disinfestation. Quaternary ammonium compounds are not effective in reducing decays of fruit, but they are widely used in food-processing plants because they are microbial biocides with high water solubility and detergent properties. They also have low mammalian toxicity and are generally noncorrosive at recommended concentrations. The efficacy of live steam depends on obtaining proper temperatures; hot water treatments that are not under pressure may hold little lethal heat energy during short exposure times. Extended exposures to heat treatments may injure the commodity. Isopropyl alcohol is less commonly used.

Sanitation by fumigation

Other inoculum-reduction treatments include chemical fumigation to in enclosed areas. Sulfur dioxide (SO_2) has been used since antiquity in food preservation, especially for dried fruit. Its use on crops, however, depends on potential phytotoxicity and crop tolerance to the treatment. In crops like grapes or raspberries, some cultivars are tolerant to SO_2 fumigation. The treatment has been a standard practice during storage or for fumigating transportation containers in California since 1928 (Jacob 1929). Because the treatment is fungitoxic to spores and aerial mycelium of fungi such as *B. cinerea*, it can inactivate inoculum on fruit surfaces and prevent nesting or decay of healthy fruit adjacent to decayed fruit (Nelson 1958). The treatment does not, however, suppress decay in fruit that was infected prior to treatment. Sulfur dioxide is more effective at high RH. The toxicant sulphurous acid is readily formed in water from gaseous SO_2 and irreversibly binds to cell proteins. Typically, a 0.1 to 0.5% (by volume) application of SO_2 for 20 to 30 minutes is made as soon as possible after harvest followed by a fumigation of 0.25% for 30 to 60 minutes every 7 to 10 days in storage. Generally, lower concentrations (0.05 to 0.1%) of SO_2 applied in more frequent intervals (<7 days) gave better control of *Botrytis* infections

(Nelson and Baker 1963; Smilanick and Henson 1992). Depending on the variety of grape, injury symptoms will begin to show when the berries have absorbed 20 to 55 ppm SO_2. Because of the high polarity of the toxicant, it is readily bound to moist surfaces. Thus, care should be taken in selecting appropriate packaging that retards moisture and accumulation of SO_2. For export marketing when grapes are in transit for extended periods, SO_2-generating pads are sometimes used. South Africa and countries in South America commonly use pads with plastic box liners when exporting to other countries. These pads contain sodium metabisulfite and allow slow release of SO_2 during transit and marketing.

Other fumigation treatments of enclosed areas include acetic acid and oxidants such as gaseous ozone, formaldehyde, ethylene oxide, and propylene oxide. The latter two fumigants have been used in the past on dried fruit (Whelton et al. 1946) and have been recently re-registered for selected dried food crops such as nuts and spices (see table 17.2). Ozone will retard the growth of fungi on the fruit surface, reduce sporulation on decayed fruit (thereby reducing inoculum in the atmosphere), and destroy offensive odors (Schomer 1948; Spalding 1966; Palou et al. in press). In California, a special local need registration for formaldehyde allows fumigation treatments to kill spores of *Penicillium* and *Geotrichum* spp. on equipment in storage, during de-greening, and in precooling facilities in citrus packinghouses, with a limit of two applications per year.

Alternatives to sulfur dioxide are needed because of concern with sulfide residues and potential phytotoxicity. Acetic acid vapors have recently been used experimentally as a fungistatic gas for the control of postharvest pathogens of apples and table grapes (Sholberg et al. 1996), as well as stone fruit. In tests with table grapes, acetic acid completely inhibited germination of *Botrytis cinerea* spores that were dried on the fruit surface, whereas decay control was similar to SO_2.

PREVENTION, SUPPRESSION, AND ERADICATION

Preventive (protective) chemical treatments are preinfection treatments that must be applied before the fruit is infected by the pathogen. These chemicals prevent the germination of fungal spores or inhibit mycelial growth. Generally, they are only effective if quiescent infections are absent, the inoculum levels of decay organisms are low, and the fruit do not have excessive mechanical injuries or insect damage. Most of the fungicides listed in tables 17.1 and 17.2 are protectants. Some of the fungicides listed also have a suppressive or even eradicant (therapeutic) action that sometimes depends on the method of fungicide application. Treatments with suppressive or eradicant action are postinfection treatments; suppressive action inhibits fungal growth as long as the chemical is present in sufficient amounts, and eradicant action irreversibly stops fungal development. The few therapeutic treatments available are either chemical or physical. Therapeutic treatments are most applicable to nonperishable crops such as grain and dried fruits. Acetic acid–propionic acid wash treatments kill organisms established in kernels and seeds of grain crops.

Because fungicides may have more than one mode of activity, preventive and suppressive treatments are discussed together in this section. Active compounds used for these treatments include simple inorganic and organic compounds, as well as materials with more complex organic structures. Fungicides are grouped into classes based on their chemical structures (see Uesugi 1998). Compounds within each class have a similar mode of action that targets either a single site or multiple sites in the biochemical pathways of the fungus. For postharvest fungicides, the carbonates, phenols, dicarboximides, phthalimids, benzimidazoles, and piperazines are the most important "older" classes; new developments include the hydroxyanilids, phenylpyrroles, and strobilurins. Although most active as postharvest treatments, some of these fungicides and others are also quite active against postharvest decays when they are applied before crop harvest. Some of these preharvest treatments, such as cyprodonil and tebuconazole maintain their activity even after fruits are passed through the postharvest washing process (Adaskaveg and Förster 2000). Therefore, preharvest treatments have become increasingly important for the export market when fruits are shipped to countries where postharvest treatments are restricted. Although antibiotics for bacterial disease control are used preharvest, they are not currently registered for postharvest use.

PREVIOUSLY AND CURRENTLY REGISTERED FUNGICIDES FOR POSTHARVEST USE

Inorganic compounds such as elemental sulfur, sulfur dioxide, and potassium sorbate have been used historically with some success for managing postharvest decays. Elemental sulfur has been used commercially with moderate success to protect peach fruit from brown rot infections. Because of potential phytotoxicity, inactivity against Rhizopus rot, and improved fungicide treatments, sulfur is rarely currently used. Potassium sorbate, a common postharvest fungicide for dried fruit or processed foods, has a broad spectrum of activity. The compound does not prevent growth unless high concentrations are used and sufficient chemical is absorbed on the processed fruit to reduce initial colonization by decay fungi such as *Rhizopus*, *Penicillium*, and *Aspergillus* species.

Some of the first treatments used to control postharvest decays of citrus fruits were alkaline solutions of borax, sodium carbonate, and sodium bicarbonate. The effectiveness of these treatments is due to accumulation of alkali in potential infection sites on the surface of the citrus fruit (Eckert and Sommer 1967). In laboratory studies, recommended concentrations of borax were lethal to conidia of *Penicillium* species after a 5-minute exposure at 43.5°C (110°F), but only weakly fungicidal at 38°C (100°F). Treatments with 6 to 8% borax that are either heated to 43.5°C (110°F) or not rinsed after treatment are effective for control Penicillium decays and stem-end rots caused by *Diplodia* and *Phomopsis* species (Eckert and Sommer 1967). Because visible residues of borax on fruit are not acceptable, commercial treatments are always rinsed with water. Borax is relatively insoluble in water, and this leads to application problems. Thus, commercial treatments usually use 4% borax and 2% boric acid at 43.5°C (110°F). Heated solutions of sodium carbonate or sodium bicarbonate are slightly more toxic; however, these treatments are less toxic to spores of *Penicillium* species as compared to borax. Still, these treatments have been shown to be effective under commercial conditions (Smilanick et al. 1997, 1999).

Sodium ortho-phenylphenate (SOPP) also has residual activity in preventing fruit decay.

After treatment, residues accumulate in potential infection sites on the fruit and prevent the development of decay from subsequent inoculation of pathogens. New wounds, however, are not protected. SOPP is quite soluble in water. At pH 10.3, the phenate ion and o-phenylphenol are in equilibrium. The fungitoxic and phytotoxic properties of the solution depend on the concentration of o-phenylphenol. Because the hydrophilic o-phenylphenate enters the fruit only at sites of injury and precipitates as o-phenylphenol, the relative concentration of the phenate ion and the undissociated o-phenylphenol in solution is not critical (Eckert and Sommer 1967). When used at high concentrations, SOPP treatments are followed by a potable water wash to prevent phytotoxicity. In citrus, a 0.5% solution at pH 11.5 to 11.8 and 43°C (109°F) is commonly used as a dip or spray treatment for management of sour rot caused by *Geotrichum citri-aurantii* and the green and blue molds, *P. digitatum* and *P. italicum*, respectively.

Biphenyl impregnated in fruit or sheet wraps has been extensively used as a citrus fruit fumigant. The treatment is unique in that it sublimates in the packed container, resulting in a fumigation treatment during shipment and storage. Biphenyl reduces the incidence of decay and prevents sporulation of *Penicillium* species on the surface of decaying fruits (Tomkins 1936). Although not currently registered, the compound was a major factor in development of world trade in citrus fruits. In the 1950s, the pthalimide captan was also widely used as pre- and postharvest protectant for stone fruit and other crops. The fungicide is still registered, but it is now rarely used as a postharvest treatment because it must be applied as a wettable powder suspension at a high rate that often leaves unacceptable visible residues on the fruit.

The benzimidazoles (benomyl, thiophanate-methyl, thiabendazole, carbendazim) were another important group of compounds used for pre- and postharvest treatments in the past (Ogawa et al. 1968; Wells and Gerdts 1971). The fungicide has a broad spectrum of activity against fungi, including *Monilinia*, *Botrytis*, *Penicillium*, *Ceratocystis*, and *Gloeosporium* species, with both protective and suppressive action. When introduced, the benzimidazoles were revolutionary compared to the previously registered protective compounds, requiring lower application rates and having a greater activity and a suppressive action. Currently, only thiabendazole is still registered for postharvest use on a variety of crops. The demethylation-inhibiting (DMI) piperazine fungicide triforine was widely used on stone fruit in the past for control of *Monilinia* spp. and is active against benzimidazole-resistant populations of the pathogens. It is not effective, however, against other decay fungi such *Botrytis cinerea* or *Rhizopus stolonifer*.

Other fungicides extensively used for postharvest decay control in the past were the dicarboximides dichloran and iprodione. Dichloran (DCNA) is still registered and is most active against *Rhizopus stolonifer*; it is ineffective against other *Rhizopus* species (Ogawa et al. 1961; Ogawa et al. 1963b; Weber and Ogawa 1965). Control of gray mold, brown rot, and Penicillium decays is unsatisfactory.

Because a complex of pathogens usually causes postharvest decays, mixtures of fungicides are commonly used in management strategies. Thus, mixtures of benomyl, thiophanate-methyl, or iprodione with dichloran were very effective against decays caused by species of *Monilinia, Botrytis, Penicillium,* and *Rhizopus* on stone fruit when applied pre- or postharvest (Ogawa and English 1991). In an attempt to increase chemical coverage and penetration into infection sites, ethanol has experimentally been added to postharvest treatments with DCNA and benomyl for peaches (Feliciano et al. 1992). When this treatment was tested on peaches in Brazil, established infections of *Colletotrichum* spp. were suppressed more effectively than by the treatment without ethanol. This suggests that other protectant fungicides may be used as suppressants by the addition of alcohol as wetting agents. In the United States, however, ethanol is a controlled substance and cannot be used in postharvest treatments.

A residue tolerance for iprodione for postharvest treatments of stone fruit was established in the United States in 1989. The fungicide is highly effective, with protective and suppressive activity against all major stone fruit decays, brown rot, gray mold, and Alternaria decays. Furthermore, when mixed with postharvest wax/oil emulsions, the fungicide is also effective against Rhizopus rot (Adaskaveg and Ogawa 1994). Thus, for a number of years it was the foundation of postharvest treatments for control of stone fruit decays. In 1996, the enactment of the federal Food Quality Protection Act required safety reviews of all existing pesticide tolerances (maximum residue limits) and a tolerance reassessment through the pesticide reregistration program. Although iprodione was determined to be eligible for reregistration, the manufacturer of the compound cancelled all high-risk uses, including postharvest registrations, in a effort to remain within the total exposure limits (i.e., "risk cup") as defined by the U.S. EPA.

NEW FUNGICIDES FOR POSTHARVEST USE

Currently, more new fungicides are being developed, introduced, and registered for use for fungal disease management than at any other time in the history of agriculture. These new tools were designed with awareness and criteria for greater environmental and human safety than older fungicides. Many of the most recently introduced fungicides offer the promise of lower application rates, greater efficacy against target pathogens, minimal effects against nontarget organisms, short persistence or nonreactivity in the environment, and greater worker and consumer safety during exposure than older fungicides. After the cancellation of the postharvest registration of iprodione, representatives of the new classes of fungicides were also evaluated for postharvest use. Some of these compounds are now available for use on selected crops or will be made available within the next few years.

In 1998, an emergency registration was obtained for postharvest use of fludioxonil on stone fruit crops. Fludioxonil belongs to a new class of fungicides, the phenylpyrroles. It is highly active against all major postharvest decays, including brown rot, gray mold, Rhizopus rot, Penicillium decays, and decays caused by species of *Gilbertella* and *Mucor* that previously could not be controlled with existing fungicides (Förster and Adaskaveg 1999). Thus, fludioxonil has the broadest spectrum of activity, is used at the lowest rate (227g per 718 kg or 8 oz per 200,000 lb of fruit), and has the lowest mammalian toxicity of any fungicide previously registered for postharvest use. The fungicide is synthetically produced but is based on the chemistry of a pyrrolnitrin. This latter compound is a naturally produced metabolite of soil bacteria in the genus *Pseudomonas*. It was first discovered in Japan (Arima et al. 1965) and was evaluated for postharvest use for the control of blue and gray mold of pome fruit in 1991 (Janisiewicz et al. 1998). Future full registrations of fludioxonil, in addition to stone fruits, are planned for citrus, kiwifruit, pome fruit, and pomegranates. Because the compound is photodegradable, preharvest uses will be restricted to seed and soil treatments of other crops.

Tebuconazole is another new fungicide that is registered for preharvest use for control of brown rot and powdery mildew of stone fruit crops. Because postharvest residues were below preharvest tolerances set for sweet cherry, the fungicide was also registered for postharvest use on sweet cherry in California. The fungicide applied at a rate of 8 oz per 25,000 lb (227 g per 11,350 kg) of fruit is very effective against brown rot and Rhizopus decays but is less effective against gray mold caused by *Botrytis cinerea*.

Two new materials, fenhexamid and azoxystrobin, represent two additional new classes of fungicides, the hydroxyanilids and the strobilurins, respectively. Like tebuconazole, these fungicides are being developed for preharvest uses on many agricultural crops, and like fludioxonil, they are considered extremely safe by standards set by the Food Quality Protection Act. Because of the low mammalian toxicity, these compounds were selected for postharvest uses. Fenhexamid will be registered for brown rot and gray mold control of stone fruit crops, and for gray mold control of pome fruits, pomegranate, and kiwifruit. Azoxystrobin is effective against Penicillium decays and will be registered for use on citrus and pome fruits. Together with registrations of fludioxonil and with planned registrations of fenhexamid and azoxystrobin on the respective crops mentioned for each fungicide, effective integrated control programs can be developed to prevent resistance from developing in target populations of decay fungi.

MANAGEMENT OF POSTHARVEST DECAYS WITH BIOLOGICAL CONTROL ORGANISMS

Concerns about the safety of chemical treatments has been the primary motivation for developing biological control methods using antagonistic organisms. These antagonistic organisms include bacteria, yeasts, and occasionally filamentous fungi. Although they are quite easily identified in the laboratory, the transition from antagonism observed in the laboratory to successful control in defined experimental packinghouse environments and ultimately to the effective implementation in commercial production agriculture has been difficult. In the process of producing large amounts of the biocontrol organism, genetic drift of the original biocontrol organism can result in loss of traits that are essential for successful commercial development. Biological control mechanisms, with their complex microbial interactions and dependency on specific environmental conditions, are still poorly understood (Larkin et al. 1998). Thus, only few biological controls have been successfully introduced on a commercial scale.

The challenges for postharvest biological decay control are different from those for control of diseases in the field. For example, colonization of the fruit surface by the antagonist before harvest is desirable for successful postharvest decay management. Thus, wounds that occur during harvesting and postharvest handling could be protected from infection by decay organisms. Consequently, preharvest treatments with biocontrols have been suggested as an application method to deliver the biocontrol agent to the fruit surface before the pathogen is introduced (Benbow and Sugar 1999). Coverage of the biocontrol agent on the crop, however, is usually poor because organisms commonly grow in localized areas on the fruit surface.

Mechanisms described for biocontrol include competition, antibiosis, parasitism, and induction of host resistance (Larkin et al. 1998). These mechanisms may reduce the amount of pathogen inoculum, protect the infection site, limit disease development after pathogen infection, or induce resistance in the host. Competition between microorganisms can be for either nutrients or space. For space competition, it has been suggested that a physical barrier of the infection site prevents the infection of the pathogen. Competition can be an effective biocontrol mechanism when the antagonist is present in sufficient quantities at the proper time and location. In antibiosis, metabolic products such as toxins, antibiotics, or enzymes are secreted by the antagonist that inhibit or kill another organism (Bull et al. 1998; Janisiewicz et al. 1991). To be an effective mechanism of biological control, these inhibitory metabolic products must be secreted at the site of interaction between pathogen and biocontrol agent. Additionally, the products must be present in sufficient amounts to be inhibitory. Parasitism occurs when the antagonist feeds on or grows within the pathogen, resulting in the direct destruction or lysis of pathogen structures. Induced resistance of the host plant occurs when the biocontrol agent induces physiological changes in the host that render it less susceptible to infection of the pathogen. These changes may include increases in activity of β-1,3 glucanases, chitinases, and peroxidases (Ippolito et al. 2000; Wilson et al. 1994) or increases in total natural phenolic compounds in the host tissue that are inhibitory to the pathogen. Mechanisms of induced host resistance have also been demonstrated with the use of natural compounds such as chitosan (Wilson et al. 1994) and methyl jasmonate (Meir et al. 2000).

For postharvest use two biological control treatments are currently registered (table 17.4). Bio-Save is a preparation of the antagonistic bacterium *Pseudomonas syringae*. It is used on citrus, cherries, pome fruits, and potatoes for control of a range of decay organisms. A combination of biological control and calcium treatments was more effective for control of postharvest Penicillium decay of apples than using the biocontrol agent alone (Janisiewicz et al. 1998). Aspire is a preparation of the yeast *Candida oleophila* and is registered on pome fruits and citrus. The commercial efficacy of Bio-Save and Aspire has been inconsistent. Like other biological controls these treatments never completely prevented fruit decay. In general, biological controls only provide a partial level of control with results that are often inconsistent (El-Gaouth 1997). Currently, they do not provide levels of control comparable to

synthetic fungicides (Mari and Guizzardi 1998). In addition, there is no curative activity, and their use is sensitive to pathogen concentration (Roberts 1994). For commercial applications, manufacturers of biocontrol treatments, as well as researchers, have suggested their use in conjunction with chemical pesticides (Droby et al. 1998). To date, biocontrols should be regarded as complementary tools for the management of postharvest decays and used together with other strategies as part of an integrated pest management program.

The use of biological controls is not necessarily the ultimate alternative to the use of synthetic chemicals (i.e., fungicides) for postharvest decay control of horticultural crops. A debate is developing on the use of these biological versus chemical methods of decay control. In the case of chemical control, a single active ingredient is used that is highly characterized (investigated) chemically and toxicologically for any adverse human health and environmental effects. In biological control, an organism or a natural product is used that is not well characterized chemically but is produced through cultivation or fermentation processes without any selective concentration or purification of the organism or its primary or secondary metabolites. Some of these metabolites are broad-spectrum antibiotics (Bull et al. 1998; Janisiewicz et al. 1991). Whether these materials or other compounds are mechanisms of action of biological controls is not clear, but known and unknown metabolites are probably present at low concentrations under conditions that allow for their production. Still, the idea of using a synthetic compound is not attractive to some people because not all aspects of the compound can be evaluated in respect to potential negative effects. It is not the intent of the authors to resolve this issue but to provide these perspectives to the reader.

MANAGEMENT OF POSTHARVEST DECAYS WITH PLANT GROWTH REGULATORS

Plant growth is regulated by naturally occurring growth regulators that act as hormones. To change the plant's physiology, synthetically produced plant growth regulators are

Table 17.4. Commercially available biological control materials and plant growth regulators (PGR) registered as postharvest treatments

Category	Organism/ Product	Year introduced	Crop	Decay organisms or function	Methods of application	Residue tolerance (ppm)*
Biocontrol	*Pseudomonas syringae* (Bio-Save)	1995	Citrus	*Penicillium digitatum, P. italicum, Geotrichum citri-aurantii*	Dip or spray	exempt
			Cherries	*Penicillium expansum, Botrytis cinerea*	Drench	exempt
			Apples, pears	*Penicillium expansum, Botrytis cinerea, Mucor piriformis*	Dip or drench	exempt
			Potatoes	*Fusarium sambucinum, Helminthosporium solani*	Dip or spray	exempt
Biocontrol	*Candida oleophila* (Aspire)	1995	Pome fruits	Decay pathogens	Any type of application	exempt
			Citrus	Decay pathogens	Any type of application	exempt
PGR	Gibberellic acid (Pro Gibb)	1955	Citrus	Delays senescence (delays onset of decay)	Storage wax	exempt
PGR	2,4-D (Citrus Fix)	1942	Citrus	Delays senescence of buttons (delays onset of decay)	Storage wax	5

Source: Adapted from Hopkins 1996; Spencer 1981.

Note: * Residue tolerances from U.S. Environmental Protection Agency 1999, 2000.

commonly applied to certain agricultural crops. These compounds may act antagonistically to the naturally occurring substances or they are complementary, synthetic derivatives of these substances. By changing the plant's physiology, selected plant growth regulators may also have an indirect effect on the fruit's susceptibility to postharvest decay caused by opportunistic (weak) pathogens.

Thus, any treatment that delays plant senescence not only delays ripening but may reduce susceptibility to pathogens that favor senescent tissues for infection. For example, postharvest treatment of lemons with gibberellic acid reduces ethylene production, delays ripening, and consequently delays the onset of sour rot caused by *Geotrichum citri-aurantii* (Coggins et al. 1965). For control of stem decays of citrus, 2,4-D treatments delay the senescence of lemon fruit buttons and thus delay the development of Alternaria stem end rot (DeWolfe et al. 1959). Diplodia and Phomopsis stem end rots of oranges (Loest et al. 1954) and Penicillium, Alternaria, and Colletotrichum decays of mandarins (Lodh et al. 1963) have been commercially controlled with postharvest fruit treatments of 2,4-D and experimentally controlled with

2,4,5-T. Similarly, treatments with ethylene biosynthesis inhibitors such as aminoethoxyvinylglycine hydrochloride (AVG) (Ju et al. 1999) or 1-methylcyclopropene (1-MCP) may offer a similar potential in the future.

APPLICATION OF POSTHARVEST TREATMENTS

TREATMENT METHODS

Methods used to apply postharvest treatments include the less-frequently used dips, flooders, foamers, brushes, fumigators, dusters, paper wraps, and box liners, as well as the currently more frequently used drenches, high-volume systems (e.g. liquid or air-nozzle sprayers), and low-volume systems such as controlled droplet (CDA) applicators. High-volume applications use from 417 to 834 l/metric ton (100 to 200 gal/ton) of fruit, whereas low volume systems use 30 to 114 l/metric ton(8 to 30 gal/ton) of fruit. It is critical that the labeled amount of fungicide is applied to the specified weight of fruit. Low- and ultra-low-volume applications are more economical and are more environmentally sound, because there is very little run-off, resulting in no disposal problems. Therefore, they are increasingly being used.

Treatments are applied either as an aqueous solution or more commonly in a wax-oil emulsion. Based on their water-oil solubility, different waxes function differently. Waxes used for postharvest treatments are derived from paraffinic oils (petroleum-based oils), vegetable oils, carnauba waxes, or shellacs. Waxes are primarily used to prevent water loss of the commodity during storage and transportation. In addition, they generally enhance fruit appearance. Most waxes, except the shellacs, allow gas exchange, so that respiration can occur with minimal water loss. Ethylene also will freely pass through nonshellac waxes. The solvents used to emulsify the waxes and oils are important factors for selecting the proper coating for the commodity. Not all additives or fruit treatments are considered food grade in different international markets. The proper postharvest treatments and additives must be selected for the intended market.

FACTORS AFFECTING TREATMENT EFFICACY

The efficacy of residual fungicides and biological controls in preventing or minimizing postharvest decays depends on several important factors. The most important factors include the activity of the treatment against any one decay organism, the spectrum of activity against the complex of postharvest decay organisms for each crop, and the preinfection (protectant) or postinfection (suppressant) activity of the treatment. The application method may also determine the efficacy of the treatment; the most efficacious treatment may perform poorly if it is applied improperly. In addition, undesirable side effects may occur, such as fruit staining or development of resistant pathogen populations.

Generally, the addition of waxes improves fruit coverage. The solubility of the chemical may be enhanced or decreased. Some methods of treatment allow for improved performance of the fungicide. As mentioned above, iprodione mixed with wax-oil emulsions significantly improved the efficacy of the fungicide against a broader spectrum of decay pathogens (Adaskaveg and Ogawa 1994; Adaskaveg et al. 1993). The efficacy

of imazalil for decay control on lemons is significantly increased when the chemical is applied at a higher temperature (Schirra et al. 1997) or in aqueous solutions without using storage or pack waxes.

High humidity decreases water loss from a crop but also makes conditions favorable for disease development, generally reducing the efficacy of chemical treatments. In addition, fungicides such as captan and the previously registered triforine undergo hydrolysis reactions, and moisture developing on the fruit may trigger the inactivation process. Storage or transportation of chemically treated crops in modified atmospheres containing low O_2, high CO_2, or N_2 has not been shown to affect the efficacy of chemical treatments. Fungicides such as iprodione can degrade in alkaline environments, similar to high-pH waxes. Furthermore, the newly registered fungicide fludioxonil is light-sensitive and may degrade in direct sunlight. In both cases, efficacy may decrease or be lost if these fungicides are placed in environments where they may degrade.

REGULATORY ASPECTS OF CHEMICAL AND BIOLOGICAL TREATMENTS

REGULATION OF PESTICIDES

In recent years there have been major regulatory changes in the United States for pesticides used on all agricultural crops, including postharvest pesticides on fruits and vegetables. In 1996, the federal Food Quality Protection Act (FQPA) amended the federal Food, Drug, and Cosmetic Act (FFDCA) and the federal Insecticide, Fungicide, and Rodenticide Act (FIFRA) to establish a new safety standard for setting tolerances of pesticides in raw and processed foods. These new safety standards directed the U.S. Environmental Protection Agency (EPA) to consider information concerning the exposure and cumulative effects of pesticides and other substances in food that have a common mode of action not only on the general population, but especially on infants and children.

Furthermore, FQPA encouraged the development and adoption of safer crop protection tools for U.S. agriculture with what began in 1994 as the "reduced-risk pesticide initiative." A pesticide is considered to be reduced risk when it broadens the adoption of integrated pest management practices or reduces the exposure risk to humans, has a lower potential toxicity to nontarget organisms, reduces the contamination of environmental resources, or promotes lower use rates and lower pesticide resistance potential. As indicated earlier in this chapter, reduced risk fungicides include the postharvest treatments azoxystrobin, fenhexamid, and fludioxonil, in addition to cyprodonil and trifloxystrobin, which are registered for preharvest use. FQPA separates reduced risk pesticides into two types: conventional reduced risk pesticides and biopesticides. The latter are distinguished from chemical pesticides by their unique modes of action, lower toxicity, target-species specificity, or natural occurrence. Examples of postharvest biopesticides are the biocontrol agents *Pseudomonas syringae* (Bio-Save) and *Candida oleophila* (Aspire).

As mandated in 1988 by congressional amendments to FIFRA, the EPA is also continuing its efforts in re-registration programs of pesticides that were registered prior to Nov. 1, 1984, when standards for government approval were less stringent. Examples of postharvest fungicides that were not re-registered are benomyl, thiophanate-methyl, and triforine. The registrants of these products found it too costly to reregister in the low-profit postharvest market. In the re-registration program, EPA makes a re-registration eligibility decision (RED) after a risk review for "unreasonable adverse effects" to human health or the environment when used according to the label. The review defines potential continued uses and restrictions of a pesticide upon any reregistration. The postharvest fungicide iprodione was approved for reregistration; however, because postharvest uses were identified as high risk, the manufacturer of the compound withdrew all its postharvest uses in 1996, and its preharvest fruit uses in 1999. The lack of an effective brown rot fungicide stimulated extensive research for finding a replacement for iprodione, which culminated with the development of fludioxonil, a reduced-risk pesticide that is highly active against all major decay fungi.

DEVELOPMENT OF NEW POSTHARVEST TREATMENTS

In contrast to preharvest treatments, postharvest treatments generally do not provide a large, high-profit market, and the agrochemi-

cal industry has been reluctant to spend time and funds on registration of new chemicals for minor crops. Interregional Research Project No. 4 (IR-4) is a federal program that was initiated in 1963 to facilitate full registration (Section 3) of pesticides for minor uses to ensure a supply of essential pest management tools for minor crop growers and food processors. This program develops residue data so that the registrants would incur minimal expenses toward minor use labels of their products that are approved by the EPA. In the past, there were relatively few effective chemicals available for postharvest treatments (Chichester and Tanner 1972). This is because the development of registered chemicals for postharvest use was aggravated by public position of stressing risks over benefits; difficulties in discovering chemicals that effectively control decay pathogens; emergence of fungicide-resistant pathogen populations (Eckert 1988; Ogawa et al. 1977; Ogawa et al. 1988); and research emphasis on nonchemical disease control strategies.

With the enactment of FQPA, development of reduced risk pesticides for minor crop uses has become much more attractive for registrants. Furthermore, consumers are much more likely to approve postharvest treatments that are rated as reduced risk and have been shown to be extremely safe and consistent in their performance. Thus, in the future we can expect a number of new treatments to become available. A special local need (Section 24C-Statewide) or an emergency registration (Section 18-Federal) is one stimulus for the development of new chemicals for use on a crop. A special local need registration exists for pesticides with existing tolerances, whereas an emergency registration sets a time-limited tolerance or action level for a specific pesticide on a crop.

USE LIMITS OF PESTICIDES

Residue tolerances must be established for all postharvest chemical treatments except those that the EPA has designated as exempt from tolerance or that the FDA has designated as Generally Regarded as Safe (GRAS). Exempt chemicals for postharvest use include chlorine solution, potassium sorbate, potassium bisulfite, and sulfur (U. S. Environmental Protection Agency 1999). Limits for application rates, however, also exist for exempt compounds. For example, there is a

25-mg/l (ppm) limit for generic hypochlorous acid used in spray or dip tanks. Exceptions exist if a label is registered for a specific hypochlorite-containing product for a given crop.

Residue tolerances that have been established for nonexempt chemicals are included in tables 17.1 and 17.2. A tolerance is the maximum residue of a chemical that is allowed to remain on the product. Pesticide registration includes evaluation of combined or aggregate effects of pesticide exposure from food, drinking water, and other nonoccupational uses, as well as the cumulative effects of pesticides that are similar in their chemistry. Pesticide manufacturers or registrants must submit a wide array of scientific studies for review before EPA will set a tolerance. Data packages are designed to identify possible harmful effects that a pesticide could have on humans (toxicity data), the amount of chemical or breakdown products likely to remain in or on food, and other possible sources of pesticide exposure, including through the home or workplace. These factors determine the "risk-cup," or amount of exposure allowed for each pesticide registration. Thus, the larger the risk-cup, the safer is the pesticide.

The risk assessment includes considerations of the amounts and types of food people eat and how widely the pesticide is used. Pesticides that are registered under FQPA can be used with "reasonable certainty of no harm" when label instructions are followed. The registered pesticide label indicates the method of application that will ensure sufficient chemical residue to provide protective, suppressive, or therapeutic activity without exceeding the established tolerance of safety.

RESISTANCE TO FUNGICIDES

DEFINITIONS AND CONCEPTS

Fungicide resistance is a genetically inherited character that allows the fungus to withstand a chemical that previously inhibited its growth. Fungicide resistance for postharvest treatments becomes evident when fruit decay develops that previously could be controlled by a specific treatment. Recent reviews on fungicide resistance are found in Hewitt (1998) and Kendall and Hollomon (1998).

Fungicides that have a single-site mode of action affect only a single step in a physiolog-

ical pathway of a fungus, resulting in the prevention of growth. All of the reduced risk materials belong in this category. In contrast, materials with a multi-site mode of action disrupt not just one but many processes that are vital for growth. If an environmental pressure such as the use of a fungicide to manage a disease is applied to a target pathogen population, the population may respond. Two of the best-described mechanisms for responses of organisms to environmental pressures are selection and mutation (Kendall and Hollomon 1998). If in a selection process the mode of action of a fungicide is only single-site, there is greater potential to select individuals from a heterogeneous population that vary in their sensitivity at this physiological site than when a fungicide acts on multiple sites. In a mutation process, a genetic change may occur at the site of action of the fungicide that may increase the survivability of the individual. Because the epidemic growth stage of most plant pathogens is asexual and haploid, mutational changes are expressed immediately, and provided the mutant is fit, its development in the fungal population is rapid (Hewitt 1998). In both selection and mutation processes, individuals that are no longer sensitive to the fungicide's site of action will be the ones that survive, grow, and become the dominant population that will be resistant to the fungicide.

The result of these processes is a shift in the pathogen population from one that was originally sensitive to one that is resistant to the fungicide. A resistant fungal population has a reduced sensitivity toward a fungicide as compared to the baseline sensitivity of the original population. This reduced sensitivity results in the loss of fungicide efficacy and eventually leads to crop loss. The shift in pathogen population may be temporary or permanent, depending on the fitness of the new population as compared to the old one. (this is discussed in more detail below under "Types of Resistance").

Resistance is much more common with fungicides that have a single-site mode of action than with those that have multiple sites of action. Thus, according to their potential to develop resistance, fungicides have also been grouped into low, moderate, and high resistance risk fungicides (Hewitt 1998). A general principle with fungicide resistance is that once a fungal population develops resistance to a fungicide of a specific class, the population will be cross-resistant to other fungicides within the same class. For example, benomyl-resistant populations of *Botrytis cinerea* are cross-resistant to thiophanate methyl, thiabendazole, and carbendazim. In addition, a fungus can have multiple resistances or resistance to fungicides that belong to different classes with different modes of action. Multiple resistances to unrelated compounds have been reported for *B. cinerea* to DCNA and benomyl (Chastagner and Ogawa 1979), and for *P. digitatum* to biphenyl, benomyl, and 2-aminobutane (Dave et al. 1980) and to imazalil, thiabendazole, and ortho-phenylphenol (Holmes and Eckert 1999).

TYPES OF RESISTANCE

Resistance in fungal populations has been described as either qualitative or quantitative (Kendall and Hollomon 1998). In qualitative resistance, a single mutation or a small number of mutations in major genes results in the sudden shift from a sensitive to a resistant population. Populations of pathogens with qualitative resistance generally remain parasitically fit and are stable populations in the absence of the fungicide. In practice, this results in the permanent presence of the resistant population. Subsequently, the efficacy of the fungicide is lost indefinitely. Examples of this type of resistance are found with the benzimidazoles for control of Penicillium decays of citrus (Eckert 1988) and brown rot decay of stone fruits (Ogawa et al. 1988).

In quantitative resistance, numerous mutations result in changes that contribute to a greater or lesser degree toward the development of a resistant population that is comprised of individuals with different degrees of sensitivity. In this type of resistance, there is no sudden shift but rather a gradual change or selection for a resistant population with the continued use of the fungicide. Because the multiple changes generally make the resistant populations less fit as compared to sensitive populations, the population will revert to sensitivity over time in the absence of the selection pressure (the fungicide). This type of resistance is typical for the DMI fungicide imazalil used on citrus for control of Penicillium decays (Holmes and Eckert 1999).

PRACTICES AND STRATEGIES TO PREVENT RESISTANCE

With these basic principles in mind, postharvest treatments must be adapted so that the potential development of resistant populations of a pathogen is minimized or avoided. Strategies must be developed and deployed that delay the development of resistance within target populations. Initially, baseline sensitivity studies of the pathogen to a new chemical should be established before the chemical is commonly used, and a monitoring program should be started for early detection of fungicide-resistant lines. For instance, monitoring in lemon packinghouses, where unrelated fungicides such as thiabendazole, 2-aminobutane, imazalil, and biphenyl are used, has made it possible to detect resistance within populations of *Penicillium* species to several of these compounds (Ogawa et al. 1983).

To help delay the development of resistant lines, a new chemical should be gradually introduced into the current chemical program, and it should not be used exclusively. Continuous exposure of any pathogen to a fungicide often results in the rapid selection of fungicide-resistant populations. Fungicides with single-site mode of action should not be used alone on a continuous basis. Most strategies emphasize rotations or mixtures between different classes of fungicides. If fungicide mixtures are employed as a strategy for resistance management, they must be used from the introduction of the single-site mode of action fungicide. The mixture should also be used in each application and at effective rates for each fungicide used in the mixture. Additionally, each material should have a similar efficacy and performance against target populations; otherwise selection of resistant populations may still occur.

Models describing the development of resistance to benomyl in target populations have indicated that once resistance is established in a population, mixtures with another material will only reduce the rate of additional selection of the resistant population (Delp 1979). The use of two effective compounds in alternation would be less costly than combinations at full dosages (Ogawa and Manji 1984). In preharvest practices or for postharvest commodities that require multiple applications, alternate applications of two or more fungicides will provide disease control and will prevent the rapid buildup of fungicide-resistant populations. Ideally, a fungicide of a different class than those registered for preharvest uses should be registered for postharvest use. This principle was applied with the development and registration of fludioxonil for management of postharvest decays of stone fruit and other commodities (Förster and Adaskaveg 1999). For this reason, fludioxonil is only used in postharvest applications.

FUNGICIDE STEWARDSHIP

Historically, efficacy and cost have been determining factors for fungicide usage, but with proper stewardship in mind a significant determining factor should also be a pathogen population's resistance potential to a given product. Maintaining a fungicide's high efficacy is in the interest not only of its users but also of its manufacturer. Resistance represents a potentially huge economic loss to the manufacturer. Thus, guidelines and recommendations are made available to users that provide strategic information for resistance management.

In addition, manufacturers are committed to resistance management through inter-company programs such as the Fungicide Resistance Action Committee (FRAC). One of the most important aspects of fungicide stewardship for a grower or user is to be aware not only of the efficacy of a fungicide but of all of its properties, including the consequences of its overuse. This may require considerable rethinking of the frequency of application and the selection of fungicides used, irrespective of the cost of the materials. As mentioned above, development of fungicide programs for disease management requires a thorough understanding of the diseases that occur on a crop and the stages of host susceptibility. Additionally, up-to-date information is needed on the fungicides currently registered, their spectrum of activity, their mode of action and class, and their persistence after application under variable environments.

Currently, fungicide guidelines that reduce the risk of resistance from developing in a target population recommend

- initiation of a fungicide disease management program that starts in the field with a multi-site mode of action fungicide,

continuing before and after harvest with single-site mode of action treatments

- use of high labeled rates of single-site mode of action fungicides

- rotation between different classes of fungicides

- limitation of the total number of applications of each fungicide class to four or less per season

CONCLUSION

Prevention of postharvest losses due to decay has been a challenge since the beginning of agriculture. Currently, the use of modern technology for selecting resistant cultivars or modifying postharvest environments has extended storage and shelf life of harvested commodities. Using reduced temperatures and controlled atmospheres to slow metabolic processes of the host, eliminating potential inoculum sources of the pathogen, and improving equipment or handling procedures to reduce potential risks of commodity injury have been critical advances in postharvest technology for maintaining high quality and reducing losses from decay. Losses of commodities, however, still occur during transportation and marketing. The agricultural economy has changed from a local to a global market, with narrow profit margins due to high standards of quality and long-distance transportation and storage costs that make any loss from decay unacceptable to the industry. Thus, integrated approaches that utilize physical and chemical methods have been developed.

Chemical management tools have evolved from simple inorganic molecules to complex organic compounds. While older compounds have a protective action and target multiple sites within the pathogen, the newer materials often have protective and suppressive action, and they target a single site within a biochemical pathway of the pathogen. In addition, the latest introductions of postharvest control treatments have characteristics that make them more acceptable to the consumer because of their extremely low toxicity to certain human populations (ethnic groups, infants, and children). Furthermore, these compounds have a low environmental impact, with soil immobility and nontoxicity to nontarget organisms including mammals, birds, and insects. Currently, chemical treatments using fungicides are the most effective means of controlling postharvest decays. They are, however, part of an integrated system that includes management practices in the field, postharvest handling and storage practices, as well as postharvest sanitation treatments and the use of biological control agents.

Although new biological control organisms will continue to be identified, their role will likely remain limited unless their efficacy can be increased. Management practices that are based on mechanisms of natural or biological systems are becoming more important in the development of new fungicides based on what has been accomplished with the reduced-risk fungicides azoxystrobin and fludioxonil. Furthermore, with the advent of computer modeling in the discovery and optimization of fungicides (Steffens and Kleier 1995), "designer" chemicals will increasingly be utilized. Thus, if target sites are chemically well-characterized, new compounds can be designed or the efficacy of existing compounds can be optimized.

With these new perspectives in fungicide development, the increased interest of chemical companies in postharvest treatments, and the regulatory emphasis on human and environmental safety, control of postharvest pathogens with chemicals and biological controls has a bright future. Growers and packers, however, face new challenges. Most of the new products do not have as much broad-spectrum toxicity against the fungal pathogens that may occur on a specific crop as compared to the older compounds. Furthermore, a higher risk for development of resistance in target populations generally exists with new fungicides because most of these compounds have only single-site modes of action. New approaches, including the concepts of "fungicide stewardship," will be needed in developing and maintaining the fungicide component in integrated disease management programs in production agriculture. Stewardship must also be regarded as interdisciplinary.

With the discovery of new highly active compounds, safety regulations should also restrict the use of specific chemical classes with similar modes of action to either human medicine or agriculture purposes of disease control. This would ensure that the development of resistance within target populations is not readily transferred between human and plant pathogens. This practice would allow for a multitude of tools to remain available against diseases of man and his food supply.

REFERENCES

Adaskaveg, J. E. 1992. Defense mechanisms in leaves and fruit of trees to fungal penetration. In R. A. Blanchette and A. R. Biggs, eds., Defense mechanisms of woody plants against fungi. New York: Springer-Verlag. 207–245.

———. 1995. Postharvest sanitation to reduce decay of perishable commodities. Perishables Handling Newsletter 82:21–25.

Adaskaveg, J. E., and H. Förster. 2000. Preharvest fungicide treatments for management of postharvest decays of stone fruit crops. Phytopathology. 90:S2.

Adaskaveg, J. E., and J. M. Ogawa. 1994. Penetration of iprodione into mesocarp fruit tissue and suppression of gray mold and brown rot of sweet cherries. Plant Dis. 78:293–296.

Adaskaveg, J. E., H. Förster, and D. F. Thompson. 2000. Identification and etiology of visible quiescent infections of Monilinia fructicola and Botrytis cinerea in sweet cherry fruit. Plant Dis. 84:328–333.

Adaskaveg, J. E., J. M Ogawa, and K. E. Conn. 1993. Iprodione-wax/oil mixtures for control of postharvest decays of fruit crops. Abstracts 6th Inter. Cong. of Plant Pathol., Montreal.

Arima, K., M. Imanaka, M. Kousaka, A. Fukuda, and G. Tamura. 1965. Studies on pyrrolnitrin, a new antibiotic. J. Antibiot. Ser. A 18:201–204.

Barkai-Golan, R., and D. J. Phillips. 1991. Postharvest heat treatment of fresh fruits and vegetables for decay control. Plant Dis. 75:1085–1089.

Benbow, J. M., and D. Sugar. 1999. Fruit surface colonization and biological control or postharvest diseases of pear by preharvest yeast application. Plant Dis. 83:839–844.

Brooks, C., and J. S. Cooley. 1928. Time-temperature relations in different types of peach rot infection. J. Agric. Res. 37:507–543.

Bull, C. T., M. L. Wadsworth, K. N. Sorensen, J. Y. Takemoto, R. K. Austin, and J. L. Smilanick. 1998. Syringomycin E produced by biological control agents controls green mold of lemons. Biol. Control 12:89–95.

Butler, E. E., J. M. Ogawa, and T. A. Shalla. 1960. Notes on Gilbertella persicaria from California. Bul. Torrey Bot. Club 87:397–401.

Chastagner, G. A., and J. M. Ogawa. 1979. DCNA-benomyl multiple tolerance in strains of *Botrytis cinerea*. Phytopathology. 69:699–702.

Chichester, D. F., and F. W. Tanner. 1972. Antimicrobial food additives. In T. E. Furia, ed., Handbook of food additives, Vol. I. 2nd ed. Cleveland, OH: CRC Press. 115–184.

Coates, L. M., G. I. Johnson, and A. W. Cooke. 1993. Postharvest disease control in mangoes using high humidity hot air and fungicide treatments. Ann. Appl. Biol. 123:441–448.

Coggins, C. W., H. Z. Hield, I. L. Eaks, L. N. Lewis, and R. M. Burns. 1965. Gibberellin research on citrus. Calif. Citrogr. 50:457–466, 468.

Dave, B. A., H. J. Kaplan, and J. F. Petrie. 1980. The isolation of *Penicillium digitatum* Sacc. strains tolerant to 2-AB, SOPP, TBZ and benomyl. Proc. Fla. State Hort. Sci. 93:344–347.

Delp, C. J. 1979. Resistance to plant disease control agents–How to cope with it. IX Inter. Cong. of Plant Protection. Washington, D.C. August 10, 1979. 12 pp.

DeWolfe, T. A., L. C. Erickson, and B. L. Brannaman. 1959. Retardation of Alternaria rot in stored lemons with 2,4-D. Proc. Am. Soc. Hortic. Sci. 74:367–371.

Droby, S., L. Cohen, A. Daus, B. Weiss, B. Horev, E. Chalutz, H. Katz, M. Keren-Tzur, and A. Shachnai. 1998. Commercial testing of Aspire: A yeast preparation for the biological control of postharvest decay of citrus. Biol. Control 12:97–101.

Eckert, J. W. 1988. Dynamics of benzimidazole-resistant *Penicillia* in the development of postharvest decays of citrus and pome fruits. In C. J. Delp, ed., Fungicide resistance in North America. St. Paul, MN: American Phytopathological Society Press. 31–35.

Eckert, J. W., and J. M. Ogawa. 1985. The chemical control of postharvest diseases: Subtropical and tropical fruits. Ann. Rev. Phytopathol. 23:421–454.

———. 1988. The chemical control of postharvest diseases: Deciduous fruits, berries, vegetables, and root/tuber crops. Ann. Rev. Phytopathol. 26:433–469.

Eckert, J. W., and N. F. Sommer. 1967. Control of diseases of fruits and vegetables by postharvest treatments. Ann. Rev. Phytopathol. 5:391–432.

El-Ghaouth, A. 1997. Biologically-based alternatives to synthetic fungicides for the control of postharvest diseases. J. Ind. Microbiol. Biotechnol. 19:160–162.

Feliciano, A., A. J. Feliciano, J. Vendrusculo, J. E. Adaskaveg, and J. M. Ogawa. 1992. Efficacy of ethanol in postharvest benomyl-DCNA treatments for control of brown rot of peach. Plant Dis. 76:226–229.

Fidler, J. C., and C. J. North. 1967. The effect of the conditions of storage on the respiration of apples: I, The effects of temperature and concentrations of carbon dioxide and oxygen on the production of carbon dioxide and uptake of oxygen. J. Hortic. Sci. 42:189–206.

Förster, H., and J. E. Adaskaveg. 1999. Fludioxonil, a new reduced risk postharvest fungicide for management of fungal decays of stone fruit. Phytopathology 89:S26.

———. 2000. Early brown rot infections in sweet cherry fruit are detected by *Monilinia*-specific DNA primers. Phytopathology. 90:171–178.

Hewitt, H. G. 1998. Fungicides in crop protection. Cambridge, UK: CAB International. 221 pp.

Holmes, G. J., and J. W. Eckert. 1999. Sensitivity of *Penicillium digitatum* and *P. italicum* to postharvest citrus fungicides in California. Phytopathology. 89:716–721.

Hopkins, W. L. 1996. Global fungicide directory. Indianapolis, IN: Ag Chem Information Services. 148 pp.

Ippolito, A., A. El-Ghaouth, C. L. Wilson, and M. Wisniewski. 2000. Control of postharvest decay of apple fruit by *Aureobasidium pullulans* and induction of defense responses. Postharvest Biol. Technol. 19:265–272.

Jacob, H.E. 1929. The use of sulfur dioxide in shipping grapes. Univ. Calif. Bull. 471, 24pp.

Janisiewicz, W. J., W. S. Conway, D. M. Glenn, and C. E. Sams. 1998. Integrating biological control and calcium treatment for controlling postharvest decay of apples. Hortscience. 33:105–109.

Janisiewicz, W. J., L. Yourman, J. Roitman, and N. Mahoney. 1991. Postharvest control of blue mold and gray mold of apples and pears by dip treatment with pyrrolnitrin, a metabolite of *Pseudomonas cepacia*. Plant Dis. 75:490–494.

Ju, Z., Y. Duan, and Z. Ju. 1999. Combinations of GA_3 and AVG delay fruit maturation, increase fruit size, and improve storage life of "Feicheng" peaches. J. Horticult. Sci Biotech. 74:579–583.

Kendall, S. J., and D. W. Hollomon. 1998. Fungicide resistance. In D. Hutson and J. Miyamoto, eds., Fungicidal activity: Chemical and biological approaches to plant protection. New York: Wiley. 87–108.

Larkin, R. P., D. P. Roberts, and J. A. Gracia-Garza. 1998. Biological control of fungal diseases. In D. Hutson and J. Miyamoto, eds., Fungicidal activity: Chemical and biological approaches to plant protection. New York: Wiley. 149–191.

Lodh, S. B., S. De, S. K. Mukherjee, and A. N. Bose. 1963. Storage of mandarin oranges. II. Effects of hormones and wax coatings. J. Food Sci. 28:519–524.

Loest, F. C., H. T. Kriel, and J. duT Deetlefs. 1954. Citrus wastage studies. Citrus Grower (S. Afr.) 243:3–8.

Lurie, S. 1998. Postharvest heat treatments. Postharvest Biol. Technol. 14:257–269.

Mari, M., and M. Guizzardi. 1998. The postharvest phase: Emerging technologies for the control of fungal diseases. Phytoparasitica. 26:59–66.

Matsumoto, T. T., and N. F. Sommer. 1967. Sensitivity of *Rhizopus stolonifer* to chilling. Phytopathology. 57:881–884.

Maxie, E. C., N. F. Sommer, and F. G. Mitchell. 1971. Infeasibility of irradiating fresh fruits and vegetables. Hortscience. 6:202–204.

Meir, S., S. Philosoph-Hadas, R. Porat, H. Davidson, S. Salim, L. Cohen, B. Weiss, and S. Droby. 2000. Methyl jasmonate induces resistance against postharvest pathogens of cut rose flowers and citrus fruits. Abstracts of the 4th International Conference on Postharvest Science, Jerusalem. March 26–31. 8.

Mitchell, F. G., R. Guilloun, and R. A. Parsons. 1972. Commercial cooling of fruit and vegetables. Univ. Calif. Agric. Exp. Sta. Manual 43. 44 pp.

Nelson, K. E. 1958. Some studies of the action of sulfur dioxide in the control of *Botrytis* rot of Tokay grapes. Amer. Soc. Hort. Sci. Proc. 71:190–198.

Nelson, K.E., and G. A.Baker. 1963. Studies on the sulfur dioxide fumigation of table grapes. Am. J. Enol. Vitic. 14:13–22.

Noble, J. P., and R. Drysdale. 1983. The role of benzoic acid and phenolic compounds in latency in fruits of *Malus* to *Venturia inaequalis*. Phytopathology. 54:92–97.

Ogawa, J. M., and H. English. 1991. Diseases of temperate tree fruit and nut crops. Oakland: Univ. Calif. Div. Ag. and Nat. Res. Publ. 3345. 461 pp.

Ogawa, J. M., and B. T. Manji. 1984. Control of postharvest diseases by chemical and physical means. In H. E. Moline, ed., Postharvest pathology of fruits and vegetables: Postharvest losses in perishable crops. Univ. Calif. Agric. Expt. Stn. Publ. NE-87 (UC Bull. 1914). 55–66.

Ogawa, J. M., J. D. Gilpatrick, and L. Chiarappa. 1977. Review of plant pathogens resistant to fungicides and bactericides. FAO Plant Protection Bull. 25:97–111.

Ogawa, J. M., S. D. Lyda, and D. J. Weber. 1961. 2,6-Dichloro-4-nitroaniline effective against *Rhizopus* fruit rot of sweet cherries. Plant Dis. Rep. 45:636–638.

Ogawa, J. M., B. T. Manji, and E. Bose. 1968. Efficacy of fungicide 1991 in reducing fruit rot of stone fruits. Plant Dis. Rep. 52:722–726.

Ogawa, J. M, J. L. Sandeno, and J. H. Mathre. 1963a. Comparisons in development and chemical control of decay-causing organisms on mechanical- and hand-harvested stone fruits. Plant Dis. Rep. 47:129–133.

Ogawa, J. M., E. Bose, B. T. Manji, and W. R. Schreader. 1972. Bruising of sweet cherries resulting in internal browning and increasing susceptibility to fungi. Phytopathology. 62:579–580.

Ogawa, J. M., B. T. Manji, J. E. Adaskaveg, and T. J. Michailides. 1988. Population dynamics of benzimidazole-resistant *Monilinia* species on stone fruit trees in California. In C. J. Delp, ed., Fungicide resistance in North America. St. Paul, MN: American Phytopathological Society Press. 36–39.

Ogawa, J. M., B. T. Manji, C. R. Heaton, J. Petrie, and R. M. Sonoda. 1983. Methods for detecting and monitoring the resistance of plant pathogens to chemicals. In G.P. Georghiou, ed., Pest resistance to pesticides. New York: Plenum. 117–162.

Ogawa, J. M., J. H. Mathre, D. J. Weber, and S. D. Lyda. 1963b. Effects of 2,6-dichloro-4-nitroaniline on *Rhizopus* species and its comparison with other fungicides on control of *Rhizopus* rot of peaches. Phytopathology. 53:950–955.

Palou, L., C. H. Crisosto, J. L. Smilanick, J. E. Adaskaveg, and J. P. Zoffoli. 2001. Effect of continuous ozone (0.3 ppm) exposure on fungal growth, fruit decay, and responses of peach and table grape fruit in cold storage. Postharvest Biol. Technol. (In press).

Phillips, D. J. 1984. Mycotoxins as a postharvest problem. In H. E. Moline, ed., Postharvest pathology of fruits and vegetables: Postharvest losses in perishable crops. Univ. of Calif. Agric. Expt. Stn. Publ. NE-87 (UC Bulletin 1914). 50–54.

Roberts, R. 1994. Integrating biological control into postharvest disease management strategies. Hortscience. 29:758–762.

Roberts, R. G., and S. T. Reymond. 1994. Chlorine dioxide for reduction of postharvest pathogen inoculum during handling of tree fruits. Appl. Environ. Microbiol. 60:2864–2868.

Schirra, M., P. Cabras, A. Angioni, G. D'hallewin, R. Ruggiu, and E. V. Minelli. 1997. Effect of heated solutions on decay control and residues of imazalil in lemons. J. Agric. Food Chem. 45:4127–4130.

Schomer, H. A. 1948. Ozone in relation to storage of apples. U.S. Dept. of Agric. Circ. 765.

Schots, A., F. M. Dewey, and R. Oliver. 1994. Modern assays for plant pathogenic fungi: Identification, detection, and quantification. Oxford, UK: CAB International. 267 pp.

Sholberg, P. L., A. G. Reynolds, and A. P. Gaunce. 1996. Fumigation of table grapes with acetic acid to prevent postharvest decay. Plant Dis. 80:1425–1428.

Smilanick, J. L., and D. J. Henson. 1992. Minimum gaseous sulphur dioxide concentrations and exposure periods to control *Botrytis cinerea*. Crop Protection 11:535-540.

Smilanick, J. L., B. E. Mackey, R. Reese, J. Usall, and D. A. Margosan. 1997. Influence of concentration of soda ash, temperature, and immersion period on the control of postharvest green mold of oranges. Plant Dis. 81:379–382.

Smilanick, J. L., D. A. Margosan, F. Mlikota, J. Usall, and I. F. Michael. 1999. Control of citrus green mold by carbonate and bicarbonate salts and the influence of commercial postharvest practices on their efficacy. Plant Dis. 83:139–145.

Smith, W. L., Jr. 1962. Reduction of postharvest brown rot and *Rhizopus* decay of eastern peaches with hot water. Plant Dis. Rep. 46:861–865.

Smith, W. L., Jr., H. E. Moline, and K. S. Johnson. 1979. Studies with *Mucor* species causing postharvest decay of fresh produce. Phytopathology. 69:865–869.

Sommer, N. F., R. J. Forlage, J. R. Buchanan, and A. A. Kader. 1981. Effect of oxygen on carbon monoxide suppression of postharvest diseases of fruit. Plant Dis. 66:357–364.

Spalding, D. H. 1966. Appearance and decay of strawberries, peaches, and lettuce treated with ozone. ARS-USDA Mktng. Res. Rep. 756.

Spencer, E. Y. 1981. Guide to the chemicals used in crop protection. 7th ed. Agriculture Canada, Publication 1093. 595 pp.

Spotts, R. A., and L. A. Cervantes. 1992. Effect of ozonated water on postharvest pathogens of pear in laboratory and packinghouse tests. Plant Dis. 76:256–259.

Spotts, R. A., and B. B. Peters. 1980. Chlorine and chlorine dioxide for control of 'd'Anjou' pear decay. Plant Dis. 64:1095–1097.

Stanghellini, M. E., and M. Aragaki. 1966. Relation of periderm formation and callose deposition to anthracnose resistance in papaya fruit. Phytopathology. 56:444–450.

Steffens, J. J., and D. A. Kleier. 1995. Computer aided discovery and optimization of fungicides. In H. Lyr, ed., Modern selective fungicides–Properties, applications, mechanisms of action. New York: Fischer Verlag. 517–542.

Thompson, J. F. 1992. Storage systems. In A. A. Kader, ed., Postharvest technology of horticultural crops. 2nd ed. Oakland: Univ. Calif. Div. Ag. and Nat. Res. Publ. 3311. 69–78.

Tomkins, R. G. 1936. Wraps for the prevention of rotting of fruit. Gt. Brit. Dept. Sci. Ind. Res., Food Invest. Board Rpt., 1935:129–131.

Uesugi, Y. 1998. Fungicide classes: Chemistry, uses and mode of action. In D. Hutson and J. Miyamoto, eds., Fungicidal activity: Chemical and biological approaches to plant protection. New York: Wiley. 23–56.

U. S. Environmental Protection Agency (EPA). 1999. EPA Compendium of Registered Pesticides, Vols. I and II. Washington, D.C.: U.S. Government Printing Office.

———. 2000. Pesticide tolerance index system (TIS) information retrieval. Version 1.0 Computer program available at the US-EPA BBS.

Weber, D. J., and J. M. Ogawa. 1965. The mode of action of 2,6-dichloro-4-nitroaniline in *Rhizopus arrhizus*. Phytopathology. 55:159–165.

Wells, J. M., and M. H. Gerdts. 1971. Pre- and postharvest benomyl treatments for control of brown rot of nectarines in California. Plant Dis. Rep. 55:69–72.

Wells, J. M., and J. M. Harvey. 1970. Combination heat and 2,6-dichloro-4-nitroaniline treatments for control of *Rhizopus* and brown rot of peaches, plums, and nectarines. Phytopathology 60:116–130.

Wells, J. M., and M. Uota. 1970. Germination and growth of five fungi in low oxygen and high-carbon dioxide atmospheres. Phytopathology. 60:50–53.

Whelton, R., H. J. Phaff, E. M. Mrak, and C. D. Fisher. 1946. Control of microbiological food spoilage by fumigation with epoxides. Part I & II. Food Ind. 18:23–25, 174–176, 318–20.

White, G. C. 1992. Handbook of chlorination and alternative disinfectants. 3rd ed. New York: Van Nostrand Reinhold. 1,308 pp.

Wilson, C. L., A. El-Ghaouth, E. Chalutz, S. Droby, C. Stevens, J. Y. Lu, V. Khan, and J. Arul. 1994. Potential of induced resistance to control postharvest diseases of fruits and vegetables. Plant Dis. 78:837–844.

18

Postharvest Diseases of Selected Commodities

Noel F. Sommer, Robert J. Fortlage,

and Donald C. Edwards

The threat of postharvest disease influences the way most fresh horticultural crops are handled. Therefore, it is necessary to understand the nature of disease organisms, the physiology of the host commodity, and how handling methods affect them both, as well as the various environmental and handling stresses that fruits and vegetables suffer.

During handling, cuts, bruises, and punctures may facilitate entrance of a pathogen into a commodity. High or low temperatures can alter physiology, increasing its susceptibility to certain pathogens.

Relative humidity (vapor pressure deficit) and atmosphere composition are other considerations in combating postharvest diseases. The presence of pathogens completes the disease triad (host-environment-pathogen). The level of disease the pathogen can cause and the number of fungus spores present determine disease incidence and severity.

THE PATHOGEN

Fungi are overwhelmingly present in postharvest diseases of fruits and vegetables. Bacteria frequently cause disease in certain vegetables, but they are generally rare in tree fruits and berries. Virus diseases may develop or intensify postharvest disease expression in certain root or tuber crops, but they do not affect fruit after harvest.

Fungal pathogens are commonly members of the class Ascomycetes and the associated Fungi Imperfecti. Phycomycetes are represented by the genus *Rhizopus* and near relatives, and by the genera *Phytophthora* and *Pythium*. Basidiomycetes, with few exceptions, are not postharvest pathogens.

Among the Ascomycetes, pathogens are usually encountered in the asexual (vegetative conidial) state in postharvest diseases. The sexual state is rarely seen in culture or in diseased commodities; in some species it is rare in nature. To use the conidial state for purposes of identification, an asexual binomial name is commonly given. Exceptions apply in certain well-known fungi, such as the stone fruit brown rot organism *Monilinia fructicola* or *Sclerotinia sclerotiorum*, which lacks a known asexual spore state.

THE INFECTION PROCESS

The propagule functioning to disperse fungi is generally a spore, but other propagules exist. In fact, most living parts of the fungus are capable of growth and developing disease under favorable conditions.

Spores of postharvest fungi exist in many sizes and shapes and may be either vegetative or sexual. Sexual spores may be part of the life cycle of the fungus; sometimes, they may serve to permit survival of the fungus during drought or winter cold. It is through the vegetative spore that diseases spread.

SPORE GERMINATION

Inactive vegetative spores are usually not dormant. If a spore is deposited in a moist, fresh wound, for example, it germinates immediately if temperature and atmospheric conditions are favorable. Besides water, certain substances are required:

- Oxygen is required for spore germination; however, low oxygen tensions suffice.

- Commonly, spores do not germinate well in the complete absence of carbon dioxide, which may be fixed during germination.

- The presence of metabolizable organic compounds in the liquid enhances germination and may sometimes be required.

- Absorption of moisture, an initial stage of germination, is usually accompanied by spore swelling. Some spores swell many-fold; others swell relatively little. As swelling becomes noticeable, oxygen consumption increases sharply and carbon dioxide evolution marks an enhanced metabolic rate.

Before germination is triggered, spores exhibit minimal metabolic activity. Germination is associated with rapid increase in the synthesis of RNA, DNA, and protein. The amount of endoplasmic reticulum and the number of mitochondria also increase.

Spores are characteristically covered by a thick spore coat. After swelling, a germ tube protrudes through the spore coat. The wall of the germ tube is continuous with and a part of the innermost layer of the spore coat. Germ tube protrusion and possibly much of the previous spore swelling depend on protein synthesis. As the germ tube lengthens through polar growth, side branches are initiated (fig. 18.1).

Spore germination is a hazardous period in the life of the fungus. During swelling, spores are susceptible to the lethal effects of gamma and ultraviolet irradiation, low and high temperatures, absence of oxygen, and exposure to toxic chemicals. Once underway, germination evidently cannot stop for long without loss of the ability to resume normal growth.

SPORE DISSEMINATION

Many spores produced in exposed structures are powdery and ideally suited to wind transport. Other spores, produced in more or less enclosed structures, may be exuded to the surface in a gelatinous or mucilaginous substance. Such spores are dispersed by rain and by windborne mist, often for long distances.

Some postharvest pathogens, primarily in the genus *Phytophthora*, produce sporangia structures that may germinate by germ tubes protruding like spores. Under favorable conditions, however, many motile spores may be formed within the sporangium instead. Upon emergence, the motile spores swim in soil water before resting. With favorable conditions, germination is by formation of a germ tube. These sporangia or motile spores are usually deposited on fruit in the orchard when soil is splashed on lower fruit by driving rain. Sprinkler irrigation systems, insects, small animals, birds, and humans assist in spore dissemination.

INITIAL FRUIT PENETRATION

Two kinds of postharvest pathogens can enter fruit through its skin. One group of decay fungi bypasses the skin through wounds: bruises, stem punctures, cuts, limb-rubbed areas, abrasions, and insect punctures. The germinating spore grows and colonizes the exposed fruit tissue. The other group forms appressoria, special structures that permit the

Figure 18.1

Sporangiospores of *Rhizopus stolonifer.* (A) ungerminated spores; (B) germinated spores. Magnification 400×.

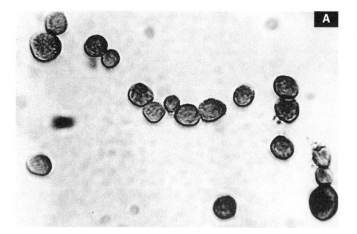
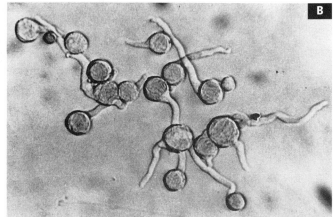

fungus to penetrate the fruit cuticle and epidermis. Commonly, these infections are initially quiescent; rot lesions do not develop until the fruit is nearly ripe.

Spores of all postharvest pathogens require high humidity or free water for several hours to germinate. Frequently, fruit surfaces are too dry to encourage germination, but spores in wounds germinate because fruit juice is present.

Stem-end rots result from infection caused when the stem is severed at harvest. *Lasiodiplodia theobromae* causes stem-end rots of citrus fruits, mango, papaya, and watermelon.

Figure 18.2

Appressoria produced on short germ tubes of sickle-shaped conidia of *Colletotrichum* sp. Magnification approximately 700×.

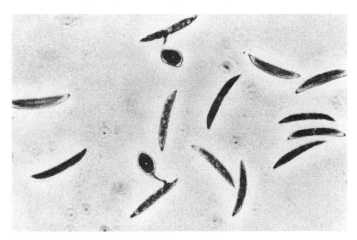

Figure 18.3

Electron micrograph of *Colletotrichum gloeosporioides* penetrating host. Magnification approximately 7,000×. (Courtesy Dr. G. Eldon Brown, University of Florida, Lake Alfred Research and Extension Center, Lake Alfred, FL)

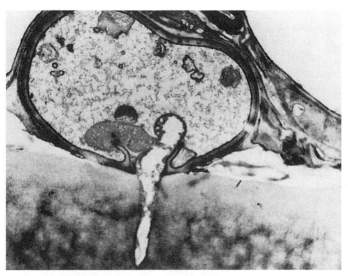

Similarly, *Thielaviopsis paradoxa* invades stem tissues of banana fingers and pineapple. Cut peduncles of apple and pear may be colonized by *Botrytis cinerea* or *Penicillium expansum*. During storage the disease progresses from the peduncle to the body of the fruit. In some fruit species, stems may be the first part of the fruit to become senescent. Senescence is likely a factor in Alternaria stem-end rots of citrus fruits.

Fungi may colonize senescent floral parts at blossoming and only much later grow into and rot the fruit proper. For example, floral parts of California Bartlett pear are infected by *Botrytis cinerea* near the end of the blossoming period. Dead styles and stamens colonized by the fungus are retained within the floral cavity near the core. Only as the fruit become senescent at the end of storage life does the fungus successfully invade the fruit flesh.

A fungus directly penetrates fruit skin according to the following scenario. Spores landing upon a fruit, when temperature and humidity are satisfactory, germinate within a few hours by sending out a germ tube. After the germ tube is well formed, a thick-walled structure, the appressorium, forms (fig. 18.2). The appressorium and germ tube adhere tightly to the fruit surface by mucilaginous material produced by the germ tube. The thickening of appressoria walls is complete except for a pore on the underside against the fruit surface, covered only by the thin germ tube wall. It is believed that enzymes are excreted through the pore to the fruit surface, including a cutinase capable of hydrolyzing the cutin that overlies the epidermis. Through the pore of the appressorium, a very fine germ-tube-like protrusion called an infection peg penetrates the cuticle where it is weakened by enzymatic action (fig. 18.3). Penetration is aided by considerable pressure from the appressorium. After penetration, the infection peg regains the normal size for mycelia of the fungus. It proceeds to branch and rebranch, thoroughly invading the fruit flesh.

Latent infections result from an interruption in infection following direct penetration. If the penetrating infection peg is unable to overcome host resistance, the infection may remain quiescent until the fruit's resistance is reduced. An example is found in *Colletotrichum gloeosporioides*, a fungus causing anthracnose of many fruits, such as apple, avocado, mango, and papaya. Fruit are commonly penetrated while developing in the orchard. Before a fruit ripens, it is highly resistant, but upon ripening it becomes susceptible. Typically, anthracnose is a rot of ripe fruit.

TISSUE INVASION AND ROTTING

Once penetration succeeds, the mycelium grows and branches, thoroughly invading the fruit flesh. The advancing mycelium excretes into the fruit tissue toxins that kill the cells. Extracellular enzymes are produced that break down constituents, degrading complex substances into low-molecular-weight compounds that can enter the fungus cell. These compounds provide the building blocks for synthesizing substances required for the fungus to grow, plus the energy for its life processes.

SPORULATION

The production of spores, the final step in the fruit rotting process, completes the vegetative spore cycle.

RESISTANCE OF FRUITS AND VEGETABLES TO INFECTION

Fruits and vegetables effectively heal wounds before harvest by walling off infections with wound periderm that appears to be made mainly of ligninlike substances (fig. 18.4). In some cases, these barriers form around wounds of fruits (apple, pear, citrus) even long after harvest.

Most fruits, while still green, are highly resistant to most pathogens and insects and do not attract animals or humans. The stomachaches that afflict young children who eat green apples are lessons well learned. Fruits become attractive to animals about the time seeds mature. Then it is advantageous for the fruit to attract animals to spread the seeds. Much of the fruit's resistance to disease is lost during ripening.

Before ripening, fruit typically contain substances, often of a phenolic nature, that are toxic to fungi. These are present in the fruit at time of infection and are generally described as preformed inhibitors. Other inhibitors, formed in response to infection, are called postinfection inhibitors or phytoalexins.

POSTHARVEST DISEASES OF DECIDUOUS FRUITS

APPLE AND PEAR (POME FRUITS)
Blue mold rot

Sexual state: Unknown
Asexual state: *Penicillium expansum* Link
= *Coremium glaucum* Link
= *Penicillium glaucum* Link

Penicillium expansum is usually considered the causal organism of blue mold rot. Studies in Poland, South Africa, and the United States show that other species of blue or blue-green *Penicillia* may cause a similar disease, but *P. expansum* is the greatest cause of loss in stored apples.

At first, blue mold lesions are light-colored and soft. As the lesion enlarges, the decayed portion can be easily separated from the surrounding sound tissue. The fungus growth on the lesion surface, at first white, becomes pale blue as sporulation occurs (fig. 18.5). Although blue mold lesions start from wound infections, the fungus in a rotting fruit can cause "nesting" by growing into sound neighboring fruit. *Penicillium expansum* produces abundant conidia that are readily airborne.

Penicillium expansum appears to be a strict wound parasite, or nearly so. Blue mold readily colonizes cuts or punctures. Less frequently, the fungus colonizes the peduncle, particularly peduncles that are thick and fleshy (figs. 18.6 and 18.7). Lenticels are also infected, usually following injury. Lenticels on some apple cultivars, such as Golden Delicious, are thin, weakly sealed, and easily ruptured by variations in pressure. Apples handled in water are especially liable to infection of the wounded lenticels, and sprays or dips in benzimidazole fungicides may be necessary to avoid excess rot.

Gray mold rot

Sexual state: *Botryotinia fuckeliana* (de Bary) Whetzl
= *Sclerotinia fuckeliana* (de Bary) Fuckel
Asexual state: *Botrytis cinerea* Pers.:Fr.
= *Haplaria grisea* Link
= *Botrytis vulgaris* Link:Fr.

Some isolates of *Botrytis cinerea* sporulate abundantly and sclerotia are absent; others develop abundant sclerotia with sparse spore production. Growth rates are highly variable.

The gray mold lesion, light brown to brown and sometimes dark brown, is characteristically less soft than blue mold lesion and has a well-defined margin, but rotted tissue cannot be neatly separated from the surrounding healthy tissue as with blue mold. Given adequate humidity, lesions become covered with gray, sporulating mycelium.

Botrytis cinerea frequently rots apple and pear in storage and may colonize stems, especially in cultivars with thick, fleshy stems such as Yellow Newtown apple or Beurre d'Anjou pear. Stem infections may grow into the fruit proper and completely colonize it (fig. 18.8). Stem punctures and other wounds in the fruit body are readily colonized.

Calyx-end rot is common in California-grown Bartlett pear (a.k.a. Williams or bon Christien) and in Packham's Triumph and Beurre Bosc pears in South Africa. Infections occur during blossoming, and colonized pistils and stamens, particularly the former, are retained within the floral tube of the fruit (fig. 18.9). Fruit rot in the orchard is rare or nonexistent. Similarly, no rot occurs in storage until the fruit starts to turn from green to yellowish green. At that time changes in fruit resistance apparently permit the mycelium to grow into the fruit proper (fig. 18.10).

Germinating spores of *B. cinerea* are unlikely to penetrate unwounded apples and pears but enter the fruit mostly through wounds or injured lenticels. This fungus can directly penetrate blossoms, however. Furthermore, the mycelium may grow and contact nearby fruit in containers, as it can penetrate the unwounded fruit surface.

Anthracnose rots

Sexual state: *Pezicula malicorticis* (Jacks.)
                  Nannf.
                  = *Neofabraea malicorticis* Jacks.
                  = *Neofabraea perennans*
                  Kienholz
Asexual state: *Cryptosporiopsis curvispora*
                  (Peck) Gremmen.
                  = *Cryptosporiopsis malicorticis*
                  (Cordl.) Nannf.
                  = *Gloeopsorium perennans*
                  Zeller & Childs.

Pezicula malicorticis causes serious losses in the U.S. Northwest (Oregon, Washington, Idaho) and British Columbia, as well as in Britain, France, and other northern European countries. The disease is present in California and most other apple-producing states without being a source of major concern. The conidia of *P. malicorticis* directly penetrate apple and pear fruit, often at the lenticel, or they invade wounds. Infections that occur during fruit development are usually latent, becoming active as the fruit ripens. Visible lesion development occurs only after a long time in storage, when fruit change from resistant to susceptible.

Bull's-eye rot of fruit and perennial canker on trees, once thought to be caused by two organisms, are now considered the asexual and sexual state of the same organism. On fruit, lesions are usually nearly round and have a light-colored center (fig. 18.11). As lesions develop, concentric circular bands of wet, cream-colored masses of conidia may be evident, forming a target pattern (fig. 18.12). Cankers form on smaller branches usually less than 5 cm (2 in) in diameter. The first indication of a canker is discolored bark extending inward to the cambium in late fall (fig. 18.13). At maturity the following spring, cankers are usually elliptical, 2.5 to 25 cm (1 to 10 in) long and 5 to 7.5 cm (2 to 3 in) across. The fungus in cankers produces spores of the asexual state. Later the sexual state develops in the canker and produces sexual spores. Fruit infections may largely result from conidia produced in the perennial cankers in the tree.

Target or lenticel spot

Sexual state: *Pezicula alba* Guthrie
Asexual state: *Phyctema vagabunda* Desmaz.
                  = *Gloeosporium album* Oster
                  = *Gloeosporium llentoidem* Peck
                  = *Gloeosporium allantosporum*
                  Fautrey
                  = *Gloeosporium diervillae*
                  Grove
                  = *Gloeosporium frigidum* Sacc.
                  = *Gloeosporium tineum* Sacc.
                  = *Trichoseptoria fructigena*
                  Maublanc

A disease similar to bull's-eye rot is caused by *Pezicula alba*, which has a comparable infection process. Infections characteristically occur at lenticels, although wounds serve as an alternative infection site. *Pezicula alba* is generally considered less damaging than *P. malicorticis*, and it occurs primarily in European apple-growing areas. Its presence in North America has been reported sporadically, but it has not been damaging and sometimes the organism may have been confused with *P. malicorticis.*

Bitter rot

Sexual state: *Glomerella cingulata* (Stonem.) Spauld. & Schrenk.
Note: There are at least 14 synonyms.

Asexual state: *Colletotrichum gloeosporioides* (Penz.) Penz. & Sacc.
= *Gloeosporium fructigenum* Berk.
Note: There are believed to be several hundred synonyms.

Bitter rot is primarily observed in the warm, humid growing areas of the southeastern United States. It probably occurs wherever apple or pear is grown near the southern limits of production. The disease is rare on apple and pear in the western United States.

The fungus has a large host range in fruit or plants of temperate, subtropical, or tropical zones. Literally hundreds of names had been published, according to the region or host upon which it was first found. A 1970 study reduced several hundred species to synonymy. The minimum temperature for development of *Colletotrichum gloeosporioides* is 3° to 5°C (37° to 41°F), but some isolates from tropical fruits have much higher minimum temperatures.

The disease may occasionally occur in immature fruit, but characteristic disease symptoms are limited to fruit that are nearly or fully grown. Disease lesions first appear as light brown, circular spots that enlarge rapidly and become sunken (fig. 18.14). Under humid conditions, conidia are produced in a semigelatinous mass, often in pinkish concentric circles. As lesions age, sporulation ceases and lesions become dark brown to black.

The fungus may sometimes produce cankers in limbs of pome fruit trees. It may grow saprophytically on lesions of other diseases, producing conidia and sometimes ascospores. Diseased fruit may overwinter as mummies in the tree or on the ground and produce conidia. The fungus can penetrate unwounded fruit, forming latent infections that become active when fruit ripen.

Elimination of cankers in the orchard and rotting fruit on the orchard floor reduces the sources of inoculum. Benzimidazole fungicides applied in the orchard generally elimi-

nate much of the inoculum and protect the fruit from infections. Postharvest handling at 0° to 3°C (32° to 37°F) should prevent disease development.

Black rot

Sexual state: = *Botryosphaeria obtusa* (Schw.) Shoem.
= *Physalospora obtusa* (Schw.) Cooke
= *Physalospora everhartii* Sacc.

Asexual state: = *Sphaeropsis malorum* Berk.
= *Sphaeropsis biformis* Peck
= *Sphaeropsis cerasina* Peck
= *Sphaeropsis druparum* (Schwein.) Cooke
= *Sphaeropsis fertilis* Peck
= *Sphaeropsis maclurae* Cooke
= *Sphaeropsis pennsylvanica* Berk. & M.A. Curtis
= *Sphaeropsis phlei* Ellis & Everh.
= *Sphaeropsis rosarum* Cooke & Ellis

The black rot organism may infect leaves (frog-eye leaf spot), produce cankers in trees, and rot fruit. Injury to trees can be highly destructive, particularly in winter-injured or neglected orchards.

Sphaeropsis malorum infects fruit usually via wounds. The rotted area is brown at first but in time usually turns dark brown to black (fig. 18.15). Concentric bands alternating brown with very dark brown or black bands often appear in the lesion. Black, conidia-producing bodies (pycnidia) usually form on cankers or on mummified fruit. Sometimes the spores are forced out in the form of coils of mucilaginous material in which they are embedded. The spores are disseminated largely by rain, splashing rain, or wind-driven mist.

Conidia of *Sphaeropsis malorum* germinate optimally at about 25° to 27°C (77° to 81°F) and slowly at 15°C (59°F). The fungus cannot grow and develop disease under usual refrigerated storage or transport conditions.

White rot

Sexual state: *Botryosphaeria dothidea* (Moug.:Fr.) Ces. & De Not.
= *Botryosphaeria berengeriana* De Not.
= *Physalospora suberumpens* Ellis & Everh.

Asexual state: *Dothiorella gregaria* Sacc. Some taxonomists consider the asexual state to be *Fusicoccum aesouli* Corda.

White rot is a relatively minor fruit rot found in the humid eastern United States on apple, pear, and peach. In California, the disease also causes a stem-end rot of avocado and citrus fruits. The disease has been found on apple in South Africa.

Rotting apples appear bleached (fig. 18.16). Cankers formed in trees may produce spores that are distributed by air, water, or wind-driven mist.

Further information on this fungus can be found in the section on Botryosphaeria rot of avocado.

Mucor rot

Pathogen: *Mucor piriformis* Fisher

Mucor rot fungus appears widely distributed in apple- and pear-growing regions of North America, Europe, Australia, and South Africa. Losses are generally not serious except in the U.S. Northwest, where the pear cultivar Beurre d'Anjou has suffered serious storage losses. The disease is not a problem in California apple or pear, although *M. piriformis* appears to be widely distributed in the state and has been observed in strawberry and feijoa. Stone fruits imported from Chile occasionally suffer losses.

At first glance, mucor rot looks like Rhizopus rot. A major difference is that *M. piriformis*, growing in high humidity, forms white, tall sporangiophores that are topped with a single black sporangium; sporangiophores of *Rhizopus* spp. are much shorter. *Mucor piriformis* rots fruit at 0°C (32°F); *Rhizopus* spp. seldom cause fruit rots at 5°C (41°F) or less. Disease symptoms on Buerre d'Anjou pear are seen as watersoaked areas of the skin, usually near the stem end or in wounds. Sporangiophores may emerge from the fruit through breaks in the skin, but often there are few sporangiophores in pome fruits until or unless the rotted fruit is nearing collapse.

Sporangiophores of *M. piriformis* may become abundant in the orchard soil near the surface. Any contact of fruit with soil on the ground or in bins risks infections in storage. Chlorine or sodium orthophenylphenate may be used in dump or flume water to minimize the buildup of sporangiospore popula-

tions. Rotted fruit should be picked up from the ground after harvest to reduce inoculum the following year.

Alternaria rot

Sexual state: Unknown
Asexual state: *Alternaria alternata* (Fr.:Fr.) Keissl.
= *Alternaria tenuis* Nees
= *Alternaria fasciculata* (Cooke & Ellis) Jones & Grout
= *Macrosporium fasciculatum* Cooke & Ellis
= *Macrosporium maydis* Cooke & Ellis

Alternaria alternata is not a vigorous pathogen but can grow at –2° to –3°C (28° to 26°F). The fungus may attack wounds and fleshy stems of cultivars such as Beurre d'Anjou pear or Yellow Newtown apple. It is one of several fungi that invade delayed sunscalded fruit in storage (fig. 18.17). Chilling injury in chilling-susceptible apple cultivars, such as Yellow Newtown, increases the susceptibility of the fruit to Alternaria rot (fig. 18.18). The fungus may contribute to "moldy core" or to "core rot."

Cladosporium rot

Sexual state: *Mycosphaerella tassiana* (De Not.) Johans.
Asexual state: *Cladosporium herbarum* (Pers.:Fr.) Link
= *Cladosporium caricicola* Corda
= *Cladosporium epiphyllum* (Pers.:Fr.) Fr.
= *Cladosporium fasciculatum* Corda
= *Cladosporium fuscatum* Link
= *Cladosporium graminum* (Pers.:Fr.) Link
= *Helminthosporium flexuosum* Corda

Cladosporium rot fungus is primarily a saprophyte, but it is a wound pathogen of mostly overripe and senescent fruit (fig. 18.19). It colonizes delayed sunburn-damaged areas in Yellow Newtown apple. Bruised or wounded tissues may be colonized, particularly in storage near the end of the fruit's life. The fungus may also be present in "moldy cores" and in "dry core rots."

Pleospora rot

Sexual state: *Pleospora* spp.
Asexual state: *Stemphylium botryosum* Wallr.
Pleospora is a weak parasite that may attack fruit that are overripe or suffering from sunburn or chilling injury. It is commonly found colonizing "delayed sunscald" on Yellow Newtown apple (fig. 18.20). This fungus may also be involved with "moldy core" or "dry core rot." Symptoms are usually nondescript, and identification requires culturing the fungus.

Phomopsis rot

Sexual state: *Diaporthe perniciosa* Marchal
Asexual state: *Phomopsis mali* Roberts
Phomopsis rot was observed as a stem-end rot in 1975–1976 in controlled atmosphere stored Yellow Newtown apples in Santa Cruz County, California (fig. 18.21). The fruit were believed to have suffered minor chilling injury. Symptoms are a brown skin and flesh discoloration along with sparse or no aerial mycelium. Differentiation from other rots without distinctive symptoms requires culturing the fungus. It is likely that orchard inoculum may come from leaf spots or pycnidia in trees or from windfall fruit that have mummified.

Pink rot

Sexual state: Unknown
Asexual state: *Trichothecium roseum*
                           (Pers.:Fr.) Link
                           = *Cephalothecium roseum* Corda
Triochothecium roseum occurred in apple and pear before the advent of modern temperature management and before effective control of scab. Often the main development was on scab lesions caused by the fungus *Venturia inaequalis* (Cke.) Aderh. It is now rare on apple and pear in this country. However, in the Himalayan foothills of northern India, which is subject to monsoon rains, *T. roseum* is common before and after harvest.

Side rot

Sexual state: Unknown
Asexual state: *Phialophora malorum* (M. N.
                           Kidd & A. Beaumont)
                           McColloch
                           Sporotrichum malorum M. N.
                           Kidd & A. Beaumont
                           = *Sporotrichum carpogenum*
                           Rühle

Incidence of side rot is evidently low except in the Beurre Bosc pear, in which significant losses have been experienced in the Pacific Northwest. Lesions are characteristically round or nearly so. There is usually no sporulation on the lesion surface (fig. 18.22). Similar lesions are sometimes produced by *Cladosporium herbarum*, which also commonly does not sporulate on the lesion surface. Isolation in pure culture is the usual means of identifying the causal fungus. *Phialophora malorum* is susceptible to sodium orthophenylphenate in a flotation fluid with sodium silicate.

Sooty blotch and fly speck

Sooty blotch and fly speck are caused by two separate organisms that are commonly found growing together superficially on apple surfaces. Sooty blotch, produced by the asexual fungus *Gloeodes pomigena* (Schw.) Colby, is observed as a thin coating of gray mycelium growing in irregular patches.

Fly speck is caused by the asexual fungus *Zygophiala jamaicensis* Mason. It is observed as small black bodies on the fruit surface reminiscent of fly crottes. Fly speck is commonly associated with blotch, which is why the two organisms are generally considered together as one condition.

Sooty blotch and fly speck are limited to warmer areas with high rainfall, but an unusual instance occurred in California's hot, dry San Joaquin Valley. When a grower sprinkled water on fruit to obtain evaporative cooling to reduce sunburn injury of Granny Smith apple, blotch and fly speck appeared.

Sooty blotch and fly speck organisms do not establish infections in the fruit, but rather grow on the surface. Nevertheless, the surface mycelium is difficult to remove and may result in unattractive fruit.

Brown rot

Sexual state: *Monilinia fructigena* Honey
                           = *Sclerotinia fructigena*
                           Aderhold ex. Sacc.
Asexual state: *Monilia fructigena* Pers. ex Fr.
Brown rot of apple, rare in North America, is common in Europe. The organism is sufficiently similar to the stone fruit brown rot fungi *Monilinia fructicola* and *M. laxa* to cause considerable trouble with identification. The rare occurrence of brown rot on ripe Bartlett pear is generally attributed to

M. laxa or *M. fructicola*. The life cycle of *M. fructigena* is similar to the life cycle of the stone fruit organisms (see the stone fruit section, below).

Phytophthora rot

Pathogens: (1) *Phytophthora cactorum* (Lebert & Cohn) Schröt.
(2) *Phytophthora syringae* (Kleb.) Kleb.

Phytophthora fruit rot is sometimes called "sprinkler rot" because rotted fruit are often found in the lower part of the tree, where irrigation water from sprinklers contacts the trees. Irrigation water can deposit sporangia or zoospores on the surface of fruit if it comes from a stream contaminated with the fungus, or if it splashes contaminated particles of soil onto the fruit. Spores readily penetrate the fruit surface to produce a rapidly developing rot (fig. 18.23). Healthy fruit in contact with diseased fruit frequently become diseased. Diseased areas are dark brown in Bartlett pear.

In the western United States, Phytophthora rot of fruit seldom warrants control measures. Presumably, metering a copper fungicide into the sprinkler irrigation water prevents infection, if needed. These fungi may produce a collar or root rot among trees in the orchard.

APRICOT, CHERRY, NECTARINE, PEACH, PLUM, AND PRUNE (STONE FRUITS)
Brown rot

Sexual state: (1) *Monilinia fructicola* (Wint.) Honey
= *Sclerotinia fructicola* (Wint.) Rehm
Asexual state: (1) *Monilia* sp.
Sexual state: (2) *Monilinia laxa* (Alderh. & Ruhl.) Honey
= *Sclerotinia laxa* Aderh. & Ruhl.
Asexual state: (2) *Monilia laxa* (Ehrenb.) Sacc. & Voglino
= *Monilia cineria* Bonord.

In California, the two species of brown rot organisms above are responsible for the most serious disease of most stone fruits. A third organism, M*onilinia fructigena* Aderh. & Ruhl., attacks and rots apple and pear as well as stone fruits in Europe.

Most active in early spring during blossoming or after fruit have started to ripen, these fungi may grow from the rotted blossom into the pedicel and into the smaller twigs or small branches, which may be girdled, causing young leaves to die. Fruit rot may occur before harvest but often occurs postharvest.

Monilinia laxa is more likely to cause blossom blight and twig blight; *M. fructicola* is more likely to cause fruit rot. The two organisms are so similar morphologically that they cannot be easily distinguished by usual microscopic methods. Similarly, *M. fructigena* is difficult to distinguish from the other two. Consequently, considerable confusion about the true identities of the brown rot organisms exists throughout the world's stone fruit—growing regions.

Airborne conidia, landing on green fruit in the orchard, may germinate and penetrate fruit skin, but they do not proceed further. These quiescent infections only become active as the fruit ripens (fig. 18.24) and provide abundant conidia for infections. These fungi nest from diseased to healthy fruit (fig. 18.25). Infections commonly develop in mechanical wounds (fig. 18.26). Stem-end rots occur in peach when fruit skin is slightly torn during harvest (fig. 18.27).

A sexual (mushroom) state may be essential to overwinter the fungus in the northern climates. It is not essential in California, where the fungus overwinters in rotten fruit mummies and diseased twigs. Conidia are produced when temperature and humidity are favorable.

Monilinia fructicola grows slowly at temperatures near 0°C (32°F) in culture. In a fruit, where the fungus must overcome resistance to infection, the minimum temperature for growth is higher. During transit at 2° to 5°C (36° to 41°F) the fungus grows slowly. Therefore, brown rot of fruit in transit usually does not occur unless fruit were not adequately cooled, transit temperature exceeded 5°C (41°F), or the fruit were overripe when loaded.

Gray mold

Sexual state: *Botryotinia fuckeliana* (de Bary) Whetzel
Asexual state: *Botrytis cinerea* Pers.:Fr.

Gray mold affects all stone fruits and many other fruits, vegetables, and ornamental species. Serious blossom blighting by *Botrytis cinerea* is not common in California on peach, plum, nectarine, and cherry, but it is occasionally serious in apricot during wet spring weather, when apricot may suffer blossom blighting and rotting of young fruit in which *B. cinerea* is joined by a close fungus relative, *Sclerotinia sclerotiorum* (Lib.) de Bary.

Gray mold causes large economic losses because it commonly develops in fruit stored at temperatures as low as 0°C (32°F). Its symptoms are similar to those of brown rot, and inexperienced observers can confuse the two.

Germinating spores of *B. cinerea* can penetrate the unbroken cuticle and epidermis of blossoms, very young fruit, and young leaves of many plants. However, most infections of stone fruits by *B. cinerea* are likely to result from contamination of harvesting and handling wounds. Contact infections occur, however, when fungal mycelia grow from a rotting fruit to nearby healthy fruit, resulting in an ever-enlarging nest of rotting fruit in the container.

Rhizopus or "whiskers" rot

Pathogens: *Rhizopus stolonifer* (Ehr.:Fr.) Vuill.
= *Rhizopus nigricans* Ehrenb.
Rhizopus arrhizus Fischer
= *Rhizopus nodosum* Namyslowski
= *Rhizopus oxyzae* Went & Prinsen Geerligs
= *Rhizopus tritici* K. Saito
Gilbertella persicaria (Eddy) Hesseltine
= *Choanephore persicaria* Eddy

At 20° to 25°C (68° to 77°F), Rhizopus rot lesions grow rapidly in ripe or near-ripe fruits. If fruit are promptly cooled to 5°C (41°F) or lower, growth of the fungus nearly stops. At lower temperatures, germinating spores and young mycelia fail to grow and the fungus is killed. It is commonly observed that peaches packed unrefrigerated quickly rot. If similar peaches are promptly cooled to near 0°C (32°F) and shipped at about 1° to 3°C (34° to 37°F), they seldom develop Rhizopus rot when removed from refrigeration.

Germinating spores of Rhizopus rot fungi initially attack fruit via wounds but spread from rotting fruit to nearby sound fruit by "nesting" (fig. 18.28). Where moist juice exists on the fruit surface, spores are capable of penetrating the fruit's cuticle and epidermis. If an infected fruit is packed in a shipping container and held at elevated tempera-

tures, the mycelium quickly grows from it to attack nearby fruit (fig. 18.29) and eventually occupies the entire container. Mycelial growth at lesions is initially white, then it changes as black sporangia are produced.

A serious problem in canned apricots due to Rhizopus rot has occurred in South Africa, Europe, and Australia, as well as in California. Severe softening of the canned fruit was usually observed 9 or more months after canning. It is now known that juice from rotting fruit contains pectolytic enzymes that remain active despite washing and the heat of canning, softening all fruit within a can. A wash containing sodium hydroxide inactivates the enzyme (fig. 18.30).

Mucor rot

Pathogens: *Mucor piriformis* Fischer
Mucor hiemalis Wehmer

The *Mucor* spp. resemble *Rhizopus* spp. in overall characteristics in culture and in rotting fruit, except that the white sporangiophores of the *Mucor* spp. often grow very long under high humidity conditions. Mucor rot has been observed in shipping containers of peaches and nectarines imported from Chile (fig. 18.31). It is not known why this fungus has not been prevalent in California, even though it has long been present and occasionally causes fruit rot of strawberry. *Mucor piriformis* is capable of rotting fruit held at –1°C (30°F).

Inoculum is produced by the fungus growing in discarded fruit in the field or on various types of organic matter in the soil. The germinating spores are believed incapable of penetrating the unbroken skin of fruit, but they may colonize harvest and handling wounds. Nesting, by growth of the fungus from a rotting fruit to nearby sound fruit in containers, occurs in a manner similar to nesting by *Rhizopus stolonifer.*

Blue mold

Sexual state: Unknown
Asexual state: *Penicillium expansum* (Lk.) Thom.

Blue mold rot is commonly found in peach, plum, and nectarine only after 1 week or more in storage at 0°C (32°F). Even after several weeks of storage, incidence of blue mold is usually much less than that of Botrytis rot. Blue mold is also common in refrigerated sweet cherries, particularly if cold, moist weather precedes harvest. It is likely that other *Penicillium* spp. occasionally are involved.

Penicillium expansum is capable of contact infection from rotted to healthy fruit to form nests in containers. However, the relatively slow growth, compared with *Monilinia fructicola* or *Botrytis cinerea,* and the short storage period for stone fruits limits nesting.

Spores of *P. expansum* are produced on fruit or on various types of organic matter on or in the soil. The readily airborne spores may be produced in quantities large enough to cover surfaces of fruit-handling equipment. To establish lesions, spores must contaminate wounds. The lesions produced are soft and watery but less so than in Rhizopus rot. At first, mycelium growing from lesions is white but turns an iridescent blue as the fungus starts to sporulate (fig. 18.32). *Penicillium expansum* readily rots fruits at 0°C (32°F), having a minimum temperature for growth of about −3.3° to −2.2°C (26° to 28°F).

Alternaria rot

Sexual state: Unknown
Asexual state: *Alternaria alternata* (Fr.) Keissler

Alternaria rot is commonly found in dark plums, where it colonizes sunburn-injured tissues (fig. 18.33), and in California sweet cherry, often colonizing an aborted fruit of a double. Alternaria rot is also common in apricot if unseasonable rain has caused the fruit to split.

Much of the Alternaria rot of sunburned fruit develops before harvest, but the lesions may enlarge after harvest (fig. 18.34). Harvest wounds are occasionally found colonized by the Alternaria rot organism. It is possible that Alternaria rot would be more common in wounds if highly vigorous competing organisms were not present.

Fruit tissues in lesions of *A. alternata* are firm and relatively dry in contrast with the soft, watery tissue found with Rhizopus or Mucor rots and, to a lesser extent, with brown rot, gray mold, or blue mold rot. *Alternaria alternata* grows at temperatures as low as 0°C (32°F), but the slow rate of growth generally limits damage unless fruits are held for excessive periods in low-temperature storage.

Cladosporium rot

Sexual state: *Mycosphaerella tassiana* (De Not.) Johans.
Asexual state: *Cladosporium herbarum* (Pers.:Fr.) Link
Note: See Cladosporium rot of apple and pear for synonyms.

Examinations of the spore flora in the air in refrigerated storage have sometimes shown *Cladosporium* to be the most prevalent fungus. Its presence is facilitated by its capability to grow at temperatures as low as 0°C (32°F). Nevertheless, stone fruits are seldom attacked unless they have been in storage more than 1 month.

KIWIFRUIT
Botrytis rot

Sexual state: *Botryotinia fuckeliana* (de Bary) Whetzel
Asexual state: *Botrytis cinerea* Pers.:Fr.
Note: See Botrytis rot of apple and pear for synonyms.

Botrytis rot in kiwifruit is a major postharvest disease in California, especially after fruit have been in storage more than 4 months at 0°C (32°F). The first indication of *B. cinerea* activity is extreme softness localized at the stylar or, more commonly, at the stem end (fig. 18.35). Occasionally the rot is centered at a wound on the side of the fruit. The fungus, generally first visible on the fruit surface as white tufts of mycelium, may spread until the lesion surface is covered. As spore production occurs, the mycelium turns gray. Occasionally, irregular-shaped sclerotia, at first gray but black at maturity, form on fruit surfaces in place of normal sporulating mycelium (fig. 18.36).

The conidia produced in storage have limited effect because the spores are unable to penetrate the fruit. However, mycelium readily penetrates sound fruit. Mycelium growing from a diseased fruit contacts surrounding fruit, enlarging a nest of rotting fruit (fig. 18.37).

The success of *B. cinerea* in rotting fruit in refrigerated storage is due to its ability to grow, albeit slowly, at −2°C (28°F). Nevertheless, fungal growth rate at 0°C (32°F) is minuscule compared with that at higher temperatures. Thus, fruit should be cooled to near 0°C (32°F) immediately after harvest.

Kiwifruit in California are relatively resistant to disease until nearly ripe. The ripening progress can be followed by measuring flesh firmness with a penetrometer that is 8 mm (5/16 in) in diameter. The firmness of kiwifruit is reduced by about half every 40 to 50 days at 0°C (32°F). Fruit commonly become susceptible when they have ripened to a firmness of 6.6 to 9 newtons (1.5 to 2 lbf). Exposure to ethylene during storage dramatically hastens softening and shortens fruit life. Fruit injury in handling also hastens incidence of rot. Widely used measures are prompt cooling and storage at 0°C (32°F), monitoring storage rooms to detect ethylene, and removing ethylene.

Surface mold, Alternaria rot, juice blotch

Sexual state: Unknown
Asexual state: *Alternaria alternata* (Fr.)
                       Keissler

Surface mold. *Alternaria alternata* grows in storage primarily on senescing fruit calyces that have not been removed by brushing (fig. 18.38). Although the fruit proper is not affected, the mycelial growth covering the calyces or other dead organic matter is unsightly. It sometimes causes the purchaser to believe, erroneously, that the fruit is rotting. Removing molding fruit from the high humidity of the transit vehicle causes the mycelium to dry and collapse, leaving little or no indication of its presence.

Alternaria rot. Fruit frequently are damaged or their disease resistance is weakened by sunburning, particularly if they are produced on young, poorly shaded vines. *Alternaria alternata* frequently colonizes sunburned fruit (fig. 18.39), necessitating its elimination by sorting.

Juice blotch. Juice blotch may result when fruit are crushed in handling. The juice contaminates other fruit and packing equipment. Juice on sound fruit provide a medium for *A. alternata* and other fungi to grow and create unsightly black blotches.

Juice blotch is most likely to occur if fruit are moved on packing lines after more than a month in storage. By then fruit are less firm and susceptible to crushing.

Dothiorella rot

Sexual state: *Botryosphaeria dothidea*
                       (Moug.:Fr.) Ces. & de Not.

Asexual state: *Dothiorella gregaria* Sacc.
                       Some taxonomists believe the
                       binomial for the asexual state
                       should be *Fusicoccum aesculi*
                       Corda.

Dothiorella gregaria, believed to be the same fungus that attacks citrus and avocado fruits in California, attacks peach fruit and causes a "white rot" of apple in southeastern states. In addition to fruit rotting, the fungus causes cankers in peach trees. There is little indication that this fungus will seriously affect California-grown kiwifruit, either as a fruit rot or as a disease in the vineyard. In storage the disease is found in overripe fruit. Generally, a portion of the fruit surface collapses (fig. 18.40). It is usually necessary to culture the organism to distinguish it from Phoma rot.

Phoma rot

Sexual state: Unknown
Asexual state: *Phoma* spp.

Phoma rot is occasionally observed on kiwifruit, especially near the end of the storage season. The rots are on the side of the fruit and are probably associated with wounds. The fruit surface characteristically becomes depressed, often like a crater (fig. 18.41), without superficial mycelium. The flesh in the area beneath the crater may be colorless but is frequently pink or purple.

Phomopsis rot

Sexual state: *Diaporthe actinidiae* Som. &
                       Ber.

Asexual state: *Phomopsis* sp.

Phomopsis rot is occasionally observed in both California-grown and New Zealand–grown kiwifruit (fig. 18.42). It appears to be found mostly in overripe fruit and does not appear to develop into a serious postharvest disease. Some leaf spots and wilted canes in the vineyard yield a *Phomopsis* spp. Affected fruit commonly are colonized at the stem end, where frothy juice often appears. That and the swarms of vinegar flies that are attracted suggest that yeasts may also be present within the lesion. Gaps or spaces in the lesion are common.

Sclerotinia rot

Sexual state: (1) *Sclerotinia sclerotiorum*
                       (Lib.) de Bary
                       = *Whetzelinia sclerotiorum*
                       (Lib.) Korf & Dumont

= *Sclerotinia libertiana* Fuckel
(2) *Sclerotinia minor* Jagger
= *Sclerotinia intermedia* Ramsey
= *Sclerotinia sativa* Drayton & Groves

Asexual state: Unknown
Sclerotinia sclerotiorum and *S. minor* grow at below 0°C (32°F) and cause serious disease losses in certain stored vegetables such as carrot and cabbage. The organisms have been observed only rarely in California kiwifruit storages.

Mucor rot
Pathogen: *Mucor piriformis* Fischer
Mucor piriformis grows below 0°C (32°F) and could become important in kiwifruit storages, but so far it has only been observed occasionally.

Blue mold
Sexual state: Unknown
Asexual state: *Penicillium expansum* (Lk.) Thom.
Penicillium expansum produces a very soft, wet rot in the commodities it attacks. Its occurrence in kiwifruit is mostly limited to overripe fruit (fig. 18.43). It is believed to infect fruit only through injuries to the skin or by acting as a secondary invader of lesions made by another fungus, such as *Botrytis cinerea*. It typically occurs as a side rot in kiwifruit. Initially white, the fungus colony turns blue as it starts to produce conidia. The fungus grows at below 0°C (32°F).

Buckshot rot
Pathogen: *Typhula* spp.
Typhula spp. appeared recently in California kiwifruit storages. The disease buckshot rot was named because of the nearly round black sclerotia produced on the fruit surface. The occasional presence of the fungus in kiwifruit storage is evidently the first instance that the disease has been reported attacking a fruit in storage (fig. 18.44). The fungus grows very slowly at 0°C (32°F); it grows much better at 15°C (59°F) than at 20°C (68°F).

STRAWBERRY
Gray mold
Sexual state: *Botryotinia fuckeliana* (de Bary) Whetzel
Asexual state: *Botrytis cinerea* Pers.:Fr.

Gray mold is the most serious postharvest disease commonly found in strawberry under modern refrigerated storage and transit conditions. The minimum temperature for growth is about −2°C (28°F). The disease is common on ripening and ripe strawberry fruit. Nesting caused by growth of the disease from rotting to sound fruit is characteristic. Infection occurs on immature green fruit, particularly under wet conditions. The fungus may attack strawberry planting stock in cold storage.

Rot may be initiated anywhere on the fruit. Affected tissue turns dull pink to brown. Fruit may become completely rotted without disintegration, and very little juice is exuded. In time, the lesion exhibits on the fruit surface white mycelium that turns gray as the fungus sporulates (figs. 18.45 and 18.46). The fungus "nests" when mycelium from a rotting fruit penetrates and colonizes adjacent fruit. Occasionally, irregularly shaped black sclerotia, 1 to 7 mm (1/25 to 1/4 in) in diameter, may form on the fruit surface if fruit are held more than a month at 0°C (32°F).

Infections are initiated in various ways. During blossoming, stamens and petals become infected when conidia land and germinate on them; from infected stamens and petals the fungus may grow into the sub-calyx area of the fruit. Infected petals may be shed and land on the fruit, and the fungus grows from them into the fruit. Sound fruit may contact the ground and become infected by fungus conidia, mycelia, or sclerotia in the soil. Conidia may also be dispersed by air currents, splashing water, or wind-driven mist.

Botrytis cinerea mycelia do not have well-developed and obvious appressoria to penetrate a sound fruit, as contrasted with *Colletotrichum* spp. Nevertheless, apparently undifferentiated mycelial strands grow from a rotting fruit to contact a sound fruit, which is penetrated without apparent pressure.

Use of plastic sheeting with slits or holes through which plants are planted greatly reduces rot in the field by preventing fruit contact with the soil, which may be heavily contaminated with *B. cinerea* conidia, sclerotia, or mycelia. Gray mold may be suppressed in strawberry by fungicidal sprays in the field.

Fruit should be moved promptly from the field to a cooler, to remove field heat. Storage should be at 0°C (32°F). Use of a modified atmosphere with 12 to 20% carbon dioxide, depending on the cultivar, suppresses fungal activity and also slows senescence of the fruit.

Leak or Rhizopus rot

Pathogen: *Rhizopus stolonifer* (Ehrenb. ex Fr.) Vuill.

Leak is found throughout the world's strawberry-growing areas. Ripe fruit in the field may be rotted, but losses occur primarily after harvest (fig. 18.47). Without good temperature management, postharvest life is shortened to as little as 1 to 3 days if *Rhizopus stolonifer* is active. A ubiquitous fungus that grows as a saprophyte on decaying organic matter, is capable of infecting (via wounds) many fruits after they are completely ripe. Sexual spores (zygospores) are produced when mated with opposite strains of the fungus.

Modern temperature management, consisting of rapid removal of field heat, and storage and transport at or near 0°C (32°F) have largely eliminated leak as a significant disease. Postharvest temperatures are usually below the minimum temperature for its growth. Furthermore, germinating sporangiospores are killed by low temperatures, and the rot after removal to ambient temperatures is usually minor.

Mucor rot

Pathogens: *Mucor piriformis* Fischer
                Mucor hiemalis Wehmer

Mucor rot resembles Rhizopus rot sufficiently to sometimes cause confusion (fig. 18.48). The most striking difference is the ability of *Mucor piriformis* to grow at low temperatures (minimum temperature for growth is below 0°C [32°F]), while the *Rhizopus* species does not. Sporangiophores are extremely long compared to those of *R. stolonifer.* Like *R. stolonifer, Mucor piriformis* causes copious leakage of juice from strawberries.

Despite an ability to grow at low temperatures, the incidence of Mucor rot in California-grown strawberries in storage or transit is usually much less than that of *Botrytis cinerea,* which is also capable of growing at below 0°C (32°F).

Anthracnose

Sexual state: *Glomerella cingulata* (Stoneman) Spauld. & Schrenk
Asexual state: *Colletotrichum gloeosporioides* (Penz.) Penz. & Sacc. With several hundred synonyms.

The sexual state of other asexual fungi listed below is unknown:

    Colletotrichum acutatum Simmonds
    Colletotrichum fragariae Brooks
    Colletotrichum dematium (Pers.) Grove

Anthracnose disease organisms may attack strawberry in the field, causing lesions on stolons, petioles, crowns, and leaves as well as fruit. Such attacks are common in Florida and other humid growing areas. Anthracnose species attack fruit primarily in the field, but the disease may develop after harvest.

Anthracnose lesions develop as tan or light brown, circular, sunken lesions on ripe or ripening fruit. As sporulation occurs (fig. 18.49), cream to salmon or pink spore masses erupt from subepidermal acervuli.

Anthracnose fungi overwinter as dormant mycelium on and in infected plants. Conidia are produced that are spread by wind, wind-driven mist, splashing water, or insects. Conidia landing on fruit germinate and form appressoria that penetrate the unwounded epidermis. Fruit lesions produce acervuli with many conidia that further spread the disease.

Leather rot

Pathogen: *Phytophthora cactorum* (Leb. & Cohn.) Schroet.

Phytophthora cactorum, a common soil pathogen with a wide host range, causes root and crown rots of many hosts. On strawberry, the fungus attacks fruit at various stages of maturity, including green fruits (fig. 18.50) on occasion.

Infected fruit that are ripe or nearly so are likely to be picked and included with sound berries. Infected fruit are likely to be noticeably lighter colored than nondiseased fruit. Diseased fruit remain quite firm and "leathery" (fig. 18.51). Diseased fruit have an unpleasant taste that permits ready identification of infected berries.

The disease is likely to be most abundant during or after excessively wet weather. In California, growing strawberry plants through plastic film prevents direct contact of fruit with the soil. For that reason, the incidence of leather rot in fruit grown on plastic is rare.

TABLE GRAPES
Botrytis rot
Sexual state: *Botryotinia fuckeliana* (de Bary) Whetzel
Asexual state: *Botrytis cinerea* Pers.:Fr.

Botrytis cinerea is the most serious rot problem of table grapes in storage and marketing (fig. 18.52). The fungus may infect berries in the vineyard, particularly when extended periods of rainy weather before harvest occur. However, wounds are common entry points for the fungus. A frequently invaded wound is at the stem where berries are joined and may be partially loosened by harvesting and handling.

Other organisms frequently found in stored grapes include *Penicillium expansum* (fig. 18.53), *Alternaria alternata,* and *Cladosporium herbarum.*

Commercial control of the diseases is obtained by inserting sulfur-dioxide-releasing generator pads into the boxes with plastic liners at the time of harvest, or by fumigating grapes in storage at 0°C (32°F) with sulfur dioxide gas. Sulfur dioxide generator pads contain sulfite salts that become hydrated within the boxes and release the gas continuously at a low rate. For storage room fumigation, an initial fumigation is applied within 12 hours of harvest, followed by weekly fumigation at a lower rate during cold storage. Older fumigation practices used rates of up to 10,000 ppm that was vented to the atmosphere after about 30 minutes. Recently, a "total utilization" method was developed, where much lower rates, as low as 200 ppm, are used without venting the gas to the atmosphere. Dosimeter tubes that measure sulfur dioxide doses are used; these record the dose that occurs within the boxes, and this dose should exceed a minimum of 100 ppm per hour. If a high disease potential is expected as a result of rain or other causes, sulfur dioxide is usually used at higher levels. Sulfur dioxide has little effect on fungal infections established in berries in the vineyard before harvest, but it prevents fruit-to-fruit contact infection and nesting of the fungus in packed boxes.

Some injury is associated with fumigation when sulfur dioxide is absorbed into wounds in the berry. The injury increases with each successive fumigation. The most obvious symptom of injury is bleaching of tissue surrounding a wound or near the point of attachment of berry to stem.

POSTHARVEST DISEASES OF SUBTROPICAL FRUITS

AVOCADO
Dothiorella rot
Sexual state: *Botryosphaeria dothidea* (Mout.:Fr.) Ces & De Not.
Asexual state: *Dothiorella gregaria* Sacc. Some taxonomists consider the asexual state to be *Fusicoccum aesculi* Corda

Dothiorella rot develops extensively only in ripened fruit. A lesion may appear on any part of the fruit as a black spot in otherwise green skin. The spots may reach 1.3 cm (0.5 in) in diameter within 3 or 4 days. The uniformly dark spots are circular, not sunken. The surface is somewhat softer than uninvaded skin. Most of the rot is limited to the skin, and if the flesh is at all affected, only the outermost part appears slightly watery. From this stage the spot spreads more rapidly, becoming soft and somewhat sunken and uneven, while a watery rot spreads slowly into the flesh. Frequently, lesions develop as a stem-end infection. These may appear earlier and penetrate more rapidly and deeply than spots in the skin (fig. 18.54).

Stem-end rot
Sexual state: *Botryosphaeria rhodina* (Cooke) Arx
Asexual state: *Lasiodiplodia theobromae* (Pat.) Griffon & Maubl. Note: See stem-end rot of citrus for synonyms of *Lasiodiplodia theobromae.*

Lasiodiplodia theobromae has a large host range and causes serious losses in warm, humid growing areas. The fungus attacks cultivated crops as a stem-end rot (such as stem-end rot of citrus, Java black rot of sweet potato, and stem-end rot of watermelon). Symptoms of Lasiodiplodia rot in avocado (fig. 18.55) are similar to those of Dothiorella rot.

Fusarium rot
Pathogen: *Fusarium* spp.
Studies in Israel showed that among 36 isolates of *Fusarium* spp. causing avocado fruit rots, 19 were *Fusarium roseum* (Lk.) Snyder & Hansen. *Fusarium moniliforme* Sheld. accounted for 8, *Fusarium solani* (Mart.) Sacc. for 5, and *Fusarium oxysporum*

Schlechtend.:Fr. for 4. The optimal temperature range for growth of these species is 20° to 30°C (68° to 86°F).

Lesions (often near the stem end) darken the skin; the invaded flesh is discolored and may become light brown to nearly black. Typically, the flesh remains firm or only slightly softer than uninvaded tissue, and there is little or no leakage of moisture from invaded areas. Tissue shrinkage often results in voids or gas pockets in invaded tissue. Light-colored mycelia may grow into these voids.

Anthracnose

Sexual state: *Gomerella cingulata* (Stonem.) Spauld. & Schrenk

Asexual state: *Colletotrichum gloeosporioides* (Penz.) Sacc.
There are several hundred synonyms not listed.

Anthracnose is a serious disease of avocado where growing conditions are humid. In the orchard the fungus may grow and sporulate on dead twigs or branches or in dead spots of leaves. Conidia of the fungus are spread by wind and wind-driven rain to fruit on the tree. Infections usually remain latent until about the time fruit soften after harvest.

Anthracnose first appears on ripening avocado fruit as circular discolorations of the skin that may appear at one location or scattered over the fruit surface. Centers of disease lesions become slightly sunken. Sporulation commonly occurs at the center of lesions, and orange to pink spore masses are exuded from the usually numerous acervuli (fig. 18.56). Fruit flesh beneath lesions is greenish-black and decayed and may be fairly firm to very soft.

Control generally includes preharvest spraying with benzimidazole fungicides and maintaining optimal temperature in storage, 7° to 13°C (45° to 55°F), depending on the cultivar.

Other diseases. In addition to diseases discussed above, those caused by *Phomopsis* spp., *Rhizopus stolonifer,* and other fungi are usually of minor extent or limited to certain growing areas.

CITRUS FRUITS
Blue and green mold rots

Sexual state: Unknown

Asexual state: Blue mold: *Penicillium italicum* Wehmer

Green mold: *Penicillium digitatum* (Pers.:Fr.) Sacc.

Blue and green mold rots, the most serious fruit diseases of citrus, occur wherever citrus fruits are grown. All species of citrus fruits are susceptible. Fruit may become diseased while still on the tree; during handling, storage, and transportation; or in the market.

The first indication of rot is usually a soft, watersoaked area at the fruit surface, usually 0.6 to 1.3 cm (¼ to ½ in) in diameter when first noticed. At favorable warm temperatures the lesion may grow to 3.8 to 5 cm (1.5 to 2 in) within 24 to 36 hours. At about that stage, colorless mycelium can be detected on the fruit surface.

At room temperatures, blue mold develops slower than green mold, but the blue mold fungus is somewhat better adapted to low temperatures. Sporulation of *Pencillium italicum* is blue, surrounded by a narrow band of white mycelium. Green mold is olive green surrounded by a broad zone of white mycelium, which in turn may be surrounded by a narrow watersoaked area (fig. 18.57). Often the blue and green molds occur as mixed infections. The slower-growing blue mold lesion appear to be "taken over" by the more vigorously growing green mold.

Most infections are believed to be the result of fungal penetration of wounds in the fruit rind, although wounds may be small. *Penicillium italicum,* however, is capable of growing from a rotting fruit contacting an adjacent sound fruit. Given sufficient time, a "nest" of rotting fruit may involve several fruit.

Minimum temperature for rotting is below 7°C (45°F) for both species.

Brown rot

Pathogen: *Phytophthora citrophthora* (R. E. Smith & E. H. Smith) Leonian.

Brown rot, caused by any of several *Phytophthora* spp. in addition to *P. citrophthora,* is found in all citrus-growing regions of the world. It affects all citrus fruit; however, in lemon the highly acidic juice causes the disease to be limited to the rind, core, and tissues between segments.

The rot, first observed as a slight surface discoloration, extends rapidly, and the lesion becomes a brownish tan or sometimes a slightly olive drab. The fruit remains firm and leathery. It has a characteristic pungent odor.

Mold on the surface of fruit is unusual, but a delicate white fungus growth may be seen during extremely wet weather (fig. 18.58).

The disease is seldom seen except during or after very wet weather. Low-hanging fruit on trees become infected as motile spores from the soil are splashed on the lower part of the tree. Ordinarily, fruit within about 1.8 m (6 ft) of the ground are infected, but strong winds may cause fruit high in the tree to become diseased. For the fungus to cause disease, fruit must be wet for a long uninterrupted period. The motile spores are capable of penetrating the fruit rind without a wound.

Harvested fruit that are believed to be infected by the brown rot fungus are commonly heat-treated. Submersion in water at 46° to 48°C (115° to 120°F) for 2 to 4 minutes kills the fungus providing it is confined to external layers of the rind. Fungus that has penetrated well below the rind survives heat treatment. Heat is especially injurious to lemons, and turgid fruit are particularly subject to injury. Thus, turgid fruit, particularly lemons picked in cool or humid weather, are allowed to wilt for 1 or 2 days before heat treatment.

Phomopsis stem-end rot
Sexual state: *Diaporthe citri* Wolf
Asexual state: *Phomopsis citri* Fawc.
Phomopsis stem-end rot is widely prevalent and results in serious losses in such humid growing areas as Florida. The optimal temperature for rotting is 23° to 24°C (73° to 75°F), and the minimum temperature is 10°C (50°F). Infections appear as a softening of the rind and a slight watersoaked appearance generally turning light brown. The decayed flesh is not discolored. Mycelium sometimes occurs on the fruit surface (fig. 18.59).

Infections may develop on the calyx (button) during the growing season and remain latent until the fruit is harvested. After harvest, lesions may develop at the calyx end and grow at the sides of fruit or at the stylar end. Susceptibility to the disease increases with increasing age of the fruit at harvest.

Phomopsis citri also causes the orchard disease melanose. In wet weather the spores may infect a few epidermal cells of the rind of very young fruit. Infections may coalesce, forming various patterns on the rind as water containing fungal spores runs over the fruit surface. The fungus is usually dead by the time the fruit mature.

Stem-end rot
Sexual state: *Botryosphaeria rhodina* (Cooke) Arx
Asexual state: *Lasiodiplodia theobromae* (Pat.) Griffon & Maubl. Note: The connection of *Botryosphaeria rhodina* as the sexual state of *Lasiodiplodia theobromae* is tenuous. See Lasiodiplodia rot of banana for synonyms.
Stem-end rot was long known as diplodia stem-end rot. Symptoms are similar to those caused by *Phomopsis citri* in that the disease normally starts at the calyx end of the fruit, but can develop from wounds at any point on the fruit. However, Lasiodiplodia rot lesions are usually a darker brown than those of Phomopsis rot. Also, the advancing margin of the Lasiodiplodia rot lesion progresses in lobes or fingers (fig. 18.60), whereas the advancing margin of Phomopsis lesions develops evenly. Lasiodiplodia rot is rapid and may decay the fruit completely in 3 or 4 days at its optimal temperature for growth, 28° to 30°C (82° to 86°F).

Alternaria stem-end rot
Sexual state: Unknown
Asexual state: *Alternaria citri* Ellis & N. Pierce.
Alternaria stem-end rot, usually a minor disease of overripe or badly stressed fruit, is found in all citrus-growing areas. Its market importance, however, increases when the rot is internal and hidden from the consumer's view.

Usually starting on the senescent calyx button after long storage on trees or in storage rooms (fig. 18.61) the fungus develops in the vascular tissues of the core and the inner tissues of the rind. Rot is not always apparent at the surface, although the central core and inner rind may be seriously rotted (fig. 18.62).

Onset depends upon the physiological condition of the buttons. Therefore, treatment with 2,4-dichlorophenoxy acetic acid (2,4-D) has been used to delay senescence of the buttons.

Anthracnose
Sexual state: *Glomerella cingulata* (Ston.) Spauld. & Shrenk
Asexual state: *Colletotrichum gloeosporioides* (Penz.) Sacc.

Anthracnose occurs on citrus fruits in the orchard, storage, or market. Conidia, produced in acervuli on the fruit rind, are at first pinkish but soon darken (fig. 18.63).

The fungus enters the fruit at mechanical injuries or senescent buttons. Drought, freeze damage, and other factors that weaken trees may increase susceptibility. Fruit such as tangerines that may be picked green and subjected to lengthy degreening treatments with ethylene are often susceptible to attack by *C. gloeosporioides*.

Harvesting at proper maturity, careful handling, avoiding long storage or excessive de-greening, and maintaining temperatures below 10°C (50°F) during postharvest handling reduces anthracnose losses.

Sour rot

Sexual state: *Galactomyces geotrichum* (E.E. Butler & L. J. Petersen) Redhead & Malloch.

Asexual state: *Geotrichum candidum* Link = *Geotrichum citri-auranti* (Ferraris) R. Cif. & F. Cif. = *Oospora lactis* (Fresen.) Sacc. = *Oospora lactis* (Fresen.) Sacc. var. *parasitica* Prit. & Port.

Sour rot disease is widely distributed throughout most citrus-growing areas, and it occurs most often in storage and transit and in ripe or overripe fruit. The rot is a soft, messy, putrid, sour-smelling, disgusting mass. Fruit flies are attracted to diseased fruit.

Geotrichum candidum is common in orchard soils where diseased fruit have rotted in previous years. Fruit that drop to the ground should be discarded because wounds can be contaminated by the sour rot fungus. The fungus can spread by contact infection from rotting to adjacent sound fruit, creating a nest of rotting fruit (fig. 18.64).

Sour rot does not develop at temperatures below about 5°C (41°F), which are appropriate for orange and mandarin. Lemon, lime, and grapefruit may be injured at temperatures below about 10°C (50°F).

Sclerotinia rot

Sexual state: *Sclerotinia sclerotiorum* (Lib.) de Bary

Asexual state: Unknown

Sclerotinia sclerotiorum produces an apothecium (a small mushroom) in the orchard from which sexual spores are forcibly ejected. The spores are readily spread by wind and may become established in fruit at the calyx or in wounds. Decaying fruit appears leathery, but mycelial growth results in a cottony look (fig. 18.65). The fungus nests by contact infection from an infected fruit to all surrounding fruit in a container.

The fungus grows in fruit that have fallen to the ground or on one of many host plants. Sclerotia form up to 1 cm (0.4 in) long from which the apothecia are produced. Alternatively, the sclerotia produce mycelium that attacks susceptible plants. Elimination of weeds and clean cultivation of the orchard floor considerably reduce incidence of the disease.

Trichoderma rot

Sexual state: *Hypocrea* sp.

Asexual state: *Trichoderma viride* Pers.:Fr. = *Trichoderma lignorum* Tode. Note: The relation between *Trichoderma viride* and *Hypocrea* sp. is tenuous.

Trichoderma rot starts in injuries contaminated with soil. The fungus may also be spread by contact between infected and sound fruit. Lemons are most likely to become diseased, particularly after long-term storage. Grapefruit and oranges are also commonly diseased. Rotting citrus fruits turn brown (fig. 18.66) and have a firm, pliable texture. A coconutlike odor is typical. Control measures involve good sanitation and avoidance of contact with soil.

Botrytis rot

Sexual state: *Botryotinia fuckeliana* (de Bary) Whetzel

Asexual state: *Botrytis cinerea* Pers.:Fr.

In cool growing areas with moist, foggy weather during spring, *Botrytis cinerea* may grow abundantly in old flower petals, which become attached to fruit surfaces and cause mycelial infection. Quiescent infections may be established in the fruit's stem end. In storage, nesting occurs in packed fruit (fig. 18.67).

POSTHARVEST DISEASES OF TROPICAL FRUITS

BANANA

Crown rot disease complex

Pathogens: The complex of organisms responsible for crown rot generally includes *Fusarium roseum*, *Lasiodiplodia theobromae*, *Thielaviopsis paradoxa*, *Verticillium theobromae*, *Nigrospora sphaerica*, *Deightoniella torulosa*, and *Colletotrichum musae*. Many other fungi are sometimes present. Inoculations with various combinations of fungi show that the greatest damage results from combinations of *T. paradoxa*, *L. theobromae*, *C. musae* and *D. torulosa*. *Fusarium roseum* makes a less serious contribution.

Crown rot in banana is largely a consequence of a technological change in which the hands are cut from the stems and packaged in fiberboard cartons rather than shipped as stems. The same organisms that now attack the hand tissue formerly attacked the cut surface of the stem in transport. The disease is characterized by the darkening of the hand and adjacent peduncle and loss of the ability of the hand to support the fruit (fig. 18.68). From the rotting hand tissue the fungi grow into the finger neck and, with time, down into the fruit proper.

Creasing of the finger stem in handling is a common injury that, along with cut surfaces of hands, permits initiation of the disease (fig. 18.69).

Anthracnose

Asexual state: *Colletotrichum musae* (Berk. & Curt.) Arx
= *Gloeosporium musarum* Cooke & Mass.
= *Myxosporium musae* Berk. & Mass.

Anthracnose disease becomes evident as fruit ripen. Symptoms include more or less circular and somewhat sunken spots in which the skin becomes black and cracks in the peel sometimes occur (fig. 18.70). Salmon-colored masses of spores, produced by acervuli within the lesion, can be seen. Disease lesions on the fruit seldom extend into the flesh, but they render bananas and plantains unmarketable. *Colletotrichum musae* is a participant, with other fungi, in crown rot disease.

In the field, *C. musae* is found growing on all aboveground parts of the banana plant, particularly on the persistent bracts, during the wet season. Spores landing on green fruit germinate if free water is present or if humidity is very high. On the germ tube an appressorium forms from which an infection hypha penetrates the fruit cuticle. Green fruit are readily penetrated, but the infection remains latent until the fruit ripens and conditions become favorable for lesion development. Infection usually takes place between adjacent epidermal cells. Only a small proportion of the appressoria give rise to disease lesions. Infections also readily occur in handling wounds or where the skin splits between carpels.

Lasiodiplodia rot

Sexual state: *Botryosphaeria rhodina* (Cooke) Arx
= *Physalospora rhodina* (Berk. & Curt.) Cooke

Asexual state: *Lasiodiplodia theobromae* (Pat.) Griffon & Maubl.
= *Botryodiplodia theobromae* Pat.
= *Diplodia theobromae* (Pat.) W. Nowell
= *Botryodiplodia gossypii* Ellis & Barth.
= *Diplodia gossypina* Cooke
= *Diplodia natalensis* Pole-Evans
= *Lasiodiplodia triflorae* Higgins
= *Diplodia tubericola* (Ellis & Everh.) Tauben.
= *Lasiodiplodia tubericola* Ellis & Everh.

Note: The connection between *Botryosphaeria rhodina* and *Lasiodiplodia* is tenuous.

In Lasiodiplodia rot, the fungus produces pycnidia on dead or dying leaves, bracts, and decaying fruit of the banana in the field. Similar dead or senescent plant materials of many other hosts also provide a substrate for colonization of the fungus and development of pycnidia. Conidia of the fungus are readily airborne. Rain, wind-driven rain, or splashing rain contribute to dispersal.

Freshly cut surfaces of banana stems, hands, and fruit provide areas moist with the juice and nutrients required for prompt germination and growth of conidia (fig. 18.71). The optimal temperature for growth of *L. theobromae* is about 27° to 30°C (81° to

86°F), and most isolates produce some growth over the range of 15° to 37°C (59° to 97°F). No growth is detected in culture after 4 weeks at 10°C (50°F).

Thielaviopsis or Ceratocystis rot

Sexual state: *Ceratocystis paradoxa* (Dade) Moreau
Ceratostomella paradoxa Dade
Endoconidiophora paradoxa (de Seyn.) Davidson
Ophiostoma paradoxum (Dade) Nannf.

Asexual state: *Thielaviopsis paradoxa* (de Seyn.) Hohnel
Note: Asexual state is considered by some to be *Chalara paradoxa* (De Seyn.) Sacc.

Thielaviopsis paradoxa has a higher growth rate in culture than most common tropical fruit-rotting fungi at optimal temperatures of 25° to 30°C (77° to 86°F). Growth occurs, but at a much reduced rate, at normal transit temperatures of 12° to 14°C (54° to 57°F) for banana.

Because of its high growth rate, *Thielaviopsis paradoxa* is a wound pathogen with considerable potential for damage. Although the growth and the extent of rotting are greater in mature fruit, *T. paradoxa* invades stem tissue and green fruit as well. A common place of entry at harvest is the cut stem or hand. Although commonly a stem-end rot (fig. 18.72), the fungus can enter and colonize wherever there is an injury. When the rot occurs in market areas on ripe or near-ripe fruit, the invaded flesh becomes soft and watersoaked.

Cigar-end rot

Two fungi, *Verticillium theobromae* and *Trachysphaera fructigena,* may cause the disease singly or in combination.
Sexual state: Unknown
Asexual states: *Verticillium theobromae* (Turc.) Mason & Hughes
= *Stachylidium theobromae* Turc.
Trachysphaera fructigena Tabor & Bunt.

Cigar-end rot causes serious losses in the West Indies, the Canary Islands, and in Africa, Iran, and India. The disease appears to be absent from banana-growing areas of Central and South America. *Verticillium theo-*

bromae is present in American growing areas, but *Trachysphaera fructigena* evidently is not.

Infection occurs in senescent floral parts from which the disease spreads to the fruit finger. Cigar-end rot in Egypt, for example, may involve half of the finger or more. The rotted portion of the banana finger is dry and tends to adhere to the fruit. The appearance of the rotted portion is like the ash of a fine cigar (fig. 18.73).

Squirter disease

Sexual state: Unknown
Asexual state: *Nigrospora sphaerica* (Sacc.) Mason
= *Trichosporum sphaerica* Sacc.

Found in Australia, primarily Queensland, and in the Cook Islands and Norfolk Island, squirter disease is associated with the Australian practice of shipping separate fruit (singles), often without refrigeration during transport. Infected green fruit are not detected at the time of packing but rot during transport and ripening.

The disease's name is descriptive of the condition of fruit in advanced stages of the disease. Fruit flesh darkens and tends to liquefy, and under pressure the liquefied flesh squirts out of the fruit.

The earliest recognizable symptom is the development of a dark center or core. Sometimes a line of dark red gumlike substance is distributed along the center. The disease may be confined to either end or may affect the entire fruit.

The fungus usually enters the freshly cut peduncle in fruit packed as singles. Spores germinate in the moist cut surface or may be drawn slightly into xylem vessels of the peduncle. Hyphae may grow in the xylem vessels through the peduncle into the fruit proper with little visible effect on the peduncle. Consequently, initial symptoms may develop in the fruit's interior with no obvious connection to the outside.

Optimal temperature for growth of *N. sphaerica* on malt agar is 22° to 25°C (72° to 77°F); maximum and minimum temperatures permitting growth are 32.5°C (90.5°F) and 5°C (41°F), respectively.

Nigrospora sphaerica is commonly found on bananas in the Western Hemisphere, but squirter disease symptoms are not. Some have proposed that the squirter disease organism is a separate species, and the name

Nigrospora musae was suggested as a name for it. That suggestion has not generally been accepted. Instead, it is believed that the Australian organism is an especially vigorous pathogenic strain of *N. sphaerica.*

Fusarium rot

Sexual state: Unknown
Asexual state: *Fusarium roseum* Link 'gibbosum' Snyder & Hans.

The Fusarium rot fungus is a major contributor to the crown rot complex. It also attacks fruit peduncles and fruit. Occasionally, infection begins in adhering floral parts and extends up the finger from the blossom end. These blossom-end rots resemble cigarend disease (fig. 18.74), but Fusarium disease development is usually comparatively slow. Also, the disease may extend from the peduncle or from the stylar end, but it is not likely to extensively invade the finger.

Injuries to fruit skin are sometimes colonized. The invaded area of the skin darkens and becomes dry, and the dead skin sometimes splits to reveal the flesh below. The flesh, colonized while the skin was still green, often appears slightly brown or pink and is invariably dry and pithy. When the fruit ripens, those portions of the flesh that have been invaded remain hard and dry.

Fusarium roseum 'gibbosum' is prevalent in plantations where it colonizes organic matter such as stigmas and styles of immature fruit, dead floral parts, and rotting fruit on the ground. Spores are effectively dispersed by splashing raindrops or wind-driven rain.

MANGO
Anthracnose

Sexual state: *Glomerella cingulata* (Stonem.) Spauld. et Schrenk
Asexual state: *Colletotrichum gloeosporioides* Penz.
Note: There are several hundred synonyms.

Fruit are infected with anthracnose rot in the young stage as well as mature, but infections in unripe fruit are mostly latent and do not start to appear until the fruit begin to ripen. Some infections enlarge and darken. Acervuli develop beneath the fruit cuticle, sometimes situated in more or less concentric circles. When acervuli develop-

ment causes the cuticle to break, salmoncolored spores in a mucilaginous liquid can be observed (fig. 18.75). Under conditions of high humidity, the lesion may largely be covered with salmon-colored spores. Sometimes, however, spore production is less obvious and lesions appear as black spots (fig. 18.76). Lesions may remain limited to the fruit skin or may invade and darken the flesh to the stone. The bitter taste associated with this organism on apples and pears is less obvious in mangos.

Isolates of *Colletotrichum gloeosporioides* from tropical sources grow at 10°C (50°F) but not at 5°C (41°F).

Stem-end rot

Sexual state: *Botryosphaeria rhodina* (Cooke) Arx
Asexual state: *Lasiodiplodia theobromae* (Pat.) Griffin & Maulb.
Note: See Lasiodiplodia rot of banana for synonyms. The connection between *Botryosphaeria rhodina* and *Lasiodiplodia theobromae* is tenuous.

Under high humidity and at 30°C (86°F), spores of the stem-end rot fungus readily infect fruit when placed on the pedicel scar or in injuries to the pedicel or injuries to the fruit peel. Sections of inoculated pedicels contain mycelium, largely in the vascular tissue and sparingly in the cortex. The fungus grows from the pedicel into a circular black lesion around the pedicel. Later the fungus commonly grows unevenly, with the advancing margin of the lesion growing much faster at certain places than at others (fig. 18.77), a characteristic of growth also observed in citrus fruits.

Other diseases. Several other diseases occasionally cause postharvest diseases of mango. In areas where limited refrigeration exists, rots caused by *Rhizopus* spp., *Aspergillus niger, Macrophomina* spp., and other fungi may be prevalent. Severe cases of sooty mold occur in high-rainfall areas of India. Most of the rot in American-grown mangos imported into the United States is due to stem-end or side rots caused by *Lasiodiplodia theobromae* or anthracnose caused by *Colletotrichum gloeosporioides.*

PAPAYA
Anthracnose
Sexual state: *Glomerella cingulata*
                 (Stonem.) Spauld. & Schr.
Asexual state: *Colletotrichum gloeosporioides*
                 (Penz.) Arx

Anthracnose is found almost everywhere papaya is grown and may be the most serious cause of loss of harvested fruit. Infections occur in the green fruit in the orchard but are latent until the fruit ripen after harvest.

The highly variable nature of *Colletotrichum gloeosporioides* is evidenced by the synonym list of around 600 binomials. Further, its host range is wide, including leaves, young twigs, and fruit of many species. Besides the destructive fruit rot of papaya, anthracnose causes rots of many other commodities. Its optimal temperature for growth on papaya fruit is between 26° and 29°C (79° and 84°F), and its minimum temperature for growth is about 9°C (48°F). Isolates from Hawaiian papaya fruit grow at 10°C (50°F) but not at 5°C (41°F).

The anthracnose disease organism is present in the field, growing on senescent leaves and petioles, fallen fruit, and other types of organic matter. Conidia are produced in an acervulus in a water-soluble matrix. Evidently conidia are not readily transported in dry wind, but spores are readily dispersed by rain and wind-driven rain. Upon landing on a fruit, the conidium may germinate to produce a germ tube, which forms an appressorium. From the appressorium a fine infection hypha penetrates the fruit cuticle. If fruit are green, the fungus usually remains quiescent. With the onset of ripening, conditions favor growth of the fungus and lesions develop. Certain stress conditions also render the fruit more susceptible, as shown by the enhanced development of anthracnose lesions following fumigation of fruit with methyl bromide for fruit fly control.

Lesions are first detectable as tiny, brown, superficial, watersoaked lesions that may enlarge to 2.5 cm (1 in) or more in diameter. Several lesions may grow together to produce a large, irregular compound lesion. Lesions become sunken but usually do not extend deep into the fruit flesh (fig. 18.78). Invaded flesh may taste bitter. On the surface, salmon-colored spore masses may form, sometimes giving the lesion a target or bull's- eye appearance. The lesions may remain a light brown to salmon color, but eventually they darken to dark brown or black. Sometimes the spore masses remain inconspicuous.

Conditions for controlling the disease by a postharvest heat treatment are almost ideal: infections are present in the fruit but have caused no damage at time of harvest; control of the disease in the orchard has been unsatisfactory; and the fruit tolerate a heat treatment that effectively inactivates latent infections in the fruit. The heat treatment consists of a preheating step of 30 minutes at 42°C (108°F) followed by 20 minutes at 49°C (120°F).

Central to the control of anthracnose and other postharvest fruit rots, as well as maintenance of the physiological condition of the fruit, is good temperature management during transportation and marketing. If fruit are quickly cooled to about 13°C (55°F), growth of the pathogen and ripening changes in the fruit are slow. If fruit are ripened quickly at 20°C (68°F) following transportation, there is little opportunity for disease to develop during normal marketing. To be avoided is slow ripening at 15° to 17°C (59° to 63°F) because the fungus can grow appreciably during the time required for ripening at those temperatures.

Phoma (Ascochyta) rot
Sexual state: *Mycosphaerella caricae* H. & P.
                 Sydow.
Asexual state: *Phoma caricae-papayae* (Tarr.)
                 Punith.
                 = *Ascochyta caricae-papayae*
                 Tarr

Phoma caricae-papayae attacks the trunk, leaves, flowers, leaf and fruit pedicels, and green and ripe fruit in the orchard, particularly during winter and early spring. Fruit still attached to trees may show a black rot extending into the fruit from the stem end or from a point of contact with a dead leaf or infected leaf stalk. An isolated lesion first appears as a small watersoaked spot that develops into a shrunken, black, circular lesion from 2.5 to 7.5 cm (1 to 3 in) in diameter. Multiple infections may coalesce to form a large rotting area. Fruit attacked in the field may wither and drop, particularly if still immature.

In the field, *Phoma caricae-papayae* frequently colonizes pedicels of senescent leaves from which the trunk of the plant is invaded. Fruit may become infected in the field during harvesting when they are injured.

After harvest, the disease lesion on a fruit is frequently observed first in the region of the stem, where rotting tissue becomes dark brown to coal black, while the invaded tissue remains relatively firm and dry (figs. 18.79 and 18.80). The surface of the lesion eventually appears roughened or pebbly from the development of pycnidia, which rupture the host epidermis.

Phomopsis rot
Sexual state: Unknown
Asexual state: *Phomopsis caricae-papayae*
                Petr. & Cif.
Phomopsis caricae-papayae invades senescent leaf petioles and other papaya trash in the field where abundant pycnidia are produced. During rainy weather, spores are carried by splashing rain, wind, or wind-driven rain.

Phomopsis rot commonly originates in the area of the peduncle or a fruit skin wound. Presumably, spores of the fungus cannot easily enter the fruit except through wounds or the freshly ruptured peduncle.

The disease can develop rapidly in ripe fruit. The invaded flesh and skin tissue is soft, watersoaked, and slightly darker than uninvaded tissue (figs. 18.81 and 18.82). Surface mycelium may be almost absent. Sometimes, however, other fungi are also present. Pycnidia of the fungus are usually produced only after the fruit is nearly completely rotten.

Lasiodiplodia rot
Sexual state: *Botryosphaeria rhodina*
                (Cooke) Arx
Asexual state: *Lasiodiplodia theobromae*
                (Pat.) Griffin & Maulb.
                Note: The connection
                between *Botryosphaeria rhodi-*
                na and *Lasiodiplodia theobro-*
                mae is tenuous.
Lasiodiplodia rot, commonly initiated near the fruit peduncle, causes a stem-end rot that may be confused with Phoma stem-end rot. Lasiodiplodia rot may usually occur at injuries to the fruit skin. The conidia ordinarily cannot penetrate uninjured skin. Fresh mechanical injuries or the fresh surface of the peduncle at the point of separation are ideal infection sites.

Lasiodiplodia theobromae is found in the orchard in dropped fruit or in senescent or decaying organic matter of various types. Spores are disseminated by wind during wet weather and especially by splashing rain. The minimum temperature for growth is 10°C (50°F) or slightly above. Growth is rapid at the optimal temperatures of 25° to 30°C (77° to 86°F).

Phytophthora rot
Pathogen: *Phytophthora nicotianae* Breda
                de Haan var. *parasitica* (Dast.)
                Waterh.
Phytophthora rot in the field can be confused with other *Phytophthora* spp. or *Pythium* spp. Fruit of any age may be infected on the tree. As the disease progresses, the fruit starts shriveling, turns dark brown, and falls to the ground, where more shriveling and mummification takes place.

Postharvest fruit rot is typically a stem-end rot. After starting near the abscission layer, the fungus progressively invades the tissue inter- and intracellularly until the entire fruit decays. The epidermis and fruit flesh, although watersoaked, remain near normal in color. Fruit surfaces are commonly covered with white mycelium that becomes encrusted (fig. 18.83).

Alternaria rot
Sexual state: Unknown
Asexual state: *Alternaria alternata* (Fr.)
                Keissler
                Note: See apple and pear for
                synonymy.
Alternaria alternata seriously affects papayas that have been stored or transported at chilling temperatures below 12°C (54°F). Such chilled fruit typically appear sound for a time, but the chilling injury increases their susceptibility to *A. alternata* and certain other fungi (fig. 18.84).

PINEAPPLE

Thielaviopsis rot, black rot, water blister
Sexual state: *Ceratocystis paradoxa* (Dade)
                C. Moreau
                = *Ceratostomella paradoxa*
                Dade
                = *Ophiostoma paradoxa*
                (Dade) Nannf.
Asexual state: *Thielaviopsis paradoxa* (de
                Seyn.) Hohnel

= *Thielaviopsis paradoxa* (de Seyn.)

Note: Some consider *Chalera paradoxa* (de Seyn.) to be the preferable name for the asexual state.

Thielaviopsis rot, well known throughout the world's pineapple-growing areas and markets, poses a major problem almost everywhere to the orderly production and marketing of fresh pineapples. For example, 3 to 5% of Puerto Rican fruit on the New York market was reported rotted in usual shipments, but in fruit harvested during or just after a rainy period, rot affected 25%.

The disease organism is found on a wide range of hosts throughout the world's temperate and tropical regions. Although the same organism attacks both pineapple and banana fruit, isolates of the organism from rotting bananas, in our experience, have not readily decayed pineapples. Similarly, isolates from rotting pineapple do not rot bananas.

The optimal temperature for growth of the fungus in culture is about 27°C (81°F); the maximum is about 37°C (99°F), and the minimum is between 0° and 5°C (32° and 41°F). However, fruit rotting is almost arrested at 10°C (50°F).

Thielaviopsis rot is usually not readily noticeable until its invasion is advanced. Rot may start at the stem end and advance through most of the flesh with little external evidence of decay (figs. 18.85, 18.86, and 18.87). The only external indication of damage is a slight darkening due to watersoaking of the skin over rotted portions of the fruit. As the flesh softens, the skin above the affected tissue readily breaks under slight pressure.

When fruit are cut, areas invaded by the fungus are soft and watery as well as deeper in color than adjacent healthy tissue. In the disease's final stages the core disintegrates with the flesh. A sweetish odor distinct from the sour odor of rot by yeasts accompanies the decay. Yeasts often cohabit the rot lesion and attract vinegar flies.

Fruitlet core rot
Sexual state: (1) *Gibberella fujikori* Sawada & Ito
Asexual state: (1) *Fusarium moniliforme* Sheld.
               (2) *Penicillium funiculosum* Thom

Fruitlet core rot, evident throughout the pineapple-growing world, is characterized by rot of individual fruitlets and is usually visible only when the fruit is cut to reveal a brown to black discoloration of the flesh. Inoculations of fruitlets with *Penicillium funiculosum* or *Fusarium moniliforme* or with a mixture of conidia of the two fungi result in typical fruitlet core rot (fig. 18.88).

The two types of symptoms are a wet rot resulting when mature, juicy fruit are infected, and dry spots that develop when fruit are inoculated before flower buds are visible and before the fruit mature and become juicy.

The two fungi appear incapable of infection unless the epidermis of the floral cavity has been broken. Insects that enter the floral cavity in search of nectar may carry the fungus spores into the cavity with them and cause the damage. Also, growth cracks in the floral cavity may provide the needed infection court.

Yeasty fermentation
Asexual state: *Saccharomyces* spp.(possibly other yeasts).

Yeasty fermentation, usually associated with overripe fruit, may start while the fruit is still on the plant or after harvesting. Evidently the organisms enter the fruit through wounds.

Fruit flesh becomes soft and bright yellow and is ruptured by large gas cavities. Production of gas forces juice from dead and dying cells out through the cracks and wounds in the shell as a frothy, sticky liquid (fig. 18.89). In time, the juice within the tissues may largely be fermented, leaving spaces within the tissues (fig. 18.90).

Marbling
Pathogen: *Erwinia carotovora* (Jones) Bergey et al.

Marbling is a bacterial disease caused by *Erwinia carotovora,* but other, unrelated bacteria may also cause nearly identical symptoms. It has been associated with fruit of low acidity such as summer fruit from warmer pineapple-growing areas.

Disease symptoms include speckled browning and abnormal hardening of internal tissues. Browning varies from bright yellowish or reddish-brown to a very dark dull brown. Hardening is most pronounced in tissues that are brown, but adjacent tissues of normal color may be abnormally crisp. These symptoms may affect the flesh of the entire fruit or be limited to a single fruitlet.

Infection is thought to take place at flowering or shortly thereafter, but the disease develops only during ripening. Large fruit seem more susceptible than small, possibly because they are generally less acid.

POSTHARVEST DISEASES OF VEGETABLES

CARROT
Bacterial soft rot
Pathogen: *Erwinia carotovora* (Jones) Bergey et al.

The bacteria that causes soft rot may affect carrot foliage in the field and may be present in soil adhering to carrot roots. Harvest bruises and insect wounds offer entrance into the root.

The soft rot bacteria decay carrot tissues rapidly at room temperature. Carrot tissues become watersoaked and the middle lamella is solubilized by the action of pectolytic enzymes of the fungus. The result is a soft, slimy, often foul-smelling, wet mass of carrot cells (fig. 18.91).

Bacterial soft rot seldom occurs at temperatures below 5°C (41°F), although the minimum temperature for growth of the bacterium is as low as 0° to 2°C (32° to 36°F). Thus, serious loss from bacterial soft rot usually occurs when carrots have not been adequately cooled or cooling has been delayed.

Gray mold rot
Sexual state: *Botryotinia fuckeliana* (de Bary) Whetzel
Asexual state: *Botrytis cinerea* Pers.:Fr
                  Note: See gray mold rot of apple and pear for synonymy.

Gray mold rot is common in carrot, particularly during storage periods of 5 months or longer. Lesions are typically covered with mycelium of the fungus, which is initially white. In time, the fungus sporulates abundantly, turning gray (fig. 18.92). Later, black, irregular sclerotia appear on the host's surface.

Conidia of *B. cinerea* are capable of directly penetrating many tissues. It appears likely, however, that most infections in storage occur from mycelium or sclerotia in adhering soil particles entering wounds made during harvesting and handling. Mycelia grow from diseased roots to nearby sound roots, like *S.*

sclerotiorum or *S. minor,* to form a nest of rotting carrots.

Carrot white rot
Sexual state: 1) *Sclerotinia sclerotiorum* (Lib.) de Bary
                  2) *Sclerotinia minor* Jagger
Asexual state: Unknown
                  See white rot of snap beans for synonymy of the organisms and figure 18.124 for illustration of apothecia of *Sclerotinia sclerotiorum.*

White rot organisms cause a soft, but not slimy, rot of carrot roots in storage at 0°C (32°F). The disease is likely to become serious if the organisms were active before harvest, when a cottony white mycelium develops on the root (fig. 18.93). In time, distinctive black, rounded sclerotia form.

The minute microconidia that are produced appear to serve as spermatia (male cells), permitting development of the perfect (ascospore) state. Ascospores play little or no part in infecting mature carrot roots in storage. Instead, mycelia in soil may produce infection hyphae that penetrate the host cuticle. More often, the mycelium in soil invades wounds caused by harvesting or handling.

Mycelia may grow from a colonized root to penetrate and infect nearby sound roots. The result is an ever-enlarging nest of rotting roots bound together by fungus mycelium.

Fusarium dry rot
Several *Fusarium* spp. attack carrot, causing a relatively dry, spongy rot in storage. The disease usually is observed only after several months in storage, generally above 8° to 10°C (46° to 50°F).

Black mold
Sexual state: Unknown
Asexual state: *Thielaviopsis basicola* (Berk. & Br.) Ferr.
                  = *Trichocladium basicola* (Berk. & Br.) Carmichael

Black mold is occasionally found on carrots (fig. 18.94). *Thielaviopsis basicola* attacks the root systems of many growing plants, particularly cucurbits, legumes, and members of the Solanaceae. This fungus produces two types of conidia. Hyaline conidia are produced internally within a conidiophore and are expelled in chains of several conidia that soon break apart as they age, and brown-walled chlamydospores are produced on hyaline conidiophores.

Black mold develops only if temperatures exceed 5°C (41°F), and possibly after the root has become senescent.

Crater rot
Sexual state: Unknown
Asexual state: *Rhizoctonia carotae* Rader
High losses due to crater rot have been experienced in carrot storages in the northeastern United States and in northern Europe. *Rhizoctonia carotae* resembles *Rhizoctonia solani,* but it is known to attack only carrot. Clamp connections on the mycelium indicate that these two fungi are basidiomycetes and may be closely related.

The pathogen grows at temperatures from below 0° to 24°C (<32° to 75°F) with an optimum of 21°C (70°F). Humidities near saturation are associated with lesion development. Inoculum is present in carrot fields, and the root may be infected at or soon after harvest.

Pits develop in the root and become large, sunken lesions resembling a crater (fig. 18.95). Avoidance of injuries to roots during harvesting and handling reduces crater rot. Losses can be further prevented by reducing storage RH below 95% and reducing temperature to 0°C (32°F).

CELERY
Bacterial soft rot
Pathogen: *Erwinia carotovora* var. *caro-tovora* (Jones) Dye
Bacterial soft rot advances down the petioles from the leaflet. Sometimes a single petiole is involved with little or no tendency for the diseased petiole to infect others (figs. 18.96 and 18.97). Contributing to bacterial soft rot are such problems as inadequate or delayed cooling or warming in transit.

Sclerotinia pink rot
Sexual state: *Sclerotinia sclerotiorum* (Lib.)
                de Bary
Asexual state: Unknown
                Note: See Sclerotinia rot of
                kiwifruit for synonymy.
Pink rot is caused by the fungus in soil, generally present where celery has been grown. Mycelium in the soil may penetrate healthy plants at the base of petioles. Affected petioles become pink and soft (fig. 18.98), either

in the field or during storage and transit. For more information on the disease, see the sections on white rot in lettuce, snap bean, and carrot.

LETTUCE
Bacterial soft rot
Pathogen: *Erwinia carotovora* var. *caro-tovora* (Jones) Dye
Bacterial soft rot may start in the field, where *Erwinia carotovora* is normally present. Leaves die as the infection progresses, often with the production of slime. The greatest potential for loss is as a postharvest disease (fig. 18.99). Commonly the disease follows delays in cooling or inadequate heat removal, giving the bacterium ideal conditions for growth. Often lettuce leaves appear to have become senescent and highly susceptible to rot before soft rot develops.

Lettuce white rot
Sexual state: (1)   *Sclerotinia sclerotiorum*
                      (Lib.) de Bary
                (2)   *Sclerotinia minor* Jagger
Asexual state: Unknown
                Note: See Sclerotinia rot of
                kiwifruit for synonyms.
These fungi cause the disease "lettuce drop" in the field. The unharvested head yellows and collapses as a consequence of infection. "White rot" is the name commonly given to postharvest rot of lettuce and many other commodities. The white mycelium grows abundantly on the rotting host and the fungus extends mycelial strands into sound lettuce heads which become infected, forming a nest of rotting heads.

The two causal organisms differ primarily in the size of the sclerotia they produce. *Sclerotinia sclerotiorum* produces relatively large sclerotia; those of *S. minor* are much smaller. Neither species produces asexual spores. Both species produce small flesh-colored flat or saucer-shaped apothecia (mushrooms), which eject ascospores into the air during a short period in spring (see fig. 18.124).

The fungus, growing saprophytically on organic matter in the soil, may contact a plant and the mycelium penetrates directly to establish infection. Ascospores landing on lettuce leaves may also establish infections, depending upon local growing and climatic conditions.

Gray mold rot

Sexual state: *Botryotinia fuckeliana* (de Bary) Whetzel

Asexual state: *Botrytis cinerea* Pers.:Fr.

Note: See gray mold rot of apple and pear for synonymy.

Gray mold rot resembles white rot in many respects. In the field or greenhouse the fungus reduces a lettuce head to a slimy rotten mass. As with white mold, the gray mold fungus has a wide and somewhat similar range of host plants.

Botrytis cinerea produces abundant conidia, causing a gray color, in contrast to *Sclerotinia* spp., which do not produce conidia. Sclerotia of *B. cinerea* function like those of *Sclerotinia* spp., but their shape is very irregular. Sexual spores are rare and appear to have little or no role in rotting host products.

Botrytis cinerea spreads by contact infections, like *Sclerotinia* spp., if diseased heads are mixed with healthy heads. Prompt cooling to near 0°C (32°F) and maintaining low temperature during transport and marketing minimizes losses.

MELONS
Sour rot

Sexual state: *Galactomyces geotrichum* (Butl. & Peter.) Redh. & Mall.

Asexual state: *Geotrichum candidum* Lk.

Note: See sour rot of citrus for synonymy.

The sour rot fungus is found in the soil and in plant debris in the soil. It can colonize wounds at any point on the fruit. On muskmelons the fungus often colonizes the fresh stem scars. In advanced stages, the rot may convert the interior of the melon to a slimy mass with a sour, disagreeable odor highly attractive to vinegar flies (*Drosophila* spp.). The disease is also found, in a lesser incidence, in watermelon, usually as a stem-end rot but sometimes at the blossom end or at wounds. The flesh of the melon may eventually liquefy (figs. 18.100, 18.101, and 18.102).

Geotrichum candidum is widely distributed in most soils, and stems are likely contaminated by soil containing propagules of the fungus. Arthrospores of the fungus are spread by wind or wind-driven mist or splashing water. Vinegar flies attracted to rotting fruit may transport spores from fruit to fruit. Melons become more susceptible to the sour rot fungus as they mature and ripen.

Rhizopus rot

Pathogen: *Rhizopus stolonifer* (Ehrenb.:Fr.) Vuill.

Note: See Rhizopus rot of stone fruits for synonymy.

Rhizopus stolonifer and possibly other *Rhizopus* spp. infect melons in injuries or at the stem scar (fig. 18.103). Stem scar infections are particularly common in cantaloupe. The resulting rot is extremely soft, and considerable liquid accumulates by the time the fungus has involved much of the fruit. At elevated temperatures the fungus colonizes the fruit very rapidly. Optimal temperature for growth of *Rhizopus stolonifer* is about 24° to 27°C (75° to 80°F); the minimum is 5°C (41°F).

Fusarium rot

Fusarium spp. infect melons in contact with the soil before harvest. Development of the disease occurs as the fruit ripen. Rotted tissues may develop into voids. Disease development is slow enough that losses in California are not excessive. Cantaloupes that have been chilled may develop mold on the fruit surface.

Fusarium roseum Lk: Fr. occasionally develops as a stem-end rot of melons (fig. 18.104).

Trichothecium rot

Sexual state: Unknown

Asexual state: *Trichothecium roseum* (Pers.:Fr.) Link

Note: See pink rot of apple and pear for synonymy.

The fungus *Trichothecium roseum* occasionally attacks melons, particularly honeydew melon (fig. 18.105). A stem-end rot is usually produced, but all deep wounds are probably infection courts. The fungus is often found in melons nearing the end of their postharvest life. Usually recognized by the rusty-red color of the sporulating fungus, the disease is sometimes confused with Fusarium stem-end rot, which may appear pink or red.

Usually of little consequence, the disease does produce a potent toxin that is strictly confined to the lesion. Chances that someone may consume the rotted lesion tissue appear minuscule.

Botrytis rot

Sexual state: *Botryotinia fuckeliana* (de Bary) Whetzel

Asexual state: *Botrytis cinerea* Pers.:Fr.
Note: See gray mold rot of apple and pear for synonymy.

Botrytis rot is found rarely as a stem-end rot of watermelon (fig. 18.106) or as invader of mechanical wounds. Botrytis rot usually follows unseasonably cool, wet weather.

Lasiodiplodia rot

Sexual state: *Botryosphaeria rhodina* (Cooke) Arx

Asexual state: *Lasiodiplodia theobromae* (Pat.) Griff. & Maubl.
Note: The connection between *Botryosphaeria rhodina* and *Lasiodiplodia theobromae* is tenuous. See Lasiodiplodia rot of banana for synonymy.

Lasiodiplodia rot is occasionally found as a postharvest rot of California watermelon (fig. 18.107). More serious losses are found in watermelon grown in humid climates, where honeydew melon and cantaloupe are also attacked. The disease is first seen as a shriveling and drying of the stem followed by browning of the area around the stem, which progressively enlarges as the disease develops. The cut flesh is noticeably softened and lightly browned. If the cut melon is exposed to the air for a few hours, the diseased areas become black (fig. 18.108). The disease develops rapidly in the fruit at temperatures of 25° to 30°C (77° to 86°F) but slowly or not at all at 10°C (50°F).

Gummy stem blight and black rot

Sexual state: *Didymella bryoniae* (Auersw.) Rehm.
= *Mycosphaerella citrullina* (C.O.Sm.) Gross.
= *Didymella melonis* Pass.
= *Mycosphaerella melonis* (Pass.) Chiu & Walker

Asexual state: *Phoma cucurbitacearum* (Fr.:Fr.) Sacc.

Gummy stem blight, found primarily in warm, humid growing areas, is a field disease whose infections sometimes cause postharvest losses in cucurbits. In the field, nodes of plants may appear oily green, and sap exudes that may partially dry to form drops of dark gum, giving the disease its name. On fruit, the first symptoms developing after harvest are dark, watersoaked spots anywhere on the fruit surface. Mature lesions are sunken, may show a pattern of concentric rings, and turn black with small pycnidia.

Anthracnose

Sexual state: *Glomerella lagenarium* F. Stevens.

Asexual state: *Colletotrichum orbiculare* (Berk. & Mont.) Arx
= *Gloeosporium orbiculare* Berk. & Mont.
= *Colletotrichum lagenarium* (Pass.) Ellis & Halst.
= *Gloeosporium lagenarium* (Pass.) Sacc.

The Anthracnose fungus can penetrate the cuticle and epidermis of leaves, stems, and fruit. With fruit, quiescent infections may be produced with no external evidence of the disease. As fruit mature and ripen the latent infections become active. Watersoaked spots appear, becoming sunken and black with the formation of acervuli. Spores are produced of a rose, brick red, or orange color as viewed en masse. The spores are distributed by water, wind-driven mist, insects, or pickers' hands.

ONION
Neck rot

Sexual state: (1)   Unknown
(2)   *Botryotinia allii* (Sawada) Yamamoto
(3)   *Botryotinia squamosa* Vien.-Bourg.
= *Sclerotinia squamosa* (Vien.-Bourg.) Dennis

Asexual state: (1)   *Botrytis aclada* Fresen.
= Botrytis *allii* Munn.
(2)   *Botrytis byssoidea* Walker
(3)   *Botrytis squamosa* Walker

Botrytis neck rot is often the most serious postharvest disease of onion. The organisms may spot leaves and cause tip dieback in the field, but the bulb is seldom affected before harvest. The disease usually progresses from the cut leaves into the inner bulb scales, which become watersoaked and light brown to dark brown. Gray fungal growth may be abundant, usually in the neck area, and sporulation may occur (fig. 18.109).

The fungus persists from one season to the next by growing on onion waste or by sclerotia in the soil. Its conidia are dispersed by wind or splashing rain.

To minimize losses from neck rot, onions should be properly cured by drying with good air circulation around them to prevent accumulation of moisture. Foliage should have matured before leaves are cut. Bulbs should not become sunburned, and every effort should be made to avoid cuts, bruises, or abrasions. Best storage is at 0°C (32°F) and 70 to 75% RH.

Bacterial soft rot

Pathogen: *Erwinia carotovora* var. *carotovora* (Jones) Dye

Bacterial soft rot may occur in high humidity and high temperatures. The bacterium gains entrance at cut leaves or by contamination of cuts or bruises in bulbs. Invaded fleshy scales become watersoaked and appear yellow or light brown (fig. 18.110). Scales may become thoroughly invaded. Bulbs with leaves cut before they are completely mature or bulbs that have not been well cured are likely to be affected.

Smudge

Sexual state: Unknown
Asexual state: *Colletotrichum circinans* (Berk.) Vogl.
= *Vermicularia circinans* Berk.
= *Colletotrichum dematium* (Pers.) Grove f. *circinans* (Berk.) Arx

Smudge is limited to the dry, outer scale leaves surrounding the bulb. Black areas on the dry scales may appear to be smeared but concentric black circles are frequently formed (fig. 18.111).

The fungus may be transported long distances on bulbs or onion sets and may persist on onion residues. Conidia of the fungus may be disseminated by wind or splashing rain in the field. The disease may be particularly prevalent in storage if rain has occurred during harvest.

Fusarium bulb rot

Sexual state: Unknown
Asexual state: *Fusarium oxysporum* Schlect.:Fr. f. sp. *cepae* (Hans.) Snyder & Hans.
Fusarium zonatum (Sherb.) Wr.

Fusarium bulb rot is caused by the *Fusarium* spp. listed above and probably by others as well. Infection may start in the field. The fungus may directly penetrate roots or bulbs or may enter via injuries caused by soil insects.

Rot may develop at anywhere on the bulb, including the basal plate (fig. 18.112). Losses are greatest in storages with high humidity or in bulbs poorly cured before storage (fig. 18.113).

Black mold

Sexual state: Unknown
Asexual state: *Aspergillus niger* v. Tiegh.

Black mold is most common in hot growing areas, but the fungus occurs wherever onions are grown. It can often be observed on dead plant debris where considerable moisture is present and the temperatures are high. The fungus grows slowly at 13°C (55°F) but the disease occurs at temperatures of 16° to 40°C (60° to 104°F).

The most common indication of the disease is the occurrence of black, powdery spore masses on the dry outer scales or between dry scales and the outermost fleshy scales. Under hot, humid conditions the fungus may cause a slow rot of bulbs. However, the main loss is usually in appearance and marketability of the bulbs.

Blue mold rot

Sexual state: Unknown
Asexual state: *Penicillium expansum* Link

Blue mold rot may be present in storage, causing a wet, soft lesion. Rotting can occur in storage at 0°C (32°F), but serious losses seldom occur if bulbs are properly cured and are not stressed by unfavorable temperatures or humidity levels. The disease is readily recognized by the blue to blue-green color of the sporulating fungus.

POTATO
Bacterial soft rot

Pathogen: *Erwinia carotovora* var. *carotovora* (Jones) Dye
E. carotovora var. *atroseptica* (Van Hall) Dye

In addition to the two main causal agents of bacterial soft rot named above, species belonging to the genera *Pseudomonas*, *Bacillus*, and *Clostridium* sometimes cause similar diseases in potato. *Erwinia carotovora* cells are motile with peritrichous flagella, rod shaped, and gram-negative. The bacteria enter tubers via lenticels or wounds.

Water on the surface of stored tubers reduces aeration and predisposes them to infection. Infections at lenticels appear watersoaked and circular at first. Rotting tissues become wet and cream or tan colored compared to sound tissues (fig. 18.114). The rotting tissues are soft and watery. With time infected tubers develop a foul odor.

Tubers for storage should be harvested when mature with minimal mechanical injuries. Ventilation prevents formation of water films or localized accumulation of elevated carbon dioxide. Tubers should not be washed before storage and when washed before marketing should be dried as soon as possible. Wash water should be changed frequently and should be treated with chlorine to lower the level of viable cells.

Ring rot
Pathogen: *Corynebacterium sepedonicum* (Spieck. & Knott.) Skapt. & Burkh.

The ring rot bacterium, *Corynebacterium sepedonicum*, is a gram-positive, nonmotile organism having predominantly wedge-shaped cells (0.4 to 0.6 by 0.8 to 1.2 μm). The bacterium is slow growing on media. The disease overwinters in infected tubers in the field or in storage. Infection occurs through tuber wounds. Contamination of knives used for cutting seed tubers or various machinery transmits the bacterial slime to tuber seed pieces. Harvesting operations in which tubers are wounded may also spread the disease (fig. 18.115).

Late blight
Pathogen: *Phytophthora infestans* (Mont.) de Bary.

Late blight (fig. 18.116) caused the potato famine in Europe in the mid-nineteenth century. It continues to threaten potatoes wherever they are grown, except in certain hot, dry, irrigated areas.

The disease causes serious losses in the field. Tubers that have been infected in the field may rot in storage (fig. 18.117). Tuber infections are most likely when wet weather occurs during harvest. Little spread occurs under good storage conditions.

Fusarium dry rot
Sexual state: Unknown
Asexual state: (1) *Fusarium solani* (Mart.) Sacc.
(2) *F. roseum* (Lk.) Snyd. & Hans.

Fusarium dry rot is found wherever potatoes are grown and handled. Although the rot lesion is characteristically dry, secondary infections by soft-rot bacteria soften the lesion and make it wet. A disagreeable odor, often associated with bacterial soft rot, may develop.

Fusarium dry rot fungi are believed to be incapable of penetrating the tuber periderm or lenticels. Some infections are associated with insect or rodent activity, and the pathogens may become secondary organisms invading lesions of other fungus diseases. Pathogens enter the tuber through mechanical injuries, usually cuts and periderm-breaking bruises inflicted during harvesting and handling.

Tubers are relatively resistant to Fusarium dry rot when harvested, but susceptibility increases during storage (fig. 18.118). Disease development is most rapid at temperatures of 15° to 20°C (59° to 68°F). Cuts in tubers result in the deposition of suberin, leading to the formation of a disease-preventing periderm barrier within several days at 20°C (68°F) and somewhat longer at lower temperatures.

Black scurf
Sexual state: *Thanatephorus cucumeris* (Frank) Donk
Partial synonymy = *Corticium areolatum* Stahel
= *Pellicularia filamentosa* (Pat.) Rogers
= *Ceratobasidium filamentosum* (Pat.) Olive
= *Hypochnus filamentosus* Pat.
= *Corticium praticola* Kotila
= *Corticium sasakii* (Shirai) Matsumoto
= *Hypochnus sasakii* Shirai
= *Pellicularia sasakii* (Shirai) Ito
= *Botryobasidium solani* (Prill. & Delacr.) Donk
= *Corticium solani* Prill. & Delacr.

Asexual state: *Rhizoctonia solani* Kühn
= *Moniliopsis solani* Kühn
= *Rhizoctonia macrosclerotia* J. Matz
= *Rhizoctonia microsclerotia* J. Matz

Black scurf occurs in the field where sclerotia on tubers or mycelium in plant debris provide the inoculum. Potato stems, roots, and stolons of the growing plants may be infected. Sclerotial development on the tuber occurs in a favorable environment (low soil temperature and high moisture level).

In the harvested tuber the disease is usually observed as black or dark brown sclerotia that are tightly appressed to the periderm (fig. 18.119), often described as "dirt that won't wash off." Sclerotia are usually a few millimeters in diameter and are irregular in size and shape. Sclerotia on tubers do not ordinarily cause them to rot, but their presence detracts from their appearance.

Other diseases. Many other diseases of tubers after harvest may be periodically or locally important.

Gangrene. Causal agents of gangrene are *Phoma exigua* Desm. var. *foveta* (Foister) Boerema and *Phoma exigua* Desm. var. *exigua* Sutton & Water. The former is the more virulent and is found in most northern European countries. Infected seed tubers provide the initial inoculum. Spores of the fungus are washed through the soil, and tuber infections may occur at that time, but infection more commonly occurs in wounds caused by harvesting or handling.

Gray mold. Botrytis cinerea Pers.:Fr. may attack foliage in the field and may occasionally cause significant rotting of tubers in storage if curing is not done properly and if storage temperature is above the optimal range.

White mold. Causal organisms of white mold, *Sclerotinia sclerotiorum* (Lib.) de Bary and *S. minor* Jagger, are active in the growing crop. Tubers are infected in the field and may continue rotting during storage.

Sclerotium rot. Sclerotium rolfsii Sacc. is primarily a field disease, but tubers freshly infected at harvest continue to rot in storage if temperatures are favorable for the fungus.

Pink rot. Pink rot of potato (fig. 18.120), caused by the pathogen *Phytophthora erythroseptica* Pethybr., is primarily a disease of tubers in the field. Basal stem decay and wilt of the top may occur. Affected tubers appear dull brown with lenticels and eyes still darker. Internal decay usually begins at the stem end; affected tissue becomes rubbery. Internal tissues gradually turn pink as decay occurs. Infected tubers may continue rotting during storage.

SNAP BEAN
Botrytis gray mold rot
Sexual state: *Botryotinia fuckeliana* (de Bary) Whetzel
Asexual state: *Botrytis cinerea* Pers.:Fr

Botrytis gray mold is frequently the most common disease of beans after harvest, and losses can be serious, particularly after an inclement growing season. Overhead irrigation may result in excessive moisture and added Botrytis rot in the field. Consequently, many infected pods may be inadvertently included among harvested pods.

Infected pods develop watersoaked lesions that may spread by mycelial contact from diseased to sound pods. Disease lesions or fungus conidia on pod surfaces provide opportunities for new infections. The fungus nests by contact infections from diseased to surrounding pods (fig. 18.121).

Snap bean white rot
Sexual state: (1) *Sclerotinia sclerotiorum* (Lib.) de Bary
(2) *Sclerotinia minor* Jagger
Asexual state: Unknown
See white rot of lettuce for added information and for synonymy of the organisms.

The two white rot pathogens are similar, differing primarily in the size of sclerotia. *Sclerotinia sclerotiorum* is larger and much more widespread than *S. minor*.

Disease after harvest results primarily from infections that have occurred before harvest (figs. 18.122 and 18.123). Disease in the field results from infections from mycelium produced by sclerotia or from ascospores produced by apothecia (small mushrooms) that grow from sclerotia under certain climatic conditions (fig. 18.124).

Anthracnose
Sexual state: *Glomerella lindemuthiana* Shear
Asexual state: *Colletotrichum lindemuthianum* (Sacc. & Magnus) Lams.-Scrib.

Anthracnose is common all over the world except in those areas with a very dry climate, such as the western United States. Typically, diseased pods develop black, sunken lesions. The center of each lesion contains a salmon-colored ooze from the many acervuli contained in the lesion (fig. 18.125).

Control where the climate favors the disease involves field application of fungicides such as benomyl. Resistant cultivars are successful against certain strains of the fungus.

SWEET POTATO
Rhizopus rot

Pathogen: *Rhizopus stolonifer* (Ehr.:Fr.) Lind.
Note: See Rhizopus rot of stone fruits for synonymy.

Rhizopus stolonifer is a ubiquitous fungus that grows on various types of organic matter. It attacks many species of ripe fruit (berries, stone fruits, tomato, papaya) and many other commodities when not refrigerated (fig. 18.126). Little pathogenic growth of the fungus occurs at temperatures below 5°C (41°F). Germinating spores of the fungus are incapable of penetrating an uninjured host.

As the disease progresses, rotting sweet potato roots soften and the flesh is partially liquefied. White mycelium protrudes through the surface at mechanical injuries, at breaks in the skin, or at lenticels. As the fungus sporulates, the aspect changes from white to black as black sporangia form. Mycelia grow from diseased to sound roots in storage but seldom penetrate well-cured roots.

Black rot

Sexual state: *Ceratocystis fimbriata* Ell. & Halst.
= *Ceratostomella fimbriata* (Ell.& Halst.) Elliott
= *Endoconidiophora fimbriata* (Ell.& Halst.) Davd.
= *Ophiostoma fimbriata* (Ellis & Halst.) Nannf.
Asexual state: *Chalara* sp.

Black rot may occur in the field and after harvest. Infected roots become black and tend to remain relatively firm. Lesions are usually associated with wounds and lenticels. The fungus produces abundant asexual spores. These include endoconidia that are single celled, thin walled, cylindrical,

smooth, colorless, and about 5 by 15 μm in size. Conidiophores are cylindrical structures often found on the surface of diseased areas.

Thick-walled, brown chlamydospores 9 to 18 by 6 to 13 μm are formed within affected tissues. Perithecia form on sprouts and roots, releasing ascospores as the cytoplasm around them dissolves.

Java black rot

Sexual state: *Botryosphaeria rhodina* (Cooke) Arx
Asexual state: *Lasiodiplodia theobromae* (Pat.) Griff. & Maubl.
Note: Refer to Lasiodiplodia rot of banana for synonymy. The connection between *Botryosphaeria rhodina* and *Lasiodiplodia theobromae* is tenuous.

The Java black rot fungus *Lasiodiplodia theobromae* was at first thought to have been imported in sweet potatoes from Java. However, the fungus was determined to already be present on many hosts, and it appears that the disease is found nearly everywhere sweet potatoes are grown.

The fungus may persist in the field in debris from sweet potato or from alternate host crops or susceptible weeds. Spores are associated with the fungus growing in the soil, although transmission via air, splashing water, wind-driven mist, and insects occurs. The fungus cannot penetrate the root periderm, but must enter through wounds in the side of the root or at the ends.

Exposure of roots to chilling temperatures renders them more susceptible to the disease. Similarly, roots become more susceptible after long-term storage (5 to 8 months or more), probably as a consequence of root senescence.

Fusarium surface rot

Sexual state: Unknown
Asexual state: *Fusarium oxysporum* Schlect.

Fusarium surface rot affects tissues immediately below the surface, seldom extending into the root more than a centimeter or so. Rotting of tissues deep within the root may indicate that secondary organisms are present. The surface epidermal layers are not affected, but some darkening may appear.

Early symptoms are circular, slightly depressed, brown lesions. Typically, they appear slightly to moderately darkened when viewed at the surface of the root. Lesions enlarge and sometimes exhibit concentric zones (fig. 18.127).

Fusarium surface rot results from soil-borne inoculum. Most infections probably occur via soil contamination of wounds caused during harvest. Exposure of roots in the field to bright sunlight for more than an hour or two enhances incidence of disease. The disease is minimized by curing to encourage wound healing.

TOMATO
Alternaria rot
Sexual state: Unknown
Asexual state: *Alternaria alternata* (Fr.:Fr.) Keissl.
Note: See Alternaria rot in apple and pear for synonyms.
Lesions of Alternaria rot in tomato are flat at the fruit surface or sunken. Usually covered by the sporulating black mycelium of the fungus (fig. 18.128), the lesion extends into the flesh where it produces a firm, dry, blackened mass of tissue thoroughly ramified by the mycelium.

Alternaria alternata is believed to produce latent infections in the developing fruit in the field by directly penetrating the cuticle. Such infections seldom develop into disease lesions unless the fruit has been chilled. Upon chilling, however, rot lesions may develop at any point on the fruit surface. Mechanical injuries provide the fungus ready access to internal fruit tissues. A circle of lesions around the stem probably results from the tendency of fruit shoulders to be abraded during handling and transport of packed fruit.

The dramatic loss of resistance after fruit have been held for extended periods at temperatures below about 13°C (55°F) and particularly below 5°C (41°F) means that such temperatures should not be maintained for more than a few days. Mature-green fruit are most sensitive to chilling injury, followed by pink fruit, then red-ripe fruit. Resistance to Alternaria rot can be lost without the appearance of other symptoms of chilling injury.

A disease with similar symptoms is caused by a fungus having *Pleospora herbarum* (Pers.:Fr.) Rabenh. as the sexual state and *Stemphylium herbarum* Simmonds as the asexual state. Ordinarily the disease organ-

ism cannot easily be distinguished with the naked eye from the Alternaria rot organism.

Buckeye rot
Pathogen: *Phytophthora nicotiana* Breda de Haan var. *parasitica* (Dastur) Waterhouse
= *Phytophthora parasitica* Dastur
Buckeye rot is generally attributed to *Phytophthora nicotiana* var. *parasitica*; however, *P. capsici* Leonin and *P. drechsleri* Tucker have also been found to cause the disease. Other *Phytophthora* spp. may sometimes be involved.

Phytophthora spp. are soil inhabitants. During warm (18° to 22°C [64° to 72°F]), wet weather, the sporangia produced give rise to motile swimming spores that infect fruit in contact with soil. Splattering rain and wind-driven mist can deposit droplets of water containing spores on fruit.

Lesions are usually not sunken. One can often see patterns beneath the skin resembling overlapping rose petals (fig. 18.129). Preventing fruit contact with the soil reduces losses.

Gray mold rot and ghost spot
Sexual state: *Botryotinia fuckeliana* (de Bary) Whetzel
Asexual state: *Botrytis cinerea* Pers.:Fr.
The fungus causing gray mold rot or ghost spot is a common postharvest pathogen of many fruits, vegetables, and flowers in addition to tomato fruit. Its optimal growth occurs at 25°C (77°F), but it can rot produce more slowly at temperatures as low as −2° or −3°C (28° to 26°F). Consequently, disease develops readily at the lowest temperature the fruit can tolerate. The fungus can penetrate the fruit skin while still in the field. The more common loci of infection are mechanical injuries or growth cracks that often occur near the stem scar.

The most common disease symptom is a "dirty white" color of mycelium over the lesion. As the fungus sporulates the color darkens to gray or gray-brown, and the appearance is typical of *Botrytis cinerea* on many other hosts. Tissues appear water-soaked when invaded.

Ghost spots develop when conidia germinate on fruit under cool, humid conditions. Germ tubes penetrate the fruit epidermis. If the fruit is subsequently exposed to hot weather, the fungus is killed, but the "ghost spot" remains (fig. 18.130).

Sour rot

Sexual state: *Galactomyces geotrichum* (Butl. & Peters.) Redh. & Mall.

Asexual state: *Geotrichum candidum* Link
                Note: See sour rot of citrus for synonymy.

Sour rot is present wherever tomatoes are grown. In California, it is particularly prevalent in processing tomato fields, where rotting fruit provide abundant inoculum that may be carried by vinegar flies and other insects. Fruit in contact with the ground can become infected, particularly if rain occurs or the soil is wet from irrigation.

Fresh market tomatoes (mature-green or pink) may also become diseased. On such fruit the lesions are watersoaked and bleached, with the surface of the lesion dull rather than shiny (fig. 18.131). The infection often starts at the stem scar from which the lesion may extend down the side of the fruit.

Mature-green fruit that have been chilled several days at 0° to 5°C (32° to 41°F) become susceptible to sour rot (fig. 18.132). Infections occur from 5° to 38°C (41° to 100°F); optimal temperature for disease development is 30°C (86°F).

Rhizopus rot

Pathogen: *Rhizopus stolonifer* (Ehrenb.:F.) Vuill.
                Note: See Rhizopus or "whiskers" rot of stone fruits for synonyms.

Rhizopus stolonifer lesions are first noted as watersoaked areas beneath the fruit skin. Often the skin ruptures after the lesion becomes large. The fungus becomes evident, first as white to gray mycelium and sporangiophores, but the aspect becomes black as the fungus sporulates (fig. 18.133). The fungus nests when mycelium from a diseased fruit penetrates nearby sound fruit (fig. 18.134).

Rhizopus rot develops most rapidly at 24° to 27°C (75° to 81°F). Pathogenic growth is slow as the temperature approaches 10°C (50°F) and essentially stops at 5°C (41°F).

Anthracnose

Sexual state: Unknown

Asexual state: *Colletotrichum coccodes* (Wallr.) Hughes
                = *Colletotrichum atramentarium* (Berk & Broome) Taubenhaus

Anthracnose caused by the above organism, and possibly by several other *Colletotrichum* spp., occurs in warm, humid growing areas. Lesions are slightly sunken and are typically about 1 to 1.5 cm (0.4 to 0.6 in) in diameter. Acervuli in the lesions produce many conidia in a slimy mass (fig. 18.135). Conidia are capable of penetrating sound fruit. However, the fruit must be ripe for the disease to become serious. On green fruit, a brown flecking indicates penetration without successful lesion development.

Other diseases. Penicillium rot is occasionally found on tomato fruit. The causal organism is commonly *Penicillium expansum* but may include other *Penicillium* spp. Penicillium rot often occurs in overripe or chilled fruit.

Fusarium rot is sometimes found causing a lesion that may be mistaken for anthracnose to the naked eye. The fungus may belong to the *Fusarium roseum* group, but definite identification has not been determined. The lesions appear to occur only on ripe fruit (fig. 18.136).

GLOSSARY

abiotic disease. A disease that results from a nonliving cause.

acervulus, pl. acervuli. A fruiting body of certain imperfect fungi. A shallow, saucer-shaped structure with a layer of conidiophores that bear conidia; found in anthracnose diseases and fungi belonging to the genus *Pezicula*.

anamorph. A fungus in the asexual state.

Anthracnose. Any disease caused by fungi that produce asexual spores in acervuli.

apothecium, pl. apothecia. Cuplike, ascus-containing fungus fruiting body. See figure 18.124.

appressorium, pl. appressoria. Bulbous or lobed swelling of a hyphal tip, often held in place by a gelatinous secretion, that forms an infection peg to penetrate plant tissues. See figures 18.2 and 18.3.

arthrospores. Spores formed by the simultaneous or random fragmentation of hyphae.

asexual state. Production of spores without previous fusion of gametes; a form of vegetative reproduction.

chlamydospore. A thick-walled spore usually resistant to adverse environmental conditions.

conidiophore. Specialized portion of mycelium on which conidia are produced.

conidium, pl. conidia. Any asexual spore except sporangiospores and chlamydospores. Produced on the conidiophore, a specialized portion of the mycelium.

fructification. Production of spores by fungi, fungus fruiting body, or spore-bearing structure.

fruiting body. A complex fungus structure containing or bearing spores, as a mushroom, perithecium, pycnidium, etc.

fungicidal. Capable of killing or inhibiting fungi.

fungicide. A chemical or physical agent that kills or inhibits fungi.

fungistatic. An agent that prevents development of fungi without destroying them; term applied to this action.

Fungi Imperfecti. A class of fungi without a known sexual state.

fungus, pl. fungi. Organisms having no chlorophyll, with reproduction by sexual or asexual spores and not by fission, and usually with mycelium having well-marked nuclei.

gamete. A reproductive cell with a haploid nucleus capable of fusion with that of a gamete of an opposite mating type.

genus, pl. genera. A taxonomic group above species and below the rank of family. The first name of a binomial such as *Monilinia fructicola*.

germ tube. The fungus hypha formed upon germination of a spore.

germination. The swelling of a spore and the protrusion of hyphae. The beginning of growth.

gram negative. A microorganism that does not retain the purple dye used in Gram's method.

haploid. Germ cell that has half the number of chromosomes present in the normal somatic cell.

host. A living organism upon which a fungus or bacterium grows and obtains sustenance.

hyaline. Colorless, transparent, or nearly so.

hypha, pl. hyphae. A single thread of fungus mycelium, a threadlike structure that increases length by growth at the tip and forms lateral branches.

imperfect fungus. One lacking any sexual reproductive stage.

in vitro. Growth occurring on nonliving substrates.

in vivo. Growth occurring on living plants or animals.

incubation period. The time between inoculation of a plant or plant tissue and the first observed disease reaction.

infect. Entrance of a pathogen into a plant where it grows and obtains sustenance.

infection court. The place where an infection may take place.

infectious. Term applied to a disease that may be communicated from one plant, or plant part, to another.

inoculate. To place inoculum in an infection court.

inoculum, pl. inocula. Infectious parts of a pathogen, such as a spore or bacterial cell, that can be transferred to healthy tissue and cause disease.

lesion. A localized spot of diseased tissue.

molds. Fungi with conspicuous mycelium or spore masses.

mummified. Dried up and shriveled, as in fruit affected by the brown rot pathogen.

mycelium, pl. mycelia. A mass of fungus hyphae.

mycology. The science dealing with fungi.

necrosis, pl. necroses. Death of plant tissues, as in rots, blights, and cankers.

parasite. An organism that lives on or in a second organism, usually causing disease in the latter.

pathogen. Any organism or factor causing disease.

pathogenic. Capable of causing disease.

pathology. The science of disease.

perfect state. Capable of sexual reproduction.

perithecium, pl. perithecia. Ascospore-producing body.

peritrichous. Having hairlike flagella all over the surface.

pycnidium, pl. pycnidia. Flasklike fruiting body containing conidia.

resistance. Ability of a host plant or plant part to suppress or retard the activity of a pathogen or other injurious factor.

saprophyte. Organism that feeds exclusively on lifeless organic matter.

sclerotium, pl. sclerotia. A resting mass of

fungus tissue, often more or less spherical, usually not bearing spores.

sexual state. The condition in which spores are produced following fusion of gametes.

sporangiophore. Hypha bearing a sporangium.

sporangiospore. A spore produced in a sporangium, as in *Rhizopus stolonifer* or *Mucor piriformis*.

sporangium, pl. **sporangia.** An organ producing nonsexual spores with a more or less spherical wall.

spore. A single- to many-celled reproductive body, in fungi and/or other lower plants.

sporulate. To produce spores.

sterilize. To remove or destroy all living organisms on or in an object or material.

substrate. The substance or object on which an organism lives and from which it obtains nourishment.

suscept. Any plant susceptible to infection by a given pathogen.

susceptible. Inability to oppose the operation of an injurious or pathogenic agent.

teliomorph. A fungus of the sexual state.

toxin. A poison formed by an organism.

viability. State of being alive.

virulent. Highly pathogenic; with a strong capacity for causing disease.

REFERENCES

Beattie, B. B., W. B. McGlasson, and N. L. Wade, eds. 1990. Postharvest diseases of horticultural produce. Vol. 1. Temperate fruit. Victoria, Australia: CSIRO. 84 pp.

Brown, G. E. 1975. Factors affecting postharvest development of *Colletotrichum gloeosporioides* in citrus fruits. Phytopathol. 65:404–409.

Caruso, F. L., and D. C. Ramsdell, eds. 1995. Compendium of blueberry and cranberry diseases. St. Paul: APS Press. 126 pp.

Clark, C. A., and J. W. Moyer. 1988. Compendium of sweet potato diseases. St. Paul: APS Press. 96 pp.

Coates, L., T. Cooke, D. Persley, B. Beattie, N. Wade, and R. Ridgway, eds. 1995. Postharvest diseases of horticultural produce. Vol. 2. Tropical fruit. Queensland, Australia: Dept. Plant Industries. 136 pp.

Davis, R. M., K. V. Subbarao, R. N. Raid, and E. A. Kurtz. 1997. Compendium of lettuce diseases. St. Paul: APS Press. 104 pp.

Eckert, J. W., and I. L. Eaks. 1989. Postharvest disorders and diseases of citrus fruits. In W. Reuther et al., eds., The citrus industry. Vol. V. Oakland: Univ. Calif. Div. Ag. and Nat. Res. 179–260.

Eckert, J. W., and J. M. Ogawa. 1985. The chemical control of postharvest diseases: Subtropical and tropical fruits. Annu. Rev. Phytopathol. 23:421–454.

———. 1988. The chemical control of postharvest diseases: Deciduous fruits, berries, vegetables and root/tuber crops. Annu. Rev. Phytopathol. 26:433–469.

Eckert, J. W., and N. F. Sommer. 1967. Control of diseases of fruits and vegetables by postharvest treatment. Annu. Rev. Phytopathol. 5:391–432.

El-Goorani, M. A., and N. F. Sommer. 1981. Fungistatic effects of modified atmospheres in fruit and vegetable storage. Hortic. Rev. 3:412–461.

Ellis, M. A., R. H. Converse, R. N. Williams, and B. Williamson, eds. 1991. Compendium of raspberry and blackberry diseases and insects. St. Paul: APS Press. 128 pp.

Farr, D. F., G. F. Bills, G. P. Chamuris, and A. Y. Rossman. 1989. Fungi on plants and plant products in the United States. St. Paul: APS Press. 1,252 pp.

Griffen, D. H. 1993. Fungal physiology. 2nd ed. New York: Wiley-Liss.

Harvey, J. M. 1978. Reduction of losses in fresh market fruits and vegetables. Annu. Rev. Phytopathol. 16:321–341.

Ismail, M. A., and G. E. Brown. 1975. Phenolic content during healing of 'Valencia' orange peel under high humidity. J. Am. Soc. Hortic. Sci. 100:249–251.

———. 1979. Postharvest wound healing in citrus fruit: Induction of phenylalanine ammonia-lyase in injured 'Valencia' orange flavedo. J. Am. Soc. Hortic. Sci. 104:126–129.

Ismail, M. A., R. L. Rouseeff, and G. E. Brown. 1978. Wound healing in citrus: Isolation and identification of 7-hydroxycoumarin (Umbelliferone) from grapefruit flavedo and its effect on *Penicillium digitatum* Sacc. HortSci. 13:358.

Jones, A. L., and H. S. Aldwinckle, eds. 1990. Compendium of apple and pear diseases. St. Paul: APS Press. 125 pp.

Jones, A. L., and T. B. Sutton. 1999. Diseases of tree fruits in the east. Michigan State Univ. Publ. NCR45.

Jones, J. B., J. P. Jones, R. E. Stall, and T. A. Zitter, eds. 1991. Compendium of tomato diseases. St. Paul: APS Press. 112 pp.

Kuc, J. 1972. Phytoalexins. Annu. Rev. Phytopathol. 10:207–232.

Kuc, J., and N. Lisker. 1978. Terpenoids and their role in wounded and infected plant storage tissue. In G. Kahl, ed., Biochemistry of wounded plant tissues. New York: de Gruyter. 203–242.

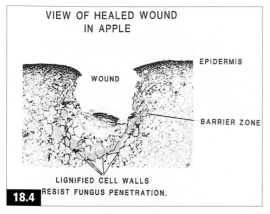

VIEW OF HEALED WOUND IN APPLE

EPIDERMIS

WOUND

BARRIER ZONE

LIGNIFIED CELL WALLS RESIST FUNGUS PENETRATION.

18.4

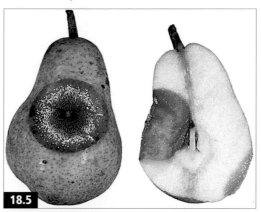

18.5

18.6

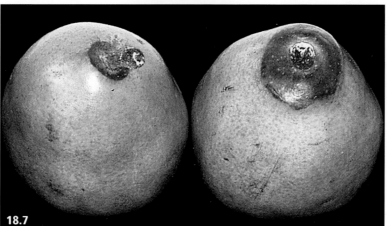

18.7

18.8

18.9

POSTHARVEST DISEASE SYMPTOMS

Figure 18.4

Wound healing in Yellow Newtown apple fruit after 6 months CA storage at 3.5°C (38°F). Note cell wall thickening near wound.

Figure 18.5

Blue mold rot (*Penicillium expansum*) in ripe Bartlett pear.

Figure 18.6

Stem-end blue mold rot (*Penicillium expansum*) of Granny Smith apple.

Figure 18.7

Stem-end blue mold rot (*Penicillium expansum*) of Beurre d'Anjou pear.

Figure 18.8

Gray mold rot (*Botrytis cinerea*) of Beurre d'Anjou pear.

Figure 18.9

Senescent style and stamen fragments in the subcalyx area of Bartlett pear.

POSTHARVEST DISEASE SYMPTOMS

Figure 18.10

Calyx-end rot (*Botrytis cinerea*) in Bartlett pear.

Figure 18.11

Bull's-eye rot (*Pezicula malicorticis*) in Yellow Newton apple.

Figure 18.12

Bull's-eye rot (*Pezicula malicorticis*) showing sporulating lesion.

Figure 18.13

Limb canker (*Pezicula malicorticis*) of apple tree. (Courtesy Dr. A. Helton, University of Idaho, Moscow)

Figure 18.14

Bitter rot (anthracnose) (*Colletotrichum gloeosporioides*) in Golden Delicious apple.

Figure 18.15

Black rot (*Botryosphaeria obtusa*) in Yellow Newtown apple.

Figure 18.16

White rot (*Botryosphaeria dothidea*) in Yellow Newtown apple.

Figure 18.17

Alternaria rot (*A. alternata*) in Yellow Newtown apple with delayed sunscald.

18.10

18.11

18.13

18.12

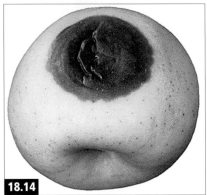

18.14

18.15

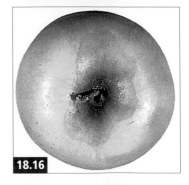

18.16

18.17

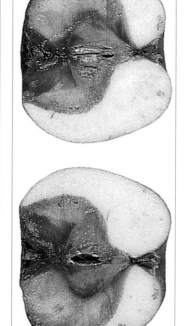

POSTHARVEST DISEASE SYMPTOMS

Figure 18.18

Alternaria stem-end rot (*A. alternata*) in chilled Yellow Newtown apple.

Figure 18.19

Cladosporium rot (*C. herbarum*) in ripe Bartlett pear.

Figure 18.20

Pleospora rot (*Stemphyllium botryosum*) in Yellow Newtown apple.

Figure 18.21

Phomopsis rot (*P. mali*) of Yellow Newtown apple.

Figure 18.22

Side rot (*Phialophora malorum*) in Beurre Bosc pear.

Figure 18.23

Phytophthora or sprinkler rot (*P. cactorum*) in Bartlett pear in the orchard.

Figure 18.24

Brown rot (*Monilinia fructicola*) in nectarines before harvest.

Figure 18.25

Brown rot (*M. fructicola*) nest of peaches.

POSTHARVEST DISEASE SYMPTOMS

Figure 18.26

Brown rot (*M. fructicola*) in fingernail wound in nectarine.

Figure 18.27

Stem-end rot (*M. fructicola*) of peach.

Figure 18.28

Nest of *Rhizopus*-infected peaches.

Figure 18.29

Rhizopus rot (*R. stolonifer*) in packed peaches.

Figure 18.30

Apricots 18 months after canning. (A) Fruit halves disintegrated because of contamination with Rhizopus-infected fruit juice with pectolytic enzymes not completely inactivated by heat of canning. (B) Treatment with 0.5 N sodium hydroxide for 2 minutes before canning inactivated pectolytic enzymes.

Figure 18.31

Mucor rot (*Mucor* spp.) in Chilean nectarines arriving in the United States.

Figure 18.32

Blue mold rot (*Penicillium expansum*) of nectarines stored at 0°C (32°F).

Figure 18.33

Alternaria rot (*A. alternata*) of plum.

Figure 18.34

Alternaria rot (*A. alternata*) of Nubiana plum.

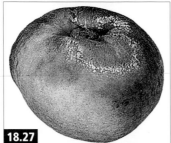

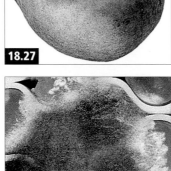

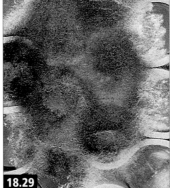

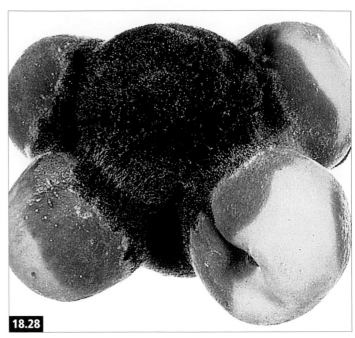

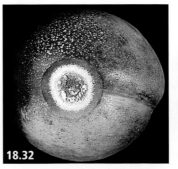

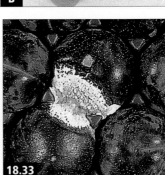

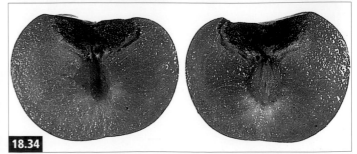

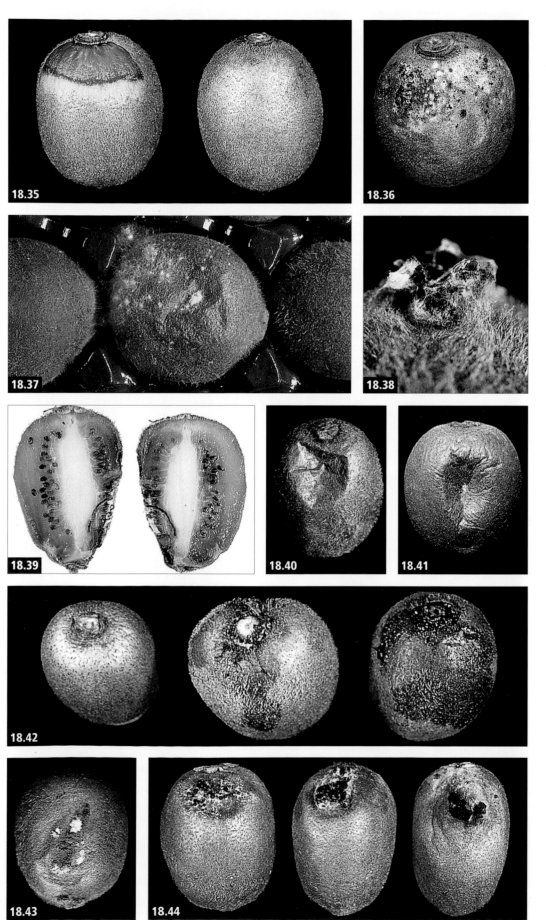

POSTHARVEST DISEASE SYMPTOMS

Figure 18.35

Botrytis stem-end rot (*Botrytis cinerea*) of kiwifruit.

Figure 18.36

Botrytis stem-end rot (*B. cinerea*) with sclerotia on the kiwifruit.

Figure 18.37

Nest of kiwifruit rotted by *Botrytis cinerea*.

Figure 18.38

Alternaria alternata growing over kiwifruit sepals.

Figure 18.39

Alternaria rot (*A. alternata*) in sunburned kiwifruit.

Figure 18.40

Dothiorella soft rot (*D. gregaria*) of kiwifruit.

Figure 18.41

Phoma rot (*Phoma* spp.) of kiwifruit.

Figure 18.42

Phomopsis stem-end rot (*Diaporthe actinidiae*) of kiwifruit with juice exudation.

Figure 18.43

Penicillium stem-end rot (*P. expansum*) of kiwifruit.

Figure 18.44

Buckshot rot (*Typhula* spp.) of stored kiwifruit.

POSTHARVEST DISEASE SYMPTOMS

Figure 18.45

Nest of strawberries rotted by *Botrytis cinerea*.

Figure 18.46

Various aspects of Botrytis rot in strawberry.

Figure 18.47

Rhizopus rot (*R. stolonifer*) of ripe or near-ripe strawberries in the field.

Figure 18.48

Mucor rot (*M. piriformis*) nesting in strawberry basket. Black sporangia have not yet formed on the sporangiophores of the fungus.

Figure 18.49

Anthracnose (*Colletotrichum acutatum*) on strawberries in the field.

Figure 18.50

Leather rot (*Phytophthora cactorum*) in green strawberries.

Figure 18.51

Leather rot (*Phytophthora cactorum*) in ripening strawberries.

Figure 18.52

Gray mold rot (*Botrytis cinerea*) in grapes stored at 0°C (32°F).

Figure 18.53

Penicillium rot (*P. expansum*) in grapes stored at 0°C (32°F).

18.45

18.46

18.47

18.48

18.49

18.50

18.51

18.52

18.53

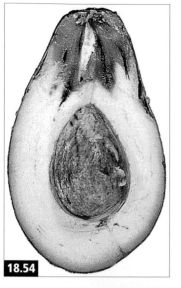

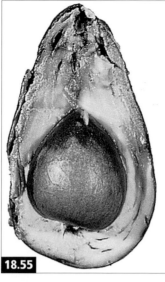

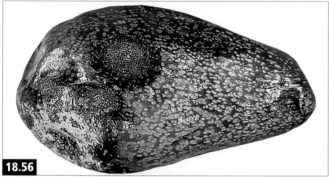

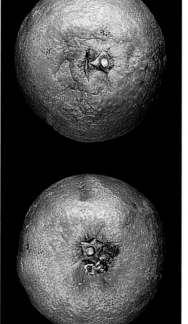

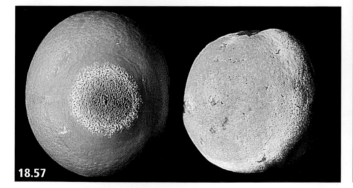

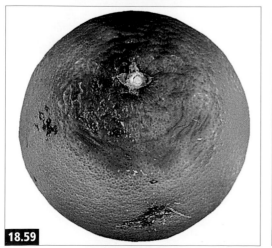

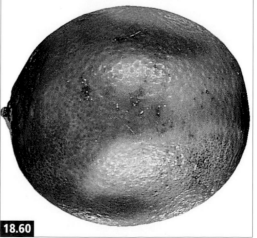

POSTHARVEST DISEASE SYMPTOMS

Figure 18.54

Dothiorella stem-end rot (*D. gregaria*) of avocado.

Figure 18.55

Lasiodioplodia stem-end rot (*L. theobromae*) of avocado.

Figure 18.56

Anthracnose (*Colletotrichum gloeosporioides*) of avocado.

Figure 18.57

Blue and green rots (*Penicillium italicum* and *P. digitatum*) in orange.

Figure 18.58

Citrus fruit brown rot (*Phytophthora citrophthora*).

Figure 18.59

Phomopsis stem-end rot (*P. citri*) of orange.

Figure 18.60

Lasiodiplodia stem-end rot (*L. theobromae*) of orange.

POSTHARVEST DISEASE SYMPTOMS

Figure 18.61

Alternaria stem-end rot (*Alternaria citri*) of grapefruit.

Figure 18.62

Alternaria rot (*A. citri*) of orange.

Figure 18.63

Anthracnose (*Colletotrichum gloeosporioides*) of Meyer lemon.

Figure 18.64

Sour rot (*Geotrichum candidum*) of navel orange.

Figure 18.65

Sclerotinia rot (*S. sclerotiorum*) of stored lemons.

Figure 18.66

Trichoderma rot (*T. viride*) of orange.

Figure 18.67

Botrytis rot (*B. cinerea*) of orange.

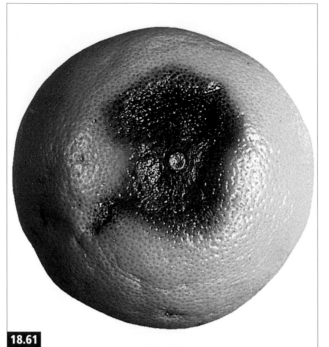

18.61

18.62

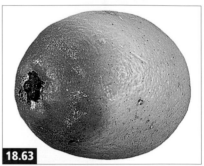

18.63

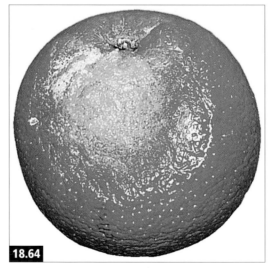

18.64

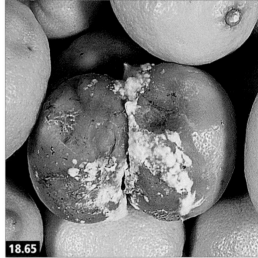

18.65

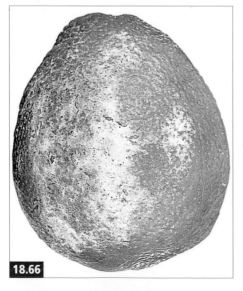

18.66

18.67

18.68

18.69

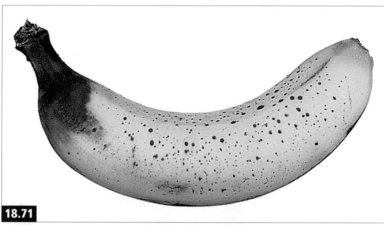

18.70

18.71

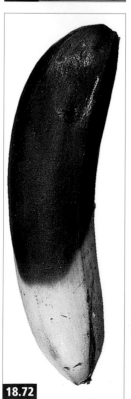

18.72

18.73

18.74

POSTHARVEST DISEASE SYMPTOMS

Figure 18.68

Crown rot of banana caused by several fungi, often in concert.

Figure 18.69

Creased stems of banana fingers and cut surfaces of crowns are loci of infection.

Figure 18.70

Anthracnose (*Colletotrichum musae*) of banana.

Figure 18.71

Lasiodiplodia stem-end rot (*L. theobromae*) of banana finger.

Figure 18.72

Thielaviopsis rot (*T. paradoxa*) of banana finger.

Figure 18.73

"Cigar-end" stylar rot disease of banana.

Figure 18.74

Fusarium blossom-end (*F. roseum*) disease of banana.

POSTHARVEST DISEASE SYMPTOMS

Figure 18.75

Anthracnose stem-end rot (*Colletotrichum gloeosporioides*) of mango.

Figure 18.76

Anthracnose (*Colletotrichum gloeosporioides*) of mango.

Figure 18.77

Lasiodiplodia rot (*L. theobromae*) of mango.

Figure 18.78

Anthracnose (*Colletotrichum gloeosporioides*) of papaya.

Figure 18.79

Black stem-end rot (*Phoma caricae-papayae*) of papaya.

Figure 18.80

Black stem-end rot (*Phoma caricae-papayae*) of papaya (cut fruit).

Figure 18.81

Phomopsis rot (*Phomopsis caricae-papayae*) of papaya.

Figure 18.82

Phomopsis rot (*Phomopsis caricae-papayae*) of papaya (cut fruit).

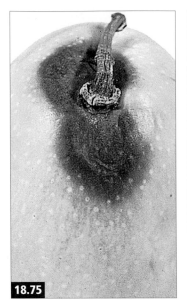

18.75

18.76

18.77

18.78

18.79

18.80

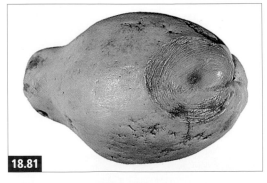

18.81

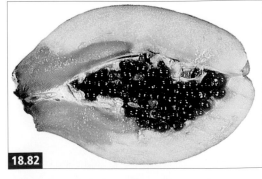

18.82

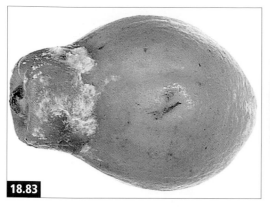

18.83

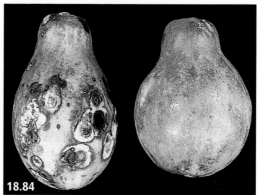

18.84

18.85

18.87

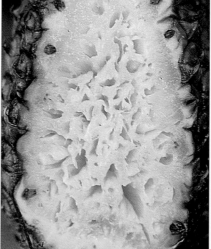

18.86

18.88

18.89

18.90

POSTHARVEST DISEASE SYMPTOMS

Figure 18.83

Phytophthora stem-end rot (*P. nicotianae* var. *parasitica*) of papaya.

Figure 18.84

Alternaria rot (*A. alternata*) of papaya following chilling.

Figure 18.85

Water blister (*Thielaviopsis paradoxa*) of pineapple (initial lesion).

Figure 18.86

Water blister of pineapple (developing rot).

Figure 18.87

Water blister of pineapple (collapsed fruit).

Figure 18.88

Fruitlet core rot of pineapple fruit caused by *Penicillium funiculosum* and/or *Fusarium moniliforme*.

Figure 18.89

Yeast fermentation of pineapple.

Figure 18.90

Yeast fermentation of pineapple (advanced stage).

POSTHARVEST DISEASE SYMPTOMS

Figure 18.91

Bacterial soft rot (*Erwinia carotovora*) of carrot.

Figure 18.92

Gray mold rot (*Botrytis cinerea*) of carrot.

Figure 18.93

White rot (*Sclerotinia minor*) of carrot.

Figure 18.94

Black mold (*Thielaviopsis basicola*) of carrot.

Figure 18.95

Crater rot (*Rhizoctonia carotae*) of carrot.

Figure 18.96

Bacterial soft rot (*Erwinia carotovora*) of celery leaflets.

Figure 18.97

Bacterial soft rot of celery petiole.

Figure 18.98

Pink rot (*Sclerotinia sclerotiorum*) of celery.

Figure 18.99

Bacterial soft rot (*Erwinia carotovora*) of lettuce.

18.91

18.92

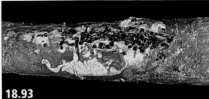

18.93

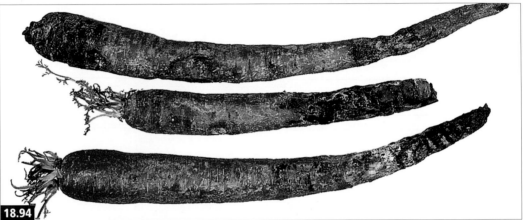

18.94

18.95

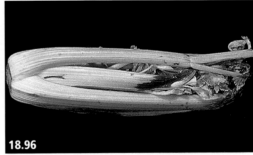

18.96

18.97

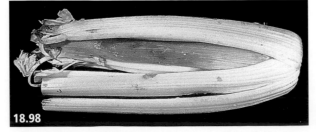

18.98

18.99

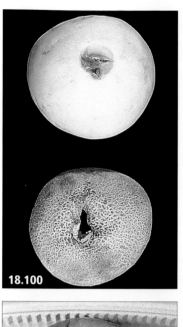

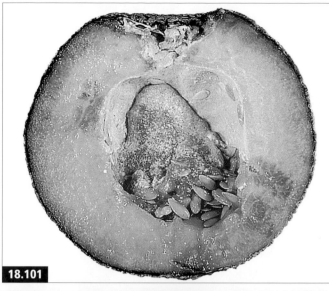

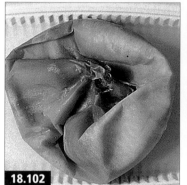

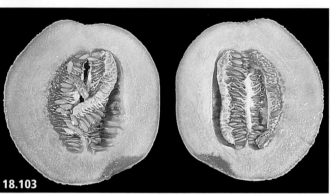

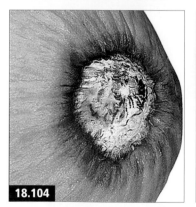

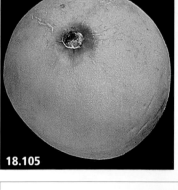

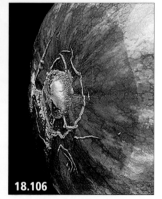

POSTHARVEST DISEASE SYMPTOMS

Figure 18.100

Sour rot (*Geotrichum candidum*) of melons.

Figure 18.101

Sour rot (*Geotrichum candidum*) development of melons.

Figure 18.102

Sour rot (*Geotrichum candidum*) with honeydew collapsed.

Figure 18.103

Rhizopus rot (*R. stonolonifer*) of cantaloupe.

Figure 18.104

Fusarium rot of melons.

Figure 18.105

Trichothecium rot (*T. roseum*) of honeydew melon.

Figure 18.106

Botrytis stem-end rot (*B. cinerea*) of watermelon.

Figure 18.107

Lasiodiplodia rot (*L. theobromae*) of watermelon.

Figure 18.108

Lasiodiplodia rot (*L. theobromae*) of watermelon after 24-hour exposure to air.

POSTHARVEST DISEASE SYMPTOMS

Figure 18.109

Neck rot (*Botrytis* spp.) of onion.

Figure 18.110

Bacterial soft rot (*Erwinia carotovora*) of onion.

Figure 18.111

Smudge (*Colletotrichum circinans*) of onion.

Figure 18.112

Fusarium basal rot (*Fusarium* spp.) of onion.

Figure 18.113

Fusarium bulb rot (*Fusarium* spp.) of onion.

Figure 18.114

Bacterial soft rot (*Erwinia carotovora*) of potato.

Figure 18.115

Bacterial ring rot of potato. (Courtesy H. Moline, USDA, Beltsville, Maryland)

Figure 18.116

Late blight (*Phytophthora infestans*) of potato (external). (Courtesy H. Moline, USDA, Beltsville, Maryland)

Figure 18.117

Late blight (*Phytophthora infestans*) of potato (internal). (Courtesy H. Moline, USDA, Beltsville, Maryland)

18.109

18.110

18.111

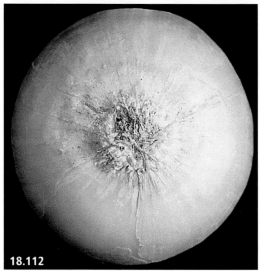

18.112

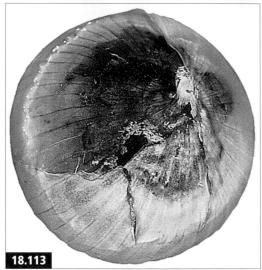

18.113

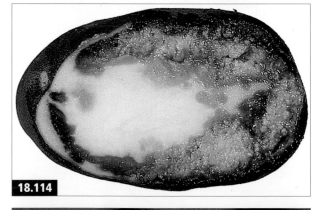

18.114

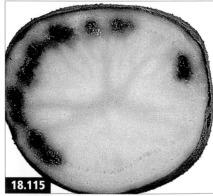

18.115

18.116

18.117

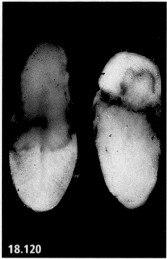

18.118

18.119

18.120

18.121

18.122

18.123

18.124

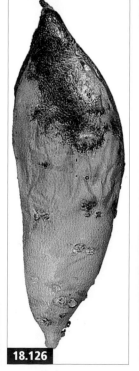

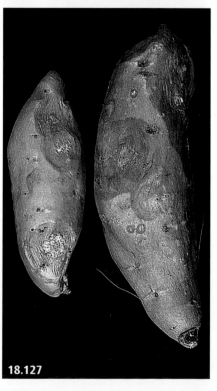

18.125

18.126

18.127

POSTHARVEST DISEASE SYMPTOMS

Figure 18.118

Fusarium dry rot (*F. solani* or *F. roseum*) of potato. (Courtesy H. Moline, USDA, Beltsville, Maryland)

Figure 18.119

Black scurf (*Rhizoctonia solani*) of potato.

Figure 18.120

Pink rot (*Phytophthora erythroseptica*) of potato. (Courtesy H. Moline, USDA, Beltsville, Maryland)

Figure 18.121

Botrytis gray mold rot (*B. cinerea*) of snap bean.

Figure 18.122

White rot (*Sclerotinia minor*) of snap bean.

Figure 18.123

White rot (*Sclerotinia minor*) of snap bean.

Figure 18.124

Apothecium of *Sclerotinia sclerotiorum*. (Courtesy J. C. Tu, Harrow Research Station, Ontario, Canada)

Figure 18.125

Anthracnose (*Colletotrichum lindemuthianum*) of snap bean. (Courtesy H. Moline, USDA, Beltsville, Maryland)

Figure 18.126

Rhizopus rot (*R. stolonifer*) of sweet potato.

Figure 18.127

Fusarium rot (*F. oxysporum*) of sweet potato.

POSTHARVEST DISEASE SYMPTOMS

Figure 18.128

Alternaria (*A. alternata*) rot of tomato.

Figure 18.129

Buckeye rot (*Phytopthora* spp.) of tomato.

Figure 18.130

Ghost spot (*Botrytis cinerea*) of tomato.

Figure 18.131

Sour rot (*Geotrichum candidum*) of tomato (early symptoms).

Figure 18.132

Sour rot (*Geotrichum candidum*) of tomato.

Figure 18.133

Rhizopus rot (*R. stolonifer*) of tomato (early stage).

Figure 18.134

Rhizopus rot (*R. stolonifer*) of tomato (late stage).

Figure 18.135

Anthracnose (*Colletotrichum* spp.) of tomato.

Figure 18.136

Fusarium rot (*Fusarium* spp.) of tomato.

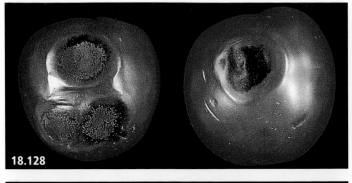

18.128

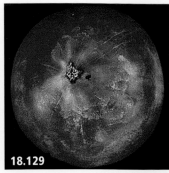

18.129

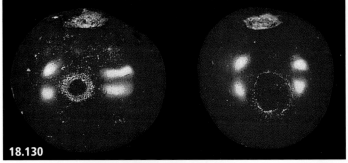

18.130

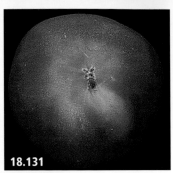

18.131

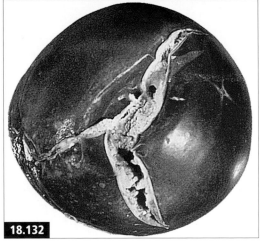

18.132

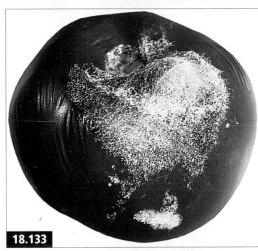

18.133

18.134

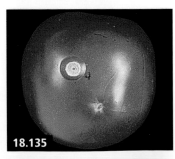

18.135

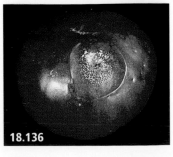

18.136

Lakshminarayana, S., N. F. Sommer, V. Polito, and R. J. Fortlage. 1987. Development of resistance to infection by *Botrytis cinerea* and *Penicillium expansum* in wounds of mature apple fruits. Phytopathology. 77:1674–1678.

Luvisi, D., H. Shorey, J. L. Smilanick, J. Thompson, B. H. Gump, and J. Knutson. 1992. Sulfur dioxide fumigation of table grapes. Oakland: Univ. Calif. Div. Ag. Nat. Res. Bull. 1932. 21 pp.

Maas, J. L., ed. 1998. Compendium of strawberry diseases. St. Paul: APS Press. 128 pp.

Matsumoto, T. T., and N. F. Sommer. 1967. Sensitivity of *Rhizopus stolonifer* to chilling. Phytopathology. 57:881–884.

Matsumoto, T. T., P. M. Buckley, N. F. Sommer, and T. A. Shalla. 1969. Chilling-induced ultrastructural changes in *Rhizopus stolonifer* sporangiospores. Phytopathology. 59:863–867.

McCracken, A. R., and J. R. Swinburne. 1980. Effect of bacteria isolated from surface of banana fruits on germination of *Colletotrichum musea* conidia. Trans. Bact. Mycol. Soc. 74:212.

Nelson, K. E. 1985. Harvesting and handling California table grapes for market. Oakland: Univ. Calif. Div. Ag. Nat. Res. Bull. 1913. 72 pp.

Offers, J. A. 1987. Citrus, diseases and defects found in the marketplace. Papendrecht, Holland: Licensed Citrus Survey and Consulting Bureau. 70 pp.

Ogawa, J. M., and H. English. 1991. Diseases of temperate zone tree fruit and nut crops. Oakland: Univ. Calif. Div. Ag. Nat. Res. Publ. 3345. 461 pp.

Ogawa, J. M., E. I. Zehr, G. W. Bird, D. F. Ritchie, K. Uriu, and J. K. Uyemoto, eds. 1995. Compendium of stone fruit diseases. St. Paul: APS Press. 128 pp.

Pearson, R. C., and A. C. Goheen, eds. 1988. Compendium of grape diseases. St. Paul: APS Press. 128 pp.

Ploetz, R. C., G. A. Zentmyer, W. T. Nishijima, K. G. Rohrbach, and H. D. Ohr, eds. 1994. Compendium of tropical fruit diseases. St. Paul: APS Press. 128 pp.

Rowe, R. C., eds. 1993. Potato health management. St. Paul: APS Press. 193 pp.

Schwartz, H. F., and S. K. Mohan, eds. 1994. Compendium of onion and garlic diseases. St. Paul: APS Press. 80 pp.

Sherf, A. F., and A. A. Mac Nab. 1986. Vegetable diseases and their control. 2d ed. New York: Wiley. 728 pp.

Slabaugh, W. R., and M. D. Grove. 1982. Postharvest diseases of bananas and their control. Plant Dis. 66:746–750.

Snowdon, A. L. 1990. A color atlas of postharvest diseases and disorders of fruits and vegetables. Vol. 1: General introduction and fruits. Boca Raton, FL: CRC Press. 302 pp.

———. 1992. A color atlas of postharvest diseases and disorders of fruits and vegetables. Vol. 2: Vegetables. Boca Raton, FL: CRC Press. 416 pp.

Sommer, N. F. 1982. Postharvest handling practices and postharvest diseases of fruit. Plant Dis. 66:357–364.

———. 1985. Role of controlled environments in suppression of postharvest diseases. Can. J. Plant Pathol. 7:331–334.

———. 1989. Manipulating the postharvest environment to enhance or maintain resistance. Phytopathology. 79:1377–1380.

Sommer, N. F., R. J. Fortlage, and D. C. Edwards. 1983. Minimizing postharvest diseases of kiwifruit. Calif. Agric. 37(1–2): 16–18.

Sommer, N. F., R. J. Fortlage, J. R. Buchanan, and A. A. Kader. 1981. Effect of oxygen on carbon monoxide suppression of postharvest diseases. Plant Dis. 65:347–349.

Stanghellini, M. E., and M. Aragaki. 1966. Relation of periderm formation and callose deposition to anthracnose resistance in papaya fruit. Phytopathology. 56:444–450.

Swinburne, T. R. 1974. The effect of store conditions on the rotting of apples, cv. Bramley's seedling, by *Nectaria galligena*. Ann. Appl. Biol. 78:39–48.

Swinburne, T. R., and A. E. Brown. 1975. The effect of carbon dioxide on the accumulation of benzoic acid in Bramley's seedling apples infected by *Nectaria galligena*. Trans. Brit. Mycol. Soc. 64:505–507.

Timmer, L. W., and L. W. Duncan. 1999. Citrus health management. St. Paul: APS Press. 221 pp.

Timmer, L. W., S. M. Garnsey, and J. H. Graham, eds. 2000. Compendium of citrus diseases. 2nd ed. St. Paul: APS Press. 128 pp.

van der Plank, J. E. 1975. Principles of plant infection. New York: Academic Press.

Uritani, I., and K. Oba. 1978. The tissue slice system as a model for studies of host-parasite relationships. In G. Kahl, ed., Biochemistry of wounded tissues. New York: de Gruyter. 287–308.

Wilson, C. L., and M. Wisniewski, eds. 1994. Biological control of postharvest disease: Theory and practice. Boca Raton: CRC Press. 182 pp.

Wilson, E. E., and J. M. Ogawa. 1979. Fungal, bacterial, and certain nonparasitic diseases of fruit and nut crops in California. Oakland: Univ. Calif. Div. Agric. Sci. 189 pp.

Zitter, T. A., D. L. Hopkins, and C. E. Thomas, eds. 1996. Compendium of cucurbit diseases. St. Paul: APS Press. 120 pp.

Postharvest Treatments for Insect Control

Elizabeth J. Mitcham,

F. Gordon Mitchell, Mary Lu Arpaia,

and Adel A. Kader

THE IMPORTANCE OF POSTHARVEST INSECT CONTROL

International markets have become increasingly important for domestically-produced perishable commodities. Under the General Agreement on Trade (GATT), many international markets have been opened for trade; however, phytosanitary restrictions continue to limit trade in many markets. Postharvest insect control is therefore critical to interstate and international trade of many fresh horticultural commodities. The ability to provide effective nondamaging insect control determines the future of many of these markets. Postharvest insect control is also essential during the storage of many horticultural products, such as tree nuts and dried fruits.

Phytosanitary restrictions are developed to protect a region's agricultural industry from the introduction of damaging insect pests. These restrictions should be based on a risk assessment, not zero risk, and should be based on scientific evidence (GATT 1966; NAFTA 1992). For example, California agriculture would be seriously affected by the introduction and establishment of Mediterranean fruit fly. For this reason, agricultural imports from growing areas that have established Mediterranean fruit fly populations are required to receive a tested and approved quarantine treatment prior to or upon entry. The absence of an approved treatment may result in the product being completely prohibited. Alternatively, if Mediterranean fruit fly were established in California, a much larger portion of the produce exported from California would be subject to quarantine treatment.

For some trading partners, import of a particular commodity may require only inspection upon arrival in the importing country. If actionable insects are found during the inspection, the product could be rejected and destroyed, sent back to the exporting country, or treated to kill the insect pests. An actionable insect is one that has potential to establish itself and cause damage to agriculture or the environment. In other cases, prior to permitting market access to the commodity from the importing country, a specific treatment is developed that will kill or sterilize the pests(s) of concern to prevent establishment.

Development of specific treatments for particular commodities and insects requires considerable experimental data. Often a very high degree of insect control is required, such as Probit 9 mortality. This means that the treatment kills 99.9968% of the insect pests or that there is only 1 survivor per 100,000 insects treated. Once the experimental data are reviewed and accepted as accurate and appropriate, there may be a requirement for a large-scale commercial test with a large number of insect pests (confirmation test). Considerable negotiation between the importing and exporting countries is common to specify the conditions under which the treatment will be conducted and verified. The time from the beginning of treatment development until the product is shipped under an acceptable treatment protocol can be 5 to 10 years, or perhaps longer.

Another important area for postharvest insect control is for stored products such as dried fruits and nuts. These products can be stored longer under ambient conditions and are therefore vulnerable to insect attack. Control of insect pests during storage must be effective but it is not required to meet the strict quarantine control mortalities. Chemical fumigation is the most common method used for stored product insect control, and controlled atmospheres (CA) with elevated

CO_2 are used to a lesser extent. A preshipment quarantine treatment may be applied at the end of storage.

TYPES OF INSECT CONTROL MEASURES

SYSTEMS APPROACH

In the systems approach to insect control, no single measure provides for complete control. Similar to integrated pest management, numerous individual steps are taken to reduce the likelihood that insect pests will be found in the packed product. Such steps may include pest control measures in the field and surrounding areas, control of the commodity's maturity at harvest, inspection during packing, special washing procedures, etc. Maturity at harvest can influence the ability of the insect pest to infest the product: for example, less-mature citrus fruits are poor hosts of various fruit flies. Treatment of the fruit with gibberellic acid may maintain the peel in an immature state longer, extending the period in which the fruit is a poor host. Under the systems approach, a specific postharvest insect control treatment is not required; however, product may be subject to inspection and certification prior to shipment.

PEST-FREE ZONES

Pest-free zones are growing areas that have been certified as free from a particular pest(s) for all or part of a growing season. Generally, a rigorous program of trapping and restriction of product movement from infested areas into the pest-free zone is required. Products exported from the pest-free zone are not subjected to a specific quarantine treatment, but inspection and certification may be required. The State of Florida has established zones certified as free of Caribbean fruit fly. Additional details on fly-free zones can be found at the USDA Animal and Plant Health Inspection Service (APHIS) website (www.aphis.usda.gov/ppq/).

INSPECTION AND CERTIFICATION

A thorough inspection of the product load may be required prior to export and/or upon arrival in the importing country. The rigor of the inspection depends on the ease of detecting the insect pest, the risk associated with establishment of the pest if imported, and the history of detection in previous ship-

ments. It is beneficial to the grower and exporter to use whatever methods are available to remove or kill insect pests that may be on their product after harvest. This could include some of the treatment methods that are discussed in this chapter. Following inspection of the load, a certificate of inspection is provided that clears the product for export. Inspection and certification may be the only requirement for import of the product or may be supplementary to an additional quarantine treatment. Detailed guidelines on the use of inspection and certification can be found at the USDA APHIS website (www.aphis.usda.gov/ppq/).

POSTHARVEST TREATMENTS

Postharvest treatments are designed to kill or sterilize the insect pest of concern with minimal damage to the commodity. Commodity response to quarantine treatments varies with the cultivar, where the commodity was grown, and the commodity maturity. These treatments are often a requirement of entry and are imposed each time the product is shipped to that destination. Most treatments are conducted prior to shipment of the commodity, but some products are treated during transport or upon arrival. Specific treatments are also important for insect management in stored products. A list of approved treatments can be found in the USDA APHIS Plant Protection and Quarantine (PPQ) Treatment Manual (website: www.aphis.usda.gov/ppq/manuals/online-manuals.htm).

POSTHARVEST TREATMENTS CURRENTLY EMPLOYED

FUMIGATION

Fumigation has been and continues to be the most common type of insect control measure, both for disinfestation and for stored product insect control. Fumigants are generally easy to use and inexpensive. However, the future of many chemical fumigants is in jeopardy because of their potential effects on the environment and human health.

Methyl bromide

Methyl bromide is the most commonly used treatment for postharvest insect control, whether as a disinfestation treatment or for stored product pests. Methyl bromide is a general biocide that is tolerated by many fresh commodities. However, the future use of methyl bromide is uncertain. Under the Montreal Protocol (a United Nations treaty) and the U.S. Clean Air Act, methyl bromide is slated to be phased out in 2005 in developed countries and in 2015 in developing countries. However, the Protocol allows exemptions for quarantine treatments and critical uses. Given the potential for loss of methyl bromide, the interest in alternatives is great.

Methyl bromide protocols exist for numerous commodity and insect pest combinations (USDA-APHIS website); an example is shown in Table 19.1. In general, at a given dose of methyl bromide, the higher the temperature, the shorter the treatment. Treatment temperature is often selected based on product tolerance and to provide the shortest possible treatment. Methyl bromide is a hazardous gas, and its use requires that strict safety procedures be followed in the design of treatment chambers and in conducting fumigations. Following fumigation, a period of ventilation is required prior to entrance into the chamber by personnel. The ventilation period required depends on the rate of off-gassing (removal of the fumigant gas from the commodity) of the product and packaging materials.

Phosphine

Phosphine is often used for control of insects in dried fruits and nuts. However, many fresh commodities are damaged by phosphine gas. Phosphine is slower to act and does not penetrate as well as methyl bromide. However, during storage of dried products, methyl bromide and phosphine are often used alternately. Phosphine is considered a potential carcinogen and its future is also questionable.

Hydrogen cyanide (HCN)

This fumigant has been used for citrus insect control, such as California red scale on California citrus shipments to Arizona. Since many commodities are damaged by HCN and the gas is very deadly, its use is limited.

HIGH- OR LOW-TEMPERATURE MODIFICATION

The benefits to the use of temperature treatments for insect disinfestation are that they are nonchemical, leaving no residues on the product. In addition, these treatments are safer for workers applying the treatments. The disadvantages include potential for product damage, higher energy costs, and potentially longer treatment times as compared to fumigation treatments. The challenge is to find a temperature and time combination that is effective for insect control while causing little to no product damage.

Cold treatment

The PPQ Treatment Manual (USDA-APHIS website) currently allows cold treatment for control of Mediterranean fruit fly (*Ceratitis capitata*), oriental citrus mite (*Eutetranychus orientalis*), Mexican fruit fly (*Anastrepha* spp.), Queensland fruit fly (*Bactrocera tryoni*), false codling moth (*Cryptophlebia leucotreta*), Natal fruit fly (*Pterandrus rosa*), melon fly (*Bactrocera cucurbitae*), pecan weevil (*Cucurlio caryae*) and lychee fruit borer (*Conopomorpha sinensis*). An example protocol for Mediterranean fruit fly on many fresh commodities is shown in table 19.2.

Cold treatment is most effective against tropical insects; however, tropical commodities are generally intolerant of this type of treatment due to their susceptibility to chilling injury. Conditioning fruit at 15° to 20°C (59° to 68°F) for 2 to 6 days allows some citrus fruits to tolerate cold treatment and has been used on Florida grapefruit during transport to Japan. Temperature control within a narrow range during conditioning treatment and subsequent cold treatment are critical to treatment success. Cold treatment is also used on Florida carambola for shipment to California and on Spanish citrus shipped to the United States.

Cold treatment is most appropriate for commodities capable of extended low-temperature storage, such as apple, pear, grape, kiwifruit, persimmon, and pomegranate. The

Table 19.1. Methyl bromide treatment schedule for a 2-hour exposure of sweet cherry to control Western cherry fruit fly (*Rhagoletis indifferens*) and codling moth (*Cydia pomonella*)

Temperature	Dosage rate g/m³ (lb/1,000ft³)
21°C (70°F) or above	32 (2.0)
16 to 20°C (60 to 68°F)	40 (2.5)
10 to 15°C (50 to 59°F)	48 (3.0)
4.5 to 9°C (40 to 49°F)	64 (4.0)

Source: USDA APHIS PPQ Treatment Manual. (www.aphis.usda.gov/ppq/manuals/online-manuals.htm)

Table 19.2. Cold treatment protocol for Mediterranean fruit fly (*Ceratitis capitata*)

Temperature	Exposure period (days)
0°C (32°F) or below	10
0.6°C (33°F) or below	11
1.1°C (34°F) or below	12
1.7°C (35°F) or below	14
2.2°C (36°F) or below	16

Source: USDA APHIS PPQ Treatment Manual (www.aphis.usda.gov/ppq/manuals/online-manuals.htm).

Note: Always refer to the latest edition of the treatment manual or the Federal Register (www.nara.gov/fedreg/) for exact treatment requirements.

10 to 16 days required for many cold treatments exceeds the potential market life of many perishable commodities, such as strawberries and cane berries. However, many of the insect pests of temperate commodities (apple and pear) are also of temperate origin, such as the codling moth (*Cydia pomonella*), and they are relatively insensitive to cold treatment.

When the cold treatment must be completed prior to transport, there are logistical problems associated with providing enough refrigerated storage capacity in production areas with heavy shipping periods. There are strict requirements for temperature monitoring in cold storage facilities to certify compliance with these treatments. Cold treatments have also been conducted during transport in marine containers.

Heat treatment (hot water, vapor heat, high-temperature forced air)

Several heat treatments are commonly used, including hot water treatment for mango and lychee and vapor heat and high-temperature forced air treatments (HTFA) for mango, papaya, and various citrus fruits. Vapor heat, one of the first postharvest insect control treatments, was developed in the 1920s. Many vapor heat treatment protocols remain in the *PPQ Treatment Manual* and are still used. Some of the older protocols having long treatment times are seldom used. HTFA is a modified version of vapor heat with greater airflow to speed heating and lower RH levels to reduce product damage. HTFA is approved for citrus from Hawaii, grapefruit and mango from Mexico, mountain papaya from Chile, and papaya from Belize and Hawaii (USDA-APHIS website).

In the HTFA treatment for papaya fruit to control tropical fruit flies, the fruit seed cavity is heated in stages, first to 41°C (105.8°F), then to 47.2°C (117°F) over about 6 hours. Hot water treatment of mangoes, also for control of tropical fruit flies, involves heating at 46.4°C (115.5°F) for about 75 minutes, depending on fruit size, variety and country of origin.

IRRADIATION

Irradiation treatment involves exposing the product to a radiation source (isotopic source using cobalt-60 or cesium-137; and E-beam or X-ray, an electrically driven machine source) until it absorbs the required dose level of gamma or X-rays. The doses tolerated by many commodities provide for insect sterilization or prevention of adult emergence, not complete kill. The potential presence of live but sterile insects in imported product requires a greater level of trust between the exporter and importer.

In 1986, the Food and Drug Administration (FDA) approved the use of radiation treatments of up to 1,000 Grays (Gy) (100 krad) on fruits and vegetables. Research has shown that the doses required for sterilization of most insects is below 300 Gy, while the doses required for effective decay control are often greater than 1,000 Gy.

While use of irradiation as a potential quarantine treatment has received much attention, to date its use has been limited. Initially, only papayas from Hawaii were approved for irradiation treatment. While approval for this treatment was granted in 1989, the treatment had not been used due to inability to select a site for an irradiation facility in Hawaii. Since 1998, papaya, lychee, rambutan, and atemoya fruit have been shipped from Hawaii to Chicago for irradiation treatment in the Chicago area. In 2000, an X-ray/E-beam irradiation facility was completed in Hawaii and is in use for tropical fruits. In addition, generic irradiation treatments for various fruit flies (regardless of the commodity) were adopted by APHIS. These dosages (0.15 to 0.25 kGy) are designed to prevent adult emergence, while most treatments in the past were designed for either insect sterilization or mortality.

Detailed guidelines on the use of irradiation as a disinfestation treatment can be found at the USDA APHIS website (www.aphis.usda.gov/ppq/). Several factors must be considered in the use of irradiation as an insect disinfestation treatment.

- Not all fruits and vegetables tolerate irradiation in the required dose range (table 19.3). Irradiation of 250 to 1,000 Gy can cause damage. Symptoms of irradiation stress include accelerated senescence, loss of green leaf color, abscission of leaves and petals of leafy vegetables and cut flowers, accelerated softening of fruit, uneven fruit ripening, and tissue browning. Irradiation stress is additive to other stresses (physical, chilling, water, etc.), which should be avoided to minimize the negative effects of ionizing radiation on fresh produce.

- APHIS or the country of destination must accept a treatment that does not provide for insect kill or Probit 9 mortality. This means that the receiving country must accept product with live (although sterilized) insects.

- Dosimetry must be considered in treatment development such that the product on the outside of a pallet will receive a higher dose to assure that the necessary minimum dose will be received at the center of the stacked pallet. The product on the outside must tolerate the higher dose. For example, the product on the outside of a pallet may have to be exposed to 500 Gy so that the product in the middle of

the pallet is exposed to 250 Gy due to absorption by the product. If pallets must be disassembled prior to treatment, the cost increases substantially.

- E-beam's lower ability to penetrate prohibits its use for pallet loads of commodities and, therefore, limits the manner in which treatments can be applied. Most products would need to be treated on a conveyor belt.

- In an area such as California or Florida that produce numerous horticultural commodities in large quantities, the logistics of radiation treatment are tremendous, depending on the percentage of product that must be treated.

- Gamma radiation plants are more costly than E-beam and X-ray facilities, and all are more economical if used essentially year-round. Fresh fruit and vegetable production is seasonal. This would require facilities to be, at a minimum, shared among commodities with somewhat different harvest schedules.

- While studies have shown that consumer acceptance of irradiated produce in the United States is increasing, serious social and public policy issues remain. Will local governments accept environmental impact statements and allow radiation facilities to be constructed in their areas? (E-beam and X-ray facilities that do not have a permanent radioactive source have fewer restrictions.) Will quarantine authorities of receiving countries accept irradiated products?

Table 19.3. Relative tolerance of fresh fruit and vegetables to irradiation doses below 1,000 Grays (100 krad)

Minimal detrimental effects	Inconsistent results	Significant detrimental effects
Apple	Apricot	Avocado
Cherry	Banana	Broccoli
Date	Cherimoya	Cauliflower
Guava	Loquat	Cucumber
Longan	Fig	Grape
Mango	Grapefruit	Green bean
Muskmelon	Kumquat	Leafy vegetables
Nectarine	Litchi	Lemon
Papaya	Orange	Lime
Peach	Passion fruit	Olive
Rambutan	Pear	Pepper
Raspberry	Pineapple	Sapodilla
Strawberry	Plum	Soursop
Tamarillo	Tangelo	Summer squash
Tomato	Tangerine	

Source: Kader 1986.

COMBINATION TREATMENTS

The most commonly used combination treatment is methyl bromide fumigation and cold treatment. Methyl bromide fumigation has been used before or after the cold treatment, depending on the protocol. The use of cold treatment allows a lower dose of and shorter treatment time with methyl bromide (USDA -APHIS website).

A combination treatment for cherimoya and limes imported into the United States from Chile involves a soapy water wash and a wax application for control of false red mite (*Brevipalpus chilensis*) of grapes (USDA -APHIS website). This treatment is similar to the systems approach in its design.

EXPERIMENTAL TREATMENTS

FUMIGATION

Alternative fumigants under study include methyl iodide, carbonyl sulfide, and sulfuryl fluoride. For example, sulfuryl fluoride is being explored as an alternative to methyl bromide for control of storage pests of walnut, including codling moth (*Cydia pomonella*), Indianmeal moth (*Plodia interpunctella*), and navel orangeworm (*Amyelois transitella*). The insect control properties of these fumigants are promising; however, the tolerance of horticultural commodities, especially perishable fruit, vegetables, and ornamentals is not fully known. None of these compounds were registered for use on perishable commodities as of August 2001. A combination of methyl bromide and CO_2 may allow reduced levels of methyl bromide to be used for insect control. However, the efficacy of the combination appears to be variable between insects and among life stages of insects. Laboratories are also exploring the use of volatile compounds, including acetaldehyde, ethyl formate, and methyl formate, for postharvest insect control on fruits and ornamentals.

TEMPERATURE

The use of heat treatments for insect disinfestation in temperate commodities and for additional tropical and subtropical commodities is under exploration, including hot air treatments for cherries and nectarines. Cold treatments are also being explored for subtropical commodities. Short-term rapid heating using radio frequency energy is also being studied for a wide range of fruits and nuts. The use of conditioning treatments, both with heat and cold, to provide better treatment tolerance is also being explored.

IRRADIATION

Additional research is under way to refine the generic radiation doses for fruit flies, based on lack of adult emergence. The response of commodities to E-beam and X-ray radiation is also being tested. Irradiation protocols are being developed for additional pests other than fruit flies, such as codling moth. Researchers are attempting to develop a rapid marker that will prove that a commodity was irradiated, which should allow for better acceptance of less-than-lethal doses.

CONTROLLED ATMOSPHERES

Treatments with high levels of CO_2 at low temperatures are being developed for several surface pests, including omnivorous leafroller (*Platynota stultana*), western flower thrips (*Frankliniella occidentalis*), and Pacific spider mite (*Tetranychus pacificus*) on harvested table grapes. This treatment requires 13 days at 0° to 2°C (32° to 36°F) and could be conducted in a marine container. Other treatments involving high CO_2 and low O_2 at high temperatures (45° to 47°C, 113° to 116.6°F) are being explored for disinfestation of codling moth in cherries and pears. These high-temperature CA treatments are generally 2 hours or shorter in duration, depending on the size of the commodity. As of August 2001, no CA treatment had been approved as a quarantine treatment.

LOOKING TOWARD THE FUTURE

The future of postharvest insect control will undoubtedly involve less use of chemicals and more physical treatments. In addition, an increased emphasis on risk assessment, nonhost status, and the systems approach in place of specific treatment requirements would be practical. Clearly, there will be no single substitute for methyl bromide fumigation for postharvest insect control. It is critical that quarantine requirements be based on science, not politics.

EXPERIMENTAL INSECT CONTROL TREATMENTS

- Alternative Fumigants
- Heat Treatment for Temperate Crops
- Cold Treatment for Subtropical Crops
- Irradiation
- Controlled Atmospheres
- Combination Treatments

REFERENCES

Aharoni, Y., J. K. Stewart, and D. G. Guadagni. 1981. Modified atmospheres to control western flower thrips on harvested strawberries. J. Econ. Entomol. 74:338–340.

Benshoter, C. A. 1987. Effects of modified atmospheres and refrigeration temperatures on survival of eggs and larvae of the Caribbean fruit fly (Diptera: Tephritidae) in laboratory diet. J. Econ. Entomol. 80: 1223–1225.

Burditt, A. K., Jr. 1982. Food irradiation as a quarantine treatment of fruits. Food Technol. 36(11): 51–54, 58–60, 62.

Carey, J. R., and R. V. Dowell. 1989 Exotic fruit fly pests and California agriculture. Calif. Agric. 43(3): 38–40.

Couey, H. M. 1989. Heat treatment for control of postharvest diseases and insect pests of fruits. HortScience. 24:198–202.

GATT. 1966. General Agreement on Tariffs and Trade, amended through 1966. Available via Internet at www.gatt.org

Hallman, G. J. 1999. Ionizing radiation quarantine treatments against tephritid fruit flies. Postharv. Biol. Technol. 16:93–106.

Kader, A. A. 1986. Potential applications of ionizing radiation in postharvest handling of fresh fruits and vegetables. Food Technol. 40(6): 117–121.

Ke, D., and A. A. Kader. 1992. Potential of controlled atmospheres for postharvest insect disinfestation of fruit and vegetables. Postharv. News and Info. 3(2): 31N–37N.

Mitcham, E. J., S. Zhou, and V. Bikoba. 1997. Controlled atmospheres for quarantine control of three pests of table grape. J. Econ. Entomol. 90(5): 1360–1370.

Mitcham, E. J., T. L. Martin, S. Zhou, and A. A. Kader. 2001. Potential of CA for postharvest insect control in fresh horticultural perishables: An update of summary tables compiled by Ke and Kader, 1992. In: CA 2001 CD-ROM. Davis: Univ. Calif. Postharv. Hort. Ser. 22.

NAFTA. 1992. North American Free Trade Agreement. (www.mac.doc.gov)

Neven, L. G., and E. J. Mitcham. 1996. CATTS (Controlled Atmosphere/Temperature Treatment System): A novel tool for the development of quarantine treatments. Amer. Entomol. 42:56–59.

Paull, R. E. 1994. Responses of tropical horticultural commodities to insect disinfestation treatments. HortScience. 29:988–996.

Paull, R. E., and J. W. Armstrong, eds. 1994. Insect pests and fresh horticultural products: Treatments and responses. Wallingford, UK: CAB International. 360 pp.

Sharp, J. L., and G. J. Hallman, eds. 1994. Quarantine treatments for pests of food plants. Boulder, CO: Westview Press. 290 p.

Sommer, N. F., and F. G. Mitchell. 1986. Gamma irradiation—A quarantine treatment for fresh fruits and vegetables? HortScience. 21:356–360.

USDA APHIS PPQ Treatment Manual. www.aphis.usda.gov/ppq/manuals/online-manuals.htm

20

Transportation

James F. Thompson

Perishable products are moved from the location of production to the consumer in a variety of transportation systems. In North America, refrigerated rail service was used extensively in the past. Now, most products are moved by refrigerated highway trucks. Most international trade into and out of North America utilizes refrigerated marine containers or ships. Airplanes that do not have temperature control capability are used for limited quantities of high-value commodities.

The goal of transportation is to move perishable products with a minimum loss of quality. Most transportation equipment controls air temperature around the product. Special equipment is sometimes available to produce a modified or controlled atmospheric (gas) composition around the product. Humidity and vibration control equipment are also used on some vehicles.

HIGHWAY TRUCKS

Large refrigerated highway vehicles are usually semi trailers. A powered tractor provides motive power and support for the front of the trailer (fig. 20.1). Trailers are sometimes placed on rail cars for a portion of the trip and then transported from a central rail location to the final destination with a highway tractor. This is sometimes referred to as a "trailer on flat car" (TOFC) shipment system. In North America refrigerated highway trailers are used for trips lasting from about 1 to 5 days.

The trailer has a self-contained engine-driven refrigeration system that provides conditioned air to the insulated load space. Most new trailers in the United States have an outside width of 2.6 m (102 in) and can be purchased in exterior lengths from 12.2 to 16.2 m (40 to 53 ft). Most new trailers are 16.2 m (53 ft) long. Interior volume ranges from 70 to 100m³ (2,500 to 3,500 ft³). The gross vehicle weight (including tractor) can not exceed 36,288 kg (80,000 lb) in the United States and individual axles also have weight limit to ensure a fairly uniform weight distribution in the vehicle. Most semi trailers have a load capacity of about 18,100 to 20,400 kg (40,000 to 45,000 lb).

Refrigeration capacity of new units is typically 12.3 to 16.4 kW (3.5 to 4.7 tons). The unit can provide heat when the trailer is operated in ambient conditions colder than the set point temperature. The refrigeration unit is located on the top front of the trailer and provides refrigerated air to an air chute in the top of the trailer (fig. 20.2). Product should be loaded and secured to provide an air space around the load (fig. 20.3). This allows the refrigerated air to flow between the load and the walls, underneath the product, and finally return to the refrigeration unit through a bulkhead in the front of the trailer. This design allows the refrigerated air to intercept this heat before it affects product temperature. Top-delivery refrigeration systems do not have enough airflow to dependably remove product heat from most commodities, and they should be cooled to transit temperature before loading.

The majority of the heat that the refrigeration unit removes comes from heat conducted across the walls and from air leaking into the trailer. If product is not center-loaded it will be warmed by contact with the walls. A series of shipments of California strawberries cooled to 1.7°C (35°F) and shipped with thermostats set at that same temperature showed average arrival temperatures in New York of 3.9°C (39°F) for center-loaded fruit and 5°C (41°F) for side-wall-loaded fruit. Warmest berries in center-loaded were 5°C (41°F) and 6.7°C (44°F) for side wall loading.

In the United States, the Refrigerated Transportation Foundation has developed a classification system for the combined insulated trailer and refrigeration unit. Vehicles using this classification system are rated for their ability to effectively transport products in four temperature categories (table 20.1). Metal placards located on the outside and inside of the trailer list the temperature range and other design characteristics of the vehicle. Most perishable fruits, vegetables, and flowers should be shipped in trailers with at least a C35 rating and equipped with an air chute and a front bulkhead.

Relative humidity is usually not controlled in refrigerated trailers. Some equipment manufacturers offer atomizers to add water to the refrigerated air. However, the added humidity reduces fiberboard strength. Acceptable levels of RH are a compromise between minimizing product moisture loss and minimizing damage caused by weakened packages. Product moisture loss can also be reduced by packing product in plastic liners, bags, or consumer-sized packages. These reduce moisture loss while allowing the refrigeration system to operate at lower humidities and protect fiberboard strength.

Atomizers will freeze if air temperature around them is close to 0°C (32°F). This may require that the thermostat be set a little warmer than the optimum for temperate fruits and cool-season vegetables. If products are not particularly sensitive to damage from moisture loss, it may be better to not use humidifiers and set the thermostat as low as possible. Humidification systems may find their best use with chilling sensitive commodities that are shipped at temperatures between 5°C (41°F) and 13°C (55°F).

Highway trailers are not airtight enough to allow their use as a gas barrier for modified or controlled atmosphere handling. Modified atmospheres can be provided through the use of semipermeable films used to form consumer packages or pallet wraps.

During transit some products can be damaged by the constant vibration caused by the road or by a shock caused by the vehicle traveling over a bump or curb. Both of these cause most damage to product loaded directly over axles. Vibration is amplified as it is transmitted through fiberboard boxes, and this causes most damage to product in topmost boxes on a load. Commercially available air ride suspension systems eliminate most of this damage. They cost more than steel spring suspensions but are becoming more commonly available on trailers. Some transport companies require that all of their new equipment have air ride suspension.

Figure 20.1

Refrigerated semi trailer.

Figure 20.2

Refrigeration and airflow system in highway trailer.

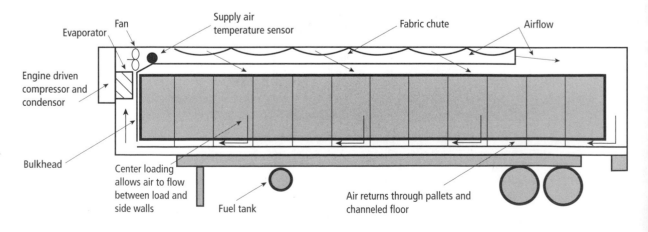

Figure 20.3

Procedures for bracing a center-loaded product in a highway trailer.

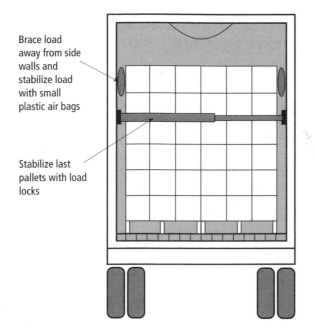

Brace load away from side walls and stabilize load with small plastic air bags

Stabilize last pallets with load locks

Table 20.1. Refrigerated Transportation Foundation temperature limit ratings for refrigerated trailers

Vehicle rating	Rated minimum temperature	Type of product protection
C65	18°C (65°F)	Controlled temperature
C35	2°C (35°F)	Fresh fruit, vegetables, flowers
F	–18°C (0°F)	Frozen foods
DF	–29°C (–20°F)	Ice cream and frozen foods

They also increase tire life and may reduce vibration damage to the trailer itself. Nearly all long-haul tractors in the United States have air ride suspension for driver comfort, so semi trailers usually have air ride suspension for the axles under the front. Vibration-sensitive product should not be loaded directly over axles suspended on steel springs. Vibration damage can also be reduced by packing product so that it is immobilized.

Over three quarters of the product that leaves California is shipped in loads with mixed products. These loads must be set up so that they are compatible. Use table 20.2 to select compatible produce. The chart divides common fruits and vegetables into four categories. Produce in the same vertical section can be safely held at the same temperature range listed on the top of the section. If produce in different temperature section are mixed, produce quality will be compromised, especially with longer transit times. The greater the difference in recommended temperature between the produces, the greater the potential for quality loss.

Dried vegetables in the top shaded area should not be mixed with any produce below them on the chart. These vegetables should be held in a 50 to 70% RH environment to prevent decay development. Most of the 0° to 2°C (32° to 36°F) vegetables are sensitive to moisture loss and should be held at more than 90% RH or packaged to minimize water loss. All the rest of the vegetables and fruit should be held at 85 to 95% RH.

Ethylene-sensitive vegetables near the top of the chart should not be mixed with ethylene-producing fruits at the bottom of the chart and luffa and ripe tomatoes. If for some reason they must be mixed, ethylene scrubbers may reduce damage. Produce items in the shaded middle of the chart are neither sensitive to ethylene nor are they ethylene producers and can be mixed with the items above or below them in the same temperature section.

Some produce items can exchange odors with other selected produces. See the notes at the bottom of table 20.2 for precautions.

Local refrigerated transport from distribution centers to stores or restaurants is nearly always a mixed load. Products may have widely differing temperature requirements and susceptibility to ethylene and odors. Most products survive this because the trip is short allowing products to be in a compromised environment for only a few hours. If frozen product is also part of the load, then the vehicles often have several compartments. Each compartment is set at a different temperature and each has a separate outside door allowing products to be carried at more ideal temperatures.

Thermostat temperature is set based on a compromise between the relative obvious damage caused by freezing temperatures and the less obvious damage caused by higher-than-recommended temperature. For products that are not chilling sensitive, thermostats are usually set below 4.4°C (40°F) but above 1.7°C (35°F). Newer refrigeration units with supply-air temperature sensing can be set lower than this because the control system is designed to ensure that the coldest temperature in the load, at the air

Table 20.2. Compatible produce for long-distance transport (Produce in the same temperature column can be safely mixed. Ethylene-sensitive vegetables should not be mixed with ethylene-producing fruits and vegetables. Dry vegetables should not be mixed with other fruits and vegetables.)

Produce	Recommended storage and transport temperatures			
	0°–2°C (32°–36°F)	4°–7°C (40°–45°F)	7°–10°C (45°–50°F)	13°–18°C (55°–65°F)
Dry vegetables	Dry onion[1,3,9], Garlic			Ginger[5], Pumpkin, Squash, winter
Ethylene-sensitive vegetables	Arugula*, Asparagus, Belgian endive, Bok choy, Broccoflower, Broccoli*, Brussels sprouts, Cabbage[1], Carrot[1,3], Cauliflower, Celery[1,3,9], Chard* — Chicory, Chinese cabbage, Collards*, Cut vegetables, Endive, Escarole, Green onion[7], Herbs (not basil), Kailon*, Kale*, Leek[8], Lettuce — Mint, Mushroom*[7], Mustard greens*, Parsley, Parsnip, Snow pea*, Spinach*, Sweet pea*, Turnip greens, Watercress	Beans, snap, etc.*[10], Cactus leaves, Cucumber*, Pepper (chili), Potato, late crop[1], Southern peas*, Tomatillo	Basil*, Chayote, Eggplant*[5], Kiwano, Long bean, Okra, Squash, summer*, Watermelon	Potato, early crop*, Tomato, mature green
Vegetables (not ethylene sensitive)	Amaranth*, Anise, Artichoke, Bean sprouts*, Beet, Celeriac, Daikon, Horseradish — Jerusalem artichoke, Kohlrabi, Lo bok, Radicchio, Radish, Rhubarb[7], Rutabaga — Salsify, Scorzonera, Shallot, Sweet corn[7], Swiss chard, Turnip, Waterchestnut		Calabaza, Haricot vert, Pepper, bell[10], Winged bean	Cassava, Jicama, Sweet potato (boniato), Taro (malanga), Yam
			Luffa*[†]	Tomato, ripe*[†]
Fruits and melons (very low ethylene producing)	Barbados cherry, Bitter melon, Blackberry*, Blueberry, Caimito, Cashew apple, Cherry, Coconut, Currant, Date, Gooseberry, Grape[6,7,8] — Longan, Loquat, Lychee, Orange, FL[4], Raspberry*, Strawberry*	Blood orange[4], Cactus pear (tuna), Jujube, Kumquat, Mandarin[4], Olive, Orange, CA, AZ[4], Pepino, Persimmon, Pomegranate, Tamarind, Tangerine[4]	Babaco, Calamondin*, Carambola, Casaba melon, Cranberry, Grapefruit[4], Juan Canary melon, Lemon[4], Lime[4], Pineapple[2,10], Pummelo[4] — Tamarillo, Tangelo[4], Ugli fruit	Breadfruit, Canistel, Grapefruit, CA, AZ[4], Jaboticaba*
Ethylene-producing fruits and melons	Apple[1,3,9], Apricot, Avocado, ripe, Cantaloupe, Cut fruits, Fig[1,7,8], Kiwifruit, Nectarine, Peach, Pear, Asian, Pear, European[1,9] — Plum, Prune, Quince	Durian, Feijoa, Guava, Honeydew melon, Persian melon	Avocado, unripe, Crenshaw melon, Custard apple, Passion fruit	Atemoya, Banana, Cherimoya, Jackfruit, Mamey sapote, Mango, Mangosteen, Papaya, Plantain, Rambutan, Sapote, Soursop*

Notes:
* Less than 14-day shelf life at recommended temperature and normal atmosphere conditions.
† Produces moderate amounts of ethylene and should be treated as an ethylene-producing fruit.
1. Odors from apples and pears are absorbed by cabbage, carrots, celery, figs, onions, and potatoes.
2. Avocado odor is absorbed by pineapple.
3. Celery absorbs odor from onions, apples, and carrots.
4. Citrus absorbs odor from strongly scented fruits and vegetables.
5. Ginger odor is absorbed by eggplant.
6. Sulfur dioxide released from pads used with table grapes will damage other produce.
7. Green onion odor is absorbed by figs, grapes, mushrooms, rhubarb, and corn.
8. Leek odor is absorbed by figs and grapes.
9. Onion odor is absorbed by apples, celery, pears, and citrus.
10. Pepper odor is absorbed by beans, pineapples, and avocados.

chute, does not drop below the thermostat set point. Good loading practices and well-cooled product in a supply-air-controlled refrigeration system allow thermostats to be set below 1.7°C (35°F). This is especially true for products with high sugar contents that freeze at temperatures significantly below 0°C (32°F) and for products that can withstand several episodes of light freezing conditions (table 20.3).

Many newer refrigeration units offer microprocessor-based controllers that automatically monitor refrigeration system operation and alert the driver to malfunctions. These systems can also transmit the information to satellites that forward the information to a company operations center. The company can then monitor vehicle location and system performance and maintain close supervision of all loads en route. This compatibility should allow companies to set thermostat set points closer to optimal conditions with less concern for load freezing.

Refrigerated trailers are used to carry a wide variety of products. The return trip to the production area may be a load of furniture, carpet, chemicals, or even pesticides. The truck owner and the company that contracts for transport of perishable foods need to ensure that residues from previous loads cannot contaminate food products with

materials that are not registered food grade materials. The guidelines for best use of refrigerated trailers are shown in fig. 20.4.

Even with these records, cargo is sometimes misdeclared. Consequently, a trailer should always be well cleaned before each load of perishables to avoid contamination by non-food grade products or foodborne pathogens. Never load a trailer that is visibly unclean or has an odor from a previous load.

Some perishables may be handled in trucks or other vehicles that are not temperature controlled. For example, products may be shipped directly to the consumer in the mail, and flowers are sometimes shipped to florists on bus services. The customer or receiver will get adequate quality only if the product had a good shelf life to begin with and the transport time is relatively short. Many products are packed in closed, insulated containers to protect them from temperature extremes, both freezing and high temperature damage are possible. Ice or gel ice products that have been well sealed in bags are sometimes added to packages to protect the products from high temperatures. Produce should be immediately placed in a temperature controlled environment after it is received.

In some rural areas refrigerated vehicles are simply not available and produce is transported in open trucks or carts. Produce must be handled very quickly because most products have very limited shelf life under these conditions. Heat gain can be minimized by shipping the product during the cool part of the day or at night and covering it to reduce heat gain from the sun. A fabric or plastic cover should be opaque and light colored to reflect solar radiation. But more importantly the cover should be supported above the product to allow a small amount of outside air to flow between the cover and the product. Ventilation removes the heat that builds up under the cover. Clean plant material can also be used as a cover.

MARINE CONTAINERS AND REFRIGERATED SHIPS

Marine containers (figs. 20.5 and 20.6) and refrigerated ships (fig. 20.7) are used to transport perishables over the ocean. Ships have longer transport times than their typical alternative of air freight; travel times are often 1 to 4 weeks. However, their cost is

Table 20.3. Susceptibility of selected fruits and vegetables to freezing injury

Injured by one light freezing	Will recover from one or two light freezing episodes	Can be lightly frozen several times without damage
Apricot	Apples	Beets without tops
Asparagus	Broccoli	Brussels sprouts
Avocado	Cabbage, new	Cabbage, mature and savoy
Banana	Carrots without tops	Dates
Beans, snap	Cauliflower	Kale
Berries, except cranberries	Celery	Kohlrabi
Cucumber	Cranberry	Parsnip
Eggplant	Grapefruit	Rutabaga
Lemons	Grapes	Salsify
Lettuce	Onion, dry	Turnip without tops
Limes	Oranges	
Okra	Parsley	
Peaches	Pears	
Peppers, sweet	Peas	
Plums	Radish without tops	
Potato	Spinach	
Squash, summer	Squash, winter	
Sweet potato		
Tomatoes		

Source: Adapted from Hardenberg et al. 1986.

much lower than air freight, and both can provide excellent temperature and environmental conditions for long-term transport. The guidelines for best use of marine containers are shown in fig. 20.4.

The important differences between using containers or refrigerated ships relate to their differences in cargo-carrying capacity. A container carries about 1,000 to 1,500 packages, and a refrigerated ship has a capacity of about 350,000 packages. Their large insulated cargo volume and built-in refrigeration allow refrigerated ships to charge less than container systems that have a large number of small, individually refrigerated units. However, containers can be transported directly to a refrigerated loading dock at the packing operation,

maintaining a continuous cold chain. Refrigerated ships are loaded from normal, open wharves and allow product to be exposed to the elements (heat, freezing temperatures, or precipitation). This is especially a problem in ships that use a common refrigeration system for two compartment levels because the refrigeration system is not operated until both compartments are loaded.

The large volume of refrigerated ships causes unique marketing problems. When 350,000 packages arrive at a port at one time, the sellers and receivers must have good marketing plan or a glut of product may depress product prices. Generally, refrigerated ships are used to transport produce that is marketed in large volumes by large

---

Figure 20.4

Guidelines for best use of refrigerated trailers, marine containers, and ships.

A. General

1. Products should be packaged in containers that are (see chapter 10 on packages for more details)
   - strong enough to withstand high humidity and vibration in transport
   - not stacked beyond the edge of a pallet
   - unitized and secured on a pallet
   - allow vertical air flow in bottom-air delivery transport (mostly marine containers and refrigerated ships)

2. Product must be cooled to proper transit temperature before loading.

3. Do not mix products that have different requirements for temperature, humidity, or are incompatible because of ethylene or odor sensitivity. Ethylene-absorbing materials can reduce ethylene damage if ethylene producers must be mixed with ethylene sensitive products.

4. Equipment must be in good condition before loading. Check to ensure that
   - Air delivery chute is in place and has no tears (mostly on trailers).
   - Door seals are in good condition.
   - Walls and ceiling are in good repair.
   - Floor and floor drains are clean.
   - Inside smells clean.

5. Vehicle should be cool before loading. Turn off refrigeration when doors are open.

6. There should be evidence that the thermostat has been calibrated.

7. Palletized loads should be well stabilized with netting or banding straps.

8. Do not allow product to touch air delivery chute; do not load above upper limit line.

9. Do not block airflow under load with crushed ice or solid load dividers.

10. Do not load product that is sensitive to vibration damage over wheel axles unless they have air ride suspension.

11. Use load bars or rear air bag to prevent rear product from shifting.

12. Place temperature monitor if needed.

B. For Highway Trailers Only

1. Check RTF classification plate to ensure that trailer was originally designed to handle the temperature conditions of the load. A C65 rating is rarely adequate for fruits, vegetables, or flowers.

2. Trailers must have a front bulkhead. Two pallets placed against the front wall can serve as a temporary bulkhead.

3. Do not load product directly on the floor unless trailer has a special deep-channel floor.

4. Load away from walls using stabilizing blocks or small inexpensive air bags.

C. For Marine Containers and Ships Only

1. Completely cover floor and pallet openings to force air through the load.

2. Produce containers should have bottom and top panel venting to allow vertical airflow. Vents must not be blocked by interior packaging materials or deck board of pallet.

3. Set fresh air exchange vent to prescribed level.

Figure 20.5

Refrigerated marine container.

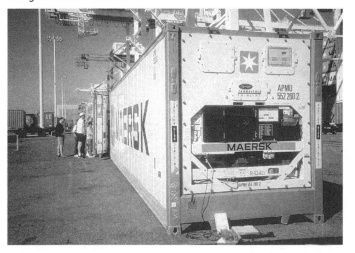

Figure 20.6

Refrigerated container ship.

Figure 20.7

Filled hold of a refrigerated ship. Floor of topmost hold is being moved into place prior to filling.

companies. Transportation contracts may be arranged for a year at a time. Bananas, table grapes, apples, and citrus fruits are commonly transported in refrigerated ships.

The key differences between refrigerated marine transport systems and refrigerated highway vehicles are that they have a bottom-delivery airflow system (fig. 20.8), they can slowly cool product during transport, and they are more gas-tight, allowing them to be used for controlled atmosphere conditions in the stowage space.

Usual outside dimensions of refrigerated marine containers are a 12.2 m (40 ft) length, 2.4 m (8 ft) width, and 2.6 to 2.9 m (8.5 to 9.5 ft) height. Interior volume ranges from 56.6 to 65.1 m^3 (2,000 to 2300 ft^3). Refrigeration capacity ranges from 8.4 to 10.2 kW (2.4 to 2.9 tons). Marine containers are usually shipped over the road so the unit plus tractor and wheeled chassis must still not exceed highway weight limits.

Typical newer refrigerated ships have a stowage volume of 10,000 to 15,000 m^3, and some are as large as 22,000 m^3. The stowage volume is typically divided into four separate holds, each with three to five cargo compartments. The compartments have a standard height of 2.2 m (7.3 ft). Ships usually have their own cranes for loading. Cargo is loaded through hatch covers over the top compartments. Compartment floors can be opened to allow product to be loaded into lower compartments. Automatic temperature control systems can maintain supply-air temperature within ± 0.1°C (about 0.2°F) of the set point temperature. In steady operating conditions, load space temperature variation is less than 2°C (about 4°F).

Most of the recommendations for best use of refrigerated highway trailers apply to containers and ships (see fig. 20.4). However, best use of the bottom-air delivery system requires loading so that the refrigerated air is forced through and around the packages and not allowed to bypass around pallet units. Bottom-delivery systems can accomplish slow product cooling if air can flow vertically through packages. Citrus from California is usually cooled in marine transport. Seven-eighths cooling is achieved in about 100 hours. Products that depend on transport cooling should be packed in boxes that have at least 3% venting on top and bottom panels, and the vents should align even if boxes

Figure 20.8

Bottom-air delivery airflow system.

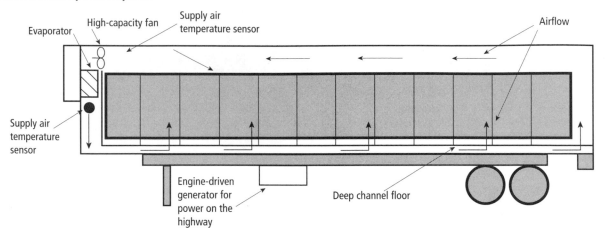

Figure 20.9

Proper loading of a marine container.

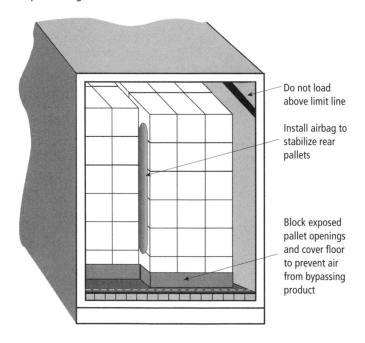

Do not load above limit line

Install airbag to stabilize rear pallets

Block exposed pallet openings and cover floor to prevent air from bypassing product

are cross-stacked. Interior packaging and pallet deck boards should not block airflow through panel vents.

The floor should be completely covered with product or solid material to force the refrigerated air around the packages and through the packages if they are vented for vertical airflow (fig. 20.9). If the product is palletized, open pallet edges should also be covered to block air from traveling horizontally through the pallet openings and escaping into an open vertical channel between pallet loads. Marine containers should not be center-loaded as is recommended for highway trail-

ers. Many have corrugated walls that guarantee airflow between the walls and product.

If the floor and pallet openings are not completely covered, refrigerated air will bypass some of the load. Product with poor airflow around it will tend to arrive too warm, especially if it had not been thoroughly cooled prior to transport. Open floor in the front of a container allows the air to flow through this open area, and product in the rear of the container will tend to be warmer because little refrigerated air reached it.

Refrigerated containers and refrigerated ships usually have built-in temperature monitoring and automatic recording of refrigeration system functioning. These data are usually only available to the shipping company, unless a special arrangement has been made before the trip. Shippers should install their own temperature recording equipment for their records. Monitors are generally installed on top of rear pallets for convenience. In this location the monitor will be exposed to air temperature that is influenced by the temperature of the air produced by the refrigeration system and heat from the product. The temperature will also be influenced by loading that allows conditioned air to bypass the rear of the load.

Containers and refrigerated ships have controlled ventilation systems to prevent high CO_2 or low O_2 concentrations and reduce ethylene levels in the load volume. Container venting is usually set at a constant level at loading (see Thompson et al. 2000 for vent settings). Refrigerated ships often have instrumentation to measure gas concentrations and vary ventilation as needed.

Controlled atmosphere transport is commercially available on some refrigerated ships. This service is becoming available in containers with new generation units specially built with onboard atmospheric modification equipment. Third-party contractors can add modified atmosphere transport service to most refrigerated containers.

Internationally marketed fruit may need to be fumigated to exclude insect pests. Refrigerated ships can slowly warm product to methyl bromide fumigation temperature just before arrival, speeding handling at the portside fumigation facility. Containers and refrigerated ships can also be used to meet quarantine requirements by holding product at low temperature. For example, USDA regulations allow control of Mediterranean fruit fly with fruit pulp temperatures of 1.1°C (34°F) for 12 days or 2.2°C (36°F) for 16 days.

RAIL CARS

Rail cars are used primarily for long-haul shipments within North America. Transportation times range from 6 to 10 days. They are mostly used to transport potatoes, citrus fruits, onions, carrots, and other less-perishable commodities. Rail shipments are often a single commodity. The cars have a stowage volume of more than 113 m³ (4,000 ft³) and can carry more than 45 metric tons (100,000 lb) of product.

The cars have their own permanently installed engine-driven generator and electric-motor-powered refrigeration system. The system incorporates a return-air temperature control system. Conditioned air is supplied to a plenum in the ceiling of the car. Air flows past the product and down wall flues and then returns underneath the load to the refrigeration unit (fig. 20.10). The units have adequate airflow and refrigeration capacity to produce slow cooling of the product if it is not packed too tightly. The cars are very air-tight when built and can be used for modified atmosphere transportation. Unintended atmospheric modification can occur if drain vents are clogged or water in them freezes in the winter.

Loading guidelines and package specifications for rail cars are generally specified by the railroad company or railroad organizations. Check with the transportation supplier for details. Very tight loads can restrict airflow and prevent air from traveling past the product. This may cause the thermostat sensor to detect warm air conditions and supply very cold air to the top of the load, causing some frozen product on the top of the load and warm product beneath. If crushed ice is applied uniformly to the top surface of the load, it blocks airflow and can prevent warm product below from being exposed to cooling conditions. If ice is used it should be applied to allow an open center area along the length of the car to permit airflow.

Several North American railroad companies offer trainload service for highway trailers that are designed to be fitted with railroad wheels. The trailers look much like conventional highway trailers, but are

Figure 20.10

Refrigerated rail car.

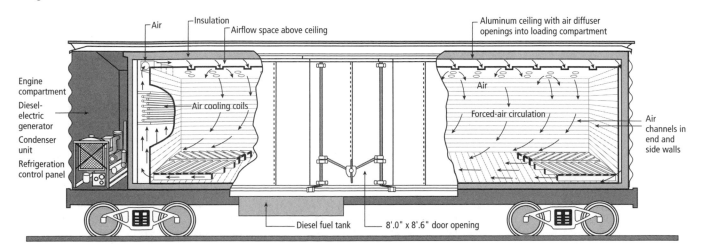

strengthened (adding about 500 kg [1,100 lb] to the vehicle tare weight) and can have special railroad wheels and couplers attached to them. They use regular over-the-road refrigeration equipment and often have air ride suspension. The vehicles can be placed on rails without the cranes or lifts needed for trailer-on-flatcar or container operations.

AIR

Air shipment is expensive and provides poor temperature control compared with refrigerated land and sea transport methods, but it often provides shorter transit times than competing methods. It is used mainly to transport highly perishable and valuable commodities to distant domestic and export markets. It is often used for early-season cherries, cut flowers, strawberries, and some tropical fruits. Products are transported in small aluminum containers that are shaped to fit inside the freight area of passenger planes or on net-covered pallets for transport in freight planes.

Most air transport containers are not refrigerated and provide minimal air circulation. Air temperature in cargo areas of planes is often set to provide safe conditions for live animals and is too warm for most perishables. At high altitudes air humidity in planes is extremely low, sometimes less than 10%, and can cause product dehydration if product is not packaged correctly or placed in a fairly airtight box. A few containers are available with CO_2 refrigerant. A battery-powered, thermostatically controlled fan moves air from the container past the refrigerant as needed to control temperature.

Improved temperature conditions can be obtained by wrapping the product to prevent warm air flowing past boxes (table 20.4). The wraps are sometimes made of reflecting materials to reduce radiant heat input when product is held on open docks or runways. Wraps should be removed if product warms to near room temperature because wraps may hold in respiration heat and cause product to heat above room temperature. Product is sometimes packed with enclosed ice, dry ice, or eutectic compounds to provide some refrigeration effect during transport. Use of dry ice must be reported to the airline because of the danger of CO_2 poising to animals or passengers. Water ice must be well enclosed to prevent water leakage and may require enough absorbent material in the box to contain an accidental leak.

Atmospheric pressure drops in flight to about 60% of sea level air pressure. Bagged product should be vented to allow pressure equalization. Bags for modified atmosphere packaging should be strong enough to withstand the low atmospheric pressure in air transport.

Travel time in the air is often in the range of 8 to 16 hours, but staging at departure and destination airports can add significantly to the total transport time. Staging areas of airlines are usually not refrigerated, and product can warm quickly, especially in hot, humid tropical climates, or it can freeze in winter conditions. Freight forwarding companies are often hired to arrange transport, handle the product at airports, and ensure best possible temperature conditions.

REFERENCES

American Society of Heating, Refrigeration, and Air Conditioning Engineers (ASHRAE). 1998. Refrigeration handbook. Atlanta, GA: ASHRAE.

Ashby, H. 1995. Protecting perishable foods during transport by truck. USDA-AMS Agr. Handb. No. 669. 88 pp.

Hardenberg, H. E., A. E. Watada, and C.Y. Wang. 1986. The commercial storage of fruits, vegetables, and florist and nursery stock. USDA-ARS Agr. Handb. No. 66. 130 pp.

Harvey, J. M., H. M Couey, C. M. Harris, and F. M. Porter. 1966. Air transport of California strawberries, factors affecting market quality in summer shipments – 1965. USDA-ARS Marketing Res. Report No. 751. 12 pp.

Table 20.4. Effect of pallet wraps on temperature rise of strawberries during air transport from San Francisco to several eastern U.S. cities

Pallet cover over open-top, tray-packed berries	Average arrival temperature
Corrugated fiberboard sheet on top of pallet	19.4°C (67°F)
Corrugated fiberboard on top and sides of pallet	12.2°C (54°F)
4 mil polyethylene film and corrugated fiberboard on top sides and under bottom boxes	8.9°C (48°F)

Source: Adapted from Harvey et al. 1966.

Note: Strawberries were originally cooled to 2.7°C (37°F); average transit time was 18 hr; average ambient temperature in the planes was about 15°C (60°F) and at airports ranged from 17°C (63°F) to 24°C (76°F).

Heap, R., M. Kierstan, and G. Ford. 1998. Food transportation. London, UK: Blackie Academic and Professional.

International Institute of Refrigeration. 1995. Guide to refrigerated transport. Paris, France: Int'l Inst. Refrig. 150 pp.

McGregor, B. M. 1989. Tropical products transport handbook. USDA Agr. Handb. No. 668. 148 pp.

Reid, M. S., and M. Serek. 1999. Guide to food transport. Controlled atmosphere. Copenhagen, Denmark: Mercantila Publishers. 153 pp.

Thompson, J. F., P. E. Brecht, R. T. Hinsch, and A. A. Kader. 2000. Marine container transport of chilled perishable produce. Oakland: Univ. Calif. Div. Ag. and Nat. Res. Publ. 21595. 32 pp.

Welby, E. M., and B. M. McGregor. 1997. Agricultural export transportation handbook. USDA Agr. Handb. No. 700. 138 pp.

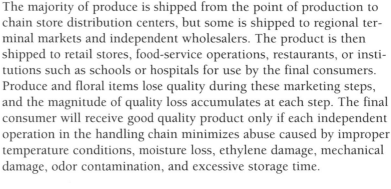

21

Handling at Destination Markets

James F. Thompson and

Carlos H. Crisosto

The majority of produce is shipped from the point of production to chain store distribution centers, but some is shipped to regional terminal markets and independent wholesalers. The product is then shipped to retail stores, food-service operations, restaurants, or institutions such as schools or hospitals for use by the final consumers. Produce and floral items lose quality during these marketing steps, and the magnitude of quality loss accumulates at each step. The final consumer will receive good quality product only if each independent operation in the handling chain minimizes abuse caused by improper temperature conditions, moisture loss, ethylene damage, mechanical damage, odor contamination, and excessive storage time.

DIRECT MARKETING

Limited amounts of horticultural crops are sold directly to the consumer through farmers' markets or local roadside stands. These markets are usually close to the point of production and offer the consumer product that is picked just before the time of sale. The short time between picking and sale to the consumer can allow the product to be harvested close to optimal eating quality. But ripe produce is often susceptible to mechanical damage and losses due to high temperature abuse and moisture loss. Most of these markets do not have refrigerated storage, and even if they do, they cannot quickly cool warm product.

Farmers must harvest each crop as it reaches optimal maturity or ripeness and without refrigerated storage, and they must sell it immediately or risk losing some product to decay, wilting or over-maturity. Refrigerated storage allows the farmer to temporarily store product during periods of large supply and market it later when customer demand exceeds supply. Storage is also vital for holding displayed commodity from one day to the next.

Non-chilling-sensitive produce can be displayed in iced displays. Crushed or flaked ice can be purchased or produced on-site from potable water. As a rough guide, a typical display requires about 20 to 24 kg/m^2 (4 to 5 lb/ft^2) of ice each day. Ice use can be minimized by protecting the display from direct sunlight and other sources of heat or drafts, and insulating the display. Design the display so that melt water does not spill into walkways. Spread ice in a thin layer over the product to provide good contact between ice and product; otherwise ice will not effectively cool the product.

Refrigerated displays are effective but can be costly to purchase. Used equipment is less expensive. Determine the potential cost effectiveness of acquiring a refrigerated display by estimating the amount of product spoilage or weight loss that might be prevented with the use of a refrigerated display. Refrigerated displays may not be cost-effective for items with a long shelf life.

If neither iced nor refrigerated displays are affordable, evaporative cooling can be used to reduce product temperature and slow moisture loss. Most vegetables and many fruits can be lightly sprayed with potable water. Direct evaporation from the product surface slows water loss and cools the product. Unit evaporative coolers can also be used to cool and humidify the air around produce displays. They are most beneficial in arid climates, although they have some value even in tropical areas. Placing product in vented plastic bags slows moisture loss but does not reduce product temperature.

Local markets should also be designed to maintain cleanliness. Each marketer should have access to potable water for product washing and preparation. Drains are needed to dispose of waste water. Daily trash removal is also needed to control odor and insect pests.

WHOLESALE AND RETAIL DISTRIBUTION

Large wholesale distribution facilities, whether independently owned or integrated with a retail chain, should strive to receive only the product that will be shipped the following day, with the exception of a few products, such as less-perishable produce (potato, onion, garlic, etc.) and mature green bananas, tomatoes, and some tree fruits that are ripened before transport to retail. In practice, few products remain at the facility for more than several days, but oversupply may occur if market conditions change or buyers overestimate store orders.

Products should be received at near their proper long-term storage temperature and then stored near that temperature. A classification of common fruits and vegetables into three categories according to their optimal temperature requirements is included in figure 12.2 (chapter 12). Air RH should be in the range of 85 to 95% except for the low temperature vegetables, which should be kept at 90 to 98% RH. The lowest temperature range, 0° to 2°C (32° to 36°F), is for the majority of the green, nonfruit vegetables and temperate fruits. If there is enough capacity in the facility, the fruits can be stored separately from the vegetables. This allows installing equipment to maintain higher humidity (90 to 98%) for the vegetables, as many of them are quite subject to wilting damage. Most floral items also need to be stored at 0° to 2°C (32° to 36°F) and 90 to 98% RH.

The two warmer temperature ranges are for the chilling-sensitive products. The highest-temperature room can also be used to ripen products that require only a warm environment for ripening. If refrigerated space is limited, low-temperature fruits and vegetables can be combined into one room, and an air-conditioned space at 20° to 25°C (68° to 77°F) temperature can be used for the highest temperature products.

Many green vegetables and most floral products are quite susceptible to ethylene damage. Ethylene must be kept away from these products. Banana ripening rooms are often the major source of ethylene contamination in a distribution facility. Most traditionally operated ripening rooms use ethylene levels near 1,000 ppm. A significant amount of the gas can leak into the area around the ripening rooms or escape as operators enter the ripening room. A well-sealed room can operate with ethylene levels of 100 ppm and still produce effective and uniform ripening. Ethylene should be vented from the ripening room to the outside after the exposure period is complete. Ripening rooms should never be vented by opening the ripening room doors. This releases the room's ethylene to adjacent storage areas within the distribution center.

The area around the ripening rooms should be vented to the outside or equipped with an ethylene scrubber. Operators often inspect product twice a day, and when they open ripening room doors ethylene is released. Even after the ethylene has been vented from the room, many of the ripened products naturally produce ethylene, causing concentrations that can damage sensitive products.

Ventilation with outside air is much less expensive to operate than scrubbers, unless outside air temperatures are very low or very high. Air around the ripening rooms should be vented to the outside once per day, preferably when air ambient temperatures are closest to the room air temperature.

Heated catalyst and potassium permanganate scrubbers are both effective in maintaining low ethylene levels. Two-thirds of the operating cost of a potassium permanganate scrubber is the cost of the reactive material. In a produce distribution warehouse, where there are many ethylene sources, the potassium permanganate system would probably be more expensive because of the large amount of ethylene that must be removed.

Propane, gasoline, and diesel-fueled vehicles emit ethylene and should not be used inside an area that contains ethylene-sensitive products. It is common practice to use battery-powered forklifts in cold storages.

Floral products are particularly sensitive to ethylene, and managers of some distribution facilities have found that even the above steps have not prevented damage to flowers. They have chosen to handle floral products with dairy or meat products, where ethylene concentrations are low, or they require that suppliers treat floral items to minimize their ethylene sensitivity.

Mechanical damage to products during the period between packing and retail display is usually caused by weak fiberboard containers or rough handling. If product arrives at the facility in crushed boxes, buyers need to work with their produce suppliers to get product packed in stronger boxes or ensure that the packed boxes are stacked and palletized correctly.

The distribution center assembles pallets of mixed products to be shipped to retail outlets. Product can easily be damaged when boxes with different footprints are stacked or heavy bags of product are placed on weak boxes. Some of this damage can be minimized by placing only strong containers of heavy produce on the bottom layers of a pallet load. Plastic foam and returnable plastic containers are often stronger than typical fiberboard boxes and reduce mechanical damage to product.

Most distribution facilities have special rooms for banana ripening that may also be used for avocados, kiwifruit, mangoes, tomatoes, stone fruits and European pears (see table 21.1 for a summary of ripening conditions for commonly ripened fruit). Pressurized or forced-air ripening rooms allow better control of ripening compared with older methods of space-stacking boxes in a warm room. The new designs force temperature-controlled air through the boxes, maintaining fairly uniform temperature of the product. Ethylene gas is added on a schedule appropriate for each product and CO_2 levels are kept below 1% by ventilating the rooms with outside air. Ripening is done with air temperature in the range of 14° to 25°C (58° to 77°F). Ripening temperature for a specific product is selected based on desired speed of ripening and the initial level of ripening. Water vapor is added to the air to keep RH above 85 to 95% in order to slow moisture loss. Some products, like stone fruits and pears, that have been treated with ethylene at the packing operation can be ripened by warming the fruit and placing it in the 13° to 18°C (55° to 65°F) area.

RETAIL MARKETS

Many larger-volume retail stores receive produce 5 to 7 days per week. This allows the potential for relatively short storage times, but some produce may be held in a back room for several days before being set out on display. Well-designed stores have two cold storage rooms, one for fruits and vegetables held near 0° to 2°C (32° to 36°F) and the other for produce that should be held at 7° to 10°C (45° to 50°F). Produce that should be held at 13° to 18°C (55° to 65°F) can be stored in the air-conditioned preparation room. The low-temperature room should be equipped with a floor drain to accommodate the meltwater from iced vegetables.

Ethylene gas can be controlled by periodically venting with outside ethylene-free air or by using ethylene scrubbers. But these precautions are rarely used, and ethylene levels in retail stores are often high enough (above 1 ppm) to damage sensitive products. Keeping products at their optimal temperature range and minimizing time in the store will reduce the damage problem.

Many flowers and floral products are particularly sensitive to ethylene. Damage can be reduced if the products are chemically treated (with either silver thiosulfate or 1-methylcyclopropene) to reduce their sensitivity to ethylene. It may also be possible in some stores to keep flowers in the meat and dairy storage areas, away from ethylene-producing fruits.

Refrigerated displays are most valuable for highly perishable products, those that are sensitive to wilting, and products that may not sell quickly. The displays are either single-layer units with air flowing up through the product or multilevel displays where the cold air flows from the top front of the display to the bottom front. Some units are designed to receive wheeled carts with already-loaded shelves. Most refrigerated displays are filled with individually placed product. Some stores are set up to receive product in returnable plastic containers, and the filled containers are placed directly on display. Hand labor is needed only to remove the last product in a box and replace empty boxes with filled ones.

Most displays can be equipped with timer-operated misting systems. These are particularly valuable for wilting-sensitive leafy vegetables and fresh herbs (table 21.2). Berries,

Table 21.1. Ripening conditions of some commonly ripened fruits

Commodity	Ripening treatment location	Ethylene concentration (ppm)	Ethylene exposure time	Ripening temperature	Predicted storage after treatment
Avocado	Shipping point (from harvest to within 2 weeks storage at 5°C)	10–100 ppm	8–48 hr, until the button pops off the stem-end of the fruit	15.5°–20°C (60°–68°F)	Less than 7 days at 5°C (41°F)
Apple (Granny Smith)	Shipping point (early- to midseason harvested up to 170 days after bloom)	10 ppm	6 days	25°C (77°F)	Less than 4 months at 0°C (32°F)
Banana	Distribution center	100–150 ppm	24–48 hr then hold until desired peel color	14°–18°C (58°–65°F)	Less than 7 days at 14°C (58°F)
Kiwifruit	Shipping point (conditioning)	100–200 ppm	12 hr	0°–5°C (32°–41°F)	3 to 6 weeks at 0°C (32°F) according to firmness
	Distribution center (unconditioned fruit stored for less than 4 weeks or with a firmness greater than 8–10 lbf)	100 ppm	12–24 hr	12°–25°C (54°–77°F)	4 days at 20°C (68°F), 12 days at 5°C (41°F)
	Distribution center (unconditioned fruit stored for more than 4 weeks or with a firmness less than 8–10 lbf)	None needed	Ripening period according to transfer/ shipping temperature, based on firmness	12°–25°C (54°–77°F)	3 days at 20°C (68°F), 8 days at 5°C (41°F)
Mango	Distribution center	100 ppm	12–24 hr	15.5°–25°C (60°–77°F)	Less than 7 days at 10°–13°C (50°–55°F) and 95% RH
Pear	Shipping point "Quality conditioning" is needed only if pears have been stored for less than 2 weeks	100 ppm	20–24 hr	20°–25°C (68°–77°F)	Storage life is not affected if fruit are cooled to 0°C (32°F) immediately after conditioning
	Distribution center	100 ppm, none if fruit have been "quality conditioned"	3–4 days, until fruit reach 10–12 lbf	13°–22°C (55°–72°F) with 90% RH	About 7 days at 0°–2°C (32°–36°F)
Stone fruits	Shipping point (treatment limits development of internal breakdown)	None for Calif. well-mature (CA WELL MAT)	48 hr	20°C (68°F)	Up to 6 weeks at 0°C (32°F), depending on cultivar and flesh firmness
	Distribution center	None for Calif. well-mature (CA WELL MAT)	Until 6–8 lbf firmness, 5–7 days at 13°C (55°F), 2–4 days at 20°C (68°F)	13°–25°C (55°–77°F) with 85–90% RH	3 days in unrefrigerated display, 6 days in refrigerated display
Tomato	Shipping point or distribution center				
	Mature-green stage	100 ppm	3–3.5 days to reach breaker stage and 5–16 days, depending on temperature, to reach full red stage	18°–20°C (65°–68°F) with 90–95% RH	7 days after reaching the red stage
	Breaker stage	None needed	5–16 days to reach full red stage (according to temperature)	15°–20°C (59°–68°F)	2 weeks if stored and ripened at a constant 15°C

Table 21.2. Produce items that benefit from misting while displayed

Artichoke	Eggplant	Radishes
Beans, snap	Endive	Rhubarb
Beets	Kale	Shallots, green
Broccoli	Leeks	Spinach
Brussels sprouts	Lettuce	Sprouts
Cabbage	Mustard greens	Squash, summer
Carrots	Onions, green	Swiss chard
Cauliflower	Parsley	Turnips
Celery	Parsnips	Watercress
Collards	Peas	
Corn	Peppers	

Note: Asparagus should be in a refrigerated display or displayed vertically oriented with cut ends on a wet absorbent pad.

mushrooms and a few other products should not be wetted and must not be placed in misted displays. Most fruits do not benefit from misting.

Most stores also display produce in nonrefrigerated displays. These are best suited to products with a long shelf life and are not sensitive to wilting, such as garlic, onion, potato, sweet potato, and melons. They are also used for products that are marketed at sale prices and will be sold quite rapidly.

Displays and storage areas must be kept clean and sanitized on a regular basis. Trimmings, waste, and other product remaining in displays are unsightly and can be sources of decay, odor, and ethylene. Flower buckets must be scrubbed and regularly disinfected to prevent buildup of fungal and bacterial organisms that infect flower stems and prevent water uptake.

Products should be grouped in displays according to their temperature and misting needs. Beyond this, produce managers may also group products by their intended use and with related items. Produce may also be grouped to produce an attractive mix of sizes, colors and shapes.

RESTAURANT AND INSTITUTIONAL FACILITIES

Restaurants often have limited facilities for handling fruits and vegetables. Refrigerated storage is usually only a single walk-in refrigerator that may be used for fresh meats, dairy products, eggs, prepared foods, and sauces in addition to fruits and vegetables. Many of these items are considered potentially hazardous because of possible bacterial contamination and by law must be held at

5°C (41°F) or below.

Most produce should be stored in the walk-in refrigerator. However, this temperature is not ideal for many items, so products should only be held a maximum of 1 or 2 days before consumption. This means that produce must be received almost every day that the restaurant is in operation. Product containers should be dated and the oldest product used first, unless a more recent batch shows signs of advanced ripening or deterioration. A few items are best held at room temperature because they are chilling sensitive and do not lose moisture rapidly. Intact dry onions, garlic, ginger, jicama, potatoes, hard rind squashes, melons, and sweet potatoes can be stored at 20°C (68°F).

Sliced and prepared produce and bean sprouts are sometimes considered potentially hazardous and must always be held at or below 5°C (41°F). Even when the thermostat is set at 5°C, temperatures near the door may be significantly above this because of frequent entry. Store these items in the colder areas of the walk-in, away from the door and close to the cold air flowing from the refrigerator coil. Measure and record temperature in all produce storage areas.

The limited space also means restaurants do not have the facilities to ripen fruits to optimal eating quality. Fruits that are harvested at a mature-green stage, like avocado, banana, mango, nectarine, pear, peach, and plum, should be received ripened and ready to use. Ripe fruits are very subject to bruising damage and have a short storage life. Supplier and chef will need to determine the best stage of ripeness for a particular establishment.

Minimize produce moisture loss by keeping a high-humidity environment around perishables. Refrigerators should be designed to maintain 85 to 95% RH. If a unit does not maintain this level of humidity or if produce loses excessive moisture in spite of a high humidity, store perishables in plastic bags or plastic boxes. The containers should have some air vents to prevent an accidentally high CO_2 environment that will damage the product. Eight to 12 small (6 mm, or ¼ in) holes in a 1-kg-capacity (2.2-lb) container is adequate.

Ethylene gas can cause damage like brown spots on head lettuce and loss of green color in vegetables. The damage is cumulative and is often caused by 24 hours of exposure to a

Table 21.3. Home storage of fruits and vegetables

Storage location	Fruits and melons	Vegetables
Store in refrigerator	Apples (more than 7 days) Apricots Blackberries Blueberries Cherries Cut fruits Figs Grapes Nashi (Asian pears) Raspberries Strawberries	Artichokes Asparagus Beets Belgian endive Broccoli Brussels sprouts Cabbage Carrots Cauliflower Celery Cut vegetables Green beans Green onions Herbs (not basil) Leafy vegetables Leeks Lettuce Lima beans Mushrooms Peas Radishes Spinach Sprouts Summer squashes Sweet corn
Ripen on the counter first, then store in the refrigerator	Avocados Kiwifruit Nectarines Peaches Pears Plums Plumcots	
Store only at room temperature	Apples (fewer than 7 days) Bananas Grapefruit Lemons Limes Mandarins Mangoes Muskmelons Oranges Papayas Persimmons Pineapple Plantain Pomegranates Watermelons	Basil (in water) Cucumbers† Dry onions* Eggplant† Garlic* Ginger Jicama Peppers† Potatoes* Pumpkins Sweet potatoes* Tomatoes Winter squashes

Notes:

*Store garlic, onions, potatoes, and sweet potatoes in a well-ventilated area in the pantry. Protect potatoes from light to avoid greening.

†Cucumbers, eggplant, and peppers can be kept in the refrigerator for 1 to 3 days if they are used soon after removal from the refrigerator.

1 ppm or higher concentration. The exposure could have occurred at several places in the handling chain. The chef will need to work with the supplier to determine source of ethylene. If the damage is caused at the restaurant, it can be minimized by regularly venting the storage areas by periodically opening doors. If ventilation is not feasible, ethylene can be controlled with potassium permanganate absorber systems.

Larger food service operations may have enough facilities to store products near their optimal temperature. They should use the temperature guidelines recommended for wholesale operations (see fig. 12.2 in chapter 12). All fresh-cut and otherwise prepared fruits and vegetables will need to be kept below 5°C (41°C).

HOME STORAGE

The home is the last step in handling produce. Most homes have adequate conditions for short-term storage of produce. However many products have a limited life because previous handling conditions and duration, and home storage conditions are rarely ideal, so most produce should be consumed within a few days of purchase.

Table 21.3 lists produce that should always be kept in the refrigerator (top group) and another group (bottom group) that should be stored on the counter because these products are damaged by refrigerator temperatures [usually 3° to 6°C (38° to 44°F)]. The middle group is fruits and fruit-type vegetables that ripen (soften and become sweeter) when held at room temperature. After they have ripened they can be stored for a few days without losing taste in the refrigerator. The bottom group is chilling-sensitive commodities that usually do not show damage until after 5 days at refrigerator temperatures followed by a day or so at room temperature. If they must be held for more than 5 days they should be placed in a cool part of the house such as a basement, or in the cooler months of the year, they can be stored in a garage if the garage does not get below chilling-damage temperatures.

The counter storage area should be away from direct sunlight to prevent produce from getting too hot. These produce items do not lose moisture rapidly, so they can be held at room temperatures for several days without shriveling. However, moisture loss can be reduced by putting them in a vented plastic bowl or a special ripening bowls that has a plastic lid. Even putting products in a paper bag will slow moisture loss. Do not put produce in sealed plastic bags on the counter because this will slow ripening due to a depletion of O_2 and accumulation of CO_2 within the bags.

Ripening in a bowl or paper bag can be speeded by placing one ripe apple with every five pieces of fruits to be ripened. The apples produce ethylene gas that speeds ripening. (Fuji or Granny Smith apples do not produce much ethylene and will not enhance ripening.)

Leafy green vegetables, carrots, and berries lose moisture quickly in the refrigerator. Put them in perforated plastic bags to retard moisture loss.

REFERENCES

Anon. 1987. Professional produce manager's manual. National-American Wholesale Grocers' Assn. and Produce Marketing Assn.

Anon. 1994. Refrigerated warehouse design (chapter 24); Retail food store refrigeration equipment (chapter 46). In ASHRAE Handbook. Atlanta: ASHRAE.

Bartsch, J. A., and R. Kline. 1992. Produce handling for direct market. Ithaca, NY: Northeast Regional Agricultural Engineering Service. 26 pp.

Kader, A., J. Thompson, and K. Sylva. 2000. Storing fresh fruits and vegetables for better taste. Oakland: Univ. Calif. Div. Ag. and Nat. Res. Publ. 21590 (poster).

Thompson, J., A. Kader, and K. Sylva. 1996. Compatibility chart for fruits and vegetables in short-term transport or storage. Oakland: Univ. Calif. Div. Ag. and Nat. Res. Publ. 21560 (poster).

Quality and Safety Factors: Definition and Evaluation for Fresh Horticultural Crops

Adel A. Kader

Quality is defined as any of the features that make something what it is, or the degree of excellence or superiority. The word *quality* is used in various ways in reference to fresh fruits and vegetables such as market quality, edible quality, dessert quality, shipping quality, table quality, nutritional quality, internal quality, and appearance quality.

The quality of fresh horticultural commodities is a combination of characteristics, attributes, and properties that give the commodity value for food (fruits and vegetables) and enjoyment (ornamentals). Producers are concerned that their commodities have good appearance and few visual defects, but for them a useful cultivar must score high on yield, disease resistance, ease of harvest, and shipping quality. To receivers and market distributors, appearance quality is most important; they are also keenly interested in firmness and long storage life. Consumers consider good-quality fruits and vegetables to be those that look good, are firm, and offer good flavor and nutritive value. Although consumers buy on the basis of appearance and feel, their satisfaction and repeat purchases are dependent upon good edible quality. Assurance of safety of the products sold is extremely important to the consumers. If the product is not safe it does not matter what its quality is; it should be eliminated from the produce distribution system.

COMPONENTS OF QUALITY

The various components of quality listed in table 22.1 are used to evaluate commodities in relation to specifications for grades and standards, selection in breeding programs, and evaluation of responses to various environmental factors and postharvest treatments. The relative importance of each quality factor depends upon the commodity and its intended use (fresh or processed). Appearance factors are the most important quality attributes of ornamental crops.

Many defects influence the appearance quality of horticultural crops. Morphological defects include sprouting of potatoes, onions, and garlic; rooting of onions; elongation of asparagus; curvature of asparagus and cut flowers; seed germination inside fruits such as lemons, tomatoes and peppers; presence of seed stems in cabbage and lettuce; doubles in cherries; floret opening in broccoli; and so on. Physical defects include shriveling and wilting; internal drying of some fruits; mechanical damage such as punctures, cuts and deep scratches, splitting and crushing, skin abrasions and scuffing, deformation (compression), and bruising; growth cracks (radial, concentric); and so on. Physiological defects include temperature-related disorders (freezing, chilling, sunburn, sunscald); puffiness of tomatoes; blossom-end rot of tomatoes; tipburn of lettuce; internal breakdown (chilling injury) of stone fruits; water core of apples; and black heart of potatoes. Pathological defects include decay caused by fungi or bacteria and virus-related blemishes, irregular ripening, and other disorders. Other defects result from damage caused by insects, birds, and hail; chemical injuries; and scars, scabs, and various blemishes (e.g., russeting, rind staining).

The texture of horticultural crops is important for eating and cooking quality and is a factor in withstanding shipping stresses. Soft fruits cannot be shipped long distances without extensive losses due to physical injuries. In many cases, this necessitates harvesting fruits at less than ideal maturity for flavor quality.

Table 22.1. Quality components of fresh fruits and vegetables

Main factors	Components
Appearance (visual)	Size: dimensions, weight, volume
	Shape and form: diameter/depth, ratio, compactness, uniformity
	Color: uniformity, intensity
	Gloss: nature of surface wax
	Defects: external, internal
	Morphological
	Physical and mechanical
	Physiological
	Pathological
	Entomological
Texture (feel)	Firmness, hardness, softness
	Crispness
	Succulence, juiciness
	Mealiness, grittiness
	Toughness, fibrousness
Flavor (taste and smell)	Sweetness
	Sourness (acidity)
	Astringency
	Bitterness
	Aroma (volatile compounds)
	Off-flavors and off-odors
Nutritional value	Carbohydrates (including dietary fiber)
	Proteins
	Lipids
	Vitamins
	Minerals
Safety	Naturally occurring toxicants
	Contaminants (chemical residues, heavy metals)
	Mycotoxins
	Microbial contamination

Evaluating flavor quality involves the perception of tastes and aromas of many compounds. Objective analytical determination of critical components must be coupled with subjective evaluations by a taste panel to yield meaningful information about flavor quality. This approach can be used to establish a minimum acceptable level. To learn consumer flavor preferences for a given commodity, large-scale testing by a representative sample of consumers is required.

Fresh fruits and vegetables play a significant role in human nutrition, especially as sources of vitamins (C, A, B_6, thiamine, niacin), minerals, and dietary fiber. Their contribution as a group is estimated at 91% of vitamin C, 48% of vitamin A, 27% of vitamin B_6, 17% of thiamine, and 15% of niacin in the U.S. diet. Fruits and vegetables also supply 26% of magnesium, 19% of iron, and 9% of the calories consumed. Legume veg-etables, potatoes, and tree nuts contribute about 5% of the per capita availability of proteins in the United States, and their proteins have a good content of essential amino acids. Other important nutrients supplied by fruits and vegetables include folacin, riboflavin, zinc, calcium, potassium, and phosphorus. Other constituents that may lower risk of cancer, heart disease, and other diseases include carotenoids, flavonoids, isoflavones, phytosterols, and other phytochemicals (phytonutrients). Postharvest losses in nutritional quality, particularly vitamin C content, can be substantial and increase with physical damage, extended storage, high temperatures, low RH, and chilling injury.

Safety factors include levels of naturally occurring toxicants in certain crops (such as glycoalkaloids in potatoes), which vary according to genotype and are routinely monitored by plant breeders so they do not exceed safe levels. Contaminants such as chemical residues and heavy metals are also monitored by various agencies to assure compliance with established maximum tolerance levels. Sanitation throughout harvesting and postharvest handling operations is essential to minimize microbial contamination; procedures that reduce the potential for growth and development of mycotoxin-producing fungi must be used. For more details about safety factors, see chapter 24.

INTERRELATIONSHIPS AMONG THE COMPONENTS OF QUALITY

It is important to define the interrelationships among each commodity's quality components and to correlate subjective and objective methods of quality evaluation. This information is essential for selecting new cultivars, choosing optimal production practices, defining optimal harvest maturity, and identifying optimal postharvest handling procedures. The point of all this effort is to provide high-quality fruits and vegetables for the consumer.

In most commodities, the rate of deterioration in nutritional quality (especially vitamin C content) is faster than that in flavor quality, which in turn is lost faster than textural quality and appearance quality. Thus, the postharvest life of a commodity based on appearance (visual) quality is often longer than its postharvest life based on maintenance of good flavor.

Quality criteria used in the U.S. standards for grades and the California Agricultural Code (tables 23.1 to 23.5 in chapter 23) emphasize appearance quality factors in most commodities. In many cases, good appearance does not necessarily mean good flavor and nutritional quality. A fruit or vegetable that is misshapen or has external blemishes may be just as tasty and nutritious as one of perfect appearance. For this reason, it is important to include quality criteria other than appearance that more accurately reflect consumer preferences. Such quality indices must be relatively easy to evaluate, and objective methods for evaluation should be developed.

FACTORS INFLUENCING QUALITY

Many pre- and postharvest factors influence the composition and quality of fresh horticultural crops. These include genetic factors (selection of cultivars and rootstocks), preharvest environmental factors (climatic conditions and cultural practices), maturity at harvest, harvesting method, and postharvest handling procedures.

Climatic conditions

Climatic factors, especially temperature and light intensity, have a strong influence on the nutritional quality of fruits and vegetables. The location and season in which plants are grown can determine their ascorbic acid, carotene, riboflavin, and thiamine content. Light is one of the most important climatic factors in determining ascorbic acid content of plant tissues. Researchers consistently find much higher ascorbic acid content in strawberries grown under high light intensity than in the same varieties grown under lower light intensity. In general, the lower the light intensity, the lower the ascorbic acid of plant tissues.

Although light does not play a direct role in the uptake and metabolism of mineral elements by plants, temperature influences the nutrient supply because transpiration increases with higher temperatures. Rainfall affects the water supply to the plant, which may influence composition of the harvested plant part.

Cultural practices

Soil type, the rootstock used for fruit trees, mulching, irrigation, and fertilization influence the water and nutrient supply to the plant, which can affect the nutritional composition of the harvested plant part. The effect of fertilizers on the vitamin content of plants is much less important than variety and climate, but their effect on mineral content is more significant. Increasing the nitrogen and/or phosphorus supply to citrus trees results in somewhat lower acidity and ascorbic acid content in citrus fruits, while increased potassium supply increases their acidity and ascorbic acid content.

Cultural practices such as pruning and thinning determine the crop load and fruit size, which can influence the nutritional composition of fruit. The use of agricultural chemicals, such as pesticides and growth regulators, does not directly influence fruit composition but may indirectly affect it due to delayed or accelerated fruit maturity.

Maturity at harvest

This is one of the main factors determining compositional quality and storage life of fruits and vegetables. All fruits, with few exceptions, reach peak eating quality when fully ripened on the tree. However, since they cannot survive the postharvest handling system, they are usually picked mature but not ripe. Tomatoes harvested green and ripened at 20°C (68°F) to table ripeness contain less ascorbic acid than those harvested at the table-ripe stage.

Harvesting method

The method of harvest can determine the variability in maturity and physical injuries and can consequently influence the nutritional composition of fruits and vegetables. Mechanical injuries such as bruising, surface abrasions, and cuts can accelerate loss of vitamin C. The incidence and severity of such injuries are influenced by the method of harvest, management of harvesting, and handling operations. Proper management to minimize physical damage to the commodity is required whether harvesting is done by hand or by machine.

Postharvest handling procedures

Delays between harvesting and cooling or processing can result in direct losses (due to water loss and decay) and indirect losses (lowering of flavor and nutritional quality). The extent of such losses is related to the condition of the commodity when picked and is strongly influenced by the temperature of the commodity, which can be several degrees higher than ambient temperatures, especially when exposed to sunlight. Temperatures higher than those that are optimum for the commodity increase the loss rate of vitamin content, especially vitamin C. In general, vegetables have more loss of ascorbic acid content in response to elevated temperatures than do fruits that are more acidic (pH 4.0 or lower), such as citrus.

Chilling injury causes accelerated losses in ascorbic acid content of sweet potatoes, pineapples, and bananas, but it does not influence ascorbic acid content of tomatoes and guavas.

METHODS FOR EVALUATING QUALITY

Quality evaluation methods can be destructive or nondestructive. They include objective scales based on instrument readings and subjective methods based on human judgment using hedonic scales.

APPEARANCE QUALITY (VISUAL)
1. Size

Dimensions: Measured with sizing rings, calipers.

Weight: Correlation is generally good between size and weight; size can also be expressed as numbers of units of commodity per unit of weight.

Volume: Determined by water displacement or by calculation from measured dimensions.

2. Shape

Ratio of dimensions: For example, diameter/depth ratio; used as index of shape in fruits.

Diagrams and models of shape: Some commodity models are used as visual aids for quality inspectors.

3. Color

Uniformity and intensity: Important appearance qualities.

Visual matching: Using color charts, guides, and dictionaries to match and describe colors of fruits and vegetables.

Light reflectance meter: Measures color on the basis of the amount of light reflected from surface of the commodity; examples include Minolta Colorimeter, Gardner and Hunter Color Difference Meters (tristimulus colorimeters), and Agtron E5W spectrophotometer.

Light transmission meter: Measures the light transmitted through the commodity; may be used to determine internal color and various disorders, such as water core of apples and black heart of potatoes.

Measurement of delayed light emission: Related to the amount of chlorophyll in the plant tissues; can be used to determine color-based maturity stages.

Determination of pigment content: Evaluates the color of horticultural crops by pigment content, i.e., chlorophylls, carotenoids (carotene, lycopene, xanthophylls), and flavonoids (anthocyanins).

4. Gloss (bloom, finish)

Wax platelets: Amount, structure, and arrangement on the fruit surface affect the gloss quality; measured using a Gloss-meter or by visual evaluation.

5. Presence of defects (external and internal)

Incidence and severity of defects are evaluated using a five-grade scoring system (1 = no symptoms, 2 = slight, 3 = moderate, 4 = severe, 5 = extreme) or a seven- or nine-point hedonic scale if more categories are needed. To reduce variability among evaluators, detailed descriptions and photographs may be used as guides in scoring a given defect. Objective evaluation of external defects using computer-aided vision techniques appears promising.

Internal defects can be evaluated by non-destructive techniques, such as light transmission and absorption characteristics of the commodity, sonic and vibration techniques associated with the mass density and elasticity of the material, X-ray transmission (which depends on mass density and mass absorption coefficient of the material), and nuclear magnetic resonance (NMR) imaging (also known as magnetic resonance imaging, or MRI), which detects the concentration of hydrogen nuclei and is sensitive to variations in the concentration of free water and oil.

TEXTURAL QUALITY
1. Yielding quality (firmness, softness)

Hand-held testers: Determine penetration force using testers such as the Magness-Taylor Pressure Tester and the Effegi penetrometer. The plunger (tip) size used depends on the fruit and varies from 3 mm ($\frac{1}{8}$ in) for cherry, grape, and strawberry; 8 mm ($\frac{5}{16}$ in) for stone fruits (other than cherries), kiwifruit, and pear; to 11 mm ($\frac{7}{16}$ in) for apple.

Stand-mounted testers: Determine penetration force using testers with a more consistent speed of punch such as the UC Fruit Firmness Tester and the Effegi penetrometer mounted on a drill stand.

Laboratory testing: Fruit firmness can be determined by measuring penetration force using an Instron Universal Testing machine or a Texture Testing system, or by measuring fruit deformation using a Deformation Tester.

It is inappropriate to use the term "pressure" in association with firmness measurements using the devices described above. While pounds-force (lbf) or kg-force (kgf) are preferred in the industry, Newton (N) is the required unit for scientific writing. The conversion factors are as follows:

$$\text{pound-force (lbf)} \times 4.448 = \text{Newton (N)}$$

$$\text{kilogram-force (kgf)} \times 9.807 = \text{Newton (N)}$$

2. Fibrousness and toughness

Shear force: Determined using an Instron or a Texture Testing system.

Resistance to cutting: Determined by using a Fibrometer.

Chemical analysis: Fiber content or lignin content.

3. Succulence and juiciness

Measurement of water content: An indicator of succulence or turgidity.

Measurement of extractable juice: An indicator of juiciness.

4. Sensory textural qualities

Sensory evaluation procedures: Evaluate grittiness, crispness, mealiness, chewiness, and oiliness.

FLAVOR QUALITY

1. Sweetness

Sugar content: Determined by chemical analysis procedures for total and reducing sugars or for individual sugars; indicator papers for quick measurement of glucose in certain commodities, such as potatoes.

Total soluble solids content: Measured using refractometers or hydrometers; can be used as indicator of sweetness because sugars are major component of soluble solids. Other constituents that contribute to total soluble solids include soluble pectins, organic acids, amino acids, and ascorbic acid.

2. Sourness (acidity)

pH (hydrogen ion concentration) of extracted juice: Determined using a pH meter or pH indicator paper.

Total titratable acidity: Determined by titrating a specific volume of the extracted juice with 0.1 N NaOH to pH 8.1, then calculating titratable acidity as citric, malic, or tartaric acid (depending on which organic acid predominates in the commodity).

3. Saltiness

Fresh vegetables and fruits: Usually not applicable.

4. Astringency

Determined by taste testing or by measuring tannin content, solubility, and degree of polymerization.

5. Bitterness

Determined by taste testing or measurement of the alkaloids or glucosides responsible for the bitter taste.

6. Aroma (odor)

Determined by sensory panels in combination with identification of volatile components responsible for specific aroma of a commodity (using gas chromatography–mass spectrometry).

7. Sensory evaluation

Human subjects: Judge and measure combined sensory characteristics (sweetness, sourness, astringency, bitterness, overall flavor intensity) of a commodity.

Laboratory panels: Detect and describe differences among samples; determine which volatile compounds are organoleptically important in a commodity.

Consumer panels: Indicate quality preferences.

NUTRITIONAL VALUE

Various analytical methods are available to determine total carbohydrates, dietary fiber, proteins and individual amino acids, lipids and individual fatty acids, vitamins, and minerals in fruits and vegetables. Several public and private laboratories have automated equipment for food analysis for use in situations where nutritional labeling is required and large numbers of samples have to be analyzed routinely.

SAFETY FACTORS

Analytical procedures, using thin-layer chromatography, gas chromatography, and high-pressure liquid chromatography, are available for determining minute quantities of the following toxic substances:

- naturally occurring toxicants, such as cyanogenic glucosides in lima beans and cassava, nitrates and nitrites in leafy vegetables, oxalates in rhubarb and spinach, thioglucosides in cruciferous vegetables, and glycoalkaloids (solanine) in potatoes

- natural contaminants, such as fungal toxins (mycotoxins), bacterial toxins and heavy metals (mercury, cadmium, lead)

- synthetic toxicants, such as environmental contaminants and pollutants, and residues of agricultural chemicals

QUALITY CONTROL AND ASSURANCE

An effective quality control and assurance system throughout the handling steps between harvest and retail display is required to provide a consistently good-quality supply of fresh horticultural crops to the consumers and to protect the reputation of a given marketing label. Quality control starts in the field with the selection of the proper time to harvest for maximum quality. Minimum acceptable flavor of fruits can be assured by determining their soluble solids content and titratable acidity (table 22.2). Careful harvesting is essential to minimize physical injuries and maintain quality. Each subsequent step after harvest has the potential to either maintain or reduce quality. Few postharvest procedures can improve the quality of individual units of the commodity.

Many attempts are currently being made to automate the separation of a given commodity into various grades and the elimination of defective units. The availability of low-cost microcomputers and solid-state imaging systems has made computer-aided video inspection on the packing line a practical reality. Solid-state video camera or light reflectance systems are used for detection of external defects, and X-ray or light transmittance systems are used for detecting internal defects. Further development of these and other systems to provide greater reliability and efficiency will be very helpful in quality control efforts.

FUTURE RESEARCH AND EXTENSION NEEDS

- Identify the important components of quality and the interrelationships among these quality factors for the various horticultural commodities and products for which such information is not available.

- Develop objective and nondestructive methods of determining quality attributes, especially those related to flavor and nutritional quality of fresh fruits and vegetables.

- Work with agencies responsible for standardization and inspection of fresh horticultural commodities to develop methods to improve the enforcement of current minimum maturity and quality standards to ensure better quality for the consumer. Also, consideration should be given to revising some of the existing quality and maturity standards with more emphasis on the eating quality of fruits and vegetables.

- Conduct consumer acceptance research aimed at relating maturity indices at harvest to the final organoleptic acceptability by the consumer.

- Continue efforts aimed at development of new genotypes with better flavor and nutritional quality in all the major fruits and vegetables and genotypes with improved appearance quality and vase life of cut flowers.

Table 22.2. Proposed minimum soluble solids content (SSC) and maximum titratable acidity (TA) for acceptable flavor quality of fruits

Fruit	Minimum SSC%	Maximum TA%
Apple	10.5–12.5 (depending on cultivar)	
Apricot	10	0.8
Blueberry	10	—
Cherry	14–16 (depending on cultivar)	
Grape	14–17.5 (depending on cultivar) or SSC:TA ratio of 20+	
Grapefruit	SSC:TA ratio of 6+	
Kiwifruit	14	—
Mandarin	SSC:TA ratio of 8+	
Mango	12–14 (depending on cultivar)	
Muskmelon	10	—
Nectarine	10	0.6
Orange	SSC:TA ratio of 8+	
Papaya	11.5	—
Peach	10	0.6
Pear	13	—
Persimmon	18	—
Pineapple	12	1.0
Plum	12	0.8
Pomegranate	17	1.4
Raspberry	8	0.8
Strawberry	7	0.8
Watermelon	10	—

- Study the effects of preharvest factors (climatic conditions, cultural practices, etc.) on quality attributes of fresh fruits, vegetables, and flowers.

- Evaluate the effects of currently used and alternative postharvest handling practices on flavor and nutritional quality (including phytonutrients contents) and safety attributes of fresh fruits and vegetables.

- Develop alternatives to currently used chemicals as part of integrated pest management strategies for control of postharvest diseases and insects of fresh horticultural crops.

- Expand the current Extension programs to reach more of the handlers, receivers, marketers, and consumers and provide them with information about proper procedures for maintaining quality and safety of fresh produce.

- Identify strategies to improve the efficiency of the distribution system for fresh fruits, ornamentals, and vegetables at the local, national, and international levels.

REFERENCES

Abbott, J. A. 1999. Quality measurement of fruits and vegetables. Postharv. Biol. Technol. 15:207–225.

Abbott, J. A., R. Lu, B. L. Upchurch, and R. Stroshine. 1997. Technologies for nondestructive quality evaluation of fruits and vegetables. Hort. Rev. 20:1–120.

Amerine, M. A., R. M. Pangborn, and E. B. Roessler. 1965. Principles of sensory evaluation of food. New York: Academic Press. 602 pp.

Arthey, V. D. 1975. Quality of horticultural products. New York: Halstead Press; Wiley. 228 pp.

Bourne, M. C. 1980. Texture evaluation of horticultural crops. HortScience. 15:51–57.

Chen, P., and Z. Sun. 1991. A review of non-destructive methods for internal quality evaluation and sorting of agricultural products. J. Agr. Eng. Res. 49:85–98.

Dull, G. G., G. S. Birth, and J. B. Magee. 1980. Nondestructive evaluation of internal quality. HortScience. 15:60–63.

Eskin, N. A. M., ed. 1989. Quality and preservation of vegetables. Boca Raton, FL: CRC Press. 313 pp.

———. 1991. Quality and preservation of fruits. Boca Raton, FL: CRC Press. 212 pp.

Francis, F. J. 1980. Color quality evaluation of horticultural crops. HortScience. 15:58–59.

Goddard, M. S., and R. H. Matthews. 1979. Contribu-

tion of fruits and vegetables to human nutrition. HortScience. 14:245–247.

Gould, W. A. 1977. Food quality assurance. Westport, CT: AVI. 314 pp.

Gunasekaran, S. 1990. Delayed light emission as a means of quality evaluation of fruits and vegetables. Crit. Rev. Food Sci. Nutr. 29:19–34.

Heintz, C. M., and A. A. Kader. 1983. Procedures for the sensory evaluation of horticultural crops. HortScience. 18:18–22.

Jen, J. J., ed. 1989. Quality factors of fruits and vegetables—Chemistry and technology. Washington, D.C.: Am. Chem Soc. Ser. 405. 410 pp.

Kader, A. A. 1983. Postharvest quality maintenance of fruits and vegetables in developing countries. In M. Lieberman, ed., Postharvest physiology and crop preservation. New York: Plenum. pp. 455–470.

Lee, S. K., and A. A. Kader. 200. Preharvest and postharvest factors influencing vitamin C content of horticultural crops. Postharv. Biol. Technol. 20:207–220.

Lipton, W. J. 1980. Interpretation of quality evaluations of horticultural crops. HortScience. 15:64–66.

O'Mahony, M. 1986. Sensory evaluation of food. Statistical methods and procedures. New York: Marcel Dekker. 487 pp.

Pattee, H. E., ed. 1985. Evaluation of quality of fruits and vegetables. Westport, CT: AVI. 410 pp.

Paull, R. E. 1999. Effect of temperature and relative humidity on fresh commodity quality. Postharv. Biol. Technol. 15:263–277.

Shewfelt, R. L. 1999. What is quality? Postharv. Biol. Technol. 15:197–200.

Shewfelt, R. L., and S. E. Prussian, eds. 1993. Postharvest handling: A systems approach. San Diego, CA: Academic Press. 358 pp.

Stevens, M. A., and M. Albright. 1980. An approach to sensory evaluation of horticultural commodities. HortScience. 15:48–50.

Tomas-Barberan, F. A., and R. J. Robins, eds. 1997. Phytochemistry of fruits and vegetables. Oxford, UK: Oxford Science. 375 pp.

U.S. Department of Agriculture (USDA). 2000. Composition of foods: Raw, processed, prepared. Available via Internet at http://www.nal.usda.gov/fnic/foodcomp

Watada, A. E. 1989. Non-destructive methods of evaluating quality of fresh fruits and vegetables. Acta Hort. 258:321–329.

Williams, A. A. 1979. The evaluation of flavour quality in fruits and fruit products. In D. G. Land and H. E. Nursten, eds., Progress in flavour research. Essex, UK: Applied Science. pp. 287–305.

23

Standardization and Inspection of Fresh Fruits and Vegetables

Adel A. Kader

Grade standards identify the degrees of quality in a commodity that are the basis of its usability and value. Such standards are valuable tools in fresh produce marketing because they

- provide a common language for trade among growers, handlers, processors, and receivers at terminal markets

- help producers and handlers do better jobs of preparing and labeling fresh horticultural commodities for market

- provide a basis for incentive payments rewarding better quality

- serve as the basis for market reporting (prices and supplies quoted by the Federal-State Market News Service in different markets can only be meaningful if they are based on products of comparable quality)

- help settle damage claims and disputes between buyers and sellers

U.S. GRADE STANDARDS

U.S. standards for fresh fruit and vegetable grades are voluntary, except when required by state and local regulations, by industry marketing orders (federal or state), or for export marketing. They are also used by many private and government procurement agencies when purchasing fresh fruits and vegetables. The USDA Agricultural Marketing Service (AMS) is responsible for developing, amending, and implementing grade standards. For more information on the AMS, access their website at http://www.ams.usda.gov.

The first U.S. grade standards were developed for potatoes in 1917. Currently there are more than 150 standards covering 80 different commodities. The quality factors used in these standards for fresh fruits, vegetables, and tree nuts are summarized in tables 23.1 to 23.5 at the end of this chapter.

The number of grades and grade names included in the U.S. standards for a given commodity vary with the number of distinct quality gradations that the industry normally recognizes and with the established usage of grade names. Currently, grades include three or more of the following: U.S. Fancy, U.S. No. 1, U.S. No. 2, U.S. No. 3, U.S. Extra No. 1, U.S. Extra Fancy, U.S. Combination, U.S. Commercial, and so on. The AMS is gradually phasing in the first four grades as uniform grades for all fresh fruits and vegetables, to represent available levels of quality.

Steps to establish or change U.S. standards include:

1. Demonstration of need, interest, and support from the industry.

2. Study of physical characteristics and quality factors, and their normal ranges for the commodity in the main production areas.

3. Consultation among all interested parties as part of data collection.

4. Development of a proposal that is practical.

5. Publication of the proposal in the *Federal Register*, and publicizing it through various means with an invitation for comments. Public hearings may be held for the same purpose.

6. Amendment of the proposal on the basis of comments received.

7. Publication of the standards in their final form in the *Federal Register* with a specified date on which they become effective (at least 30 days after publication date).

APPLYING THE STANDARDS

USDA inspectors are located at most shipping points and at terminal markets. In many cases cooperative agreements between the USDA and the states are in place to allow federal-state grading by USDA-licensed state inspectors. Some inspectors are full-time employees, while others are seasonal employees hired during the peak production season in a given location.

METHODS OF INSPECTION

1. Continuous inspection. One or more inspectors are assigned to a packinghouse. They make frequent quality checks on the commodity along the packing lines and examine samples of the packed product to determine whether it meets the U.S. grade specifications for which it is being packed. The inspector gives oral and/or written reports to management so that they can correct problems.

2. Inspection on a sample basis. Representative samples of a prescribed number of boxes out of a given lot are randomly selected and inspected to determine the quality and condition of the commodity according to grade specifications. Automatic sampling systems are used for some commodities that are handled in bulk bins or trailers, such as tomatoes, grapes, and cling peaches destined for processing. When inspection is completed, certificates are issued by the inspector on the basis of the applicable official standards. USDA inspectors can also inspect quality or condition based on a state grade or other specifications agreed upon by the parties involved. The cost of inspection is paid by the party requesting the service.

Each grade allows for a percentage of individual units within a lot that do not meet the standard. This reflects the practical limitations in sorting perishable products accurately into grades within a limited time. Tolerances, or the number of defects allowed, are more restrictive in U.S. No. 1 grade than in U.S. No. 2. The penalty for noncompliance with the U.S. grade specified on a given container may be rejection, resorting and repacking, or reclassification to a lower grade.

To ensure uniformity of inspection, inspectors are trained to apply the standards; visual aids (color charts, models, diagrams, photographs, and the like) are used whenever possible; objective methods for determining quality and maturity are used whenever feasible and practical; and good working environments with proper lighting are provided.

Recently, the Fresh Products Branch of the AMS equipped inspectors in designated market offices with digital cameras and enhanced computer technology for taking and transmitting images of produce or containers. AMS is offering the images to applicants over the Internet as an additional resource in its fresh fruits and vegetable inspection service. Inspectors also use the imaging to confer with produce quality experts working in USDA headquarters in Washington, D.C.

CALIFORNIA STANDARDS

California is one of the few states that has quality standards for horticultural crops produced within the state. The standards for fresh fruits and vegetables in the California Agricultural Code (summarized in tables 23.1 to 23.5) are mandatory minimum standards enforced by the California Department of Food and Agriculture (CDFA) Division of Inspection Services, Fruit and Vegetable Quality Control, through each county agricultural commissioner's staff. The cost of this inspection is paid by taxpayers. Noncompliance results in destruction of the commodity or its resorting and repacking to meet minimum requirements.

Steps for establishing new standards or revising existing ones are similar to those mentioned above for U.S. grade standards, except that they are carried out at the state level by the same agency responsible for inspection. Uniformity of inspection is assured by methods similar to those mentioned above for U.S. grade standards.

INDUSTRY STANDARDS

Some industries establish their own quality standards or specifications for a given commodity; examples include apricots, clingstone peaches, processing tomatoes, and walnuts. The standards are established by agreement between producers and processors, who pay application costs. Inspection is performed by independent agencies such as the California Dried Fruit Association and the Federal-State Inspection Service.

Some companies, cooperatives, and other organizations have quality grades that are applied by their quality-control personnel. Examples include quality grades for bananas, papayas, pineapples, and fresh-cut (lightly processed) fruits and vegetables.

INTERNATIONAL STANDARDS

International standards for fruits and vegetables were defined by the European Economic Commission (EEC) in 1954. Many standards have since been introduced, mainly under the Organization for Economic Cooperation and Development (OECD) scheme drawn up for this purpose. The first European International Standards were promulgated in 1961 for apples and pears, and now there are standards for about 40 commodities. Each includes three quality classes with appropriate tolerances: Extra class = superior quality; Class I = good quality; and Class II = marketable quality. Class I covers the bulk of produce entering into international trade. These standards or their equivalents are mandatory in the European Union (EU)

countries for imported and exported fresh fruits and vegetables. Inspection and certification is done by exporting and/or importing EU countries.

REFERENCES

California Department of Food and Agriculture. 1983. Fruit and vegetable quality control standardization. Extracts from the Administrative Code of California. Sacramento: CDFA. 154 pp.

Organization for Economic Cooperation and Development. Various dates. International standardization of fruits and vegetables. Paris: OECD.

U.S. Department of Agriculture (USDA). 1998. U.S. Standards and inspection instructions for fresh fruits and vegetables and other special products. Washington, D.C.: USDA Agric. Marketing Serv., Fruit and Vegetable Programs, Fresh Products Branch. 9 pp.

———. Various dates. U.S. standards for grades of fresh fruits and vegetables. Washington, D.C.: USDA Agric. Marketing Serv., Fruit and Vegetable Programs, Fresh Products Branch. Available via Internet at http://www.ams.usda.gov/standards

Table 23.1. Quality factors for fresh fruits in the U.S. Standards for Grades (U.S.) and the California Food and Agricultural Code (CA)

Fruit	Standard (date*)	Quality factors		
Apple	US (1976)	Maturity, color (color charts) related to grade, firmness, shape, and size, and freedom from decay, internal browning, internal breakdown, scald, scab, bitter pit, Jonathan spot, freezing injury, water core, bruises, russeting, scars, insect damage, and other defects		
	CA (1990)	Maturity as determined by starch staining pattern and/or soluble solids content [SSC] and firmness tests		
		Cultivar	SSC (%)	Firmness (lbf)
		Red Delicious	11.0	18
		Golden Delicious	12.0	18
		Jonathan	12.0	19
		Rome	12.5	21
		Newtown Pippin	11.0	23
		McIntosh	11.5	19
		Gravenstein	10.5	—
		Size, color, flesh condition, and freedom from defect (such as scald, spot, internal breakdown, water core, bruises, sunburn, russeting) and decay		
Apricot	US (1994)	Maturity, size, and shape, and freedom from defect and decay		
	CA (1983)	Maturity (>¾ of external surface area has a color equal to No. 3 yellowish green of the CDFA standard color chart or at least ½ has attained No. 4 yellow), and freedom from insect injury, decay, and mechanical damage		
Avocado	US (1957)	For Florida avocados: maturity, shape, texture, skin and flesh color, and freedom from decay, anthracnose, freezing injury, bruises, russeting, scars, sunburn, mechanical damage, and other defects		
	CA (1990)	Maturity (18.4 to 21.9% dry weight of the flesh, depending on cultivar), size, and freedom from defect, insect damage, freezing injury, rancidity, and decay		
Blueberry	US (1995)	Maturity, color, size, and freedom from defect and decay		
Cherry, sweet	US (1971)	Maturity, color, size, shape, and freedom from cracks, hail damage, russeting, scars, insect damage, and decay		
	CA (1983)	Maturity (entire surface with at least a solid light red color and/or 14 to 16% soluble solids, depending on the cultivar), and freedom from bird pecks, insect injury, shriveling, growth cracks, other defects, and decay		
Citrus				
Grapefruit	US (1950)	California and Arizona: maturity, color, firmness, size, shape, skin thickness, smoothness, and freedom from defect and decay		
	US (1997)	Florida: maturity, color (color charts), firmness, size, smoothness, shape, and freedom from discoloration, defect, and decay		
	US (1969)	Texas and other states: maturity, color, firmness, size, shape, smoothness, and freedom from discoloration, defect, and decay		
	CA (1983)	Maturity (minimum soluble solids:acid ratio of 5.5 or 6 [desert areas] and >⅔ of fruit surface showing yellow color—0.9 GY 6.40/5.7 Munsell color), and freedom from decay, freezing damage, scars, pitting, rind staining, and insect damage		
Lemon	US (1964)	Maturity (28 or 30% minimum juice content by volume, depending on grade), firmness, shape, color, size, smoothness, and freedom from discoloration, defect, and decay		
	CA (1983)	Maturity (30% or more juice by volume), size uniformity, and freedom from decay, freezing damage, drying, mechanical damage, rind stains, red blotch, shriveling, and other defects		
Lime	US (1958)	Color, shape, firmness, smoothness, and freedom from stylar end breakdown, bruises, dryness, other defects, and decay		
	CA (1983)	Maturity, and freedom from defect (freezing injury, drying, mechanical damage) and decay		
Orange	US (1957)	California and Arizona: maturity, color, firmness, smoothness, size, and freedom from defect and decay		
	US (1997)	Florida: maturity, color (color charts), firmness, size, shape, and freedom from discoloration, defect, and decay (used also for tangelos)		
	US (1969)	Texas and other states: maturity, color, firmness, shape, size, and freedom from discoloration, defect, and decay		
	CA (1983)	Maturity as indicated by soluble solids:acid ratio of 8 or higher and orange color on 25% of the fruit (7.5 Y 6/6 Munsell color) or soluble solids:acid ratio of 10 or higher and orange color on 25% of fruit (2.5 GY 5/6 Munsell color), size uniformity, and freedom from defect and decay		

Table 23.1. Cont.

Fruit	Standard (date*)	Quality factors
Tangerine and mandarin	US (1948)	States other than Florida: maturity, firmness, color, size, and freedom from defect and decay
	US (1997)	Florida: maturity, color (color charts), firmness, size, shape, and freedom from decay and defect
	CA (1983)	Maturity (yellow, orange, or red color on 75% of fruit surface and soluble solids:acid ratio of 6.5 or higher), size uniformity, and freedom from defect and decay
Cranberry	US (1971)	Maturity, firmness, color, and freedom from bruises, freezing injury, scars, sunscald, insect damage, and decay
Date	CA (1983)	Freedom from insect damage, decay, black scald, fermentation, and other defects
Dewberry, blackberry	US (1928)	Maturity, color, and freedom from calyxes, decay, shriveling, mechanical damage, insect damage, and other defects
	CA (1983)	Maturity and freedom from decay and damage due to frost, bruising, insects, or other causes
Grape, table European *Vinifera* type	US (1991)	Maturity (as determined by percent soluble solids as set forth by the producing states); for states other than California and Arizona, and countries exporting to U.S.:

<table>
<tr><td></td><td>Minimum SCC</td></tr>
<tr><td>Cultivar</td><td>(%)</td></tr>
<tr><td>Muscat...................................... 17.5
Cardinal, Ribier, Olivette
Blanche, Emperor, Perlette,
Rish Baba, Red Malaga, and
Similar cultivars......................... 15.5
All other cultivars...................... 16.5</td><td></td></tr>
</table>

Fruit	Standard (date*)	Quality factors
		Color, uniformity, firmness, berry size, and freedom from shriveling, shattering, sunburn, waterberry, shot berries, dried berries, other defects, and decay; bunches: fairly well filled but not excessively tight; stems: not dry and brittle, and at least yellowish-green in color
	CA (1983)	Maturity (minimum percent soluble solids of 14 to 17.5, depending on cultivar and production area, or soluble solids:acid ratio of 20 or higher, or a combination of a minimum soluble solids:acid ratio and percent soluble solids), and freedom from decay, freezing injury, sunburned or dried berries, and insect damage (same for Arizona)
American bunch type	US (1983)	Maturity (juiciness, ease of separation of skin from pulp), color, firmness, compactness, and freedom from defect and decay
Kiwifruit	US (1986)	Maturity (more than 6.5% soluble solids), firmness, cleanness, and freedom from growth cracks, insect injury, broken skin, bruises, scars, sunscald, freezing injury, internal breakdown, and decay
Nectarine	US (1966)	Maturity, color (depending on variety), shape, and size, and freedom from growth cracks, insect damage, scars, bruises, russeting, split pits, other defects, and decay
	CA (1983)	Maturity (surface ground color, fruit shape), and freedom from insect injury, split pits, mechanical damage, and decay
Olive	CA (1983)	Freedom from insect injury, especially scale
Peach	US (1995)	Maturity (shape, size, ground color), and freedom from decay and defect (split pit, hail injury, insect damage, growth cracks)
	CA (1983)	Maturity (skin and flesh color, and fullness of shoulders and suture) and freedom from defect and decay
Pear		
Winter	US (1955)	Maturity (color, firmness), size, and freedom from internal breakdown, black end, russeting, other defects, and decay
Summer and fall	US (1955)	Maturity (color, firmness), shape, size, and freedom from defect and decay
	CA (1983)	Maturity (Bartlett: average firmness test of <23 lb, and/or soluble solids content 13%, and/or yellowish-green color on CDFA color chart), and freedom from insect damage, mechanical damage, decay, and other defects
Persimmon	CA (1983)	Maturity as indicated by surface color: Hachiya: blossom end's color is orange or reddish color equal to or darker than Munsell color 6.7 YR 5.93/12.7 on at least 1/3 of the fruit's length with the remaining 2/3 a green color equal to or lighter than Munsell color 2.5 GY 5/6; other cultivars: yellowish-green color equal to or lighter than Munsell color 10 Y 6/6; freedom from growth cracks, mechanical damage, decay, and other defects

Table 23.1. Cont.

Fruit	Standard (date*)	Quality factors
Pineapple	US (1990)	Maturity, firmness, uniformity of size and shape, and freedom from decay, sunburn, sunscald, bruising, internal breakdown, gummosis, insect damage, and cracks; tops: color, length, and straightness
Plum and fresh prune	US (1973)	Maturity, color, shape, size, and freedom from decay, sunscald, split pits, hail damage, mechanical damage, scars, russeting, and other defects
	CA (1983)	Maturity as indicating by surface color (minimum color requirements are described for 56 cultivars), and freedom from decay, insect damage, bruises, sunburn, hail damage, gum spot, growth cracks, and other defects
Pomegranate	CA (1983)	Maturity (<1.85% acid content in juice and red juice color equal to or darker than Munsell color 5 R 5/12), and freedom from sunburn, growth cracks, cuts or bruises, and decay
Quince	CA (1983)	Maturity and freedom from insect damage, mechanical damage, and decay
Raspberry	US (1931)	Maturity, color, shape, and freedom from defect and decay
	CA (1983)	Maturity and freedom from decay and damage due to insects, sun, frost, bruising, or other causes
Strawberry	US (1965)	Maturity (>½ or >¾ of surface showing red or pink color, depending on grade), firmness, attached calyx, size, and freedom from defect and decay
	CA (1983)	Maturity (>⅔ of fruit surface showing a pink or red color) and freedom from defect and decay

Note: *Date when standard was issued or revised.

Table 23.2. Quality factors for fresh vegetables in the U.S. Standards for Grades (U.S.) and the California Food and Agricultural Code (CA)

Vegetable	Standard (date*)	Quality factors
Anise, sweet	US (1973)	Firmness, tenderness, trimming, blanching, and freedom from decay and damage caused by growth cracks, pithy branches, wilting, freezing, seedstems, insects, and mechanical means
Artichoke	US (1969)	Stem length, shape, overmaturity, uniformity of size, compactness, and freedom from decay and defects
	CA (1983)	Freedom from decay, insect damage, and freezing injury
Asparagus	US (1966)	Freshness (turgidity), trimming, straightness, and freedom from damage and decay, diameter of stalks, percent green color
	CA (1983)	Turgidity, straightness, percent showing white color, stalk diameter, and freedom from decay, mechanical damage, and insect injury
Bean, lima	US (1938)	Uniformity, maturity, freshness, shape, and freedom from damage (defect) and decay
Bean, snap	US (1990)	Uniformity, size, maturity (not overmature = woody or fibrous), firmness (not wilted or flabby), and freedom from defect and decay
Beet, bunched, or topped	US (1955)	Root shape, trimming of rootlets, firmness (turgidity), smoothness, cleanness, minimum size (diameter), and freedom from defect
Beet, greens	US (1959)	Freshness, cleanness, tenderness, and freedom from decay, other kinds of leaves, discoloration, insects, mechanical injury, and freezing injury
Broccoli	US (1943)	Color, maturity, stalk diameter and length, compactness, base cut, and freedom from defects and decay
	CA (1983)	Freedom from decay and damage due to overmaturity, insects, or other causes
Brussels sprouts	US (1954)	Color, maturity (firmness), no seedstems, size (diameter and length), and freedom from defect and decay
	CA (1983)	Freedom from decay, from burst, soft, or spongy heads, and from insect damage
Cabbage	US (1945)	Uniformity, solidity (maturity or firmness), no seedstems, trimming, color, and freedom from defect and decay
	CA (1983)	Conform to U.S. commercial grade or better
Cantaloupe	US (1968)	Soluble solids (>9%), uniformity of size, shape, ground color and netting; maturity and turgidity; and freedom from "wet slip," sunscald, and other defects
	CA (1983)	Maturity (soluble solids >8%) and freedom from insect injury, bruises, sunburn, growth cracks, and decay
Carrot, bunched	US (1954)	Shape, color, cleanness, smoothness, freedom from defects, freshness, length of tops, and root diameter
	CA (1983)	Number, size, and weight per bunch, freshness, and freedom from defect and decay (tops)
Carrot, topped	US (1965)	Uniformity, turgidity, color, shape, size, cleanness, smoothness, and freedom from defect (growth cracks, pithiness, woodiness, internal discoloration)
	CA (1983)	Freedom from defect (growth cracks, doubles, mechanical injury, green discoloration, objectionable flavor or odor) and decay
Carrots with short rimmed tops	US (1954)	Roots: firmness, color, smoothness; freedom from defect (sunburn, pithiness, woodiness, internal discoloration, and insect and mechanical injuries) and decay; leaves (cut to <4 inches),: freedom from yellowing or other discoloration, disease, insects, and seedstems
Cauliflower	US (1968)	Curd cleanness, compactness, white color, size (diameter), freshness and trimming of jacket leaves, and freedom from defect and decay
	CA (1983)	Freedom from insect injury, decay, freezing injury, and sunburn
Celery	US (1959)	Stalk form, compactness, color, trimming, length of stalk and midribs, width and thickness of midribs, no seedstems, and freedom from defect and decay
	CA (1983)	Freedom from pink rot and other decay, blackheart, seedstems, pithy condition, and insect damage
Collard green and broccoli greens	US (1953)	Freshness, tenderness, cleanness, and freedom from seedstems, discoloration, freezing injury, insects, and diseases
Corn, sweet	US (1992)	Uniformity of color and size, freshness, plump and milky kernels, cob length, and freedom from insect injury, discoloration, and other defects, coverage with fresh husks
	CA (1983)	Milky, plump, well-developed kernels, and freedom from insect injury, mechanical damage, and decay

Table 23.2. Cont.

Vegetable	Standard (date*)	Quality factors
Cucumber	US (1958)	Color, shape, turgidity, maturity, size (diameter and length), and freedom from defect and decay
Cucumber, greenhouse	US (1985)	Freshness, shape, firmness, color, size (length of 11 inches or longer), and freedom from decay, cuts, bruises, scars, insect injury and other defects
Dandelion greens	US (1955)	Freshness, cleanness, tenderness; freedom from damage caused by seed stems, discoloration, freezing, diseases, insects, and mechanical injury
Eggplant	US (1953)	Color, turgidity, shape, size; freedom from defect and decay
Endive, escarole or chicory	US (1964)	Freshness, trimming, color (blanching), no seed stems, and freedom from defect and decay
Garlic	US (1944) CA (1983)	Maturity, curing, compactness, well-filled cloves, bulb size, and freedom from defect Size (bulb diameter)
Honeydew and honey ball melons	US (1967) CA (1983)	Maturity, firmness, shape, and freedom from decay and defect (sunburn, bruising, hail spots, and mechanical injuries) Maturity, soluble solids (>10%), and freedom from decay, sunscald, bruises, and growth cracks; honey ball melons should be netted and should have pink flesh
Horseradish roots	US (1936)	Uniformity of shape and size, firmness, smoothness, and freedom from hollow heart, other defects, and decay
Kale	US (1934)	Uniformity of growth and color, trimming, freshness, and freedom from defect and decay
Lettuce, crisp-head	US (1975) CA (1983)	Turgidity, color, maturity (firmness), trimming (number of wrapper leaves), and freedom from tip burn, other physiological disorders, mechanical damage, seedstems, other defects, and decay Freedom from insect damage, decay, seedstems, tip burn, freezing injury, broken midribs, and bursting; for sectioned, chopped, or shredded lettuce: same as intact heads plus freedom from discoloration and excessive moisture
Lettuce, greenhouse leaf	US (1964)	Well-developed, well-trimmed, and freedom from coarse stems, bleached or discolored leaves, wilting, freezing, insects, and decay
Lettuce, romaine	US (1960)	Freshness, trimming, and freedom from decay and damage caused by seedstems, broken, bruised, or discolored leaves, tip burn, and wilting
Melon, casaba and Persian	CA (1983)	Maturity and freedom from growth cracks, decay, mechanical injury, and sunburn
Mushroom	US (1966) CA(1983)	Maturity, shape, trimming, size, and freedom from open veils, disease, spots, insect injury, and decay Freedom from insect injury
Mustard greens and turnip greens	US (1953)	Fresh, tenderness, cleanness, and freedom from damage caused by seedstems, discoloration, freezing, disease, insects, or mechanical means; roots (if attached): firmness and freedom from damage
Okra	US (1928)	Freshness, uniformity of shape and color, and freedom from defect and decay
Onion, dry Creole Bermuda- Granex Grano Other cultivars	US (1943) US (1995) US (1995)	Maturity, firmness, shape, size (diameter), and freedom from decay, wet sunscald, doubles, bottlenecks, sprouting, and other defects
Onion, dry	CA (1983)	Freedom from insect injury, decay, sunscald, freezing injury, sprouting, and other defects
Onion, green	US (1947)	Turgidity, color, form, cleanness, bulb trimming, no seedstems, and freedom from defect and decay
Onion sets	US (1940)	Maturity, firmness, size, and freedom from decay and damage caused by tops, sprouting, freezing, mold, moisture, dirt, disease, insects, or mechanical means
Parsley	US (1930)	Freshness, green color, and freedom from defects, seedstems, and decay

Table 23.2. Cont.

Vegetable	Standard (date*)	Quality factors
Parsnip	US (1945)	Turgidity, trimming, cleanness, smoothness, shape, size (diameter), and freedom from defects and decay,
Pea, fresh	US (1942)	Maturity, size, shape, freshness, and freedom from defects and decay
	CA (1983)	Maturity, and freedom from mechanical damage, insect damage, decay, yellowing, and shriveling
Pea, Southern (Cowpea)	US (1956)	Maturity, pod shape, and freedom from discoloration and other defects
Pepper, sweet	US (1989)	Maturity, color, shape, size, firmness, and freedom from defects (sunburn, sunscald, freezing injury, hail, scars, insects, mechanical damage) and decay
	CA (1983)	Freedom from insect damage, bacterial spot, and decay
Potato	US (1991)	Uniformity, maturity, firmness, cleanness, shape, size, intactness of the skin, and freedom from sprouts, scabs, growth cracks, hollow heart, blackheart, greening, and other defects
	CA (1983)	A minimum equivalent of U.S. No. 2 grade; maturity is described in terms of extent of skin missing or feathered
Radish (topped)	US (1968)	Tenderness, cleanness, smoothness, shape, size, and freedom from pithiness and other defects
Rhubarb	US (1966)	Color, freshness, straightness, trimming, cleanness, stalk diameter and length, and freedom from defect
Shallot, bunched	US (1946)	Firmness, form, tenderness, trimming, cleanness, and freedom from decay and damage caused by seed stems, disease, insects, mechanical and other means; tops: freshness, green color, and no mechanical damage
Spinach, bunches	US (1987)	Freshness, cleanness, trimming, and freedom from decay and damage caused by coarse stalks or seedstems, discoloration, insects, and mechanical means
Spinach, leaves	US (1946)	Color, turgidity, cleanness, trimming, and freedom from seedstems, coarse stalks, and other defects
Squash, summer	US (1984)	Immaturity, tenderness, shape, firmness, and freedom from decay, cuts, bruises, scars, and other defects
Squash, winter and pumpkin	US (1983)	Maturity, firmness, uniformity of size, and freedom from discoloration, cracking, dry rot, insect damage, and other defects
Sweet potato	US (1963)	Firmness, smoothness, cleanness, shape, size, and freedom from mechanical damage, growth cracks, internal breakdown, insect damage, other defects, and decay
	CA (1983)	Freedom from decay, mechanical damage, insect injury, growth cracks, and freezing injury
Tomato	US (1991)	Maturity and ripeness (color chart), firmness, shape, size, and freedom from defect (puffiness, freezing injury, sunscald, scars, catfaces, growth cracks, insect injury, and other defects) and decay
	CA (1983)	Freedom from insect and freezing damage, sunburn, mechanical damage, blossom-end rot, catfaces, growth cracks, and other defects
Tomato, greenhouse	US (1966)	Maturity, firmness, shape, size, and freedom from decay, sunscald, freezing injury, bruises, cuts, shriveling, puffiness, catfaces, growth cracks, scars, disease, and insects
Turnip and rutabaga	US (1955)	Uniformity of root color, size, and shape, trimming, freshness, and freedom from defects (cuts, growth cracks, pithiness, woodiness, water core, dry rot)
Watermelon	US (1978)	Maturity and ripeness (optional internal quality criteria: soluble solids content = >10% very good, >8% good), shape, uniformity of size (weight), and freedom from anthracnose, decay, sunscald, and whiteheart
	CA (1983)	Maturity (arils around the seeds have been absorbed and flesh color is >75% red), and freedom from decay, sunburn, flesh discoloration, and mechanical damage

Note: *Date when standard was issued or revised.

Table 23.3. Quality factors for processing fruits in the U.S. Standards for Grades (US) and the California Food and Agricultural Code (CA)

Fruit	Standard (date*)	Quality factors
Apple	US (1961)	Ripeness (not overripe, mealy, or soft), and freedom from decay, worm holes, freezing injury, internal breakdown, and other defects that would cause a loss of >5% (U.S. No. 1) or >12% (U.S. No. 2) by weight
Berries	US (1947)	Color, and freedom from caps (calyxes), decay, and defect (dried, undeveloped and immature berries, crushing, shriveling, sunscald, insect damage, and mechanical injury)
Blueberry	US (1950)	Freedom from other kinds of berries, clusters, large stems, leaves and other foreign material, and freedom from damage caused by decay, shriveling, dirt, overmaturity, or other means
Cherry, red sour	US (1941)	Color uniformity, and freedom from decay, pulled pits, attached stems, hail marks, windwhips, scars, sun scald, shriveling, disease, and insect damage
Cherry, sweet for canning or freezing	US (1946)	Maturity, shape; freedom from decay, worms, pulled pits, doubles, insect and bird damage, and mechanical injury; freedom from damage caused by freezing softness, shriveling, cracks and skin breaks, scars, and sunscald; tolerance is 7% (U.S. No. 1) or 12% (U.S. No. 2) by count
Cherry, sweet for sulfur brining	US (1940)	Maturity (ease of pit separation), firmness, shape, and freedom from decay and defect (bruises, bird and insect damage, skin breaks, russeting, shriveling, scars, sunscald, and lim rubs)
Cranberry, red sour	US (1957)	Maturity, color, firmness, size, and freedom from defect (insect damage, bruises, scars, sunscald, freezing injury, and mechanical injury) and decay
Currant	US (1952)	Color, stem attached, and freedom from decay and damage caused by crushing, drying, shriveling, insects, and mechanical means
Grape, American type for processing and freezing	US (1943)	Maturity (>15.5% soluble solids), color; freedom from shattered, split, crushed, or wet berries, and freedom from decay; freedom from damage caused by freezing, heat, sunburn, disease, insects, or other means
Grape, juice (European or vinifera type)	US (1939)	Maturity (>16 to 18% soluble solids, depending on cultivar); freedom from crushed, split, wet, waterberry and redberry; freedom from defect (insect, disease, mechanical injury, sunburn, and freezing damage)
	CA (1983)	Maturity (minimum soluble solids content of 14 to 17.5%, depending on cultivar or soluble solids:acid ratio of 20 or higher), and freedom from decay, freezing injury, waterberry, redberry, and other defects
Grape for processing and freezing	US (1977)	Maturity (>15.5% soluble solids content), and freedom from decay and defect (dried berries, discoloration, sunburn, insect damage, and immature berries)
Peach, freestone for canning or pulping	US (1966)	Maturity, color (not greener than yellowish green), shape, firmness, and freedom from decay, worms and worm holes, split pits, scab, bacterial spot, insects, and bruises; grade is based on the severity of defects with 10% tolerance
Pear for processing	US (1970)	Maturity, color (less than yellowish green), shape, firmness, and freedom from scald, hard end, black end, internal breakdown, decay, worms and worm holes, scars, sunburn, bruises, and other defects; grade is based on the severity of defects with 10% tolerance
Raspberry	US (1952)	Color, and freedom from decay and defect (dried berries, crushing, shriveling, sunscald, scars, bird and insect damage, discoloration, or mechanical injury)
Strawberry, growers' stock for manufacture	US (1935)	Color, size and cap removal, freedom from decay and defect (crushed, split, dried or undeveloped berries, sunscald, and bird or insect damage)
Strawberry, washed and sorted for freezing	US (1935)	Color, cleanness, size, cap removal, and freedom from decay and defect (crushed, split, dried or undeveloped berries, bird and insect damage, mechanical injury)

Note: *Date when standard was issued or revised.

Table 23.4. Quality factors for processing vegetables in the U.S. Standards for Grades (US)

Vegetable	Standard (date*)	Quality factors
Asparagus, green	US (1972)	Freshness, shape, green color, size (spear length), and freedom from defect (freezing damage, dirt, disease, insect injury, and mechanical injuries) and decay
Bean, shelled lima	US (1953)	Tenderness, green color, freedom from decay and from injury caused by discoloration, shriveling, sunscald, freezing, heating, disease, insects, or other means
Bean, snap	US (1985)	Freshness, tenderness, shape, size, and freedom from decay and from damage caused by scars, rust, disease, insects, bruises, punctures, broken ends, or other means
Beet	US (1945)	Firmness, tenderness, shape, size, and freedom from soft rot, cull material, growth cracks, internal discoloration, white zoning, rodent damage, disease, insects, and mechanical injury
Broccoli	US (1959)	Freshness, tenderness, green color, compactness, trimming, and freedom from decay and damage caused by discoloration, freezing, pithiness, scars, dirt, or mechanical means
Cabbage	US (1944)	Firmness, trimming, and freedom from soft rot, seedstems, and damage caused by bursting, discoloration, freezing, disease, birds, insects, or mechanical or other means
Carrot	US (1984)	Firmness, color, shape, size (root length), smoothness, not woody, and freedom from soft rot, cull material, and from damage caused by growth cracks, sunburn, green core, pithy core, water core, internal discoloration, disease, or mechanical means
Cauliflower	US (1959)	Freshness, compactness, color, and freedom from jacket leaves, stalks, and other cull material, decay, and damage caused by discoloration, bruising, fuzziness, enlarged bracts, dirt, freezing, hail, or mechanical means
Corn, sweet	US (1962)	Maturity, freshness, and freedom from damage by freezing, insects, birds, disease, cross-pollination, or fermentation
Cucumber, pickling	US (1936)	Color, shape, freshness, firmness, maturity, and freedom from decay and from damage caused by dirt, freezing, sunburn, disease, insects, or mechanical or other means
Mushroom	US (1964)	Freshness, firmness, shape, and freedom from decay, disease spots, and insects, and from damage caused by insects, bruising, discoloration, or feathering
Okra	US (1965)	Freshness, tenderness, color, shape, and freedom from decay, insects, and damage caused by scars, bruises, cuts, punctures, discoloration, dirt, or other means
Onion	US (1944)	Maturity, firmness, and freedom from decay, sprouts, bottlenecks, scallions, seedstems, sunscald, roots, insects, and mechanical injury
Pea, fresh shelled for canning/ freezing	US (1946)	Tenderness, succulence, color, and freedom from decay, scald, rust, shriveling, heating, disease, and insects
Pea, Southern	US (1965)	Pods: maturity, freshness, and freedom from decay; seeds: freedom from scars, insects, decay, discoloration, splits, cracked skin, and other defects
Pepper, sweet	US (1948)	Firmness, color, shape, and freedom from decay, insects, and damage by any means that results in 5 to 20% trimming (by weight), depending on grade
Potato	US (1983)	Shape, smoothness, size, specific gravity, glucose content, fry color, and freedom from decay and defect (freezing injury, blackheart, sprouts)
Potato for chipping	US (1978)	Firmness, cleanness, shape; freedom from defect (freezing, blackheart decay, insect injury, and mechanical injury); size; optional tests for specific gravity and fry color are included
Spinach	US (1956)	Freshness; freedom from decay, grass weeds, and other foreign material; freedom from damage caused by seedstems, discoloration, coarse stalks, insects, dirt, or mechanical means
Sweet potato for canning/ freezing	US (1959)	Firmness, shape, color, size, and freedom from decay and defect (freezing injury, scald, cork, internal discoloration, bruises, cuts, growth cracks, pithiness, stringiness, and insect injury)
Sweet potato for dicing/ pulping	US (1951)	Firmness, shape, size, and freedom from decay and defect (scald, freezing injury, cork, internal discoloration, pithiness, growth cracks, insect damage, and stringiness)

Table 23.4. Cont.

Vegetable	Standard (date*)	Quality factors
Tomato	US (1983)	Firmness, ripeness (color as determined by a photoelectric instrument), and freedom from insect damage, freezing, mechanical damage, decay, growth cracks, sunscald, gray wall, and blossom-end rot
Tomato, green	US (1950)	Firmness, color (green), and freedom from decay and defect (growth cracks, scars, catfaces, sunscald, disease, insects, or mechanical damage)
Tomato, Italian type for canning	US (1957)	Firmness, color uniformity, and freedom from decay and defect (growth cracks, sunscald, freezing, disease, insects, or mechanical injury)

Note: *Date when standard was issued or revised.

Table 23.5. Quality factors for tree nuts in the U.S. Standards for grades (US) and the California Food and Agricultural Code (CA)

Nut	Standard (date*)	Quality factors
Almond, shelled	US (1997)	Similar varietal characteristics (shape, appearance), size (count per ounce), degree of dryness, cleanness (freedom from dust, particles, and foreign materials), and freedom from decay and defect (rancidity, insect injury, doubles, split or broken kernels, shriveling, brown spot, or gumminess)
Almond, in-shell	US (1997)	Shell: similar varietal characteristics (shape, hardness), cleanness (freedom from loose extraneous and foreign materials), size (thickness), brightness and uniformity of color, and freedom from discoloration, insect infestation, adhering hulls, and broken shells; kernel: degree of dryness, and freedom from decay and defect (rancidity, insect damage, shriveling, brown spot, gumminess, and skin discoloration)
Brazil nut, in-shell	US (1966)	Shell: degree of dryness, cleanness (freedom from dirt, extraneous, and adhering foreign materials), size (diameter), and freedom from damage caused by splits, breaks, punctures, oil stains, and mold; kernel: degree of development (must fill more than 50% of the shell capacity) and freedom from decay and defect (rancidity, insect damage, and discoloration)
Filbert, in-shell	US (1970)	Shell: shape, size (diameter), cleanness, brightness, and freedom from defect (blanks, broken or split shells, stains, and adhering husk); kernel: degree of dryness (less than 10% moisture content), development (must fill more than 50% of the shell capacity), shape, and freedom from decay and defect (insect injury, shriveling, rancidity, and discoloration)
Mixed nuts, in-shell	US (1981)	Each species of nut must conform to a minimum size and grade (same quality criteria used for that species); grade of the mix is also determined by percent allowable for each component (almonds, brazils, filberts, pecans, walnuts)
Pecan, shelled	US (1969)	Degree of dryness, degree of development (amount of meat in proportion to width and length), color (plastic models for color standards are available), color uniformity, size (number of halves per pound or diameter of pieces), cleanness (freedom from dust, dirt, and adhering material), and freedom from decay and defect (shriveling, insect damage, internal discoloration, dark spots, skin discoloration, and rancidity)
Pecan, in-shell	US (1976)	Shell: color uniformity, size (number of nuts per pound), cleanness, and freedom from decay and defect (insect damage, dark stains, split or cracked shells, and broken shells); kernel: same for shelled pecans (above)
Pistachio, shelled	US (1990)	Degree of dryness; freedom from foreign material and damage caused by mold, insects, spotting, rancidity, and other defects; and size (whole kernels, broken kernels)
Pistachio, in-shell	US (1992)	Freedom from foreign material, loose kernels, shell pieces, particles and dust, and blanks; freedom from nonsplit shells, shells not split on suture, adhering hull material, and staining; nut size; degree of kernel dryness and freedom from defects
Walnut, shelled	US (1968)	Color (USDA color chart), degree of dryness, cleanness (freedom from shells, dirt, dust, and foreign material), size (diameter of halves or pieces), and freedom from decay and defect (insect injury, rancidity, shriveling, and meat discoloration
Walnut, in-shell	US (1976)	Shell: dryness, cleanness, brightness, freedom from decay and defect (splits, discoloration, broken shells, perforated shells, and adhering hulls), and size (diameter); kernel: same as for shelled walnuts (above)
	CA (1983)	Shell: dryness, size, and freedom from blanks, decay, and defect (insect damage, adhering hulls, and perforations affecting more than $1/8$ of the surface); kernel: size and freedom from decay and defect (insect damage, shriveling, and rancidity)

Note: *Date when standard was issued or revised.

24

Safety Factors

Linda J. Harris, Devon Zagory,

and James R. Gorny

Fresh fruits and vegetables are inherently some of the most nutritious and safest foods available. Fruits and vegetables have been implicated as vectors of foodborne illness far less often than other food groups such as poultry, meat, or fish. Yet fruits and vegetables have been implicated in foodborne illness, and the risks are real, if small. Recent attention focused on foodborne illness has raised awareness of the potential of fresh fruits and vegetables to serve as vectors for human pathogens and has led to an increase in monitoring as well as in expectations that, where possible, the produce industry reduce risks to consumers.

Foodborne illness associated with fresh produce has been relatively unusual, and outbreaks have been irregularly distributed in time and location as well as incompletely reported. Therefore, it has not been possible to define relative risks associated with produce in the form of a quantitative risk assessment. There is disagreement as to whether foodborne illness associated with fresh produce is on the rise, or if it is only being tracked and reported more effectively. Produce distribution has become more global and less local, and there has been consolidation in all sectors of the industry resulting in ever-larger operating units. It can be argued that this has led to larger outbreaks of illnesses and outbreaks involving pathogens not typically seen in temperate locations. Much better techniques in tracking the sources of outbreaks and in information exchange among public health agencies has resulted in regulatory scrutiny and industry response in the form of heightened attention to all aspects of food safety associated with the growing, harvesting, processing, and distribution of fresh produce.

Data from the U.S. Centers for Disease Control and Prevention (CDC) indicate that the number of foodborne illness outbreaks linked to fresh produce, and the number of persons affected in these outbreaks, has increased in the recent past (table 24.1). A number of reasons have been proposed for this increased association of foodborne illness with fresh produce. Since the early 1970s a significant increase in the consumption of fresh produce has been observed in the United States, presumably due, in part, to active promotion of fruits and vegetables as an important part of a healthful diet. During this same time, there has been a trend toward greater consumption of foods away from home and an increase in the popularity of salad bars (buffets). Greater volumes of fruits and vegetables are being shipped from central locations and distributed over much greater geographical areas to many more people. This, coupled with increased global trade, potentially increases human exposure to a wide variety of foodborne pathogens and also increases the chances that an outbreak will be detected.

The perishability of produce and the complexity of the distribution system have made it difficult to effectively investigate produce-related outbreaks. It has been particularly difficult to trace outbreaks back to their origin because of the complexity of the distribution system and the practice of commingling produce in packinghouses. Epidemiological investigations often take weeks before a link is made between reported illnesses and a produce item, by which time there is nothing left to test. However, improvements in outbreak investigations and pathogen detection methods have contributed to an increase in recognition of the role of fresh produce in human illness.

While produce quality can be judged by outward appearance on such criteria as color, texture, and aroma, food safety cannot. Casual inspection of produce cannot determine if it is, in fact, safe and

Table 24.1. U.S. foodborne outbreaks associated with the consumption of fresh fruits and vegetables

	Percent of total reported		
	1982–1987	1988–1991	1992–1997
Cases in outbreaks*	4	8	14
Outbreaks†	2	4	2.3

Source: Adapted from National Advisory Committee on Microbiological Criteria for Foods 1999; U.S. Centers for Disease Control and Prevention 2000.

*A case is defined as an ill person associated with an outbreak.

†An outbreak is defined as two or more cases of a similar illness resulting from the ingestion of a common food.

wholesome to consume. Chlorinated water, ozone, organic acid washes, ultraviolet light, "antibacterial" packaging materials, and irradiation may all have a place in produce sanitation. However, once fruits and vegetables have been contaminated with bacterial or viral pathogens or with parasites, none of these methods will assure the safety of the product. It is possible to reduce the numbers of pathogens on produce by washing in sanitized water, but it is not currently possible to eliminate them through any of the above means. The only treatment currently available that will completely eliminate vegetative pathogens (bacteria and fungi, excluding spores) from fruits and vegetables is thorough cooking. Consequently, management of growing and handling conditions is paramount in preventing the contamination of fresh produce with human pathogens. Preventing contamination of fresh fruits and vegetables with these pathogens, dangerous levels of chemical residues, or physical contaminants is the best way to assure that foods are wholesome and safe for human consumption.

BIOLOGICAL SAFETY ISSUES

OVERVIEW

A wide variety of fresh fruits and vegetables has been associated with diseases caused by microbial pathogens (table 24.2). The contamination of fresh produce with pathogenic microorganisms can and has occurred during growth, harvesting, distribution, and final preparation. Unlike fresh meats or other grocery items for which a kill step is expected, fruits and vegetables are often consumed raw with minimal preparation on the part of the consumer.

As raw agricultural products, fresh produce may be expected to harbor a wide variety of microorganisms, including the occasional pathogen. For the most part, produce is grown outdoors, where animals, birds, and insects may be in fields or orchards and can transmit human pathogens to produce prior to or during harvest. The key to addressing biological contamination is to focus on reducing the risk of contamination of raw produce where possible all along the food chain.

FOODBORNE PATHOGENS ASSOCIATED WITH FRESH PRODUCE

Numerous pathogenic microorganisms have been isolated from a variety of fresh fruits and vegetables (table 24.3). Not all of the microorganisms listed in this table have been linked to produce-associated illnesses, but this does not mean that such outbreaks have not occurred. Under the right conditions, all of these microorganisms have the potential to cause produce-associated illness.

Bacteria, viruses, and parasites have all been linked to outbreaks of illness associated with fresh produce (see table 24.2). These microorganisms are physiologically diverse, but they do share some common features (table 24.4). The foodborne pathogens that are frequently associated with fresh produce mostly originate from enteric environments, that is, they are found in the intestinal tract and fecal material of humans or animals. Exceptions include *Clostridium botulinum*, which is usually isolated from soils, water, and decaying plant or animal material, and *Listeria monocytogenes*, which can be readily isolated from human and animal feces but also from many other environments including soil, agricultural irrigation sources, decaying plant residue on equipment or bins, cull piles, packing sheds, and food processing facilities.

The illnesses caused by the organisms listed in table 24.4 can sometimes be severe, especially in susceptible individuals such as young children, the elderly, and persons with suppressed immune systems. In addition, the infective dose (minimum number of organisms necessary to cause illness) is very low in many cases. A low infective dose means that the microorganism need only contaminate the food and survive, without reproducing, through the time of consumption. While

Table 24.2. Selected fruits and vegetables that have been implicated in outbreaks of foodborne illness (point of contamination was often undetermined, but growing, harvesting, processing and handling practices have all played a role in one or more outbreaks)

Pathogen	Implicated produce
BACTERIA	
Bacillus cereus	Seed sprouts
Clostridium botulinum	Cabbage coleslaw; dried, chopped garlic in oil
Enterotoxigenic *E. coli* spp.	Carrots
E. coli O157:H7	Unpasteurized apple juice, spring mix, alfalfa sprouts, cabbage coleslaw
Listeria monocytogenes	Cabbage, coleslaw
Salmonella spp.	Cantaloupe, mangoes, tomatoes, seed sprouts, watermelon, unpasteurized orange juice, unpasteurized apple juice
Shigella spp.	Green onion, parsley, lettuce
Vibrio cholerae	Coconut milk
PARASITES	
Cyclospora cayentanensis	Raspberries, mesclun lettuce, basil, basil-containing products
Cryptosporidium parvum	Unpasteurized apple juice
VIRUSES	
Hepatitis A virus	Lettuce, frozen raspberries, frozen strawberries, tomatoes
Norwalk/Norwalk-like virus	Melon, green salad, celery

Source: Adapted from National Advisory Committee on Microbiological Criteria for Foods 1999.

Table 24.3. Examples of pathogens that have been isolated from raw produce (international studies have been included)

Pathogen	Produce
Aeromonas spp.	Alfalfa sprouts, asparagus, broccoli, cauliflower, lettuce, pepper, spinach
Bacillus cereus	Cress sprouts, mustard sprouts, bean sprouts, soybean sprouts
Campylobacter jejuni	Mushrooms
E. coli O157:H7	Cabbage, celery, cilantro, lettuce, pineapple
Listeria monocytogenes	Bean sprouts, cabbage, cucumber, potatoes, radish, tomatoes
Salmonella spp.	Artichoke, bean sprouts, beet leaves, cabbage, cantaloupe, cauliflower, celery, eggplant, endive, fennel, lettuce, mustard cress, parsley, pepper, radish, spinach
Shigella spp.	Parsley, cilantro
Staphylococcus spp.	Carrots, lettuce, parsley, radish, seed sprouts
Vibrio cholerae	Cabbage

Source: Adapted from Beuchat 1996.

enhancing the likelihood of illness, temperature abuse and multiplication of bacteria is *not always necessary* for foodborne illness to occur. In fact, pathogenic parasites and viruses are unable to multiply outside of a human or animal host.

CONTAMINATION OF PRODUCE

Produce can become contaminated with microbial pathogens by a wide variety of mechanisms (fig. 24.1). Contamination leading to foodborne illness has occurred during production, harvest, and processing, and at retail outlets, in foodservice establishments, and in the home kitchen. Contamination at any point in the food handling chain can be exacerbated by improper handling and storage of produce prior to consumption. Contamination with fecal material can occur directly or indirectly via animals or insects, soil, water, dirty equipment, and human handling. For example, fruit flies have been shown to transfer *Escherichia coli* O157:H7 to wounded apples under laboratory conditions (Janisiewicz et al. 1999). This may have implications during harvesting and in packing sheds or processing facilities where damaged produce is inevitable and flies may be difficult to control. Humans and animals can shed foodborne pathogens in the absence of the signs of illness. While domestic animals may be separated from fruit- and vegetable-growing operations, wild animals and birds can be controlled only to a limited extent. Human hygiene and hand washing all along the food chain are critical in reducing or eliminating this source of fecal pathogens.

SURVIVAL AND MULTIPLICATION OF PATHOGENS ON RAW PRODUCE

The survival and/or growth of pathogens on fresh produce is influenced by the organism, produce item, and conditions of storage. In general, pathogens will survive on the outer surface of fresh fruits or vegetables, especially if the humidity is high. Growth on intact surfaces is not common because foodborne pathogens do not produce the enzymes necessary to break down the protective outer barriers on most produce. This restricts the availability of nutrients and moisture. The survival of microorganisms is often enhanced at refrigerated temperatures even though these conditions reduce or eliminate the ability of the organisms to multiply.

Table 24.4. Characteristics of selected microbial pathogens linked to outbreaks of produce-associated illness

Microorganism	Incubation period	Symptoms	Infectious dose (number of cells)	Source
BACTERIA				
Clostridium botulinum	12–36 hours	Nausea, vomiting, fatigue, dizziness, dryness of mouth and throat, muscle paralysis, difficulty swallowing, double or blurred vision, drooping eyelids, and breathing difficulties	Intoxication; growth and toxin production in food	Soil, lakes, streams, decaying vegetation
E. coli O157:H7	2–5 days	Watery diarrhea, often containing blood; abdominal pain; can lead to hemolytic uremic syndrome and kidney failure especially in children and elderly	10 to 1,000	Animal feces, especially cattle, deer, and human; cross-contamination from raw meat or poultry
Salmonella spp.	18–36 hours	Abdominal pain, diarrhea, chills, fever, nausea, vomiting	10 to 100,000	Animal and human feces; cross-contamination from raw meat, poultry, or eggs
Shigella spp.	1–3 days	Abdominal pain, diarrhea, fever, vomiting	~10	Human feces
Listeria monocytogenes	1 day to 5 or more weeks	Flulike illness in healthy adults; may lead to spontaneous abortion or stillbirth in pregnant women; severe septicemia and meningitis in neonates and immunocompromised adults; mortality may be 20 to 40%	Unknown; dependent upon health of individual	Soil; animal and human feces; silage
PARASITES				
Cryptosporidium spp.	1–12 days	Profuse watery diarrhea, abdominal pain, anorexia, vomiting	~30	Human and animal feces
Cyclospora spp.	1–11 days	Watery diarrhea, nausea, anorexia, abdominal cramps (duration 7 to 40 days)	Unknown, probably low	Human feces, others?
VIRUSES				
Hepatitis A	25–30 days	Fever, malaise, anorexia, nausea, abdominal pain, jaundice, dark urine	10 to 50	Human feces and urine
Norwalk/Norwalk-like virus	12–48 hours	Vomiting diarrhea, malaise, fever, nausea, abdominal cramps	Unknown, probably low	Human feces

The survival of foodborne pathogens on produce is significantly enhanced once the protective epidermal barrier has been broken either by physical damage, such as punctures or bruising, or by degradation by plant pathogens (bacteria or fungi). These conditions can also promote the multiplication of pathogens, especially at higher temperatures. Various enteric pathogens have been shown to multiply on the surface of cut melons, on shredded lettuce, and on chopped parsley, and even under acidic conditions such as on chopped tomatoes and wounded apple tissue. Temperature control becomes critical for preventing bacterial reproduction on any cut produce item.

While the acidity of most fruit juices prevents the multiplication of pathogens, survival in these products is much better than had previously been assumed. In addition, pathogen survival is substantially enhanced at refrigerated temperatures. Pathogen viability decreases with increasing temperature but juice shelf life also decreases due to the rapid growth of yeasts and other spoilage organisms at the higher temperatures.

USE OF WASH SOLUTIONS TO REDUCE MICROBIAL POPULATIONS ON THE SURFACE OF RAW PRODUCE

If produce becomes contaminated by human pathogens there is no currently

Figure 24.1

Mechanisms by which produce can become contaminated with pathogenic microorganisms. (Adapted from Beuchat 1996)

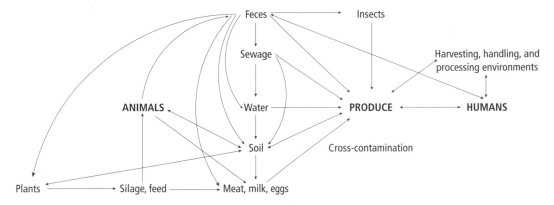

available process, other than thorough cooking, that can completely eliminate them. This is why prevention of contamination is the first line of defense in any produce operation. Fresh fruits and vegetables are often washed in preparation for processing or marketing. While washing can remove some of the surface microorganisms, it cannot remove them all. Reductions observed over a wide range of conditions are typically from 0 to 1,000-fold. However, improperly sanitized wash water can itself become a major source of contamination and should be carefully monitored and controlled (table 24.5). In table 24.5, cut cantaloupe cubes inoculated with Log 7.0 CFU/piece (10^7 CFU/piece) *Salmonella* spp. were soaked for 1 minute in 50 ml of clean unchlorinated or chlorinated (50 ppm, pH 7.0) water. The population remaining on the cantaloupe was unchanged after soaking regardless of the soak solution. However, the population of *Salmonella* in the unchlorinated water increased to Log 5.8 CFU/ml ($6.3×10^5$ CFU/ml) while the population in the chlorinated water remained at <Log 1 (<10 CFU/ml). Additional uninoculated cantaloupe cubes were subsequently soaked for 1 minute in each of the soak solutions. Those soaked in the chlorinated water remained *Salmonella*-free while those soaked in the unchlorinated water picked up substantial levels of *Salmonella* (Log 4.2 CFU/piece; $1.6×10^4$ CFU/piece.

Bacteria have been shown to preferentially adhere to stomata and cut surfaces of lettuce, making them less accessible to wash treatments. This phenomenon is likely true for

many other fruits and vegetables and may be one reason that washing fails to eliminate pathogens.

Infiltration of wash water into intact fruit has been demonstrated in a number of fruits and vegetables and is thought to have contributed to an outbreak of salmonellosis associated with fresh-market tomatoes. Wash water contaminated with microorganisms, including pathogens, can infiltrate the intercellular spaces through pores when conditions are right. Internal gas pressures and surface hydrophobicity usually prevent uptake of water. However, when produce temperature is much higher than the water temperature, the pressure difference created may be sufficient to draw water into the fruit. The temperature of the wash water may need to be increased and monitored to prevent this. Addition of detergents to the water appears to enhance infiltration, likely due to reduced surface tension. Under some circumstances, wash water may enter an intact fruit through the stem scar or other openings, such as the blossom or stem end of an apple. Conditions that reduce infiltration of plant pathogens should also reduce infiltration of human pathogens.

Chlorinated or otherwise sanitized wash water is usually no more or only slightly more effective than plain water at removing surface bacteria. However, wash water in tanks, recirculated water, or water that is reused in a spray wash system can become contaminated with pathogens if contaminated produce coming in from the field is washed in that water. Pathogens can survive for relatively long times in water and can

Table 24.5. Effect of sanitized wash water on cross-contamination

	Log Total CFU Salmonella		
Treatment	Inoculated cantaloupe	Wash water	Uninoculated cantaloupe
Untreated	7.0	<1	<1
			<1
Unchlorinated water (1 min soak)	7.1	5.8	4.2
			4.2
50 ppm HOCl (1 min soak)	6.9	<1	<1
			<1

subsequently contaminate clean produce that passes through that water. Wash water sanitation is important to ensure that pathogens introduced into the water are killed and cannot act as a source of contamination thereafter (see table 24.5).

Many wash water sanitation compounds have been evaluated for fresh produce, including sodium hypochlorite, calcium hypochlorite, acidified calcium hypochlorite, sodium bromide, chlorine dioxide, chlorine gas, hydrogen peroxide, ozone, organic acids, and ultraviolet light. With the exception of ultraviolet light, which is a purely physical method, all of the chemicals are oxidizing compounds. They are biocidal (they kill microorganisms) through their oxidizing capacity.

Not all chemical sanitizers have been approved for contact with intact fresh fruits and vegetables, and the approval may differ for cut produce. A number of compounds are currently awaiting approval by the U.S. Food and Drug Administration (FDA). When approved, chemical sanitizers must always be used in accordance with their label. Always check with the chemical supply company and the local regulatory authority before changing wash water sanitation procedures.

A number of methods are used to measure the concentrations of sanitizing compounds in water. Regardless of the sanitizing and monitoring system used, it is critical that the efficacy of the method be validated. Effective concentrations and other important factors, such as pH and temperature, must be maintained in the wash water. These properties need to be measured on a routine basis and appropriate records should be kept. Where possible, continuous monitoring is preferable to spot-checking.

FRESH-CUT FRUITS AND VEGETABLES

All fresh-cut produce has been injured through peeling, cutting, slicing, or shredding. These same operations can transfer pathogenic microorganisms from the surface of the intact fruit or vegetable to the internal tissues. Injured cells and released cell fluids provide a nourishing environment for microbial growth. Maintaining low temperatures throughout distribution is critical to maintaining the quality of fresh-cut fruits and vegetables. Low temperatures reduce enzymatic reactions and greatly slow the multiplication of spoilage organisms. Low temperatures also prevent the multiplication of most foodborne pathogens with the exception of *Listeria monocytogenes* and a few others that are capable of growing, albeit slowly, at refrigerated temperatures.

The emphasis should be on preventing contamination by pathogens. The best way to prevent the introduction of pathogens into fresh-cut produce is by employing Good Agricultural Practices (GAPs), Good Manufacturing Practices (GMPs), developing and using Standard Sanitation Operating Procedures (SSOPs) and, where appropriate, implementing an effective Hazard Analysis Critical Control Point (HACCP) plan. These programs, discussed below (see the references for additional materials), work together to identify potential points of contamination and ensure that potential hazards are monitored and controlled to enhance safety. Food safety does not begin and end at the doors of the processing facility. The processor must work with suppliers and customers to ensure that safe practices are maintained throughout the production, distribution, and marketing chain.

A vigorous population of nonpathogenic bacteria is another barrier to reduce the risk of foodborne illness from fresh-cut products. The nonpathogenic bacteria do not necessarily prevent the growth of pathogens but they do provide indicators of temperature abuse and age of the produce by causing detectable spoilage. Most pathogens do not cause produce to spoil even at relatively high populations. In the absence of spoilage, high levels of pathogens may be achieved and the item may be consumed because it is not perceived as spoiled. For this reason, specifications requiring very low microbial counts may, in some cases, compromise produce safety.

INFLUENCE OF MODIFIED ATMOSPHERE PACKAGING

Fresh produce packaged in gas-permeable films can modify its own atmosphere, creating more-favorable conditions for storage. Three mutually interacting processes determine the course of this modification: respiration by the fruit or vegetable, gas diffusion through the produce item, and gas transmission through the film. As a result of produce respiration the O_2 concentration in the package is decreased and the CO_2 concentration is increased. This subject has been reviewed in depth (see Kader et al. 1989).

Each fruit or vegetable has a particular temperature and atmosphere at which its postharvest potential will be maximized. Knowing these conditions for a given fruit or vegetable is a necessary prerequisite for successful modified atmosphere packaging (MAP). The reduced O_2 and elevated CO_2 typically found within produce MAP have been shown to slow the growth of some, but not all, bacteria. In fact, suppression of the growth of some aerobic spoilage bacteria is one of the chief ways in which MAP can forestall spoilage and increase shelf life. However, it has also been shown that MAP has little effect in slowing the growth of most human pathogenic bacteria and may even stimulate the growth of some (see Farber 1991). This combination of a reduction in sensory spoilage organisms and an increase in time for pathogens to grow has been a cause of concern with fresh-cut produce in MAP. While disease outbreaks associated with MAP produce have been few, some have occurred, and concerns continue.

Growth and toxin production by *Clostridium botulinum* is of particular concern among regulatory personnel because of the increased potential for botulism. This bacterium is found in soil and around agricultural environments and may be detected as a "hitchhiker" on many kinds of produce. It normally only grows and produces toxin in the absence, or near absence, of oxygen, conditions that have not previously been associated with fresh produce. Produce stored in MAP, if exposed to warm temperatures, can rapidly deplete the O_2 in the package through increased respiratory demand. These warm, low-O_2 conditions, together with extended shelf life from suppression of spoilage bacteria, could provide an opportunity for *C. botulinum* to grow

and produce toxin before the food is unacceptable. While this has not been found to happen in commercial practice, most fresh mushrooms, for example, are not packaged in MAP because they are known to sometimes carry spores of *C. botulinum*.

Under low-O_2 conditions, the normally large populations of spoilage bacteria reproduce and cause significant sensory spoilage before *C. botulinum* can produce toxin. Such challenge studies should be performed to demonstrate produce safety before new products are packaged in MAP.

SEED SPROUTS

Over the past several years sprouts have become a common fresh produce item linked to foodborne illness, especially from *Salmonella* infections. A scientific advisory group to the FDA has recognized sprouts as a special problem. This is because bacterial pathogens that may be present at very low levels on sprout seeds at the time of sprouting can multiply to very high levels during the 3- to 10-day sprouting process.

Most sprout outbreaks have been due to seed that is contaminated with a bacterial pathogen before the sprouting process begins. Many pathogens can survive for months under the dry conditions used for seed storage. Although contaminated alfalfa seed have been identified as the source in many outbreaks, clover, radish, and mung bean sprouts have also been associated with outbreaks. Any type of sprout seed may potentially be contaminated with bacterial pathogens before it is sprouted.

The FDA has published guidelines for sprout processors to reduce the potential for foodborne illness related to these products. The guidelines include treatment of seeds in a sanitizer solution (currently a special allowance for 20,000 ppm chlorine) prior to sprouting, as well as testing of the sprout wash water for *Salmonella*, *E. coli* O157:H7, and *Listeria monocytogenes* prior to harvest.

The FDA has published two related documents entitled *Guidance for Industry: Reducing Microbial Food Safety Hazards for Sprouted Seeds* and *Guidance for Industry: Sampling and Microbial Testing of Spent Irrigation Water During Sprout Production*. These guidelines are intended to provide recommendations to suppliers of seeds for sprouting and to sprout producers about how to reduce microbial food safety hazards found in the production of raw sprouts. They are further intended to help ensure that sprouts are not a cause of food-borne illness and that sprout producers comply with the food safety provisions of the Federal Food, Drug, and Cosmetic Act. The FDA and California State Department of Health Services, Food and Drug Branch, have recently produced a video that provides an overview of safe sprout production (see the references).

CHEMICAL SAFETY ISSUES

Examples of chemical hazards that may contaminate produce during production, handling, or storage include insecticides, fungicides, herbicides, rodentacides, machine lubricants from forklifts or packing lines, heavy metals (lead, mercury, arsenic), industrial toxins, and compounds used to clean and sanitize equipment. Allowable levels of contaminants, such as chemical residues and heavy metals, are established by the FDA and the U.S. Environmental Protection Agency (EPA) to assure compliance with established maximum tolerance levels. Despite consumer concerns about pesticide residues on fresh produce, there is no evidence that maximum tolerance levels present a health hazard or cause illness.

Chemical residues well above maximum residue levels can and have been harmful to human health. In the United States and abroad consumers have become ill from consumption of food illegally contaminated with pesticide residues. For example, in 1985, several hundred consumers in the United States and Canada became ill from residues of the insecticide aldicarb that was illegally applied to watermelons (Green et al. 1987). An evaluation of the chemical contamination risks for growing, harvesting, and packing fresh fruits and vegetables and establishment of adequate controls and record keeping should prevent this type of hazard.

Naturally occurring toxins in certain crops (such as glycoalkaloids in potatoes) vary according to genotype and are routinely monitored by plant breeders to ensure that they do not exceed safe levels.

FOOD SAFETY AND PRODUCE DERIVED FROM BIOTECHNOLOGY

Advanced breeding techniques that use biotechnology have the potential to substantially benefit the produce industry by reducing the primary causes of deterioration, enhancing nutritional quality, and reducing the use of agricultural inputs. Currently, there is no scientific evidence to indicate that food crops derived from biotechnology present any unique food safety risk. The regulation of foods derived from biotechnology on a nation-by-nation basis are diverse and currently in flux. However, in the United States there is currently stringent regulatory oversight to assure production of safe and wholesome food products including foods derived from biotechnology. The FDA is the primary federal agency responsible for ensuring the safety of commercial food products. FDA policy with respect to foods derived from new plant varieties was published in the *Federal Register* (vol. 67, no. 104) on Friday, May 29, 1992. This policy established that the key factors in reviewing safety concerns are based on the characteristics of the food, irrespective of the method by which it is developed. It also put forth extensive procedures to review the safety of foods derived from biotechnology, including tests that assure that

- no new food allergens are introduced
- increased levels of natural toxicants are not observed
- important nutrients are not reduced
- unapproved additives are not present in the food

Current FDA policy does not require the labeling of foods derived from biotechnology, as long as they are "substantially equivalent" (nutritionally and compositionally) to the conventionally grown crop from which they are derived. FDA Commissioner Jane E. Henney stated in a press release on May 3, 2000, that "the FDA's scientific review continues to show that all bioengineered foods sold in the United States today are as safe as their non-bioengineered counterparts."

PHYSICAL SAFETY ISSUES

Examples of physical hazards associated with produce during production, handling, or storage include

- fasteners (staples, nails, screws, bolts)
- aluminum beverage cans that may be

subsequently shredded in processing equipment

- glass
- pieces of plastic
- wood splinters

While physical hazards do not reproduce and spread the way some pathogens can, they can still be significant problems and must be excluded from food through vigilance and comprehensive prevention programs. Maintenance of harvest and packinghouse equipment and exclusion of certain materials (e.g., glass) from facilities are steps that can be used to reduce these risks.

CONTROLLING HAZARDS IN PRODUCE

While pathogens cannot be eliminated from fresh produce, there are a number of things that can be done to significantly reduce the risk of contamination. All potential paths of contamination (see fig. 24.1) should be systematically evaluated at each farm and in each processing or food service facility, and, where possible, preventative measures should be implemented. To this end, numerous produce-related associations and individual companies have developed documents to help implement food safety programs from production through processing and retail distribution.

GOOD AGRICULTURAL PRACTICES

The U.S. FDA published *Guidance for Industry: Guide to Minimize Microbial Food Safety Hazards for Fresh Fruits and Vegetables* in 1998. Although this document carries no regulatory or legal weight, due diligence requires producers to take prudent steps to prevent contamination of their crops. This document gives guidance on those prudent steps. A number of retail chains have begun to require independent third-party audits of fresh fruit and vegetable producers based, in part, on this document.

The guide identifies eight principles of food safety within the realms of growing, harvesting, and transporting fresh produce and suggests that the reader "use the general recommendations in this guide to develop the most appropriate good agricultural and management practices for your operation." The application of the principals is aimed at preventing contamination of fresh produce with human pathogens. The eight principles are listed below followed by areas of implementation (U.S. FDA 1998, p. 8).

Basic Principles of Good Agricultural Practices (GAPs)

1. Prevention of microbial contamination of fresh produce is favored over reliance on corrective actions once contamination has occurred.

2. To minimize microbial food safety hazards in fresh produce, growers or packers should use GAPs in those areas over which they have a degree of control while not increasing other risks to the food supply or the environment.

3. Anything that comes in contact with fresh produce has the potential of contaminating it. For most foodborne pathogens associated with produce, the major source of contamination is associated with human or animal feces.

4. Whenever water comes in contact with fresh produce, its source and quality dictate the potential for contamination.

5. Practices using manure or municipal biosolid wastes should be closely managed to minimize the potential for microbial contamination of fresh produce.

6. Worker hygiene and sanitation practices during production, harvesting, sorting, packing, and transport play a critical role in minimizing the potential for microbial contamination of fresh produce.

7. Follow all applicable local, state, and federal laws and regulations, or corresponding or similar laws, regulations, or standards for operators outside the United States for agricultural practices.

8. Accountability at all levels of the agricultural environment (farms, packing facility, distribution center, and transport operation) is important to a successful food safety program. There must be qualified personnel and effective monitoring to ensure that all elements of the program function correctly and to help track produce back through the distribution channels to the producer.

LAND USE

The safety of food grown on any given parcel of land is influenced not only by the current agricultural practices but also by former land use practices. Heavy metals and pesticide residues may persist in soils for long periods of time. Soil should be tested to assure that unsafe levels of these compounds are not present. Former land use should also be investigated and documented to assure that the production land was not formerly used for hazardous waste disposal or for industrial purposes that may have left toxic residues. If production land was previously used for agricultural purposes, pesticide use records should be reviewed to assure that proper pesticide management practices were followed. Production acreage should not have recently been used as a feedlot or for animal grazing, since pathogens associated with fecal contamination of the soil may persist.

FERTILIZERS

Improperly composted or uncomposted manure is a potential source of human pathogens. Human pathogens may persist in animal manure for weeks or even months. *E. coli* O157:H7 has been found to survive in uncomposted dairy manure incorporated into soil for up to 250 days (Suslow 1999). Proper composting via thermal treatment will reduce the risk of potential foodborne illness. However, the persistence of many human pathogens in untreated agricultural soils is currently unknown. Inorganic fertilizers should be certified to be free of heavy metals and other chemical contaminants.

IRRIGATION WATER

Irrigation water is another potential vector by which contaminants may be brought in contact with fruits and vegetables. Deep well water is less likely to be contaminated with human pathogens than surface water supplies. However, all irrigation water sources should be periodically tested for contamination by pesticides and human pathogens. Presence of *E. coli* is a useful indicator for fecal contamination and possible presence of human pathogens. Inexpensive test kits for generic *E. coli* are available from several vendors. Overhead irrigation systems are more likely than flood, furrow, or drip irrigation to spread contamination since contaminated water is applied directly to the edible portions of fruits and vegetables. Water used to mix or spray agricultural chemicals should be tested prior to use.

PESTICIDE USAGE

All pesticide usage should be done in strict accordance with manufacturer recommendations as well as federal, state, and local ordinances. Monitoring and documentation of proper pesticide usage should be done to prevent unsafe or illegal residues from contaminating fruits and vegetables. All pesticide applications should be documented and proper records of application should be available and reviewed by management on a regular basis. Appropriately trained and licensed individuals should make pesticide use recommendations and perform applications.

HARVEST OPERATIONS

During harvest operations field personnel may contaminate fresh fruits and vegetables by simply touching them with an unclean hand or knife blade. Portable field latrines equipped with hand wash stations must be available and used by all harvest crew members. Training, monitoring, and enforcement of fieldworker hygiene practices, such as washing hands after using the latrine, are necessary to reduce the risk of human pathogen contamination. Once harvested, produce should not be placed on soil before being placed in clean and sanitary field containers. Field harvesting tools should be clean and sanitary and should not be placed directly in contact with soil. Field containers should be cleaned and sanitized on a regular basis and should be kept free of contaminants such as mud, industrial lubricants, metal fasteners, or wood splinters. Plastic bins and containers are recommended as they are easier to clean and sanitize than wooden bins.

SANITARY POSTHARVEST HANDLING OF PRODUCE

Depending upon the item, produce may be field-packed in containers that will go all the way to the destination market, or it may be temporarily placed in bulk bins, baskets, or bags that will be transported to a packing shed. Employees, equipment, cold storage facilities, packaging materials, and any water that will contact the harvested produce must be kept clean and sanitary to prevent contamination.

EMPLOYEE HYGIENE

Gloves, hairnets, and clean smocks are commonly worn by packinghouse employees in export-oriented packing sheds. The cleanliness and personal hygiene of employees handling produce at all stages of production and handling must be managed to minimize the risk of contamination. Adequate toilet facilities and hand-washing stations must be provided and used properly to prevent cont-amination of produce by packinghouse employees. Shoe or boot cleaning stations may also be in place to reduce the amount of field dirt and contamination that enters the packing shed from field operations. Employee training regarding sanitary food handling practices should be done when an employee is hired, before beginning work, and on a regular basis thereafter. All such training should be documented and kept on file.

EQUIPMENT

Food contact surfaces on conveyor belts, dump tanks, etc., should be cleaned and san-itized on a regularly scheduled basis with approved cleaning compounds. A 200 ppm hypochlorite (bleach) solution is an example of a food contact surface sanitizer. Sanitizers should be used only after thorough cleaning with abrasion to remove organic material such as dirt or plant materials. Use of steam to clean equipment should be avoided since steam may actually "cook on" organic mate-rials, which may aid in the formation of biofilms that render equipment very difficult to sanitize. Steam or high-pressure water may also create bacterial aerosols and actual-ly help spread contamination throughout the packinghouse.

COLD STORAGE FACILITIES

Cold storage facilities, and in particular refrigeration coils, refrigeration drip pans, forced-air cooling fans, drain tiles, walls, and floors should be cleaned and sanitized on a regular basis. The human pathogen *Listeria monocytogenes* can multiply at refrigerated temperatures in moist conditions and may contaminate produce if condensation from refrigeration units or ceilings drips onto the produce. A relatively common environmen-tal pathogen, *L. monocytogenes* may get on walls, in drains, and into cooling systems. Comprehensive sanitation programs that tar-get these areas are instrumental in prevent-ing the establishment of this pathogen.

PACKAGING MATERIALS

All packaging materials should be made of food-contact-grade materials to assure that toxic compounds in the packaging materials do not leach out of the package and into the produce. Toxic chemical residues may be pre-sent in some packaging materials due to use of recycled base materials. Packages such as boxes and plastic bags should be stored in an enclosed storage area to protect them from insects, rodents, dust, dirt, and other poten-tial sources of contamination. Plastic field bins and totes are preferred to wooden con-tainers since plastic surfaces are more amenable to cleaning and sanitizing. Field bins should be cleaned and sanitized after every use. Wooden containers or field totes are almost impossible to sanitize since they have a porous surface, and wooden or metal fasteners, such as nails from wooden contain-ers, may accidentally be introduced into pro-duce. Cardboard field bins, if reused, should be visually inspected for cleanliness and lined with a clean plastic bag before reuse to pre-vent the risk of cross-contamination.

PRODUCE WASH AND HYDROCOOLING WATER

All water that comes in contact with produce for washing, hydrocooling, or vacuum cool-ing must be potable. To achieve this, water should contain between 2 and 7 ppm free chlorine and have a pH between 6 and 7. Alternatively, an oxidation-reduction poten-tial greater than 650 mV using any oxidative sanitizer will ensure that bacteria in the water are killed on contact. Chlorine use prevents cross-contamination of produce in the washing or hydrocooling system but does not sterilize the produce. Rinsing pro-duce with potable water reduces the number of microorganisms present on the produce but does not remove all bacteria.

REFRIGERATED TRANSPORT

Produce is best shipped in temperature-con-trolled refrigerated trucks. Maintaining most perishables below 5°C (41°F) (except for tropical fruit) extends shelf life and signifi-cantly reduces the growth rate of microbes, including human pathogens. Cut produce, including tropical fruits, should always be stored below 5°C. Trucks used during trans-portation should be cleaned and sanitized on a regular basis. Trucks that have been used to transport live animals, animal products, or toxic materials should not be used to transport produce.

RECALL AND TRACEBACK PLANS

Recall of product is the last line of defense in a food safety emergency. This action may be initiated by the company, performed on a voluntary basis, or done at the request of the FDA because of a suspected hazard in the product.

The FDA has defined three recall classifications:

Class I: An emergency situation involving removal of products from the market which could lead to an immediate or long-term, life-threatening situation and involve a direct cause-effect relationship, e.g., *Clostridium botulinum* in the product.

Class II: A priority situation in which the consequences may be immediate or long-term and may be potentially life-threatening or hazardous to health, e.g., *Salmonella* in food.

Class III: A routine situation in which life-threatening consequences (if any) are remote or nonexistent. The products are recalled because of adulteration (filth in produce relating to aesthetic quality) or misbranding (label violation), and the product does not involve a health hazard.
Every food provider should develop a recall plan and organizational structure that enable it to remove product from the marketplace quickly and efficiently. A traceback plan is different from a recall plan in that it helps to trace product back to the farm or field from which it was harvested.

SANITATION IN PACKINGHOUSES AND PROCESSING FACILITIES

Good sanitation practices should focus on those places and practices where contamination of food is most likely. Contamination with pathogens is most likely to occur from

- contaminated wash water in the processing facility

- drip or splash from contaminated floors, drains, overhead pipes, or cooling systems

- unsanitary food-contact surfaces

- handling by workers practicing poor personal hygiene

Cleanliness of all work surfaces and equipment is an important quality assurance and produce shelf life issue. Produce contaminated with high populations of bacteria is likely to become decayed and/or slimy sooner than similarly handled cleaner produce. However, the chief environmental safety issue is the possible presence of *Listeria monocytogenes* within the packing or processing facility. *Listeria* is a common environmental contaminant that thrives in cold, wet environments such as those encountered in fresh-cut processing facilities. Once established, *Listeria* can be difficult to eliminate, and good sanitation programs coupled with environmental testing are necessary to prevent its establishment. *Listeria* spp. are most likely to be found in drains, refrigeration drip pans, and any place where cold water accumulates and stands. A comprehensive environmental sanitation program may include specific swab tests for *Listeria* spp. and vigorous cleaning and sanitation of all areas where *Listeria* is likely to be found. Following sanitation schedules and procedures, and documenting that they have been followed, are important to controlling this pathogen.

SUMMARY

While produce quality can be judged by outward appearance on such criteria as color, texture, and aroma, food safety cannot. Casual inspection of produce cannot determine if it is, in fact, safe and wholesome to consume. While quality may vary depending on the product, price, availability, and market conditions, safety does not. Safety must be held to the same high standard no matter what the market. For this reason, programs that combine quality and safety are not recommended since the appropriate priorities of safety and quality may be lost. Safety programs should stand alone and operate independently of quality programs and must be held to a higher, and constant, standard. Management of growing, harvesting, packing, and processing conditions are paramount in preventing the contamination of fresh produce by physical hazards, harmful chemicals, and human pathogens. Since pathogens cannot be completely removed from contaminated produce, food safety assurance relies on preventing the contamination.

REFERENCES

Beuchat, L. R. 1992. Surface disinfection of raw produce. Dairy Food Environ. San. 12:6–9.

————. 1996. Pathogenic microorganisms associated with fresh produce. J. Food Prot. 59:204–216.

Bartz, J. A. 1999. Washing fresh fruits and vegetables: Lessons from treatment of tomatoes and potatoes with water. Dairy Food Environ. San. 19:853–864.

Farber, J. M. 1991. Relative effect of CO_2 on the growth of foodborne microorganisms. J. Food Prot. 54:58–70.

Green, M. A., M. A. Heumann, H. M. Wehr, L. R. Foster, L. P. Williams, J. A. Polder, C. L. Morgan, S. L. Wagner, L. A. Wanke, and J. M. Witt. 1987. An outbreak of watermelon-borne pesticide toxicity. Amer. J. Pub. Health 77:1431–1434.

Henney, J. E. 2000. Press release. May 3. Washington, D.C.

Janisiewicz, W. J., W. S. Conway, M. W. Brown, G. M. Sapers, P. Fratamico, and R. L. Buchanan. 1999. Fate of *Escherichia coli* O157:H7 on fresh-cut apple tissue and its potential for transmission by fruit flies. Appl. Environ. Microbiol. 65:1–5.

Kader, A. A., D. Zagory, and E. L. Kerbel. 1989. Modified atmosphere packaging of fruits and vegetables. CRC Critical Rev. Food Sci. Nut. 28:1–30.

National Advisory Committee on Microbiological Criteria for Foods. 1999. Microbiological safety evaluations and recommendations on fresh produce. Food Control 10:117–143.

Nguyen-the, C., and F. Carlin. 1994. The microbiology of minimally processed fresh fruits and vegetables. Crit. Rev. Food Sci. Nutr. 34:371–401.

Rushing, J. W., F. J. Angulo, and L. R. Beuchat. 1996. Implementation of a HACCP program in a commercial fresh-market tomato packinghouse: A model for the industry. Dairy Food Environ. San. 16:549–553.

Suslow, T. 1999. Addressing animal manure management issues for fresh vegetable production. Perishables Handling 98: 7–9.

Tauxe, R., H. Kruese, C. Hedberg, M. Potter, J. Madden, and K. Wachsmuth. 1997. Microbial hazards and emerging issues associated with produce. A preliminary report to the National Advisory Committee on Microbiologic Criteria for Foods. J. Food Prot. 60:1400–1408.

U.S. Centers for Disease Control and Prevention (CDC). 2000. Surveillance for foodborne disease outbreaks—United States, 1993–1997. MMWR 49(SS01): 1–51.

U.S. Food and Drug Administration (U.S. FDA). 1998. Guide to minimize microbial food safety hazards for fresh fruits and vegetables. Washington, D.C.: FDA. Also available via Internet at http://www.foodsafety.gov/~dms/prodguid.html

FOOD SAFETY RESOURCES

Beuchat, L. R. 1998. Surface decontamination of fruits and vegetables eaten raw: A review. Geneva, Switzerland: World Health Organization. Available from World Health Organization Publications, Distribution and Sales, 1211 Geneva 27 Switzerland, Fax (+41 22) 791 48 57; E-mail orders: publications@who.ch

Current Good Manufacturing Practices (GMPs) regulations, 21 CFR 100–169. Available from Superintendent of Documents, U.S. Government Printing Office, Washington, D.C. 20402, (202) 783-3238.

Guidance for industry: Reducing microbial food safety hazards for sprouted seeds. Federal Register 64, no. 207 (October 27, 1999). Guidance for industry: Sampling and microbial testing of spent irrigation water during sprout production. Federal Register 64, no. 207 (October 27, 1999). Both can be obtained from Center for Food Safety and Applied Nutrition, Food and Drug Administration, 200 C St. SW, Washington, D.C. 20204, (202) 205-4200, or via the World Wide Web at http://vm.cfsan.fda.gov/~lrd/hhsprout.html and http://vm.cfsan.fda.gov/~mow/sprouts2.html

Guide to minimize microbial food safety hazards for fresh fruits and vegetables. U.S. FDA, 1998. Currently available in English, Spanish, Portuguese, and French. Available from Food Safety Initiative Staff, HFS-32, U.S. Food and Drug Administration, Center for Food Safety and Applied Nutrition, 200 C Street SW, Washington, D.C. 20204 (202) 260-8920 or via the World Wide Web at http://www.foodsafety.gov/~dms/prodguid.html

Safer processing of sprouts (video). The California Department of Health Services, Food and Drug Branch, and the FDA jointly developed this video in cooperation with the CDC, university researchers, and industry representatives to assist the industry in producing a safer product. The video may also be useful for retailers, regulators, and anyone working with the industry that wants to better understand the product and current recommendations for best production practices. For an order form see http://vm.cfsan.fda.gov/~dms/sprouvid.html. Additional videos for processed vegetables and unpasteurized juices are currently under production.

Gorny, J. R., ed. 2001. Food safety guidelines for the fresh-cut produce industry. 4th ed. : Alexandria, VA: International Fresh-Cut Produce Association. Available from International Fresh-Cut Produce Association, 1600 Duke Street, Suite

440, Alexandria, VA 22314, (703) 299-6282 or via the World Wide Web at www.fresh-cuts.org

WEBSITES OF INTEREST

American Food Safety Institute: www.americanfood-safety.com

California Department of Food and Agriculture, Food Safety Issues: www.cdfa.ca.gov/foodsafety

Centers for Disease Control and Prevention: www.cdc.gov/health/foodill.htm

Davis Fresh Technologies: www.davisfreshtech.com

Food Safety Guide:
http://vm.cfsan.fda.gov/~dms/prodguid.html

Gateway to Government Food Safety Information: www.foodsafety.gov

HACCP information can be found at the Seafood Network Information (http://seafood.ucdavis.edu/) and Food Safe Program (http://foodsafe.ucdavis.edu/homepage.html) sites.

International Fresh-Cut Produce Association: www.fresh-cuts.org

North Carolina State University Cooperative Extension Service: www.ces.ncsu.edu/depts/food-sci/agentinfo

Partnership For Food Safety Education: www.fight-bac.org

Produce Marketing Association: www.pma.com

Restaurant Association of Maryland, Fruit and Vegetable Food Safety Tips: www.marylandrestaurants.com/fruitveg-safety.html

University of California, Davis: www.ucdavis.edu From this website select "Public Service" to access the following home pages: Center for Consumer Research, FoodSafe Program, Fruit and Nut Research and Information Center, Postharvest Outreach Program, Seafood Network Information, Vegetable Research and Information Center

USDA Food and Nutrition Information Center: www.nal.usda.gov/fnic

USDA Food Safety and Inspection Service Q&A about HACCP:
www.fsis.usda.gov/OA/haccpq&a.htm

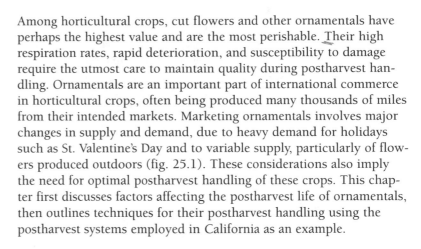

25

Postharvest Handling Systems: Ornamental Crops

Michael S. Reid

Among horticultural crops, cut flowers and other ornamentals have perhaps the highest value and are the most perishable. Their high respiration rates, rapid deterioration, and susceptibility to damage require the utmost care to maintain quality during postharvest handling. Ornamentals are an important part of international commerce in horticultural crops, often being produced many thousands of miles from their intended markets. Marketing ornamentals involves major changes in supply and demand, due to heavy demand for holidays such as St. Valentine's Day and to variable supply, particularly of flowers produced outdoors (fig. 25.1). These considerations also imply the need for optimal postharvest handling of these crops. This chapter first discusses factors affecting the postharvest life of ornamentals, then outlines techniques for their postharvest handling using the postharvest systems employed in California as an example.

FACTORS AFFECTING POSTHARVEST QUALITY OF ORNAMENTALS

Although the factors affecting the postharvest life of ornamentals are largely similar to those for fruits and vegetables, they differ in some important aspects.

FLOWER MATURITY

Minimum harvest maturity for a cut-flower crop is the stage at which harvested buds can be subsequently opened fully and have satisfactory display life after distribution. Many flowers are best cut in the bud stage and opened after storage, transport, or distribution. Although this technique is seldom used, it has many advantages, including reduced growing time for single-harvest crops, increased product packing density, simplified temperature management, reduced susceptibility to mechanical damage, and reduced desiccation. Many flowers are presently harvested when the buds are starting to open (rose, gladiolus); others are normally fully open or nearly so (chrysanthemum, carnation). Flowers for local markets are generally harvested more open than those intended for storage or long-distance transport.

FOOD SUPPLY

The high respiration rate and rapid development of flower buds and flowers indicate the need for a substantial carbohydrate supply to the flowers after harvest. Starch and sugar stored in the stem, leaves, and petals provide much of the food needed for cut-flower opening and maintenance. These carbohydrate levels are highest when plants are grown in high light conditions and with proper cultural management. Carbohydrate levels are generally highest in the late afternoon, after a full day of sunlight. However, flowers are preferably harvested in the early morning, because temperatures are low, plant water content is high, and a whole day is available for processing the cut flowers.

The quality and vase life of many cut flowers can be improved by pulsing them after harvest with a solution containing sugar. The cut flowers are allowed to stand in solution for a short period, usually less than 24 hours, and often at low temperature. The most dramatic example of the effect of added carbohydrate is in spikes of tuberose and gladiolus: flowers open further up the spike, are bigger, and have a longer vase life after overnight treatment with a solution containing 20% sucrose and a biocide to inhibit bacterial growth (fig. 25.2).

Figure 25.1

Variability in the flower market and the effect of holiday periods is shown by the changing wholesale value of red roses and gypsophila through the year.

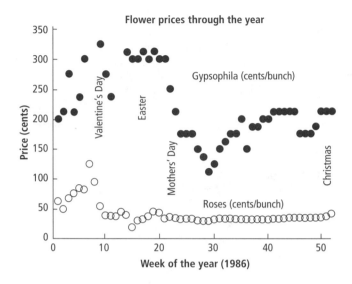

Figure 25.2

Pulsing gladiolus flowers overnight in a cold room with a 20% sucrose solution enhances flower size, opening, and vase life.

Sugar is also an essential part of the solution used to open bud-cut flowers before distribution, and of the vase "preservatives" used at the retail and domestic level. Red roses opened without sugar in the vase solution turn a dark purple (blueing). Blackening of leaves of cut flowers of *Protea nerifolia* can be prevented by maintaining the flowers in high light intensity. The problem is induced by the low carbohydrate status of the harvested inflorescence.

TEMPERATURE

The rapid respiration of cut flowers, an indicator of their rate of growth and senescence, generates heat as a by-product. As in all biological systems, the respiration of cut flowers increases logarithmically with increasing temperature (fig. 25.3). For example, a flower held at 29°C (85°F) is likely to respire up to 45 times faster than a flower held at 2°C (36°F). The rate of senescence can be reduced dramatically by cooling the flowers (fig. 25.4). When cut flowers were stored at a range of temperatures for a five-day transportation simulation, the subsequent vase life of the flowers under standard conditions was directly related to their respiration rate. Since reducing the temperature markedly reduces respiration, rapid cooling and proper temperature maintenance are essential to maintaining quality and satisfactory vase life of cut flowers.

The optimal temperature for storage of most common cut flowers is 0°C (32°F), just above freezing (see fig. 25.4). In contrast, some tropical crops, such as anthurium, bird-of-paradise, some orchids, and ginger, are injured at temperatures below 10°C (50°F). Symptoms of this chilling injury include darkening of the petals (fig. 25.5), watersoaking of the petals (which look transparent), and, in severe cases, collapse and drying of leaves and petals (fig. 25.6).

WATER SUPPLY

Cut flowers, especially those with leafy stems, have a large surface area, so they lose water and wilt rapidly. They should be stored at above 95% RH, particularly during long-term storage. Water loss is dramatically reduced at low temperatures, another reason for prompt cooling.

Even after flowers have lost considerable water (for example, during transportation or

Figure 25.3

Respiration and heat production of cut carnations increases logarithmically with increasing temperature.

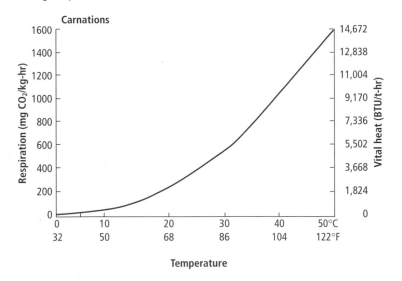

Figure 25.4

Optimal storage temperature for most cut flowers is 0°C (32°F). Flowers kept at 10°C (50°F) deteriorate 3 times faster than those kept at 0°C (32°F).

Figure 25.5

Anthurium flowers are sensitive to low temperatures. After storage of 7 days at 0°C (32°F), the flowers show symptoms typical of chilling injury.

Figure 25.6

Poinsettias, like many other tropical plants, are very sensitive to low-temperature storage. Exposure for 1 week to temperatures below 7°C (45°F) causes complete collapse of the plants.

storage), they can be fully rehydrated using proper techniques. Cut flowers absorb solutions without difficulty, providing there is no obstruction to water flow in the stems. Air embolism, bacterial plugging, and poor water quality can reduce solution uptake.

Air embolism. Air embolism occurs when small bubbles of air (emboli) are drawn into the stem at the time of cutting. These bubbles cannot move far up the stem, so the upward movement of solution to the flower is restricted. Emboli can be removed by recutting the stems under water (removing about 2.5 cm or 1 in), by ensuring that the rehydration solution is acid (pH 3 or 4), or by placing the stems in ice water, or in a vase solution heated to 41°C (105°F) (warm, but not hot).

Bacterial plugging. The quality and vase life of flowers are improved by supplying them with sugar after harvest. Sugar can, however, also act as food for the growth of detrimental fungi and bacteria in the water; this growth can be enhanced by organic materials from the cut stem. Substances produced by the bacteria and the bacteria themselves can plug the water-conducting system (fig. 25.7). For this reason, buckets should be cleaned and disinfected regularly, and flower-holding solutions should contain germicides to prevent the growth of microorganisms. Acidic vase solutions not only improve water flow by overcoming embolism but also inhibit bacterial growth.

Water quality. Hard water contains minerals that make the water alkaline. Alkaline

Figure 25.7

Scanning electron micrograph of a cross-section of the stem of the rose flower ([A] before; [B] after) held for 3 days in deionized water. The fine tubes conducting water to the flower have become plugged with bacterial slime and fungal hyphae.

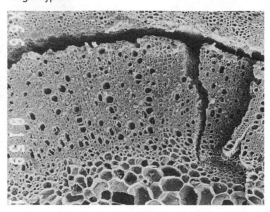

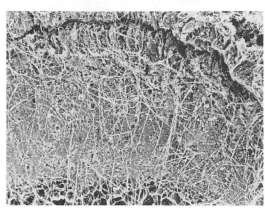

water does not move readily through stems and can substantially reduce vase life. Such water can be treated by removing the minerals using a deionizer or by making the water acidic. Commercial flower preservatives may not contain enough acid to acidify some very alkaline waters; in that case, more acid should be added. Recent studies have explained why the presence of low salt concentrations in the water may improve flow in cut flower stems.

Certain ions found in tap water are toxic to some cut flowers. Sodium (Na^+), present in high concentrations in soft water, is, for example, toxic to carnations and roses. Fluoride (F^-) is very toxic to gerbera, gladiolus, freesia, and some rose varieties; fluoridated drinking water contains enough F^- (about 1 to 2 ppm) to damage these cut flowers.

ETHYLENE

Some flowers, especially carnations and some rose cultivars, perish rapidly if exposed to minute concentrations of ethylene. In carnations and sweet peas, ethylene produced by the flower induces the normal processes of senescence. In many compound flowers, such as snapdragon and delphinium, ethylene causes flower abscission, or shattering (fig. 25.8). Ethylene is produced in large quantities by some ripening fruits and during combustion of organic materials (e.g., gasoline, firewood, tobacco). Levels of ethylene in the air above 0.1 ppm in the vicinity of many cut flowers can cause damage. Storage and handling areas should be designed not only to minimize contamina-

Figure 25.8

Effect of ethylene and MCP on flower loss in snapdragon.

Control   1-MCP

1ppm ethylene, 3 days

tion with ethylene but to have enough ventilation to remove any ethylene that does occur. Treatment with the anionic thiosulfate complex of silver (STS) reduces the effects of ethylene (exogenous or endogenous) in some flowers (fig. 25.9); 1-methylcyclopropene (1-MCP), another inhibitor of ethylene action, provides similar protection. Finally, refrigeration greatly reduces ethylene production and the sensitivity of the product to ethylene.

GROWTH TROPISMS

Certain responses of cut flowers to environmental stimuli (tropisms) can result in quality loss. Most important are geotropism (bending away from gravity) and phototropism

Figure 25.9

Effect of STS pretreatment on opening of cut Lovely Girl roses.

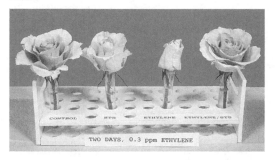

Figure 25.10

Laying snapdragons on the bench results in unsightly curvature.

(bending toward light). Geotropism often reduces quality in spike-flower crops like gladiolus and snapdragon, because the flowers and spike bend upward when stored horizontally (fig. 25.10). These flowers should be handled upright whenever possible, although proper storage temperatures will greatly reduce bending.

MECHANICAL DAMAGE

Bruising and breakage of cut flowers should be avoided. Flowers with torn petals, broken stems, or other obvious injuries are undesirable for aesthetic reasons. Disease organisms can more easily infect plants through injured areas. In fact, most diseases are caused by infection of injured areas. Finally, respiration and ethylene evolution are generally higher in injured plants, further reducing storage and vase life.

DISEASE

Flowers are susceptible to disease, not only because their petals are fragile, but also because the sugar solution secreted by their nectaries is an excellent nutrient supply for even mild pathogens. To make matters worse, transfer from cold storage to warmer handling areas can result in condensation of water on the harvested flowers. The ubiquitous spores of gray mold (*Botrytis cinerea*) can germinate wherever free moisture is present. In the humid environment of the flower head, it can even grow (albeit more slowly) at temperatures near freezing. Proper greenhouse hygiene, temperature control, and minimizing condensation on flowers all reduce losses caused by gray mold. Some fungicides, such as vinclozolin, iprodione, and a copper complex (Phyton-27) have been approved for use on cut flowers and are effective against gray mold.

POSTHARVEST MANAGEMENT TECHNIQUES USED FOR CUT FLOWERS

Systems for harvesting and marketing cut flowers vary according to individual crops, growers, production areas, and marketing systems. All involve a series of steps—harvesting, grading, bunching, sleeving, storage, packing, precooling, transportation, and retail marketing, not necessarily in that order. Management systems should be selected that maximize postharvest life of the flowers, a goal that usually requires prompt precooling and proper temperature management throughout the marketing chain. Increasingly, producers are trying to reduce the number of steps in the postharvest handling system. For example, some field flower growers cut, grade, bunch, and pack their product in the field (fig. 25.11). The packed boxes are then taken directly to the precooler. Such systems, where appropriate, reduce damage to the flowers and may decrease labor costs.

Figure 25.11

Harvesting, grading, bunching, and packing are all carried out under shade in this chrysanthemum bouquet operation.

Figure 25.12

These specialized shears hold a rose stem after it has been cut, simplifying the task of harvesting.

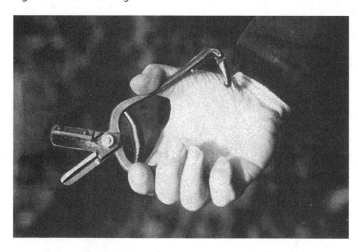

HARVESTING

Harvesting is normally done by hand, using shears or a sharp knife. Simple mechanical aids are used for some crops; examples are the hook-shaped comma that permits chrysanthemum harvest without stooping and rose shears that grip the flower stem after it has been cut, allowing it to be withdrawn single-handedly from the bush (fig. 25.12). At no time should harvested flowers be placed on the ground, where they may become contaminated with disease organisms. Ideally, harvesting, grading, and packing should all be done dry, without the use of chemical solutions or water. If this is not possible, clean buckets containing clean

water and a biocide should be used. With hard water and for difficult-to-rehydrate flowers, clean water containing sufficient citric acid to bring the pH to 3.5 and an appropriate biocide should be used instead.

GRADING

The designation of grade standards for cut flowers is controversial. There are no U.S. federal or state standards for cut flowers or other ornamentals. Large producers, wholesalers, or retailers may have internal grade standards, but they are highly variable. The Society of American Florists has promulgated "recommended" grade standards for some major crops, but they cannot be enforced. Stem length, still the major quality standard for many flowers, may bear little relationship to flower quality, vase life, or usefulness. Straightness of stems, stem strength, flower size, vase life, freedom from defects, maturity, uniformity, and foliage quality are among the factors that should also be used in cut flower grading. Bunch weight, proposed 60 years ago as a grade standard for New York flower producers, is now being applied by the Holland flower auction for certain field flowers. Mechanical grading systems, including semiautomatic systems using vision technology to determine stem length and bud maturity, should be carefully designed to ensure efficiency and avoid damaging flowers.

BUNCHING

Flowers are normally bunched, except for standard chrysanthemums, anthuriums, orchids, and some other specialty flowers. The number of flowers in the bunch varies according to growing area, market, and species. Groups of 5, 10, 12, and 25 are common for single-stemmed flowers. Spray-type flowers are bunched by the number of open flowers, by weight, or by bunch size.

Bunches are held together by string, paper-covered wire, or elastic bands and are frequently sleeved soon after harvest (fig. 25.13) to separate them, protect the flower heads, prevent tangling, and identify the grower or shipper. Materials used for sleeving include paper (waxed or unwaxed) and polyethylene (perforated, unperforated, and blister). Sleeves can be preformed (although variable bunch size can be a problem), or they can be formed around each bunch, using tape, heat sealing (polyethylene), or staples.

Figure 25.13

Thin plastic sleeves are pulled over the flower bunches to provide protection and keep flowers separate in the box.

Damage through multiple handlings can be reduced if grading, sizing, and even bunching are done in the field or greenhouse. Flowers should be graded and bunched before being treated with chemicals or being placed in storage. When the flowers are badly wilted, or when labor is not available for grading and bunching, flowers should be rehydrated under refrigeration until these operations can be carried out.

CHEMICAL SOLUTIONS

The diverse solutions in which flowers may be placed after harvest have specific purposes.

Rehydration. Wilted flowers should be rehydrated in a cooler, using deionized water containing a germicide. Wetting agents (0.01 to 0.1%) can be added, and the water should be acidified with citric acid, hydroxyquinoline citrate (HQC), or aluminum sulfate to a pH near 3.5. No sugar should be added to the solution. The solution should contain a biocide.

Pulsing. The term *pulsing* means placing freshly harvested flowers for a short time in a solution formulated to extend their storage and vase life. Sucrose is the main ingredient of pulsing solutions, and the proper concentration ranges from 2 to 20%, depending on the crop. Some cut flowers are also fumigated with 1-MCP or pulsed with STS to reduce the adverse effects of ethylene. Flowers can be pulsed with STS for short periods at warm temperatures (e.g., 10 minutes at 21°C [70°F]) or long periods at cool temperatures (e.g., 20 hours at 2°C [36°F]). Short pulses (10 seconds) in solutions of silver nitrate are valuable for some crops. Chinese asters and maidenhair fern respond well to solutions containing 1,000 ppm silver nitrate. Other flowers (e.g., gerberas) respond well to 100 to 200 ppm. The function of the silver nitrate is not fully understood. In some cases it seems to function strictly as a germicide (e.g., chrysanthemums). The residual silver nitrate solution should be rinsed from the stems before packing.

Bud opening. Bud-cut flowers are opened in bud-opening solutions before they are sold. These solutions contain a germicide and sugar. Foliage of some flowers (especially roses) can be damaged if the sugar concentration is too high. Buds should be opened at relatively warm temperatures (21° to 27°C [70° to 80°F]), moderate humidities (60 to 80% RH), and reasonably high light intensities (100 to 200 foot-candles).

Tinting. Artificial coloration of flowers (tinting) is done in two ways: through the stem (for carnations) and by dipping the flower heads (for other flowers, principally daisies). In tinting carnations, proprietary dye solutions (combinations of food-type dyes with adjuvants designed to increase uptake of the solution) are mixed in a bucket and warmed to about 41°C (105°F). The carnations to be tinted (usually White Sim) are allowed to dry somewhat (overnight in the packing shed at 18°C [65°F]) to increase their rate of solution uptake. Dyeing is stopped before the flowers reach the desired color, because dye still in the stem is flushed into the flower by the vase solution. Tinting by dipping is done with proprietary tinting solutions containing aniline dyes dissolved in organic solvents such as rubbing alcohol (isopropanol). The heads of the flowers are dipped in a vat of the dye, shaken to remove surplus solution, and placed on a rack to dry before storage or packing.

PACKING

Most boxes for cut flowers are long and flat (fig. 25.14). This design restricts the depth of the flowers in the box, which may in turn reduce physical damage. Flower heads can be placed at both ends of the container for better use of space. Layers of whole sheets of newspaper have traditionally been used to prevent

Figure 25.14

Boxes commonly used in California for packing cut flowers are long and flat. From left: a regular fiberboard box designed for a bottom air delivery trailer, a box constructed from polyurethane-sprayed fiberboard, and a polystyrene "casket."

Figure 25.15

Cleats (padded with newspaper and foam) to hold the product are stapled to each side of the box.

the layers of flowers from injuring each other, but using small pieces of newspaper to protect only the flower heads allows for more efficient cooling of flowers after packing.

It is critical that boxes be packed in ways that minimize transport damage. Most packers anchor the product by filling the box so that the contents are firmly held. To avoid longitudinal movement, packers use one or more "cleats," normally foam- or newspaper-covered wood pieces that are placed over the product, pushed down, and stapled into each side of the box (fig. 25.15). Padded metal straps, high-density polyethylene blocks, and cardboard tubes can also be used as cleats. To ensure that the flowers do not move, it is useful to place a bottom cleat in the box, so that the bottom of the box cannot bulge and allow movement. The heads of the flowers are normally placed 7 to 12 cm (3 to 5 in) from the end of the box to eliminate the danger of petal bruising should the contents of the box shift, and to provide a plenum to distribute precooling air to the contents of the box. With some flowers, a staggered arrangement of the bunches anchors with the product.

Gladioli, snapdragons, and some other species are often packed in vertical hampers to prevent unsightly geotropic curvature. Cubic hampers are used for upright storage of daisies and other flowers. Specialty flowers such as anthurium, orchids, ginger, and bird-of-paradise are packed in various ways to minimize friction damage during transport (fig. 25.16). Flower heads may be individually protected by paper or polyethylene sleeves. Cushioning materials, such as shredded paper, paper wool, or wood wool may be distributed between the packed flowers to further reduce damage. For some tropical flowers (anthurium, heliconia) these materials may be moistened to reduce water loss. The bases of high-value flowers, such as orchids, may be sealed in small vials containing water or a preservative solution to ensure that they do not dehydrate during transportation. This is particularly useful for chilling-sensitive flowers, because the higher handling temperatures that they require are apt to accelerate desiccation.

COOLING

By far the most important part of maintaining the quality of harvested flowers is cooling them as soon as possible after harvest

Figure 25.16

High-value flowers are often specially packed. Here, cattleya orchids with individual tubes containing floral preservative are separated and cushioned from vibration with shredded newsprint.

Figure 25.17

In forced-air precooling of flowers, cool air is sucked or blown through the packed boxes.

and maintaining optimal temperatures during marketing. Most flowers should be held at 0 to 2°C (32 to 36°F). Chilling-sensitive flowers (anthurium, bird-of-paradise, ginger, tropical orchids) should be held at temperatures above 10°C (50°F).

Individually, flowers cool (and warm) rather rapidly (half-cooling times of a few minutes). Individual flowers brought out of cool storage into a warmer packing area warm quickly and develop condensation before packing. The simplest way to keep packed flowers adequately cooled and dry is to pack them in the cool room. Although this method is not always popular with packers and may slow packing somewhat, it ensures a cooled, dry product.

Packed boxes of cut flowers do not cool well if they are merely placed in a refrigerated room. The rapid respiration of the flowers and the insulating properties of a packed box result in heat buildup in packed flower boxes unless steps are taken to ensure temperature reduction. Forced-air cooling of boxes with end holes or closable flaps is the most common and effective method for cooling cut

flowers. Cool air is sucked or blown through the boxes (fig. 25.17). Care must be taken to pack in such a way that air can flow through the box and not be blocked by the packing material (fig. 25.18). In general, packers use less paper when packing flowers for precooling. The half-cooling time for forced-air cooling ranges from 10 to 40 minutes, depending on product and packaging.

If the packages are to remain in a cool environment after precooling, vents may be left open to assist removal of the heat of respiration. Flowers that are to be transported at ambient temperatures can be packed in polyethylene caskets, foam-sprayed boxes, or boxes with the vents resealed. Using ice in the box is only effective if it is placed so that it intercepts heat entering the carton (i.e., it must surround the product), and care must be taken to ensure that the ice does not melt onto the flowers or cause freezing damage.

Tropical flowers shipped in a mixed load need special care. The flowers should be packed in plenty of insulating material (an insulated box packed with shredded newsprint, for example). These flowers should not be cooled. If they are to be shipped by truck, they should be placed in the middle of the load, away from direct exposure to cooling air.

STORAGE

Although it would be advantageous to store cut flowers for short periods (up to a month), few cut flowers maintain their quality during cool storage. With the exception of bud-cut

Figure 25.18

Proper precooling of cut flowers requires that the flowers be packed so that airflow through the box is not impeded (as shown by the arrows). Upper box is correctly packed; in the lower box, paper protection prevents satisfactory airflow.

carnations, which can be stored for several months, flowers should not be stored for more than 2 to 3 weeks.

The factors limiting storage life of flowers are not fully understood, but development of *Botrytis* is commonly observed. Flowers intended for storage should be of premium quality and should be pretreated appropriately (STS, 1-MCP, sugar, growth regulators), depending on the species. Flowers to be stored for any length of time should be held dry (packed, and preferably wrapped, after cooling, in polyethylene sheet to minimize water loss). Flowers can be stored for a few days with their bases in preservative solution, but vase life is reduced if the storage period is longer. The storage room should be at 0°C (32°F) for most cut flowers, and the temperature should vary from this by no more than 0.2°C (0.5°F). To avoid larger temperature fluctuations, which are likely to cause condensation on the flowers and consequent growth of *Botrytis* and other pathogens, access to the storage room should be limited. The RH in the room should be at least 95%, and ethylene-producing commodities (especially fruits) should be rigorously excluded.

In general, controlled or modified atmosphere storage has not proved to be useful with cut flowers. Carnations last well in CA conditions, but they also last well in regular storage. The vase life of daffodils falls less rapidly during storage in a nitrogen atmos-

phere, and some flowers (lily) can withstand high CO_2 concentrations in storage, which may be useful for inhibiting pathogen growth. After storage, flowers should be rehydrated in the cool room, using water containing a biocide, and preferably at pH 3.5.

RETAIL HANDLING

To ensure maximum quality and satisfaction to the customer, it is vital that retail florists handle flowers properly on receipt. On arrival, packed boxes should be unpacked or placed immediately in a cool room. After unpacking, flower bunches should be recut and placed in a cool room (in a warmed rehydration solution) for at least a few hours. If flowers are to remain in the store for several days, they should be displayed in a clean bucket or vase containing a vase preservative.

The florist's cool room should be held at 0° to 2°C (32° to 36°F). Chilling-sensitive flowers (anthurium, bird-of-paradise, ginger, tropical orchids) should be held at temperatures above 10°C (50°F). Cut flowers that are held during storage with their bases in water (wet storage) are usually kept only 1 to 3 days, while those stored dry are often kept for longer periods.

Vase solutions. Vase solutions normally contain a low concentration of sugar (0.5 to 2.0%), a chemical to keep the pH of the solution low, and a germicide. Flowers are kept in this solution indefinitely. This treatment is usually provided by the retail seller or the customer. Preservatives are supplied to the trade in bulk or may be purchased in individual sachets intended for 1 quart (0.9 L) of solution. A typical solution contains 1 to 2% sugar, a biocide (HQC or a mixture of quaternary ammonium compounds) and 300 ppm citric acid. Various other chemicals are sometimes incorporated, including aluminum sulfate (which forms flocculent aluminum hydroxide to trap particles in the water and also lowers the pH), slow-release chlorine compounds (good biocides, but sometimes toxic) and 6-benzyl adenine (a plant growth regulator that increases the vase life of some flowers). These chemicals should be used at recommended strengths.

Flowers in the supermarket. Flowers intended for supermarket sales require special care throughout the distribution chain. First, only flowers that have adequate vase life to remain saleable for the intended period of time should be used; thus some short-lived flowers such as iris, tulip, and narcissus should be avoided. Flowers with common postharvest problems (roses, gypsophila) are a danger, too. It is particularly important that ethylene-sensitive flowers (notably carnations, some rose cultivars, and gypsophila) be treated with STS or 1-MCP if they are to be marketed through produce stores.

DRYING

A number of cut flower crops are sold in both fresh and dried forms. For some, such as statice and strawflower, the drying process can be as simple as hanging bunches upside down (to keep the stems straight) in a warm, dry location. Others, such as gypsophila and silver-dollar eucalyptus, which become too brittle if dried this way, are placed in a freshly prepared solution of 30% glycerol. The glycerol moves through the plant in the course of a day or so. Dye can be added to the glycerol to color the dried foliage or flowers. After the glycerol treatment, the plants are hung upside down in a warm dry environment to dry. Materials dried in this way remain supple for years. Improved methods of drying and preservation are still being developed. Plastic impregnation of whole plants and freeze-drying cut flowers are two such techniques.

REFERENCES

Armitage, A. M. 1993. Bedding plants–Prolonging shelf performance, postproduction care and handling. Batavia, IL: Ball. 71 pp.

Berkholst, C. E. M., et al. 1986. Snijbloemen: Kwaliteitsbehoud in de afzetketen. Wageningen, Neth.: Sprenger Instituut. 222 pp.

Blessington, T. M., and P. C. Collins. 1993. Foliage plants–Prolonging quality, postproduction care and handling. Batavia, IL: Ball. 203 pp.

Carow, B. 1981. Frischhalten von Schnittblumen. Stuttgart: Verlag Eugen Ulmer. 144 pp.

Goszcynska, D. M., and R. M. Rudnicki. 1988. Storage of cut flowers. Hortic. Rev. 10:35–62.

Halevy, A. H., and S. Mayak. 1979. Senescence and postharvest physiology of cut flowers. Part 1. Hortic. Rev. 1:204–36.

———. 1981. Senescence and postharvest physiology of cut flowers. Part 2. Hortic. Rev. 3:61–143.

Hardenburg, R. E., A. E. Watada, and C. Y. Wang. 1986. The commercial storage of fruits, vegetables, and florist and nursery stocks. USDA Handb. 66. 130 pp.

Mayak, S., and A. H. Halevy. 1980. Flower senescence. In K. V. Thimann, ed., Senescence in plants. Boca Raton, FL: CRC Press. 131–56.

Nell, T. A. 1993. Flowering potted plants–Prolonging shelf performance, postproduction care and handling. Batavia, IL: Ball. 96 pp.

Nell, T. A., J. E. Barrett, and R. T. Leonard. 1997. Production factors affecting postproduction quality of flowering potted plants. HortScience. 32:817–819.

Nowak, J., and R. M. Rudnick. 1990. Postharvest handling and storage of cut flowers, florist greens, and potted plants. Portland, OR: Timber Press. 210 pp.

Reid, M. S. 1997. Produce facts on alstroemeria, anthurium, antirrhinum, carnation and chrysanthemum. Available via Internet at University of California Postharvest Technology Center Website (http://postharvest.ucdavis.edu).

Rij, R. E, J. F. Thompson, and D. S. Farnham. 1979. Handling, pre-cooling and temperature management of cut flower crops for truck transportation. Oakland, CA: USDA, Western Region AAT-W-5. 26 pp. (Issued jointly as Univ. Calif. Div. Ag. Sci. Leaflet 21058.)

Sacalis, J. N., and J. L. Seals. 1993. Cut flowers–prolonging freshness, postproduction care and handling. Batavia, IL: Ball. 110 pp.

Salunkhe, D. K., N. R. Bhat, and B. B. Desai. 1990. Postharvest biotechnology of flowers and ornamental plants. New York: Springer-Verlag. 192 pp.

Staby, G. 1994. Flower and plant care manual. Alexandria, VA: Society of American Florists. 180 pp.

Van Doorn, W. G. 1996. Water relations of cut flowers. Hort. Rev. 18:1–85.

Van Gorsel, R. 1994. Postharvest technology of imported and trans-shipped tropical floricultural commodities. HortScience. 29:979–981.

26

Postharvest Handling Systems: Fresh Herbs

Marita I. Cantwell and Michael S. Reid

Culinary herbs are leafy plant materials used in small amounts to contribute aroma and flavor to foods. Common examples are parsley, mint, basil, sage, and oregano. They have always been an important component of the human diet, providing variation in flavor of staple foods and being a vital part of some food preservation techniques. The major commercial form of culinary herbs is still the shelf-stable dried product, since it can be transported easily and stored more than 1 year under proper conditions. However, because of their superior flavor, fresh culinary herbs are increasingly in demand.

Distribution of fresh herbs through the major wholesale, retail, and foodservice channels has grown in recent years, but marketing them has not been wholly successful. Many herbs are extremely perishable; and although they are of high value, the small quantities sold mean that even a small inventory takes time to turn over. Moreover, marketing strategies have tended to use a uniform technology for all herbs. Because herbs are diverse in botanical origin and physiological state, postharvest conditions that are adequate for one fresh herb may be completely inappropriate for others. For example, herb bouquets on sale at retail markets may combine an extremely perishable herb such as chervil with long-lasting herbs such as rosemary and thyme. Another example is the storage temperature incompatibility of most culinary herbs with the chilling-sensitive basil.

Fresh herbs are complex structures consisting of stem, leaf, flower, and sometimes root tissues. Most herbs are harvested as soft or semiwoody leafy stems. Sometimes, as in the case of dill, oregano, and basil, the herb includes immature or mature flowers. Many salad herbs harvested as developing leaves (sorrel, arugula) or intact plants (mache) consist of leaves at different physiological stages. This diversity means that postharvest requirements may differ for each product, although in practice the herbs are handled much like leafy vegetables.

HARVEST AND HANDLING

Herbs for fresh market may be grown in the field or in protective enclosures such as tunnels or greenhouses. Production for fresh market is typically on a small scale, and harvesting, grading, trimming, bunching, and packing are all manual operations (fig. 26.1). Harvest in the field is commonly done with scissors, shears, or knives, and the herbs may be bunched at the time of harvest or taken in bulk to a packing station where they may be trimmed before bunching. If the herbs are very susceptible to water loss, bunching in the field may significantly reduce their fresh appearance and decrease shelf life. Herbs may be packed in bulk, in bunches, or placed directly in polyethylene pouches or rigid plastic boxes designed as point-of-purchase packages.

As with other leafy greens, quality components of fresh culinary herbs are largely visual: an appearance of freshness; uniformity of size, form, and color; and lack of defects such as adhering soil, damaged leaves, yellowing, and evidence of decay. There is no doubt that the most important quality components of culinary herbs are the expected aroma and flavor, but these usually are not considered in legal standards for fresh produce. There are presently neither legal standards for the quality of fresh herbs nor standard unit weights or volumes for marketing them.

Figure 26.1

Basil, brought to a packing station in bulk plastic containers, is trimmed, graded, and bunched.

Table 26.1. Effect of temperature and ethylene on visual quality of fresh culinary herbs after 10 days. Relative humidity was 90–95%.

Herb	Scientific name	Visual quality score after 10 days at indicated temperature*		
		0°C (32°F)	10°C (50°F)	20°C (68°F)
Basil	*Ocimum basilicum*	2	8	7
Chervil	*Anthriscus cerefolium*	8	6†	1
Chives	*Allium schoenoprasum*	9	6	3
Cilantro	*Coriandrum sativum*	9	4†	1
Dill	*Anethum graveolens*	9	6†	2
Epazote	*Chenopodium ambrosioides*	9	7†	5
Mache	*Valerianella locusta*	8	5	2
Marjoram	*Origanum majorana*	9	8†	1
Mint	*Mentha* spp.	9	6†	2
Mitsuba	*Cryptotaenia japonica*	9	7†	4
Rosemary	*Rosmarinus officinalis*	9	9	7
Sage	*Salvia officinalis*	9	8	—
Shiso	*Perilla frutescens*	6	8†	3
Tarragon	*Artemisia dracunculus*	8	6	—
Thyme	*Thymus vulgaris*	9	8	7

Source: Adapted from Cantwell and Reid 1990.

Notes:

*Quality score: 9 = excellent; 7 = good, minor defects; 5 = fair, moderate defects, limit of salability; 3 = poor, major defects; 1 = unusable.

†Herbs that showed reduced visual quality at 10°C (50°F) when exposed to 5–10 ppm ethylene.

POSTHARVEST TECHNOLOGY

The small size of many herb operations permits attention to producing and preparing a quality product. However, the perishability of fresh herbs means that maintaining quality during marketing requires careful postharvest handling. There are six principal considerations.

Temperature. As with other perishable products, temperature is the overriding factor affecting the life of cut herbs. Herbs such as cilantro, chervil, mitsuba, dill, mint, and mache are susceptible to yellowing. Most of these same herbs, along with large-leafed basil and shiso, are prone to rapid water loss. For most herbs, the optimal storage conditions are 0°C (32°F) and high RH (95 to 98%). The visual quality of herbs stored for 10 days at different temperatures is shown in table 26.1. Over a 10-day simulated marketing period, many herbs still have acceptable quality if held in the range of 0° to 10°C (32° to 50°F), but storing at 20°C (68°F) greatly limits shelf life. In commercial handling, where temperature fluctuations frequently occur, shelf life may be considerably less than indicated in table 26.1.

Basil and shiso show reduced quality when stored at 0°C (32°F) because these two herbs are chilling sensitive (fig. 26.2; table 26.1). This sensitivity presents practical problems, since basil is a major component of most mixed herb shipments. Herb shipments containing basil are often held at an intermediate temperature, between 5° and 10°C (41° and 50°F), but this may still induce chilling in basil while substantially increasing the rate of deterioration of other herbs. The time required for chilling symptoms to appear on basil stored at different temperatures is shown in figure 26.3. Chilling symptoms on basil include discolored leaves, discolored stems, and loss of aroma.

Because of the small scale of herb production, facilities for temperature management are often rudimentary or lacking. If herbs are harvested cool in the early morning, the need for cooling is reduced. Some producers send their product to distributors of major-volume leafy vegetables, who have the necessary facilities for better cooling and handling. However, due to smaller volumes and lower priority, herbs may be subject to temperature stress and moisture loss while waiting to be cooled. Some fresh herbs (parsley, watercress, and mint) can be hydrocooled and iced. Others, with tender leaves, are better forced-air or room-cooled. Some growers have successfully vacuum-cooled fresh herbs.

Fresh herbs are sometimes transported in mixed loads with other leafy greens in

Figure 26.2

Typical symptoms of chilling injury on basil stored 10 days at 5°C (41°F).

Figure 26.4

Herbs packed in a cold room in perforated polyethylene bags are placed in a waxed carton.

Figure 26.3

Development of chilling injury on basil stored at different temperatures. An injury score of 2 is sufficient to reduce marketability.

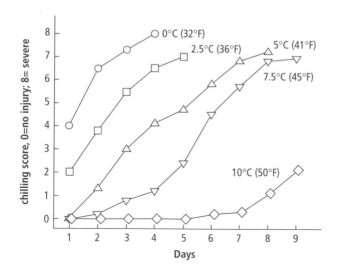

refrigerated trailers or containers. More commonly, because of their specialty nature and the small volumes marketed, they are shipped by air to reduce transit time. Gel ice packs are routinely used with air shipments to reduce heat buildup.

Moisture loss. Many fresh herbs lose water rapidly through their leaves, and wilting after harvest can be significant. Protection from the sun and wind and rapid transport to a protected packing and cooling area are necessary. Storing herbs at low temperatures greatly reduces the rate of water loss. Some handlers place the cut stems of the

herbs (e.g., mint, basil) in water, achieving a partial rehydration. However, microorganisms may grow rapidly in the water, and the technique is not practical for transport and storage. Another option for some markets is to handle the herbs in pots with roots and growing substrate.

A preferred way to reduce water loss is to package the cut herbs in plastic films (fig. 26.4). For herbs packaged this way, it is necessary to maintain constant temperatures to reduce condensation inside the bag and the consequent risk of microbial growth. The bags may be partially ventilated with perforations or may be constructed of a polymer that is partially permeable to water vapor. Paper sheets inside film packages can be useful to absorb free moisture yet maintain high-humidity conditions. These paper liners are most useful for herbs marketed in bulk quantities since they obscure the visibility of the product. The packing areas, cold rooms, and transport vehicles should be kept above 95% RH where practicable.

Physical injury. Most fresh herbs are very susceptible to damage during postharvest handling. Mishandling can result in extensive discoloration of the tender leaves of mint, basil, and cilantro and may provide microbial infection sites. Careful handling

Figure 26.5

Yellowing of mitsuba and abscission of mint leaves are caused by exposure to 5 ppm ethylene.

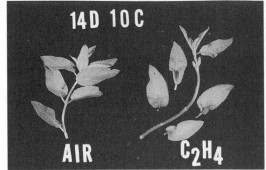

Figure 26.6

Quality of cilantro stored at different temperatures in air (left) or controlled atmospheres (right). (Loaiza and Cantwell 1997)

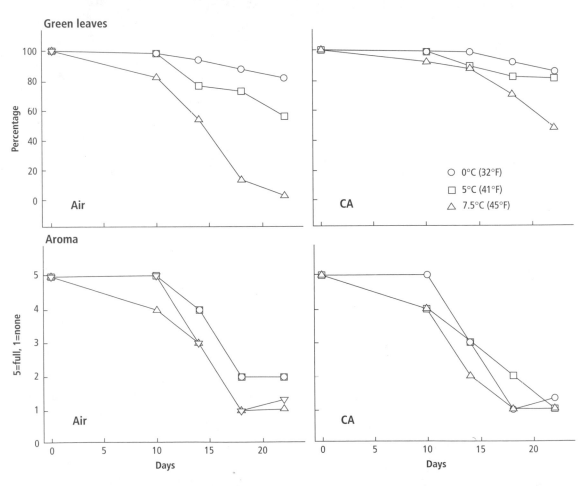

and packaging are means of minimizing damage. For retail marketing of delicate herbs, rigid clear plastic containers are sometimes used.

Ethylene. Like other leafy tissues, herbs are detrimentally affected by ethylene, with such symptoms as yellowing and leaf drop (fig. 26.5) and epinasty (petiole bending). Sensitivity to ethylene varies among herb species (see table 26.1); many show epinasty, but yellowing and abscission are usually observed only in the most sensitive species. Ethylene effects can be minimized by maintaining proper storage temperatures.

Figure 26.7

Postharvest defects of cilantro: freeze damage (left), decay (middle), and CO_2 injury (right).

Atmosphere modification (CA, MA).
Fresh herbs respond positively to reduced O_2 and increased CO_2 concentrations. The benefit of high CO_2 atmospheres to the shelf life of herbs can be illustrated with data on cilantro (fig. 26.6). Cilantro stored in air at 0°C (32°F) has good visual quality for 18 to 22 days. Good visual quality can be maintained only 14 and 7 days at 5°C (41°F) and 7.5°C (45°F), respectively. A high CO_2 atmosphere (5 to 10%) can double the shelf life of cilantro stored at 7.5°C (45°F), principally by maintaining the leaves green. However, a 10% CO_2 atmosphere may eventually cause damage to the leaves (fig. 26.7). It is important to note that the high CO_2 atmospheres were not effective in extending the period of typical cilantro aroma (see fig 26.6). Modified atmospheres do not reduce the chilling sensitivity of basil.

Although the main benefit obtained with packaging in plastic films is probably reduced water loss, film packages may also provide herbs with beneficial modified atmospheres. If herbs are marketed in lighted displays, however, beneficial CO_2 levels may be reduced because of photosynthesis. Lighted marketing conditions may also stimulate detrimental photo-oxidative reactions, and the quality of packaged herbs may be less when they are stored in the light than in the dark.

Pathogens. The shelf life of fresh herbs, especially those packaged in film bags, is frequently terminated by the growth of pathogens such as gray mold and bacterial soft rot. Control measures include preharvest and handling hygiene (including use of chlorinated water), storing at the proper tempera-

ture, and avoidance of injury and water condensation during marketing. Because harvest and postharvest handling operations typically involve large work crews and many hands, fresh herbs are particularly susceptible to contamination by human pathogens such as *Shigella* and *Salmonella* spp. (Nguyen-the and Carlin 1994). See Chapter 24 regarding these food safety issues.

REFERENCES

Aharoni, N., A. Reuveni, and O. Dvir. 1989. Modified atmospheres in film packages delay senescence and decay of green vegetables and herbs. Acta Hortic. 258:37–45.

Åpeland, J. 1971. Factors affecting respiration and colour during storage of parsley. Acta Hortic. 20:43–52.

Cantwell, M., and M. S. Reid. 1990. Postharvest physiology and handling of fresh culinary herbs. J. Herbs, Spices, Medicinal Plants 1:93–127.

Hruschka, H. W., and C.-Y. Wang. 1979. Storage and shelf life of packaged watercress, parsley, and mint. USDA Mktg. Res. Rept. 1101. 19 pp.

Ishi, K., and M. Okubo. 1984. The keeping quality of Chinese chive (*Allium tuberosum* Rottler) by low temperature and seal-packaging with polyethylene bag. J. Jap. Soc. Hort. Sci. 53:87–95.

Lange, D. D. and A. C. Cameron. 1994. Postharvest shelf life of sweet basil (*Ocimum basilicum*). HortScience. 29: 102–103.

Lipton, W. J. 1987. Senescence of leafy vegetables. HortScience. 22:854–859.

Loaiza, J., and M. Cantwell. 1997. Postharvest physiology and quality of cilantro (*Coriandrum sativum* L.). HortScience. 31:104–107.

Nguyen-the, C., and F. Carlin. 1994. The microbiology of minimally processed fresh fruits and vegetables. Crit. Rev. Food Sci. and Nutr. 34: 371–401.

Philosoph-Hadas, S., D. Jacob, S. Meir, and N. Aharoni. 1993. Mode of action of CO_2 in delaying senescence of chervil leaves. Acta Hortic. 343:117–122.

Simon, J. E., A. F. Chadwick, and L. E. Craker. 1984. Herbs: An indexed bibliography, 1971–1980. Hamden, CT: Shoe String Press. 770 pp.

Small, E. 1997. Culinary Herbs. Ottawa, Canada: National Research Council Press. 710 pp.

Yamauchi, N., and A. E. Watada. 1993. Pigment changes in parsley leaves during storage in controlled or ethylene containing atmosphere. J. Food Sci. 58: 616–618, 637.

27

Postharvest Handling Systems: Pome Fruits

Elizabeth J. Mitcham and

F. Gordon Mitchell

California is a major producer of apple and European pear fruit for U.S. and export markets. California produces a more limited volume of Asian pears and quince. The apple industry in California has increased greatly over the past 10 years. New plantings in the Central Valley, in addition to older plantings along the coast and in the foothills, have increased California's production such that it is generally the third largest producing state in the United States after Washington and Michigan. Granny Smith, Gala, and Fuji have dominated these new plantings. The pear industry in California is more than 90% Bartlett pear, with approximately half the volume diverted to processing, while the fresh volume remains substantial.

I. Apple

MATURITY

Maturity changes in apples include skin color, seed color, flesh firmness, soluble solids content, starch content, titratable acid content, respiratory rate, ethylene production, and production of other flavor and odor constituents. Suggested maturity indices have included all of these as well as time (days from full bloom), accumulated heat units (e.g., degree-days above 7°C [45°F]), fruit size, and various combinations of these.

Most possible maturity indices have limitations. For example, ethylene production and respiratory rate changes may occur too late or are too variable to be useful for timing of harvest. Some other changes are too subtle to be effective. Most commercial applications of maturity indices use days from full bloom as the rough guide, with fine tuning using mostly starch index and flesh firmness. Maturity standards must take into account whether fruit is to be stored (and for how long) or shipped immediately. The goal is to determine a reliable index that can predict the best harvest date for long-term controlled atmosphere storage or for short-term air storage. Harvesting immature fruit results in smaller fruit with greater susceptibility to water loss and storage disorders, such as bitter pit and scald. Overmature fruit have greater susceptibility to softening, mealiness, decay, and controlled atmosphere–related disorders.

Many apple-growing regions have developed maturity standards that guide the start of harvest. These standards may be voluntary or legally binding. In California, Granny Smith apples are released for harvest by county according to the California Granny Smith Starch Iodine Chart. Inspectors rate a 30-apple sample from the orchard for starch degradation pattern, and the average for the orchard must meet the minimum level of starch degradation (a score of 2.5 on the Starch Iodine Chart). A similar starch score is used to determine the start of Granny Smith harvest in New Zealand. While starch disappearance is also used to guide the start of Granny Smith harvest in Washington, the scale is quite different from that used in California and New Zealand. Unfortunately, the starch disappearance charts used in different apple growing regions are not uniform. Scales can range from 0 to 6, 0 to 7, or 0 to 10. The higher number always indicates more starch disappearance.

Starch is a useful index of harvest maturity because starch is degraded into sugars as the apple matures and begins to ripen. The starch content is measured at the center cross-section of the apple. Starch disappearance is visualized by staining the cut surface of the apple with an iodine–potassium iodide solution (see recipe above). The iodine solution turns starch black and provides a view of the starch clearance pattern. Starch charts can be used to determine a starch score based on the percentage of the cross-section clear of staining.

HANDLING

Apple fruit are hand-picked into bags, gently transferred into field bins, and transported to the packing facility. Care must be taken during harvest of apples to avoid impact bruising. Solid-framed, padded picking bags into which pickers can place, rather than drop, harvested apples are ideal. Careful training and supervision of picking crews is essential to avoid excessive fruit bruising.

Apples, which are often stored in field bins to await later packing, may be drenched with a scald inhibitor and fungicide and are sometimes treated with calcium chloride for bitter pit control before storage.

Apple fruit can benefit from precooling prior to cold storage. Precooling can be accomplished by forced-air cooling or hydro-cooling. While hydrocooling is faster than forced-air cooling, there is the potential to spread disease organisms from diseased fruit or soil to healthy fruit. If liquid chlorine is used to reduce the potential for disease spread, care must be taken to change the drench water daily to prevent a buildup of sodium, resulting in fruit burn. The same is true for dump tank water systems.

Fruit to be packed are unloaded from bins using water flotation dumps, washed, pre-

sized to eliminate undersized fruit, and sorted. Sorting is generally by a combination of electronic and human sorters. Electronic sorters for color and size are very common in larger packinghouses. Electronic sorting for defects is used to a very limited extent but will likely increase in the near future. Most defect sorting is by human sorters at this time. Fruit is often segregated into several fresh grades, as well as processing grades and by-product outlets.

Most fresh-market apples are tray-packed, while lower grades may be volume filled. Bagging of apple fruit is used more frequently, not only for smaller apples but also for large fruit destined for consumer warehouse stores. Bruising continues to be a concern for bagged fruit, but newer machine designs have reduced this problem somewhat.

Packed cartons are segregated by fruit size and stacked onto unitized pallets. Final cooling of these pallet loads occurs after packing. Cooling may involve forced-air or static room cooling. Most apple cartons used today are not ventilated or are inadequately ventilated, a situation that seriously impedes cooling after packing.

A flow diagram for the handling of apples (fig. 27.1) shows the steps involved in fresh-market handling.

STORAGE

TEMPERATURE MANAGEMENT

As a general rule, pome fruits respond best to rapid cooling and storage at as low a temperature as possible without danger of freezing or low-temperature injury. Many apple varieties can be safely stored at 0°C (32°F); however, there are exceptions, some of which are shown in table 27.1. Low-temperature injury can be manifested as internal flesh browning. Specific recommendations for varieties may require confirmation under localized conditions. For example, California-grown Yellow Newtown apples are recommended to be stored at 3.5° to 4.5°C (38° to 40°F), while Yellow Newtown apples from other growing areas have been safely stored at 0°C (32°F).

Cooling delays are associated with shortened storage life, flesh softening, increased physiological disorders, and diseases. Thus, fruit should be cooled as soon after harvest as possible. However, for fruit with severe

Figure 27.1

Postharvest handling of apples and Bartlett pears.

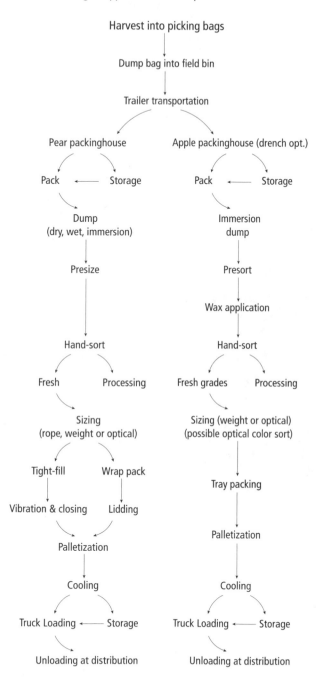

this delay becomes even more important.

While some apples are packed immediately after harvest, the majority are stored for a period of time in either regular cold storage or in controlled atmosphere storage. During the packing operation, fruit can warm considerably, and recooling after packing is important to maintain fruit quality. The lack of adequate ventilation on most apple boxes greatly slows the recooling process after packing. Boxes should be designed with at least 5% venting on all four sides to allow for air passage when boxes are cross-stacked.

CONTROLLED ATMOSPHERE STORAGE

Controlled atmospheres (CA) are often used when apples are stored longer than 3 months; however, Gala apples have been shown to benefit from CA storage for as short as 1.5 months, and this has been used commercially. Most fruits are CA-stored in field bins, often with unsorted, field-run fruit. However, some apples are dumped, sorted, sized, and refilled into their bins before moving to CA storage. Maintenance of the proper levels of O_2 and CO_2 for each variety is critical to the success of CA storage as is proper harvest maturity. In general, fruit harvested earlier in the harvest season is a better candidate for long-term CA storage than later harvested fruit. Fruit with advanced maturity can be more susceptible to developing CA-related storage disorders. General CA recommendations by variety are shown in table 27.1.

RELATIVE HUMIDITY

Warm fruits lose water faster than cold fruits, especially if the RH of the air is less than 95%. Loss of water is loss of saleable weight and can result in fruit shrivel and loss of gloss. Water loss is cumulative from the time the fruits are harvested until consumption. Visible shrivel is not apparent until a critical amount of water is lost, usually 3 to 5% of the original weight. Rapid transport of fruits from the field to the cooler, rapid cooling and maintenance of 90 to 95% RH in storage are essential to reduce fruit water loss. Use of polyethylene film bin liners can reduce water loss in storage, particularly at less than 95% RH. The liner should have an open or highly perforated bottom if apples will be drenched or hydrocooled. The disadvantage of the liner is a significant reduction in

watercore, cooling delays can reduce the severity of the disorder. The importance of cooling speed can vary with cultivar. For example, Gala apples often soften significantly in storage, and rapid cooling can greatly extend the storage life. When antioxidants such as diphenylamine are used to delay storage scald development, a cooling delay of up to 1 day may be required for maximum scald control. Rapid cooling of the fruit after

Table 27.1. Recommended temperatures and CA for storage of selected apple varieties

Cultivar	Temperature °C (°F)	O$_2$ concentration	CO$_2$ concentration	Storage life (months)
Braeburn	0–1 (32–34)	1.5–2.0	≤1.0	6–9
Fuji	0–1 (32–34)	1–2	≤1.0	7–9
Gala	0–1 (32–34)	1.5–2.0	1–2	4–6
Golden Delicious	0–1 (32–33)	1–2	1–3	7–10
Granny Smith	0–2 (32–35)	1–2	≤1.0	7–11
McIntosh	2–3 (36–38)	1.5–2.5	1.5–4	5–8
Mutsu	0–1 (32–34)	1.5–2.0	1–3	6–9
Pink Lady	0 (32)	1.5–2.0	1	9
Red Delicious	0–1 (32–34)	1–2	1–3	7–11

Source: Kupferman 1997.

Figure 27.2

Storage scald on Granny Smith apples.

cooling rate if room cooling or forced-air cooling are used. A RH of 90 to 95% is considered ideal for apples, but it is difficult to maintain using large-surface cooling coils alone. Supplemental humidification, especially with fog spray nozzles, is now widely used. Unfortunately, the water added in such a system increases the coil defrost problem.

PHYSIOLOGICAL DISORDERS

STORAGE SCALD
A number of physiological disorders can cause serious losses in apples, especially after prolonged storage. Apple storage scald is most severe on less mature fruits, but on highly susceptible varieties, such as Granny Smith and Red Delicious, scald occurs even on fruit harvested at optimal maturity. Fruit grown in hot, dry climates, such as California, are more susceptible. The disorder appears as brown patches on the fruit surface that may not develop until the fruit are warmed after storage (fig. 27.2). In very severe cases, browning will be seen on the fruit in cold storage and will intensify when the fruit are warmed after storage. Thorough cooling and maintaining a low storage temperature are important in scald control. Often, an antioxidant such as diphenylamine (DPA) is applied as a bin drench (fig. 27.3), according to label directions, before storage to reduce the incidence and severity of scald. Controlled atmosphere storage can significantly delay scald development. Very low O$_2$ concentrations in CA storage (less than 1%) can provide control similar to DPA in some cases, although care must be taken to avoid low-O$_2$ injury to the fruit.

WATER CORE
The watersoaking symptoms of this disorder result from flooding of the intercellular spaces by a fruit solution high in sorbitol. Water core is often associated with low calcium levels in the fruit and is most severe in fruit from young, vigorous, lightly cropped trees. Red Delicious, Fuji, and Granny Smith are sensitive varieties. Fruit of advanced maturity are generally more severely affected. Fruit grown at high temperatures, especially if exposed to afternoon sun and cold nights, appear most prone to water core. Mild water core may disappear during short-term cold storage, but if severe, tissue may turn brown during storage. If harvested fruit is found to contain water core, fruit should be marketed as soon as possible, and fruit should not be stored for long periods. Cooling delays of 1 to 2 days may reduce the severity of water core symptoms and allow fruit to be stored in regular storage. Under no circumstances should water cored fruit be placed in controlled atmosphere storage. In California, water core has not been a significant problem in Fuji apples; however, in Washington, water core can be severe in Fuji.

BITTER PIT
Bitter pit symptoms are expressed by the development of dark, dry, pithy spots or pits near or below the fruit surface, especially at the calyx end (fig. 27.4). Bitter pit is another common cause of losses in apples. Lenticel blotch pit is a similar disorder with very shallow pits on Granny Smith apple. Tree

Figure 27.3

Apple fruit in bins being drenched with diphenylamine (DPA) just after harvest for control of storage scald.

Figure 27.4

Symptoms of bitter pit on Granny Smith apples. Pits can be of various sizes on the surface of the apple or just under the skin of the apple.

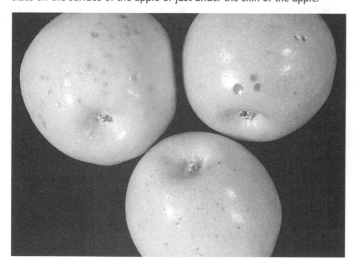

conditions that are associated with bitter pit include extreme vigor, light crop, and low fruit calcium content. Early-harvested, early-maturity fruit are most susceptible. The symptoms are sometimes visible at harvest but commonly develop during the first month of storage. Rapid cooling and high RH can reduce the development of symptoms during storage. Multiple calcium sprays on fruit before harvest, following label instructions, are the most effective treatment for bitter pit. Calcium application and rapid cooling following harvest may further reduce the problem. Application of pressure or vacuum, while dipping in calcium solution, has increased penetration into the flesh and

reduced bitter pit incidence in susceptible cultivars, but there may be an added risk of internal browning from waterlogged tissues, and decay.

INTERNAL BROWNING

Development of internal browning as a result of CO_2 injury has been reported in Fuji, Cox's Orange Pippin, Braeburn, and Jonathan apples. Brown discoloration occurs in the flesh, usually originating in or near the core, and the discoloration is firm, but moist. The brown areas have well-defined margins and may include dry cavities resulting from desiccation. Symptoms can range from a small spot of brown flesh up to nearly the entire apple's flesh being affected in severe cases. Symptoms develop during the first month of CA storage. This disorder is associated with later harvested, large fruit stored with high CO_2 concentrations; however, the incidence varies season to season and orchard to orchard. Because internal browning is not visible externally, the disorder may be detected by buyers and consumers. Nondestructive means of detection are currently under development.

BRUISING

Apples can be damaged by bruising. Roller bruising affects the surface of the fruit, resulting in brown areas or bands where rolling, rubbing, or vibration occurred, and damage is usually not apparent below the surface. Impact bruising affects the flesh, resulting in brown spots under the skin that may or may not be visible from the surface. The browning results from oxidation of phenols in the presence of an oxidizing enzyme, primarily polyphenoloxidase. Climatic differences in severity are apparently related to the level of phenols or the relative activity of the browning enzyme. Any mechanical injury can stimulate ethylene production, and thus speed fruit ripening and deterioration. Any surface damage can result in inoculation by fruit rot organisms and more rapid water loss from the fruit.

PATHOLOGICAL DISORDERS

GRAY MOLD AND BLUE MOLD

Gray mold (*Botrytis cinerea*) and blue mold (*Penicillium expansum*) can develop in wounds or at the stem or calyx end on apple fruit. Gray mold can be quite common after long-term storage. Sanitation in the field and packinghouse (especially water systems) and careful handling of fruit to avoid wounding are the best means of control. Temperature management will slow the growth of fungi that might be present. Fungicides added to the drench water can provide some protection; however some fungi have developed resistance to commonly used thiabendizole-type fungicides.

MUCOR ROT

Mucor rot has become a serious problem in Granny Smith and Fuji apples grown in California. The fungus, *Mucor piriformis*, lives in the orchard soil. Fruit collected from the orchard floor and soil attached to field bins transport the fungus into drench systems and dump tanks where healthy fruit can become infected. *Mucor* causes a very soft rot that can liquify fruit tissue, resulting in dripping of spores onto healthy fruit in the bin and thereby spreading of the disease in storage. While low temperatures greatly reduce its growth, this fungus continues to grow slowly at 0°C (32°F). Chlorine and most fungicides are not very effective against this organism. The best means of control are orchard sanitation and preventing orchard soil from entering the drench or dump tank water. Leave ground fruit in the orchard, place field bins on trailers in the field, and use a fresh water prerinse of loaded bins and trailer prior to the drench system.

CORE ROT

Core rot is a disease in which one or more of a series of fungi attack the core area of the apple fruit. The disease may begin in the field or during storage. Inoculation of the core area can occur during petal fall in the orchard or during drenching of the fruit. The disease may be restricted to the seed cavity or may involve the apple flesh and eventually the entire apple. Core rots present a marketing challenge because they are not visible externally, unless very severe. Efforts are under way to develop nondestructive methods for sorting out defective fruit during packing.

II. European Pears

MATURITY

In California, Bartlett pear standards utilize an index combining firmness and soluble solids content that is modified by fruit diameter and color (table 27.2). Most legal standards define minimum maturity; California also imposes a maximum maturity standard (minimum flesh firmness level) for Bartlett pears destined for processing. Many other pear varieties are picked on the basis of flesh firmness (table 27.3).

MATURITY VERSUS RIPENING

European pears that are allowed to ripen on the tree typically develop poor texture and lack juiciness and the characteristic flavor of the cultivar. Thus, they are harvested when physiologically mature but still quite firm, then ripened before processing or consumption. Freshly harvested Bartlett pears, especially those harvested early in the season, ripen slowly and unevenly and with poor

Table 27.2. California minimal maturity standards for Bartlett pears,

| Minimum soluble solids | Fruit firmness (lbf) | |
	Diameter 2⅜ to 2½ in (60.3 to 63.5 mm)	Diameter 2½ in (63.5 mm) and larger
< 10%	19.0 (84.5)	20.0 (89.0)
10%	20.0 (89.0)	21.0 (93.4)
11%	20.5 (91.2)	21.5 (95.6)
12%	21.0 (93.4)	22.0 (97.9)
13%	No maximum	No maximum

Source: California Pear Advisory Board, 1999

Table 27.3. Pear harvest firmness recommendations

| Cultivar | Flesh firmness (lbf)* | | |
	Maximum	Optimum	Minimum
Anjou	15	13	11
Bartlett	19	17	15
Bosc	16	13	11
Comice	13	11	9
Hardy	11	10	9
Kiefer	15	13–14	12
Seckel	18	16	14
Winter Nelis	15	12.5	11

Note: *Measurements with penetrometer using an 8-mm (⁵⁄₁₆-in) tip.
Conversions: lbf × 2.2 = kgf; lbf × 4.448 = Newtons.

eating quality if placed immediately into a ripening environment (13° to 24°C [55 to 75°F]) without prior time in cold storage. d'Anjou pears can remain hard and unripe for 50 to 60 days under the same conditions. However, the same fruit ripen quickly and uniformly if first stored at low temperature (–1° to 5°C [30° to 41°F]) for 2 to 8 weeks, depending upon the cultivar and harvest maturity. During cold storage, ethylene precursors form within the pear tissue so that when fruit are placed at ripening temperature they increase their ethylene production and ripen uniformly. Most European pears can be ripened without cold storage by applying ethylene to the freshly harvested fruit.

A conditioning treatment is often used for Bartlett and d'Anjou pears that are transported to market without sufficient cold storage to allow for uniform ripening. For Bartletts, warm, freshly harvested fruit are packed, exposed to 100 ppm ethylene at 20° to 25°C (68° to 77°F) for 24 hours, then forced-air cooled and transported to market. The fruit lose little flesh firmness during the treatment and do not become more subject to handling damage, but they ripen quickly and uniformly when warmed during marketing. d'Anjou pears are treated with ethylene for 2 to 3 days at near 20°C (68°F), with a goal of softening the pears to 11 to 12 lbf firmness prior to recooling and shipping, and these pears are not intended for long storage.

HANDLING

European pears are harvested by hand into large field bins. Most California Bartlett pears are packed within 2 days of harvest. However, in other growing areas, pears are frequently stored for weeks to many months in regular cold storage or CA storage prior to packing. Upon packing, fruit are dumped from bins using either dry, wet, or water immersion dumps. A presizer is used to remove undersized fruit to processing outlets. Most sorting is by hand, particularly for defects. Fruit are segregated into fresh and processing grades based on shape and surface defects. Frequently, over 50% of California Bartlett pears are diverted from a fresh-market sorting line to processing. (Some Bartlett pears are grown strictly for processing.) Sizing is accomplished with weight sizers or optical sizers.

Pears may be wrap-packed, tight-fill packed, tray-packed, or bagged. A large percentage of fresh-market California Bartlett pears are tight-fill packed into corrugated cartons. The remaining California Bartletts and most other pear varieties are either wrap-packed or tray-packed into corrugated cartons, although some wood cartons are still in use, especially internationally. A number of single or double layer cartons, some corrugated and others of reusable plastic, are gaining popularity. The corrugated cartons used in California are well ventilated to allow for efficient forced-air cooling; however, pear boxes from many other growing areas are not vented and may include a plastic liner. Bagging of pears is also increasing in popularity, particularly for marketing at consumer warehouse stores. A flow diagram for the handling of Bartlett pears (see fig. 27.1) shows the steps involved in fresh-market handling and processing of this commodity.

STORAGE

TEMPERATURE MANAGEMENT
Because pears are so sensitive to temperature effects, immediately following packing, California Bartlett pears are forced-air cooled prior to shipment or storage. For Bartlett pears, the recommended storage temperature is 0° to 0.5°C (32° to 33°F) for up to 1 month of storage and −1°C (30°F) if fruit is to be stored more than one month. The lowest safe storage temperature is related to the extent of temperature fluctuation in the room and the soluble solids content of the fruit. Freezing injury usually occurs first in the smallest pears with the lowest soluble solids content. Therefore, knowledge of the soluble solids content of the core tissue is necessary to predict the lowest safe storage temperature. Bartlett pears in California should be cooled to near storage temperature within 24 hours of harvest. The estimated maximum storage life for European pear cultivars stored in air at −1°C (30°F) is shown in table 27.4.

REDUCING WATER LOSS
Maintaining 95% RH in the storage room will prevent loss of moisture from the fruit during storage. In many commercial cold storages, this level of humidity is difficult to maintain. Humidity should be considered in the design and sizing of the refrigeration system. Even with the best system design, supplemental humidity may need to be added. During bin storage, wetting of wooden bins prior to loading, use of plastic bins, and use of plastic bin liners can help to reduce water loss from fruit during storage.

Using perforated polyethylene bags or liners within the packed box to maintain humidity, once a common practice, is now seldom done in California because it interferes with temperature management; but it is still used in many other growing areas. Wax or plastic coatings on corrugated containers are extensively used, when polyethylene liners are not used, to provide a moisture barrier for pear fruit.

CONTROLLED ATMOSPHERE STORAGE
Pears are second only to apples in the volume stored under controlled atmospheres (CA). Although CA storage of pears is used to a limited extent in California for pears, it is very common in other growing areas. Maintenance of the proper concentrations of O_2 and CO_2 are critical to successful CA storage, as is storage of fruit at the proper stage of maturity. Overmature fruit harvested toward the end of the harvest season is more subject to low O_2 or high CO_2 damage. Fruit harvested early in the harvest season is the best candidate for long-term CA storage. Recommended atmospheres for pear varieties are shown in table 27.4.

PHYSIOLOGICAL DISORDERS

SUPERFICIAL SCALD
D'Anjou, Packham, and Bartlett pears are susceptible to superficial scald, a disorder similar to storage scald in apples. Superficial scald develops during cold storage, but it may not become visible until after fruit are transferred to warm temperatures for ripening. It is a surface disorder that results in browning of the skin. Unlike storage scald in apples, pear susceptibility to superficial scald appears to be less dependent on fruit maturity. Treatment of freshly harvested fruit with an antioxidant such as Ethoxyquin can greatly reduce the incidence of superficial scald. Ethoxyquin is not registered (neither is DPA) for use on pears in California, but it is used in other growing

Table 27.4. Storage conditions and estimated maximum postharvest life of European and Asian pears stored in air or CA at −1° to 0°C (30° to 32°F)

Cultivar	Months in air	% O_2	% CO_2	Months in CA
European Pears				
Anjou	6–7	1–2.5	0–0.5	7–8
Bartlett	2–3	1–2	0–3	3–5
Bosc	3–4	1–2.5	0.5–1.5	4–8
Comice	4	1.5–4	0.5–4	5–6
Forelle	4–5	1.5	0–1.5	6–7
Hardy	2–4	2–3	3–5	4–6
Packham	5–6	1.5–1.8	1.5–2.5	7–9
Asian Pears				
Chojuro	2–3	2	1–2	3–4
Kosui	2–3	1–2	0–2	3–4
Tsu Li	2–4	1–2	0–3	3–5
20th Century	3–4	1–3	0–1	5
Ya Li*	2–3	4–5	0–3	3–4

Source: Adapted from Hardenburg et al. 1986; Richardson and Kupferman 1997.

*Ya Li may show chilling injury at temperatures less than 5°C (41°F).

areas. CA storage can reduce the incidence of storage scald, especially when low O_2 is used (1%), but with low O_2, care should be taken to keep CO_2 below 0.5%. More research is needed to address fruit tolerance issues as well as the effectiveness of low-O_2 CA for scald control.

SENESCENT SCALD

Senescent scald develops in pears that have been stored beyond their potential postharvest life. The background color of scalded fruit changes from green to yellow during storage, and the fruit lose their capacity to ripen. Fruit from very early or late harvest, fruit suffering delayed cooling, and fruit held at too high a storage temperature are all more susceptible to senescent scald. Symptoms begin on the fruit surface but can progress into the flesh during ripening. Proper harvest maturity, good temperature management, and avoiding storing for too long are all important in minimizing senescent scald of pears. CA storage can delay fruit senescence and thus delay development of senescent scald.

CORK SPOT

Similar to bitter pit in apples, cork spot occurs in tissue low in calcium or with a high potassium to calcium ratio. Patches of corky brown tissue, usually near the calyx end, occur either just beneath the skin or deep in the flesh. Affected pears become partly yellow and exhibit premature ripening. d'Anjou and Packham's Triumph pears are highly susceptible. Preharvest calcium sprays are beneficial, especially late in the season, but there is a risk of phytotoxicity.

CARBON DIOXIDE OR LOW OXYGEN INJURY

Pears are much less tolerant of CO_2 than apples. CO_2 injury is evident as browning in the core area. Cavities may develop in the browned tissues. The susceptibility of pears to CO_2 injury increases with advanced maturity, delayed storage, and low O_2 levels in the atmosphere. Fruit grown in a cool season have a greater susceptibility to CO_2 injury. Low-O_2 injury can occur at less than 1% O_2 and is manifested as a browning in the core area that may be tinged with pink. Fruit with advanced maturity are more susceptible.

CORE BREAKDOWN

Pears can suffer losses due to various core breakdown problems (sometimes called internal breakdown or brown core). Symptoms are browning and softening of the tissue in and around the fruit core (fig. 27.5). Late-harvested fruit are most susceptible to these disorders. Symptoms may develop in storage or during subsequent ripening. A range of similar symptoms associated with specific cultivars or handling and storage treatments have been described by various researchers.

Flesh breakdown that develops during storage and ripening can be reduced by good postharvest temperature management. This should include rapid movement from tree to cooler, rapid cooling, and maintenance of the proper low storage temperature. When climatic, cultural, or maturity factors predispose the fruit to breakdown, the problem may be unavoidable; however, storage life may be lengthened and intensity of the disorder can be reduced by good temperature management. In some tests with watery breakdown of Bartlett pears in California, late-season, high-maturity pears developed no breakdown if fruit was cooled within 1

Figure 27.5

Core breakdown in senescent Bartlett pear.

day of harvest, as compared with 5% breakdown with 2-day cooling, and over 10% breakdown with 3-day cooling. In other tests with similar pears that were cooled within 1 day of harvest, stored for 5 weeks, and then ripened, breakdown incidence was 0% after –1°C (30°F) storage, almost 2% after 0.5°C (33°F) storage, and about 3.5% after 2°C (36°F) storage.

PREMATURE RIPENING

Reports from cool-climate production areas describe a fruit breakdown in Bartlett pears caused by premature ripening. In Oregon, this problem can develop when daytime high temperatures do not exceed 21°C (70°F) and night temperatures are no higher than 7°C (45°F) for 2 days, 10°C (50°F) for 9 days, or 13°C (55°F) for 21 days. Affected fruit develop a pink color around the calyx (blossom) end. Night temperatures above 13°C (55°F) or day temperatures above 32°C (90°F) prevent premature ripening. Bartlett pears grown in cool climates also tend to soften faster and be more susceptible to core breakdown. By contrast, d'Anjou pears grown in

warm climates have a higher incidence of mealy breakdown of the flesh.

WATERY BREAKDOWN

A problem associated especially with Bartlett pear, watery breakdown involves soft, watery breakdown in portions of the fruit, usually without brown discoloration during its early stages. This enzymatic softening can affect any part of the fruit and probably results from severe physiological stress on the fruit. In some seasons in California, this disorder is responsible for loss of as much as 10% of fruit destined for processing. Fast cooling and low storage temperatures (avoiding freezing) are effective in minimizing the problem.

ROLLER BRUISING (BRUSH BURN, FRICTION DISCOLORATION)

This can occur whenever fruit have freedom to move, rub, or vibrate against a hard surface such as bins or package surfaces, packing belts, or other fruit. Pears are generally more susceptible to roller bruising damage than are apples. As Bartlett and d'Anjou pears remain in storage, they normally become more susceptible to roller bruising injury, and if packaging does not occur within 1 to 2 weeks of harvest, special care must be taken to avoid fruit injury. Data on other cultivars are unavailable. Avoidance of roller bruising involves prevention of fruit rubbing or vibrating at all stages of handling from harvest to market.

IMPACT BRUISING

Injuries can follow any impact of the fruit, and incidence and severity of bruises increase with increasing height of drop of the fruit. Impact bruising is important because of its effect on fruit appearance, and because the injury induces higher respiratory activity of the tissue, thus reducing storage life. Some reports suggest that symptoms are less apparent in fruits that are impacted after CA storage.

III. Asian Pears

Asian pears comprise a large group of pears (*Pyrus serotina*) that are crisp in texture and ready to eat at harvest. Asian pears do not markedly change texture after harvest or storage, as do European pears. Hundreds of varieties are grown in Asia, and more than 25 varieties are known in California. Asian pears are also called apple pears, salad pears, Nashi (Japanese for "pear"), and Oriental, Chinese, or Japanese pear.

In the United States, most Asian pears are grown in California, Oregon, and Washington. There are three basic types: round or flat fruit with green to yellow skin (20th Century), round or flat fruit with bronze-colored skin and a light bronze russet (Shinko, Hosui), and pear-shaped fruit with green or russet skin (Ya Li).

HARVEST

Asian pears are harvested in California from mid-July through September, depending on the variety. Harvest maturity is usually assessed by background color and soluble solids content. Studies have shown that background color is best related to the storage potential of the fruit. Other studies have indicated that a combination of starch index, firmness, and soluble solids would provide a good index. Immature fruit usually have poor flavor quality, while overmature fruit are more susceptible to surface marking and storage disorders.

Asian pears appear highly susceptible to surface damage from abrasions to the skin. Careful handling from harvest to consumption is necessary to avoid unsightly fruit damage manifested as brown or black markings on the fruit surface. Pickers may need to wear rubber or soft cotton gloves to avoid fruit injury. Padded, clean picking buckets are essential, and efforts must be made to avoid stem punctures. Often, stems are clipped close to the stem bowl. It is recommended that Asian pears be handled similarly to tree-ripened soft fruit. Handling steps should be minimized. Often, fruit are sized by eye and packed by hand. Many Asian pears are wrapped, using paper or plastic materials, and some are packed into plastic trays or foam punnets. Fruit are often packed one or two layers deep, and padding is added to the box to secure the fruit and prevent movement during transit which would result in roller bruising.

STORAGE

Some varieties can be stored 1 to 3 months at 0°C (32°F) without problems. Hosui and Shinko became spongy and developed storage rots after 2½ months of storage at 0°C (32°F). After 4 months of storage, core browning developed.

Based on limited studies, it appears that the magnitude of CA benefits for Asian pears is cultivar-specific and is generally less than that for European pears and apples. CA may extend the storage life of some Asian pear cultivars by about 25% relative to storage in air. Recommended CA conditions are shown in table 27.4. Care should be taken to avoid low-O_2 or high-CO_2 injury, as described below.

Some cultivars, such as 20th Century, Kosui, and Niitaka, produce very little ethylene and are nonclimacteric. Other cultivars, such as Tsu Li, Ya Li, Chojuro, Shinsui, Kikusui, and Hosui, are climacteric and produce between 1 and 14 µl/kg·hr ethylene. Exposure of climacteric cultivars to greater than 1 ppm ethylene accelerates loss of green color and slightly increases softening at 20°C (68°F). The effects of ethylene at 0°C (32°F) are minimal.

PHYSIOLOGICAL DISORDERS

LOW OXYGEN INJURY
Discolored surface depressions resulting from exposing 20th Century pears to 1% O_2 for 4 months at 0°C (32°F) and from exposing Ya Li and Tsu Li pears to 1% O_2 for 2 months, 2% O_2 for 4 months, or 3% O_2 for 6 months at 0°C (32°F).

HIGH CARBON DIOXIDE INJURY
Exposure to high levels of CO_2 can cause fruit to develop core or medial flesh browning, which may contain cavities in severe cases as a result of tissue death and subsequent desiccation. Ya Li pears may exhibit CO_2 injury after exposure to 5% CO_2 for 6 weeks at 0°C (32°F).

FLESH SPOT DECAY OR INTERNAL BROWNING

Development of brown to dark brown water-soaked areas in the core and flesh occurs either on the tree or during the first 2 to 6 weeks of cold storage. Despite the name often used to describe this disorder, it is physiological in nature and not due to the activity of decay organisms. Harvesting fruit prior to the overmature stage is the best means of controlling this disorder. Studies with Ya Li and Seuri pears indicated that fruit should be harvested before they change color from green to light green-yellow. Fruit harvested yellow developed internal browning within 1 month of cold storage. Results from New Zealand indicate that delaying fruit cooling 36 to 48 hours after harvest reduces the incidence of flesh spot decay.

WATER CORE

A disorder similar to that which occurs in apples, glassy, water-soaked areas develop in the flesh either on the tree or, less commonly, in storage. The affected tissue may be clear and sweet-tasting or brown and bitter-tasting. Avoid harvesting overripe fruit to avoid this disorder.

CORE BROWNING

The core of the fruit turns brown or black and senesces while the flesh remains healthy. Overmature fruit are more prone; therefore harvest maturity is the best means of control. This disorder may be related to flesh spot decay.

REFERENCES

Blanpied, G. D. 1990. Controlled atmosphere storage of apples and pears. In M. Calderon and R. Barkai-Golan, eds., Food preservation by modified atmospheres. Boca Raton, FL: CRC Press. 265–299.

California Pear Advisory Board. 1999. Annual report. Sacramento, CA: California Pear Advisory Board. p. 39.

Cappellini, R. A., M. J. Ceponis, and G. W. Lightner. 1987. Disorders in apple and pear shipments to the New York market, 1972–1984. Plant Dis. 71:852–56.

Crisosto, C. H., D. Garner, G. M. Crisosto, G. S. Sibbett, and K. R. Day. 1994. Early harvest prevents internal browning in Asian pears. Calif. Agric. 48:17–19.

Ferguson, I. B., and C. B. Watkins. 1989. Bitter pit in apple fruit. Hort. Rev. 11:289–355.

Hardenburg, R. E., A. E. Watada, and C. Y. Wang. 1986. The commercial storage of fruits, vegetables, and florist and nursery stock. USDA Agric. Handb. 66. 130 pp.

Kingston, C. M. 1992. Maturity indices for apple and pear. Hort. Rev. 13:407–432.

Kupferman, E. 1997. Controlled atmosphere storage of apples. In Proc. Seventh International Controlled Atmosphere Research Conference. Vol. 2. Davis: Univ. Calif. Postharv. Hort. Ser. 16. 1–30.

Larsen, F. E., S. S. Higgins, M. E. Patterson, V. K. Jandhyala, and W. Nichols. 1993. Quality, maturity, and storage of Asian pears grown in Central Washington. J. Prod. Agric. 6:247–252.

Meheriuk, M., R. K. Prange, P. D. Lidster, and S. W. Porritt. 1994. Postharvest disorders of apples and pears. Ottawa: Agric. Canada. Publ. 1737/E. 67 pp.

Richardson, D. G., and E. Kupferman. 1997. Controlled atmosphere storage of pears. In Proc. Seventh International Controlled Atmosphere Research Conference. Vol. 2. Davis: Univ. Calif. Postharv. Hort. Ser. 16. 31–35.

Tyler, R. H., W. C. Micke, D. S. Brown, and F. G. Mitchell. 1983. Commercial apple growing in California. Oakland: Univ. Calif. Div. Ag. and Nat. Res. Leaflet 2456. 20 pp.

White, A. G., D. Cranwell, B. Drewitt, et al. 1990. Nashi: Asian Pear in New Zealand. Wellington: DSIR Publishing.

Wrolstad, R. E., P. B. Lombard, and D. G. Richardson. 1991. The pear. In N. A. M. Eskin, ed., Quality and preservation of fruits. Boca Raton, FL: CRC Press. 67–96.

28

Postharvest Handling Systems: Stone Fruits

I. Peach, Nectarine, and Plum

Carlos H. Crisosto and F. Gordon Mitchell

California is a major producer and shipper of peach, nectarine, and plum fruits in the United States. Fresh peaches, nectarines, and plums constitute the bulk of the California fresh tree-fruit industry, and current shipments approach 60 million 11.4-kilogram (25-lb) packages of more than 450 cultivars of these three kinds of stone fruits. In the San Joaquin Valley, harvest of early cultivars starts in late April, and harvest of late cultivars of nectarines, peaches, and plums is completed in early October. In recent years, a large increase in the production of white-flesh peach and nectarine cultivars has occurred. Peach, nectarine, and plum exports are mainly to Canada, Taiwan, Hong Kong, Mexico, Brazil, and New Zealand.

DETERIORATION PROBLEMS

Stone fruits are characteristically soft-fleshed and highly perishable, and they have a limited market life potential. Because of the large number of cultivars spanning a 5-month harvest season, long-term storage has historically not been a concern in California. Potential opportunities for export marketing, combined with the desire to store some late-season cultivars to extend the marketing season, has increased interest in procedures to extend postharvest life.

INTERNAL BREAKDOWN

The major physiological cause of deterioration in peach, nectarine, and plum is a low-temperature or chilling injury problem generically called internal breakdown (IB). The disorder can manifest itself as dry, mealy, woolly, or hard-textured fruit (not juicy); flesh or pit cavity browning; and flesh translucency usually radiating through the flesh from the pit (fig. 28.1). An intense red color in the flesh ("bleeding"), usually radiating from the pit, may be associated with this problem in some cultivars. In all of the cases, flavor is lost before visual symptoms are evident. However, there is large variability in internal breakdown susceptibility among peach, nectarine, and plum cultivars. In general, peach cultivars are more susceptible than nectarine and plum.

In susceptible cultivars, internal breakdown symptoms develop faster and more intensely when fruit are stored at temperatures between about 2° and 7°C (36° and 45°F) than when similar fruit are stored at 0°C (32°F) or below but above the freezing point (fig. 28.2). At the shipping point, fruit should be cooled and held near or below 0°C (32°F) if possible. During transportation, if IB-susceptible cultivars are exposed to approximately 5°C (41°F), it can significantly reduce their postharvest life.

Several treatments to delay and limit development of this disorder have been tested. Among them, preripening fruit before storage is used commercially within the United States. The success of CA treatment (6% O_2 with 17% CO_2) in ameliorating chilling injury depends on cultivar market life, fruit maturity, shipping time, and fruit size.

FRUIT DECAY ORGANISMS

Postharvest loss of stone fruits to decay-causing fungi is considered

Figure 28.1

Internal breakdown symptoms in peaches and nectarines include mealiness (left) and browning (right).

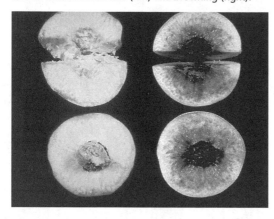

Figure 28.2

Storage temperature influences incidence and severity of internal breakdown in stone fruits.

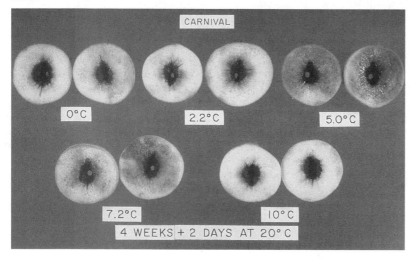

MECHANICAL INJURIES

Stone fruits are susceptible to mechanical injuries including impact, compression, and abrasion (or vibration) bruising. Careful handling during harvesting, hauling, and packing operations to minimize such injuries is important because the injuries result in reduced appearance quality, accelerated physiological activity, potentially more inoculation by fruit decay organisms, and greater water loss. Incidence of impact and compression bruising has become a greater concern as a large part of the California industry is harvesting fruit at more advanced maturity (softer) to maximize fruit flavor quality. Several surveys carried out in the Fresno area indicated that most impact bruising damage occurs during the packinghouse operation and during long transportation from orchard to packinghouse. Critical impact bruising thresholds (the minimum fruit firmness measured at the weakest point to tolerate impact abuse) has been developed for many peach, nectarine, and plum cultivars. Plums are less susceptible to impact bruising than peaches and nectarines.

Abrasion damage can occur at any time during postharvest handling. Protection against abrasion damage involves procedures to reduce vibrations during transport and handling by immobilizing the fruit. These procedures include installing air suspension systems on axles of field and highway trucks, using plastic film liners inside field bins, installing special top pads on bins before transport, avoiding abrasion on the packing line, and using packing procedures that immobilize the fruit within the shipping container before they are transported to market. Using half-size plastic bins can also reduce abrasion and impact damage.

If abrasion damage occurs during harvesting on fruit that have heavy metal contaminants, such as iron, copper and/or aluminum, on their skin, a dark discoloration is formed on the surface of peaches and nectarines (fig. 28.3). These dark or brown spots or stripes (inking or skin discoloration) are a cosmetic problem and a reason to cull the fruit. Heavy metal contaminants on the surface of fruit can occur as a consequence of foliar nutrients and/or fungicides applied within 15 days or 7 days before harvest, respectively. Light brown spots or stripes are also produced on the surface of white-flesh

the greatest deterioration problem. Worldwide, the most important pathogen of fresh stone fruits is Botrytis rot, caused by the fungus *Botrytis cinerea*. In California an even greater cause of loss is brown rot, caused by the fungus *Monilinia fructicola*. Good orchard sanitation practices and fungicide applications are essential to reduce these problems. It is common to use a postharvest fungicidal treatment against these diseases. An EPA-approved fungicide is often incorporated into a fruit wax for uniformity of application. Careful handling to minimize fruit injury, sanitation of packinghouse equipment, and rapid, thorough cooling to 0°C (32°F) as soon after harvest as possible are also important for effective disease suppression.

Figure 28.3

Surface discoloration (inking) on peaches and nectarines.

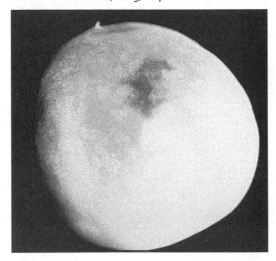

peaches and nectarines as a consequence of abrasion occurring mainly during harvesting and hauling operations.

WATER LOSS

Loss of approximately 5 to 8% of the fruit's water content, based on weight at harvest, can cause visual shrivel in peaches and nectarines. While there is a large variability in susceptibility to water loss among cultivars, all stone fruits must be protected to assure the best postharvest life. Fruit waxes that are commonly used as carriers for postharvest fungicides can reduce the rate of water loss when brushing is not excessive. Mineral oil waxes can potentially control the rate of water loss better than vegetable oil and edible coatings. Because fruit shrivel results from cumulative water loss throughout handling, it is important to maintain low water deficit conditions throughout harvesting, packing, storage, transport, and distribution. Short intervals between harvest and cooling, efficient waxing with gentle brushing, and fast cooling followed by storage under constant low temperature and high RH are the main ways of limiting water loss. Because of injuries caused on peaches when they are brushed to remove the trichomes ("fuzz"), this treatment increases water loss.

TEMPERATURE MANAGEMENT

Cooling requirements depend in part upon the scheduling of the packing operation. Fruit

can be cooled in field bins using forced-air cooling or hydrocooling. Forced-air cooling in side-vented bins can be by either the tunnel or the serpentine method (see chapter 11 for more details on cooling). Hydrocooling is normally done by a conveyor-type hydrocooler. Cooling of packed fruit is normally by forced air, using either the tunnel or cold wall method. Fruit in field bins can be cooled to intermediate temperatures (5° to 10°C [41° to 50°F]) provided that packing and subsequent forced-air cooling will occur the next day. If packing is to be delayed beyond the next day, fruit should be thoroughly cooled in the bins to near 0°C (32°F). With IB-susceptible cultivars, fast cooling within 8 hours and maintaining fruit temperature near 0°C (32°F) are recommended.

Fruit in packed containers should be cooled to near 0°C (32°F). Even fruit that were thoroughly cooled in the bins will warm substantially during packing and should be thoroughly recooled after packing. Forced-air cooling is normally indicated after packing. An exception to the need for cooling after packing would be a system that handles completely cooled fruit and provides protection against rewarming during packing.

Stone fruit storage and overseas shipments should be at or below 0°C (32°F). Maintaining these low temperatures requires knowledge of the freezing point of the fruit and knowledge of the temperature fluctuations in the storage system. Temperatures during truck transportation within the United States, Canada, and Mexico should be kept below 2°C (36°F). Holding stone fruits at these low temperatures minimizes the losses associated with rotting organisms, excessive softening, and water loss, as well as the deterioration resulting from internal breakdown in susceptible cultivars.

HARVEST MATURITY

The maturity at which stone fruits are harvested greatly influences their ultimate flavor and market life. Harvest maturity controls the fruit's flavor components, physiological deterioration problems, susceptibility to mechanical injuries, resistance to moisture loss, susceptibility to invasion by rot organisms, market life, and ability to ripen. Stone fruits that are harvested too soon (immature) may fail to ripen properly or may ripen abnormally. Immature fruit typically soften slowly and irregularly, never reaching the desired melting texture of fully matured fruit. The green ground color of fruit picked immature may never fully disappear.

Because immature fruit lack a fully developed surface cuticle, they are more susceptible to water loss than properly matured fruit. Immature and low-maturity fruit have lower soluble solids content and higher acids than properly matured fruit. This contributes to inadequate flavor development. Low-maturity fruit are more susceptible to abrasion and the development of internal breakdown symptoms than properly matured fruit.

Overmature fruit have a shortened postharvest life, primarily because they are already approaching a senescent stage at harvest. Such fruit have partially ripened, and the resulting flesh softening renders them highly susceptible to mechanical injury and microbial invasion. By the time these fruit reach the consumer they may have become overripe, with poor eating quality including off-flavors and mealy texture.

The optimal maturity for stone fruit harvest must be defined for each cultivar. The highest maturity at which a cultivar can be successfully harvested is limited by postharvest handling and temperature management procedures. Although maturity selection is more critical for distant markets than for local markets, this does not necessarily mean lower maturity for distant markets. When stone fruits are harvested at low maturity to reduce senescent breakdown problems during long-distance marketing, they become more susceptible to losses from internal breakdown.

Because of the availability of new cultivars that adapt well to harvesting at a more mature stage (softer), and because of the increase in demand for high-quality, less-firm fruit and the use of more sophisticated packinghouse equipment, a larger proportion of California stone fruits are being picked in a more advanced maturity stage.

FIELD HANDLING

Peaches, nectarines, and plums are hand-picked into bags, baskets, or totes (fig. 28.4). The fruit are dumped in bins carried by trailers in the orchard. Totes are placed directly inside the bins; baskets are placed on modified trailers. Fruit picked at advanced matu-

Figure 28.4

Postharvest handling of fresh-shipped stone fruits.

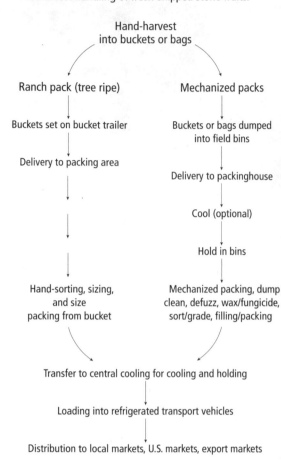

rity stages, and white-flesh peaches or nectarines, are generally picked and placed into baskets or totes. The fruit are hauled for short distances by trailers, but if the distance is longer than 10 km (6 mi), the bins are loaded on a truck for transportation to packinghouses.

Physical wounding of stone fruits can occur at any time from harvest until consumption. Good worker supervision helps assure adequate protection against impact bruising and mechanical damage during picking, handling, and transport of fruit. Protection against roller bruising may require modifications of transport equipment and procedures, as described in chapter 8. If severe injuries are encountered, consider using a top bin pad that maintains a slight tension across the top fruit. It is also helpful to grade farm roads to reduce roughness, avoid rough roads during transport, and establish strict speed limits for trucks operating between orchards and packinghouses.

Figure 28.5

Washing of stone fruits at the packinghouse.

PACKING AND HANDLING

Stone fruits are transported from orchard to packinghouse and cooler as soon as possible after harvest (see fig. 28.4). Fruit should be shaded during any delay between harvest and transport. Stone fruits are often cooled as soon as they arrive from the orchard, then packed cool the next day. Some fruits are packed upon arrival from the orchard and cooled immediately after packing. At the packinghouse the fruit are dumped (almost exclusively using dry bin dumps) and cleaned. Here trash is removed and fruit may be washed with detergent. Peaches are normally wet-brushed (fig. 28.5) to remove the trichomes (fuzz), which are single-cell extensions of epidermal cells. Waxing and fungicide treatment may follow. Water-emulsifiable waxes are normally used, and fungicides may be incorporated into the wax. Waxes are applied cold. In most of the "ranch pack" operations, peaches are not normally brushed.

Sorting is done to eliminate fruit with visual defects and sometimes to divert fruit of high surface color to a high-maturity pack (fig. 28.6). Attention to details of sorting line efficiency, as described in chapter 8, is especially important with stone fruits, where a range of fruit colors, sizes, and shapes can be encountered. Sizing segregates fruit by either weight or dimension. Sorting and sizing equipment must be flexible to efficiently handle a broad range of fruit sizes and volumes.

Most plums and some yellow-flesh peaches and nectarines are volume-fill packed,

Figure 28.6

Sorting plums by skin color at the packinghouse.

with the fruit automatically filled by weight into shipping containers. Most white-flesh peaches and "tree ripe" stone fruits are packed into one-layer (tray) boxes (flat). Yellow-flesh peaches, nectarines, and a few plums are packed into two-layer (trays) boxes. In some cases, mechanical place-packing units use hand-assisted fillers where the operator can control the belt speed to match the flow of fruit into plastic trays.

Limited volumes of stone fruits are "ranch packed" (tree ripe) at the point of production. In a typical ranch pack operation, fruit are picked into buckets or totes that are carried by trailer to the packing area. These packers work directly from the buckets to select, grade, size, and pack fruit into plastic trays (fig. 28.7). In these operations, the fruit are not washed, brushed, waxed, or fungicide treated. In other operations, fruit are picked into buckets or totes but then dumped into a smooth-operating, low-volume packing line for washing, brushing, waxing, sorting, and packaging. In both types of operations, because of the lesser amount of fruit handling, a higher maturity standard can be used, and growers can benefit from increased fruit size, red color, and greater yield. High-quality fruit can also be produced by managing the orchard factors (fruit thinning, girdling, fertilization, irrigation) properly and picking firm fruit. In this

Figure 28.7

Ranch packing system for stone fruits.

case, ripening at the shipping or retailer points is essential to assure good flavor quality for consumers.

RIPENING PROTOCOL FOR RECEIVERS

Stone fruits are usually harvested when they reach a minimum (or higher) maturity, but they are not completely ripe ("ready to eat") when harvested. Initiation of the ripening process must occur before consumption to satisfy consumers. This section describes a protocol designed to properly ripen peaches, nectarines, and plums at destination handling facilities.

1. **Check the initial fruit ripeness.** Flesh firmness is the best indicator of stone fruit ripeness and a good predictor of potential shelf life. A penetrometer with an 8 mm ($\frac{5}{16}$ in) tip is a quick and simple device for determining fruit firmness. Either a hand-held or drill-press-mounted instrument can be used. The drill-press-mounted instrument is recommended for greater accuracy and consistency. Fruit that reach 27 to 36 N (6 to 8 lbf) are considered "ready-to-buy." Fruit that reach approximately 9 to 13.5 N (2 to 3 lbf) flesh firmness are considered ripe ("ready-to-eat").

2. **Communicate with the merchandisers.** Find out the anticipated consumption schedule (fruit rotation schedule) and establish the ripening protocol accordingly. For IB-susceptible cultivars, fruits should be ripened down to 27 N (6 lbf) and then stored at 0°C (32°F) to avoid mealiness and flesh browning, if necessary.

3. **Determine the rate of softening.** Although the rate of fruit softening varies among cultivars, it is controlled by temperature. A high rate of softening is achieved at 20° to 25°C (68° to 77°F), and a low rate of softening is maintained by using lower temperatures. Temperatures higher than 25°C (77°F) will reduce the softening rate, induce off-flavors, and cause irregular ripening.

4. **Maintain fruit ripening conditions.** Most cultivars that are harvested at (or higher than) the California Well Mature stage do not need treatment with ethylene to ripen properly; the fruit softening rate depends only on temperature. Some plum, nectarine, and peach cultivars need exogenous ethylene treatment (10 to 100 ppm for 1 to 2 days) to initiate ripening. Maintain adequate air circulation within the ripening room to assure uniform fruit temperature. High RH (90 to 95%) around the fruit during the ripening process to prevent fruit shrivel is also necessary. In a tightly sealed room, it is important to assure that CO_2 produced by the fruit does not accumulate above 1% in the room. This can be done by continuously introducing fresh air or by periodically opening the room for an air change. For more details about management of fruit ripening, see chapters 16 and 21.

REFERENCES

Ceponis, M. J., R. A. Cappellini, J. M. Wells, and G. W. Lightner. 1987. Disorders in plum, peach and nectarine shipments to the New York market, 1972–1985. Plant Dis. 71:947–952.

Crisosto, C. H. 1994. Stone fruit maturity indices: A descriptive review. Postharv. News and Info. 5:65N–68N.

Crisosto, C. H., F. G. Mitchell, and S. Johnson. 1995. Factors in fresh market stone fruit quality. Postharv. News and Info. 6:17N–21N.

LaRue, J. H., and R. S. Johnson, eds. 1989. Peaches, plums, and nectarines: Growing and handling for fresh market. Oakland: Univ. Calif. Div. Ag. and Nat. Res. Publ. 3331. 246 pp. (See especially chapters 22 to 29.)

Lill, R. E., E. M. O'Donoghue, and G. A. King. 1989. Postharvest physiology of peaches and nectarines. Hortic. Rev. 11:413–452.

Mitchell, F. G. 1987. Influence of cooling and temperature maintenance on the quality of California grown stone fruit. Internat. J. Refrig. 10:77–81.

II. Apricots

Carlos H. Crisosto

Most of the apricots grown in the United States are grown in California, with much smaller amounts grown in Washington and Utah. The greatest hazard in handling or shipping apricots is decay, mainly brown rot and Rhizopus rot. Quick cooling of apricots to temperatures of 4°C (39°F) or lower and holding them as near 0°C (32°F) as possible will retard ripening (softening) and decay.

QUALITY AND MATURITY INDICES

Quality indices include fruit size, shape, and freedom from defects (including gel breakdown and pit burn) and decay. High consumer acceptance is attained for fruit with high (>10%) soluble solids content (SSC) and moderate acidity (0.7 to 1.0%). Apricots with 9 to 13.5 N (2 to 3 lbf) flesh firmness are considered "ready to eat." Most apricot cultivars soften very quickly, making them very susceptible to bruising and subsequent decay.

In California, apricot harvest date is determined by changes in skin ground color from green to yellow. The exact yellowish-green color depends on the cultivar and shipping distance. Apricots should be picked when still firm because of their high bruising susceptibility when fully ripe and soft.

PREPARATION FOR MARKET

Apricots are always harvested by hand, usually into picking bags or plastic totes. Apricots are generally handled in half bins or totes and hand-packed. In some cases, apricots are dry-dumped over a soft packing line. Apricots are packed in single- or two-layer trays, or volume-filled (about 10 kg, or 22 lb net). Apricots should be uniform in size, and not more than 5% by count of the apricots in each container may vary more than 6 mm (¼ in) when measured through the widest portion of the cross-section.

STORAGE CONDITIONS

TEMPERATURE AND RELATIVE HUMIDITY

Apricots are seldom stored in large quantities, although they keep well for 1 to 2 weeks, or possibly even 3 to 4 weeks, depending on the cultivar, at –0.5°C to 0°C (31° to 32°F) and a RH of 90 to 95% RH. Susceptibility of cultivars to freezing injury depends on SSC, which may vary from 10 to 14%. The highest freezing point is –1.0°C (30°F).

Chilling-sensitive cultivars develop chilling injury symptoms (gel breakdown, flesh browning, loss of flavor) more rapidly at 5°C (41°F) than at 0°C (32°F). Storage at 0°C (32°F) is necessary to minimize incidence and severity of chilling injury on susceptible cultivars.

CONTROLLED ATMOSPHERES (CA)

The major benefits of CA during storage or shipment are to retain fruit firmness and ground color. CA conditions of 2 to 3% O_2 and 2 to 3% CO_2 are suggested for moderate commercial benefits; the extent of benefits depends on the cultivar. Exposure to less than 1% O_2 may result in development of off-flavors; exposure to greater than 5% CO_2 for longer than 2 weeks can cause flesh browning and loss of flavor. The addition of 5 to 10% CO_2 as a fungistat during transport (less than 2 weeks), may improve the potential for benefit from CA. Prestorage treatment with 20% CO_2 for 2 days may reduce incidence of decay during subsequent transport or storage in CA or air.

ETHYLENE PRODUCTION AND EFFECTS

Ethylene production rates increase with ripening and storage temperature (<0.1 µl/kg·hr at 0°C [32°F] to 4 to 6 µl/kg·hr at 20°C [68°F]) for firm-ripe apricots and higher ethylene production rates for soft-ripe apricots. Exposure to ethylene at high temperatures may reduce ripening variability, as indicated by softening and color changes from green to yellow.

DISORDERS AND DISEASES

PHYSIOLOGICAL DISORDERS

Gel breakdown or chilling injury. This physiological problem is characterized in the earlier stages by the formation of water-soaked areas that subsequently turn brown. Breakdown of tissue is sometimes accompanied by sponginess and gel formation. Fruit stored at 2.2° to 7.6°C (36° to 46°F) have short market life and lose flavor.

Pit Burn. Flesh tissue around the stone softens and turns brown when the apricots are exposed to temperatures above 38°C (101°F) before harvest. This heat injury increases with higher temperatures and longer durations of exposure.

POSTHARVEST DISEASES

Brown rot. This fungal disease, caused by *Monilinia fructicola*, is the most important postharvest disease of apricot. Infection begins during flowering. Fruit rots may occur before harvest, but they often occur postharvest. Orchard sanitation to minimize infection sources, preharvest fungicide application, and prompt cooling after harvest are among the control strategies.

Rhizopus rot. A fungal disease caused by *Rhizopus stolonifer*, Rhizopus rot occurs frequently in ripe or near-ripe apricot fruits held at 20° to 25°C (68° to 77°F). Cooling the fruit and keeping them below 5°C (41°F) is very effective against this fungus.

REFERENCES

Andrich, G., and R. Fiorentini. 1986. Effects of controlled atmosphere on the storage of new apricot cultivars. J. Sci. Food Agr. 37:1203–1208.

Brecht, J. K., A. A. Kader, C. M. Heintz, and R. C. Norona. 1982. Controlled atmosphere and ethylene effects on quality of California canning apricots and clingstone peaches. J. Food Sci. 47:432–436.

Trurter, A. B., J. C. Combrink, and L. J. von Mollendorff. 1994. Controlled-atmosphere storage of apricots and nectarines. Decid. Fruit Grower 44:421–427.

III. Sweet Cherry

Elizabeth J. Mitcham and Carlos H. Crisosto

Approximately 20,000 hectares (50,000 acres) of sweet cherries are grown in the United States, mainly in California, Washington, and Oregon, producing about 150,000 metric tons (165,000 tons) annually. Approximately 45% of this production is sold fresh, while 35% is brined, and the remaining 20% is canned. The numerous varieties include red sweet cherries, such as Bing and Lambert; yellow sweet cherries, such as Rainier and Royal Anne; and the more heat-tolerant varieties Brooks, Tulare, King, and Garnet.

HARVEST MATURITY AND QUALITY

Being a nonclimacteric fruit, cherries must be harvested fully ripe for good flavor quality. While skin color may become darker red after harvest, sugar content does not increase. Therefore, fruit quality is at its optimum at harvest. Skin color and fruit soluble solids content are the main criteria used to judge fruit maturity and readiness for harvest. The U.S. Standards for maturity state that the fruit should be mature, having reached a stage of growth that will ensure the proper completion of ripening. In addition, the fruit should be "fairly well colored," meaning that at least 95% of the fruit surface should show the characteristic color for mature cherries of that variety. Minimum maturity in California requires that the entire cherry surface have a minimum of light red color and the fruit contain 14 to 16% soluble solids, depending on the variety. The red mahogany stage is recommended for harvest of Brooks, Garnet, Tulare, and King cherries. Fruit eating quality is related to the soluble solids content, titratable acidity, the ratio of soluble solids to titratable acidity, and fruit firmness. Visual quality factors include fruit size, color and luster, and absence of defects such as pitting, cracking, shriveling, decay, and misshapen fruit (doubles, spurs). Green, fleshy stems are often associated with freshness and quality.

HARVEST

Sweet cherries are hand-harvested, with stems attached, into small buckets that are transferred into field boxes or one-half-size field bins (100 by 120 by 60 cm, or 40 by 48 by 24 in). If field boxes are used they are loaded into one-half-size field bins for transport to the packing facility. Most harvesting is completed by midday to reduce the heat load in the fruit at harvest. Fruit maturity on a given tree can be quite variable. Some orchards are color-picked, with the ripest fruit being removed at an earlier pick and the bulk of the fruit removed in a second harvest, while other orchards are harvested in a single pick.

Supervision of the picking crew is essential to prevent excessive fruit injury. Pickers should grasp stems rather than fruit and remove clusters from the limb with an upward motion to leave the fruiting spur intact. Fruit should be gently placed (not thrown) into the picking bucket to avoid impact damage. Padding the bottom of the bucket can reduce fruit injury. Fruit should be kept in the shade while awaiting transport to the packinghouse.

PACKINGHOUSE OPERATIONS

If field bins are used, automatic bin dumps are employed. Water bath bin dumps are often used to reduce impact damage, but dry dumps are also used. If field boxes are used, these must be dumped by hand onto a conveyor. A flighted, inclined conveyor lifts fruit out of the water and conveys it under an air stream to remove leaves from fruit on the bar conveyor.

Because sweet cherries are sold based on size, the clusters must be cut in order to size the individual fruit. As fruit move by on a conveyor, tines catch the clusters of fruit and cause the stems to move into a saw, which cuts the stems to separate the fruit. The individual fruit then move through an eliminator, which consists of pairs of smooth, counter-rotating rollers that slope downward so that fruit slides between the rollers. The gap between the pairs of rollers increases slightly in the direction of fruit travel. Under-sized fruit pass between the rollers and are sent to processing outlets or discarded. The remaining fruit are transferred to sorting tables on conveyor belts or in water flumes. Sprays of water are used to lubricate the fruit on the cluster cutter and eliminator.

Fruit are conveyed over a flat belt past sorters who remove damaged, misshapen, or immature fruit. The fruit are then hydrocooled in an inline unit that removes heat by showering fruit with cold water or by moving immersed fruit through a cold water bath. Following hydrocooling, fruit are sized using a set of parallel rollers similar to the eliminator used to remove undersized fruit. Just prior to box filling, a fungicide may be sprayed onto the fruit as it passes by on conveyors. Fruit are generally bulk-filled into containers.

PACKAGING

Sweet cherries are commonly packaged into fiberboard boxes with plastic fold-over liners. Absorbent pads are used to prevent water from accumulating in the bottom of the liner. Boxes are hand-filled by being held under a belt carrying the fruit. The box may pass over a vibrating point in the belt to settle the fruit. Box weight and grade are checked and the box is sealed. Standard box size in the Pacific Northwest is 5.4 or 9.0 kg (12 or 20 lb). In California, the standard is 8.1 kg (18 lb). Fruit may also be packaged into clamshell containers, bags, or other types of consumer packages.

MODIFIED ATMOSPHERE PACKAGING

Fruit in boxes and in consumer packages can be exposed to a modified atmosphere (MA). Fruit can be sealed into MA bags within the fiberboard box. Bag sealing may involve pulling a vacuum on (partially evacuating) the bag of fruit and introducing a premixed atmosphere. Alternatively, the bag can simply be sealed, allowing the fruit to modify the atmosphere through respiration. Consumer packages can be covered with specialized films to achieve the desired package atmosphere. The goal is to achieve 10 to 15% CO_2 and 3 to 10% O_2 within the bag or package. MA packaging has been demonstrated to reduce the growth of decay organisms, reduce the rate of fruit deterioration, maintain soluble solids and titratable acidity concentrations, and maintain green stems. These benefits are more noticeable when

fruit are shipped under less than ideal temperatures (>2°C, or 36°F) and the time to destination exceeds 1 week. Flavor volatiles may be reduced following several weeks of CA storage, resulting in fruit of good visual quality but poor eating quality.

COOLING

Cherries should be cooled as rapidly as possible after harvest, and the temperature should be held as close as possible to 0°C (32°F) with 90 to 95% RH during handling, storing, and shipping.

The fruit may be hydrocooled at a field location when the orchard is far from the packinghouse and temperatures are high, or it may be hydrocooled upon arrival at the packinghouse. Cherries are commonly hydrocooled during the packing process. Shower coolers quickly remove heat from the fruit, with an 11°C (20°F) reduction in fruit temperature over a 7- to 10-minute period. Immersion coolers are slower to cool fruit and are often used for fruit that has previously been cooled. Chlorine (100 ppm) should be added to the water system to prevent cross-contamination with disease organisms in the water. Following packaging, cherries may be forced-air cooled to bring their temperature down to 0°C (32°F). Forced-air cooling is more common than hydrocooling in cherry growing areas in Europe. Forced-air cooling can result in fruit and stem desiccation, and the use of plastic liners within boxes reduces the efficacy of this method.

FUMIGATION

Sweet cherries exported to certain markets are currently fumigated with methyl bromide. For codling moth disinfestation, a 2-hour methyl bromide fumigation is required. The concentration of methyl bromide varies with the temperature. In general, methyl bromide treatment, especially at higher temperatures, results in increased fruit pitting and stem browning. In some operations, fruit for export to markets that require methyl bromide move through the packing line and are diverted to bin fillers after the sorting operation. In this way, only fruit of export quality are fumigated. Following fumigation and ventilation, fruit are transferred into a screened packinghouse for additional sorting, cooling, and final packaging. Fruit are generally packaged into unvented boxes to prevent reinfestation. Boxes with screened vents are also used.

SHIPPING

AIR
Thirty to forty percent of sweet cherries grown in the United States are shipped to export markets. A significant number of these, particularly from the early production areas in California, are shipped by air. While air shipment gets product to market rapidly, the breaks in the cold chain create challenges for maintenance of fruit quality. To combat these breaks in the cold chain, many air-shipped fruit are placed under reflective pallet covers with dry cold packs placed over the top of the pallet. This system appears to maintain cooler temperatures around the fruit during transit.

SEA CONTAINER
The remaining product that is exported overseas is shipped in marine containers, which should be set at 0°C (32°F) to maintain product quality during transit. Many marine containers have bottom airflow, and boxes should be designed with bottom and top vents at the corners to allow air to flow within the box and around the plastic liner. Because of the air restriction caused by the plastic liner, fruit must be thoroughly cooled prior to container loading. Containers can provide a MA for the product with similar benefits to those provided by MA packaging. The difference is that the atmosphere is lost when the container is opened at destination. In export markets where temperature management is very difficult, removal of the MA may be beneficial to prevent product damage caused by harmful atmospheres.

DOMESTIC TRUCK
Nearly all domestic cherries are shipped by truck. Most highway trucks have horizontal air delivery systems, and boxes should have side venting for this shipping method. Thermostats should be set at 0°C (32°F), but product temperature is most important and should be monitored. The loading pattern should allow for good air circulation within the truck.

PHYSICAL AND PHYSIOLOGICAL DISORDERS

Postharvest life of sweet cherries is closely related to the respiration rate of the fruit, which increases with higher temperature and fruit injury. Fruit injury also greatly increases the incidence of fruit decay. Studies have shown that approximately 30% of fruit is damaged upon arrival at the packinghouse, due to harvesting and field operations. Damage from packinghouse operations was shown to vary from 4 to 46%, depending on the operation. In packinghouse operations, damage can be minimized by eliminating drops onto rough surfaces, slowing fruit speed in cluster cutters, operating cluster cutters at high capacities, and reducing the water drop height within shower hydrocoolers to less than 20 cm (8 in).

PITTING
An unsightly indentation in the surface of the fruit is caused by the collapse and death of cells under the skin. These pits appear to result from impact injuries. Firm fruit and cold fruit appear to be more susceptible to pitting damage, but there is less agreement on the relationship between firmness and pitting. Pitting has been shown to be greater when fruit flesh is less than 10°C (50°F) during handling. Hydrocooled fruit may be more susceptible to pitting than air-cooled fruit. Darker-skinned, more mature cherries appear to be less susceptible to pitting injury.

BRUISING
Bruising can result from compression, vibration, or impact damage. Firm fruit and cold fruit are less susceptible to compression bruising damage. Vibration damage is not influenced by fruit temperature. Compression bruising is evidenced by a flat area on the fruit surface and soft tissue beneath that can be detected by feel. While bruises are quite apparent on yellow-skinned cherries, they may not be visible on black-skinned cherries.

CRACKING
Fruit cracking is usually associated with preharvest rains. Splits and cracks in the skin develop around the stem, on the shoulders, or at the blossom end. Cracked cherries are

much more susceptible to decay and deterioration and should be removed during sorting.

PATHOLOGICAL DISORDERS

Brown rot, caused by *Monilinia fructicola*, often begins in the orchard and manifests itself in the postharvest environment. The disease can also occur as a result of wounding and inoculation during postharvest handling. Preharvest and postharvest control measures are necessary. Gray mold, caused by *Botrytis cinerea*, can also be significant. This fungus is particularly troublesome because it cannot be completely controlled with temperature management, since it continues to grow even at 0°C (32°F). Blue mold, caused by *Penicillium expansum*, is an opportunistic pathogen that infects wounds on fruit. Rhizopus rot, caused by *Rhizopus stolonifer*, occurs when fruit are exposed to temperatures of 5°C (41°F) or higher.

Proper temperature management, including rapid cooling to and maintenance at 0°C (32°F), can completely control Rhizopus rot and can significantly reduce brown rot and gray mold. Eliminating injured and diseased fruit from the packed box is important to prevent spreading (nesting) of the disease. Preharvest and postharvest fungicide treatments are often beneficial.

REFERENCES

Brown, G. K., and G. Kollar. 1996. Harvesting and handling sour and sweet cherries for processing. In A. D. Webster and N. E. Looney, eds., Cherries: Crop physiology, production and uses. Wallingford, UK: CAB International. 443–469.

Crisosto, C. H., D. Garner, J. Doyle, and K. R. Day. 1993. Relationship between fruit respiration, bruising susceptibility, and temperature in sweet cherries. HortScience. 28:132–135.

Couey, H. M., and T. R. Wright. 1974. Impact bruising of sweet cherries related to temperature and fruit ripeness. HortScience. 9:586.

Facteau, T. J., and K. E. Rowe. 1979. Factors associated with surface pitting of sweet cherry. J. Amer. Soc. Hort. Sci. 104:706–710.

Looney, N. E., A. D. Webster, and E. M. Kupferman. 1996. Harvest and handling of sweet cherries for the fresh market. In A. D. Webster and N. E. Looney, eds., Cherries: Crop physiology, production and uses. Wallingford, UK: CAB International. 411–441.

Mitchell, F. G., G. Mayer, and A. A. Kader. 1980. Injuries cause deterioration of sweet cherries. Calif. Agric. 34(4): 14–15.

Thompson, J. T., J. Grant, E. M. Kupferman, and J. Knutson. 1997. Reducing sweet cherry damage in postharvest operations. HortTechnology. 7:134–138.

29

Postharvest Handling Systems: Small Fruits

I. Table Grapes

Carlos H. Crisosto and F. Gordon Mitchell

The table grape is a nonclimacteric fruit with a relatively low rate of physiological activity. The fruit is subject to significant water loss following harvest, which can result in stem drying and browning, berry shatter, and even wilting and shriveling of the berries. Gray mold, caused by the fungus *Botrytis cinerea*, requires constant attention and treatment during storage and handling. The bloom (natural wax) on the grape berry's surface is a primary appearance quality factor. Rough handling and rubbing destroys this bloom, giving the skin a shine rather than the more desirable luster appearance.

CULTIVARS

In California the major cultivars are Thompson Seedless (Sultanina) and Flame Seedless, marketed mostly during the summer months up to 8 to 10 weeks after harvest. Perlette is still important in the early production area of the Coachella Valley in California, and other seedless cultivars such as Ruby Seedless, Sugraone (Superior Seedless), and Crimson Seedless make up the bulk of the remaining production. There is also increasing production of recently introduced seedless cultivars including Autumn Royal and Princess. The seeded Redglobe cultivar is important for export in the mid-to-late season. Little is known about the specific postharvest requirements of these new cultivars.

MATURITY

Grapes are harvested when mature, based upon the soluble solids concentration of the berries. Titratable acidity and sugar to acid ratio are also used as maturity indices (e.g., Thompson Seedless, 18:1 sugar to acid ratio). The minimum requirements vary with cultivar and growing area. Cultivars other than green-colored ones also have minimum color requirements based on the percentage of berries in the cluster that show a certain minimum color intensity and coverage. Detailed information on maturity requirements according to cultivars is presented in chapter 23.

WATER LOSS

In general, cumulative water loss during postharvest handling results in weight loss, stem browning, berry shatter, and shriveling of berries. In all of the cultivars studied, there is a high correlation between cluster water loss and stem browning. The high rate of respiration of stems may also be a contributor to stem browning, as the respiration rate of the stems may be 15 times or more that of berries. When water loss reaches 2.0% or more for Perlette, Flame Seedless, Thompson Seedless, Ruby Seedless, and Fantasy Seedless, stems will show symptoms of browning approximately 7 days later in storage. A survey indicated that water loss ranged from 0.5 to 2.1%, based on the initial weight (measured at harvest) within the 8-hour period before cooling. The magnitude of the losses was directly related to the length of exposure, temperature during the delay before cooling, and type of box material. Even a few hours delay at high temperatures can cause severe drying and browning of cluster stems (fig. 29.1), especially on the hottest

Figure 29.1

Cumulative water loss during postharvest handling results in weight loss, stem browning, and shriveling of berries. Even a few hours delay at high temperatures can cause severe stem drying and browning, especially on the hottest days.

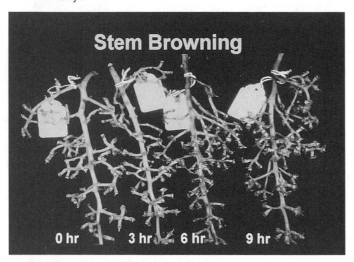

Figure 29.2

Sulfur dioxide dosimeter tubes used to monitor sulfur dioxide concentration during fumigation of table grapes.

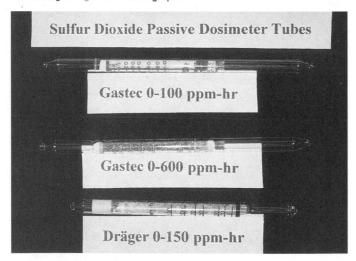

days. Water loss during storage can reach 2.5% for plain packed grapes. (A box is "plain packed" when fruit is placed cluster to cluster in the box until the appropriate net weight is attained.) The use of cluster bags reduced water losses in all three of the containers tested during their postharvest life (Crisosto et al. 2001).

Because stems and fruit are susceptible to deterioration from water loss, grapes are normally forced-air cooled as soon as possible after harvest. Grapes do not tolerate the wetting associated with hydrocooling.

FRUIT ROTS

GRAY MOLD

Control of the fungus *Botrytis cinerea*, which causes Botrytis rot (gray mold), requires constant attention and treatment during storage and handling. Gray mold is the most aggressive postharvest disease of table grapes because of its ability to develop at temperatures as low as –0.5°C (31°F) and move by mycelial growth from berry to berry. Botrytis rot can be identified by the characteristic "slipskin" condition that develops, and later, by "nests" of decayed berries encased in white mycelium. Botrytis rot of grapes is not sufficiently avoided by fast cooling alone. It is standard practice in California and other production areas to fumigate with sulfur dioxide (SO_2) immediately after packing, followed by lower-dose SO_2 treatments weekly during storage. An exception is for grapes produced in the Coachella Valley, which are marketed soon after harvest. Formulas for calculating SO_2 fumigation dosages are available in the publications by Nelson (1985) and Luvisi et al. (1992). Because of the recent increased interest in the export market, there is a greater demand for the use of SO_2-generating pads, especially for long-distance transport. Sodium metabisulfite is incorporated into the pads, allowing the release of SO_2 during transit and marketing.

One problem associated with SO_2 fumigation of grapes is the constant potential for injury to the fruit and stems. Injured tissue first shows bleaching of color, followed by sunken areas where accelerated water loss has occurred. These injuries first appear on the berry where some other injury has occurred, such as a harvest wound, transit injury, or breakage at the cap stem attachment. Symptoms may also be seen around the cap stem, and slowly spread over the berry. Careful attention to SO_2 treatment procedures is necessary to minimize this damage. Another problem with SO_2 fumigation of grapes is the level of sulfite residue remaining at time of final sale. Sulfur dioxide was once included on the list of "Generally Recognized As Safe" (GRAS) chemicals, for which no registration is required. Heavy usage of sulfites in some other foods has caused a change in regulation, because some people are highly allergic to sulfites. Sulfite residues in grapes are currently limited to less than 10 ppm, and there are limits on the number of repeat SO₂ fumigations allowed, depending upon cultivar.

Recently it has been demonstrated that the amount of sulfur dioxide gas needed to kill *Botrytis* spores, or to inactivate exposed mycelium, is dependent on the SO₂ concentration and fumigation time. A cumulative concentration, calculated as the product of the concentration and exposure time, called the CT product, describes the sulfur dioxide exposure needed to kill *Botrytis cinerea*. A CT of at least 100 ppm-hour is the minimum required to kill spores and mycelium of *Botrytis* at 0°C (32°F), or approximately 30 ppm-hour at 20°C (68°F). The CT-100 dose can be obtained with an average concentration of either 100 ppm for 1 hour, 200 ppm for ½ hour, 50 ppm for 2 hours, or an equivalent combination of concentration and time. This finding was the basis for the development of the total utilization system.

The total utilization system differs from the traditional system in that there is no

Figure 29.3

A slow-release SO₂-generating pad combined with a perforated or microperforated polyethylene box liner is used to control decay and reduce water loss during transport of table grapes.

excess SO₂ fumigant at the end of the fumigation treatment, reducing both air pollution and sulfite residues in the fruit. It can be used with forced-air cooling for initial fumigation and in cold storage for subsequent periodic treatments. Total utilization typically uses about half as much SO₂ as the traditional method, and improves the uniformity and effectiveness of the SO₂ fumigant.

THE TOTAL UTILIZATION SYSTEM

Initial fumigation. The first fumigation is done in conjunction with forced-air cooling. The forced air flows through the boxes and ensures good penetration of SO₂, even to the center boxes within a pallet. In most combinations of boxes and packs, this system produces over 80% penetration, measured as percentage of the room air CT product.

Passive fumigation. This fumigation process is applied every 7 to 10 days. After SO₂ application in the room, fans should run at high speed for over 3 hours so that nearly all of the SO₂ is absorbed by the fruit, packaging materials, and room surfaces. At the end of fumigation, the concentration of SO₂ in the room air should be less than 2 to 5 ppm and no venting or scrubbing is needed. In this system, each cold storage room should be calibrated to determine the amount of SO₂ to use. Center boxes within a pallet have lower SO₂ exposures than corner boxes, and pallets closest to the SO₂ inlet have higher fumigant exposures than those farthest away. To check fumigant penetration and distribution, inexpensive SO₂ dosimeter tubes are available. These dosimeters were originally designed for human safety monitoring. There is a large difference in SO₂ penetration according to box materials. For example, SO₂ penetration is higher in EPS boxes than wood-end and corrugated boxes; SO₂ penetration in corrugated is lower than in wood-end boxes.

Dosimeters designated for SO₂ fumigation doses at marked levels from 0 to 150, 0 to 100, and 600 ppm-hr are available (fig. 29.2). These dosimeter tubes work well for measuring the SO₂ CT product inside packed grape boxes. The glass dosimeter tubes are placed in the center of the boxes inside tissue wraps of cluster bags if these are present, and usually in boxes located in the center of the pallets. After fumigation the tubes are removed promptly and the ppm-hr exposure to SO₂ is

Figure 29.4

Postharvest handling of table grapes.

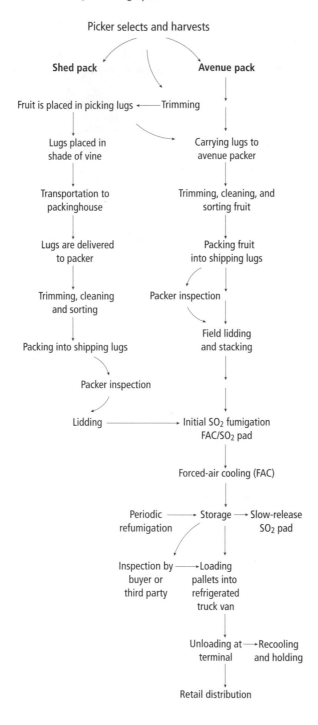

protected from decay but not exposed to fumigant levels that might cause excessive residues and bleaching. For details on this work see Luvisi et al. (1992).

Shipment. During ocean shipment for periods longer than 10 days or long retail handling in which SO_2 fumigation cannot be applied, the use of SO_2-generating pads in combination with a box plastic liner is advised (fig. 29.3). These SO_2-generating pads have sodium metabisulfite incorporated into them to allow a constant and slow release of SO_2 during shipment and marketing.

The slow-release SO_2-generating pad combined with a perforated polyethylene box liner (6.5 mm hole, 7.5 cm center, or ¼ in. hole, 3 in. center) reduces water loss and assures *Botrytis* control without enhancing bleaching (SO_2 phytotoxicity). This perforated box liner did not interfere with the initial fumigation (Crisosto et al. 1994, 2000). Initial fumigation is essential to control *Botrytis* during long-term storage or shipping. Water loss in the field and during storage or shipping was greatly reduced using this perforated box liner.

PACKING SYSTEMS

The two methods of handling table grapes in California are summarized in figure 29.4. Most California table grapes are packed in the field. In contrast to South Africa and Chile, few grapes are shed-packed.

FIELD PACKING

The most common field-packing system is the "avenue pack" (fig. 29.5). The fruit is picked and placed into shallow plastic picking lugs. Usually, the picker also trims the cluster. The picking lug is then transferred a short distance to the packer, who works at a small, shaded portable stand in the avenue between vineyard blocks. It is common for the packer and several pickers to work as a crew. Packing materials are located at the packing stand, which also shades the packer. With many packing stands around the vineyard, supervision is more difficult than in a packing shed. Lidding is done in the field. Substandard fruit can be accumulated in field lugs for transport to wineries or other processors.

SHED PACKING

Shed-packed fruit is harvested by pickers and placed in field lugs without trimming,

recorded. The tubes should be read promptly because some tubes can overestimate the dosage if examining their color reaction is delayed. A dose of 100 ppm-hr is the minimum adequate dose. This allows the operator to adjust the amount of fumigant applied to ensure that most boxes are adequately

Figure 29.5

"Avenue pack" of table grapes.

Table 29.1. Advantages and disadvantages of corrugated, wood-end, and foam boxes for packing table grapes

Attribute	Type of box*		
	Wood-end	Corrugated	Foam
Historical identification as a grape box	+++	+	+
Pleasant appearance and presentation of the fruit	+++	+	+++
Most favorable cost	+	+++	++
Ease of storing empty boxes and logistics of transporting to field	+	+++	+
Ease of packing	+	+	+
Protection against heat gain in the field	+	+	++
Protection of the fruit against shatter	+	+	+++
Protection of the fruit against bruising and splitting	+	+	+++
Stability of columns of boxes prior to palletizing	+++	+	++
Firmness of interlock for stability in the pallet	+	+	+++
Penetrability of palletized boxes to circulating air and fumigant	++	+	+++
Light weight during handling and shipping	+	++	+++
Conservation of storage and shipping space due to small bulk of box	+++	+++	+
Stability of the box in multiple pallet stacks	+++	+	++
Stability of the box during high humidity cold storage	+++	−	+++
Perceived (by receivers and public):			
Environmental acceptability	++	+++	−
Potential for recycling	+++	+++	+++
Actual amount of recycling (1995)	−	++	−

Source: Luvisi et al. 1995.

Note: *+++ = most or best attribute; ++ = intermediate attribute; + = average attribute; − = negative attribute.

then placed in the shade of the vines to await transport to the shed. At the packing shed the field lugs are distributed to packers, who select, trim, and pack the fruit. Generally, there are two grades packed simultaneously by each packer to facilitate quality selection. In some operations, trimming, color sorting, and a first quality sorting may have occurred in the field.

In all of the systems, grapes are nearly always packed on a scale to facilitate packing to a precise net weight, whether field or shed packed. In general, mid- and late-season grapes are packed in plastic bags or wrapped in paper. For early-season grapes, bulk pack is mainly used. In all cases, packed lugs are subject to quality inspection and weight checking.

PACKAGES

Three types of box materials are available to the California industry: corrugated, the technical Kraft Veneer (TKV, or wooded end), and the expanded polystyrene (EPS or Styrofoam). The advantages and disadvantages of these materials are summarized in table 29.1. In general, the use of corrugated has increased, especially for earlier cultivars and for fruit that are not stored for long periods. The use of the EPS container is becoming more popular for late cultivars and long storage periods. The choice of box material is often influenced by factors other than maintaining the quality of packed fruit (see table 29.1), such as the preferences of the receiver, environmental issues (recycling), cost, cold storage humidity conditions, storage length period, box weight, etc. TKV and foam boxes are mainly used for long storage periods because they maintain their structural integrity in high humidity conditions better than corrugated boxes. Approximately 0.1% of the total grape production is packed in clamshell containers.

Mainly due to retail pressures to use the 100 by 120 cm (40 by 48 in) pallet, the range of box sizes is diversified well beyond the standard LA lug (14 by 17.5 in). Since 1994, the MUM (35 by 40 cm, or 14 by 16 in), metric (40 by 50 cm, or 16 by 20 in), and shoe box (30 by 50 cm, or 12 by 20 in) containers have been used by the California industry. Detailed studies of the relationship between pack volume and packing height in the box versus grape quality have been carried out for the different box materials and sizes (see Luvisi et al. 1995).

In the last 6 years, different inner-packaging styles have been developed, driven mainly by retailers. During the 1998–99 season, approximately 70% of the total crop was packed using plastic cluster bags, approximately 20% was plain-packed (no inner packaging), and approximately 7% was wrap-packed. Plastic cluster bags provide consumer-size units and reduce the drop of loose berries onto produce department floors. The use of the plastic cluster bags has been reported to greatly reduce fruit damage during marketing (Luvisi 1992). Recent work has developed a cluster bag ventilation system (patented in 2000) that restricts water loss and slows drying and shriveling of the fruit and stems (Crisosto et al. 2001). In general, the use of the plastic cluster bag has been increasing, especially for mid- and late-season cultivars and for fruit that will be stored for long periods.

PALLETIZATION

After packing and lidding, grapes are palletized on disposable or recycled pallets. Some strapping in the field before loading is necessary in grapes packed in shoe-box boxes. Often, loaded pallets coming from the field pass through a "pallet squeeze," a device that straightens and tightens the stacks of containers. These pallet loads are unitized, usually by strapping or netting.

In shed-packing operations, some palletizing glue is used. This glue bonds the corrugated containers vertically on the pallet so that only horizontal strapping is required.

COOLING AND STORAGE OPERATIONS

After palletization is complete, the pallets are moved either to a fumigation chamber for immediate SO_2 treatment, to a forced-air cooler combined with fumigation, or to a forced-air cooler, where fumigation is done at the end of the day's packing. In any case, cooling must start as soon as possible and SO_2 applied within 12 hours of harvest. Many grape forced-air coolers in California are designed to achieve seven-eighths cooling in 6 hours or less. After cooling is completed, the pallets are moved to a storage room to await transport. Ideally the storage room operates at −1° to 0°C (30° to 32°F) and 90 to 95% RH, with a moderate airflow 20 to 40 cfm per ton of stored grapes. The constant low temperature, high RH and moderate airflow are important to limit the rate of water loss from fruit stems. Fruit should be stored at −0.5° to 0°C (31° to 32°F) pulp temperature throughout its postharvest life.

Fumigation is commonly repeated every 7 to 10 days during storage. Grapes should be regularly monitored during storage for physiological deterioration, fruit rot, SO_2 injury, and stem drying. When grapes are loaded for transport or shipment they may receive an additional SO_2 fumigation before loading to assure a longer market life because fumigation is seldom available in receiving markets. Unless SO_2 fumigation is available, the receiver must order grapes for immediate needs and must complete distribution and marketing within a reasonable time after arrival. An exception would be when SO_2-generating pads are placed in the container before shipment.

REFERENCES

Cappellini, R. A., M. J. Ceponis, and G. W. Lightner. 1986. Disorders in table grape shipments to the New York market, 1972–1984. Plant Dis. 70:1075–1079.

Crisosto, C. H., J. L. Smilanick, and N. K. Dokoozlian. 2001. Illustrating the importance of water loss during cooling delays for California table grapes. Calif. Agric. 55(1): 39–42.

Crisosto, C. H., J. L. Smilanick, N. K. Dokoozlian, and D. A. Luvisi. 1994. Maintaining table grape post-harvest quality for long distant markets. In Proc. International Symposium on Table Grape Production, June 28 and 29, 1994. Davis, CA: American Society for Enology and Viticulture. 195–199.

Crisosto, C. H., J. Smilanick, and A. A. Gardea. 1995. Uso de dióxido de azufre para controlar botrytis durante el manejo postcosecha de uva de mesa. Horticultura Mexicana 3(1): 33–40.

Harvey, J. M., and W. T. Pentzer. 1960. Market diseases of grapes and other small fruits. USDA Handb. 189. 37 pp.

Luvisi, D. A., H. Shorey, J. Smilanick, J. Thompson, B. Gump, and J. Knutson. 1992. Sulfur dioxide fumigation of table grapes. Oakland: Univ. Calif. Div. Ag. and Nat. Res. Bulletin 1932. 21 pp.

Luvisi, D., H. Shorey, J. Thompson, T. Hinsch, and D. Slaughter. 1995. Packaging California grapes. Oakland: Univ. Calif. Div. Ag. and Nat. Res. Publ. 1934. 16 pp.

Nelson, K. E. 1985. Harvesting and handling California table grapes for market. Oakland: Univ. Calif. Div. Ag. and Nat. Res. Bull. 1913. 72 pp.

Reynaud, E., and P. Ribereau-Gayon. 1971. The grape. In A. C. Hulme, ed., The biochemistry of fruits and their products. Vol. 2. New York: Academic Press. 172–206.

Ryall, A. L., and W. T. Pentzer. 1982. Handling, transportation and storage of fruits and vegetables. Vol. 2. Fruits and tree nuts. 2nd ed. Westport, CT: AVI. 257–262, 529–542.

Winkler, A. J., J. A. Cook, W. M. Kliewer, and L. A. Lider. 1974. General viticulture. Berkeley: Univ. Calif. Press. 710 pp.

II. Strawberries and Cane Berries

Elizabeth J. Mitcham and F. Gordon Mitchell

STRAWBERRIES

Strawberries are one of the most perishable fresh fruits, yet worldwide they are being successfully marketed in increasing volume. Because much of the marketing is at a great distance from the growing sites, effective handling procedures are required to prevent excessive deterioration. In California, a large fresh strawberry industry exists based largely on fruit delivery to markets up to 5,000 km (2,000 to 3,000 mi) away. California strawberry production begins in late winter in the south and continues through late fall in the north. Fresh market is the primary outlet, followed by freezing and jam manufacturing markets.

The primary market for fresh strawberries extends across the United States and into southern Canada. Some fruit are exported overseas by air to markets such as Japan and Australia. Most domestic shipments move by surface transport (highway trucks). Surface-transported strawberries must have a 5- to 7-day market life when shipped to cities in the eastern United States.

HARVESTING AND FIELD HANDLING

Strawberry growers and harvesters play a key role in determining fruit quality and deterioration throughout marketing and distribution. Important factors in the field include preharvest disease control and field sanitation; maturity selection; avoiding injuries while harvesting and packing fruit into crates; grading to eliminate injured, diseased, and defective fruit; protection from warming; and prompt movement from field to cooler (fig. 29.6).

Field disease control may include fungicidal treatment (when needed and following label instructions on the proper use of the chemical), removal of all diseased fruit from the plants at each harvest, and avoiding fruit contact with damp soil.

Harvest should be as frequent as needed to avoid overmature fruit, and any overripe berries encountered should be diverted to processing or discarded. Fruit color should be within a fairly narrow range when harvested (normally at least three-fourths red color) so that all berries will perform and respond to handling conditions in a similar manner. Fruit should be sorted carefully to remove even small lesions and all injuries (cuts, finger bruises, torn or removed calyx, etc.), and care should be taken to avoid wounding the fruit during the harvest and packing operations. Harvesting, grading, and packing are done simultaneously by the pickers in the field. The crates are usually placed into a picking cart or stand, which keeps the crate off the ground and facilitates easy packing of the crate by the picker during harvest (fig. 29.7).

Figure 29.6

Postharvest handling of strawberries.

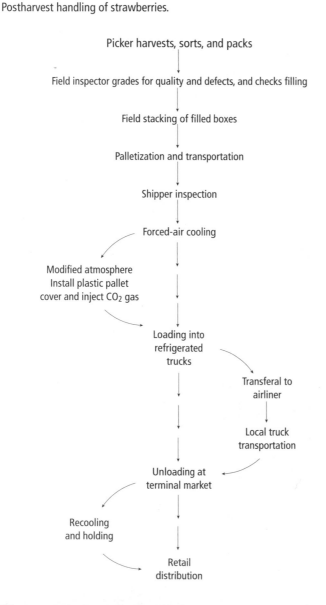

Figure 29.7

Figure 29.7

A strawberry harvest operation. Picking from only half the bed on each side of the picker eliminates excessive reaching and simplifies selection, sorting, and placement of the berries into trays on the wheeled cart.

Figure 29.8

Package design holding eight 0.45-kg (1-lb) clamshell baskets made of clear thermoformed plastic.

Figure 29.9

Common package design holding 4 to 5 kg (9 to 11 lb) of fruit in twelve 0.55-l (1-dry-pt) polypropylene mesh baskets.

Strawberries in California are shipped in corrugated fiberboard crates holding a variety of baskets. Clear, thermoformed plastic clamshell baskets in quart and pint (0.9 and 0.5 l) sizes are most commonly used (fig. 29.8). This basket can be placed directly on the retail display without repackaging or adding a cover. Open mesh plastic baskets are also used (fig. 29.9), and the crate holds 4 to 5 kg (9 to 11 lb) of fruit. These plastic baskets can be capped or uncapped, but the crate is not lidded for subsequent handling.

Trays are grouped together in a two-high stack with wires at both ends of the trays. The wires protrude above the top tray and fit into slots in the tray stacked above. Fiberboard tie sheets are placed on top of the sixth, tenth, and top tray layers. Wires and tie sheets stabilize the pallet load and protect the fruit so that pallet wraps or strapping are typically not necessary.

Berries are picked with the caps (calyxes) attached. Some special "stem-grade" fruit are picked by pinching the stem about 5 cm (2 in) from the calyx. The fruit must be loosely held in the hand without squeezing. Any squeezing of the fruit will cause bruising injury and discoloration. Strawberries should be placed into the crate, not dropped into it. While the crates need to be well filled for marketing, they should not be filled so full that berries will be crushed when crates are subsequently stacked. Often, berries are cut by the basket rim on the open-mesh baskets as a result of being packed over the tops of the baskets. Other overpackaged fruit are injured by abrasion against the corrugated crate. Injuries from this type of packing cause both a direct loss from physical damage as well as an increased risk of Botrytis rot (gray mold), which can spread to non-damaged berries. The lidded baskets eliminate this type of damage.

Any berries that show rot, sunburn, insect feeding, irregular shape, softness, or over-ripeness should be removed from the plant and discarded. If the calyx has been accidentally removed during picking, that fruit should also be discarded.

PREPARING HARVESTED STRAWBERRIES FOR MARKET
Temperature management

Good temperature management, including rapid cooling and maintenance of low pulp temperatures, is the single most important factor in protecting strawberries to minimize decay and over-ripening and to maximize their postharvest life. Both exposure temperature and duration are important to the amount of deterioration that will occur. Because of this relationship, constant circulation of delivery trucks to and from the field is required, and frequent, small loads must be delivered to the cooler. Arriving fruit should be moved promptly to the cooler without delays for inspection or other processing.

Because of the harmful effects of cooling delays it is recommended that cooling begin within approximately 1 hour of picking to prevent decreases in the percentage of marketable fruit (fig. 29.10). All strawberries in California are forced-air cooled in order to remove field heat and cool the fruit to storage temperature as soon as possible. Forced-air cooling is a specific method of cold air management in which pallets of fruit are positioned so that cold air must pass through package openings and around individual berries (see chapter 11 for more details). The design of the strawberry crates commonly

used in California, with their large side ventilation openings, allows large volumes of air to move across packed berries with only modest air pressure differences, and berry cooling can be quite rapid. Approximate airflow characteristics needed for cooling strawberries in standard corrugated crates with mesh baskets are shown in table 29.2. Cooling rates for mesh baskets and pint-sized clamshells are similar, but quart clamshells take slightly longer to cool. The amount of venting on the clamshell baskets should be close to 13% of the side surfaces.

Seven-eighths cooling time is the time required to cool the berries seven-eighths of the difference between their initial temperature and the temperature of the cold air. A 24°C (75°F) strawberry in –1°C (30°F) air would be seven-eighths cool when it reaches 2°C (about 36°F). These values are based upon the warmest fruit in the pallet (on the downstream position, inside the tunnel). The pallet furthest from the fan will usually contain the warmest fruit.

Refrigerated holding

Depending upon their handling and harvest maturity, strawberries may have a market life of from 1 to 2 weeks. Strawberry coolers and holding facilities should be maintained as close to 0°C (32°F) as possible with minimum temperature fluctuation. The highest freezing point that has been reported for strawberries is –0.8°C (30.6°F), but this is dependent upon their soluble solids concentration (higher soluble solids would result in a lower freezing point). Maintaining high RH in storage areas minimizes water loss.

Carbon dioxide treatment

Many strawberries are shipped with elevated CO_2 treatment, sometimes called modified atmosphere transport. This may be useful in slowing the physiological activity of the berries, thus slowing their rate of deterioration. It can also reduce the spread and development of rots (especially gray mold). This effect on gray mold development is measurable at transport temperatures above about 2°C (36°F), when growth and spread of the rot are more active. The value of CO_2 treatment will depend upon the transport time and temperature and upon the decay potential of the fruit. The greatest benefit would be expected for berries harvested

Figure 29.10

Strawberries should be cooled as soon as possible after harvest. Delays beyond 1 hour reduce the percentage of marketable fruit.

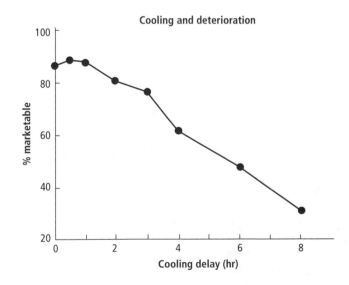

Cooling and deterioration

Table 29.2. Airflow characteristics and time to cool strawberries

Hours to seven-eighths cool warmest fruit	1.5	2	3	4
Airflow (l.sec/kg or cfm/lb berries)	2.0	1.4	0.8	0.5
Static air pressure (in) across pallet*	0.40	0.20	0.08	0.04

Note: *Static air pressure in × 0.2487= kPa.

Figure 29.11

A pallet bag is placed over a pallet of thoroughly cooled strawberries; the bag will be sealed to the plastic pallet liner under the fruit.

Figure 29.12

With the plastic bag and pallet liner carefully taped together, a nozzle pierces the bag to exhaust some air and then replace it with CO_2. Tape will cover the point of injection in order to maintain the gas seal.

after periods of cool, moist, or foggy weather when free water might collect on berries in the field, and when gray mold spread would be expected to be most severe.

The standard method of CO_2 treatment is to completely enclose pallet loads of cold berries in sealed plastic bags, pull a slight vacuum, then add CO_2 to create a 12 to 15% CO_2 atmosphere within the pallet bag and around the fruit (figs. 29.11 and 29.12). It is important that the fruit be thoroughly cold, since the plastic pallet cover will impede further cooling, and condensation can form on the plastic when the berries are not fully cooled. All preparation and treatment should occur inside the refrigerated holding rooms, and pallet covers should be applied just prior to truck loading.

TRANSPORTING FRESH STRAWBERRIES TO MARKET

The period during which fresh strawberries are in transport is a major portion of their total postharvest life. Thus, assuring good conditions, including maintenance of low fruit temperatures during loading and transport, is critical to the successful marketing of fresh strawberries.

Refrigerated loading. Refrigerated holding rooms or loading docks for strawberries should be equipped with sealed loading doors that allow berries to remain under refrigeration while they are loaded into transport vehicles.

The transport vehicle should be thoroughly inspected to assure that the refrigeration system and airflow system are operating properly, and that the vehicle and any contents have been completely cooled before strawberry loading begins.

Experience suggests that damage to strawberries during transport is least when fruit are placed near the front and center of the vehicle. There the tractor axle under the load is often equipped with air-ride suspension (for driver comfort), whereas the rear trailer axles most commonly have normal spring suspension.

Past studies have shown that considerable heating of fruit can occur through trailer walls, with strawberries loaded against side walls warming substantially during cross-country transport. To resolve this problem, the California strawberry industry pioneered the use of "center loading" of pallets, using dunnage blocks along the sides (between walls and pallets) to stabilize the load. This assures space for airflow between the trailer wall and the pallets.

Setting the thermostat. The vehicle thermostat should be set at as low a temperature as possible without danger of freezing the strawberries. If the freezing point is near –0.8°C (30.6°F) (depending on the soluble solids concentration of the fruit) and if the accuracy of the thermostatic control equipment of the vehicle is ± 1.5°C (± 3°F), then a 1°C (34°F) setting may be possible. Any produce to be loaded with strawberries should be compatible.

Air transport. Most strawberries transported by air are destined for export markets. Strawberries for air transport should receive the same preparation and protection as strawberries for surface transport. Berries should be cooled and loaded as discussed above, and the fruit should be kept cold as long as possible before being staged for air cargo handling. Efforts should be made to minimize exposure of strawberries (pallets or air freight containers) to open sunlight, especially on hot tarmac staging areas. Reflective pallet covers with cold packs may help reduce warming. Tests have shown that even though strawberries may warm during transport, it is better to provide cooling whenever possible than not to cool at all. Upon arrival at destination, pallet covers should be removed and the fruit should be promptly returned to refrigeration, using forced-air (if possible) to recool them to near 0°C (32°F) as quickly as possible.

CARE OF STRAWBERRIES IN DISTRIBUTION AND RETAILING

Receiving. Cold fruit in refrigerated vehicles should be unloaded directly into the 0°C (32°F) warehouse whenever possible. If the fruit are warmer than about 2°C (36°F) they will benefit from being recooled and held at 0°C (32°F). If inspection indicates the presence of decay and reconditioning is required, the fruit should still be placed at cold temperatures and small lots removed momentarily for sorting.

Handling bagged pallets. Strawberries in pallets that have been bagged and treated with CO_2 will warm somewhat during transport as a result of their own respiration. The bags do not allow cold air to flow through

the boxes within the pallets to remove this respiratory heat. Because of handling abuses to the pallets during loading and unloading, it is likely that some of the plastic pallet covers will have been punctured, and the high-CO_2 atmosphere will have been lost. It is thus important that these plastic bags be removed and the berries cooled thoroughly upon receipt.

Refrigerated holding. Strawberries in distribution should be held between 0° and 2°C (32° and 36°F) under reasonably high (90 to 95%) RH. In low-RH air, the fruit will readily lose water, and this can quickly cause those fruit to shrivel and appear old. Loss of vitamin C has also been observed when strawberries were stored under low-RH air. Because fruit shrivel is a result of cumulative water loss, fruit that appear sound may be approaching the stage at which shrivel becomes visible at the time of arrival in the distribution market.

Loading for retail store distribution. Because of their high perishability, strawberries should receive special handling in their final distribution. They should not be allowed to accumulate in warm staging areas where individual store orders may be assembled. They should not be loaded into warm distribution vehicles awaiting assembly of other products to complete the load. Whenever possible, strawberries should be kept cold until the delivery trucks are ready to depart, and only then placed into the load for final movement to retail outlets.

Many of the principles that apply to handling of strawberries also apply to handling of other fresh caneberry fruit.

RASPBERRIES AND BLACKBERRIES

Raspberries and blackberries grow on cane bushes and are harvested by hand for the fresh market when fully ripe. Some processed raspberries are machine-harvested. Raspberries should be fully red, and blackberries should be fully black and easily separate from the core at harvest. Harvesting will be necessary at least every other day to prevent development of overripe fruit. Berries should be handled gently as they bruise very easily. Fruit should be harvested into small containers and quickly transported to a shaded area for packing.

- Strawberries and caneberries are among the most perishable fruits.

- Berry deterioration is caused by injuries from harvesting and handling, decay, and self-destruction (natural senescence).

- Reducing berry losses must start with picker training and supervision.

- Careful grading at harvest to prevent packing of overripe, injured, or diseased berries is essential to prevent the spread of disease to healthy berries.

- Good temperature management is the single most important factor in minimizing deterioration and maximizing strawberry and caneberry shelf life.

The packing process involves sorting for defects, including damaged, decayed, and overripe fruit; placing fruit into small cardboard or plastic baskets; and overwrapping with cellophane or into clamshell thermoformed plastic containers.

Rapid cooling is very critical for the highly perishable raspberry and blackberry. Fruit should be cooled to 0°C (32°F) using forced-air cooling. Avoid leaving the fruit on the forced-air cooler beyond the time needed to seven-eighths cool to prevent excessive dehydration. Carbon dioxide treatment, as described for strawberries, can equally benefit these berries by reducing decay development.

BLUEBERRIES

Blueberries are more hardy than raspberries and blackberries but still require careful and expedited handling for successful marketing. Blueberries for fresh market are nearly always harvested by hand, while those destined for processing may be machine-harvested. The ability to resist damage during the harvesting operation varies considerably among varieties. Berries of some varieties may be picked soon after they turn dark blue, but those of other varieties are not ripe even though they have turned dark blue. These fruit should be picked after they develop good flavor, but while they are still firm enough for successful marketing.

As with the other berries discussed, handling of blueberries should be minimized to

reduce fruit damage. Fruit should be harvested into small containers or directly into small cardboard or plastic baskets. Baskets are overwrapped with cellophane. Cool fruit to 0°C (32°F) as quickly as possible after harvest.

CRANBERRIES

Cranberries grow on small shrubs. Berries can be harvested by hand, with a scoop (tines and bucket), or by machine in a dry field. Alternatively, the field can be flooded (bog) and a machine can be used to water-rake the fruit off the shrubs. The berries float and are collected onto elevators using wind to move the fruit. The water system results in greater yields but causes more injury to the fruit. Water-harvested fruit also have greater decay, particularly if held in the water up to 24 hours. If the fruit are held in the water more than 12 hours, a physiological breakdown of the berry occurs after harvest. The incidence of this breakdown increases with increasing time in the water and varies with variety.

Rapid cooling after harvest is important to maintain fruit quality. However, cranberries are chilling-sensitive and should not be stored at temperatures lower than 3°C (37.4°F).

REFERENCES

Ceponis, M. J., and A. W. Stretch. 1983. Berry color, water-immersion time, rot and physiological breakdown of cold-stored cranberry fruits. HortScience. 18:484–485.

Dale, A. E., J. Hanson, D. E. Yarborough, R. J. NcNeil, et al. 1994. Mechanical harvesting of berry crops. Hort. Rev. 16:255–382.

Kader, A. A. 1991. Quality and its maintenance in relation to the postharvest physiology of strawberry. In A. Dale and J. J. Luby, eds., The strawberry in the 21st century. Portland, OR: Timber Press. 145–152.

Manning, K. 1993. Soft fruits. In G. B. Seymour et al., eds., Biochemistry of fruit ripening. London: Chapman and Hall. 347–377.

Miller, W. R., and R. E. McDonald. 1988. Fruit quality of rabbiteye blueberries as influenced by weekly harvests, cultivars, and storage duration. HortScience. 23:182–184.

Mitchell, F. G., E. Mitcham, J. E. Thompson, and N. Welch. 1996. Handling strawberries for fresh market. Oakland: Univ. Calif. Div. Ag. and Nat. Res. Publ. 2442. 14 pp.

Morris, J. R., and W. A. Sistrunk. 1991. The strawberry. In N. A. M. Eskin, ed., Quality and preservation of fruits. Boca Raton, FL: CRC Press. 181–206.

Perkins-Veazie, P. 1995. Growth and ripening of strawberry fruit. Hort. Rev. 17:267–297.

Perkins-Veazie, P., J. K. Collins, J. R. Clark, and J. Magee. 1994. Postharvest quality of southern highbush blueberries. Proc. Fla. State Hort. Soc. 107:269–271.

Robbins, J. A., and J. K. Fellman. 1993. Postharvest physiology, storage and handling of red raspberry. Postharv. News and Info. 4:53N–59N.

Talbot, M. T., J. K. Brecht, and S. A. Sargent. 1995. Cooling performance evaluation of strawberry containers. Proc. Fla. State Hort. Soc. 108:258–268.

III. Kiwifruit

Carlos H. Crisosto and
F. Gordon Mitchell

Approximately 2,100 hectares (5,200 acres) of kiwifruit (mainly of the cultivar Hayward) are grown in California, producing approximately 10 million to 11 million 3.6-kilogram (8-lb) boxes of fruit valued at $40 million to $43 million. Most production in California is located in the San Joaquin and Sacramento Valleys. Because of the hot, dry climate, kiwifruit must be well irrigated, but under these conditions kiwifruit develop very high soluble solids and mature earlier than in cooler climates. Most of the production area is relatively free of heavy winds so wind scarring of fruit is minimal, but care must be taken to avoid sunburn.

PHYSIOLOGY

Kiwifruit have an initial high starch content throughout the flesh before ripening, which is converted into sugars with time in storage. Consequently, soluble solids content (SSC) increases sharply in kiwifruit after harvest and may more than double during the first 1 to 2 months of storage. Meanwhile, titratable acidity decreases by as much as 50%. Most dramatic of all changes during early storage is a reduction in flesh firmness. Flesh firmness typically declines by 30 to 50% during each month of air storage at 0°C (32°F), until the fruit is fully ripe (near 0.9 kgf using an 8-mm tip, or 2 lbf using $\frac{5}{16}$ in tip).

Both CO_2 and ethylene production rates show a slight increase during early storage, then level out and continue at a fairly constant rate through a long period of storage. This small early peak in respiratory activity may be a response to handling injuries (even the wound caused by picking the fruit), rather than a real acceleration in physiological activity related to ripening. Respiration and ethylene production increase concurrently, and respiration almost always peaks 1 or 2 days before ethylene production. The climacteric rise of both ethylene-treated and control kiwifruit began only when fruit soften to 0.7 kilogram-force(1.5 lbf), well after the fruit were soft enough to eat (1.3 kgf, or 2.9 lbf) (Ritenour et al. 1999).

MATURITY

Kiwifruit exhibit little change in appearance and density as they approach maturity. They reach nearly full size well in advance of maturity. Although there is a large size variation among fruit on the vine, this appears unrelated to fruit maturity. For these reasons, any maturity standard must be based upon a single harvest of all fruit.

Several physical and chemical characteristics have been studied in kiwifruit, such as surface color, flesh color, soluble solids content (SSC, which is mostly sugars), titratable acidity (TA), SSC:TA ratio, starch disappearance, seed color change, and flesh firmness. Surface and flesh color change little over long periods of fruit development. Acid composition changes somewhat with development, but the level of titratable acids declines only after a period of storage. Starch disappearance is not easily measurable until ripening begins. Soluble solids content (SSC) taken at harvest time has been the most widely used maturity index to predict a minimum quality and storage performance. A 6.5% minimum SSC level of freshly harvested kiwifruit is commonly used as a standard in California. Because much of the starch conversion into sugars occurs after harvest, an initial SSC measure is valid only if taken immediately after harvesting the fruit because monitoring the SSC of fruit after packing or after any other delays that allow for starch conversion to sugar will result in higher SSC levels.

PHYSICAL INJURY

Flesh softening occurs as fruit mature on the vine, thus, flesh firmness measurements may help growers determine how late they can delay the harvest. The variability among fruit from a given location increases as softening on the vine continues and especially when flesh firmness drops below about 6.5 kilogram-force (14 lbf). Thus, there would be softer fruit if harvest occurs when flesh firmness drops below this level. To avoid possible increased fruit-handling injuries, harvest should not be delayed beyond an average flesh firmness of 6.5 kilogram-force (14 lbf).

When firm fruit (>6 kgf, or 13 lbf) are impacted, a light, whitish bruise results. The white color results from failure to convert starch to sugar in the injured cells. When the fruit softens to about 3 kilogram-force(6.6 lbf), a translucent bruise results. This injured flesh no longer contains starch. At intermediate firmness between about 6 and 3 kilogram-force(13.2 and 6.6 lbf), no visual bruising symptoms appear. Kiwifruit injured at above 6 kilogram-force (13.2 lbf) flesh firmness do not show a physiological response in either elevated CO_2 or ethylene production. However, below that firmness, there is a sharp increase in ethylene production, which persists for more than 2 weeks after injury. This is another reason for completing fruit harvest before flesh firmness drops below this level.

Vibration bruising of kiwifruit usually results in only minor signs of surface injury, but it could cause severe internal flesh injury. Such injury occurs when fruit soften to about 2.3 kilogram-force(5 lbf). Concurrent with the injury is a sharp increase in ethylene production that persists for at least a week. Opportunity for vibration bruising can be expected during transport from storage to distribution market. This provides a compelling reason for attempting to market kiwifruit at firmness above 2.3 kilogram-force(5 lbf).

Late-harvested kiwifruit retain their flesh firmness during storage better than early harvested fruit. Even though these fruit are less firm at harvest, they will emerge after 4 to 6 months storage at 0°C (32°F) firmer than earlier harvested fruit. Thus, growers should not rush to harvest fruit destined for long storage. A good rule for kiwifruit storage management appears to be "last harvested-last marketed." It is also known that kiwifruit with higher SSC levels will store better than ones with lower levels. Thus, late harvest is recommended to assure high kiwifruit quality after a long storage period in California.

QUALITY

In-store consumer acceptance tests were carried out for three consecutive seasons on the relationship between ripe soluble solids concentrations (RSSC) and/or ripe titratable acidity (RTA) on Hayward kiwifruit (Crisosto and Crisosto 2000). Based on test results, kiwifruit with RSSC that ranged from 11.6% to greater than 13.5% were always liked by consumers but with different degrees of liking. A 12.5% RSSC is being proposed as a minimum quality index for early-marketed Hayward kiwifruit. RTA played a significant role in consumer acceptance only on kiwifruit that had RSSC less than 11.6% with RTA equal to or greater than 1.17% ("sour"). These data confirm recommendations from New Zealand that ripe Hayward kiwifruit should have a minimum of 12.5% SSC to achieve consumer acceptance.

TOTAL SOLIDS

A total solids determination measured at any time during kiwifruit postharvest life should predict final SSC (ripeness) and fruit quality. Percent total solids is the dry weight (after removing all of the water) divided by the fresh weight of the tissue. There is a highly significant correlation between total solids and ripe soluble solids concentration (Slaughter and Crisosto 1998). To speed up the total solids determination, the drying period has been shortened to about 52 minutes by using a microwave oven. However, this procedure is a destructive method and it involves careful fruit sample preparation for its determination, making it difficult to use.

NIR SPECTROSCOPY

A nondestructive quick method, NIR spectroscopy technique, can be used to determine fructose, glucose, SSC, and total solids (dry weight). The reliable measurement of total solids any time during postharvest handling can be used as a quality index. Because of its potential for high-speed measurements, this optical technique may be suitable for quality segregation in the packing shed (Slaughter and Crisosto 1998).

ORCHARD FACTORS

Fruit from certain vineyards maintain firmness during storage better than fruit from other vineyards consistently from year to year. A survey of numerous vineyards within California production areas indicated that vineyard nutrition was related to these differences (Johnson et al. 1997). Fruit with a high nitrogen concentration softened more rapidly in storage. Potassium and calcium were also positively correlated with fruit storage ability but less consistently than nitrogen.

POSTHARVEST DISEASE MANAGEMENT

Most kiwifruit decay problems are a result of infection by *Botrytis cinerea*. This organism grows and spreads slowly at 0°C (32°F) storage, but because of the long storage duration (up to 6 months) for kiwifruit, it is a major cause of fruit loss. Botrytis rot can invade the fruit directly, but it also enters through wounds, invades dead floral parts or other organic matter on the fruit, and spreads from infected fruit to healthy surrounding fruit (nesting). It is important to maintain cleanliness in the vineyard, to avoid fruit injuries during handling, to brush the fruit to remove dead floral parts and other material on the fruit surface, to avoid contamination (such as juice from soft fruit), to cool the fruit rapidly, and to maintain a constant 0°C (32°F) storage temperature. A preharvest fungicide spray is recommended when high Botrytis pressure is detected. Because Botrytis rot is associated with soft fruit, any practices that maintain flesh firmness during storage decrease the fruit-rotting problem.

During long-term storage, some individual kiwifruit become rotted, particularly from *Botrytis cinerea*. Fruit infected with Botrytis rot produce ethylene at a higher rate, and this can affect flesh softening of healthy fruit. Even a single soft, decayed kiwifruit in the center of a flat can speed the softening of surrounding fruit. Fruit farthest from the rot can soften more rapidly than fruit in trays that are rot-free.

Sampling fruit from vineyards 4 months after fruit set and recording the incidence of *Botrytis* colonization in sepals or stem ends is being used as a field-monitoring method to predict the incidence of kiwifruit Botrytis gray mold after 3 to 5 months in cold storage. Spraying a registered fungicide 1 or 2 weeks before harvest significantly reduced postharvest gray mold after 5 months storage only in vineyards with more than 6% gray mold prediction. Preharvest spray is not recommended when predicted gray mold is below 6%. This prediction method can be successfully used by growers to make decisions about preharvest sprays, sorting, repacking, and shipping (Michailides and Morgan 1996).

STORAGE CONDITIONS

Kiwifruit flesh softening is rapid during the first few weeks of storage, even at 0°C (32°F). Exposure to ethylene during any cooling delay can substantially accelerate flesh softening during subsequent 0°C (32°F) air storage. The greater the ethylene concentration and the longer the cooling delay period, the greater the effect on fruit softening. Advancing fruit maturity also increases the sensitivity of freshly harvested fruit to ethylene exposure. The best protection of kiwifruit quality after harvest is forced-air cooling to near storage temperature within 6 hours of harvest, avoiding any ethylene exposure, and storing at 0°C (32°F). We determined a reduction in Botrytis incidence on kiwifruit exposed to an ethylene–free delayed cooling period under specific environmental conditions ("curing"). A period of 48 hours curing at 15°C, 95% RH, and high air velocity (2 m/sec, or 6.5 ft/sec) around the kiwifruit are recommended for a successful treatment.

Avoid ethylene exposure during harvest, transport, and storage. Even very low ethylene levels (5 to 10 ppb) will induce fruit softening. Continuous ventilation during air storage helps to assure low ethylene levels. This method works very well in the Central Valley of California when the outside air is ethylene-free. However, during burning days, we found that air ethylene levels increase to dangerous concentrations.

Extensive studies have been conducted on the potential benefits of controlled atmosphere (CA) storage for kiwifruit. The major benefits are a delay in flesh softening during storage and reduction in Botrytis rot problems. Best results have been obtained in a CA atmosphere of about 5% CO_2 and 2% O_2. The fruit must be promptly cooled and placed under the CA conditions as soon after harvest as possible. There is little benefit if delays in establishing CA conditions exceed 1 week. Upon removal from CA conditions the fruit are firmer and retain a greater market life. Ethylene accumulation must be avoided in CA storage as in air storage. Ethylene at low levels causes rapid flesh softening, and a fruit injury problem involving an ethylene-CO_2 interaction has been identified. Avoiding ethylene contamination is therefore important, and only CA generating equipment that does not produce ethylene should be used. It is possible that ethylene scrubbing may be required during CA storage, but no reliable commercial CA storage system for kiwifruit has been critically monitored.

We found that the rate of kiwifruit softening stored in either ethylene-free air or 5% CO_2 and 2% O_2 at 0°C (32°F) during the 16-week cold storage period was related to fruit size and storage conditions. However, soluble solids concentration accumulation was independent of the fruit size and the storage conditions. Under both storage conditions, large size (approximately 101 g, or 3.5 oz) fruit had a slower rate of softening than medium (93 g, or 3.2 oz) and small size (81 g, or 2.9 oz) kiwifruit. Air-stored kiwifruit softened approximately 2.5 times faster than CA-stored fruit. Kiwifruit are more susceptible to physical damage during packaging when they soften below 1.8 kilogram-force(4 lbf). Under air conditions large, medium, and small kiwifruit reached 1.8 kilogram-force fruit firmness by 11, 12, and 13 weeks, respectively. Large, medium, and small kiwifruit under CA conditions reached 1.8 kilogram-force fruit firmness by 25, 35, and 57 weeks, respectively. Thus, the length of the bin cold storage period prior to packaging depends on fruit size and storage conditions.

KIWIFRUIT RIPENING

For general information on management of fruit ripening, see chapters 16 and 21. Kiwifruit ripening protocols for packers, shippers, buyers, handlers, and receivers have been developed. The *Ripening Guidelines for Kiwifruit Handlers and Receivers* are available from the California Kiwifruit Commission or from Dr. Carlos Crisosto.

REFERENCES

Arpaia, M. L., F. G. Mitchell, and A. A. Kader. 1994. Postharvest physiology and causes of deterioration. In J. H. Hasey, R. S. Johnson, J. A. Grant, and W. O. Reil, eds., Kiwifruit growing and handling. Oakland: Univ. Calif. Div. Ag. and Nat. Res. Publ. 3344. 88–93.

Arpaia, M. L., F. G. Mitchell, A. A. Kader, and G. Mayer. 1985. Effects of 2% O_2 and varying concentrations of CO_2 with or without C_2H_4 on the storage performance of kiwifruit. J. Amer. Soc. Hort. Sci. 110:200–203.

Crisosto, C. H., and G. M. Crisosto. 2001. Understanding Consumer acceptance of early harvested 'Hayward' Kiwifruit. Postharv. Biol. Technol. 22:205–213.

Crisosto, C. H., D. Garner, G. M. Crisosto, and R. Kaprielian. 1997. Kiwifruit preconditioning protocol. Acta Hort. 444: 555–560.

Crisosto, C. H., D. Garner, R. S. Johnson, and J. P. Zoffoli. 1992. Maturity indices for kiwifruit. Sacramento: California Kiwifruit Commission.

Johnson, R. S., F. G. Mitchell, C. H. Crisosto, W. H. Olson, and G. Costa. 1997. Nitrogen influence storage life. Acta Hort. 444:285–290.

Michailides, T., and D. P. Morgan. 1996. New technique predicts gray mold in stored kiwifruit. Calif. Agric. 50(3): 34–40.

Mitchell, F. G. 1990. Postharvest physiology and technology of kiwifruit. Acta Hort. 282: 291–307.

Mitchell, F. G., G. Mayer, W. Biasi, and D. Golli. 1990. Estimating kiwifruit maturity through total solids measurements. Sacramento: California Kiwifruit Commission.

Ritenour, M. A., C. H. Crisosto, D. T. Garner, G. W. Cheng, and J. P. Zoffoli. 1999. Temperature, length of cold storage and maturity influence the ripening rate of ethylene-preconditioned kiwifruit. Postharv. Biol. and Technol. 15:107–115.

Slaughter, D. C., and C. H. Crisosto. 1998. Nondestructive internal quality assessment of kiwifruit using near-infrared spectroscopy. Semin. in Food Anal. 3:131–140.

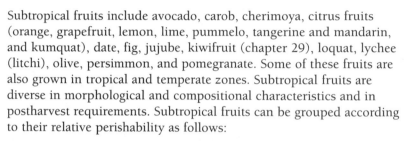

30

Postharvest Handling Systems: Subtropical Fruits

Adel A. Kader and Mary Lu Arpaia

Subtropical fruits include avocado, carob, cherimoya, citrus fruits (orange, grapefruit, lemon, lime, pummelo, tangerine and mandarin, and kumquat), date, fig, jujube, kiwifruit (chapter 29), loquat, lychee (litchi), olive, persimmon, and pomegranate. Some of these fruits are also grown in tropical and temperate zones. Subtropical fruits are diverse in morphological and compositional characteristics and in postharvest requirements. Subtropical fruits can be grouped according to their relative perishability as follows:

- Highly perishable: fresh fig, loquat, lychee

- Moderately perishable: avocado, cherimoya, olive, persimmon

- Less perishable: citrus fruits, carob (dry), dried fig, date, jujube (Chinese date), kiwifruit, pomegranate

This chapter relates the general characteristics of subtropical fruits to their postharvest biology and handling requirements. The emphasis is on avocado and citrus fruits, the fruits that are most important commercially. Commercial production of citrus fruits in the United States is limited to Arizona, California, Florida, and Texas. The United States produces about 30% of the world's production of lemons and 40% of its oranges. Florida is the leading U.S. producer of citrus fruits, most of which (greater than 80%) is processed. Most California citrus fruits are marketed fresh. California accounts for almost all U.S. production of dates, figs, kiwifruit, olives, persimmons, and pomegranates.

Three strains of cultivated avocados are Mexican (e.g., Bacon), Guatemalan (e.g., Hass, Reed), West Indian (e.g., Pallock, Waldin), and their hybrids (e.g., Fuerte). Fresh avocados from Florida (West Indian cultivars) are available from July through February; California avocados (Mexican, Guatemalan, and Mexican-Guatemalan hybrid cultivars) are available year-round. California produces more than 80% of U.S. avocados.

MORPHOLOGICAL AND COMPOSITIONAL CHARACTERISTICS

Avocados are one-seeded berries. Cultivars vary in size. Usually pear shaped, avocados can also be round or oval. The flesh has more energy value than meat of equal weight and is a good source of niacin and thiamine. Of all tree fruits, avocados and olives are the highest in protein and fat content. In California, minimum maturity of the avocado is defined in terms of dry weight, which is highly correlated with oil content of the flesh. The minimum dry weight required before harvesting can begin varies with cultivar from 19 to 25%. Florida-grown cultivars have lower oil content and are harvested on the basis of a calendar date (days after full bloom).

Citrus fruits rank first in their contribution of vitamin C to human nutrition in the United States. Botanically, citrus fruits are classified as a hesperidium, a specialized berry. The rind has two components: the pigmented part, called the *flavedo* (epidermis and several subepidermal layers), and the whitish part, called the *albedo*. The juicy part consists of segments filled with juice sacs. Minimum maturity requirements of citrus fruits are based on juice content (lemon and lime) or soluble solids content, titratable acidity, and the ratio of the two (orange, grapefruit, and tangerine).

POSTHARVEST PHYSIOLOGY

Avocados have a relatively high respiration and ethylene production rate, and both rates exhibit a climacteric pattern. Citrus fruits are nonclimacteric and have low respiration and ethylene production rates. Postharvest compositional changes in citrus fruits are minimal, whereas avocados undergo many changes in composition, texture, and flavor associated with ripening.

Avocados do not ripen on the tree. The exact nature of the ripening inhibitor is not known, but it continues to exert its effect for about 24 hours after harvest. Avocado ripening can be hastened by exposure to 10 ppm ethylene at 15°C to 18°C (59° to 65°F) and 85 to 90% RH for 12 to 48 hours; temperatures up to 25°C (77°F) can be used if faster ripening is desired. On the other hand, removal or exclusion of ethylene from the storage environment helps extend the storage life of avocados by delaying softening, onset of chilling injury, and decay incidence.

Cold nights followed by warm days, are necessary for loss of green color and appearance of yellow or orange color in citrus fruits. This is why citrus fruits remain green after attaining full maturity and good eating quality in tropical areas. Occasionally, re-greening of Valencia oranges occurs in certain production areas after they have reached full orange color.

Degreening of citrus fruits essentially results in removal of chlorophyll from the flavedo but does not influence composition of the fruits' edible portion. The need for and duration of de-greening treatments depend on the cultivar and the fruit's condition at harvest—the amount of chlorophyll to be removed. Lemons are usually de-greened at 16°C (60.8°F) with or without added ethylene; higher temperatures may be used for faster de-greening. Recommended conditions for de-greening California oranges and grapefruits are:

- Temperature: 20° to 25°C (68° to 77°F)
- Relative humidity: 90%
- Ethylene concentration: 5 to 10 ppm
- Air circulation: One room volume per minute
- Ventilation: One to two air changes per hour, or sufficient changes to keep CO_2 below 0.1%

In Florida, a temperature of 27° to 29°C (80.6° to 84.2°F) and an ethylene concentration of 1 to 5 ppm are recommended.

PHYSIOLOGICAL DISORDERS

CHILLING INJURY

Susceptibility to chilling injury varies in subtropical fruits according to species and cultivar. For example, grapefruit, lemon, and lime are much more susceptible to chilling injury than orange and mandarin. Orange cultivars grown in Florida are reportedly less sensitive to chilling injury than those grown in California and Arizona. Date, fig, kiwifruit, and Hachiya persimmon are not sensitive to chilling injury. Fuyu persimmon, pomegranate, olive, and other subtropical fruits are chilling sensitive. Ripe avocado fruit tolerate lower temperatures than unripe fruit, without danger of chilling injury. Symptoms of chilling injury on selected subtropical fruits are summarized in table 30.1 and illustrated in figures 30.1 and 30.2 for orange and grapefruit, respectively.

OTHER DISORDERS

Fruits exposed to temperatures below their freezing point before or after harvest can suffer serious injuries; for example, injured citrus fruits become dry and useless and have to be separated in the packinghouse by flotation or X-ray techniques. High-temperature disorders resulting from preharvest exposure

Table 30.1. Chilling injury symptoms on avocados and selected citrus fruits

Fruit	Minimum safe temperature* °C	°F	Symptoms
Avocado	5–10	41–50	Skin pitting, scalding, and blackening are the main external symptoms. Internal symptoms include grayish-brown discoloration of flesh, discoloration of the vascular tissue, softening, and development of off-flavors.
Grapefruit	10–13	50–55	Pitting, scald, watery breakdown.
Lemon	10–13	50–55	Pitting, membranous stain, red blotch.
Lime	10–13	50–55	Pitting, accelerated decay.
Mandarin	5–8	41–46	Pitting, brown discoloration.
Orange	3–5	37–41	Pitting, brown stain.

Note: *Varies with cultivar, maturity-ripeness stage, and duration of storage.

Figure 30.1

Chilling injury symptoms on oranges. Darker areas are brown.

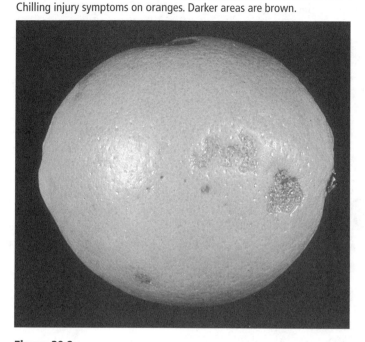

Figure 30.2

Scoring system for severity of chilling injury on grapefruit. Darker areas are brown.

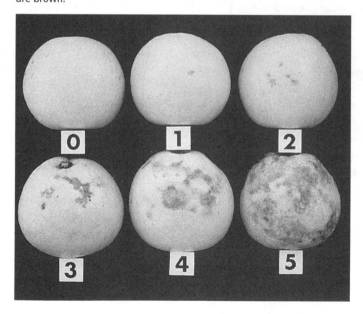

to sun can result in sunburned avocado and citrus fruits. Exposure of avocado to temperatures above 25°C (77°F) may cause uneven softening, skin discoloration, flesh darkening, and off-flavors.

Citrus fruit peel disorders, other than chilling injury, include

- oil spotting, or oleocellosis (breaking of oil cells, causing the oil to extrude and damage surrounding tissue) (fig. 30.3)

- rind staining of navel orange (an indication of peel overmaturity that can be controlled by preharvest application of gibberellin)

- stem-end rind breakdown of orange

- stylar-end breakdown of lime

- shriveling and peel injury around the stem end indicating aging

PATHOLOGICAL BREAKDOWN

AVOCADO

Avocado fruit can be affected by one or more pathogens. *Dothiorella gregaria* (probably the asexual state of *Botryosphaeria ribis*) is a postharvest rot of California avocados. Anthracnose occurs particularly in humid areas such as Florida. Neither organism is usually a serious problem in California unless the weather has been unusually wet at or near harvest time. Stem-end rots (*Diplodia natalesis, Phomopsis citri*) can also be serious in Florida and other humid growing areas.

Control methods include good orchard sanitation, effective preharvest fungicide application, careful handling to minimize physical injuries, prompt cooling to optimal temperature for the cultivar, and maintaining that temperature during marketing.

CITRUS FRUITS

Postharvest diseases also limit the postharvest life of citrus fruits. Blue mold (*Penicillium italicum*) and green mold (*Penicillium digitatum*) occur in citrus fruits in all production areas. In humid areas, stem-end rots (*Diplodia* spp. and *Phomopsis* spp.) and anthracnose (*Colletotrichum gloeosporioides*) are common. Sour rot (*Geotrichum candidum*) affects lemons during long-term storage, especially during wet seasons. Phytophthora brown rot occurs in California following cool, wet weather, and can be controlled by heat treatment. Alternaria stem-end rot (*Alternaria citri*) usually follows senescence of the calyx of the fruit.

Citrus diseases can be controlled by the following procedures:

Reduce the pathogen population in the environment. Use an effective preharvest disease control program to reduce postharvest incidence of stem-end rots and anthracnose. Use chlorine (e.g., sodium hypochlorite) in wash water. Regularly disinfest field

Figure 30.3

Scoring system for severity of oil spotting on lemons. Darker areas are green.

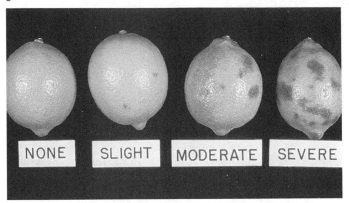

| NONE | SLIGHT | MODERATE | SEVERE |

Figure 30.4

Harvesting avocados.

containers, packinghouse equipment, and storage facilities using a fog of 1% formaldehyde solution or quaternary ammonium products. Circulate *Penicillium* spore–laden air through filters in a special box-dumping room for stored lemons.

Maintain fruit resistance to infection. Minimize mechanical injuries during harvesting and postharvest handling. Use proper temperature and relative humidity management throughout postharvest handling. Use 2,4-D treatment (200 ppm) on lemon to maintain vitality of button tissue and reduce development of stem-end rots, or use gibberellic acid (50 ppm in the storage wax) as an alternative to 2,4-D; this treatment decreases the incidence of *Geotrichum* during storage. Use postharvest fungicides, such as sodium orthophenylphenol (SOPP), thiabendizole (TBZ), sec. butylamine, and imazalil. New fungicides are continually being evaluated. Choosing a fungicide depends upon whether it has been approved for use and whether it has been accepted by importing countries. Judicious use of fungicides is a valuable component of a disease control program, as resistance to fungicides can develop quickly.

ALTERNATIVES TO POSTHARVEST FUNGICIDES

Without the use of available postharvest fungicides or replacements, storage life of citrus fruits would be significantly reduced. Consequently, exports of fresh citrus fruits, other than air-shipped, would be curtailed, and postharvest losses would increase in both domestic and export marketing. Short-term and long-term options and alternatives to currently used postharvest fungicides are listed below.

SHORT-TERM ALTERNATIVES

More careful handling during harvest and postharvest operations to reduce mechanical injuries will reduce fungal infection and losses due to decay. This, coupled with providing the optimal temperature and RH and expedited handling during all marketing steps, can extend the postharvest life of citrus fruits about 2 to 3 weeks, depending on the cultivar.

Controlled or modified atmospheres (including carbon monoxide) can be used during transport and temporary storage. CO at 5 to 10%, added to 5% O_2, provides adequate fungistatic control of many fungi causing citrus postharvest diseases. Use of CO requires strict safety precautions to protect transport and storage facility workers. Also, the cost of maintaining such atmosphere is greater than the cost of treatments with postharvest fungicides. More research is needed, however, to evaluate the fruit tolerance to

Figure 30.5

Postharvest handling of avocados.

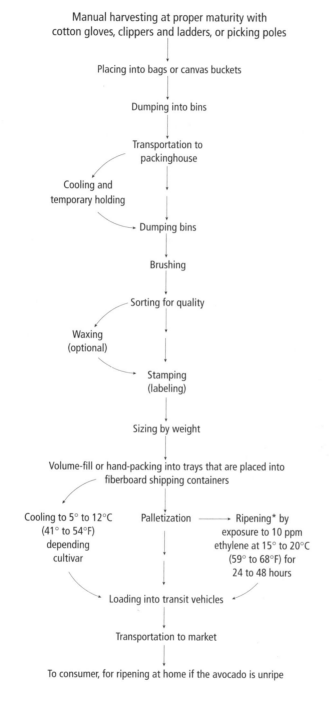

Manual harvesting at proper maturity with cotton gloves, clippers and ladders, or picking poles

Placing into bags or canvas buckets

Dumping into bins

Transportation to packinghouse

Cooling and temporary holding

Dumping bins

Brushing

Sorting for quality

Waxing (optional)

Stamping (labeling)

Sizing by weight

Volume-fill or hand-packing into trays that are placed into fiberboard shipping containers

Cooling to 5° to 12°C (41° to 54°F) depending cultivar

Palletization

Ripening* by exposure to 10 ppm ethylene at 15° to 20°C (59° to 68°F) for 24 to 48 hours

Loading into transit vehicles

Transportation to market

To consumer, for ripening at home if the avocado is unripe

*Ripening treatment may be applied at wholesale markets or distribution centers instead of at shipping point.

fungistatic CA with elevated CO_2 and with or without elevated O_2 (to reduce the negative effects of high CO_2 on fruit flavor).

Biological control (use of antagonistic microorganisms) has been and continues to be investigated to identify the most effective agents. Some biocontrol products have been approved for use and are used either alone or in combination with postharvest fungicides (at lower concentrations) for effective decay control.

LONG-TERM ALTERNATIVES

Treatment with ionizing radiation for decay control has been suggested. However, the dose needed to effectively control decay is between 1.5 and 2.0 kGy (150 and 200 krad). Such doses can result in rind injuries and increase fruit softening. Furthermore, the currently approved upper limit for irradiating fresh fruit is 1 kGy (100 krad). Combining irradiation with heat treatments may reduce the dose required and consequently the resulting detrimental effects.

Heat treatments, such as dipping citrus fruits in 44°C (111°F) water for 2 to 4 minutes, have been tested as possible means to kill fungal spores on fruit surfaces and reduce decay. The limiting factor for heat treatments is the narrow margin between the time and temperature combinations that reduce decay and those that cause fruit injury.

Breeding new citrus cultivars whose fruit show resistance to decay-causing fungi is a long-term option.

POSTHARVEST HANDLING PROCEDURES

HARVESTING

Research into mechanical harvesting of citrus fruits (especially in Florida for processing fruit) has been extensive, but no satisfactory system is available. Chemicals that promote abscission will probably be part of any mechanical harvesting system. Several harvest aids, such as mobile ladders and picker platforms, have been tested, but few are in commercial use. California avocados (fig. 30.4) and citrus fruits are harvested with hand clippers. Some Florida citrus fruits are snap-picked (twist and pull method), but this may increase their susceptibility to decay. Some Florida processing oranges and grapefruit are picked and dropped on the ground. This practice is detrimental to the fruit even though they are processed within a day or two after harvest.

Figure 30.6

Dry-dumping of avocados at the packinghouse.

Figure 30.7

Sizing avocados by weight.

PACKINGHOUSE OPERATIONS

A flow diagram of the postharvest handling system used for avocados in California is shown in figure 30.5. The dumping, sizing, and packing operations are illustrated in figures 30.6, 30.7, and 30.8, respectively.

The packinghouse operations for citrus fruits are summarized in figure 30.9. Figures 30.10 to 30.14 illustrate surface drying, quality sorting, hand-packing, pattern-packing, and packing in bags. Lemons are usually

Figure 30.8

Packing avocados from an accumulation bin.

sorted into four color classes (dark green, light green, silver, and yellow) by electronic sorters based on their light reflectance (fig. 30.15). Some orange and tangelo cultivars are colored with a certified food dye in Florida, but this treatment is not allowed in California. Cooling methods include hydrocooling, forced-air cooling, and room cooling. Attention to proper and fast cooling for citrus fruits is badly needed in most citrus-handling facilities.

Seal packaging (wrapping with various types of plastic film) of individual citrus fruit has been extensively tested and is currently used by a few shippers. The treatment reduces water loss and maintains the vitality of the peel because of the high RH maintained around the fruit. It also prevents the spread of decay from fruit to fruit. For decay control, fruit must be treated with fungicides before wrapping. While seal packaging of individual fruit may allow short-term holding of citrus fruits without refrigeration, it must be combined with refrigeration for long-term storage to maintain good quality and reduce losses.

Citrus fruits produced in certain areas must be treated for insect control before shipment to some markets. The main disinfestation method in use was once fumigation with ethylene dibromide (EDB) against fruit flies. Since EDB was completely withdrawn from the EPA's list of approved chemicals in 1987, cold treatments or fumigation with methyl bromide or phosphine have been used. These treatments result in some phytotoxicity. Losses due to phytotoxicity as a result of cold treatment can be mitigated by conditioning the fruit for 1 week at 16°C

Figure 30.9

Postharvest handling of citrus fruits.

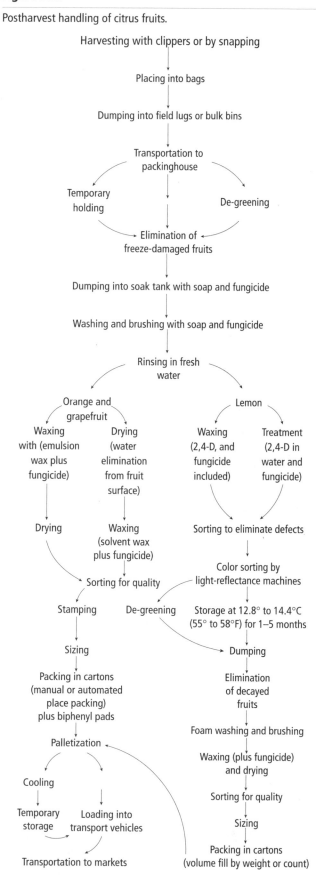

Harvesting with clippers or by snapping
↓
Placing into bags
↓
Dumping into field lugs or bulk bins
↓
Transportation to packinghouse
↓
Temporary holding | De-greening
↓
Elimination of freeze-damaged fruits
↓
Dumping into soak tank with soap and fungicide
↓
Washing and brushing with soap and fungicide
↓
Rinsing in fresh water

Orange and grapefruit
- Waxing with (emulsion wax plus fungicide)
- Drying (water elimination from fruit surface)

Lemon
- Waxing (2,4-D, and fungicide included)
- Treatment (2,4-D in water and fungicide)

Drying → Waxing (solvent wax plus fungicide) → Sorting for quality

Sorting to eliminate defects
↓
Color sorting by light-reflectance machines
↓
Stamping | De-greening | Storage at 12.8° to 14.4°C (55° to 58°F) for 1–5 months
↓
Sizing
↓
Dumping
↓
Packing in cartons (manual or automated place packing) plus biphenyl pads
↓
Elimination of decayed fruits
↓
Palletization
↓
Foam washing and brushing
↓
Cooling
↓
Waxing (plus fungicide) and drying
↓
Temporary storage | Loading into transport vehicles
↓
Sorting for quality
↓
Transportation to markets
↓
Sizing
↓
Packing in cartons (volume fill by weight or count)

Figure 30.10

Drying citrus fruits with warm air to remove surface moisture.

Figure 30.11

Quality sorting of citrus fruits.

(61°F). Other alternatives to chemical fumigation being evaluated include heat treatments, irradiation, and controlled atmospheres. The citrus cultivar and the stage of maturity at harvest influence the fruit's response to the quarantine treatment.

QUALITY AND STORAGE LIFE OF CITRUS FRUITS

The composition and quality of citrus fruits at harvest and the fruits' potential for storage are influenced by many pre- and postharvest factors. Preharvest factors include rootstock and cultivar, fruit maturity at harvest, harvesting season, tree condition (vigor), weather conditions (temperature, RH, rain), and cultural practices (fertilization, irrigation, pest control, growth regulators). Harvesting methods influence the uniformity among fruit and the extent of mechanical injuries due to rough handling.

Postharvest factors that influence the postharvest life span of citrus fruits include delays between harvest, packing, and cooling;

Figure 30.12

Rapid packing of citrus fruits.

Figure 30.13

Automated pattern packing system for oranges.

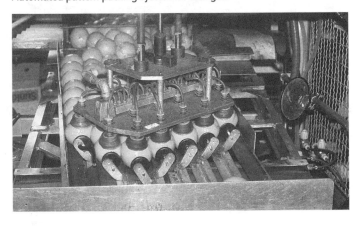

Figure 30.14

Packing citrus fruits into consumer packages.

de-greening conditions; fungicidal treatments; waxing; seal-packaging; growth-regulator treatments; temperature and RH management; and presence of ethylene and other volatiles in storage. Scrubbing ethylene from the storage environment can be useful in retarding fruit senescence and decay incidence.

STORAGE

Some citrus cultivars may be left on the tree for up to 5 months after attaining legal maturity. Depending on the cultivar, avocados will remain attached to the tree for 3 to 12 weeks after maturity before excessive abscission begins. The duration of "on-tree storage" depends on the cultivar. Quality of citrus fruits and avocado may, however, deteriorate during on-tree storage. For successful postharvest storage, maintain the conditions summarized in table 30.2. These recommendations also apply for optimal transport and temporary storage conditions.

REFERENCES

Baldwin, E. A. 1993. Citrus fruits. In G. B. Seymour et al., eds., Biochemistry of fruit ripening. London: Chapman and Hall. 107–149.

Biale, J. B., and R. E. Young. 1971. The avocado pear. In A. C. Hulme, ed., The biochemistry of fruits and their products. Vol. 2. New York: Academic Press. 2–64.

Bower, J. P., and J. G. Cutting. 1988. Avocado fruit development and ripening physiology. Hort. Rev. 10:229–271.

Ceponis, M. J., R. A. Cappellini, and G. W. Lightner. 1986. Disorders in citrus shipments to the New York market, 1972–1984. Plant Dis. 70:1162–1165.

Davies, F. S., and L. G. Albrigo. 1994. Fruit quality, harvesting and postharvest technology. In F. S. Davies and G. Albrigo, Citrus. Wallingford, UK: CAB International. 202–224.

Dezman, D. J., S. Nagy, and G. E. Brown. 1986. Postharvest fungal decay control chemicals: Treatments and residues in citrus fruits. Residue Rev. 97:37–92.

Eaks, I. L. 1977. Physiology of degreening: Summary and discussion of related topics. Proc. Int. Soc. Citricult. 1:223–226.

Eckert, J. W., and I. L. Eaks. 1989. Postharvest disorders and diseases of citrus fruits. In W. Reuther et al., eds., The citrus industry. Vol. 5. Oakland: Univ. Calif. Div. Ag. and Nat. Res. Publ. 3326. 179–260.

Figure 30.15

Light reflectance machine used for sorting lemons by color.

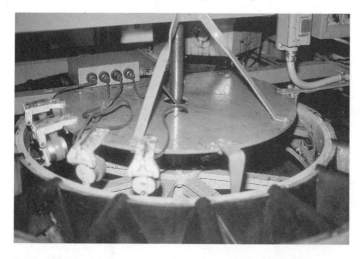

Table 30.2. Optimal storage conditions for avocado and selected citrus fruits

Commodity	Temperature		Approximate storage life*	Modified atmospheres if used[†]	
	°C	°F	(weeks)	% O_2	% CO_2
Avocado, unripe[‡]	5–12	41–54	2–4	2–5	3–10
Avocado, ripe[‡]	2–4	36–40	1–2	—	—
Grapefruit	12–14	54–57	4–8	3–10	5–10
Kumquat	4–8	39–46	2–4	5–10	0–5
Lemon[§]	12–14	54–57	16–24	5–10	0–10
Lime	10–12	50–54	6–8	5–10	0–10
Mandarin	5–8	41–46	2–4	5–10	0–5
Orange	4–8	39–46	4–8	5–10	0–5
Pummelo	8–10	46–50	8–12	5–10	5–10

Notes:

*Under optimal temperature and 90 to 95%RH.

[†]MA use on citrus is limited; 5 to 10% CO may be added to MA for decay control during transport to export markets.

[‡]Response to temperature and MA is dependent upon cultivar.

[§]Storage life for dark-green lemons; for other stages: light-green, 8–16 weeks; silver, 4–8 weeks; yellow, 3–4 weeks.

Grierson, W., and T. T. Hatton. 1977. Factors involved in storage of citrus fruits: A new evaluation. Proc. Int. Soc. Citricult. 1:227–231.

Grierson, W., W. M. Miller, and W. F. Wardowski. 1978. Packingline machinery for Florida citrus packinghouses. Univ. Fla. Coop. Ext. Serv. Bull. 803. 30 pp.

Lee, S. K., R. E. Young, P. M. Shiffman, and C. W. Coggins Jr. 1983. Maturity studies of avocado fruit based on picking dates and dry weight. J. Am. Soc. Hort. Sci. 108:390–394.

Lindsey, P. J., S. S. Briggs, K. Moulton, and A. A. Kader. 1989. Postharvest fungicides on citrus: Issues and alternatives. In Chemical use in food processing and postharvest handling: Issues and alternatives. Davis: Univ. Calif. Ag. Issues Ctr. 23–38.

Nagy, S., and J. A. Attaway, eds. 1980. Citrus nutrition and quality. Symposium Series 143. Washington: American Chemical Society. 456 pp.

Nagy, S., and P. E. Shaw. 1980. Tropical and subtropical fruits: Composition, properties, and uses. Westport, CT: AVI. 570 pp.

Schirra, M., ed. 1999. Advances in postharvest diseases and disorders control of citrus fruit. Trivandrum, India: Research Signpost. 161 pp.

Seymour, G. B., and G. A. Tucker. 1993. Avocado. In G. B. Seymour et al., eds., Biochemistry of fruit ripening. London: Chapman and Hall. 53–81.

Smoot, J. J., L. G. Houck, and H. B. Johnson. 1971. Market diseases of citrus and other subtropical fruits. USDA Handb. 398. 115 pp.

Ting, S. V., and J. A. Attaway. 1971. Citrus fruits. In A. C. Hulme, ed., The biochemistry of fruits and their products. Vol. 2. New York: Academic Press. 107–171.

Ting, S. V., and R. L. Rousett. 1986. Citrus fruits and their products—Analysis and technology. New York: Marcel Dekker. 312 pp.

Wardowski, W. F., S. Nagy, and W. Grierson, eds. 1986. Fresh citrus fruits. Westport, CT: AVI. 571 pp.

31

Postharvest Handling Systems: Tropical Fruits

Adel A. Kader, Noel F. Sommer,

and Mary Lu Arpaia

The most important tropical fruit by far in temperate-zone markets is the banana; its per capita consumption in the United States is higher than any other tropical fruit. Its popularity among consumers is enhanced by a reasonable price, high fruit quality, and year-round availability. The other fresh tropical fruits commonly found in the temperate-zone markets are mango, papaya, and pineapple. The influx of many people from tropical countries into the United States has increased the demand for these and other tropical fruits.

BANANAS

The banana is a large herbaceous plant. The underground tuberous stem, or "bulb," gives rise to leaves and the fruit bunch. The above-ground trunk of the banana tree is a pseudostem consisting of tightly oppressed leaf bases. The pseudostem dies after a fruit bunch has been harvested and is replaced by one of the young pseudostems that have emerged. The banana inflorescence has three types of flowers. The fruit originate from female flowers that are produced first, followed by hermaphrodites, and lastly male flowers.

The postharvest operations required for bananas include transportation to packinghouse, dehanding, washing to remove dirt and latex, disease control, packaging, transportation to market, ripening, and retail sale (fig. 31.1).

HARVESTING

Bunches are examined about 3 months before harvest. Those that have completed their female (fruit-producing) stage have their buds removed to prevent further floral development. One or two apical hands are also removed at this time to promote development of the remainder. Removed buds may be consumed or discarded.

Each bunch is covered with a polyethylene bag. The top of the bag is secured to the stalk and a colored ribbon is attached to the end of the bag or inflorescence. Different colored ribbons are used each week as a ready record of bunch age. The polyethylene bag protects the bananas from leaf scarring and keeps dust off, and the bag may be impregnated with an insecticide.

During fruit development, the bunches may require props to support the inflorescence weight (fig. 31.2). Sometimes a pseudostem is provided with twine guys from the crown to the bases of nearby pseudostems.

Bananas are harvested green at about the 75% mature stage and are ripened in market areas. More-mature fingers often split and tend to be mealy. The maturity at which bananas are harvested depends on the time required to get them to market. Fruit shipped from Central America to Europe are usually harvested less mature than those shipped to North America. Of course, a penalty of lost finger weight is paid when bunches are harvested before fingers are fully developed. Expanded use of controlled atmospheres (CA) during transport to prevent ripening has facilitated harvesting more-mature bananas regardless of the intended market. The sizes and shapes of finger sections at various stages of maturity are illustrated in figure 31.3.

At harvest, crews pass through the plantation, usually at 3- or 4-day intervals, selecting bunches for harvest. Colored ribbons provide information regarding age. The diameter (caliper) of fruit is also monitored.

Harvesting and field handling vary with location. However, harvesting is usually a two-person operation. The cutter makes a cut with a

Figure 31.1

Postharvest handling of bananas.

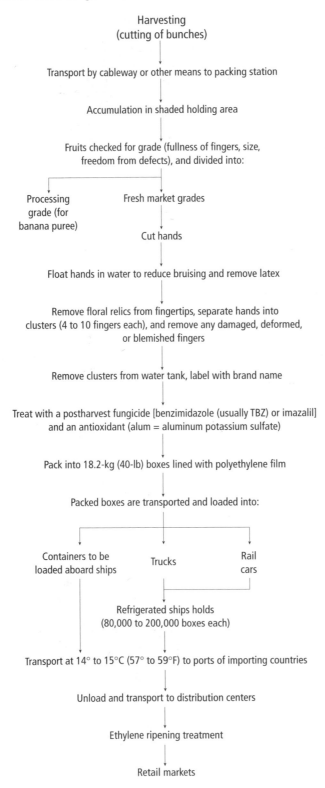

Harvesting
(cutting of bunches)

↓

Transport by cableway or other means to packing station

↓

Accumulation in shaded holding area

↓

Fruits checked for grade (fullness of fingers, size, freedom from defects), and divided into:

Processing grade (for banana puree)   Fresh market grades

↓

Cut hands

↓

Float hands in water to reduce bruising and remove latex

↓

Remove floral relics from fingertips, separate hands into clusters (4 to 10 fingers each), and remove any damaged, deformed, or blemished fingers

↓

Remove clusters from water tank, label with brand name

↓

Treat with a postharvest fungicide [benzimidazole (usually TBZ) or imazalil] and an antioxidant (alum = aluminum potassium sulfate)

↓

Pack into 18.2-kg (40-lb) boxes lined with polyethylene film

↓

Packed boxes are transported and loaded into:

Containers to be loaded aboard ships   Trucks   Rail cars

↓

Refrigerated ships holds (80,000 to 200,000 boxes each)

↓

Transport at 14° to 15°C (57° to 59°F) to ports of importing countries

↓

Unload and transport to distribution centers

↓

Ethylene ripening treatment

↓

Retail markets

Figure 31.2

Banana fruit bunches are enclosed in plastic bags, and guys are run to prevent pseudostems from falling.

bunch firmly. The cutter then severs the bunch from the pseudostem, just below the basal hands.

TRANSPORTATION TO PACKING STATION

Many banana plantations are equipped with a system of cableways. A backer carries a cut bunch to the nearest cableway, where the bunch is attached by its base to a roller on the cable. The bunches are separated by spacer bars to prevent contact. A train of up to 75 or 150 bunches of 30 to 60 kg (66 to 132 lb) each forms and is pulled along the cableway to the packing area by a small tractor. A tractor that hangs from the cable has the advantage of not requiring roadways or bridges to cross drainage ditches. Bananas waiting to be packed are held in the shade to prevent sunburn (fig. 31.4).

In Queensland and New South Wales, Australia, winter temperatures limit banana production to sunny, northern exposures, often on very steep hills. Some cableway systems have been installed that use gravity as the means of locomotion. Otherwise, bunches accumulate at roadways that have been bulldozed across the slopes. Bunches are placed on small trucks or on trailers, usually two or three bunches deep. Padding is placed on the vehicle bed and between bunches to limit damage. Low tire pressure and low speed during transit are important injury avoidance measures.

In some operations, the hands are separated from the bunch soon after harvest and placed on a padded vehicle bed with pads between the hands to minimize abrasion damage during transport to the packinghouse.

machete, partially severing the pseudostem at about its midpoint. A backer, positioned under the bunch, catches and braces the

Figure 31.3

Changes in size and shape of banana fruit sections at various stages of maturity.

Maturity stages of banana fruit

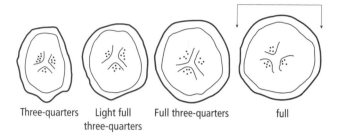

Three-quarters   Light full three-quarters   Full three-quarters   full

Figure 31.5

Inspecting banana stems.

Figure 31.4

A train of banana stems moves to the packinghouse by aerial tramway.

Figure 31.6

A worker cuts the hands from banana stems. Hands are placed in water to clean the fruit and avoid latex stains.

PACKINGHOUSE OPERATIONS

Upon entering the packinghouse, bananas are checked for finger fullness and length, and for blemishes from leaf rub, insect activity, pathogens, and handling bruises (fig. 31.5). Those not meeting a fresh fruit grade can be ripened and processed as puree, sold on secondary markets, or discarded.

Most bananas from the American tropics are shipped as hands in fiberboard cartons. Hands are removed from the stalk by the dehander, using a sharp curved knife (fig. 31.6). When the hand is cut away, latex flows from the wound. If the latex is allowed to coat the surface of the fingers, the resulting stains seriously detract from appearance.

Consequently, the hands are immediately placed in water to coagulate the exuded latex and reduce staining (fig. 31.7). Dust and dirt are also removed at the same time.

Selectors at the dehanding tank remove dead floral parts still adhering to the end of the fingers and sort out undersized, damaged, deformed, and blemished fingers. Large hands are divided into smaller clusters to facilitate packing and provide a convenient unit for the consumer. The bananas may be floated in a second water tank for an additional 10 or 15 minutes to permit latex exudation.

The water tank is a potential source of serious disease problems. Fungus spores on dead floral parts may accumulate in the

Figure 31.7

Bananas are floated in water to coagulate latex exuding from cut surfaces. Withered floral parts are removed from the fingers.

Figure 31.8

Banana hands emerge from a plastic enclosure where the fruit were sprayed with TBZ for disease control and alum as an added precaution against latex staining.

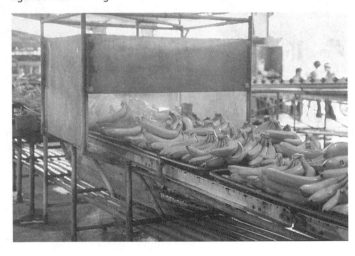

water and contaminate the cut surfaces of freshly cut hands. In particular, crown rot results from inoculation by mixed spores of *Colletotrichum musae*, *Fusarium roseum*, *Nigrospora sphaerica*, *Thielaviopsis paradoxa*, *Botryodiplodia theobromae*, and other fungi.

Sodium hypochlorite in the water is used to kill spores and reduce the likelihood of inoculation. Chlorine concentrations must be carefully monitored and maintained between 75 and 125 ppm. Indicator papers provide a simple and inexpensive means of checking chlorine concentrations. Water should be changed regularly to minimize accumulation of organic materials.

As an alternative, a fungicide (such as TBZ, or imazalil), is added to the wash water instead of as a separate treatment. Relatively large amounts of the fungicide are usually required because the wash water becomes dirty and must be changed from time to time. The fungicide must be included with fresh water.

A fungicide may be applied as a separate dip or spray treatment after washing (fig. 31.8). Alum (aluminum potassium sulfate), which may be applied at the same time, serves as an antioxidant to prevent subsequent latex exudations from darkening and staining the fruit surface. This is the last step before packing. The shorter the time between dehanding and fungicide application the better the subsequent decay control.

PACKING AND SHIPPING

In the American tropics, clusters of fruit are packed in corrugated fiberboard cartons lined with a thin polyethylene film to prevent scuffing of the fingers against the fiberboard (fig. 31.9). The cartons are then moved to conveyors for loading onto trucks or trains for transportation to the wharf. Cartons move by conveyer to an elevated conveyer, or gantry, that lowers them into the hold of a ship. The ship's refrigeration system cools the fruit and holds it at 13° to 14°C (56° to 58°F).

Marine shipment of bananas in refrigerated containers on container ships (fig. 31.10) continues to increase. Faster loading and unloading is an economic advantage for container ships. Containers allow the use of pallets and forklift trucks, and are adaptable to intermodal transportation. Fruit cartons can be loaded into containers at the packing stations and remain untouched until arrival in the destination market area. Presumably, the reduced handling minimizes fruit bruising.

HANDLING IN MARKET AREAS

Banana fingers should be green upon arrival in market areas to provide time for ripening and distribution to retail markets. Facilities for controlled ripening are often provided at produce distribution centers (figs. 31.11 and 31.12). Ripening is triggered by an ethylene exposure of about 24 to 48 hours; by scheduling temperatures and time, distributors have a measure of control over the stage of fruit ripeness delivered to retail stores.

Figure 31.9

Bananas are hand-packed into polyethylene-lined corrugated paper containers.

Figure 31.10

Containerized marine transport of bananas.

Figure 31.11

Ripening rooms for bananas.

Figure 31.12

Forced-air ripening of bananas on pallets.

RIPENING FACILITIES

Ripening rooms require close temperature control. The rooms must be well insulated and provided with heating and refrigeration. Vigorous air circulation is required to thoroughly disperse the ethylene and remove respiration heat from the fruit. Rooms are made as nearly airtight as possible to contain the ethylene. High RH (90 to 95%) is essential in ripening rooms to avoid fruit dehydration. Moisture may be introduced automatically in the form of steam or a mist or spray of water at ambient temperature. In the absence of a special system, walls and floors may be wetted before closing the room to initiate ripening.

The size of ripening rooms varies with the volume of bananas handled. In modern facilities, bananas are normally handled in palletized cartons. Valuable floor space is better used if rooms are designed for pallets stacked two or three high on supporting framework.

To provide ripening bananas continuously to retail stores, a minimum of three ripening rooms is required; bananas can be delivered from each room on two successive days. Fruit leaving on the second day are somewhat riper than the first day's output. With six or more ripening rooms, all fruit from a single room can be delivered the same day.

CONTROLLED RIPENING

The objective of controlled ripening is to provide retail stores with bananas at a stage of ripeness desired by consumers. The state of ripeness is judged primarily by skin color, using a color number scale of 1 to 7 common in the industry:

1: The finger is hard and completely green
2: Green but with a trace of yellow
3: About half green and half yellow
4: More yellow than green
5: Yellow with green tips
6: Fully yellow
7: Yellow with brown spots

Fruit should be ripened at least to color number 3 before delivery to retail stores, or ripening may not continue normally. Generally, the fruit are not riper than number 4 when shipped from the distribution center to the retail store, as fruit may suffer handling injury if too ripe.

Ripening is initiated by releasing ethylene into the ripening room for 24 to 48 hours with fruit pulp temperatures at 15.5° to 18°C (60° to 65°F) and RH maintained at 90 to 95%. Enough gas is used to provide a concentration, by volume, of 100 to 150 ppm in air with vigorous air circulation, such as forced-air systems to ensure uniform distribution of ethylene. The rate of ripening can be controlled by selecting the appropriate temperature/time schedule.

The ripening rate varies to some extent between lots. Cloudy conditions or low temperature during growth may slow the rate of ripening. Temperature conditions during handling and transit may also affect the ripening rate. In particular, if elevated temperatures have caused some bananas to ripen en route, as shown by slight color changes or slight softening of pulp, the rate of ripening will be much faster than average. The more mature the fingers at harvest the better the quality when bananas ripen.

RESPONSE TO CONTROLLED ATMOSPHERES (CA)

A CA of 2 to 5% O_2 and 2 to 5% CO_2 delays ripening and reduces respiration and ethyl-ene production rates. Exposure to less than 1% O_2 and/or more than 7% CO_2 may cause undesirable texture and flavor. Postharvest life of mature green bananas ranges from 4 to 6 weeks in CA versus 2 to 4 weeks in air at 14°C (58°F). CA of 2% O_2 and 10% CO_2 can be used to delay ripening of partially ripe (color 3 or 4) bananas for up to 1 week at 14°C (58°F).

CHILLING INJURY

Symptoms of chilling injury include surface discoloration, dull or smoky color, subepidermal tissues with dark-brown streaks, failure to ripen, and, in severe cases, flesh browning. Fruit exposed for extended periods to temperatures below about 13°C (56°F) for a few hours to a few days (depending on cultivar, maturity, and temperature) exhibit symptoms of chilling injury. For example, moderate chilling injury will result from exposing mature-green bananas to 1 hour at 10°C (50°F), 5 hours at 11.7°C (53°F), 24 hours at 12.2°C (54°F), or 72 hours at 12.8°C (55°F). Chilled fruit are more sensitive to mechanical injury.

PAPAYAS

Handling and market preparation requirements are influenced greatly by mechanical injury and the susceptibility of papayas to certain diseases. The most important of these is anthracnose, caused by the fungus *Colleto-trichum gloeosporioides* (see chapter 18). Infections occur on the tree by direct fungal penetration during fruit development, but disease development does not proceed at this fruit stage because of the almost-complete immunity of the fruit flesh. Infections remain latent until the start of ripening, when the climacteric rise in respiration rate occurs. Fungicidal sprays in the orchard during fruit development are essential for disease control. These sprays do not eliminate the need for postharvest treatments, but they reduce disease pressure considerably. Control requires the use of postharvest treatments.

The major postharvest diseases other than anthracnose are stem-end rots caused by *Ascochyta caricae-papayae* (see chapter 18), *Phomopsis caricae-papayae*, and *Phytophthora nicotiana* var. *parasitica*. Although characteristically colonizing the fruit stem, these fungi often invade wounds caused by handling.

The presence of the Mediterranean, oriental, and melon fruit flies in Hawaii necessitates postharvest treatment to eliminate flies before shipment to noninfested areas. Fruit handling after disinfestation must be in screened areas. Packages must be sealed to avoid reinfestation, unless they can be loaded into marine containers with screen protection.

HARVESTING

Papaya trees normally have a single trunk, and the lowermost fruits are the oldest. The fruit's position on the tree, therefore, provides an indication of relative maturity (fig. 31.13).

Fruit are generally picked when they exhibit a slight overall loss of green color, with some hint of yellow at the blossom end. Riper fruit are categorized as one-quarter, one-half, and three-quarters yellow. A minimum soluble solids content of 11.5% is required by the Hawaiian grade standards, and this requires the fruit to be at the color-break stage at harvest.

Figure 31.13

Developing papayas are produced successively on the stem. The lowest fruit are the oldest. Size is variable despite position.

For long-distance shipment, fruit are generally harvested at color break or between color break and one-quarter color. To obtain maximum fruit life for long-distance surface shipments to mainland North America, the fruit are better harvested at the color-break stage. It is difficult to distinguish immature-green from mature-green fruit. Immature-green fruit will not ripen after long-distance refrigerated transport.

Fruit can be harvested by pickers standing on the ground while trees are small. As the fruit-bearing area progresses higher, a long-handled suction cup (plumber's friend) is positioned by the picker over the stylar end of the fruit. A twist of the handle breaks the fruit's pedicel, and the picker catches the fruit if it falls from the suction cup. Fruit are placed in a pail and transferred to field boxes or bins. Motorized picking platforms are used in large plantings as a harvesting aid. Bins or other containers on the picking aid machines permit accumulation of the fruit.

TRANSPORTATION AND PACKINGHOUSE OPERATIONS

Bulk bins and 18.2-kg (40-lb) field boxes are used. Extreme care is essential when filling bins or boxes and transporting them to the packinghouse, because of the fruit's sensitivity to abrasion injury. Compression or impact bruises of partially ripe papayas result from careless handling or transportation over rough roads. Sometimes equally serious is abrasion damage to the fruit's tender skin. The damaged skin area generally does not de-green as the fruit ripens.

Preliminary operations include washing and sorting to separate cull and ripe fruit (which are diverted to processing) from those suitable for fresh marketing. This reduces the volume of fruit that must be heat-treated for insect disinfestation and handled within the insect-proof packinghouse.

Papayas are sorted for off-size or defective fruit as they move on conveyor belts. Fruit can be sized by automatic weight sizers (fig. 31.14). Uniformly sized fruit are hand-packed into shipping cartons (fig. 31.15).

For decay control, bins or pallet loads of fruit may be submerged for 20 minutes in water at 46° to 50°C (114.8° to 122°F) with vigorous circulation. As the fruit warms, the water may cool several degrees.

Figure 31.14

A weight sizer is often used with papayas.

Figure 31.15

Workers hand-pack papayas.

The heat treatment operation, in some cases, has been integrated into the packing-house line. At the speed at which fruit moves through such lines, the dwell period in the hot water (without excessively long tanks) would necessarily be very brief. To obtain enough heat penetration for fungicidal effectiveness, the temperature of the water must be very high, 54°C (129°F) for 3 minutes. Positive movement of fruit through such a spray to insure that each fruit is exposed to an exact dwell time is essential. Longer periods lead to heat injury. In papayas, heat injury results in scald, failure to degreen, and increased decay incidence.

HEAT TREATMENTS FOR INSECT CONTROL

There are currently two approved heat treatments for papayas: vapor heat and forced hot air. The vapor treatment is administered in a room with accurate temperature control and adequate air circulation. Fruit temperature is raised until the center of the fruit reaches 47.2°C (117°F) at a RH of 60 to 95%, depending upon treatment.

In 1989 a multistage high-temperature forced-air heat treatment was approved as follows:

Temperature		
Air °C(°F)	Seed cavity °C (°F)	Time (hr)
43 (109.4)	41 (105.8)	2
45 (113.0)	44 (111.2)	2
46.5 (115.7)	46 (114.8)	2
49 (120.2)	47.2 (117.0)	Final

After reaching the final stage, the fruit is then immediately cooled in water at 20° to 25°C (68° to 77°F).

TRANSPORTATION AND STORAGE TEMPERATURES

Optimal temperatures: 0°C (50°F) for mature-green to one-half yellow papayas for less than 21 days; 7°C (45°F) for ripe (greater than one-half yellow) papayas.

Chilling injury. Symptoms include blotchy coloration, uneven ripening, skin scald, hard core (hard areas in the flesh around the vascular bundles), air pockets in the tissues, and increased susceptibility to decay. Increased Alternaria rot is observed in mature-green papayas kept for 4 days at 2°C (36°F), 6 days at 5°C (41°F), 10 days at 7.5°C (45°F), or 14 days at 10°C (50°F). Susceptibility to chilling injury varies among cultivars and is greater in mature-green than ripe papayas: 10 versus 17 days at 2°C (36°F); 20 versus 26 days at 7.5°C (45°F).

Heat Injury. Exposure of papayas to temperatures above 30°C (86°F) for longer than 10 days or to temperature-time combinations beyond those needed for decay and/or insect control results in heat injury (uneven ripening, blotchy ripening, poor color, abnormal softening, surface pitting, accelerated decay). Quick cooling to below

30°C (86°F) after heat treatments minimizes heat injury.

RESPONSES TO ETHYLENE TREATMENT

Exposure to 100 ppm ethylene at 20° to 25°C (68° to 77°F) and 90 to 95% RH for 24 to 48 hours results in faster and more uniform ripening (skin yellowing and flesh softening, but little or no improvement in flavor) of papayas picked at color break to one-quarter yellow stage. Commercial treatments are not recommended due to the rapid softening.

RESPONSES TO CONTROLLED ATMOSPHERES

Optimal CA for papaya is 3 to 5% and 5 to 8% CO_2. Benefits of CA include delayed ripening and firmness retention. Postharvest life potential at 13°C (55°F) is 2 to 3 weeks in air and 2 to 4 weeks in CA, depending on cultivar and ripeness stage at harvest.

MANGOES

The essential operations that prepare mangoes for market include harvesting, transport to packing area, washing, sorting, sizing, packaging, and transport to market. Frequently, mangoes must be treated for quarantine purposes before fruits can enter certain markets.

HARVESTING

Mangoes picked for nearby markets may exhibit color changes from dark green to light green or even to light yellow. Fruit are harvested while still dark green if destined for distant markets requiring several days of transit. Harvesting mangoes prior to external color change makes it difficult to discriminate between fruit that can ripen to an acceptable quality and those that cannot. In general, dark-green mature fruit are harvested when the cheeks have filled out. In some cultivars, maturity is indicated by the degree to which the shoulder extends out at the stem end.

Maturity indices commonly used with other fruits have not proven adaptable to dark-green mangoes. Changes in flesh texture (softening), total soluble solids, and acidity have not proved useful, as they occur mostly after the proper harvest time for distant markets. Studies suggest that flesh color, starch

Figure 31.16

A picking pole of the type commonly used for mangoes or fruits of other large trees.

content, and specific gravity might be useful indices for some cultivars.

Mangoes are harvested by hand if the pickers can reach them. The fruit is twisted sharply sideways or upward to break the pedicel. To avoid stem punctures, any long pedicels are trimmed flush with the stem end of the fruit. Fruit on high branches are harvested with a picking pole that has a cloth bag and cutting knife at the top (fig. 31.16). When the pedicel is severed, the fruit drops into the bag, the picking end of the pole is gently lowered to the ground, and fruit placed into 18- to 23-kg (40- to 50-lb) field boxes or bins for transport to the packinghouse.

PACKINGHOUSE OPERATIONS

Mature-green mangoes exude a large amount of latex from the cut stem. This is washed off with water in a tank. The water may contain imazalil (or thiabendizole, TBZ) fungicide, mainly to control anthracnose caused by the fungus *Colletotrichum gloeosporioides*. Disease control is more effective if the solution is heated (fig. 31.17). Postharvest fungicidal treatments commonly consist of water at 50°C ± 2°C (122°F ± 4°F) containing 0.1% TBZ. Mangoes are submersed in the water for 5 to 10 minutes (depending on fruit size).

Sorters generally examine fruit as it passes on moving belts. Fruit that are judged immature, overmature, or undersized, and those exhibiting limb-rub or other defects, are diverted. Sizing is usually by eye, but use of weight sizers and other mechanical sizers is increasing.

Figure 31.17

Mango packinghouse. A hot water–thiabendazole bath is in the background, and a sorting table is in the foreground.

Although various containers are used, mangoes are commonly packed in one or more layers in fiberboard boxes that may have individual compartments for separating fruit.

Postharvest quarantine treatment for fruit flies

A hot water dip treatment is now registered for mangoes. The duration and temperature of the treatment varies with the country of origin, cultivar, and fruit size. Pulp temperature prior to treatment for all mangoes must be a minimum of 21.1°C (70°F). Generally the fruit are treated in hot water at 46.4°C (115.5°F) for 65 to 90 minutes. The fruit pulp temperature must be between 45.6°C (114°F) and 46.1°C (115°F) for 10 minutes during this treatment. Exceeding the time or temperature combinations recommended for decay or insect control causes heat injury (skin scald, blotchy coloration, uneven ripening).

MANAGEMENT OF RIPENING

Mature-green mangoes may be subjected to an ethylene treatment of 100 ppm for 24 to 48 hours at 20°C (68°F) and 90 to 95% RH to promote faster and more uniform ripening. The treatment may be applied at the shipping point if transit time to market is less than 5 days, or at the destination if transit times are longer.

TRANSPORTATION AND STORAGE TEMPERATURES

Mature-green mangoes should be handled at 13°C (55°F) while partially ripe mangoes can be kept at 10°C (50°F). At lower temperatures mangoes are susceptible to chilling injury, depending on the duration. Symptoms include uneven ripening, poor color and flavor, surface pitting, grayish scald-like skin discoloration, increased susceptibility to decay, and, in severe cases, flesh browning. Chilling injury incidence and severity depend on cultivar, ripeness stage (riper mangoes are less susceptible), and the temperature and duration of exposure.

PINEAPPLES

The pineapple fruit is a multiple structure that evolved, presumably, from a racemose inflorescence. Berrylike fruitlets, generally 100 to 200 in number, and their subtending leafy bracts are together fused to the core, a continuation of the fibrous peduncle. The fruitlets are in a regular spinal pattern on the fruit axis, the pattern usually consisting of two distinct spirals, one turning to the left and the other to the right.

The short, shootlike, leaf-bearing growth at the top of the fruit is called the crown. The crown is a continuation of the original meristem of the plant's main axis extending through the fruit. Crowns are frequently used as planting materials after they are removed from fruits that are to be processed. Slips and suckers developing from axillary buds at the base of the leaves below the fruit and base of the plant, respectively, are other choices for planting materials.

The most widely grown pineapple cultivar is the Smooth Cayenne, although other cultivars are locally important, especially for the fresh market. These include Red Spanish, Queen, Singapore, Spanish, Selangor Green, Sarawak, and Maritius. Postharvest handling operations are shown in figure 31.18.

MATURITY

During maturation pineapples increase in weight, flesh soluble solids, and acidity. During ripening, the carotenoid pigments and soluble solids of the flesh increase dramatically, and the fruit attains its maximum aesthetic and eating quality. During ripening, the shell of the pineapple loses chlorophyll rapidly, starting at the fruit base, in a process similar to that of de-greening citrus fruit.

Approximately 110 days elapse between the end of flowering and ripeness. Ethephon may be sprayed about 6 days before planned harvest to stimulate color change from green to yellow.

Harvest time is often when the base of the fruit has changed from green to yellow or light brown. Fruit may be harvested for fresh market before striking color changes have occurred, since acceptable quality may develop before color changes occur in the shell. Since the pineapple fruit has no accumulation of starch, there is no reserve for major postharvest quality improvements. As a non-climacteric fruit, obvious compositional changes after harvest are mostly limited to de-greening and a decrease in acidity. A minimum soluble solids of 12% and a maximum acidity of 1% will assure minimum flavor acceptability by most consumers.

Figure 31.18

Postharvest handling of pineapples.

Harvest by hand
↓
Place on conveyor belt
↓
Automatically load into gondolas
↓
Transport to packinghouse or processing plant
↓
Dump
↓
Trim stem to 1 to 3 cm (0.4 to 1.2 in)
↓
Wax spray or dip with a benzimidazole fungicide
↓
Sort for quality
↓
Sizing
↓
Hand-pack into cartons
(dividers may be used to reduce fruit-to-fruit contct)
↓
Cooling
↓
Load into transit vehicle
↓
Air or marine transport to market

HARVESTING

Pickers select fruit by color, size, or both, and twist it from the stalk. In small operations the harvested fruit may be field-packed or placed in sacks or baskets that are carried to the end of the row for pickup.

Large-scale production avoids hand-carrying by using a harvesting aid consisting of a belt extending across a number of rows (fig. 31.19). Fruit picked by hand are placed on the belt and carried to the machine where the fruit are accumulated in a bin or truck gondola. The machine and belts, usually mounted on a truck chassis, move slowly across the field at a speed determined by the pickers.

PACKINGHOUSE OPERATIONS

At the packinghouse fruit are sorted to eliminate defective fruit. Sizing may be by eye or weight sizers (fig. 31.20).

A fungicidal treatment consists of a Bayleton dip or spray, commonly applied before packing, to control black rot disease caused by the fungus *Thielaviopsis paradoxa*.

Figure 31.19

A pineapple harvest aid in use.

Figure 31.20

Weight-sizing pineapples.

Figure 31.21

Wax is applied to pineapple fruits.

Figure 31.22

Workers hand-pack pineapple fruits.

Uncontrolled, the disease is serious in fruit after harvest. The fungus enters the fruit via wounds or at the broken stem and grows rapidly throughout the fruit flesh. A very watery soft rot is produced. In the past, the stem surface was smeared with a paste containing a fungicide such as benzoic acid to prevent water blister.

Figure 31.23

Pineapples are packed in shipping cartons.

Figure 31.24

A refrigerated marine container is loaded with pineapples.

Pineapples are sometimes waxed to reduce chilling disorders (fig. 31.21) and improve appearance. Fruit are packed into single-layer or full telescoping cartons (figs. 31.22 and 31.23) with inside dimensions of approximately 30.5 cm wide by 45 cm long by 3l cm deep (12 by 17.7 by 12.2 in). From 8 to 14 or 16 fruit of uniform size are placed in each container.

HOLDING AND TRANSIT

Pineapple fruit are shipped in marine containers (fig. 31.24) or by air. Postharvest life potential is 2 to 4 weeks in air and 4 to 6 weeks in CA (3 to 5% O_2 + 5 to 8% CO_2) at 10°C (50°F), depending on cultivar and ripeness stage. The pineapple is a chilling-sensitive tropical fruit. Chilling injury symptoms include darkening of flesh tissues, particularly around the central cylinder. This condition was called endogenous brown spot (EBS) and was frequently seen in pineapples (fig. 31.25) harvested in the winter. Temperatures below about 6°C (42.8°F) may result in chilling injury.

Figure 31.25

Endogenous brown spot disorder of pineapples.

Waxing pineapples to encourage accumulation of up to 5 to 8% CO_2 is effective in reducing chilling injury symptoms. A heat treatment at 35°C (95°F) for 1 day has been shown to ameliorate EBS symptoms in pineapples transported at 7°C (45°F) by inhibiting activity of polyphenol oxidase and consequently tissue browning. However, this treatment is difficult to apply on a commercial scale and more expensive than waxing.

REFERENCES

Alvarez, A. M., and W. T. Nishijima. 1987. Postharvest diseases of papaya. Plant Dis. 71:681–686.

Barmore, C. R. 1974. Ripening mangoes with ethylene and ethephon. Proc. Fla. State Hortic. Soc. 87:331–334.

Cappellini, R. A., M. J. Ceponis, and G. W. Lightner. 1988a. Disorders in apricot and papaya shipments to the New York market, 1972–1985. Plant Dis. 72:366–370.

———. 1988b. Disorders in avocado, mango, and pineapple shipments to the New York market, 1972–1985. Plant Dis. 72:270–274.

Chen, N. M., and R. E. Paull. 1986. Development and prevention of chilling injury in papaya fruit. J. Am. Soc. Hortic. Sci. 111:639–643.

Couey, H. M. 1989. Heat treatment for control of postharvest diseases and insect pests of fruits. HortScience. 24:198–202.

Dodd, J. C., D. Prusky, and P. Jeffries. 1997. Fruit diseases. In R. E. Litz, ed., The mango: Botany, production and uses. Wallingford, UK: CAB International. 257–280.

Eckert, J. W., and J. M. Ogawa. 1985. The chemical control of postharvest diseases: Subtropical and tropical fruits. Annu. Rev. Phytopathol. 23:421–454.

Gomez-Lim, M. A. 1997. Postharvest physiology. In R. E. Litz, ed., The mango: Botany, production and uses. Wallingford, UK: CAB International. 425–445.

Hassan, A., and E. B. Pantastico, eds. 1990. Banana: Fruit development, postharvest physiology, handling and marketing in ASEAN. Kuala Lumpur: ASEAN Food Handling Bureau. 147 pp.

John, P., and J. Marchal. 1995. Ripening and biochemistry of the fruit. In S. Gowen, ed., Bananas and plantains. London: Chapman and Hall. 434–467.

Johnson, G. I., J. L. Sharpe, D. L. Milne, and S. A. Oosthuyse. 1997. Postharvest technology and quarantine treatments. In R. E. Litz, ed., The mango: Botany, production and uses. Wallingford, UK: CAB International. 447–507.

Marriott, J. 1980. Bananas—Physiology and biochemistry of storage and ripening for optimum quality. Crit. Rev. Food Sci. Nutr. 13:41–88.

Mendoza, D. B., Jr., and R. B. Wills, eds. 1984. Mango: Fruit development; postharvest physiology, and marketing in ASEAN. Kuala Lampur: ASEAN Food Handling Bureau. 111 pp.

Mitra, S., ed. 1997. Postharvest physiology and storage of tropical and subtropical fruits. Wallingford, UK: CAB International. 423 pp.

Nagy, S., and P. E. Shaw. 1980. Tropical and subtropical fruits—Composition, properties and uses. Westport, CT: AVI. 570 pp.

Nagy, S., P. E. Shaw, and W. F. Wardowski, eds. 1990. Fruits of tropical and subtropical origin: Composition, properties and uses. Lake Alfred, FL: Florida Science Source. 391 pp.

Nakasone, H. Y., and R. E. Paull. 1998. Tropical fruits. Wallingford, UK: CAB International. 445 pp.

Pantastico, E. B. 1975. Postharvest physiology, handling and utilization of tropical and subtropical fruits and vegetables. Westport, CT: AVI. 560 pp.

Paull, R. E. 1993. Pineapple and papaya. In G. B. Seymour et al., eds., Biochemistry of fruit ripening. London: Chapman and Hall. 291–323.

Paull, R. E., and N. J. Chen. 1989. Waxing and plastic wraps influence water loss from papaya fruit during storage and ripening. J. Am. Soc. Hortic. Sci. 114:937–942.

Paull, R. E., and K. G. Rohrbach. 1985. Symptoms development of chilling injury in pineapple fruit. J. Am. Soc. Hortic. Sci. 110:100–105.

Paull, R. E., W. Nishijima, M. Reyes, and C. Cavaletto. 1997. Postharvest handling and losses during marketing of papaya. Postharv. Biol. Technol. 11:165–179.

Seymour, G. B. 1993. Banana. In G. B. Seymour et al., eds., Biochemistry of fruit ripening. London: Chapman and Hall. 83–106.

Shaw, P. E., H. T. Chan Jr., and S. Nagy, eds. 1998. Tropical and subtropical fruits. Auburndale, FL: AgScience. 569 pp.

Slabaugh, W. R., and M. D. Grove. 1982. Postharvest diseases of bananas and their control. Plant Dis. 66:746–750.

Thompson, A. K., and O. J. Burden. 1995. Harvesting and fruit care. In S. Gowen, ed., Bananas and plantains. London: Chapman and Hall. 403–433.

Wainwright, H., and M. B. Burbage. 1989. Physiological disorders in mango (*Mangifera indica* L.) fruit. J. Hort. Sci. 64:125–135.

Yahia, E. M. 1998. Modified and controlled atmospheres for tropical fruits. Hort. Rev. 22:123–183.

Yon, R. M., ed. 1994. Papaya: Fruit development, postharvest physiology, handling and marketing in ASEAN. Kuala Lumpur: ASEAN Food Handling Bureau. 144 pp.

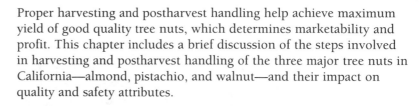

32

Postharvest Handling Systems: Tree Nuts

Adel A. Kader and James F. Thompson

Proper harvesting and postharvest handling help achieve maximum yield of good quality tree nuts, which determines marketability and profit. This chapter includes a brief discussion of the steps involved in harvesting and postharvest handling of the three major tree nuts in California—almond, pistachio, and walnut—and their impact on quality and safety attributes.

HARVESTING

WHEN TO HARVEST

Tree nuts should be harvested as soon as possible after maturation to avoid quality loss and to minimize problems involving fungal attack and infestation with insects, especially the navel orangeworm. The following indices determine optimal harvesting dates.

- Almond: Dehiscence (splitting) of the hull; separation of hull from shell; decrease in fruit removal force (development of abscission zone); drying of hull, shell, and kernel.

- Pistachio: Ease of hull separation from the shell; shell dehiscence (splitting) and color; decrease in fruit removal force; kernel dry weight and crude fat content.

- Walnut: Ease of hull removal (hullability); packing tissue browning (when the packing tissue between and around the kernel halves has just turned brown). Often the hull matures later than the kernel in warm-climate zones.

Uneven maturation presents a problem in once-over harvesting. Nuts on the tree's periphery usually mature earlier than those at the center. The use of ethephon to overcome this problem or accelerate maturation has produced mixed results that vary with the species. Although ethephon application accelerates almond maturation, it does not improve uniformity of maturity, and it may reduce yield and induce tree gummosis. Consequently, ethephon is not used on almond. Studies with pistachio indicate that ethephon application does not influence nut maturation. Ethephon applied at the time of "packing-tissue browning" allows walnut harvest within 7 to 10 days (instead of 15 to 20 days without ethephon). This treatment is used commercially in walnut harvesting, but applications to stressed trees can cause defoliation.

HARVESTING PROCEDURES

Harvest season in California depends on the cultivar, production area, and cultural practices in use. Almonds are harvested between July and October. Pistachios (primarily the Kerman cultivar) are harvested during September. Walnuts are harvested during September and October. Delayed harvest results in darker color walnut kernels, which have lower commercial value. Orchard floor management is important in preparing for the harvest of almond and walnut, since the nuts are knocked to the ground during harvesting. Once harvest begins, it is important to pick up, hull, and dry nuts as rapidly as possible. Prolonged exposure to damp ground increases mold incidence and shell staining.

Most harvest operations are mechanized. Almonds and walnuts are usually knocked or shaken to the ground by mechanical tree shakers (fig. 32.1), raked into windrows, and then picked up with sweepers

Figure 32.1

Tree shaker used for harvesting almonds and walnuts.

Figure 32.2

Walnut sweeper.

Figure 32.3

Shake-catch system used for harvesting pistachios.

(fig. 32.2). Pistachios are harvested with a shake-catch mechanical harvester (fig. 32.3), and the nuts are placed in bins 1.2 by 1.2 by 0.6 m (4 by 4 by 2 ft). Pistachios should not be shaken to the ground because of their open shells and high moisture content, relative to almonds and walnuts, at harvest.

Hand-harvesting (knocking) is used on young trees and where steep terrain makes mechanical harvesting difficult. Tarps are spread under the tree. Pickers then hit the scaffold branches with a maul (rubber mallet) or tap the branches with poles to knock the nuts loose, and the nuts are collected on the tarps.

POSTHARVEST HANDLING

A comparison of the handling systems for almond, walnut, and pistachio is illustrated in figure 32.4. Several considerations should be noted.

Almonds are picked up from the orchard floor as soon as they are dry to avoid exposure to adverse weather conditions, especially rain, and to minimize fungal infection and insect damage. Exposure of almonds to wet and hot conditions results in concealed damage, an internal disorder characterized by a rust-brown to black discoloration of the nut meat and, in extreme cases, an unpalatable off-flavor.

Walnuts must be picked up, hulled, and dried as soon as possible after harvest. Walnuts left on the ground can deteriorate (the kernels darken) rapidly, especially at high ambient temperatures, such as 3 hours at 32°C (90°F).

Pistachios should be hulled and dried soon after harvest to minimize shell staining and decay. Noticeable shell staining may occur if unhulled pistachios are held for 16 hours at 40°C (104°F). They can be held for up to 24 hours at 25°C (77°F) without staining under California's dry environment. Unhulled pistachios can be kept cool with airflow through them induced by highway transport. If temporary storage of fresh pistachios at the dehydration plant is necessary, they should be cooled and held before hulling at 0°C (32°F) and a RH lower than 70%. Sorting before cold storage to remove defective nuts, leaves, twigs, and other foreign materials (which are usually much more susceptible to decay) minimizes losses during cold storage.

Figure 32.4

Postharvest handling of almonds, walnuts, and pistachio nuts.

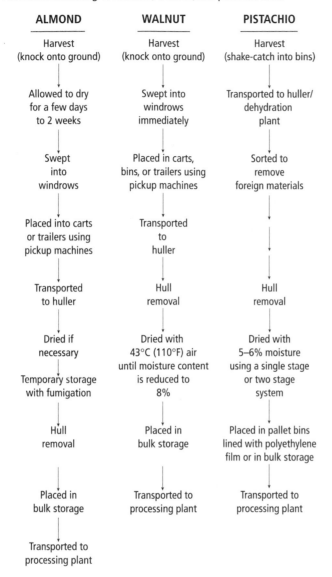

ALMOND	WALNUT	PISTACHIO
Harvest (knock onto ground)	Harvest (knock onto ground)	Harvest (shake-catch into bins)
Allowed to dry for a few days to 2 weeks	Swept into windrows immediately	Transported to huller/ dehydration plant
Swept into windrows	Placed in carts, bins, or trailers using pickup machines	Sorted to remove foreign materials
Placed into carts or trailers using pickup machines	Transported to huller	
Transported to huller	Hull removal	Hull removal
Dried if necessary	Dried with 43°C (110°F) air until moisture content is reduced to 8%	Dried with 5–6% moisture using a single stage or two stage system
Temporary storage with fumigation		
Hull removal	Placed in bulk storage	Placed in pallet bins lined with polyethylene film or in bulk storage
Placed in bulk storage	Transported to processing plant	Transported to processing plant
Transported to processing plant		

Fumigation with methyl bromide or hydrogen phosphide is often used to control insects in stored nuts. It may have to be repeated periodically depending on duration of storage and if temperature exceeds 15°C (59°F).

DRYING

Water is present in plant tissues in three forms: bound water, which is bound with other constituents by strong chemical forces; adsorbed water, which is held by molecular attraction to adsorbing substances; and absorbed water, which is held loosely in the extracellular spaces by the weak forces of capillary action. The absorbed and adsorbed water constitute the "free water," most of which is removed during drying. Bound water is not removed except at very high temperatures that would also decompose some organic matter.

The moisture content of almonds, pistachios, and walnuts at harvest ranges between 5 and 15%, 40 and 50%, and 10 and 30% (fresh weight basis), respectively. To improve stability and ensure the safety of the nuts, they should be dried to 5 to 8% moisture (water activity of 0.50 to 0.65) as soon after harvest as possible.

Almonds

Most of the final drying of almonds occurs naturally while they are on the orchard floor. However, when rainy conditions prevail during harvesting, heated-air drying may be used to complete their dehydration.

Wet almonds must be dried before hulling either in a batch dryer with a maximum temperature of 49° to 54°C (120° to 130°F) or in continuous-flow dryers with a maximum temperature of 82°C (180°F). Many batch dryers are specially designed 5-ton-capacity wagons with a perforated floor that allows heated air to be distributed underneath the nuts (fig. 32.5). The continuous-flow dryers are either horizontal belt dryers or cross-flow grain dryers. Dried almonds hull more easily than those with wet hulls, allowing hulling equipment to operate at maximum capacity.

Walnuts

Walnuts are dried to 8% (range of 3 to 13%) final moisture in batch dryers with air temperature not exceeding 43°C (110°F), as higher temperatures induce rancidity. Drying times may be as little as 4 hours for nuts that come from the field nearly dry to 2 days for extremely wet nuts. At 20% RH, increasing air temperature from 32° to 43°C (90° to 110°F) reduces the drying time by one-third (from 25 to 17 hours). The wagon dryers used for almonds are also suited to walnut drying, although most walnuts are dried in self-unloading bin dryers. Another system, unique to walnuts, is the pot hole dryer (see fig. 32.5). Metal pallet bins with expanded metal floors are filled with nuts and placed on top of an underground air plenum that supplies heated air to the bins.

Figure 32.5

Three types of dehydrators used for drying nut crops.

Wagon dehydrator

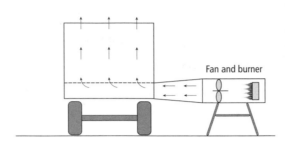

"Pot hole" dehydrator

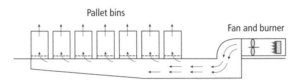

Stationary bin dehydrator

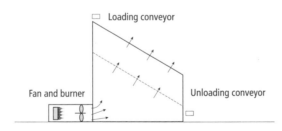

Pistachios

Most recently installed pistachio drying facilities use a two-stage process. The hulled nuts are first dried for about 3 hours in a column dryer originally designed for grain (fig. 32.6). Air temperature is kept below about 82°C (180°F) to prevent the shell from opening too far and allowing the kernel to separate from the shell. Nuts are dried to 12 to 13% moisture and then transferred to a flat-bottomed grain bin, where they continue to dry to 5 to 6% moisture with unheated air or air heated to less than 49°C (120°F). The final drying stage requires 24 to 48 hours. The nuts are left in the bins for storage until they are processed. This system increases the drying capacity of the expensive grain dryers and uses much less energy than complete drying in a high-temperature dryer. Some operations use a rotating drum dryer in place of the grain dryer for the first stage of drying. A few smaller operations use a self-unloading bin dryer for single-stage drying to the final moisture. Air temperatures in these are kept below 60° to 66°C (140° to 150°F), and drying time is about 8 hours.

Post-drying storage

Dried nuts are usually stored in bins, silos, or other bulk-storage containers for a few weeks or several months before final processing and preparation for market. Optimal

Figure 32.6

Cross-flow dryers used for pistachios.

storage conditions of 10°C (50°F) or lower and less than 65% RH must be maintained to minimize deterioration during storage. Protection against insects is also essential during the storage period.

PREPARATION FOR MARKET

QUALITY AND SAFETY
Appearance. Important factors in a nut's appearance include freedom from defect (shell staining, insect damage, mold, adhering hull, kernel discoloration, shriveling), kernel size relative to nut size (percent edible portion), and cleanliness.

Texture. Crispness, chewiness, and other textural quality attributes are influenced by the degree and uniformity of dryness within the nut and among nuts.

Flavor and nutritive value. Sweetness, oiliness, and roasted flavor are usually related to good overall flavor, while rancidity is a major factor in poor quality. Nuts are a healthy snack if eaten in moderation since they are good sources of protein, essential fatty acids, fiber, vitamin E, and minerals. Recent studies have shown an association between frequent nut consumption and reduced risk of coronary heart disease.

Safety. Aflatoxin contamination must be avoided by protecting the nuts against growth of *Aspergillus flavus* before harvest, and after harvest before drying. Aflatoxin is a highly carcinogenic secondary metabolite produced by the fungi *Aspergillus flavus* and *A. parasiti-*

cus. Often *A. flavus* is used in a collective sense to also include *A. parasiticus.*

Most prone to contamination in the orchard are the nuts poorly protected by hulls. In pistachios, early-split nuts vary between 1 and 5% from tree to tree. Early splits are nuts in which the shell splits before the hull has dehisced. Consequently, the entire pericarp splits to expose the kernel to the elements. In one study, about 20% of each 50-nut sample of early-split but otherwise good nuts were found to be contaminated with aflatoxin, as opposed to no contamination in regular-split nuts, where the shell had split within the loose, intact hull. Thin-shelled almonds are much more likely to be contaminated than thick-shelled nuts, probably because of poorer shell seal. Walnuts are less likely to develop aflatoxin than other nuts, but all nut species exposed to moist soil on the orchard floor risk infection.

All nuts infested with navel orangeworm or certain other insects risk aflatoxin contamination, but these nuts pose less of a danger because they are easier to eliminate by sorting, and consumers are likely to reject them. In contrast, nuts not infested with insects cannot usually be eliminated by normal sorting methods. Furthermore, consumers will not necessarily reject them.

Early-split pistachio nuts not infected in the orchard may become infected during transport and handling. High humidities and temperatures within bulk bins provide ideal conditions for the infection of early-split nuts. The incidence of nuts with aflatoxin increases and aflatoxin levels rise dramatically, until nuts are mycologically stabilized by drying or refrigeration after removal of field heat. Fungicidal sprays in the orchard at the time of early splitting might materially reduce the potential for aflatoxin development, but no tests have demonstrated their effectiveness.

PROCESSING OPERATIONS
Processing nuts involves the following steps:

1. Sorting to eliminate defects—nuts with adhering hulls, stained and moldy nuts—using manual or light-reflectance electronic sorting techniques.

2. Separating empty or partially empty pistachios and walnuts by an air stream.

3. Cracking shells to extract kernels when desired.

4. Sorting to eliminate nuts with possible aflatoxin contamination, a labor-intensive and costly, but necessary, operation. Efforts to develop reliable automated sorters continue.

5. Sizing in-shell or shelled nuts into size categories using mesh screens.

6. Sorting by color of shell or kernel; chemical bleaching of in-shell almonds and walnuts to improve shell color.

7. Treatment with antioxidants to slow rancidity resulting from oxidation of fatty acids.

8. Blanching, slicing, or chopping almond kernels.

9. Salting, flavoring, and roasting.

10. Packaging of in-shell nuts, shelled nuts, and nutmeats (broken kernels) in various types and sizes of package to protect against insects, provide an effective moisture barrier, exclude oxygen to slow rancidity development, and exclude light to minimize color deterioration of some nuts.

11. Rejected nut meats may be pressed for oil extraction. Shells are often used for charcoal production or as boiler fuel.

MARKETING AND UTILIZATION

Tree nuts are marketed in the shell or as shelled intact kernels or kernel pieces for use as snacks or in confectionery and bakery products. U.S. per capita consumption in 1999 was 260 grams (9.2 oz) for almonds, 80 grams (2.8 oz) for pistachios, and 170 grams (6.0 oz) for walnuts. Export marketing accounts for as much as 75% of the U.S. production for some tree nuts.

STORAGE OF NUTS

Maintenance of quality and storage life of nuts depends on their moisture content, the RH and temperature in storage, the exclusion of O_2, and effective insect control. The role each factor plays in determining the storage potential of nuts is discussed below.

MOISTURE CONTENT

The relationship between air RH and water activity (a_w) of nuts can be expressed as

$$a_w = 0.01 \times RH$$

when RH is in equilibrium with nut moisture content. Water activity is important in relation to the nuts' susceptibility to fungal attack, including mycotoxin-producing organisms (*Aspergillus flavus* and *A. parasiticus*). FDA regulations for tree nuts define a "safe moisture level" (moisture content that does not support fungal growth) as an a_w that does not exceed 0.70 at 25°C (77°F).

The relationship between a_w and moisture content of selected tree nut meats at 21°C (70°F) is shown below.

Nut	Moisture content (%) at a_w = 0.2 to 0.8
Almond	3.0–8.7
Pistachio	2.2–8.2
Walnut	2.8–7.0

EFFECT OF RELATIVE HUMIDITY

The equilibrium moisture content of a product as a function of RH at a given temperature can be illustrated by the sorption curve as shown in figure 32.7 for walnuts and pistachios. This curve has three sections:

• The lowest section of the curve, representing bound water, is concave to the RH axis. As the moisture content rises, the water molecules form a monolayer that coincides with about 20% RH.

• As the RH increases, water molecules form successive layers of diminishing bond to the adsorbing substance of the commodity. This section of the curve (30 to 70% RH) is almost linear.

• Above 75% RH, the product absorbs water to saturation (large increases in moisture

Figure 32.7

Relationship between pistachio and walnut moisture content (on a fresh-weight basis) and ambient relative humidity at indicated temperatures.

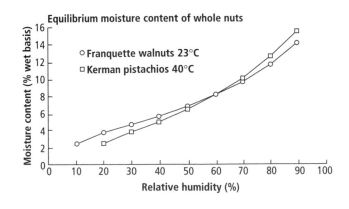

content result from small increases in RH). This encourages deterioration and attack by microorganisms.

The equilibrium moisture of almonds ranges from 3.8 to 5% at 30% RH and from 8.1 to 10% at 75% RH.

EFFECT OF TEMPERATURE

The relationship between moisture content and equilibrium relative humidity (ERH) is temperature-dependent. Between 20% and 80% ERH for any given moisture content, ERH rises about 3% for every 10°C (18°F) rise in temperature. For any given RH, air contains more water vapor at a high temperature than at a low temperature.

Lower temperatures reduce or stop insect activity and retard mold growth and deterioration, including lipid oxidation (which results in rancidity). Temperatures between 0° and 10°C (32° and 50°F) are recommended for tree nuts, depending on expected storage duration. Duration can exceed 1 year; the lower the temperature, the longer the storage life.

Walnuts must be protected from freezing (−10°C, or 14°F) because they have a very short shelf life after thawing.

EFFECT OF OXYGEN LEVEL

Low O_2 (less than 0.5%) is a beneficial supplement to proper temperature management to maintain flavor quality and insect control.

Table 32.1. Options and alternatives for insect disinfestation of nuts

	Chemical fumigation		Alternatives	
	Methyl bromide	Phosphine	Controlled atmospheres (low O_2 and/or high CO_2)	Irradiation
Treatment duration	Up to 24 hr + purging/aeration	2–3 days + purging aeration	3–7 days at 27°C (80.6°F) + purging/aeration	Relatively short exposures
Effectiveness	100% kill rate	100% kill rate	Can achieve 100% kill rate	Some larvae may survive up to 2 months
Protection against reinfestation	NONE, unless combined with nightly pyrethrin or vapona fogs; requires refumigation	Same as methyl bromide	NONE; requires repeated treatments or continuous maintenance of the CA	Some limited protection due to possible presence of sterilized males. Reirradiation is not allowed
Quality of product	Good	Good	Good—may be better than with chemical fumigation	Good at recommended levels
Health effects				
Workers	Highly toxic if directly exposed	Highly toxic if directly exposed	Presumed safer than chemical fumigation	Safe if facility properly shielded and operated
Consumers	Leaves residues that may be carcinogenic	No known hazards	No known hazards	No known hazards
Environmental effect	May involve release of toxic chemicals into environment	May involve release of toxic chemicals into environment	Does not pollute environment; somewhat energy-intensive	Some versions involve transportation and use of radioactive materials
Approximate annual cost of treatment (cents/lb)				
Almonds	—	<0.1¢	(low O_2 generator) 0.1–0.25¢	0.36–1.0¢
Walnuts	<0.1¢	—	0.25–0.5¢	0.42–3.6¢
Other considerations	Requires little capital investment; same per-lb cost for small and large operators; could be withdrawn from market or require Prop. 65 identification in California	Has some potential insect resistance problems; same per-lb cost for any size operator	Longer kill times could require construction of additional treatment chambers at some locations or alteration of existing storage facilities to make them gas-tight	Requires large capital investment; more economical for large operators; requires labeling of treated product with potential for consumer resistance

Source: Lindsey et al. 1989.

Exclusion of O_2 is usually done by vacuum packaging or by packaging in nitrogen.

In-shell almonds and walnuts have about twice the storage life of shelled nuts. The shell is an effective package to protect the kernel from physical damage. The pellicle of a walnut kernel acts as a protective barrier and contains antioxidants.

INSECT CONTROL

Fumigation treatments using methyl bromide or phosphine may be used on stored nuts before final processing and packaging. Alternative treatments include controlled atmospheres (especially O_2 levels below 0.5% and CO_2 levels above 40%) and irradiation (table 32.1). Temperatures near freezing or between 40° and 50°C (104° and 120°F) are also effective for insect control as mentioned above; their use as fumigation substitutes is expected to increase in the future. The use of insect-proof packaging is essential to prevent reinfestation.

REFERENCES

Beuchat, L. R. 1978. Relationship of water activity to moisture content in tree nuts. J. Food Sci. 43:754–758.

Ferguson, L., A. Kader, and J. Thompson. 1995. Harvesting, transporting, processing and grading. In Pistachio production. Davis: Univ. Calif. Davis Fruit and Nut Res. and Info. Ctr. 110–114.

Guadagni, D. G., E. L. Soderstrom, and C. L. Storey. 1978. Effects of controlled atmosphere on flavor stability of almonds. J. Food Sci. 43:1077–1080.

Kader, A. A. 1996. In-plant storage. In W. C. Micke, ed., Almond production manual. Oakland: Univ. Calif. Div. Ag. and Nat. Res. Publ. 3364. 274–277.

Kader, A. A., C. M. Heintz, J. M. Labavitch, and H. L. Rae. 1982. Studies related to the description and evaluation of pistachio nut quality. J. Am. Soc. Hortic. Sci. 107:812–816.

Labavitch, J. M. 1978. Relationship of almond maturation and quality to manipulations performed during and after harvest. In W. C. Micke, ed., Almond orchard management. Oakland: Univ. Calif. Div. Agric. Sci. Publ. 4092. 146–150.

Labavitch, J. M., C. M. Heintz, H. L. Rae, and A. A. Kader. 1982. Physiological and compositional changes associated with maturation of 'Kerman' pistachio nuts. J. Am. Soc. Hortic. Sci. 107:688–692.

Lindsey, P. J., S. S. Briggs, A. A. Kader, and K. Moulton. 1989. Methyl bromide on dried fruits and nuts: Issues and alternatives. In Chemical use in food processing and postharvest handling: Issues and alternatives. Davis: Univ. Calif. Ag. Issues Ctr. 41–50.

Olson, W. H., J. M. Labavitch, G. C. Martin, and R. H. Beede. 1998. Maturation, harvesting, and nut quality. In D. E. Ramos, ed., Walnut production manual. Oakland: Univ. Calif. Div. Ag. and Nat. Res. Publ. 3373. 273–276.

Santerre, C. R., ed. 1994. Pecan technology. New York: Chapman and Hall. 164 pp.

Schirra, M. 1997. Postharvest technology and utilization of almonds. Hort. Rev. 20:267–311.

Soderstrom, E. L., and D. G. Brandl. 1990. Controlled atmospheres for preservation of tree nuts and dried fruits. In M. Calderon and R. Barkai-Golan, eds., Food preservation by modified atmospheres. Boca Raton, FL: CRC Press. 83–92.

Thompson, J. F. 1981. Reducing energy costs of walnut dehydration. Oakland: Univ. Calif. Div. Agric. Sci. Leaflet 21257. 11 pp.

Thompson, J. F., T. R. Rumsey, and J. A. Grant. 1998. Dehydration. In D. E. Ramos, ed., Walnut production manual. Oakland: Univ. Calif. Div. Ag. and Nat. Res. Publ. 3373. 277–284.

Thompson, J. F., T. R. Rumsey, and M. Spinoglio. 1997. Maintaining quality of bulk-handled, unhulled pistachio nuts. Appl. Eng. Agr. 13:65–70.

Wells, A. W., and H. R. Barber. 1959. Extending the market life of packaged shelled nuts. USDA Mktg. Res. Rept. 329. 14 pp.

Woodroof, J. G. 1979. Tree nuts: Production, processing, products. Westport, CT: AVI. 712 pp.

33

Postharvest Handling Systems: Fruit Vegetables

Marita I. Cantwell and

Robert F. Kasmire

The fruit vegetables comprise two important groups that can be distinguished by the stage of maturity at harvest:

Immature fruit vegetables:

- Legumes: snap, lima, and other beans; snow pea, sugar snap and garden peas
- Cucurbits: cucumber, soft-rind or summer squashes, chayote, bitter melon, luffa
- Solanaceous vegetables: eggplant, tomatillo
- Others such as okra and sweet corn

Mature fruit vegetables:

- Cucurbits: cantaloupe, honeydew, and other muskmelons; watermelon, pumpkin, hard-rind or winter squashes
- Solanaceous vegetables: mature-green and vine-ripe tomatoes, mature-green and ripe peppers

With the exceptions of peas and broad beans, fruit vegetables are warm-season crops. All the fruit vegetables, with the exception of sweet corn and peas, are subject to chilling injury. Fruit vegetables are not generally adaptable to long-term storage. Exceptions are the hard-rind (winter) squashes and pumpkin. This chapter discusses the general postharvest requirements and handling systems for this group of commodities. The immature fruit vegetables and the mature fruit vegetables are discussed separately.

IMMATURE FRUIT VEGETABLES

MATURITY INDICES

The harvest index for most immature fruit vegetables is based principally on size and color. Immature soft-rind squashes, for example, may be harvested at several sizes or stages of development, depending upon market needs. Fruit that are too developed when harvested are of inferior quality and show undesirable seed development and color changes after harvest. Because these vegetables are growing very rapidly during the harvest period, frequent harvests are necessary to ensure harvesting at the desired stage of maturity. These vegetables are subject to very rapid compositional changes (e.g., conversion of sugars to starch in peas and sweet corn, fiber development in okra, increases in sugars and acids in tomatillo), high rates of water loss and loss of firmness (beans and summer squash), and rapid color changes (cucumbers, bitter melon). It is, therefore, extremely important to cool them as soon as possible after harvest to minimize undesirable postharvest changes.

HARVEST

Most of the immature fruit vegetables are harvested by hand, with pickling cucumbers, snap beans, peas, and sweet corn being the major exceptions. Mechanically harvested peas and sweet corn may be harvested at night when product temperatures are the coolest. Most of these vegetables have very tender skins that are easily damaged during harvest and handling. Special care must be taken in all handling operations to prevent product damage and the associated loss of visual appearance, increased water loss, and increased decay. Minimizing handling transfers of the vegetables is key to reducing physical damage.

Figure 33.1

Postharvest handling of immature fruit vegetables such as summer squash, eggplant, and cucumbers.

Hand-harvest, eliminating defective fruit that cannot be packed
Place fruit into clean plastic trays or buckets

Field-packed | Packinghouse

Fruit often not cleaned; can be immersed in clean water, spray-washed, or wiped with a moist clean cloth

Stack trays on trailer for transport to packing area

De-stack trays and spray-wash or immerse in clean water

Select and classify by size, maturity and defects

Apply protective wax or vegetable oil by spray application or wiping fruit

Pack by weight or count into shipping containers

Select and classify by size, maturity, and defects

Palletize and transport to cooler

Pack by weight or count into shipping containers

Cool to 7° to 13°C (45° to 55°F)

Cool to 7° to 13°C (45° to 55°F)

Transport to distribution center (7° to 13°C, or 45° to 55°F)

Transport to distribution center (7° to 13°C, or 45° to 55°F)

FIELD-PACKING

Field-pack operations, in which grading, sorting, sizing, packing, and palletizing are carried out in the field, minimize product handling (fig. 33.1). Mobile packing facilities are commonly towed through the fields for eggplant, cucumber, and summer squashes. An alternative is to have small shaded packing tables at the ends of rows. The vegetable is harvested into buckets or plastic containers and then packed directly from the harvest container. This reduces product damage and, therefore, increases pack-out yields. Handling costs are also reduced in field-pack operations. One difficulty with field packing is the need for increased supervision to maintain consistent quality in the packed product. Proper product sanitation and cleaning is difficult

with field-packing. A further consideration with field-pack operations is that small lots of packed product need to be frequently transported to a cooling facility to minimize water loss and other undesirable postharvest changes.

PACKINGHOUSE OPERATIONS

A centralized packinghouse can result in packed product of a more consistent quality (fig. 33.1). However, the product is usually harvested and dumped into a larger container or bin, which in turn is dumped onto the packing line. If a water dump is used, and the water recycled, chlorine or another sanitizing agent needs to be used to avoid microbial contamination. In small operations, product may be packed directly from the harvest container, as is done with field packing. If plastic crates are used, the product can be batch-washed or dipped to remove surface dust on the product. Although using a packinghouse may permit a more uniform pack-out, it also usually causes more damage. In addition to dumping injuries, dirt on conveyors or tables can easily scratch and damage the skin of the tender immature vegetables. However, a packinghouse operation permits use of fungicides and waxes or other treatments that are not easily accomplished in the field.

After an initial sorting or selection for removing cull and other defective fruit, food-grade waxes, but more commonly vegetable oils, may be applied to cucumber, eggplant, and summer squash. Postharvest fungicides are seldom applied to these vegetables. Application of wax and postharvest fungicides must be indicated on each shipping container. European cucumbers are frequently shrink-wrapped rather than waxed.

Fruit may be classified by quality into two or more grades according to U.S. standards, California grade standards, or a shipper's own more rigorous grade standards. The immature fruit vegetables are often sized manually, although weight and diverging roller sizers are sometimes used for cucumbers. Okra, cucumber, and legumes are commonly weight- or volume-filled into shipping containers. They and the other immature fruit vegetables may be place-packed into shipping containers by count. Those that are place-packed are often sized during the same operation.

Table 33.1. Typical chilling injury symptoms on fruit vegetables

Immature fruit vegetables	Product Symptoms
Beans (snap, lima, and long)	Darkening or dullness if stored below 5°C (41°F); rusty brown lesions if stored at 5°–7.5°C (41°–46°F); discoloration of seeds; increased susceptibility to decay; surface pitting
Cucurbits (bittermelon, luffa, fuzzy melon, cucumber, summer squash)	Surface pitting followed by brown or black lesions; water-soaked areas; increased susceptibility to decay
Eggplant	Brown, discolored areas; surface pitting leading to large sunken areas; calyx discoloration; seed and flesh browning; off-odors
Okra	Darkening and discoloration; pitting, water-soaked areas; increased susceptibility to decay
Mature fruit vegetables	
Cantaloupe	Only slightly sensitive to chilling injury; if stored below 2°C (36°F) for extended periods, fruit can show surface browning and increased decay after removal from storage.
Honeydew and other melons	Failure to ripen normally; water-soaked areas, increased susceptibility to decay; dull or bronzed surface
Peppers	Surface pitting leading to large sunken areas; seed browning; calyx discoloration; water-soaked tissue; increased susceptibility to decay, especially Alternaria
Tomatoes	Short exposures (4–6 days) to <10°C (50°F) results in poor flavor; longer exposures cause failure to ripen normally, pitting, shriveling, softening, and increased susceptibility to decay, with Alternaria rot a diagnostic symptom
Watermelon	Surface pitting and sunken areas that dehydrate upon removal from storage; off-flavors; internal brown discolored areas on the rind
Winter squash and pumpkins	Weakening of tissue, especially on the stem end with increased susceptibility to decay, particularly Alternaria rot.

Source: Adapted from Hardenburg et al. 1986; Kader 1996; Zong et al. 1992.

Figure 33.2

Shelf life of cucumber in relation to storage temperature.

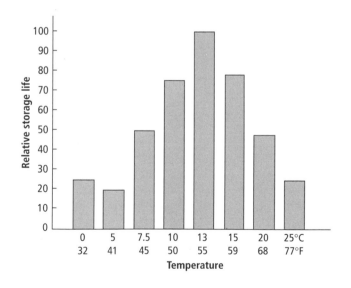

Except for sweet corn, the immature fruit vegetables are often handled in low-volume operations, where palletizing is not common because of lack of forklifts. In these cases, the products are palletized at a centralized cooling facility or as they are loaded for transport. Palletizing is usually done after hydrocooling or package-ice cooling, but before forced-air cooling. In field-pack operations, palletizing is generally done in the field.

COOLING

Various methods are used for cooling the immature fruit vegetables. Forced-air cooling can be used for virtually all these vegetables, but it is most commonly used for beans, cucumbers, peas, and soft-rind squashes. Evaporative cooling is used to a limited extent on green beans, squashes, and eggplant, all particularly sensitive to water loss. Beans, sweet corn, and okra are hydrocooled before or after packing, depending on whether they are packed in a waxed or other water-resistant container. Sweet corn is routinely liquid-iced; less commonly, package ice is used as a supplement to hydrocooling. After cooling, the packed product is then temporarily stored in a cold room or transported to a centralized short-term storage facility at a distributor.

STORAGE AND TRANSPORT TEMPERATURES

Peas and sweet corn are the only immature

Figure 33.3

External (left) and internal (right) quality of bitter melon in relation to storage temperature.

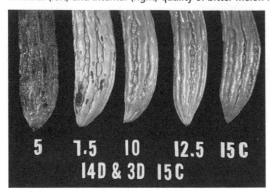 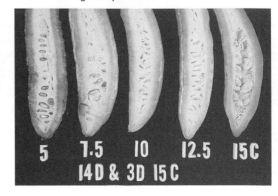

Figure 33.4

Rusty brown lesions as symptoms of chilling injury on green beans. About 7 days at 5°C (41°F) are required to obtain these symptoms.

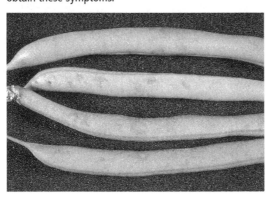

Figure 33.5

Typical external symptoms of chilling injury on egg-plants include discoloration of the calyx, pitting, and discoloration of the fruit surface.

fruit vegetables that should be stored at 0°C (32°F) and 95% RH. A temperature of 5°C (41°F) results in a more rapid loss of sugar than storage at 0°C (32°F), even in super-sweet corn varieties (see fig 33.7). All the other immature fruit vegetables are chilling sensitive (table 33.1). Chilling injury occurs when they are stored below the recommend-ed temperature; chilling injury is cumula-tive, and its severity depends on the temper-ature and the duration of exposure (figs. 33.2, 33.3, 33.4, 33.5, 33.6). The chilling susceptibility of cucumbers, summer squash, eggplants, and green beans may vary greatly depending on the variety. For example, table 33.2 shows that Chinese eggplants are more tolerant to storage at chilling temperatures than are globe or Japanese-type eggplants.

The optimal product temperatures with 90 to 95% RH for short-term storage and transport of the chilling-sensitive immature fruit vegetables are

Figure 33.6

External and internal symptoms of chilling injury on globe eggplant stored 2 weeks at different temperatures.

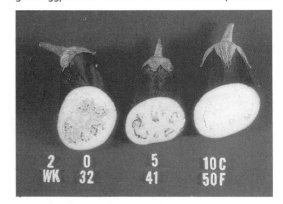

- Eggplant, cucumber, soft-rind squashes, okra: 10° to 12.5°C (50° to 55°F).

- Lima beans, snap beans: 5°to 8°C (41° to 46°F)

Table 33.2. Time required for development of visible chilling injury symptoms in three types of eggplant

Temperature		Days required for any chilling symptom to appear		
°C	°F	Globe type cv. Black Bell	Japanese type cv. Millionaire	Chinese type cv. unknown
0	32	1–2	—	2–3
2.5	36	4–5	5–6	5–6
5	41	6–7	8–9	10–12
7.5	45	12	12–14	15–16
10	50	No symptoms	No symptoms	No symptoms

Note: The calyx is more sensitive than the fruit surface to chilling injury.

Table 33.3. CA and MA requirements and recommendations for fruit vegetables

	Temperature		Atmosphere	
Product	Optimum °C (°F)	Range °C (°F)	% O₂	% CO₂
Immature fruit vegetables				
Beans, green, snap	8 (46)	5–10 (41–50)	2–3	4–7
Beans, processing	8 (46)	5–10 (41–50)	8–10	20–30
Cucumber, fresh	12 (54)	8–12 (46–54)	1–4	0
Cucumber, pickling	4 (40)	1–4 (33–40)	3–5	3–5
Okra	10 (50)	7–12 (45–54)	Air	4–10
Peas, sugar	0 (32)	0–5 (32–41)	2–3	2–3
Sweet corn	0 (32)	0–5 (32–41)	2–4	5–10
Mature fruit vegetables				
Cantaloupe	3 (38)	2–7 (36–46)	3–5	10–20
Pepper, bell	8 (46)	5–12 (41–54)	2–5	2–5
Pepper, chili	8 (46)	5–12 (41–54)	3–5	0–5
Pepper, processing	5 (41)	5–10 (41–50)	3–5	10–20
Tomato, mature-green	12 (54)	12–20 (54–68)	3–5	2–3
Tomato, partially ripe	10 (50)	10–15 (50–59)	3–5	3–5

Source: Adapted from Saltveit 1997.

Chilling injury is a serious practical problem for these products, compounded by the fact that they are frequently shipped in small volumes as part of mixed loads. One option to reduce exposure may be the use of pallet blankets to maintain a slightly warmer temperature around the product. Another practical option is minimizing the period of exposure. In many cases, as illustrated in table 33.2, short periods at chilling temperatures are not sufficient to cause permanent physiological damage that leads to chilling symptoms.

Figure 33.7

Yellowing of cucumbers due to ethylene exposure. The fruit were exposed to air (fruit on right) or 10 ppm ethylene (2 fruit on left) for 2 days and then stored an additional 5 days at 12.5°C (55°F).

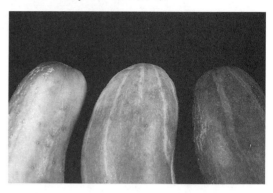

ETHYLENE EXPOSURE

Because the immature fruit vegetables are often produced in small volumes, most are transported in trucks as part of mixed-load shipments. Often these vegetables are shipped with commodities that produce ethylene, and many of them, such as cucumber, bitter melon, and eggplant are very sensitive to ethylene exposure. Ethylene exposure can favor decay development, cause discoloration and abscission of the calyx in eggplant, and de-greening of cucumbers (fig. 33.7).

MODIFIED ATMOSPHERES

Modified atmospheres (MA) are seldom used commercially for immature fruit vegetables, although extension of shelf life can be demonstrated for some (see table 33.3). Short-term holding of green beans under high-CO₂ atmospheres is beneficial to reduce brown discoloration before processing, and MA help retain green color in immature fruit vegetables such as cucumber and bitter melon. Figure 33.8 shows changes in the visual quality of sweet corn and sugar content of the kernels in relation to air or CA storage.

MATURE FRUIT VEGETABLES

MATURITY INDICES

The harvest index for mature fruit vegetables depends on several characteristics, and proper harvest maturity is the key to adequate shelf life and good quality of the ripened fruit. The principal harvest indices for cantaloupe

Figure 33.8

Quality changes in hybrid sweet corn stored at 0°C (32°F) or 5°C (41°F) in air or at 5°C in a controlled atmosphere of 3% O_2 and 10% CO_2.

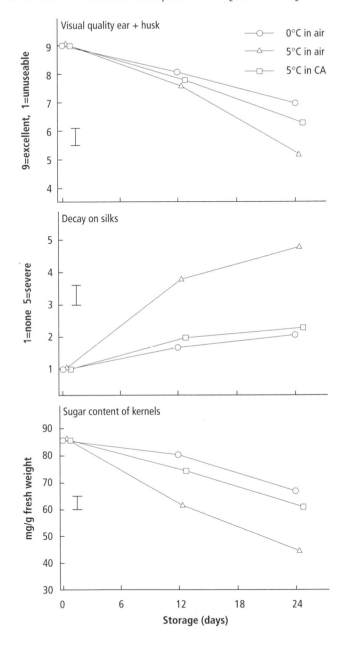

Figure 33.9

Cantaloupe exhibiting harvest indices: well-developed abscission zone, well-developed net, ground color change from green to green-yellow, and death of the subtending leaf.

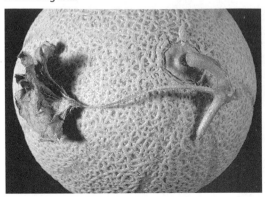

for melons. Cantaloupes are sometimes harvested with "sack" crews, who empty the melons into bulk trailers, but most cantaloupes are field-packed, and physical handling is minimized. Mixed melons, which are very susceptible to mechanical injury, may also be field-packed but are often harvested and placed in lined low-volume trailers for transport to a packing facility. Mature-green tomatoes are usually hand-harvested into buckets that are emptied into field bins or gondolas. Almost all fresh-market tomatoes grown in California are bush-type, and the plants are typically harvested once, sometimes twice. At the time of harvest, 5 to 10% of the tomatoes should have external pink or yellow color fruit, which are harvested and packed as vine-ripes. Harvesting the field at this time maximizes the proportion of mature-green fruit that can ripen into acceptable eating quality and minimizes immature fruit. Immature fruit respond to ethylene during ripening but have poor eating quality because of low levels of sugars and acids. Some varieties of cluster tomatoes may lack uniform quality at the ripe stage due to differences in maturity of fruit on the raceme. The least mature fruit on the raceme should have some external red color (breaker stage at minimum) at time of harvest.

Many of the mature fruit vegetables are hauled to packinghouses, storage, or loading facilities in bulk bins (hard-rind squashes, peppers, pink tomatoes), gondolas (mature-green tomatoes and peppers), or bulk field trailers or trucks (muskmelons, hard-rind squashes).

are green-yellow surface color, a well-developed net, and the formation of the abscission zone (fig. 33.9). Maturity and ripeness classes for honeydew melon are described in table 33.4 and those for tomato are shown in table 33.5.

HARVEST

The mature fruit vegetables are harvested by hand. Some harvest aids may be used, including pickup machines and conveyors

Table 33.4. Maturity and ripeness classes for honeydew melons*

Class	Characteristics	Internal ethylene (ppm)	Pulp firmness† (kgf)	Soluble solids (%)
0 = Immature	Greenish external color; peel fuzzy/hairy; no aroma; may be harvested by mistake	—	—	—
1 = Mature, unripe	External color white with greenish aspect; peel slightly fuzzy/hairy; no aroma; melon splits when cut, pulp is crisp; minimum commercial harvest maturity; minimum 10% soluble solids	0.8	3.1	10
2 = Mature, ripening	External color white with trace of green; peel not fuzzy, slightly waxy; slight to noticeable aroma; melon splits when cut, flesh crisp; harvest for long-distance markets	5.2	2.1	11–12
3 = Ripe	External color creamy white to pale yellow; peel waxy; noticeable aroma; stem may begin to separate from fruit; flesh firm, when sliced does not split; ideal eating; harvest for local markets	27.1	1.5	12–14
4 = Overripe	External color yellow; soft at blossom end; very aromatic; fruit is separated from stem; flesh soft, somewhat water soaked in appearance	29.4	1.1	14–15

Source: Adapted from Pratt 1971.

Notes: * Values averaged from 5 honeydew varieties.   † Firmness measured using a 1.1-cm diameter probe on a 5-kgf penetrometer.

Table 33.5. Maturity and ripeness classes for fresh-market tomatoes

Class	USDA classification	Description
Immature	—	Seed cut by a sharp knife on slicing the fruit; no jellylike material in any of the locules; fruit is more than 10 days from breaker stage
Mature-green A	1	Seed fully developed and not cut on slicing fruit; jellylike material in at least one locule; fruit is 6 to 10 days from breaker stage; minimum harvest maturity
Mature-green B	1	Jellylike material well developed in locules but fruit still completely green; fruit is 2 to 5 days from breaker stage
Mature-green C	1	Internal red coloration at the blossom end, but no external color change; fruit is 1 to 2 days from breaker stage
Breaker	2	First external pink or yellow color at the blossom end
Turning	3	More than 10% but not more than 30% of the surface, in the aggregate, shows a definite change in color from green to tannish-yellow, pink, red, or a combination thereof
Pink	4	More than 30% but not more than 60% of the surface, in the aggregate, shows pink or red color
Light red	5	More than 60% of the surface, in the aggregate, shows pinkish-red or red, but less than 90% of the surface shows red color
Red	6	More than 90% of the surface, in the aggregate, shows red color
Full red	—	Fruit has developed full final red color; fruit is more aromatic and softer than at red stage

Harvesting at night or near daybreak, when products are the coolest, is sometimes used for cantaloupes. Night harvest may reduce the time and costs of cooling products, may result in better and more uniform cooling and helps maintain product quality. Fluorescent lights attached to mobile packing units have permitted successful night harvesting of cantaloupe in California, although this is not done frequently due to difficulties in scheduling harvest crews.

FIELD VERSUS SHED PACKING

For many of the mature fruit vegetables, grading, sorting, sizing, packing, and palletizing are carried out in the field (figs. 33.10 and 33.11). The products are then transported to a central cooling facility.

Figure 33.10

Mobile packing unit for field-packing of cantaloupe.

Figure 33.11

Postharvest handling of melons.

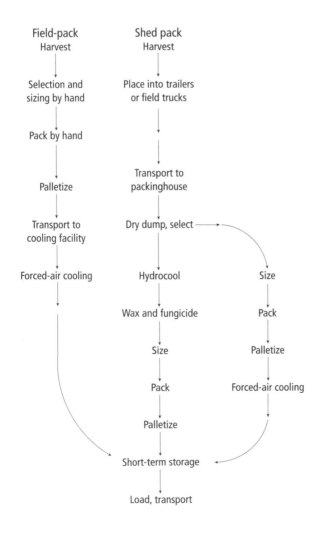

Field-pack
Harvest

↓

Selection and
sizing by hand

↓

Pack by hand

↓

Palletize

↓

Transport to
cooling facility

↓

Forced-air cooling

Shed pack
Harvest

↓

Place into trailers
or field trucks

↓

Transport to
packinghouse

↓

Dry dump, select ⟶

↓   ↓

Hydrocool   Size

↓   ↓

Wax and fungicide   Pack

↓   ↓

Size   Palletize

↓   ↓

Pack   Forced-air cooling

↓

Palletize

↓

Short-term storage

↓

Load, transport

Figure 33.12

Harvest of greenhouse vine-ripe tomatoes into plastic crates on a railed cart.

Figure 33.13

Hand-packed single-layer carton of vine-ripe tomatoes. Some fruit were turned to show undesirable variation in ripeness within a single tray.

Mobile packing facilities are commonly towed through the fields for cantaloupe, honeydew melon, peppers, and vine-ripe tomatoes. Handling costs are reduced in field-pack operations, and there is much less handling of products than in packinghouses. In melons, for example, field packing means less rolling, dumping, and dropping and thus helps reduce the "shaker" problem, in which the seed cavity loosens from the peri-carp wall. It also reduces scuffing of the net, which reduces water loss, a major cause of firmness loss in cantaloupes. One difficulty with field packing, however, is the need for increased supervision to maintain consistent quality in the packed product. Field packing is not generally used for commodities that require classification for both color and size, such as tomato. For greenhouse tomatoes or

Figure 33.14

Postharvest handling of vine-ripe tomatoes.

Hand-harvest
Eliminating defective fruit that cannot be packed

Place fruit into clean vented plastic trays
Stack trays for transport to packing area

De-stack trays and wash tomatoes
Submerge briefly into a clean water followed by a chlorinated water bath
or shower wash or use a moist clean cloth to remove dust

Drain tray and allow tomatoes to air-dry before further handling

Packer selects fruit from tray and classifies individual fruits
(3 sizes, 2 or 3 color stages)
Places fruit into shipping container (single- or double-layer carton box;
paper or foam pad between layers)

Open containers move by gravity conveyer and are palletized after
inspection for final quality and closing of carton

Pallets are strapped vertically to reduce vibration injury during transport

Palletized fruit are cooled to 10° to 13°C (50° to 55°F) as soon as possible
(at packing site) and transported (trailer or air)

field-grown vine-ripes, a simplified handling system is used that minimizes handling steps much as is done in field packing operations (figs. 33.12, 33.13, 33.14). To keep cluster tomatoes on their stems, handling should be minimal. The racemes are cut carefully and placed in crates, from which they are subsequently sorted and packed.

PACKINGHOUSE OPERATIONS

Loaded field vehicles should be parked in shade to prevent product warming and sunburn damage. Products may be unloaded by hand (some muskmelons, watermelon), dry-dumped onto sloping, padded ramps (cantaloupe, honeydew melon, sweet peppers,) or onto moving conveyor belts (mature-green tomatoes), or wet-dumped into tanks of moving water to reduce physical injury (melons, tomatoes, peppers) (figs. 33.15, 33.16). Considerable mechanical damage occurs in dry-dumping operations; bruising, scratching, abrading, and splitting are common examples. The water temperature in wet-dump tanks for tomatoes should be

slightly warmer (5°C, or 10°F) than the product temperature to prevent uptake of water and decay-causing organisms into the fruits. The dump tank water needs to be chlorinated (see chapter 17). An operation may have two tanks separated by a clean water spray to improve overall handling sanitation.

PRESIZING, SORTING, AND SELECTION

For many commodities, fruit below a certain size are eliminated manually or mechanically by a presizing belt or chain. Undersize fruit are diverted to a cull conveyor or used for processing. The purpose of the sorting step is to eliminate cull, overripe, misshapen, and otherwise defective fruit and separate products by color, maturity, and ripeness classes (e.g., tomato and muskmelons) (see fig. 33.15). Electronic color sorters are used in some tomato operations to separate mature-green and vine-ripe tomatoes or to classify fruit at different ripening stages. Fruit are sorted by quality into two or more grades according to U.S. standards, California grade standards, or a shipper's own grade standards.

WAXING

Food-grade waxes may be applied to cantaloupes and tomato, although their use is decreasing. Waxing replaces some of the natural waxes removed in the washing and cleaning operations, reduces water loss, and may improve appearance. Fungicides (e.g., SOPP and potassium sorbate) may be added to the wax. Application of wax and postharvest fungicides must be indicated on each shipping container.

SIZING

After sorting for defects and color differences, the fruit vegetables are segregated into several size categories. Sizing may be done manually (hard-rind squashes, watermelon), by diverging bar sizers (peppers, melons), volumetric sizers (cantaloupe melons), or by belt or weight sizers (tomatoes) (fig. 33.17).

PACKING AND PALLETIZING

Mature-green and pink tomatoes, and bell and chili peppers, are commonly weight- or volume-filled into shipping containers. All other fruit-type vegetables (and many tomatoes and peppers) are place-packed into shipping containers by count, bulk bins

Figure 33.15

Postharvest handling of mature-green tomatoes.

Harvested at mature-green stage into buckets
that are dumped into trailer-mounted gondola;
1 harvest per field typical;
begin to harvest when about 5–10% of fruit show external color

↓

Gondola to packinghouse (minutes to hours),
held temporarily in shaded area
Tomatoes flumed into dump tank (chlorinated water,
100–150 ppm active chlorine, sometimes heated)
Follow with clean water wash

↓

Presize and presort to remove very small and defective fruit;
Fruit with external color are diverted to vine-ripe packingline

↓

Fruit sometimes sprayed with wax emulsion (mineral or vegetable waxes)
that is uniformly distributed by sponge rollers;
Wax may contain fungicide (SOPP or potassium sorbate, typically)

↓

Fruits classified for size by belt sizer, weight or electronic sizer;
volume fill 25-lb (11.4-kg) carton

↓

Palletize, strap, store (cool to 13°C, or 55°F)	Palletize and move to degreening room

↓ ↓

Transport (7° to 13°C, or 45° to 55°F)	Treat with ethylene gas (100 ppm, 16° to 20°C) (60 to 68°F), high RH for 3–4 days; then drop temperature for storage)

↓ ↓

De-green with ethylene at distribution center	Remove from degreening room, Repack fruit for uniformity if necessary

↓ ↓

Depalletize, empty cartons onto repack line or repack directly from volume-fill cartons into final place-pack carton	Palletize; may cool to 10° to 13°C (50° to 55°)

↓ ↓

If fruit near ripe, may recool to 10° to 13°C (50° to 55°F)	Transport to distribution center 7° to 13°C (45° to 55°F)

(hard-rind squashes, pumpkin, muskmelons, watermelon) or bulk trucks (watermelon). Fruit-type vegetables that are place-packed (greenhouse tomatoes and peppers) are often sized during the same operation. Packed shipping containers of most fruit vegetables in large-volume operations are palletized for shipment.

Figure 33.16

Mature-green tomatoes being flumed from a gondola.

COOLING

Various methods are used for cooling the mature fruit vegetables. Forced-air cooling is used for melons, peppers, and tomatoes. Delays from harvest to forced-air cooling can result in excessive water loss. Forced-air evaporative cooling is sometimes used in small-volume operations (e.g., cherry tomato), but it could also be used for any of the chilling-sensitive mature fruit vegetables. Hydrocooling is sometimes used before grading, sizing, and packing of cantaloupe and other melons that are not field packed. Hydrocooling cycles are rarely long enough during hot weather, when product temperatures are the highest. This can be remedied if, after packing and palletizing, enough time is allowed in the cold room to cool the product to recommended temperatures before loading for transport to markets. Package icing and liquid-icing are still used to a limited extent for cooling cantaloupes.

STORAGE AND TRANSPORT CONDITIONS

Good temperature management can effectively control the rate of ripening of mature fruit vegetables, as illustrated for tomato in table 33.6 and fig. 33.18. Most mature fruit vegetables are sensitive to chilling injury when held below the recommended storage temperature (fig. 33.19; see table 33.2). Chilling injury is cumulative, and its severity depends on the temperature and the duration of exposure. In the case of tomato, exposure to chilling temperatures below 10°C (50°F) results in lack of color development, decreased flavor, and increased decay

Figure 33.17

Belt sizing of mature-green tomatoes.

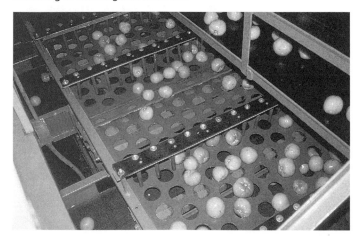

Figure 33.18

Effect of temperature on ripening physiology of tomatoes.

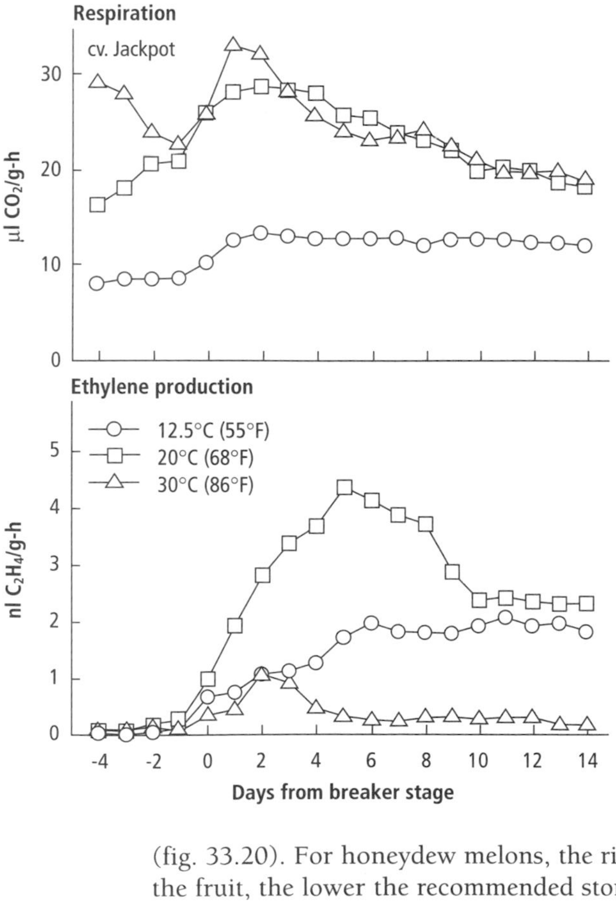

(fig. 33.20). For honeydew melons, the riper the fruit, the lower the recommended storage temperature. Class 1 fruits (see table 33.4) can be stored at 10°C (50°F), Class 2 fruits can be stored at 7° to 10°C (45° to 50°F), and Class 3 fruits can be stored at 5° to 7°C (41° to 45°F) without causing chilling injury.

The optimal temperatures for short-term storage and transport are

- Mature-green tomatoes, pumpkin, and hard-rind squashes: 12.5° to 15°C (55° to 60°F).

- Partially to fully ripe tomatoes, muskmelons (except cantaloupe): 10° to 12.5°C (50° to 55°F).

- Honeydew melons that are ripening: 5° to 7.5°C (41° to 45°F).

- Watermelon: 7° to 10°C (45° to 50°F) for short periods; 10° to 15°C (50° to 59°F) for longer than 1 week.

- Bell and chili peppers: 5° to 7.5°C (41° to 45°F). Storage below 7.5°C will cause chilling injury after about 10 days.

- Cantaloupe: 2.5° to 5°C (36° to 41°F).

The optimal RH range is 85 to 90% for tomato and muskmelons (except cantaloupe), 90 to 95% for cantaloupe, and 60 to 70% for pumpkin and hard-rind squashes.

ETHYLENE SENSITIVITY

Among the mature fruit vegetables, watermelon is detrimentally affected by ethylene, resulting in softening of the whole fruit, flesh mealiness, and rind separation. Exposure of winter squash and pumpkin to ethylene may cause abscission of the stem and de-greening, but they are much less sensitive than watermelon. Watermelon should not be shipped with cantaloupe and honeydew melons since they produce large amounts of ethylene gas.

MODIFIED ATMOSPHERES

Modified atmospheres are not frequently used for these commodities, although there are increasing numbers of commercial shipments (marine containers) of cantaloupes under MA for long-distance markets (see table 33.3). The potential for commercial shipments of tomatoes to long-distance markets under MA is also being reevaluated, and consumer packaging of vine-ripe tomatoes may also involve the use of MA. For tomatoes held at recommended temperatures, O_2 levels of 3 to 5% slow ripening, with CO_2 levels held below 5% to avoid injury.

For melons, especially cantaloupes, recommended atmospheres under normal storage conditions are 3 to 5% O_2 and 10 to 15%

Figure 33.19

Effect of temperature on quality and ripening of tomatoes.

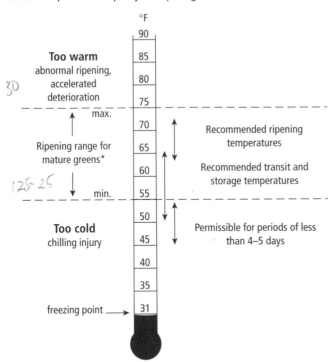

*The ripening range is somewhat wider (2° or 3°F) on each side for partially-ripened fruit.

Figure 33.20

Chilling injury (Alternaria decay) on tomatoes held 10 days at 5°C (41°F).

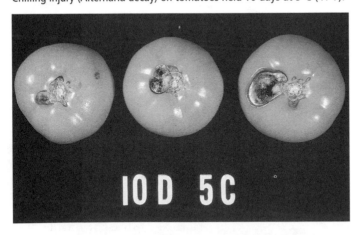

Table 33.6. Effect of temperature on average ripening rate of mature-green, breaker, turning, and pink tomatoes (conventional varieties)

Ripeness stage	°C	12.5	15	17.5	20	22.5	25
	°F	55	59	64	68	72	77
		Days to table-ripe stage at indicated temperature					
Mature-green		18	15	12	10	8	7
Breaker		16	13	10	8	6	5
Turning		13	10	8	6	4	3
Pink		10	8	6	4	3	2

Source: Kader 1986.

CO_2. The main benefit for long-distance shipment of melons is derived from CO_2, which retards decay development (fungistatic effect) on the stem end and on the rind and also retards color change and softening. Cantaloupes in carton boxes may also be packed in plastic film bags in which the atmosphere has been partially evacuated for rapid accumulation of CO_2 from fruit respiration. The plastic film also stops water loss and maintains melon firmness, and it can be maintained through to late stages of marketing. Melons should be thoroughly cooled before using this bagging technique.

RIPENING MATURE FRUIT VEGETABLES

Ripening tomatoes

For uniform and controlled ripening, ethylene is often applied to mature-green tomatoes. Satisfactory ripening occurs at 12.5° to 25°C (55° to 77°F); the higher the temperature, the faster the rate of ripening (see table 33.6). Above 30°C (86°F), red color (lycopene synthesis) development of tomato is inhibited due to inhibition of ethylene production (see fig. 33.18). Fruit ripened above 25°C (68°F) will be less firm than those ripened at 15° to 20°C (59° to 68°F). An ethylene concentration of about 100 ppm is commonly used. Tomatoes are usually held at 20°C (68°F) with high humidity and treated for up to 3 days. If tomatoes are at a minimum color stage of "breaker" (table 33.7), ethylene treatment will not further accelerate ripening, since the fruit are producing their own ethylene.

Ethylene treatments may be done at the shipping point or at destination markets, although ripening uniformity and final fruit quality are generally considered best if the treatment is applied at the shipping point soon after harvest. Tomatoes may be ethylene-treated before or after packing, but most are treated after packing. An advantage of treating before packing is that the warmer conditions favor development of any decay on the fruit, so infected fruit can be eliminated before final pack-out. Packing after ethylene treatment also permits a more uniform pack-out. Most mature-green tomatoes produced in California are packed and then ethylene-treated, and lack of uniformity in color development, or "checkerboarding," often requires a repacking operation.

Table 33.7. Typical color changes during the ripening of tomato fruit

Stage of development/color	USDA classification	L*	a*	b*	Chroma	Hue
Mature-green	1	62.7	−16.0	34.4	37.9	115.0
Breaker	2	55.8	−3.5	33.0	33.2	83.9
Pink-orange	4	49.6	16.6	30.9	35.0	61.8
Orange-red	5	46.2	24.3	27.0	36.3	48.0
Red; table-ripe	6	41.8	26.4	23.1	35.1	41.3
Dark red; overripe	—	39.6	27.5	20.7	34.4	37.0

Note: L* indicates lightness (high value) to darkness; a* changes from green (negative value) to red. Chroma and hue are calculated from a* and b* values and indicate intensity and color, respectively. The lower the hue value, the redder the tomato.

Figure 33.21

Ripening physiology of conventional, *rin*, and hybrid tomatoes.

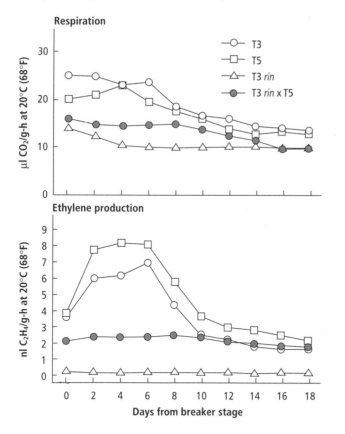

Decay is a common problem during tomato handling. Most of the decay-causing organisms are related to physical injury (*Geotrichum, Rhizopus*, etc.), but chill damage increases decay due to *Alternaria*. Prestorage conditioning treatments, heat treatments, intermittent warming, and irradiation have all been tested for effectiveness in reducing chilling injury or decay. Although such treatments may be partly effective, the best option is to avoid prolonged exposure to chilling temperatures.

Ripening mutant genes (*rin*, ripening inhibitor; *nor*, nonripening) have been used for development of long shelf life (LSL) varieties, especially for the vine-ripe greenhouse tomato industry. Figure 33.21 illustrates the effect of the *rin* mutation on tomato ripening physiology. Crosses between a *rin* and a normal ripening line result in intermediate ripening behaviors, and therefore longer shelf life. LSL varieties typically maintain firmness (table 33.8) longer than conventional varieties during storage, but they may have reduced flavor quality. Also, red color development (table 33.7) may be reduced in LSL varieties with reduced ethylene production rates.

Tomatoes have also served as a model crop for the molecular modification of genes associated with ethylene production and cell wall enzyme activity. Resulting transgenic varieties with modified ripening and storage characteristics may provide more options in postharvest handling regimes for fresh-market tomatoes.

Ripening melons

Honeydew melons (usually maturity class 1 or 2) and other melons are sometimes held in ethylene up to 24 hours to improve aroma, induce softening, and for other ripening-associated changes. However, sugar content does not increase with this treatment since melons do not have starch reserves that can be converted into sugars. Since cantaloupes and melons are harvested partially ripe and produce their own ethylene, temperature conditioning (15° to 20°C, or 59° to 68°F) without ethylene treatment will ensure good eating quality. If cantaloupes have been stored for 2 weeks or longer, decay will occur rapidly during ripening, and they will need to be marketed quickly. Melons have also been the subject of molecular studies, with focus on genetic manipulation of ethylene-producing capabilities and cell wall enzyme activities to slow the rate of ripening and pulp softening.

Ripening peppers

Bell peppers are nonclimacteric fruits, and ethylene does not enhance ripening of partially colored peppers (fig. 33.22). Holding bell and chili peppers at warm temperatures

Table 33.8. Typical textural characteristics of tomatoes and their relationship to objective firmness and slice integrity measurements

Firmness class	Description based on resistance to finger pressure	Firmness* (mm compression)	Description based on slicing characteristics	Slice integrity† (% weight loss)
Very firm	Fruit yields only slightly to considerable pressure	0.5–1.0	No loss of juice or seed when sliced	0–2
Firm	Fruit yields only slightly to moderate pressure	1.0–1.5	A few drops of juice or seed may be lost when sliced	2–5
Moderately firm	Fruit yields to moderate finger pressure	1.5–2.0	Some juice and seed are lost when sliced	5–8
Moderately soft	Fruit yields readily to moderate finger pressure	2.0–2.5	Some juice and seed are lost when sliced	5–8
Soft	Fruit yields to slight finger pressure	2.5–3.0	Considerable juice and seed are lost when sliced	8–10
Very soft	Fruit yields very readily to slight finger pressure	>3.0	Much of the juice and seed is lost when sliced	>10

Notes:

*Measured by placing a 500-g (17.6-oz) weight for 10 seconds on the equator of the fruit.

†Measured by weighing fruit before and after slicing into 0.8-cm-wide (0.3-in-wide) slices and draining.

Figure 33.22

Ripening of bell pepper fruit stored in air or 100 ppm ethylene at 20°C (68°F).

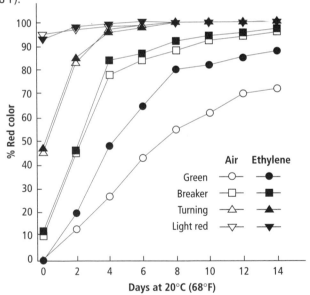

of 20° to 25°C (68° to 77°F) with greater than 90% RH is the best way to complete ripening or color changes. Chili peppers are climacteric in their ripening behavior, and they may respond to ethylene treatment, although the response appears to be highly variety dependent.

REFERENCES

Bartz, J. A. 1988. Potential for postharvest disease in tomato fruit infiltrated with chlorinated water. Plant Dis. 72:9–13.

Cantwell, M., J. Flores-Minutti, and A. Trejo-González. 1992. Developmental changes and postharvest physiology of tomatillo fruits (*Physalis ixocarpa* Brot.). Scientia Hortic. 50:59–70.

Davies, J. N., and G. E. Hobson. 1981. The constituents of tomato fruit—The influence of environment, nutrition and genotype. Crit. Rev. Food. Sci. Nutr. 15:205–280.

Fallik, E., N. Temkin-Gorodeiski, S. Grinberg, and H. Davidson. 1995. Prolonged low-temperature storage of eggplants in polyethylene bags. Postharv. Biol. Technol. 5:83–89.

Hardenburg, R. E., A. E. Watada, and C. Y. Wang. 1986. The commercial storage of fruits, vegetables, and florist and nursery stocks. USDA Agric. Handb. No. 66. 130 pp.

Hobson, G. E., and D. Grierson. 1993. Tomato. In G. B. Seymour, J. E. Taylor, and G. A. Tucker, eds., Biochemistry of fruit ripening. London: Chapman and Hall. 405–442.

Kader, A. A. 1986. Effect of postharvest handling procedures on tomato quality. Acta Hort. 190:209–221.

———. 1996. Maturity, ripening, and quality relationships of fruit-vegetables. Acta Hort. 434:249–255.

Kader, A. A., L. L. Morris, M. A. Stevens, and M. Albright-Holton. 1978. Composition and flavor quality of fresh market tomatoes as influenced by some postharvest handling procedures. J. Am. Soc. Hortic. Sci. 103:6–13.

Kasmire, R. F. 1973. Precooling, refrigeration, and postharvest handling of tomatoes and cantaloupes. In ASHRAE Symp. LO-73-7, 19–20.

Washington, D.C.: Am. Soc. Heating, Refrig. Air Cond. Eng. 19–20.

Leshuk, J. A., and M. E. Saltveit Jr. 1990. Controlled atmosphere storage requirements and recommendation for vegetables. In M. Calderon and R. Barkai-Golan, eds., Food preservation by modified atmospheres. Boca Raton, FL: CRC Press. 315–352.

McColloch, L. P., H. T. Cook, and W. R. Wright. 1968. Market diseases of tomatoes, peppers and eggplants. USDA Handb. 28. 74 pp.

Mercado, J. A., M. A. Quesada, V. Valpuesta, M. Reid, and M. Cantwell. 1995. Storage of bell peppers in controlled atmospheres at chilling and nonchilling temperatures. Acta Hort. 412:134–142.

Miller, C. H., and T. C. Whener. 1989. Green beans. In N. A. M. Eskin, ed., Quality and preservation of vegetables. Boca Raton, FL: CRC Press. 245–264.

Moretti, C. L., S. A. Sargent, C. A. Sims, and R. Puschmann. 1997. Flavor alteration in tomato fruit due to internal bruising. Proc. Fla. State Hort. Soc. 110:195–197.

Pratt, H. K. 1971. Melons. In A. C. Hulme, ed., The biochemistry of fruits and their products. Vol. 2. New York: Academic Press. 207–232.

Ryall, A. L., and W. J. Lipton. 1979. Handling, transportation and storage of fruits and vegetables. Vol. 1. Vegetables and melons. 2nd ed. Westport, CT: AVI. 587 pp.

Saltveit, M. E. 1997. A summary of CA and MA requirements and recommendations for harvested vegetables. In M. E. Saltveit, ed., Proc. 7th Internat. CA Res. Conf. Vol 4. Vegetables and ornamentals. Davis: Univ. Calif. Postharv. Outreach Prog. 98–117.

Seymour, G. B., and W. B. McGlasson. 1993. Melon. In G. B. Seymour, J. E. Taylor, and G. A. Tucker, eds., Biochemistry of fruit ripening. London: Chapman and Hall. 273–290.

Sistrunk, W. A., A. R. Gonzales, and K. J. Moore. 1989. Green beans. In N. A. M. Eskin, ed., Quality and preservation of vegetables. Boca Raton, FL: CRC Press. 185–215.

Snowdon, A. L. 1992. Color atlas of post-harvest diseases and disorders of fruits and vegetables. Vol. 2. Vegetables. Boca Raton, FL: CRC Press. 416 pp.

Stevens, M. A. 1985. Tomato flavor: Effects of genotype, cultural practices, and maturity at picking. In H. E. Pattee, ed., Evaluation of quality of fruits and vegetables. Westport, CT: AVI. 367–386.

Wiley, R. C., F. D. Schales, and K. A. Corey. 1989. Sweet corn. In N. A. M. Eskin, ed., Quality and preservation of vegetables. Boca Raton, FL: CRC Press. 121–157.

Zong, R. J., M. Cantwell, L. Morris, and V. Rubatzky. 1992. Postharvest studies on four fruit-type Chinese vegetables. Acta Hort. 318:345–354.

34

Postharvest Handling Systems: Flower, Leafy, and Stem Vegetables

Marita I. Cantwell and

Robert F. Kasmire

The leafy, stem, and floral vegetables are represented by the following commodities:

- Leafy vegetables: lettuce, cabbage, Chinese cabbage, Brussels sprouts, rhubarb, celery, spinach, chard, kale, endive, escarole, green onion, witloof chicory (Belgian endive), radicchio, sprouts, and other leafy greens.
- Stem vegetables: asparagus, kohlrabi, fennel, and cactus stems (nopalitos).
- Floral vegetables: artichoke, broccoli, and cauliflower
- Mushrooms

These commodities are generally characterized as very perishable, with high respiration and water loss rates. Cabbages are a notable exception and may be stored for long periods. The respiration rates of leafy vegetables can vary greatly, as illustrated by data on specialty salad greens and full size lettuce heads (fig. 34.1). The visual appearance (i.e., freshness and characteristic green color) is closely related to the nutritional value of the leafy vegetables, as is shown by data on broccoli stored at 3 temperatures (fig. 34.2). The rapid changes in chlorophyll are closely associated with the visual symptom of yellowing of the beads (florets); shelf life is limited to 2 days at 20°C (68°F), 10 days at 10°C (50°F) and more than 30 days at 0°C (32°F). The vitamin C (ascorbic acid) and carotenoid (about 80% is pro-vitamin A or beta-carotene) concentrations closely follow the changes in chlorophyll concentration. Therefore, in the case of broccoli, a fresh green appearance is also a good indicator of nutritive value.

Postharvest quality and shelf life of the leafy vegetables can be greatly impacted by production practices and variety selection. For example, broccoli varieties can vary by more than 50% in their potential shelf life (table 34.1). High levels of nitrogen fertilization generally reduce shelf life and can affect composition (table 34.2). Another consideration is that postharvest bacterial decay may be higher with excessive applications of nitrogen fertilizer to broccoli and other leafy green vegetables.

The maximum shelf life or the best retention of quality for most of the commodities in this chapter is achieved with storage temperatures near 0°C (32°F) and high RH. For example, the shelf life of broccoli is substantially reduced if it is stored above the optimal temperature (fig. 34.3). Although broccoli is commonly liquid-iced to keep it near 0°C, note that with a long-shelf-life variety stored at 5°C (41°F) and high RH, yellowing does not occur until 21 days.

HARVESTING

The determination of horticultural maturity varies with commodity, but in general size is the principal criterion. For nonheading lettuces, the number of leaves can be used as a harvest index. For iceberg lettuce and cabbages, the solidity of the head determines harvest maturity. Maturity classes for iceberg lettuce are shown in table 34.3 and figure 34.4.

Virtually all leafy vegetables are cut by hand, but harvesting aids may be used with some (Brussels sprouts, celery, and parsley) (figs. 34.5, 34.6). Mechanical harvesting systems have been developed for iceberg lettuce, celery, cabbage, Brussels sprouts, and cauliflower, but

Figure 34.1

Variation in respiration rates of immature and mature leafy vegetables.

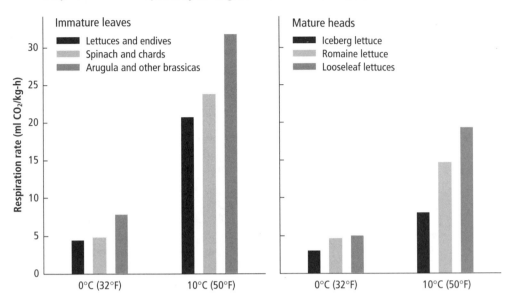

Respiration rates of speciality salad greens and full-size lettuces

Figure 34.2

Composition of broccoli in relation to storage temperature.

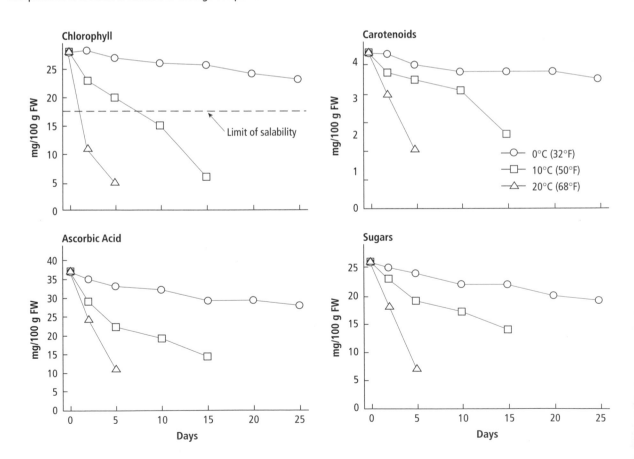

they are not used in California. However, mechanical harvesting is used for spinach and for the small leafy greens for salad mixes.

Stem vegetables are also hand-harvested. A limited amount of asparagus has been experimentally machine-harvested, but damage to the roots with subsequent yield reductions has restricted its use. Asparagus is generally hand-cut when spears are 23 cm (9 in) above the soil surface (figs. 34.7, 34.8). The maturity of floral vegetables is determined by head size and development. All floral vegetables are hand-harvested, but harvest aids (conveyors) are sometimes used for broccoli.

FIELD PACKING

Field packing is used with most of the leafy vegetables (figs. 34.9, 34.10, 34.11). The products are selected for maturity and quality, and then cut, trimmed, wrapped or sleeved, packed in cartons or crates, transported to cooling facilities, cooled, put into temporary cold storage prior to loading or loaded directly, and transported to market. Field packing generally provides greater marketable yields because of reduced mechanical damage. Wrapped and unwrapped lettuce, celery, cauliflower, broccoli, and spinach are mostly field-packed, though the latter three are still taken to packinghouses by a few shippers.

Small celery stalks may be trimmed and packed as hearts in the field or field-packed in bulk containers after harvesting and transported to packinghouses for trimming, sorting, prepackaging, and packing as celery hearts. Wrapped lettuce and cauliflower are hand-selected, cut, and trimmed and then placed on mobile field units, where they are wrapped and packed into cartons (figs. 34.12, 34.13). They are then palletized and transported to the cooling facilities for cooling and subsequent handling.

Rough handling in field packing is a major cause of lettuce and cauliflower marketing losses. For cauliflower it is critical to harvest, trim, and overwrap the head without touching or damaging the very tender curds. Much of the postharvest discoloration and decay on cauliflower can be traced to rough handling at harvest.

Keeping the commodity clean is a problem in field-packing operations, particularly when fields are muddy. Product should be placed not on the ground after cutting but on a clean belt or packing area. Sprays of chlorinated water are sometimes used to clean the heads of cauliflower and other leafy vegetables packed in the field. Careful trimming of outer leaves and periodic cleaning of the harvest knives should minimize dirt in the packing area.

Table 34.1. Shelf life range of commercial broccoli varieties harvested from a fall trial in the Salinas Valley (heads stored in air with >95% RH)

Variety	Days shelf-life at 7.5°C (45°F)
Packman	17–19
Liberty	12–17
Patriot	12–17
Green Valiant	13–15
Pinnacle	11–13
Brigadier	9–12
Majestic	9–14

Table 34.2. Differences in shelf life and composition (dry weight basis) of broccoli florets (cv. Legacy) due to varying nitrogen fertilization (Cantwell and LeStrange, unpublished data)

Total nitrogen applied (lb)*	Head size (cm)	Shelf life: Days at 5°C (41°F)	Dry weight (%)	Sugar (mg/g DW)	Chlorophyll (mg/g DW)	Carotenoids (mg/g DW)	Ascorbic acid (mg/g DW)
60	8.1	35	15.6	186	1.1	0.4	10.5
120	10.2	32	15.2	161	1.3	0.4	8.2
180	10.8	29	14.4	146	1.8	0.6	9.0
240	10.9	30	14.0	151	1.5	0.5	9.9
300	11.3	26	13.8	136	1.9	0.6	7.5
LSD.05	1.2	3	0.2	19	0.2	0.1	0.9

Note: * Nitrogen applied in 1 preplant and 2 sidedress applications.

Figure 34.3

Shelf life of broccoli in relation to storage temperature.

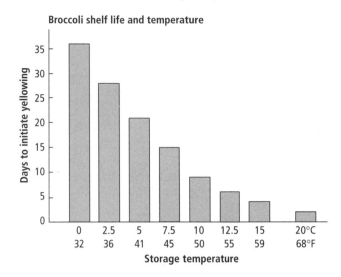

Broccoli shelf life and temperature

Table 34.3. Solidity (maturity) classes of iceberg (crisphead) lettuce

Solidity class	Postharvest characteristics
1 = Soft, no head formation	Immature; leaves very tender and susceptible to physical damage; has a higher respiration rate than more mature lettuce; sweet flavor, unacceptable for market
2 = Fairly firm, slight head formation	Higher respiration rate than mature lettuce; flavor is sweet, no bitterness
3 = Firm, good head formation; optimal density	Maximum storage life; flavor good and characteristic with little bitterness
4 = Hard, maximum density but no split ribs	More susceptible to russet spotting, pink rib, and other physiological disorders than mature lettuce; less storage life than mature lettuce; less sweet than mature lettuce and has bitter notes
5 = Extra hard; split midribs common; extreme internal pressure	Has minimum storage and shelf life remaining; most difficult to vacuum-cool; poor flavor and bitter taste

Source: Adapted from Kader et al. 1973.

Figure 34.4

Three stages of maturity of iceberg lettuce. These correspond to classes 2, 3, and 4 in table 34.3.

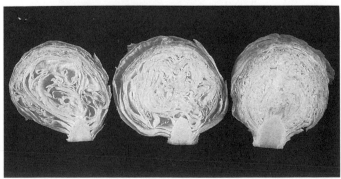

PACKINGHOUSE OPERATIONS

Brussels sprouts, witloof chicory, and mush-rooms are typically taken to a packinghouse, as can be many of the other vegetables considered here. The vegetables are selected, cut, placed in bulk containers, and then transported to packinghouses for all subsequent handling operations. Compared with field-packing, packinghouse handling requires more energy and generally results in more physical damage to the product, reducing marketable yields. For example, white *Agaricus* mushrooms are very easily bruised by careless dumping from the harvest containers onto the packing line, and the rough handling results in damaged areas that brown rapidly. In addition, waste management of discarded green material is also a consideration for packinghouse operations.

Packinghouse operations needed to prepare these products for market include

- trimming and cleaning with chlorinated water (desirable concentrations vary from 50 to 200 ppm active [total available] chlorine) (mushrooms are not washed)

- sorting and grading to eliminate defective products

- sizing, in some cases (sizing is usually subjective and done by hand)

- wrapping or tying individual units (cauliflower, broccoli), or in some cases, prepackaging (Brussels sprouts, broccoli and cauliflower heads, punnets of mushrooms, bagged celery and romaine hearts)

- packing in carton shipping containers (often wax-impregnated) or wood crates

COOLING

Delays between harvest and cooling should be avoided, especially during warm weather, since water loss rates will be high. Small leafy items, such as the specialty salad greens and young spinach, are particularly prone to rapid dehydration. Cooling delays will reduce total shelf life and quality of all these vegetables. In asparagus, for example, delays before cooling are the main cause of toughening of the spears (fig. 34.14).

Different cooling methods may be applied to the same commodity. The most common cooling methods in commercial use are

Figure 34.5

Harvesting and trimming celery.

Figure 34.6

Sizing and field-packing celery on a small field cart.

Figure 34.7

Harvesting asparagus spears.

Figure 34.8

Plastic crates for asparagus harvest.

- Vacuum cooling for crisphead lettuce, leaf lettuce, spinach, cauliflower, Chinese cabbage, bok choy, cabbage and other leafy vegetables, and mushrooms.

- Hydro-Vac cooling (vacuum cooling with injection of water prior to vacuum cycle) for celery and many other leafy vegetables.

- Hydrocooling for artichoke, asparagus, leaf lettuce, celery, spinach, some green onions, leek, and many other leafy vegetables.

- Package-icing and liquid-icing for broccoli, spinach, parsley, green onions, and Brussels sprouts.

- Room cooling, primarily for artichoke and cabbage, and for the other leafy vegetables in some operations (not generally recommended for this group of vegetables because it is too slow).

- Forced-air cooling (sometimes with initial spraying of water), primarily for cauliflower and to a limited extent for other leafy and stem vegetables, including sprouts. Forced-air cooling is increasingly used for leafy vegetables and mushrooms. In the case of mushrooms, humidified forced-air is used to reduce the drying effect of the rapid air movement.

Figure 34.9

Postharvest handling of leafy vegetables, such as lettuce, celery, and green onions.

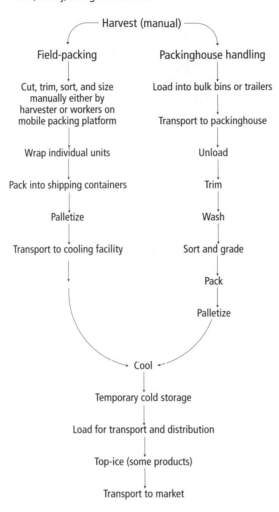

Figure 34.10

Postharvest handling of asparagus.

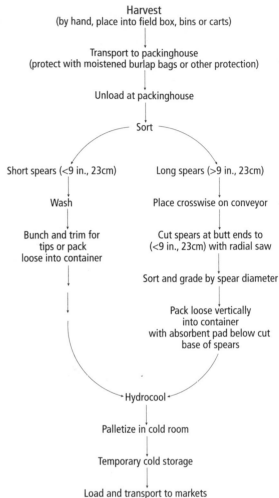

RECOMMENDED STORAGE CONDITIONS

Most of these vegetables are grown throughout the year in different production areas.
For this reason, long-term storage is usually
not required. These products are frequently
loaded into refrigerated trailers and marine
containers immediately after cooling. In general, these products respond best to storage
temperatures of 0° to 1°C (32° to 34°F).

Freezing must be avoided; it can occur at
temperatures slightly below 0°C (32°F),
since sugar content is generally low in most
of these vegetables. Slight freeze damage of
lettuce may occur in the field and results in
the cuticle separating from the leaf. This may
lead to increased discoloration and decay
during storage. Freeze damage during stor-

age typically appears as dark translucent
spots, which result in water-soaked areas and
loss of crisp texture once the leaf tissue
thaws. Symptoms of freezing injury on other
leafy vegetables are similar.

For temporary storage, temperatures of 0°
to 2°C (32° to 36°F) and a 90 to 95% RH are
recommended. For marketing within 1 to 2
weeks, however, storage temperatures consistently below 5°C (41°F) are generally adequate to maintain good quality. Figure 34.15
shows the result of storing gai-lan, a close relative of broccoli. The most perishable part,
the flower bud, deteriorated much more at
5°C (41°F) than it did at 0°C (32°F) during
the 3-week storage. A few commodities in
this group of vegetables are chilling-sensitive.
The optimal storage temperature of asparagus
is 2.5°C (36°F). A few other commodities are
very chilling-sensitive, such as tropical leafy

Figure 34.11

Postharvest handling of floral vegetables (artichoke, broccoli, and cauliflower).

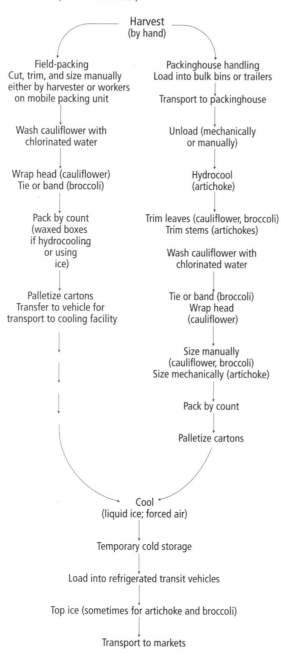

Figure 34.12

Field-packing of crisphead lettuce into cartons with plastic film liners.

Figure 34.13

Field-packing of cauliflower on a mobile packing unit.

greens (e.g., water spinach) and nopalitos, or cactus stems, and their recommended storage temperatures are 7.5° to 10°C (45° to 50°F). See the appendix for optimal storage conditions for specific products.

A high RH (>90%) is required for optimal storage of this group of vegetables. For some, such as mushrooms and sprouts, it is especially important to avoid water condensation on the product surface since it will greatly favor bacterial decay. For most of the other vegetables in this group, free moisture on the surface is not problematic, as long as the product is kept cold.

Long-term storage is not recommended, except for cabbage, Chinese cabbage, and celery. In long-term storage, temperatures near ideal are needed to maximize storage life and reduce losses. For long-term storage, air circulation should be minimized to that required for proper temperature control to avoid excessive water loss.

Figure 34.14

Toughening of asparagus spears due to cooling delays. Spears were held at 25°C (77°F) with high RH for indicated durations, then cooled and stored at 2.5° C or 5° C for 14 days.

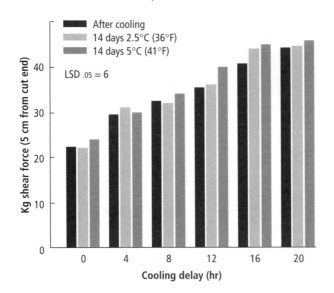

Figure 34.15

Visual appearance of gai-lan after storing 3 weeks at 0°C (32°F) or 5°C (41°F).

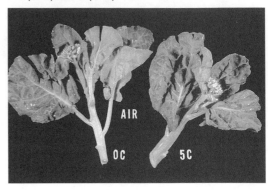

Figure 34.16

Russet spotting on crisphead lettuce: no symptoms are apparent on leaf at left; severe symptoms appear at right.

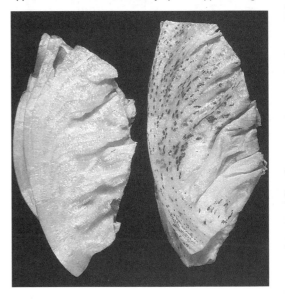

Exposure to ethylene should be avoided during storage and throughout the handling system. Ethylene induces russet spotting disorder in lettuce (fig. 34.16) and decreases the shelf life of all green, leafy vegetables. Factors to take into consideration are variety susceptibility, ethylene concentration, length of exposure, and storage temperature. There is considerable variation among varieties of iceberg lettuce in their susceptibility to ethylene and the development of russet spotting. Although varieties of Chinese cabbage also vary in the development of black speck, a similar defect, this physiological disorder is not induced by ethylene (fig. 34.17). Ethylene increases yellowing in the dark leafy greens and causes abscission of leaves in cauliflower and cabbage heads. Figure 34.18 shows the effect of ethylene on several quality characteristics of gai-lan. See chapter 16 for more information on undesirable effects of ethylene and how to avoid injury.

Exposure to light causes undesirable greening and bitterness in witloof chicory (Belgian endive). This can be controlled by using solid carton boxes with no vents to avoid light, packaging the product in dark paper, and at retail, using light-proof containers or keeping endive temperature at 5°C (41°F) or lower. At low temperature the desired yellow-white color of the heads is maintained since chlorophyll synthesis is inhibited.

Modified Atmospheres. Most leafy, stem, and floral commodities respond favorably to modified atmospheres, although this technique is used on a limited scale commercially for these intact vegetables (for fresh-cut vegetables, see chapter 36). Atmosphere recommendations for selected commodities are shown in table 34.4. Low-O_2 atmospheres (2 to 3% O_2) favor longer shelf life in all products except asparagus and mushrooms. The recommendations for CO_2 modification are more variable. Asparagus tips, easily damaged during sorting and packing, develop less decay when stored under high-CO_2 atmospheres (fig. 34.19). Spinach and broccoli, although benefited by modified atmospheres, can easily develop off-odors if O_2 concentrations drop too low.

Figure 34.17

Black speck development in 16 Chinese cabbage varieties. Heads were stored at 5°C (41°F) for 14 days. The higher the bar, the more severely affected by black speck.

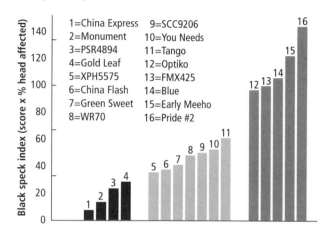

1=China Express
2=Monument
3=PSR4894
4=Gold Leaf
5=XPH5575
6=China Flash
7=Green Sweet
8=WR70
9=SCC9206
10=You Needs
11=Tango
12=Optiko
13=FMX425
14=Blue
15=Early Meeho
16=Pride #2

Table 34.4. CA and MA recommendations for leafy, stem, and floral vegetables

Product	Temperature range °C	°F	% O_2	% CO_2
Artichoke	0–5	32–41	2–3	2–3
Asparagus	1–5	34–41	21	10–14
Broccoli	0–5	32–41	2	5–10
Cabbage	0–5	32–41	2–3	3–4
Cauliflower	0–5	32–41	2–3	3–4
Celery	0–5	32–41	1–4	3–5
Chinese cabbage	0–5	32–41	1–2	0–5
Green onions	0–5	32–41	2–3	0–5
Leeks	0–5	32–41	1–2	2–5
Lettuce	0–5	32–41	1–3	0
Mushrooms	0–5	32–41	21	10–15
Spinach	0–5	32–41	7–10	5–10
Witloof Chicory	0–5	32–41	3–4	4–5

Source: Adapted from Saltveit 1997.

REFERENCES

Cantwell, M., A. Rodriguez-Felix, and F. Robles-Contreras. 1992. Postharvest physiology of prickly pear cactus stems. Scientia Hortic. 50:1–9.

Everaarts, A. P. 1994. Nitrogen fertilization and head rot in broccoli. Netherlands J. Agric. Sci. 42(3): 195–201.

Hansen, M., C. E. Olsen, L. Poll, and M. I. Cantwell. 1993. Volatile constituents and sensory quality of cooked broccoli florets after aerobic and anaerobic storage. Acta Hort. 343:105–111.

Hernández-Rivera, L., R. Mullen, and M. Cantwell. 1992. Textural changes of asparagus in relation to delays in cooling and storage conditions. HortTechnology. 2:378–381.

Jimenez, M., F. Laemmlen, X. Nie, V. Rubatzky, and M. I. Cantwell. 1998. Chinese cabbage cultivars vary in susceptibility to postharvest development of black speck. Acta Hort. 467:363–370.

Kader, A. A., W. J. Lipton, and L. L. Morris. 1973. Systems for scoring quality of harvested lettuce. HortScience. 8:408–409.

King, G. A., P. L. Hurst, D. E. Irving, and R. E. Lill. 1993. Recent advances in the postharvest physiology, storage and handling of green asparagus. Postharv. News and Info. 4:85N–89N.

King, G. A., and S. C. Morris. 1994. Physiological changes of broccoli during early postharvest senescence and through the preharvest-postharvest continuum. J. Amer. Soc. Hort. Sci. 19:270–275.

———. 1995. Early compositional changes during postharvest senescence of broccoli. J. Amer. Soc. Hort. Sci. 119:1000–1005.

Lipton, W. J. 1987. Senescence of leafy vegetables. HortScience. 22:854–859.

———. 1990. Postharvest biology of fresh asparagus. Hortic. Reviews 12:69–155.

Lipton, W. J., and E. J. Ryder. 1989. Lettuce. In N. A. M. Eskin, ed., Quality and preservation of vegetables. Boca Raton, FL: CRC Press. 217–244.

Lipton, W. J., J. K. Stewart, and T. W. Whitaker. 1972. An illustrated guide to the identification of some market disorders of head lettuce. USDA Market Res. Rept. 950. 7 pp.

Moline, H. E., and W. J. Lipton. 1987. Market diseases of beets, chicory, endive, escarole, globe artichokes, lettuce, rhubarb, spinach, and sweet potatoes. USDA Handb. No. 155. 86 pp.

Morris, L. L., A. A. Kader, and J. A. Klaustermeyer. 1974. Postharvest handling of lettuce. ASHRAE Trans. 80:341–349.

Nestle, M. 1997. Broccoli sprouts as inducers of carcinogen-detoxifying enzyme systems: Clinical, dietary, and policy implications. Proc. Natl. Acad. Sci. 94:11149–11151.

Nilsson, T. 1993. Influence of the time of harvest on keepability and carbohydrate composition during long-term storage of winter white cabbage. J. Hort. Sci. 68:71–78.

Noble, R. and K. S. Burton. 1993. Postharvest storage and handling of mushrooms: Physiology and technology. Postharv. News and Info. 4:125N–129N.

Paull, R. E. 1992. Postharvest senescence and physiology of leafy vegetables. Postharv. News and Info. 3:11N–20N.

Figure 34.18

Effect of a low concentration of ethylene on green color, flower bud deterioration, and pithiness of gai-lan stored at 5°C (41°F).

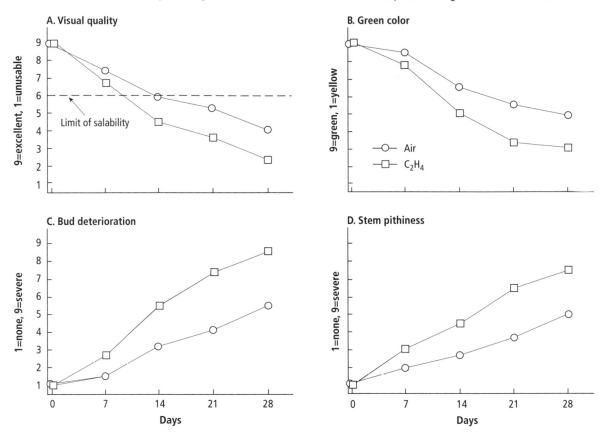

Perkins-Veazie, P. M., V. M. Russo, and J. K. Collins. 1992. Postharvest changes during storage of packaged radicchio. J. Food Quality 15:111–118.

Phan, C. T., ed. 1985. Postharvest handling of vegetables. Acta Hort. 157. 313 pp.

Pritchard, M. K., and R. F. Becker. 1989. Cabbage. In N. A. M. Eskin, ed., Quality and preservation of vegetables. Boca Raton, FL: CRC Press. 265–284.

Ramsey, G. B., and M. A. Smith. 1961. Market diseases of cabbage, cauliflower, turnips, cucumbers, melons and related crops. USDA Handb. 184. 49 pp.

Roy, S., R. C. Anantheswaran, and R. B. Beelman. 1996. Modified atmosphere and modified humidity packaging of fresh mushrooms. J. Food Sci. 61:391–397.

Ryall, A. L., and W. J. Lipton. 1979. Handling, transportation and storage of fruits and vegetables. Vol. 1, Vegetables and melons. 2nd ed. Westport, CT: AVI. 587 pp.

Saltveit, M. E. 1997. A summary of CA and MA requirements and recommendations for harvested vegetables. In M. E. Saltveit, ed., Proc. 7th Internat. CA Res. Conf, Vol 4: Vegetables and ornamentals. Davis, CA. 98–117.

Toivonen, P. M. A., B. J. Zebarth, and P. A. Bowen. 1994. Effect of nitrogen fertilization on head size, vitamin C content and storage life of broccoli (*Brassica oleracea* var. *Italica*). Can. J. Plant Sci. 74:607–610.

Varoquaux, P., G. Albagnac, C. N., The, F. Varoquaux, 1996. Modified atmosphere packaging of fresh bean sprouts. J. Sci. Food. Agric. 70:224–230.

Yamauchi, N., and A. E. Watada. 1991. Regulated chlorophyll degradation in spinach leaves during storage. J. Amer. Soc. Hort. Sci. 116:58–62.

Zong, R. J., L. L. Morris, M. J. Ahrens, V. Rubatzky, and M. I. Cantwell. 1998. Postharvest physiology and quality of Gai-lan (*Brassica oleracea* var. *alboglabra*) and Choi-sum (*Brassica rapa* subsp. *parachinensis*). Acta Hort. 467:349–356.

Figure 34.19

Changes in the quality of the tips and stems of asparagus held in air or high-CO_2 atmospheres.

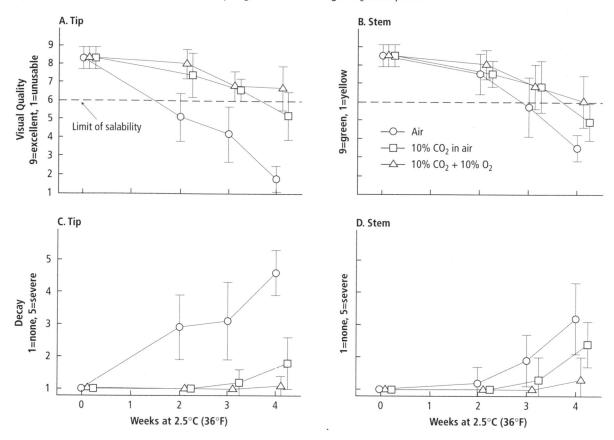

35

Postharvest Handling Systems: Underground Vegetables (Roots, Tubers, and Bulbs)

Marita I. Cantwell and

Robert F. Kasmire

The edible portions of this group of vegetables develop mostly underground and include several botanical structures:

- Roots: beet, carrot, celeriac, radish, horseradish, parsnip, turnip, sweet potato, cassava, jicama
- Tubers: potato, Jerusalem artichoke, yam
- Bulbs: onion, garlic, shallot, and related *Allium* spp.
- Others: ginger rhizomes, taro (dasheen) corms

Vegetables in this group can also be divided into two subgroups based on their postharvest temperature requirements:

- Temperate-zone underground vegetables: beet, carrot, celeriac, radish, horseradish, parsnip, turnip, potato, onion, garlic, shallot, daikon, salsify, water chestnut
- Subtropical and tropical underground vegetables: sweet potato, yam, cassava, ginger, taro, jicama, malanga

The commodities in the underground vegetables grouping have several common characteristics. They are all storage organs, principally of carbohydrates; they generally have low respiration rates (depending on the stage of development); they are considered relatively nonperishable, especially if the tops are removed; they continue growth after harvest (rooting and sprouting); and they can generally be stored for relatively long periods.

HARVESTING

Maturity indices vary with commodity. Many of these products may be harvested and marketed at various stages of development (e.g., "new" or immature potatoes versus mature potatoes; "baby" or immature carrots versus mature carrots). For some mature root and bulb crops, practices such as irrigation management and rolling or chemical destruction of vegetation, may be used to speed up maturation. Criteria commonly used to harvest at the mature stage include:

- Carrot: size, length of root
- Radish: days from planting, size
- Potato: drying of foliage, setting of skins
- Cassava and taro: drying of foliage begins
- Garlic and onion: drying and bending over of tops (fig. 35.1)
- Sweet potato: days from planting, size

Both mechanical and manual harvesting are used for this group of vegetables. Most roots and tubers are harvested mechanically and transported in bulk to packinghouses or processing facilities. Garlic and onion for fresh market are mechanically undercut, hand-harvested and hand-trimmed, cured in the field, and field-packed or transported to packinghouses. Some fresh-market onions are now mechanically harvested in California. Sweet potato harvesting is still done mostly by hand, although vine cutting and lifting are done mechanically. Vine-killing chemicals may be used on potatoes before mechanical harvesting. Carrots are undercut, lifted by their tops, and de-topped during mechanical harvest.

Figure 35.1

Onion tops bent over, indicating bulbs are mature and ready for harvest.

Figure 35.2

Stem end of undamaged potatoes (left) and internal discoloration caused by dropping (right).

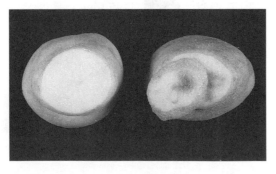

Table 35.1. Effect of temperature on the wound healing process of potatoes

Temperature		Days necessary to form	
°C	°F	Suberin	Periderm
25	77	1	2
15	59	2	3
10	50	3	6
5	41	5–8	10
2	36	7–8	Not formed

Source: Adapted from Burton 1982.

Physical damage during the harvesting operations can be extensive and is a major cause of postharvest losses. The mechanical detopping of carrots often causes nicks on the stem end, and if the carrots are cold and turgid, it can cause longitudinal cracking. Internal bruising of potatoes is a common defect and can occur at all points of handling. The stem end is particularly sensitive to discoloration after dropping injury because it contains high levels of the phenolic compounds that form the undesirable blue-black pigments in the bruised areas (fig. 35.2).

POSTHARVEST PROCEDURES

CURING

One of the simplest and most effective ways to reduce water loss and decay during storage of root, tuber, and bulb crops is curing after harvest. In roots and tubers, curing refers to the process of wound healing, with the development and suberization of new epidermal tissue (wound periderm). The effect of temperature on wound healing in the potato is shown in table 35.1. The type of wound also affects periderm formation: abrasions result in the formation of deep, irregular periderm, cuts result in a thin periderm, and compressions and impacts may entirely prevent periderm formation. Figure 35.3 shows wound healing on sweet potatoes in a commercial storage. In this case, warm product was loaded into the storage and held without ventilation for 1 week to favor curing, then temperature was dropped to storage conditions. Figure 35.4 shows that curing damaged areas of jicama roots prevented development of decay during storage.

In bulb crops, curing refers to the process of drying of the neck tissues and of the outer leaves to form dry scales. Some water loss takes place during curing. Removing decayed bulbs before curing and storage ensures a greater percentage of usable product after storage. When onions and garlic are cured in the field, they are undercut, then hand-pulled. Sometimes the roots and tops are trimmed and the bulbs then are allowed to dry in field racks or bins from 2 to 7 days or longer, depending on ambient conditions (fig. 35.5). Sometimes they are pulled and cured before trimming. Curing may be done in windrows with the tops covering the bulbs to prevent sunburn. Where

Figure 35.3

Cured areas of sweet potatoes damaged during harvest.

Figure 35.4

Jicama stored at 12.5°C (55°F). Roots were undamaged (left), damaged and not cured (middle), or damaged and cured (right) at 30°C (86°F).

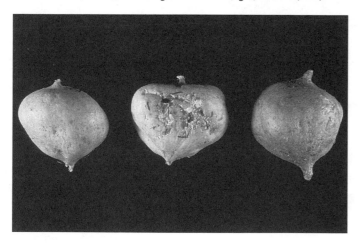

Figure 35.5

Mechanically undercut and manually trimmed garlic for fresh market being field-cured in California. Bins are covered with garlic stalks to prevent sun injury.

ambient conditions are unfavorable, curing may be done in rooms with warm forced air. Onions develop the best scale color if cured at temperatures of 25° to 32°C (77° to 90°F) (fig. 35.6).

Recommended conditions for curing vary according to commodity, as shown in table 35.2.

PREPARATION FOR MARKET

Onion, garlic, potato, and sweet potato are often stored after curing and before preparation for market (cleaning, grading, sizing, and packing). These products, and other root crops such as carrot and turnip, may be stored from 3 to 10 months in mechanically refrigerated or ventilated storages (see discussion on storage conditions and table 35.3).

The following operations are commonly used to prepare root, tuber, and bulb crops for market (figs. 35.7, 35.8, 35.9).

Bin or flume dumping. Water is usually used for roots and tubers to reduce physical injury, but it is not used for bulb crops.

Cleaning. Dry-brushing or washing and partial drying, removing excess moisture.

Sorting. Eliminating defective product and plant debris.

Decay control. Chlorination of wash and flume waters provides sanitation for carrot and potato. Postharvest fungicides are used to a limited extent on some of these commodities, such as sweet potato.

Sizing. Sizing is done mechanically or by hand. Mechanical sizers are generally diverging rollers or weight sizers. Modified volumetric sizers are used for potatoes. Carrots present special sizing problems, since they must be sized by diameter (diverging rollers) and by length (manually or with a gravity-length sizer).

Grading. Separating into quality grades.

Packing. Packing into consumer units (bags, trays) and then into master shipping containers; or bulk-pack into shipping containers (bags, boxes, and bins). Plastic bags are used for products such as carrots that are very susceptible to water loss.

Loading into transit vehicles. Bulk transport to processing plants is sometimes used for onion, potato, and radish. For fresh market, most packed product is palletized, but it may be transferred to slip-sheets for transport in refrigerated trailers. For railcar shipments, sacks are often manually stacked.

Figure 35.6

Curing of field-packed onions under vertical air fans before shipment.

Table 35.2. Conditions for curing root, tuber, and bulb crops

Commodity	Temperature °C	°F	Relative humidity (%)	Duration (days)
Cassava	30–40	86–104	90–95	2–5
Onion and garlic	30–45	86–113	60–75	1–4
Potato	15–20	59–68	85–90	5–10
Sweet potato	30–32	86–90	85–90	4–7
Yam	32–40	90–104	90–100	1–4

Table 35.3. Storage conditions recommended for tropical root-type vegetables (compiled from various sources)

Commodity	Temperature °C	°F	Relative humidity (%)	Storage period
Cassava	5–8	41–46	80–90	2–4 weeks
	0–5	32–41	85–95	<6 months
Ginger	12–14	54–57	65–75	<6 months
Jicama	12–15	54–59	65–75	3 months
Sweet potato	12–14	54–57	85–90	<6 months
Taro	13–15	55–59	85–90	<4 months
Yam	13–15	55–59	Near 100	<6 months
	27–30	80–86	60–70	3–5 weeks

COOLING

All temperate-zone root crops except potato, onion, and garlic can be hydrocooled. In areas where potatoes are harvested during hot weather, they may also be hydrocooled as part of the washing operation. These same potatoes are typically room-cooled after packing to 13° to 16°C (55° to 60°F) before shipment. Tropical root crops, onion, and garlic are also occasionally room-cooled before shipment to market.

Potatoes and onions destined for storage are cooled (after curing) during the early phase of the storage period with cool air forced through storage piles or bins. Cooling may be done with cold ambient air or by mechanical refrigeration.

STORAGE CONDITIONS

NONREFRIGERATED STORAGE METHODS

Some growers occasionally store mature potatoes in the ground for several weeks before harvest. Ground storage is also used for several of the tropical and subtropical roots, including cassava and jicama. Pits, trenches, and clamps are used for short-term, small-scale storage of potatoes, sweet potatoes and other root crops.

Ventilated storage in cellars and warehouses is used for potatoes, sweet potatoes, garlic, and onions. Newer facilities with temperature and RH controls provide forced-air circulation through bulk piles of potatoes or onions, or through and around stacks of bulk bins. Ventilation with cool night air can maintain adequate storage temperatures in many areas.

STORAGE OF SUBTROPICAL AND TROPICAL-ZONE ROOT VEGETABLES

The general storage recommendations for tropical-zone root vegetables, summarized in table 35.3, show that most of these products are chilling-sensitive. Symptoms of chilling injury include excessive decay, internal discoloration, and textural changes (e.g., "hard core" in cooked sweet potato and cassava). Figure 35.10 illustrates internal discoloration of jicama due to storage at chilling temperature. Some of these root crops (e.g., cassava) are successfully field-stored but deteriorate rapidly if harvested and held under ambient conditions. Evaporatively

Figure 35.7

Harvest and postharvest handling of root and tuber vegetables.

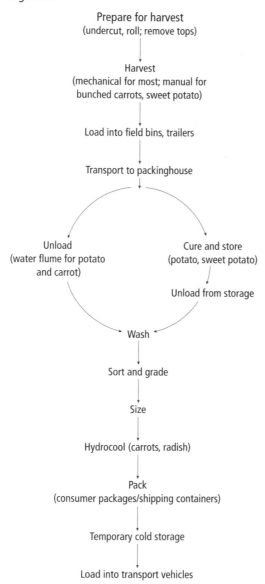

Figure 35.8

Harvest and postharvest operations for bulb vegetables.

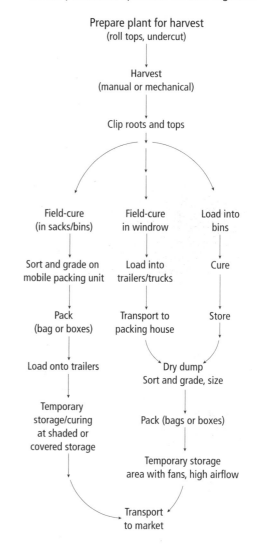

cooled storages, which maintain high humidity and moderate temperature (13° to 16°C; 55° to 60°F) conditions, are used in California for storing sweet potatoes for about 9 months.

STORAGE OF TEMPERATE-ZONE ROOT VEGETABLES

In California, temperate-zone root vegetables are not usually stored. When they are stored, the following conditions should be maintained: 0°C (32°F), 95 to 98% RH, and adequate air circulation to remove vital heat from the product and prevent CO_2 accumu-

lation. Beets, carrots, and radishes with tops have a shelf life of only 2 to 3 weeks, whereas these products can be stored several months if the tops are removed.

Potatoes can be stored up to 10 months under proper conditions. Most long-term potato storage facilities are in the northern United States. Because of the long storage periods, weight loss control is very important. In recent years, potato storages have been designed with an "air envelope" exterior to minimize temperature and humidity fluctuations and reduce weight loss.

For fresh market, potatoes should be stored under the following conditions: 4° to 7°C (39° to 45°F), 95 to 98% RH, enough air circulation to prevent O_2 depletion and CO_2 accumulation (about 23 l/min per 45 kg, or

Figure 35.9

Mobile packing unit for field-packing of onions.

Figure 35.10

Internal browning of jicama root stored 3 weeks at 10°C (50°F) (top) and 12.5°C (55°F) (bottom). The internal discoloration is one symptom of chilling injury in jicama. The roots did not differ in external appearance.

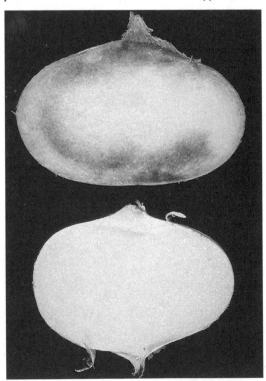

0.8 ft³/min per 100 lb of potatoes). For processing (e.g., chipping), the proper conditions are 8° to 12°C (46° to 54°F), 95 to 98% RH, adequate ventilation, and exclusion of light. This higher temperature storage retards undesirable sweetening of the potatoes and consequent dark color of the processed products. Sugars and respiration rates increase

during low-temperature storage, indicating that potatoes are slightly chilling-sensitive. Seed potatoes are best kept at 0° to 2°C (32° to 36°F), 95 to 98% RH, with adequate ventilation. In this case low-temperature sweetening is not important, and decay and water loss are minimized. Some potato varieties are more susceptible to chilling injury than others. Storage at very low temperatures over extended periods can result in mahogany browning and cavitation of internal tissues (fig. 35.11). Decay, especially soft rot due to *Erwinia* bacteria, can be a problem during long-term storage of potatoes. Proper curing of harvest wounds on the potatoes before storage is important to reduce decay.

Garlic should be kept at 0°C (32°F) or slightly below for long-term storage (6 to 7 months); 20° to 30°C (68° to 86°F) can be used for storage up to 1 to 2 months. Garlic cloves sprout most rapidly at intermediate temperatures (4° to 18°C; 40° to 64°F). Ventilation of about 1 m³ of air per minute per cubic meter (1 ft³ of air per minute per cubic foot) of garlic with 70% RH is adequate.

Onions vary in their storage capability. The more pungent types with high soluble solids contents store longer, whereas mild onions with low soluble solids content are not usually stored for more than 1 to 3 months. Storage temperatures should be either 0° to 5°C (32° to 41°F) or 20° to 30°C (68° to 86°F), as intermediate temperatures favor sprouting. Relative humidity should be maintained at 65 to 70%, and ventilation rate should be from 0.5 to 1 m³ of air per minute for each cubic meter (0.5 to 1 ft³ of air per minute per cubic foot) of onions. Avoid light exposure to prevent greening. With long-term storage, the disorder watery scales or translucency can develop (fig. 35.12). Decays associated with *Aspergillus* (black mold), but especially *Botrytis* (gray mold), increase during storage.

MODIFIED ATMOSPHERES

The use of controlled or modified atmospheres for this group of commodities is very limited. Potatoes and carrots are not benefited by MA. Limited commercial CA storage (3% O_2 and 5 to 10% CO_2) of mild types of onions has been used recently. Besides retarding decay, high-CO_2 atmospheres retard sprout development in onions and garlic (fig. 35.13).

Figure 35.11

Severe chilling injury on potato, showing internal browning and cavitation. Potatoes (cv. Yellow Finn) were stored 6 months at about 2°C (36°F).

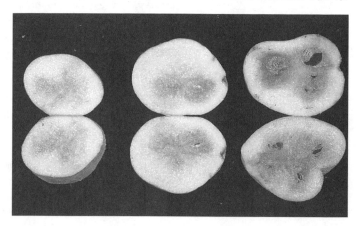

Figure 35.12

Watery or translucent scales of stored onions.

Figure 35.13

Longitudinal sections of garlic cloves stored 5 months at 0°C (32°F) to show sprout development in air (2 cloves on left) or a 0.5% O_2 and 10% CO_2 atmosphere (2 cloves on the right).

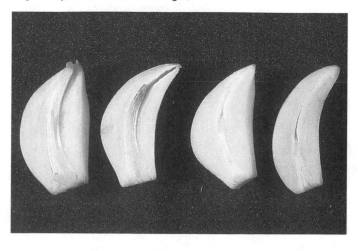

ETHYLENE

Most of the products considered in this chapter produce very low amounts of ethylene and are not detrimentally affected by exposure to ethylene. There are important exceptions. Ethylene affects potatoes by stimulating sprout initiation (reduces dormancy), but it then inhibits sprout growth. Ethylene also increases darkening of cooked potatoes. Ethylene is very detrimental to carrots. Low concentrations of ethylene (>0.1 ppm) in the storage environment of carrots and parsnips increase respiration rates and induce the formation of isocoumarin, the compound responsible for bitterness (fig. 35.14). Cut carrot segments, dropped carrots, freshly harvested carrots, and immature carrots develop very high concentrations of isocoumarin and bitterness.

SPECIAL CONSIDERATIONS

SPROUT INHIBITION

For long-term storage, onions and potatoes (not seed potatoes) are generally sprayed with maleic hydrazide (MH) a few weeks before harvest to inhibit sprouting during storage. Aerosol applications of chlorpropham (CIPC; 3-chloro-isopropyl-N-phenyl carbamate) are often circulated around stored potatoes to further inhibit sprouting. In recent years there have been attempts to look for alternative, naturally occurring sprout inhibitors. Irradiation (0.03 to 0.15 kGy, or 3 to 15 krad, doses) has also been shown to effectively inhibit sprout growth in onions, garlic, and potatoes and is used in some countries.

GLYCOALKALOIDS IN POTATOES

Light needs to be excluded during potato storage and handling to avoid greening. Greening is due to chlorophyll synthesis and is closely associated with accumulation of toxic glycoalkaloids, principally solanine and chaconine. At low levels, these compounds contribute to flavor, but 20 mg alkaloids per 100 g (0.002 oz per 10 oz) (fresh weight) potato is considered the upper safe limit for human consumption. Alkaloids are not destroyed by cooking, and it is therefore necessary to avoid their accumulation. To keep toxic alkaloid content low, potatoes should be stored at low temperatures (<7.5°C, or <45°F), kept away from light, marketed in opaque plastic or paper bags,

Figure 35.14

Development of isocoumarin in young carrots stored at low temperature with low concentrations of ethylene.

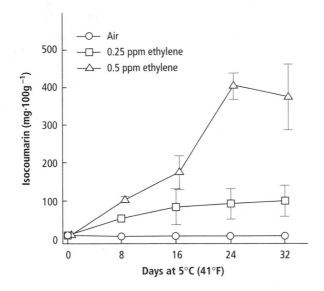

and rotated frequently on retail displays. Bruising also stimulates the formation of these alkaloids and is another reason to reduce physical injury to potatoes.

PUNGENCY IN ONIONS

Storage onions have high soluble solids content and high pungency, while the short-shelf-life mild or "sweet" types have low soluble solids and low pungency. Pungency is due to the enzymatic release of sulfur-containing aroma compounds. Onion fields are sometimes sampled for analysis of pyruvate (a by-product of the pungency reactions) to distinguish different qualities of mild onions. Onions with a pyruvate concentration below 5 μm/g fresh weight are considered mild or sweet.

REFERENCES

Booth, R. H., and R. L. Shaw. 1981. Principles of potato storage. Lima: Internat. Potato Center (CIP). 105 pp.

Brewster, J. L., and H. D. Rabinowitch. 1990. Onions and allied crops. 3 vols. Boca Raton, FL: CRC Press.

Brice, J. B., L. Currah, A. Malins, and R. Bancroft. 1997. Onion storage in the tropics: A practical guide to methods of storage and their selection. Chatham, UK: Natural Resources Institute. 119 pp.

Burton, W. G. 1982. Postharvest physiology of food crops. New York: Longman.

———. 1989. The potato. New York: Longman/Wiley. 742 pp.

Cantwell, M., W. Orozco, V. Rubatzky, and L. Hernández. 1992. Postharvest handling and storage of jicama roots. Acta Hort. 318:333–343.

Clark, C. A. 1992. Postharvest diseases of sweet potatoes and their control. Postharv. News and Info. 3:75N–79N.

Cooke, R. D., J. E. Rickard, and A. K. Thompson. 1988. The storage of tropical root and tuber crops—Cassava, yam and edible aroids. Exper. Agric. 24:457–470.

Dale, M. F. B., D. W. Griffiths, and H. Bain. 1998. Effect of bruising on the total glycoalkaloid and chlorogenic acid content of potato (*Solanum tuberosum*) tubers of 5 cultivars. J. Sci. Food Agric. 77:499–505.

Kopsell, D. E., and W. M. Randle. 1997. Onion cultivars differ in pungency and bulb quality changes during storage. HortScience. 32:1260–1263.

Lafuente, M. T., G. López-Gálvez, M. Cantwell, and S. F. Yang. 1996. Factors influencing ethylene-induced isocoumarin formation and increased respiration in carrots. J. Amer. Soc. Hort. Sci. 121:537–542.

Mazza, G. 1989. Carrots. In N. A. M. Eskin, ed., Quality and preservation of vegetables. Boca Raton, FL: CRC Press. 75–119.

McGee, R. S. 1995. Current state of the art in potato storages. ASAE Publ. I-95:538–545.

Prange, R. K., W. Kalt, B. J. Daniels-Lake, C. L. Liew, R. T. Page, J. R. Walsh, P. Dean, and R. Coffin. 1998. Using ethylene as a sprout control agent in stored 'Russet Burbank' potatoes. J. Amer. Soc. Hort. Sci. 123:463–469.

Purcell, A. E., W. M. Walter Jr., and L. G. Wilson. 1989. Sweet potatoes. In N. A. M. Eskin, ed., Quality and preservation of vegetables. Boca Raton, FL: CRC Press. 285–304.

Rastorski, A., A. van Es, et al. 1987. Storage of potatoes. Post-harvest behavior, store design, storage practices, handling. Wageningen, Netherlands: Pudoc. 453 pp.

Ravi, V., and J. Aked. 1996. Review on tropical root and tuber crops. II. Physiological disorders in freshly stored roots and tubers. Crit. Rev. Food Sci. Nutr. 36:711–731.

Ravi, V., and C. Balagopalan. 1996. Review on tropical root and tuber crop. I. Storage methods and quality changes. Crit. Rev. Food Sci. Nutr. 36:661–710.

Salunkhe, D. K., B. B. Desai, and J. K. Chavan. 1989. Potatoes. In N. A. M. Eskin, ed., Quality and preservation of vegetables. Boca Raton, FL: CRC Press. 1–52.

Schouten, S. P. 1987. Bulbs and tubers. In J. Weichmann, ed., Postharvest physiology of vegetables. New York: Marcel Dekker. 555–581.

Shibairo, S. I., M. K. Upadhyaya, and P. M. A. Toivonen. 1997. Postharvest moisture loss characteristics of carrot (*Daucus carota* L.) cultivars during short-term storage. Scientia Hortic. 71:1–12.

Smith, O. 1977. Potatoes: Production, storing, processing. 2nd ed. Westport, CT: AVI. 776 pp.

Smith, W. L., Jr., and J. B. Wilson. 1978. Market diseases of potatoes. USDA Handb. 479. 99 pp.

Smittle, D. A. 1989. Controlled atmosphere storage of Vidalia onions. In J. K. Fellman, ed., Proc. International Controlled Atmosphere Research Conf. Vol. 2. Moscow: University of Idaho. 171–177.

Snowdon, A. L. 1992. Color atlas of post-harvest diseases and disorders of fruits and vegetables. Vol. 2. Vegetables. Boca Raton, FL: CRC Press. 416 pp.

Stoll, K., and J. Weichmann. 1987. Root vegetables. In J. Weichmann, ed., Postharvest physiology of vegetables. New York: Marcel Dekker. 541–553.

Thompson, J. F., and R. W. Scheuerman. 1993. Curing and storing California sweet potatoes. Davis: Univ. Calif. Dept. Biol. Agric. Engr. 4 pp.

Wilson, L. G., C. W. Averre, J. V. Baird, E. O. Beasley, A. R. Bonanno, E. A. Estes, and K. A. Sorensen. 1989. Growing and marketing quality sweet potatoes. Raleigh: N.C. State Univ Agric. Ext. Publ. AG-09. 28 pp.

Wright, P. J., and D. G. Grant. 1997. Effects of cultural practices at harvest on onion bulb quality and incidence of rots in storage. N.Z. J. Crop Hort. Sci. 25:353–358.

36

Postharvest Handling Systems: Fresh-Cut Fruits and Vegetables

Marita I. Cantwell and

Trevor V. Suslow

Fresh-cut products have grown rapidly during the past decade, extending from the foodservice sector to the retail shelf. These fruit and vegetable products are prepared and handled to maintain their fresh state while providing convenience to the user. Although more expensive than bulk produce on a weight basis, successful fresh-cut products are often more cost-effective for the user due to reduced waste. Minimal processing at central facilities greatly reduces the number of on-site employees involved in fresh produce preparation. Preparation of fresh-cut products involves cleaning, washing, trimming, coring, slicing, shredding, and other related steps, many of which increase the perishability of the produce items. Other terms used to refer to fresh-cut products are "minimally processed," "lightly processed," "partially processed," "fresh-processed," and "pre-prepared."

Examples of fresh-cut vegetables include peeled and sliced potatoes, shredded lettuce and cabbage, mixed salads (fig. 36.1), washed and trimmed spinach, peeled "baby" carrots (fig. 36.2), cauliflower and broccoli florets, and cleaned and diced onions. These products are expected to have a shelf life of 10 to 14 days and represent about 70% of the total volume of fresh-cut items available commercially. Other important vegetable items include peeled garlic (fig. 36.3), fresh salsas, vegetable snacks such as carrots and celery, sliced mushrooms, sliced and diced tomatoes and peppers, and fresh or microwaveable vegetable trays. More recent fresh-cut vegetable products include salad meals or "home replacement meals," which contain meat and other food items. Fresh-cut fruit products include peeled and cored pineapple; peeled citrus fruits (fig. 36.4) and segments; apple (fig. 36.5), peach, mango, and melon (fig. 36.6) slices; and fruit salads. Expected shelf life of these products is generally much less than for the vegetable products. Sales of fresh-cut products currently account for 8 to 10% of the fresh fruits and vegetables marketed through foodservice and retail channels in the United States (see chapter 2).

Whereas most food processing techniques stabilize the products and lengthen their storage and shelf life, minimal processing of fruits and vegetables increases their perishability. Although the industry began as a salvage operation to utilize off-grade and second-harvest products, it was soon recognized that high-quality raw materials were required because of the increased perishability caused by product preparation. Due to this and the need for strict sanitation, preparation and handling of these products require an integration of production, postharvest, and food science technologies and marketing expertise.

FRESH-CUT PRODUCT PHYSIOLOGY AND SHELF LIFE IMPLICATIONS

Fresh-cut products generally have higher respiration rates than the corresponding intact products. Higher respiration rates indicate a more active metabolism and, usually, a faster deterioration rate. Higher respiration rates can also result in more rapid loss of acids, sugars, and other components that determine flavor quality and nutritive value. The increased O_2 demand due to the higher respiration rates of fresh-cut products dictates that packaging films with sufficient permeability to O_2 are required to prevent fermentation and off-odors. The physical damage or wounding caused by preparation increases respiration and ethylene production within minutes, with associated increases in rates of other biochemical reactions responsible for changes in color

Figure 36.1

Retail bags of mixed "European" salads from five major U.S. processors.

Figure 36.2

Baby carrots packaged in polyethylene bags for retail. Carrots in the bag on the left show surface drying or "white blush."

Figure 36.3

Peeled garlic cloves for foodservice distribution.

Figure 36.4

Peeled orange fruits in rigid tray with film lid.

Figure 36.5

Apple slices treated with browning inhibitors in a Fresh-Hold package.

Figure 36.6

Melon chunks and single-serving mixed fruit salads.

Figure 36.7

Figure 36.7

Respiration rates of intact and shredded cabbage stored at four temperatures.

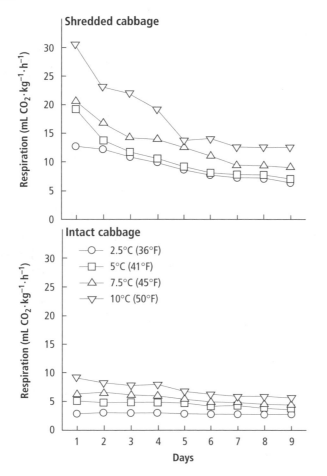

Table 36.1. Respiration rates of leaves of kale stored at various temperatures for 5 days

Product	Respiration rate (mL $CO_2 \cdot kg^{-1} \cdot h^{-1}$)			
	0°C (32°F)	5°C (41°F)	10°C (50°F)	15°C (59°F)
Full-sized leaves	8	12	29	33
Small leaves	14	21	42	57
Chopped (pieces 2 by 2 cm [¾ by ¾ in] cut from full-sized leaves)	15	23	46	53
Shredded (pieces 0.3 cm [¹⁄₁₀ in] from full-sized leaves)	17	18	59	68

(including browning), flavor, texture, and nutritional quality (sugar, acid, and vitamin content). The degree of processing and the quality of the equipment (e.g., blade sharpness), significantly affect the wounding response.

Strict temperature control is required to minimize the increased respiration and meta-bolic rates of fresh-cut products. The importance of temperature is illustrated with data from intact and shredded cabbage stored at different temperatures (fig. 36.7). Young leaf tissue will have higher respiration rates than mature fully developed leaves, as shown with data on kale leaves (table 36.1). Salad-size (2 by 2 cm, or 0.75 by 0.75 in) pieces from mature kale leaves have respiration rates almost double those of the intact leaves, but their rates are similar to rates of the small leaves; shredding mature leaves approximately doubled respiration rates. Different parts of a vegetable may have very different respiration rates, as illustrated by data from broccoli (fig. 36.8). These differences in respiration rates have implications for the quality and shelf life of mixed medleys and salad mixes. The quality of an entire fresh-cut item is only as good as that of its most perishable component. In mixed salads, it is important to ensure that components included for their color or flavor qualities be as fresh as possible and similar in shelf life to the major salad components. These considerations also apply to a product such as cleaned and washed spinach, where differences in leaf age or physical damage may yield a mixed product of variable perishability.

The greater the degree of processing, the higher the induced rates of respiration. Intact garlic bulbs have relatively low respiration rates, but they have high respiration rates when cloves are peeled or chopped, especially if stored at temperatures above 5°C (41°F) (table 36.2). The respiration rates of iceberg and romaine lettuces cut as salad pieces (2 to 3 cm 2 to 3 cm, or 0.75 to 1.18 in by 0.75 to 1.18 in) are only 20 to 40% higher than rates of the respective intact heads. The respiration rates of shredded lettuce and shredded cabbage are 200 to 300% greater than those of the intact heads and remain high throughout the storage period (see fig. 36.7). The relationship between respiration rates and changes in quality at different temperatures can be illustrated by data from intact and sliced mushrooms (fig. 36.9). Respiration rates and deterioration rates can be minimized by quickly cooling the product and storing at 5°C (41°F) or below.

Cutting carrots into large segments does not significantly change their respiration rate but does make them much more sensitive to ethylene. Unpeeled segments and slices of

Figure 36.8

Respiration rates of florets, stem, and intact heads of broccoli.

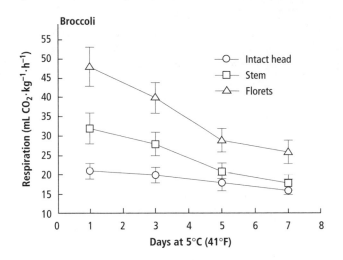

Table 36.2. Average respiration rates of intact, peeled, and chopped cloves of garlic stored 5 days

	Respiration rate (mL $CO_2 \cdot kg^{-1} \cdot h^{-1}$)			
Product	**0°C (32°F)**	**5°C (41°F)**	**10°C (50°F)**	**15°C (59°F)**
Intact	4	6	8	13
Peeled cloves	12	20	57	69
Chopped cloves	16	22	73	107

carrots exposed to ethylene become bitter due to the synthesis of isocoumarin. However, the peeled "baby" carrots do not produce this bitter compound even if exposed to ethylene because isocoumarin is formed predominantly in the peel tissue.

Fresh-cut fruits generally have a more complicated physiology than fresh-cut vegetables. Stage of ripeness at the time of processing may alter the physiological responses to cutting. In cantaloupe, respiration rates of pieces from fruit harvested at different stages of ripeness are similar. Ethylene production rates, however, are much higher in pieces from riper fruits (three-quarters or full slip) than less-ripe fruit (one-quarter to one-half slip). Piece size also greatly affects the physiological response of the fresh-cut fruits. Cantaloupe cut into very small pieces (0.2 mm, or 0.008 in) had a large increase in ethylene production, whereas large pieces (1 by 2 cm, or 0.4 by 0.75 in) were not different in their physiology from the intact fruit. Respiration rates of sliced peaches, bananas, kiwifruit and tomatoes average about 65% higher than rates of the corresponding intact fruits at 0° to 10°C (32° to 50°F). Examples of fruit responses to fresh-cut processing are summarized in table 36.3.

Although temperature is the principal controlling factor for respiration rates, modified

Table 36.3. Responses of fresh-cut fruit pieces compared to the physiology of intact fruit (compiled by Cantwell, 1998)

Fruit	Stage of ripeness	Temperature °C	Temperature °F	Piece size	Respiration compared to intact fruit	Ethylene production compared to intact fruit
Apple	Ripe	2	36	Wedge	Increase	—
Banana	Unripe	20	68	0.4 cm	—	Increase 4×
	Ripening	20	68	0.4 cm	Increase	Increase
	Ripe	20	68	4 cm	Same	Same
Cantaloupe	Ripe	20	68	0.2 mm	—	Increase 10×
	Ripening	2	36	2- × 1-cm cylinder	Same	Same
	Ripe	2	36	2- × 1-cm cylinder	Same	Same
	Ripening	10	50	2- × 1-cm cylinder	Same	Same
	Ripening	20	68	2- × 1-cm cylinder	Increase 2×	Same
Kiwifruit	Ripe	20	68	1 cm	Increase	Increase 8×
Pear	Ripening	2	36	1-cm wedge	Same	Same
		20	68	1-cm wedge	Increase	Reduced
Strawberry	Ripe	2	36	Quarters	Same	Same, none
		20	68	Quarters	Increase	Increase 4×
Tomato	Mature-green	20	68	1 cm	Same	Same
	MG+C_2H_4	20	68	0.7-cm slice	Same	Same
	Ripening	20	68	1 cm	Increase	Increase 5×
	Ripe	20	68	1 cm	Same	Same

atmospheres also reduce rates (see chapter 14). Limited information is available regarding respiration rates of fresh-cut products under CA or MA. An atmosphere of 5% O_2 and 5% CO_2 only slightly reduces the respiration rates of fresh-cut carrots, leek, and onion but slightly increases the rates of cut potatoes (Mattila et al. 1995). CA of 1 to 2% O_2 and

Figure 36.9

Respiration rates, color, and firmness changes of intact and sliced white mushrooms.

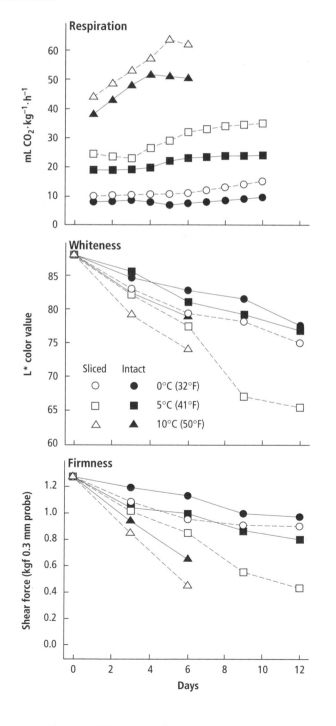

10% CO_2 reduced respiration rates of minimally processed strawberries, peaches, and honeydew by 25 to 50% at 5°C (41°F) (Qi and Watada 1997). These same atmospheres also substantially reduced ethylene production rates and softening of the fruit tissues.

Control of the wound response is the key to providing a fresh-cut product of good quality. Low temperatures minimize differences in respiration and ethylene production rates between the fresh-cut and the intact product. Low temperatures are also essential to retard microbial growth and decay on cut surfaces. Variety, production conditions, stage of maturity, piece size, and storage conditions all contribute to variations in fresh-cut product physiology.

PROCEDURES FOR PREPARATION OF FRESH-CUT VEGETABLES

Minimally processed products may be prepared at the source of production or at regional and local processors, depending on the perishability of the fresh-cut item relative to the intact product and also depending on the quality required for the designated use of the product. Table 36.4 summarizes some considerations in relation to location of processing. Fresh-cut vegetable processing has shifted from destination (local) to source processors and regional processors as improvements in equipment, MA packaging, and temperature management have become available. Other than pineapple, processing and marketing of fresh-cut fruits has remained at the regional or local level.

An example of the operations involved in minimal processing is shown in figure 36.10 for lettuce. In the past, fresh-cut lettuce operations often salvaged lettuce remaining in the fields after harvesting for fresh market. It is now recognized that first-cut lettuce of optimal maturity (heads firm but not hard) should be used to maximize processed product quality. After reception and emptying of the bins, all operations are done in cold, clean processing rooms. After trimming and coring, piece size may be reduced with rotating knives or by tearing into salad-size pieces. Damage to cells near cut surfaces influences the shelf life and quality of the product. For example, shredded lettuce cut by a sharp knife with a slicing motion has a storage life

Table 36.4. Advantages, disadvantages, and requirements of fresh-cut vegetable products prepared at different locations

Location of processing	Characteristics and requirements
Source of production	Raw product processed fresh when it is of the highest quality.
	Processed product requires a minimum of 14 days postprocessing shelf-life.
	Good temperature management critical. Economy of scale.
	Avoid long-distance transport of unusable product.
	Vacuum- and gas-flushing common; differentially permeable films.
Regional	Raw product processed when of good quality, typically 3 to 7 days after harvest.
	Reduced need to maximize shelf life; about 7 days postprocessing life required.
	Good temperature management vital.
	Several deliveries weekly to end users.
	Can better respond to short-term demands.
	Vacuum- and gas-flushing common; differentially permeable films.
Local	Raw product quality may vary greatly, since processed 7 to 14 days after harvest.
	Relatively short postprocessing life required or expected.
	Good temperature management required but is often deficient.
	Small quantities processed and delivered.
	More labor-intensive; discard large amounts of unusable product.
	Simpler and less costly packaging; less use of vacuum or gas-flushing techniques.

approximately twice that of lettuce cut with a chopping action. Shelf life of lettuce is less if a dull knife is used rather than a sharp knife. A common defect in commercially prepared fresh-cut melon is translucency or glassiness, and dull cutting equipment increases the incidence of this defect.

Washing the product immediately after cutting removes sugars and other nutrients at the cut surfaces that favor microbial growth and tissue discoloration. The wash water is usually cold and also provides rapid cooling. Because of differences in composition, some products such as cabbage are known as "dirty" products because they release a lot more organic nutrients with processing. It is therefore desirable to maintain separate processing lines or thoroughly clean the line before another product follows cabbage. Cut tissues take up water in the washing and

cooling flumes, as illustrated with spinach (fig. 36.11); continuous and strict water disinfection is necessary. Excess moisture must be completely removed after washing and cooling. Centrifugation is generally used after passing the product over a vibration screen. Forced-air tunnels over the conveyor line are used to a limited extent but may be especially useful for fresh-cut fruits and other delicate products. Ideally, the drying process should remove at least the same amount of moisture that the product retained during processing. For lettuce products, removal of slightly more moisture (i.e., slight desiccation of the product) may favor longer postprocessing life.

Centrifuged or dried product may be packaged in preformed bags on equipment that pulls a vacuum before sealing (i.e., shredded lettuce in 3- to 5-lb foodservice bags) or into bags on a form-fill-seal machine in which product is fed through a stainless steel tube around which the bag is formed from plastic film rollstock (e.g., retail salad mixes). In this case, nitrogen gas flushing is often done immediately before sealing to drop O_2 levels and rapidly establish a beneficial MA in the bag. The sealed bags then drop onto a conveyor that passes through a metal detector and are boxed and palletized, usually in an adjoining cold room. To ensure a continuous cold chain, palletized product should be loaded onto only precooled trailers at enclosed refrigerated docks.

Peeled "baby" carrots are a popular fresh-cut vegetable that is also produced in large volumes. Varieties and growing conditions have been developed to produce a long, tender carrot that is mechanically harvested and cut into 5-cm (2-in) segments. The segments may be stored to ensure a continuous supply to the processing line. Segments are flumed through course and fine abraders to remove the peel and cell wall debris, and then washed, diameter-sized, and packaged. Low-density polyethylene (LDPE) bags with a few small holes are often used to maintain high humidity and free moisture (no MA), and the bags are packed into waxed carton boxes that may be top-iced to ensure low temperature during distribution. Whitening of the carrot surface is a common defect that results from drying of the abraded and exposed cell walls (see fig. 36.2). Edible surface coatings, hygroscopic coatings, free water in the bag, and other treatments have been applied to

Figure 36.10

Preparation of minimally processed lettuce products.

Harvest
Lettuce from first harvest results in better quality
Overmature lettuce browns more after cutting
Trim outer leaves

↓

Field-pack and local transport
Plastic bins or totes or fiberboard boxes or bins; avoid wood bins
Transport on flatbed trucks; if distances far,
transport in refrigerated trailer

↓

Vacuum or forced-air cooling
Field temperatures and delay determine need to cool
Vacuum and forced-air cooling most common

↓

Reception, dump, trim, and core
Revision and selection of heads on conveyer
Further trimming to remove damaged outer leaves
Manual removal of stem tissue with manual coring device

↓

Chop/shred/tear
Continuous-feed cutter for salad pieces (3 by 3 cm, or 1.75 by 1.75 in.)
or shreds (<0.5 cm, or 0.2 in.)
Manual cutting for some lettuce types (romaine)
Very sharp knives reduce damage and subsequent browning

↓

Wash and Cool
Water containing disinfectant, usually chlorine
Residence time from 15–30 sec
Processing aids to reduce browning may be used

↓

Centrifugation or other drying technique
Vibration screen removes large volumes of water
Centrifugation and air tunnels remove moisture so surface of product dry
Basket centrifuges of different sizes depending on product

↓

Combine different products for salad mixes
Or "color" items may be added after washing and centrifuged together

↓

Package in plastic film bags
Centrifuged product dumped onto conveyor feeding filler
Manual or automated form-fill-seal machines
Vacuum- or gas-flushing with nitrogen
Check for leakers in pressurized water chamber

↓

Box, palletize, and store temporarily
Pass bags through a metal detector
On conveyor to boxing and palletizing area
Temporary storage <5°C (<41°F); 0°C (32°F) is optimum

↓

Transport to food service outlets and/or retail markets
Pre-cooled clean trucks; thermostat at <5°C (41°F)
Load at enclosed docks to maintain cold chain

reduce this visual defect. However, a dry surface is less conducive to microbial growth, and the pieces can later be rehydrated rapidly in chilled water.

Items such as broccoli and cauliflower florets require manual trimming of the florets. The broccoli stems may also be trimmed and cut for slaw and various other products. In the handling of cauliflower it is extremely important to reduce damage to the florets during harvest as well as during the fresh-cut operations. Bacterial growth on the damaged surface can be a common defect, especially if the cauliflower florets are not held at low temperature (fig. 36.12).

Key factors for maintaining quality and shelf life of fresh-cut vegetable products include

- using high-quality raw product
- using strict sanitation procedures
- minimizing mechanical damage by using sharp knives
- rinsing and sanitizing cut surfaces
- drying to remove excess water
- packaging with an appropriate atmosphere
- scrupulous control of product temperature at 0° to 5°C (32° to 41°F) during storage, transportation, and handling

Selected characteristics, including respiration rates, quality defects, and beneficial atmospheres of many fresh-cut vegetables are summarized in table 36.5.

PROCEDURES FOR PREPARATION OF FRESH-CUT FRUITS

Preparation of fresh-cut fruit products is complicated by the inherent nature of fruits in which softening and other ripening processes continue after harvest. Finding the right balance between flavor quality and firmness is a key consideration for adequate shelf life of fresh-cut fruit products. Fruit varieties often change during the production season, and fruits may be stored for relatively long periods of time before processing. Defining the best storage and conditioning procedures for fruits destined for fresh-cut processing is also a challenge.

For fresh-cut fruit products, the exterior

Table 36.5. Physiology and storage characteristics of fresh-cut products (all products should be stored at 0° to 5°C)

Commodity	Fresh-cut product	Respiration rate in air at 5°C (mL $CO_2 \cdot kg^{-1} \cdot h^{-1}$)*	Common quality defects (other than microbial growth)	Beneficial atmosphere[†] % O_2	% CO_2
Apple	Sliced	3–7 (2°C)	Browning	<1	—
Asparagus tips	Trimmed spears	40	Browning, softening	10–20	10–15
Beans, snap	Cut	15–18	Browning	2–5	3–12
Beets	Cubed	6	Leakage; color loss	5	5
Broccoli	Florets	20–35	Yellowing, off-odors	3–10	5–10
Cabbage	Shredded	13–20	Browning	3–7	5–15
Carrots	Sticks, shredded	7–10; 12–15	Surface drying ("white blush"); leakage	0.5–5	10
Cauliflower	Florets	—	Discoloration; off-odors	5–10	<5[‡]
Celery	Sticks	2–3 (2.5°C)	Browning, surface drying	—	—
Cucumber	Sliced	5	Leakage	—	—
Garlic	Peeled clove	20	Sprout growth, discoloration	3	5–10[‡]
Jicama	Sticks	5–10	Browning; texture loss	3	10[‡]
Kiwifruit	Sliced	1–3 (0°C)	Juice leakage; texture loss	2–4	5–10
Leek	Sliced, 2 mm	25	Discoloration	5	5
Lettuce, iceberg	Chopped, shredded,	6; 10	Browning of cut edges	<0.5–3	10–15
Lettuce, other than iceberg	Chopped	10–13	Browning of cut edges	1–3	5–10
Melons					
Cantaloupe	Cubed	5–8	Leakage; softening; glassiness (translucency)	3–5	5–15
Honeydew	Cubed	2–4	Leakage; softening; glassiness (translucency)	2–3	5–15
Watermelon	Cubed	2–4	Leakage; softening	3–5	5–15
Mushrooms	Sliced, 5 mm	20–45	Browning	3	10**
Onion, bulb	Sliced, diced	8–12	Texture, juice loss, discoloration	2–5	10–15
Onion, green	Chopped	25–30	Discoloration, growth; leakage	—	—
Orange	Peeled; sectioned	3	Juice leakage; off-flavors	14–21	7–10
Peach	Sliced	6	Browning; mechanical damage	1–2	5–12
Pear	Sliced	6–8 (2.5°C)	Browning; mechanical damage	0.5	<10
Peppers	Sliced; diced	3; 6	Texture loss, browning	3	5–10
Persimmon	Sliced	—	Glassiness (translucency); darkening	2	12
Pineapple	Cubed	3–7	Leakage; discoloration	3	10[‡]
Pomegranate	Arils	2	Color loss; juice leakage	21	15–20
Potato	Sticks, peeled	4–8	Browning, drying	1–3	6–9
Rutabaga	Cubed	10	Discoloration, drying	5	5
Spinach	Cleaned, cut	6–12	Off-odors; rapid deterioration of small pieces	1–3	8–10
Squash, summer	Cubed, sliced 5 mm	12–24	Browning; leakage	1	—
Strawberry	Sliced; topped	12; 6	Loss of texture, juice, color	1–2	5–10
Tomato	Sliced	3	Leakage	3	3

* *Source:* Adapted from Mirjami et al. 1995; Watada et al. 1996; Avena-Bustillos et al. 1997.

[†] *Source:* Adapted from Gorny 1997.

[‡] Unpublished results.

** Not used because of *C. botulinum* risk.

Figure 36.11

Water uptake by young (6-cm, [2.4-in]) and mature (12-cm [4.7-in]) spinach leaves in relation to the number of cut surfaces.

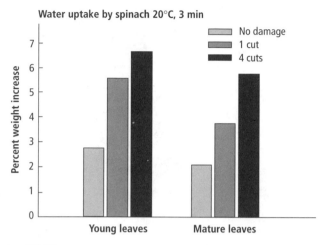

Water uptake by spinach 20°C, 3 min

Legend:
- No damage
- 1 cut
- 4 cuts

Y-axis: Percent weight increase
X-axis: Young leaves, Mature leaves

Figure 36.12

Cauliflower florets with physical damage stored at 6 days in air at 3 temperatures.

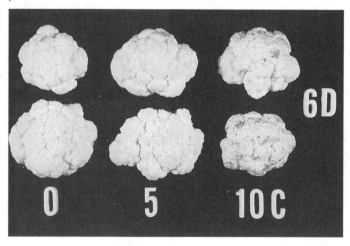

Figure 36.13

Control of browning of Granny Smith apple slices by CA (3% O_2 and 10% CO_2) and a 1% ascorbic acid dip. Pieces were stored at 5°C (41°F) for 2 weeks.

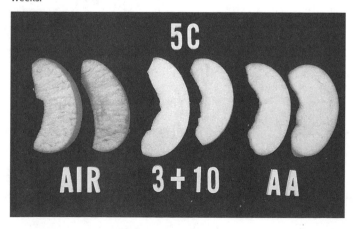

surface is disinfected; the fruit is peeled and sectioned and then may be passed through a cutter/slicer or cut by hand. Depending on the fruit, pieces may be washed after cutting. Fresh-cut soft fruit products need to be packaged in rigid containers, usually lidded with a plastic film that permits development of MA. Loss of fluid from juicy fresh-cut fruits, such as melons, during storage and handling is a common problem.

Selected characteristics, including respiration rates, quality defects, and beneficial atmospheres, of many fresh-cut fruits are summarized in table 36.5.

The visual acceptance and shelf life of many fruit products depend on the use of treatments to retard browning beyond that achieved by the use of low temperatures and MA. Many fruits (e.g., apples, peach, pear) and some vegetables (e.g., potatoes, artichoke) have high levels of preformed phenolic compounds. After cutting, very rapid surface browning will occur. Apple and peach varieties can vary greatly in browning potential of cut pieces. The use of antioxidants such as ascorbic acid (fig. 36.13) and erythorbic acid, and the use of acidifying and chelating agents (citric acid, EDTA), and combinations of these can be useful to reduce browning problems. Rinses with hypochlorite solutions may also retard browning. Sulfites are very effective but are not permitted for use on fresh-cut products in the United States. Calcium salts can also be useful to retard browning and maintain firmness, as illustrated with data on pear slices (fig. 36.14). Dips into calcium chloride or calcium lactate (which has less effect on flavor) solutions have been shown to improve the firmness of fresh-cut cantaloupe. Table 36.6 lists requirements and considerations for efficient minimal processing of vegetables and fruits.

STORAGE TEMPERATURE

Low temperatures are necessary to reduce respiration rates, retard microbial growth, and retard deterioration such as browning and softening in fresh-cut products, as illustrated for diced onions, broccoli florets, and peppers (figs. 36.15, 36.16, and 36.17, see color well). In general, all fresh-cut items should be stored at 0° to 5°C (32° to 41°F) to maintain their quality, safety, and shelf life.

Figure 36.14

Effects of a 1% calcium chloride dip on color and firmness of pear slices. (Gorny and Kader in Cantwell 1998)

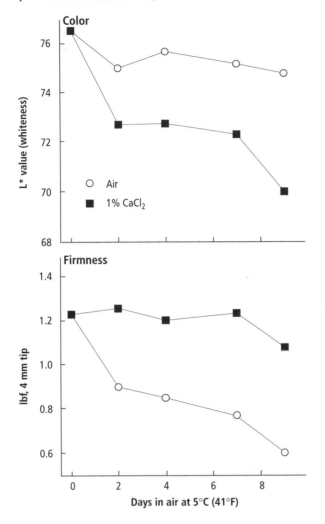

Figure 36.15

Diced yellow onion quality after 20 days in air storage at three temperatures.

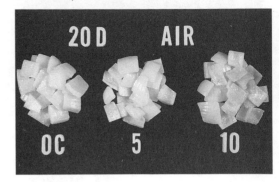

Figure 36.16

Visual quality of broccoli florets stored in air or CA (2% O_2 and 10% CO_2) at three temperatures.

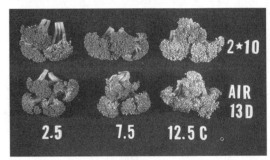

Figure 36.17

Visual quality of diced green peppers stored in air or CA (2% O_2 + 10% CO_2) at three temperatures. The pepper dices kept in air (lower right) have yellowed more than those kept in CA.

The recommendation to store fresh-cut products as close to 0°C (32°F) as possible also generally applies to items prepared from chilling-sensitive produce such as peppers, honeydew melons, jicama, and tomatoes. Fresh-cut products are usually taken directly from cold rooms and used without transfer to warmer temperatures, conditions that favor the development of chilling injury symptoms on intact sensitive products. For fresh-cut squash, cucumber, and watermelon, there are reports that storage at 2° to 3°C (36° to 38°F) may result in better shelf life than storage at 0°C (32°F). However, for chilling-sensitive commodities in general, low temperatures retard the rate of deterioration of the fresh-cut products more than they induce chilling injury. In addition, microbial safety concerns dictate that fresh-

cut products always need to be kept as cold as possible.

MODIFIED ATMOSPHERES AND PACKAGING

For many fresh-cut products, MA packaging is a necessary supplement to low-temperature storage to further reduce rates of deterioration.

Table 36.6. Basic requirements for preparation of minimally processed fruits and vegetables

- High-quality raw material
  Variety selection
  Production practices
  Harvest and storage conditions

- Strict hygiene and good manufacturing practices
  Use of HACCP principles
  Sanitation of processing line, product, and workers

- Low temperatures during processing

- Careful cleaning and/or washing before and after peeling
  Good-quality water (sensory, microbiological, pH)

- Use of mild processing aids in wash water for disinfection or prevention of browning and texture loss
  Chlorine, ozone, other disinfectants
  Antioxidant chemicals such as ascorbic acid, citric acid, etc.
  Calcium salts to reduce textural changes

- Minimize damage during peeling, cutting, slicing, and shredding operations
  Sharp knives and blades on cutters
  Elimination of defective and damaged pieces

- Gentle draining, spin- or air-drying to remove excess moisture

- Correct packaging materials and packaging methods
  Selection of plastic films to ensure adequate O_2 levels to avoid fermentation

- Correct temperature during distribution and handling
  Keep all minimally processed products at 0°–5°C (32°–41°F)

Source: Adapted from Ahvenainen 1996.

Film packaging also reduces water loss from the cut surfaces. There have been many recent improvements and innovations in plastic films and packaging equipment specifically aimed at fresh-cut products.

There are many examples of the benefit of MA on fresh-cut products. Data from fresh-cut cantaloupe stored at 5°C (41°F) provides one example of the potential benefit of CA on shelf life (fig. 36.18). In this case, high-CO_2 atmospheres in air or low O_2 gave similar results and were effective in retarding microbial growth, softening, color change, and off-odors.

For fresh-cut lettuce, discoloration of the cut surfaces is a major quality defect (fig. 36.19). Cutting stimulates enzymes involved in phenolic metabolism, which in turn leads to the formation of undesirable brown pigments. To ensure that packaged salad products have no brown edges, very low O_2 (<0.5%) and high CO_2 (>7%) atmospheres are used in commercial production (fig. 36.20). These conditions may lead to increases in acetaldehyde and ethanol con-

centrations, indicating a shift from aerobic to anaerobic or fermentative metabolism. These changes are greater in the iceberg salads than in romaine salads and are correlated with the development of off-odors.

Although modified atmosphere packaging (MAP) maintains visual quality by retarding browning, off-odors increase and lettuce crispness decreases during storage of the salad products. With current packaging technology, it is possible to have product of good visual quality even at temperature-abuse conditions (fig. 36.21). Although product stored at temperatures of 20°C (68°F) are unlikely, short periods near 10°C (50°F) can readily occur. The visual quality of the product is only slightly reduced by holding at 10°C (50°F), but atmosphere composition, production of fermentative volatiles, and off-odor development are notably different from product stored at 0°C (32°F). These data underscore the importance of low-temperature storage in conjunction with appropriate MAP conditions. Figures 36.16 and 36.17 for diced peppers and broccoli florets also show that CA cannot compensate for improper storage temperatures.

In the case of lettuce, the atmospheres effective in retarding cut edge browning are very different from the atmospheres recommended for intact lettuce heads (lettuce heads develop the disorder brown stain when exposed to CO_2 >2%). Green bell peppers provide another example in which MA conditions beneficial for the fresh-cut product differ substantially from those recommended for the intact product. As more research is conducted on fresh-cut products we can expect more examples in which temperature and atmosphere requirements will be very different from those recommended for the intact products. Current recommendations of beneficial atmospheres for fresh-cut fruits and vegetables and their common quality defects are summarized in table 36.5.

Packaging technology is indispensable for most fresh-cut products. The selection of the plastic film packaging material involves achieving a balance between the O_2 demand of the product (O_2 consumption by respiration) and the permeability of the film to O_2 and CO_2. In practice, films are often selected on the basis of the O_2 transmission rate (OTR, expressed in units of ml/m²-day-atm). Several product factors need to be considered in

Figure 36.18

Changes in quality of cantaloupe melon pieces stored at 5°C (41°F) under different controlled atmospheres. (Portela et al. 1997)

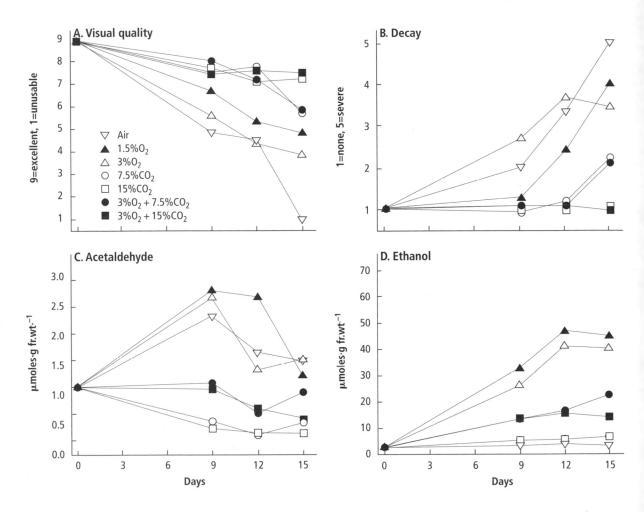

Figure 36.19

Romaine lettuce pieces showing none (1), moderate (3), and severe (5) browning.

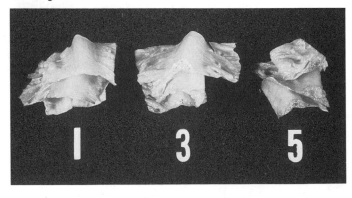

selecting film packaging: the rate of respiration of the product, the specific cut, the quantity of product, and the desirable equilibrium concentrations of O_2 and CO_2. Plastic film characteristics that need to be considered include the permeability of a given thickness of the plastic film to O_2, CO_2, and water at a given temperature; total surface area of the sealed package; and the free volume inside the package.

Many types of films are commercially available for fresh-cut packaging, including polyethylene (PE), polypropylene (PP), blends of PE and ethylene vinyl acetate (EVA), and coextruded polymers or laminates of several plastics. Besides the permeability characteristics described above, films must also satisfy other requirements (see Zagory 1995). They must have strength and be resistant to tears and splits (oriented PP or polystyrene), punctures (low-density PE), and stretching (oriented PP or polyethylene terephthalate); slip to work on bagging machines (acrylic coatings or stearate additives); have flex resistance, clarity, printability, and in some cases, resealability

Figure 36.20

Average changes in gas composition, fermentative volatiles, visual quality, and off-odors of commercial iceberg (garden salad) and romaine (Caesar salad) from five major processors. (López-Gálvez et al. 1997)

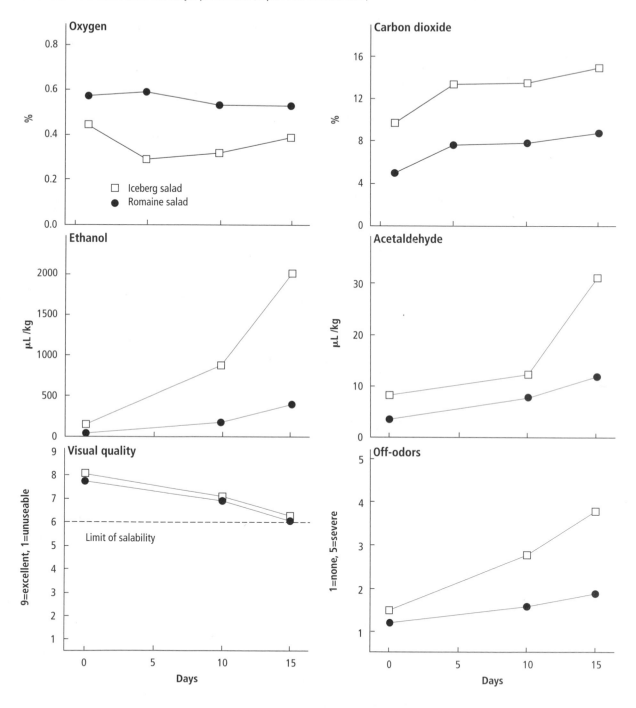

(ziplock or sticky seals). Consumer tactile appeal and ease of opening are also important considerations.

Film selection is a compromise between the strengths and weakness of the different materials. Many currently used films are coextrusion or laminates of several kinds of plastics, each providing a specific benefit.

Recent advancements in controlling the chain length of plastic polymers have resulted in high-OTR films with superior strength, good clarity, and rapid sealing. Rapid sealing is extremely important for high-volume form-fill-seal packaging equipment. Bags are usually checked periodically on the processing line for seal integrity (in a water-filled

Figure 36.21

Effect of storage temperature on the gas composition and quality parameters of iceberg lettuce salads packaged for retail market.

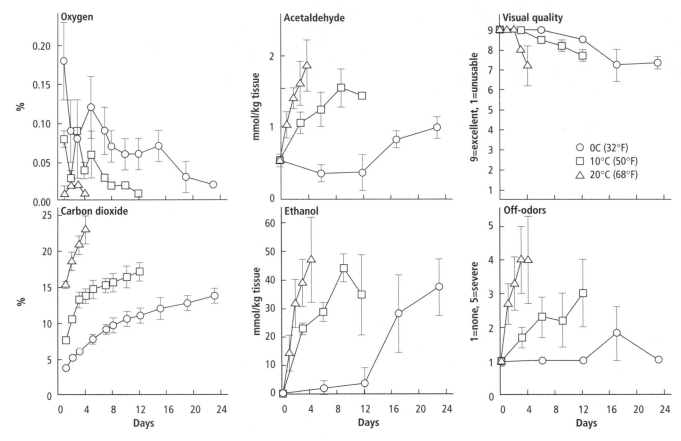

Table 36.7. Effect of small holes on atmospheres in bags of commercial products stored at 0° or 10°C (32° or 50°F) (holes made in bags with a hot, 25-gauge needle)

Product	Bag perforations and storage temperature	Day 3		Day 5	
		% O_2	% CO_2	% O_2	% CO_2
Coleslaw, retail bag	Control 0°C	1.36	3.7	1.07	3.8
	Control 10°C	0.22	5.5	0.28	5.1
	1 hole 10°C	3.07	7.0	4.97	6.7
	4 holes 10°C	10.73	4.3	15.70	3.9
Garden salad, foodservice bag	Control 0°C	0.69	9.2	0.17	8.8
	Control 10°C	0.06	12.1	0.02	11.7
	1 hole 10°C	1.46	10.0	1.03	9.7
	4 holes 10°C	4.80	8.0	10.57	7.8
Garden salad with romaine, retail bag	Control 0°C	0.03	10.6	0.03	10.6
	Control 10°C	0.01	12.8	0.01	13.0
	1 hole 10°C	3.97	10.6	4.80	10.0
	4 holes 10°C	6.60	5.2	14.97	5.0

pressurized chamber) or "leakers." There can be considerable variability in O_2 concentrations even in well-sealed salad bags, perhaps due to slight variations in film uniformity during the manufacturing process. An example of how a poor seal or pinholes in a package affect O_2 concentrations but have less effect on CO_2 concentrations is shown in table 36.7.

Other packaging options include rigid impermeable trays covered with a permeable film or membrane patch. Microperforated films (e.g., FreshHold, fig. 36.5) provide very small holes (40 to 200 µm) and allow elevated levels of O_2 in combination with intermediate CO_2 concentrations. With temperature fluctuations, the permeability of most common films changes very little in comparison to the dramatic increases in respiration rates (O_2 demand) at warmer temperatures. With lack of O_2, anaerobic metabolism occurs resulting in off-odors and other quality problems (see fig. 36.21). "Intelligent films," "customized films," or "sense and respond," film technologies are being developed to address the problems caused by fluctuating temperatures. One company employs "temperature-activated pores" that open and allow an increase in the OTR under temperature abuse situations. Antifog films capable of dispersing water droplets to avoid condensation, incorporation of antimicrobials in films, and use of time-temperature indicators on or incorporated into plastic films are also under development.

MICROBIOLOGICAL ASPECTS

The major groups of microorganisms involved in the spoilage or contamination of fresh produce are bacteria and fungi, although viruses (e.g., *Hepatitis*) and parasites (e.g., *Giardia*) can also be of concern. With minimally processed products, the increase in cut-damaged surfaces and availability of cell nutrients provide conditions that favor microbial growth. Furthermore, the increased handling during preparation of these convenience products provides greater opportunity for contamination with human pathogens such as *E. coli*, *Listeria*, *Yersinia*, and *Salmonella* spp.

There are several microbiological concerns specific to fresh-cut products: they are generally consumed raw with no critical kill step for pathogens, temperature abuse may occur during distribution and display, and some microorganisms of concern may grow under low temperatures and modified atmospheres. Because of these potential hazards, the microbiological quality and safety of minimally processed fruits and vegetables is a high priority. The International Fresh-Cut Produce Association has prepared specific voluntary guidelines for fresh fruit and vegetable processing operations to maintain high levels of microbiological safety (Gorny, 2001). Implementation of Good Manufacturing Practices (GMP) and Hazard Analysis Critical Control Points (HACCP) principles are integral parts of these guidelines.

Microbial growth on minimally processed products is controlled principally by good sanitation and temperature management. Sanitation of all equipment and use of chlorinated water are standard approaches. The large volumes of water used in fresh-cut operations are often recycled, filtered, and sanitized with chlorinated compounds (hypochlorites, chlorine gas, chlorine dioxide). The purpose of adding chlorine to the process water is to disinfect the water, which in turn prevents contamination of the produce. Vigorous washing of product with clean water may remove as many organisms as washing with chlorinated water. Continuous monitoring of active chlorine levels is considered essential to ensure consistent

microbial quality of the washed product. Moisture increases microbial growth; therefore, removal of excess water after washing and cooling by centrifugation or other methods is critical. Combination treatments may be very effective. For example, irradiation of cut lettuce after a chlorinated water wash resulted in very low microbial levels during storage. The use of common sanitizers, potentiators of common disinfectants, and alternative sanitizers (e.g., ozone) are active areas of research (see chapter 24).

Clean product can become recontaminated, especially after passing through operations where debris can accumulate, such as cutters and package-filling equipment. This problem is illustrated by data in table 36.8, in which swabs from the package filler showed a much higher bacterial count than swabs from equipment in the immediately preceding operations.

Changes in the environmental conditions surrounding a product can result in significant changes in the microflora. The risk of pathogenic bacteria may increase with film packaging (high RH and very low O_2); with packaging of products of low salt content and high cellular pH (i.e., most vegetables); and with storage of packaged products at too high temperatures (>5°C, or 41°F). Food pathogens such as *Clostridium*, *Yersinia*, *Salmonella*, and *E. coli* spp. can potentially develop on minimally processed fruits and vegetables under such conditions. Low temperatures during and after processing generally retard microbial growth but may select for psychrotropic organisms such as pseudomonads and *Listeria* spp.

Microorganisms differ in their sensitivity to modified atmospheres. Low O_2 (1%) atmospheres generally have little effect on the

Table 36.8. Total microorganisms at various steps of a fresh, process lettuce line

Operation	Number/in²
Bin dump	92,000
Coring belt	210
Cutter	2,290
Transfer belt	40
Cooling water	5
Centrifuge	10
Package filler	3,350

Source: Adapted from Hurst 1990.

growth of fungi and bacteria. Concentrations of CO_2 at 5 to 10% are usually needed to have an effect on microbial growth. High CO_2 concentrations may indirectly affect microbial growth by retarding deterioration (softening, compositional changes) of the product. High CO_2 atmospheres may have a direct affect by lowering cellular pH and affecting the metabolism of the microorganisms. This is illustrated with data on fresh-cut melon stored under various O_2 and CO_2 concentrations (table 36.9). Gram-negative bacteria are very sensitive to CO_2. Anaerobic bacteria and lactic acid bacteria are quite resistant to CO_2. Fungi are generally very sensitive to CO_2 whereas yeasts are relatively resistant.

Packaging materials modify the humidity and atmosphere composition surrounding minimally processed products and may modify the microbial profile. MA may cause changes in the composition of the microflora on fresh-cut products. For example, the growth of common spoilage bacteria may be suppressed by MAP, but a pathogenic organism such as *Listeria monocytogenes*, which can grow at low temperatures under modified atmospheres, may not be.

Another issue is that MA can extend the visual shelf life of fresh-cut products by suppressing common spoilage organisms. An organism such as *Listeria*, which presents no spoilage symptoms, could develop to high levels by the end of visual shelf life in the MA-packaged product.

Products that are visually acceptable to consumers may have high microbial populations. The total microbial population, however, has no direct bearing on the safety of the product, as discussed in chapter 24. Tests for specific pathogens are needed to evaluate the microbial risk of a given product. This is one reason that no microbiological standards for fresh-cut products exist in the United States; food safety is based on prevention strategies through GMP and HACCP programs.

Although there is no kill step, several "hurdles" are used to maintain the microbiological quality of fresh-cut products. The combinations of cleanly cut products, strict sanitation procedures, appropriate MAP, and low temperatures during processing and distribution not only favor high sensory quality of the fresh-cut products but also serve to minimize microbiological risks (see table 36.6).

RAW MATERIAL QUALITY

Consumers expect minimally processed products to be visually acceptable and appealing. Fresh-cut products must have a fresh appearance, be of consistent quality throughout the package, and be reasonably free of defects. Field defects such as tip burn on lettuce can reduce the quality of the processed product because the cut defective brown tissue will become distributed throughout the processed product. On melons, common defects such as sunburn and groundspot areas can seriously reduce the quality of the fresh-cut product. Pieces from sunburned and groundspot areas consistently have lower sugar content, less orange color, less aroma, and less firmness than pieces from sound areas. Areas of produce items with tipburn, sunburn, and other defects such as bruises should be removed from the processing line before cutting and mixing with good-quality product.

Improvements in processing equipment, packaging materials, and preparation procedures have greatly advanced the fresh-cut fruit and vegetable industry. Products of high visual quality are being produced, but in the future more emphasis will be placed on the aroma, flavor, and other sensory characteristics as well as on the nutritional qualities of fresh-cut products. This will be an even greater challenge for fresh-cut fruits, which inherently have more rapid quality losses than most fresh-cut vegetable products.

The development of varieties for different growing areas with "trait targeting" for fresh-cut quality will be increasingly important.

Table 36.9. Microbial counts of fresh-cut cantaloupe stored in air or CA at 10°C (50°F) and 5°C (41°F)

Atmosphere	Day 0 APC/g*	Day 6 at 10 °C APC/g	Day 12 at 5 °C APC/g	Day 12 at 5 °C Lactobacilli/g
Air	1.0×10^2	1.0×10^8	1.8×10^{11}	4.0×10^7
1.5% O_2		3.9×10^7	1.9×10^9	9.9×10^6
3% O_2		1.3×10^7	1.1×10^{10}	1.4×10^7
7.5% CO_2		8.1×10^6	7.6×10^6	1.8×10^5
15% CO_2		1.6×10^6	1.1×10^6	6.5×10^5
3% O_2 + 7.5% CO_2		8.1×10^5	3.8×10^5	1.5×10^5
3% O_2 + 15% CO_2		1.0×10^5	2.7×10^5	4.2×10^4

Source: Portela et al. 1997.
Note: *APC = Aerobic plate count at 29°C (84°F).

Until now, varieties for the fresh market have been evaluated for their potential usefulness in fresh-cut operations. But there are specific requirements for fresh-cut varieties that need to be met: varieties of lettuce, potato, apples, peaches, etc., with low browning potential; fruits with high sugar contents and firm texture; and varieties that facilitate cleaning, trimming, and cutting operations. Varieties with desirable characteristics will help in the production of fresh-cut products of consistently high quality throughout the year.

Many fresh-cut products are also handled in an "interrupted chain" in which the product may be stored before processing or may be processed to different degrees at different locations. Because of this variation in time and point of processing, it would be useful to be able to evaluate the quality of the raw material and predict the shelf life of the processed product. For example, iceberg lettuce can be stored 1 week but not more than 2 weeks before there are significant differences in browning rate of the salad product (fig 36.22). In the case of romaine lettuce, only 1 week storage was required to reduce the quality of the salad product. The variable quality of fresh-cut products such as peppers, tomatoes, jicama, squash, and beans may be related to preprocessing storage at chilling temperatures. Much California-grown product goes to regional processors, and a better understanding of the impact of preprocessing conditions on fresh-cut product quality and shelf life is needed.

GRADES AND STANDARDS

There are no U.S. grade standards for fresh-cut fruits and vegetables separate from those applied to the original raw product. There are, however, quality and inspection guidelines to facilitate marketing (USDA 1994). In addition, industry guidelines for safe production and handling of fresh-cut products, and expectations by regulatory agencies, place them in the food processing realm (U.S. FDA 1998; Gorny 2001).

The terms *quality* and *shelf life*, as they apply to fresh-cut products, are not consistently defined or applied. Useful criteria, which encompass sensory, nutritional, and microbiological qualities, are needed for more accurate determinations of the shelf life of fresh-cut fruits and vegetables.

REFERENCES

Ahvenainen, R. 1996. New approaches in improving the shelf life of minimally processed fruit and vegetables. Trends Food Sci. Technol. 7:179–187.

Anon. 1997. Fresh-cut produce handling guidelines. 2nd ed. Alexandria, VA: International Fresh-cut Produce Assoc.; Newark, DE: Produce Marketing Assn. 31 pp.

Avena-Bustillos, R. J., J. M. Krochta, and M. E. Saltveit. 1997. Water vapor resistance of red delicious apples and celery sticks coated with edible caseinate-acetylate monoglyceride films. J. Food Sci. 62:351–354.

Barmore, C. R. 1987. Packing technology for fresh and minimally processed fruits and vegetables. J. Food Quality 10:207–217.

Barth, M. M., and H. Zhuang. 1996. Packaging design affects antioxidant vitamin retention and quality of broccoli florets during postharvest storage. Postharv. Biol. Technol. 9:141–150.

Beauchat, L. R. 1996. Pathogenic microorganisms associated with fresh produce. J. Food Protection 59:204–216.

Bennik, M. H. J., H. W. Peppelenbos, C. Nguyen-the, F. Carlin, E. J. Smid, and L. G. M. Gorris. 1996. Microbiology of minimally processed, modified-atmosphere packaged chicory endive. Postharv. Biol. Technol. 9:209–221.

Bolin, H. R., and C. C. Huxsoll. 1989. Storage stability of minimally processed fruit. J. Food Processing and Preservation 13:281–292.

Figure 36.22

Browning of cut edges of lettuce in relation to storage of the intact heads. Intact heads and the fresh-cut products were stored at 5°C (41°F).

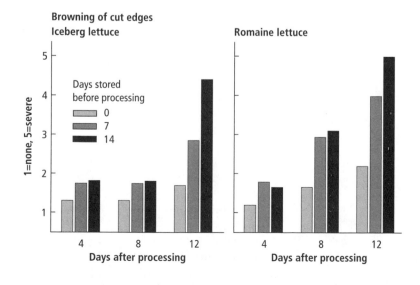

————. 1991. Effect of preparation procedures and storage parameters on quality retention of salad-cut lettuce. J. Food Sci. 56:60–62, 67.

Brackett, R. E. 1992. Shelf stability and safety of fresh produce as influenced by sanitation and disinfection. J. Food Protection 55:808–814.

Brecht, J. 1995. Physiology of lightly processed fruits and vegetables. HortScience. 30:18–22.

Cameron, A. C., P. C. Talasila, and D. W. Joles. 1995. Predicting film permeability needs for modified-atmosphere packaging of lightly processed fruits and vegetables. HortScience. 30:25–34.

Cantwell, M. 1997. Physiological responses of fresh-cut produce. In Proc. Australasian Postharvest Horticulture Conf. Hawkesbury, NSW, Australia: Univ. Western Sydney. pp. 178–191.

Cantwell, M. (compiler). 1998. Fresh-cut products. Maintaining quality and safety. Davis: Univ. Calif. Postharv. Hort. Ser. 10.

Gorny, J. R. 1997. A summary of CA and MA requirements and recommendations for fresh-cut (minimally processed) fruits and vegetables. In J. Gorny, ed., Proc. Seventh Intl. Controlled Atmosphere Research Conference. Vol 5. Davis: Univ. Calif. Postharv. Hort. Ser. 19. 30–66.

Gorny, J. R., ed. 2001. Food safety guidelines for the fresh-cut produce industry. 4th ed. Alexandria, VA: International Fresh-cut Produce Association. 220 pp.

Gorny, J. R., B. Hess-Pierce, and A. A. Kader. 1998. Effects of fruit ripeness and storage temperature on the deterioration rate of fresh-cut peach and nectarine slices. HortScience. 33:110–113.

Hägg, M. U. Häkkinen, J. Kumpulainen, E. Hurme, and R. Ahvenainen. 1994. Effects of preparation procedures and packaging on nutrient retention in different vegetables. Proc. 6th Intl. Symp. Postharvest Treatment of Fruit and Vegetables. Luxembourg: Office for Official Publications of the European Communities. 6 pp.

Hobson, G. E., and W. G. Tucker, eds. 1996. Lightly-processed horticultural products. Postharv. Biol. Technol. 9:113–245.

Hotchkiss, J. H., and M. J. Banco. 1992. Influence of new packaging technologies on the growth of microorganisms in produce. J. Food Protection 55:815–820.

Izumi, H., and A. E. Watada. 1995. Calcium treatment to maintain quality of zucchini squash slices. J. Food Sci. 60:789–793.

Kader, A. A., D. Zagory, and E. L. Kerbel. 1989. Modified atmosphere packaging of fruits and vegetables. Crit. Rev. Food Sci. Nutr. 28:1–30.

Kim, D. M., N. L. Smith, and C. Y. Lee. 1993. Quality of minimally processed apple slices from select-ed cultivars. J. Food Sci. 58:1115–1117, 1175.

King, A. D., Jr., and H. R. Bolin. 1989. Physiological and microbiological storage stability of minimally processed fruits and vegetables. Food Technol. 43:132–135, 139.

Klein, B. P. 1987. Nutritional consequences of minimal processing of fruits and vegetables. J. Food Quality 10:179–193.

Laurila, E., R. Kervinen, and R. Ahvenainen. 1998. The inhibition of enzymatic browning in minimally processed vegetables and fruits. Postharv. News and Info. 9:53N–66N.

Loaiza-Velarde, J. G., F. A. Tomás-Barderán, and M. E. Saltveit. 1997. Effect of intensity and duration of heat-shock treatments on wound-induced phenolic metabolism in iceberg lettuce. J. Amer. Soc. Hort. Sci. 122:873–877.

López-Gálvez, G., M. Saltveit, and M. Cantwell. 1996. The visual quality of minimally processed lettuces stored in air or controlled atmosphere with emphasis on romaine and iceberg types. Postharv. Biol. Technol. 8:179–190.

López-Gálvez, G., R. El-Bassuoni, X. Nie, and M. Cantwell. 1997. Quality of red and green fresh-cut peppers stored in controlled atmospheres. In J. Gorny, ed., Proc. 7th Intl. Controlled Atmosphere Research Conf. Vol 5. Davis: Univ. Calif. Postharv. Hort. Ser. 19. 152–157.

López-Gálvez, G., G. Peiser, X. Nie, and M. Cantwell. 1997. Quality changes in packaged salad products during storage. Zeitschrift Lebensmittel Untersch. Forsch. A 205:64–72.

Mattila, M., R. Ahvenainen, E. Hurme, and L. Hyvonen. 1995. Respiration rates of some minimally processed vegetables. In J. DeBaerdemaeker et al., eds., Postharvest treatment of fruit and vegetables, proc. of workshop, September 14–15, 1993, Leuven, Belgium. Luxembourg: Commission of the European Communities. 135–145.

McDonald, R. E., L. A. Risse, and C. R. Barmore. 1990. Bagging chopped lettuce in selected permeability films. HortScience. 25:671–673.

Nguyen-the, C., and F. Carlin. 1994. The microbiology of minimally processed fresh fruits and vegetables. Crit. Rev. Food Sci. and Nutr. 34:371–401.

Portela, S., X. Nie, T. Suslow, and M. Cantwell. 1997. Changes in sensory quality and fermentative volatile concentrations of minimally processed cantaloupe stored in controlled atmospheres. In Proc. 7th Intl. Controlled Atmosphere Research Conf. Vol. 5. Davis: Univ. Calif. Postharv. Hort. Ser. 19. 123–129.

Rosen, J. C., and A. A. Kader. 1989. Postharvest physiology and quality maintenance of sliced pear and strawberry fruits. J. Food Sci. 54:656–659.

Saltveit, M. E. 1997. Physical and physiological changes in minimally processed fruits and vegetables. In F. A. Tomás-Barberán and R. J. Robins, eds., Phytochemistry of fruit and vegetables. Oxford, UK: Clarendon Press. pp. 205–220.

Sapers, G. M., and R. L. Miller. 1998. Browning inhibition in fresh-cut pears. J. Food Sci. 63:342–346.

Toivonen, P. M. A. 1997. Non-ethylene, non-respiratory volatiles in harvested fruits and vegetables: their occurrence, biological activity and control. Postharv. Biol. Technol. 12:109–125.

Tomás-Barberán, F. A., J. Loaiza-Velarde, A. Bonfanti, and M. E. Saltveit. 1997. Early wound- and ethylene-induced changes in phenylpropanoid metabolism in harvested lettuce. J. Amer. Soc. Hort. Sci. 122:399–404.

U.S. Department of Agriculture (USDA). 1994. Fresh-cut produce. Shipping point and market inspection instruction. Washington, D.C.: USDA Agriculture Marketing Service, Fruit and Vegetable Division. 47 pp.

U.S. Food and Drug Administration (U.S. FDA). 1998. Guidance for Industry. Guide to minimize microbial food safety hazards for fresh fruits and vegetables. Washington, D.C.: U.S. FDA, Center for Food Safety and Applied Nutrition.43 pp.

Varoquaux, P., and R. C. Wiley. 1994. Biological and biochemical changes in minimally processed refrigerated fruits and vegetables. In R. C. Wiley, ed., Minimally processed refrigerated fruits and vegetables. New York: Chapman and Hall. 226–268.

Varoquaux, P., J. Mazollier, and G. Albagnac. 1996. The influence of raw material characteristics on the storage life of fresh-cut butterhead lettuce. Postharv. Biol. Technol. 9:127–139.

Vankerschaver, K., F. Willocx, C. Smout, M. Hendrickx, and P. Tobback. 1996. Modeling and prediction of visual shelf life of minimally processed endive. J. Food Sci. 61:1094–1098.

Watada, A. E., K. Abe, and N. Yamuchi. 1990. Physiological activities of partially processed fruits and vegetables. Food Technol. 44(5): 116, 118, 120–22.

Watada, A. E., N. P. Ko, and D. A. Minott. 1996. Factors affecting quality of fresh-cut horticultural products. Postharv. Biol. Technol. 9:115–125.

Wiley, R. C., ed. 1994. Minimally processed refrigerated fruits and vegetables New York: Chapman and Hall. 368 pp.

Wright, K. P., and A. A. Kader. 1997. Effect of controlled-atmosphere storage on the quality and carotenoid content of sliced persimmons and peaches. Postharv. Biol. Technol. 10:89–97.

———. 1997. Effect of slicing and controlled-atmosphere storage on the ascorbate content and quality of strawberries and persimmons. Postharv. Biol. Technol. 10:39–48.

Zagory, D. 1995. Principles and practice of modified atmosphere packaging of horticultural commodities. In J. M. Farber and K. L. Dodds, eds., Principles of modified-atmosphere and sous vide product packaging. Lancaster, PA: Technomic. 175–206.

37

Processing of Horticultural Crops

Diane M. Barrett

PRINCIPLES OF HORTICULTURAL CROP PRESERVATION

Food production in the United States, which includes the agricultural sector, food processing, and marketing operations and all the related support industries, generates about 20% of the gross national product and utilizes approximately 25% of the total U.S. work force. If one considers the activities of food production, manufacturing, marketing, restaurants, and institutions combined, the food industry is easily the largest industrial enterprise in the United States (Potter and Hotchkiss 1995). This chapter will highlight the principles of horticultural crop preservation, the importance of raw material quality, unit operations common to most types of processing, and specific technologies applicable to the processing of horticultural crops. Finally, the advantages and disadvantages of the various available technologies will be discussed, followed by general comments on packaging and quality control.

WHY PROCESS?

Ideally, most consumers would like to obtain and consume fruits and vegetables when they are the freshest, that is, when their color, texture, flavor and nutritional value are optimal. Unfortunately, most of us do not have the ability to maintain gardens or greenhouses that would supply us year-round with the variety of horticultural products that we desire. In addition, many fruits and vegetables are only produced in certain growing regions, for example, pineapple, papayas, blueberries, sweet corn, and artichokes. Some horticultural products are only produced during a very short window of time, and once that passes fresh products are no longer available. For these reasons and those listed below, processing and later consumption of horticultural products is often a desirable alternative to fresh consumption. Processing serves to

- extend seasonality
- allow consumption in regions distant from the growing region
- prolong shelf life and arrest deterioration during storage
- improve the nutritional quality and digestibility of some foods
- enhance the microbiological and chemical safety of food
- allow for long-term storage in case of drought or famine
- increase the variety and desirability of food (add value)
- improve quality control and provide a more consistent product
- permit extraction of key nutrients for health or medicinal uses
- enhance convenience and reduce preparation time prior to consumption
- in some cases, reduce cost of fresh horticultural product

MAJOR CAUSES OF FOOD DETERIORATION

Horticultural crops naturally mature and ripen, then begin the processes of senescence and deterioration. Over time, undesirable changes in color, texture, aroma and flavor, nutritional value, and safety may render fresh fruits and vegetables unacceptable for consumption. These changes are caused by physical, chemical, and biological factors that include (Potter and Hotchkiss 1995):

- survival and growth of bacteria, yeasts and molds
- chemical reactions, including those catalyzed by enzymes
- insect, parasite, and rodent infestation
- poor control of temperature
- loss or gain of moisture
- reactions with O_2
- reactions with light
- physical abuse or stress
- time

In most cases these factors do not act in isolation; depending on the commodity and environmental conditions, they may act in concert to increase the rate of horticultural crop deterioration. Because of the number of deteriorative factors and the diverse nature of fruit and vegetable crops, many different types of preservation technologies are utilized to prevent deterioration.

PRINCIPLES OF FOOD PRESERVATION

Food preservation may be targeted to either the short or long term. Short-term preservation is applicable to horticultural commodities that are consumed relatively soon after harvest. For them, the best preservation technique involves keeping the product alive and respiring. The fresh-cut horticultural product industry operates along this principle, and prepared salads, carrots, broccoli, cauliflower, and melon products are stored under refrigerated MA conditions that maintain metabolism. If it is not possible to maintain horticultural products in a live state, short-term preservation will be enhanced if the commodity is cleaned, covered, and cooled as quickly as possible. Short-term preservation does not involve destruction of microorganisms or enzymes; deteriorative reactions will therefore proceed, often at a faster rate due to the stresses imparted during harvesting and handling operations.

This chapter focuses primarily on long-term preservation, which normally encompasses inactivation or control of microorganisms and enzymes and reduction or elimination of chemical reactions that cause food deterioration. Microorganisms may be controlled through the use of heat, cold, dehydration, acid, sugar, salt, smoke, atmos-

pheric composition, and radiation (Potter and Hotchkiss 1995). Mild heat treatments in the range of 82° to 93°C (180° to 200°F) are commonly used to kill bacteria in low-acid foods (pH ≥4.6), but to ensure spore destruction, temperatures of 121°C (250°F) wet heat for 15 minutes or longer are required. High-acid foods (pH <4.6) require less heat, and often a treatment of 93°C (200°F) for 15 minutes will ensure commercial sterility. If the water activity (a_w), or water available for microbial growth, is 0.85 or below, no thermal process is required regardless of pH. Design of the thermal process that will ensure the safety of a specific product is best carried out by "process authorities," individuals with experience and knowledge of bacteriology and thermal process design.

Refrigeration and freezing slow microbial growth and may kill a small fraction of microorganisms present in or on a fruit or vegetable, but they do not kill all bacteria. If all of the water in a product exists in a solid state, growth of microorganisms will be prevented, but growth will resume at the same or perhaps at an even more rapid rate when thawed. Dehydration serves to remove water required for growth from microbial cells and preserve horticultural crops against microbial deterioration. In a similar fashion, sugar and salt act as preservatives because they cause osmotic dehydration of microbial cells and eventual death. Control of moisture or RH of storage environments is an important consideration in terms of desired maintenance of horticultural crop life and undesired preservation of microorganisms.

Natural fruit and vegetable acids reduce the severity of thermal treatment required because acids denature bacterial proteins and may result in cell death. Atmospheric composition is often utilized in storage and packaging of horticultural products to prevent deterioration. Oxygen is a prerequisite for not only many microbial but also chemical deterioration reactions; therefore, replacement with either nitrogen or carbon dioxide may be beneficial. Finally, microorganisms are inactivated to various degrees by X-rays, microwaves, ultraviolet light, and ionizing and other forms of radiation.

SHELF LIFE DATING

Shelf life is generally defined as the time it

takes for a product to become unacceptable for consumption or unsalable. Unacceptability may be determined by the processor, the retailer, or the consumer and may be the subject of controversy. Quantification of shelf life depends on measurement of sensory, instrumental, or microbial deterioration, and the rate is affected by the type of product, processing method, packaging, and storage conditions. Many processors use code dates that include the "pack date," "display date," "sell by date," "best used by date," or "expiration date" (Potter and Hotchkiss 1995).

RAW MATERIAL QUALITY

Earlier chapters in this book were devoted to a discussion of fruit and vegetable maturity (chapter 6) and quality (chapters 22 and 23) factors specific to crops destined for fresh consumption. Although in many cases these same factors may be required of horticultural crops destined for processing, this is not always true. Standards of raw material quality for processed products may be less stringent because the fruit or vegetable may ultimately be peeled, sliced, pureed, or even juiced. Factors important to raw material quality are discussed below.

VARIETY SELECTION

The selection of horticultural varieties for the specific purpose of processing is somewhat different from selection strictly for the fresh market. Varieties destined for processing may be exposed to mechanical harvesting, rigorous cleaning and peeling operations, size reduction operations, and high- or low-temperature preservation methods. Different varieties of the same commodity may deviate in, for example, the ease with which peels are removed, the textural integrity after thermal processing, the color stability during drying, the flavor retention during frozen storage, or the nutritional value after irradiation. Some varieties may be best targeted for one specific processed product rather than another. For example, one tomato variety may be an excellent raw material for ketchup products because of its color and flavor, yet this same variety may be a poor raw material for diced tomato products because of poor textural integrity. It is important to evaluate not only the raw product but also the intended processed

product when selecting processing varieties. Many large processors of horticultural products provide seed to growers with whom they contract.

MATURITY

Desired harvest maturity depends on a number of factors, including the variety, growing region, weather conditions, soil type, ease of harvest, hand versus mechanical harvest, and final processed product. Fruits generally increase in both sugar and acid content as they mature, but other factors such as textural integrity and appearance may decline with maturity. If the fruit is intended for juice processing, where texture and appearance are of little significance, harvesting should be carried out when the fruit is fully ripe. If textural integrity and appearance are important, for example, in the case of a "fruit half" or sliced product, the fruit might be harvested at an earlier maturity stage. Vegetables differ from fruits in that they often become tougher and more lignified as they mature, and this may adversely affect the ease of processing, flavor, and texture of the processed product. In addition, the starch content may increase in some seed and many root and tuber crops, and this may be either desirable or undesirable in the processed vegetable. Determination of optimal harvest maturity is not always straightforward, and many factors other than quality may affect this decision.

QUALITY SPECIFICATIONS

Quality standards or specifications for processed horticultural products may be set by the federal or state government, trade organizations, or by private companies. Government standards may be mandatory when they deal with the protection of consumer health and prevention of consumer deception. Other government standards such as the U.S. Grade Standards are voluntary and are aimed at helping producers, brokers, wholesalers, processors, retailers, and consumers in the marketing of processed food products.

Examples of voluntary standards are the USDA Standards for Processed Fruit and Vegetable Products, which were put in place to identify the degree of quality in a product and thereby assist in establishing usability or value. Fresh fruit, nut, and vegetable grade standards were established in 1917, and

Table 37.1. Relative importance of factors involved in USDA standards for processed fruit and vegetable products

Product	Absence of defects	Color	Flavor	Character	Consistency	Uniformity	Texture	Tenderness and maturity	Clearness of liquor
Apples	20	20	—	40	—	20 size	—	—	—
Apple butter	20	20	20	20 finish	20	—	—	—	—
Apple juice	20	20	60	—	—	—	—	—	—
Apple sauce	20	20	20	20 finish	20	—	—	—	—
Apricots	30	20	—	30	—	20 size	—	—	—
Asparagus	30	20	—	—	—	—	—	40	10
Beans, green and wax	35	15	—	—	—	—	—	40	10
Beans, dried	40	—	—	40	20	—	—	—	—
Beans, lima	25	35	—	30	—	—	—	—	10
Beets	30	25	—	—	—	15	30	—	—
Berries	30	20	—	30	—	20	—	—	—
Blueberries	40	20	—	40	—	—	—	—	—
Carrots	30	25	—	—	shape	15 size	30	—	—
Cherries, sweet	30	30	—	20	—	20 size	—	—	—
Cherries, sour	30	20	—	30	—	20 pits	—	—	—
Chili sauce	20	20	20	20	20	—	—	—	—
Corn, cream	20	10	20	—	20	—	—	30	—
Corn, whole	20	10	20	—	—	—	—	40	10
Cranberry sauce	20	20	20	—	40	—	—	—	—
Figs, kadota	30	20	—	35	—	15 size	—	—	—
Frozen apples	20	20	—	40	—	20	—	—	—
Fruit cocktail	20	20	—	20	—	20	—	—	20
Fruit jelly	—	20	40	—	40	—	—	—	—
Fruit preserv. (jam)	20	20	40	—	20	—	—	—	—
Fruit salad	30	20	—	30	—	20 size	—	—	—
Grapefruit	20	20	—	20	—	(Wholeness, 20)	(Drained wt., 20)		
Grapefruit juice	40	20	40	—	—	—	—	—	—
Grape juice	20	40	40	—	—	—	—	—	—
Lemon juice	35	35	30	—	—	—	—	—	—

standards for canned, frozen, and dried fruit and vegetables, and other related products such as preserves were created in 1928. Grading is typically based on a scoring system for four quality factors: color, size uniformity, absence of defects, and character (maturity, texture, and tenderness). The relative importance of various factors to specific processed horticultural crops was itemized by Kramer and Twigg (1970) and is illustrated in table 37.1. Definitions of each factor are included with the standard for that particular commodity.

Trade organizations are generally established by the members of a particular industry on a voluntary basis to assure at least minimum acceptable quality and to prevent a lowering of quality standards. Company standards for quality often go beyond the minimum quality standards suggested by the government.

UNIT OPERATIONS

Because the number and type of processing technologies applied to fruit and vegetable products is immense and diverse, food scientists attempt to organize them into the common "unit" operations that comprise them (see fig. 37.1). For example, heating or thermal processing is common to the manufacture of orange juice, strawberry-kiwi jam, canned peaches, french fried potatoes, cucumber pickles, and sun-dried tomatoes. Some processing systems are composed of very few unit operations, while other systems include a multitude of operations. In order to successfully preserve fruit and vegetable products it is essential to select and combine the most appropriate unit operations into more complex processing systems. A number of the unit operations most commonly used in

Table 37.1. Cont.

Product	Absence of defects	Color	Flavor	Character	Consistency	Uniformity	Texture	Tenderness and maturity	Clearness of liquor
Mushroom	30	30	—	20	—	20 size	—	—	—
Olives, green	30	30	—	20	—	20 size	—	—	—
Olives, ripe	10	15	30	25	—	20	—	—	—
Orange juice	20	40	40	—	—	—	—	—	—
Orange juice concen.	20	40	40	—	—	—	—	—	—
Orange marmalade	20	20	40	—	20	—	—	—	—
Okra	20	15	15	—	—	10	—	35	5
Peaches	30	20	—	30	—	20 size	—	—	—
Peanut butter	30	20	30	—	20	—	—	—	—
Pears	30	20	—	30	—	20 size	—	—	—
Peas	30	10	—	—	—	—	—	50	10
Peas, field	40	20	—	40	—	—	—	—	—
Cucumber, pickled	30	20	—	—	—	20	30	—	—
Pimientos	40	30	—	10	—	20 size	—	—	—
Pineapples	30	20	—	30	—	20	—	—	—
Pineapple juice	40	20	40	—	—	—	—	—	—
Plums	30	20	—	30	—	20 size	—	—	—
Potatoes, peeled	40	20	—	—	—	20	20	—	—
Prunes, dried	30	20	—	35	—	15	—	—	—
Pumpkins and squash	30	20	—	20 finish	30	—	—	—	—
Raspberries	20	25	—	35	—	20 size	—	—	—
Sauerkraut	10	15	45	15 crisp	—	—	—	—	15
Sauerkraut, bulk	10	15	45	15 crisp	—	—	—	—	15
Spinach	40	30	—	30	—	—	—	—	—
Sweet potatoes	40	20	—	20	—	20 size	—	—	—
Tomatoes	30	30	—	—	—	(Wholeness, 20)	(Drained wt., 20)	—	—
Tomato juice	15	30	40	—	15	—	—	—	—
Tomato paste	40	60	—	—	—	—	—	—	—
Tomato pulp, puree	50	50	—	—	—	—	—	—	—
Tomato sauce/catsup	25	25	25	—	25	—	—	—	—

Source: Kramer and Twigg 1970.

horticultural product processing are described below.

MATERIALS HANDLING

Materials handling operations encompass those that move horticultural crops from the farm to the packinghouse or processing plant and subsequently through the various unit operations in the plant. Included are hand and mechanical harvesting, refrigerated trucking, container transportation, and conveying of fruits and vegetables through the processing system. Throughout all materials handling operations it is essential to maintain sanitary conditions, minimize product loss (quantitative and qualitative), maintain raw material quality, minimize microbial contamination and growth, and schedule transfers and deliveries so that process delays are minimized.

Figure 37.1

Unit operations in horticultural crop processing.

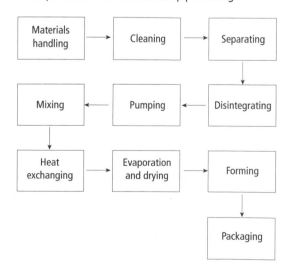

CLEANING AND SANITIZING

Cleaning and sanitizing of food materials and processing equipment surfaces is important to ensure that safe products are produced. However, fruits and vegetables that are peeled are seldom washed before peeling. Cleaning may range from simple removal of soil and other foreign matter on pome fruits by dumping in a tank, to vigorous scrubbing of pineapples with high-pressure water jets. Cleaning may be accomplished with brushes, high-velocity air, steam, water, vacuum, magnetic attraction of metal contaminants, mechanical separation, etc. (Potter and Hotchkiss 1995). The use of clean water is also essential; municipal or local water may require additional treatments such as controlled chemical flocculation of suspended matter, sand filtration, carbon purification, microfiltration, and de-aeration to ensure its cleanliness. Processing equipment and the floors and walls of processing facilities may be cleaned with acid or alkaline detergents, or a combination of both. The use of a sanitizer such as chlorine is recommended after cleaning of food or processing equipment to ensure safety.

SEPARATING

Examples of separating operations include physical sorting or removal of defective, immature, or broken products, shells from nuts, or peels from vegetables; filtration of a liquid and solid material; or pressing of juice from fruit pulp. Recently, new methods of sorting such as magnetic resonance imaging and near-infrared radiation have been developed for horticultural products.

PEELING

Peeling may be accomplished by using steam, water, or lye (sodium hydroxide) solutions, or by using mechanical means such as abrasive peelers. With some horticultural commodities it may be advantageous to use a combination of a hot water scald and steam under pressure. Many processors claim that the quality of fruits peeled with lye is preferable because of the relatively small amount of peel removed and the high flesh retention in the final product. However, the disposal of lye is cause for environmental concern, and a number of companies are moving away from this technique. Mechanical peeling is appropriate for relatively hard commodities, such as apples, pears, and potatoes.

DISINTEGRATING

Disintegrating operations subdivide large sections of food materials into smaller ones. These include chopping, grinding, pulping, homogenizing, dicing, and other size-reduction operations. In some cases the disintegration process may involve heat generation, so it may be desirable to cool the food material and the equipment used.

PUMPING

Pumping of liquids and solids from one location in a processing plant to another is essential, and many different types of pumps are available for this use. For solid materials, the size and fragility of the pieces to be pumped dictate the choice of pump. For example, very gentle pumps have been designed for aseptic particulate processing systems in which particulate integrity is critical to quality. Some of the pump types available include the cam and piston pump, external or internal gear pumps, two- to four-lobe pumps, single- to triple-screw pumps, swinging or sliding vane pumps, sine pumps, and shuttle block pumps (Potter and Hotchkiss 1995).

MIXING

Mixing operations may involve integration of various solids, liquids, liquids with solids, gases with solids, gases with liquids, and so forth. As with pumps, many different types of mixing devices are available, and the choice depends on the physical materials to be mixed.

HEAT EXCHANGING

Heat exchanging includes heating and cooling products, two of the most common unit operations in the processing of horticultural products. In many fruit and vegetable products, thermal treatment is required for destruction of microorganisms and their spores, inactivation of enzymes, flavor or taste enhancement, texture or viscosity adjustments, and numerous other reasons. However, most horticultural commodities are temperature-sensitive, and extended periods at high temperatures may cause undesirable changes in texture, color, flavor, and nutritional quality. To optimize quality, it is vital to heat and cool fruits and vegetables as quickly as possible. Many different types of heating systems exist, and selection should be based on the specific commodity, end product, and the capital and operating costs required. Products may be heated or cooled using direct or indirect heat. Direct steam injection, frying, dehydrating, and direct flame contact are used with certain fruit and vegetables products. Common indirect heating devices include plate heat exchangers, scraped surface heat exchangers, tubular heat exchangers, steam jacketed kettles, and retorts.

Cooling may be accomplished with some of the same equipment used in heating (for example, most of the heat exchangers), but refrigerant or cold water is the indirect contact medium rather than water or steam. Many fruits and vegetables are refrigerated or frozen as a form of preservation. Freezing may be carried out with air blast freezers, contact freezers, or cryogenic (liquid nitrogen or solid carbon dioxide) freezers.

EVAPORATION AND DRYING
Although evaporation and drying involve the removal of water, evaporation typically involves a 2- to 3-fold concentration whereas drying often takes food to 2 to 3% total moisture. During evaporation, flavor volatiles may also be recovered or removed, according to their desirability. In many cases evaporation is carried out under vacuum so that the water boils off at a lower temperature and thermal damage is avoided. Dehydration is discussed in more detail below.

FORMING
Forming involves pressure extrusion through dies of various shapes, or using molds into which hot products are placed and subsequently cool and harden into the mold pattern. For example, tablet-forming machines are used to produce fruit juice tablets from dehydrated fruit juice crystals.

PACKAGING
Food is packaged in order to protect it from microbial contamination, dirt, insect attack, light, moisture or flavor adsorption or loss, and physical abuse (Potter and Hotchkiss 1995). A wide variety of packaging materials exists, including tin cans, glass and plastic bottles, paper, plastic films, metallic films, and laminates comprised of layers of different materials. Special machines designed to fill and close packages are used to ensure adequate seals.

FERMENTATION

Fermentation is one of the oldest forms of food preservation, and in many less developed countries it still serves as a major preservation method. Fermentation results when microorganisms grow in an anaerobic environment, consuming organic materials for their own metabolism. Although microbial growth is usually undesirable, in this case the select microorganisms that are encouraged to grow produce acids, alcohols, and other compounds that have a pleasing flavor and extend keeping quality. In order to ensure the desired product characteristics, it is common to control the following in a food fermentation: level of acid, level of alcohol, use of starters, temperature, level of O_2, and amount of salt. Some examples of fermented fruit and vegetable products include (Potter and Hotchkiss 1995): dill and sour pickles (cucumbers); green olives; sauerkraut (cabbage); kimchi (mixed vegetables and Chinese cabbage); poi (taro); tempeh (soybeans); soy sauce (soybeans); and wine (grapes and other fruit).

Many types of fermentation, or pickling, involve the use of a salt brine. This is done to create an environment in which the desired type of microorganism will grow. For example, in the manufacture of cucumber pickles, a 5 to 8% salt brine is used to encourage the growth of salt-tolerant lactic acid bacteria and to prevent growth of proteolytic and spoilage microorganisms. Following fermentation, the salt is washed out and vinegar (a preservative that replaces lactic acid) and calcium chloride (a firming agent) are added; the pickles are then typically pasteurized to increase shelf life. Fresh-packed pickles may be produced by adding vinegar directly to the cucumber and then pasteurizing.

Fermentation can actually improve the nutritional quality of foods in three ways: microorganisms synthesize complex vitamins (riboflavin, vitamin B_{12}, and vitamin C) and other growth factors; microorganisms may allow for better digestion of food materials by enzymatically breaking them down into simpler compounds; and microorganisms may physically (e.g., penetration of mold mycelia) or enzymatically liberate nutrients that are sequestered in indigestible plant parts or cells (Potter and Hotchkiss 1995).

JUICE PROCESSING

Although juices may be prepared from either fruits or vegetables, the leading juice commodity by far is oranges. Many of the unit operations involved are common to all juice processing, and these may include extraction, clarification, de-aeration, liquefaction, pasteurization, concentration (optional), essence add-back, canning or bottling, and possibly freezing. Apple juice production is illustrated in figure 37.2. Juice extraction typically involves grinding the product, but in the case of oranges and other commodities with bitter peels, the extraction process may require special machinery to extract juice without inclusion of these flavor components. After extraction, the pulp is often clarified using enzymes followed by filters, presses, or high-speed centrifuges. It may be necessary to treat the pulp with digestive enzymes in order to hydrolyze pectic materials and increase the efficiency of clarification

(McLellan 1996). In the production of cloudy beverages it is desirable to allow a certain amount of suspended pulp to remain in the juice; many consumers believe that this imparts a more natural quality to the juice.

Removal of entrapped air, or de-aeration, will prevent undesirable oxidation reactions and improve the quality of the juice produced. Special equipment has been designed to efficiently remove air from liquid systems. Pasteurization of juices is often advisable in order to decrease microbial population and inactivate endogenous enzymes. Fruits and vegetables contain the enzyme pectinesterase, which catalyzes pectin demethylation, eventual gelling or settling out of the insoluble pulp. In orange juice production, a temperature of 90°C (194°F) is required for inactivation of this enzyme, while destruction of normal microbial populations requires a process of only 74°C (165°F) (Desrosier and Desrosier, 1977).

Production of high-quality low-acid vegetable juices is a challenge because much of the desirable flavor of the fresh vegetable resides in the pulp, and extraction of these components is difficult. Under normal juice processing conditions, low-acid juices must

Figure 37.2

Apple juice production.

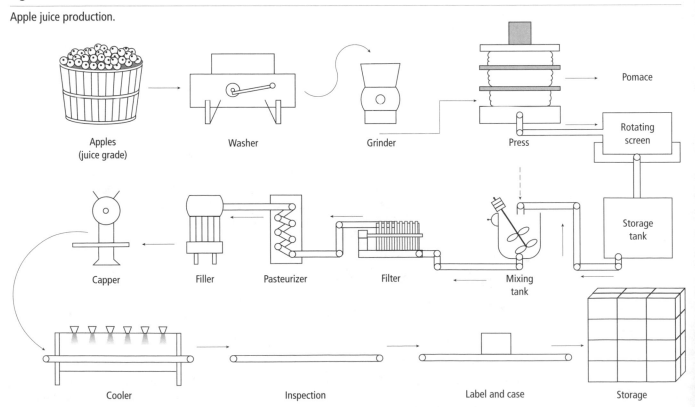

Apples (juice grade) → Washer → Grinder → Press → Pomace / Rotating screen → Storage tank

Capper ← Filler ← Pasteurizer ← Filter ← Mixing tank

Cooler → Inspection → Label and case → Storage

be sterilized at 121°C (250°F) to inactivate enzymes and kill microorganisms and their spores. This process also results in a loss of most of the flavor components typical of the vegetable commodity. It is possible to recover fruit and vegetable essences that are volatilized during pasteurization or concentration and add them to enhance flavor. Concentration is typically carried out in a vacuum evaporator to allow for water removal at lower temperatures and prevent thermal damage to the juice. Membrane filtration methods such as reverse osmosis and ultrafiltration have replaced vacuum evaporators in many juice concentration operations because of the improved quality that these technologies produce. However, membrane filtration systems may be significantly more expensive than vacuum evaporation.

Fruit nectars are another type of beverage that are usually produced from fruit that is low in flavor as a single-strength clarified juice or strongly flavored (McLellan 1996). Fruit nectars consist of a blend of sugar syrup, fruit acid, and high-solids (concentrated) juice or puree.

The nutritional composition of fruit juices and juice blends may be very high, in particular with regard to vitamin C. Apple juice is typically low in vitamin C and is therefore fortified (Potter and Hotchkiss 1995). Consumption of high-quality juices as nutritional beverages is on the rise, and technologies such as aseptic processing that preserve vitamin content are increasingly favored.

DEHYDRATION AND CONCENTRATION

Dehydration refers to the nearly complete removal of water from crops under controlled conditions, causing minimum (or ideally no) changes in food properties (Potter and Hotchkiss 1995). Dehydration of horticultural crops has been practiced for centuries; in environments where drying is slow, microbial growth may be avoided by using either smoke and salt as further preservatives. Although dehydration serves as a form of preservation, it is also useful in reducing weight and bulk, which may lower shipping and container costs. Dried fruit and vegetables products generally contain 1 to 5% moisture, and their shelf life at room temperature may be 1 year or longer. Horticultural commodities may be dried whole (e.g., grape, berries, apricot, plum), in sliced form (e.g., banana, mango, mushroom, papaya, peppers, kiwifruit, tomato), as leather, or as a powder (Ratti and Mujumdar 1996).

The principle of dehydration technology is to get heat into the product and get moisture out. Factors that affect the rate of dehydration include product surface area, temperature, air velocity and humidity, atmospheric pressure and vacuum, evaporation rate and temperature, and overall drying time and temperature. Moisture loss does not occur at a constant rate; the largest percentage of the total moisture in a product is removed in a relatively short time, but it may take as long or longer to remove the remaining moisture. The laws of heat and mass transfer dictate these rates of drying. Figure 37.3 shows a typical drying curve for carrots, illustrating 90% of the moisture is lost in 4 hours, but another 4 hours is required to remove the remaining 10%. In fact, under practical conditions, moisture contents never go to zero. Instead, final moisture levels in most products are below 5%.

Many different methods are available for horticultural crop dehydration, but the most important are sun drying, tunnel or cabinet drying, drum drying, spray drying, and

Figure 37.3

Drying curve for diced carrots.

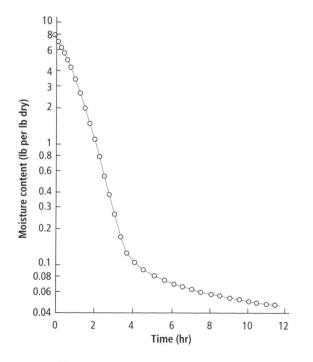

Figure 37.4

Counterflow tunnel drier construction.

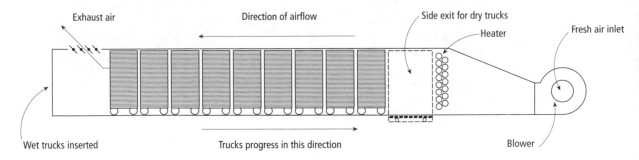

Table 37.2. Common drier types used for liquid and solid foods

Drier	Usual food type
Air convection driers	
Kiln	Pieces
Cabinet, tray, or pan	Pieces, purees, liquids
Tunnel	Pieces
Continuous conveyor belt	Purees, liquids
Belt trough	Pieces
Air lift	Small pieces, granules
Fluidized bed	Small pieces, granules
Spray	Liquids, purees
Drum or roller driers	
Atmospheric	Purees, liquids
Vacuum	Purees, liquids
Vacuum driers	
Vacuum shelf	Pieces, purees, liquids
Vacuum belt	Purees, liquids
Freeze-driers	Pieces, liquids

Source: Potter and Hotchkiss 1995.

freeze drying. Figure 37.4 depicts a counterflow tunnel dryer, which utilizes convection to convey the hottest and driest air to the nearly dry product (Potter and Hotchkiss 1995). Menon and Mujumdar (1987) classified the over 200 types of commercial dryers available based on

- mode of processing (batch, continuous, semibatch)

- operating pressure (vacuum, atmospheric, high pressure)

- mode of heat transfer (conduction, convection, radiation, dielectric, or a combination)

- adiabatic or nonadiabatic (with or without heat exchangers immersed within the dryer)

- physical form of product (granular, whole, continuous liquid etc.)

- physical state of product (stationary, moving, conveyed, vibrated, fluidized, dispersed, atomized, stirred, etc.)

Potter and Hotchkiss (1995) summarized commonly used drier types and the type of food usually dried with each (table 37.2).

Concentration also involves moisture removal, but not to the same extent as used in dehydration. Examples of concentrated fruit and vegetable products include juice and nectar concentrates, jams and jellies, tomato paste, and fruit purees. Concentration is typically not considered a form of preservation, unless the concentrate is high in sugar ($\geq$70% sucrose) or salt (18 to 25% solution) (Potter and Hotchkiss 1995). These concentrated sugar or salt solutions exert high osmotic pressure, which either draws water from microbial cells or prevents normal diffusion of water into these cells, therefore imparting a preservative effect. Concentration may be carried out with a variety of methods, including solar concentration, open kettles, flash evaporators, thin-film evaporators, vacuum evaporators, freeze concentration, or ultrafiltration and reverse osmosis.

Jellies, jams, preserves, and marmalades are all sugar-acid-pectin gels or low-methoxyl-pectin-calcium gels, but the fruit component is incorporated differently in each (Baker et al. 1996). Only strained fruit is used in jellies, while jams are made with either crushed or ground fruit pieces. Preserves consist of whole fruit or large pieces of fruit; marmalades are clear jellies in which fruit slices or shreds are suspended. Processing of these types of fruit concentrates involves blending together the four components (fruit, sugar, acid, and pectin); evaporating to a desired solids or sugar level to reduce moisture available for microbial growth; and pasteurization of the product to destroy microorganisms.

REFRIGERATION AND FREEZING

Refrigeration and freezing technologies are used to slow the growth of microorganisms and to reduce the rate of metabolism and enzyme-catalyzed deteriorative reactions in horticultural crops. Refrigeration typically refers to the use of storage temperatures above freezing, for example, from 16°C (60°F) to –2°C (28°F) (Potter and Hotchkiss 1995). Commercial and household refrigerators generally operate at 4.5° to 7°C (40° to 44.5°F). Although pure water freezes at 0°C (32°F), most foods do not freeze until about –2°C (28°F), because they contain solids that depress the freezing point. Temperatures of –18°C (0°F) or lower are recommended for maintenance of quality during frozen storage.

Ideally, the refrigeration of fruits and vegetable starts at the time of harvest and continues through each point of the postharvest chain up to the point of consumption. Refrigeration is a relatively gentle form of preservation, and although it maintains horticultural crops in the most "freshlike" state, it cannot compete with more intense preservation methods such as thermal processing or dehydration in prevention of deterioration. Rapid cooling of fruits and vegetables following harvest and throughout storage significantly decreases respiration rates and maintains flavor, color, texture, and nutritional value. For further information on cooling technologies and refrigerated transport and storage, see chapters 11, 12, and 20.

Freezing serves as a method of preservation because water activity can be lowered to levels that prevent microbial activity and reduce the rates of chemical reactions. There are three steps in the freezing process:

- cooling from initial temperature to freezing point (specific heat removal)
- phase change from liquid to solid state (latent heat of fusion)
- lowering temperature from freezing point to desired storage temperature (specific heat removal)

By undergoing a change in state from liquid to solid, water is essentially unavailable to microorganisms for growth or for chemical reactions. As water is solidified during freezing, the concentration of solutes in the remaining liquid becomes greater, and adverse quality reactions may take place unless the freezing rate is fairly rapid. In addition, fast freezing favors the formation of small ice crystals, whereas slow freezing allows large ice crystals to form within and between plant cells. These large crystals may cause physical rupture and separation of cells, which results in undesirable textural changes perceivable on thawing of the product.

There are three basic freezing methods used commercially: freezing in air, freezing by indirect contact with the refrigerant, and freezing by direct immersion in a refrigerating medium. Potter and Hotchkiss (1995) subdivided these three classifications further into specific methods (table 37.3). Still-air "sharp" freezing simply involves placing food in an insulated cold room at –23° to –30°C (–9° to –22°F). Air blast freezers operate at –30° to –45°C (–22° to –49°F), with forced-air velocities of 10 to 15 m/sec (2,000 to 3,000 ft/min) (Potter and Hotchkiss 1995). A special class of air blast freezer called fluidized bed utilizes air velocities that just exceed the free-fall velocity of the food particles. When this high-velocity air is blown up through a wire mesh belt, the particles have a dancing-boiling motion and increased contact with the cold air.

Indirect contact freezers consist of a number of metal plates through which refrigerant is circulated. Flat packages of food are placed between these cold plates, and the plates are pressed together to increase contact. Immersion freezing consists of dipping product into a refrigerant or spraying the refrigerant directly on the product. Two types of refrigerants are available: low freezing point liquids (sugars, sodium chloride and glycerol), which are chilled by indirect contact with another refrigerant; and cryogenic liquids

Table 37.3. Commercial freezing methods

Air freezing	Indirect contact freezing	Immersion freezing
Still-air "sharp" freezer	Single plate	Heat exchange fluid
Blast freezer	Double plate	Compressed gas
Fluidized-bed freezer	Pressure plate	Refrigerant spray
	Slush freezer	

Source: Potter and Hotchkiss 1995.

(compressed nitrogen or carbon dioxide), which cool by evaporating (Potter and Hotchkiss 1995).

THERMAL PROCESSING

Thermal processing is one of the most common forms of food preservation because it efficiently reduces microbial population, destroys natural enzymes, and renders horticultural products more palatable. Varying degrees of heat may be used in fruit and vegetable processing and preservation, and it is important to define thermal processing terms. *Sterilization* refers to the complete destruction of microorganisms and their spores; conditions typically cited are 121°C (250°F) wet heat for 15 minutes or its equivalent. Sterilization of all parts of a container of food requires knowledge of heat penetration rates, whether heating is conductive or convection, and the cold point in the food mass.

In order to maintain the best quality of thermally processed horticultural crops, processors seek to apply the mildest heat treatment that guarantees freedom from pathogens and toxins, and also prevent spoilage. In order to do this, it is vital to know the time-temperature combination required to inactivate the most heat-resistant pathogens and spoilage organisms in a specific food. The most heat-resistant pathogen in foods, especially those processed and stored under low-acid anaerobic conditions, is *Clostridium botulinum*. Determining the process time and lethality to be used requires experience and basic training in the principles of process engineering; particularly in the case of low-acid or acidified foods, process authorities must be consulted to assist with system design.

It is fortunate that many foods do not need to be completely sterile to be safe and have good shelf life. In practice, it is common to produce a food that is *commercially sterile*, meaning that all of the pathogenic and toxin-forming organisms have been destroyed, as well as other organisms that would grow and produce spoilage under normal conditions of handling and storage. Most canned and bottled fruits and vegetables are produced under conditions of commercial sterility and have a shelf life of 2 years or longer (Potter and Hotchkiss 1995). Deterioration that occurs in these products after long-term storage is more commonly caused by texture, flavor, or color deterioration rather than microbial growth.

Pasteurization has been discussed earlier in this chapter in conjunction with the preservation of juices. This is a low-temperature (typically below boiling point of water) process that aims to destroy pathogenic organisms associated with the specific product and to extend shelf life by reducing total microbial population and enzyme activity. Pasteurized horticultural products still contain living microorganisms capable of growth, but this technology is often combined with refrigeration, which assists in extending shelf life. Blanching is a specific type of pasteurization that is often used prior to freezing; its primary purpose is inactivation of natural deteriorative enzymes. Blanching treatments typically result in destruction of some microorganisms, but this should not be the target in setting blanch conditions.

Thermal processing essentially involves either heating foods in their final containers or heating foods prior to packaging and then packaging under sterile conditions. Heating foods in their final containers is common practice and requires little technical sophistication, but product quality may not be as good as the second method due to the slowness of heat penetration. The conventional still retort is in reality very similar to a pressure cooker (fig. 37.5) in which crates containing cans or bottles are placed and exposed to steam under pressure. Equipment available for heating foods in their containers includes still retorts, agitating retorts, hydrostatic cookers, direct flame sterilizers, and in-package pasteurizers. Heating foods prior to packaging may be accomplished by batch pasteurization, high-temperature short-time (HTST) pasteurization, aseptic processing, hot-fill processing, or microwave heating (Potter and Hotchkiss 1995).

IRRADIATION, MICROWAVE, AND OHMIC PROCESSING

Irradiation and microwave processing utilize radiant energy, which changes foods as it is absorbed, while ohmic processing raises the temperature of foods by passing an electrical current through them. These three processes are less common than the ones discussed above, but they have unique applications.

Figure 37.5

Conventional still retort.

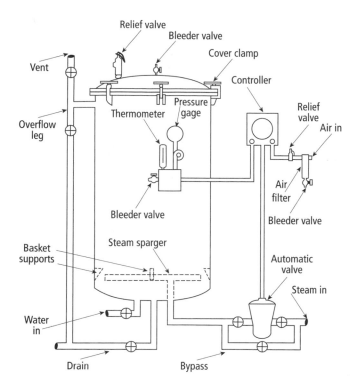

Irradiation has been under intense investigation since 1945, and in 1983 it was approved by the U.S. Food and Drug Administration (FDA) as a means for controlling microorganisms on spices. In 1986 the FDA widened the allowed uses to include irradiation of fruits and vegetables at doses up to 1 kGy for the following purposes: as an alternative to chemical fumigation to control insects; to inhibit sprouting in crops such as potatoes; and to delay fruit ripening by a mechanism not well understood (Potter and Hotchkiss 1995; Willemot et al. 1996).

Radiant energy, or ionizing radiation, is radiation of high enough energy to expulse electrons from atoms and to ionize molecules. Ionizing radiation has penetrating power but does not produce radioactivity in treated products, nor does it produce significant heat. Irradiation causes direct effects due to the interaction of high-energy rays and microbial cells or food molecules, and it also causes indirect effects due to the production of free radicals, which also interact with microorganisms and food components (Potter and Hotchkiss 1995). The literature reports varying success in terms of radiation treatments on fruits, and successful application depends on a number of factors, including commodity and cultivar, radiation dose, degree of maturity, physiological status of fruit, temperature and atmosphere during and after treatment, pre- and postharvest practices, cost, and susceptibility of microorganisms to be controlled (Willemot et al. 1996). The food industry is concerned that consumers may believe irradiated products are unsafe. However, studies by numerous federal agencies have unanimously concluded that irradiation does not result in an unsafe product, particularly at the low doses used for pasteurization, sprout inhibition, and insect control.

Microwave energy occurs as alternating electric current at frequencies of either 915 or 2,450 megahertz, which means the electric field reverses 915 or 2,450 million times per second. Water and other molecules in food are dipolar, that is, they have distinct positive and negative ends that oscillate to align themselves with alternating microwave current. These high-speed oscillations cause friction, which heats the food. Microwave heating differs from conventional thermal processing because microwaves are able to penetrate uniformly several centimeters into the food. Heat is generated quickly and evenly throughout the mass, and the steam generated heats adjacent areas by conduction. Microwave heating applications to fruit and vegetable products include concentrating heat-sensitive solutions at low temperatures; cooking large pieces without high temperature gradients; uniform dehydration; rapid enzyme inactivation; combination with freeze drying to accelerate final moisture removal; heating temperature-sensitive products; and controlled thawing of frozen products (Potter and Hotchkiss 1995).

Ohmic processing involves the continuous processing of particulates in a conducting solution by passing them through a series of low-frequency alternating electric currents of 50 to 60 hertz. One major advantage of this process is that there is no significant gradient in the temperature of particulates from inside to outside. Both the particulate and the carrier liquid heat quickly and are then cooled using technology similar to the heating by alternating electrical currents and are packaged aseptically.

ADVANTAGES AND DISADVANTAGES OF VARIOUS PRESERVATION METHODS

Growers of horticultural crops who are considering "adding value" by processing may view the selection of a preservation method as an overwhelming task with no clear choice. However, there are advantages and disadvantages to all of the preservation methods discussed above; summarizing these (table 37.4) should provide some assistance.

PACKAGING

Processed products must be protected from physical damage, chemical attack, and contamination from biological agents such as microorganisms, insects, and rodents. The goal of the packaging industry is to create containers which form a *hermetic seal*, a complete seal against the ingress of gases, vapors, and microorganisms. Jelen (1985) summarized the functional requirements of packaging materials (table 37.5). There are a limited number of packaging materials used for processed fruit and vegetable products, including metal (steel and aluminum), glass,

Table 37.4. Advantages and disadvantages to selected preservation methods

Preservation method	Advantages	Disadvantages
Fermentation	Relatively inexpensive Simple technology Shelf-stable long time May improve nutrition May improve aroma & flavor of product Acidity prevents microbial growth	Strong flavor and acidity may limit consumption May have light-catalyzed oxidation if in glass Acid may affect thickener selection
Juice processing	Relatively simple technology May be used with poorer-quality fruits and vegetables Improved palatability and digestability of product Canned or bottled products may be quite stable	May require refrigeration May require relatively high capital investment for equipment Settling of pectic materials over time may be unsightly
Dehydration	Relatively low-cost, simple technology Long shelf life if stored properly Reduction in product weight and volume	Potential for nutrient and quality losses May incur insect and microbial damage (esp. sun drying) Economic value of product may be low
Concentration	Very shelf-stable product Reduction in product weight and volume May be used with poorer-quality fruit and vegetables	May require relatively high capital investment for equipment May result in loss of flavor essences unless these are added back
Refrigeration	Relatively simple technology Maintains most "freshlike" appearance Film wraps may prevent dehydration	Equipment required may be relatively costly Short shelf life Merely slows down biological processes Horticultural crops have different optimums
Freezing	Typically carried out within hours of harvest Quality typically very good Nutritional content typically very high	May require relatively high capital investment for equipment Storage life limited Quality degradation may be ongoing unless water is in solid state
Thermal processing	Relatively simple technology Efficient destruction of microorganism and enzymes Long-term storage May increase palatability and digestibility	May require relatively high capital investment for equipment May adversely affect quality and nutritional content
Irradiation	Efficient destruction of microorganisms and enzymes At low doses is preferable to other technologies for certain applications	May require relatively high capital investment for equipment Relatively new and unfamiliar to consumers
Microwave processing	Rapid heating process Quality typically very good Nutritional content may be preferable	May require relatively high capital investment for equipment May be difficult to make continuous process Short shelf life
Ohmic processing	Rapid and uniform heating of both particulates and carrier liquids	May require relatively high capital investment for equipment

paper, paperboard, fiberboard, wood, cloth, plastics, laminates, and edible films. The choice of packaging material depends on many things, including the type of food, storage length and conditions, abuse the product may experience, functional properties of the package, and package cost and attractiveness. Thermally processed fruits and vegetables are typically packed in steel cans or glass, but aseptically processed products are often packed in laminates of metal sheet and foil, plastic, and paper. Dehydrated horticultural products must be packaged in containers that prevent exposure to oxygen and light, both of which accelerate quality degradation. Laminates of foil, plastic, and paperboard are often used for dried fruits and vegetables. Frozen products are commonly packaged in plastic pouches, laminated cartons, or composite cans.

QUALITY ASSURANCE IN FOOD PROCESSING

Quality assurance is an organized system of establishing, monitoring, and evaluating safety and quality-related activities associated with processing of a specific product. The quality assurance department of a processing facility may be involved in inspection; laboratory analysis for physical, chemical, and organoleptic quality; sanitation and microbiological testing: and research and development. Quality "assurance" and quality "control" are often used interchangeably, but quality control relates more specifically to the assessment of raw material and finished product quality and safety using standard procedures. In order to determine whether observed variations in product specifications are in an acceptable range, quality control charts are established that document the mean, range, and normal frequency of observed measurements. Statistical quality control charts may be used to monitor specific measurements of size, color, texture, flavor, ingredient composition, nutrients, microbial counts, process temperatures, can seams, and many other variables. Quality assurance is the key to any processing operation, as it allows for control of product safety and quality.

REFERENCES

Baker, R. A., N. Berry, and Y. H. Hui. 1996. Fruit preserves and jams. In L. P. Somogyi, H. S., Ramaswamny, and Y. H. Hui, eds., Processing fruits: Science and technology. Vol. 1: Biology, principles and applications. Lancaster, PA: Technomic. 117–133.

Desrosier, N. W., and J. N. Desrosier. 1977. Technology of food preservation. 4th ed. Westport, CT: AVI.

Jelen, P. 1985. Food packaging technology. In P. Jelen, Introduction to food processing. Reston, VA: Reston Publishing. 249–266.

Kramer, A., and B. A. Twigg. 1970. Quality control for the food industry. 3rd ed. Vol. 1. Westport, CT: AVI.

McLellan, M. R. 1996. Juice processing. In: L. P. Somogyi, H. S. Ramaswamny, and Y. H. Hui, eds., Processing fruits: Science and technology. Vol. 1. Biology, principles and applications. Lancaster, PA: Technomic. 67–94.

Menon, A. S., and A. S. Mujumdar. 1987. Drying of solids: Principles, classification, and selection of dryers. In: A. S. Mujumdar, ed., Handbook of industrial drying. New York: Marcel Dekker. 3–46.

Potter, N. N., and J. H. Hotchkiss. 1995. Food science. 5th ed. New York: Chapman and Hall.

Ratti, C., and A. S. Mujumdar. 1996. Drying of fruits. In: L. P. Somogyi, H. S. Ramaswamny, and Y. H. Hui, eds., Processing fruits: Science and technology. Vol. 1. Biology, principles and applications. Lancaster, PA: Technomic. 185–220.

Ronsivalli, L. J., and E. R. Vieira. 1992. Elementary food science. 3rd ed. New York: Van Nostrand Reinhold.

Willemot, C., M. Marcotte., and L. Deschenes. 1996. Ionizing radiation for preservation of fruits. In: L. P. Somogyi, H. S. Ramaswamny, and Y. H. Hui, eds., Processing fruits: Science and technology. Vol. 1. Biology, principles and applications. Lancaster, PA: Technomic. 221–260.

Table 37.5. Functional requirements of packaging materials

Functional property	Specific factors
Gas permeability	O_2, CO_2, N_2, H_2O vapor
Protection against environmental factors	Light, odor, microorganisms, moisture
Mechanical properties	Weight, elasticity, heat-sealability, mechanical sealability, strength (tensile, tear, impact, bursting)
Reactivity with food	Grease, acid, water, color
Marketing-related properties	Attractiveness, printability, cost
Convenience	Disposability, repeated use, resealability, secondary use

Source: Jelen 1985.

Making the Link: Extension of Postharvest Technology

38

I. Identifying Scale-Appropriate Postharvest Technology

Lisa Kitinoja

An enormous amount of technically useful postharvest technology is described in this text and others, and new or improved technologies for handling and marketing horticultural crops are regularly reported on in technical and trade journals. Throughout the United States and abroad, postharvest extension specialists at universities and experiment stations recommend potentially useful postharvest handling practices in newsletters, extension publications, videos, and via the Internet. Extension agents, farm advisors, postharvest consultants, and development personnel must be able to identify technologies that are cost-effective, feasible, and appropriate for their clients, as well as acceptable to consumers. With all the technologies from which to choose, how can extension professionals help their clientele determine whether any given postharvest technology will solve the problem at hand?

It is all too common to hear of technical fixes that caused more problems than they solved. During the 1970s and 1980s, postharvest losses were often attributed to a lack of storage facilities on-farm and in wholesale markets. Yet when large-scale commercial storage and marketing facilities were constructed they sat unused in many parts of the world. A recent example comes from Senegal, where a new wholesale market built opposite the existing market sits empty, due mostly to vendors' unwillingness to pay higher fees for their booth spaces (fig. 38.1). Various studies have determined the lack of use of new postharvest structures in Africa, India, and Latin America to be due to high fees, inconvenient locations, management problems, or perceived security issues. In the United States, consumers are often willing to pay more for produce that has been handled in what is considered a safe manner, perhaps by avoiding the use of pesticides or by participating in a food safety program at the farm and packinghouse level. By becoming aware of a host of possible factors that can affect whether a postharvest technology is cost-effective, feasible, and/or appropriate for the client, as well as culturally acceptable to consumers, extension professionals can assist in the transfer of information that is truly useful to their clientele.

FACTORS AFFECTING THE ADOPTION OF IMPROVED POSTHARVEST TECHNOLOGY

Casual observation or conversation with postharvest handlers and marketers of horticultural crops provides a host of examples of typical practices and conditions that contribute to postharvest losses and quality problems in the United States and in other countries. The technical literature on postharvest handling of horticultural commodities documents the improper use of various practices and the use of traditional practices that appear to be counterproductive. Physical postharvest losses of fruits, vegetables, and cut flowers can reach 20 to 30% in the United States and often are as high as 50% in developing countries. In addition, the loss of product value as quality declines during postharvest handling, storage, and distribution is an important source of economic losses.

Figure 38.1

Large modern wholesale market in Senegal sitting empty due to high cost of space rentals.

COMMON PRACTICES AND CONDITIONS AFFECTING POSTHARVEST LOSSES, QUALITY, AND FOOD SAFETY

Many handlers unknowingly contribute to postharvest losses by using common practices or by not using certain practices known to reduce losses and help maintain produce quality and safety (table 38.1). The practices and conditions listed in table 38.1 have been observed within many horticultural production and fresh marketing systems in both the United States and abroad. Each example is considered an improper practice since it has definite negative effects on fresh produce, leading either to increased waste and losses, quicker quality deterioration, or food safety problems.

Most of these improper practices and conditions cannot be labeled "technical problems," and they cannot be solved by initiating new research projects or simply by extending

Table 38.1. Common practices and conditions affecting postharvest losses, quality, and food safety

Preharvest	Inadequate planning regarding planting and harvesting dates, or growing cultivars that mature when market prices are lowest.
	Production of cultivars with high yields but short postharvest life or susceptibility to postharvest pests and diseases.
	Use of poor-quality planting materials.
	Overfertilization of vegetables with nitrogen.
	Lack of use of calcium sprays for pome fruits.
	Inappropriate irrigation practices.
	Poor orchard and field sanitation, leading to latent infections and insect damage.
	Lack of pruning, propping limbs, and/or thinning fruits, leading to small fruits with nonuniform maturation.
	Lack of pest management (spraying for insect or fungal control, bagging produce susceptible to insect or bird damage).
Harvest	Harvesting at improper maturity, leading to increased severity of storage disorders, lower eating quality (poor flavor and/or texture), failure to ripen, or excessive softening.
	Use of rough and/or unsanitary field containers.
	Harvesting during the hot hours of the day.
	Rough handling, dropping or throwing produce, fingernail punctures.
	Leaving long or sharp stems on harvested produce.
	Long exposure to direct sun after harvesting.
	Overpacking of field containers.
Curing	Lack of curing or improper curing of root and tuber crops before postharvest handling, packing, and storage.
	Improper drying of bulb crops.
Packinghouse operations	Lack of proper sorting.
	Lack of cleaning, washing, or sanitation.
	Rough handling.
	Improper trimming.
	Misuse of postharvest treatments (overwaxing, inadequate chlorine in wash water, misuse of hot water dips for pest management).
	Use of inappropriate chemicals or misuse of registered compounds.
	Long delays without cooling.
	Lack of accepted and/or implemented quality grades or standards for commodities.
	Lack of quality inspection.

existing well-proven technical information. Often, postharvest losses take time to develop, and the specific cause of quality problems may not be fully understood by produce handlers along the chain. Other times, the handler may deliberately choose not to use a practice known to protect produce because of its cost or because consumers perceive the practice as undesirable. On occasion, a lack of reliable supplies, market information, or other infrastructural problems may make changes in handling impractical. Postharvest losses and changes in quality affect both the volume and perceived value of produce as it moves from the field to its final destination market, and any changes in practices will also have an affect. Part of any potential technical solution, therefore, is a consideration of the socioeconomic, cultural, and institutional constraints facing growers, handlers, and marketers when they attempt to make changes in the way they handle and market horticultural crops.

INFORMATION GAPS

Continuing use of some of these common improper practices can be traced to information gaps. Unless careful records have been kept of the handling practices undergone by a commodity between harvest and the final destination market, it can be difficult to determine the source of losses and quality problems. Even when moving through the simple direct marketing channel from grower or shipper to retailer, many people are involved in the postharvest handling, transport, and marketing of fresh produce. The effects of poor postharvest handling practices are cumulative, and losses can be caused by one or more events. However, identifying appropriate technical solutions can be impossible if the final handler is just guessing

Table 38.1. Cont.

Packing and packaging materials	Use of flimsy or rough packing containers. Lack of liners in rough baskets or wooden crates. Overuse of packing materials intended to cushion produce, causing interference with ventilation. Containers designed without adequate ventilation. Overloading containers. Use of containers that are too large to provide adequate product protection. Misuse of films for MA packaging, overreliance on MAP versus appropriate temperature management.
Cooling	In developing countries, general lack of the use of any methods of cooling during packing, transport, storage, or marketing of fruits or vegetables. In developed countries, use of inappropriate cooling methods, misuse of cooling methods, overcooling (chilling injury, freezing). Inadequate monitoring of temperature and chlorine levels in hydrocooler water.
Storage	In developing countries, general lack of storage facilities on the farm or at wholesale or retail markets, lack of ventilation and cooling in existing onfarm facilities. Poor sanitation and inadequate management of temperature and RH in larger-scale storages. Overloading of cold stores. Stacking produce too high for container strength. Mixing lots of produce with different temperature or RH requirements. Lack of regular inspections for pest problems and temperature or RH management.
Transportation	Overloading vehicles. Use of bulk transport or poor quality packages, leading to compression damage. Lack of adequate ventilation during transport. Lack of air suspension on transport vehicles. Rough handling during loading. Lack of cooling during delays. Ethylene damage or chilling injury resulting from transporting mixed loads.
Destination handling	Rough handling during unloading. Lack of sorting, poor sanitation, improper disposal of culls. Improper de-greening of citrus crops and misuse of ripening practices. Lack of protection from direct sun during direct marketing. In developing countries and farmers' markets, open horticultural markets exposed to sun, wind, dust, and rain. In developed countries, overcooling in supermarket displays of chilling-injury-susceptible produce.

Table 38.2. Factors affecting implementation of changes in postharvest technology

Identifying sources of losses	Difficulties in determining amount and causes of losses.
	Lack of knowledge on where and when losses occur and who is responsible.
	Availability of training in loss assessment methods.
	Availability of resources (money, time, equipment) for loss assessment.
Identifying possible solutions (technology)	Availability of scale-appropriate applied research results.
	Profitability of postharvest technology (costs versus benefits).
	Governmental regulations, price supports or controls, incentives or disincentives.
	Existing fee structures for common packages, transport, or storage services.
	Technology fit with existing commodity system (handling methods, distribution system).
	Technology fit with local skills of handlers.
Social or cultural norms	Cultural barriers (gender, religion, local customs, traditions, beliefs).
	Competing responsibilities (i.e., political or community service, social obligations, water bearing, child care, home-based tasks).
	Consumer demands and preferences (definitions of quality vary by culture).
	Requirement of produce "losses" for other purposes (i.e., animal feed, poverty relief programs).
	Environmental concerns (a new technology may create pollution).
	Effects on jobs (may displace workers).
	Existence of grades and standards for produce.
Postharvest equipment, supplies, fuel	Local availability of supplies, spare parts, tools, and equipment.
	Costs and availability of materials, fuel, maintenance.
	Reliability of supply of raw materials (for packing or processing).
	Reliability of supply of electricity.
	Environmental concerns (Is the technology reusable/recyclable?).
Credit and loans	Availability of credit.
	Collateral requirements for loans.
	Prevailing market interest rates.
Postharvest labor/who will implement the change?	Seasonal changes in labor supply, labor shortages.
	Availability of trained personnel for harvesting, packing, and skilled jobs.
	Motivation and loyalty of the workforce.
Training, education, and instruction	Availability of resources for postharvest educational and training programs.
	Availability of effective extension services (public or private).
	Differences in access to existing services (women vs. men, small- vs. large-scale enterprises, farmers' marketing cash crops vs. food crops).
	Extension workers' level of knowledge and skills in postharvest handling and/or food processing.
Supporting infrastructure	Existence and condition of roads.
	Availability of cooling, storage, transportation, and marketing facilities.
	Location and size of existing facilities.
	Access to existing facilities.
	Institutional linkages (i.e., communication between research and extension).
	Government investment priorities in facilitating services.
	Budget for maintenance of existing infrastructure.
	Availability of staff with management skills for running facilities.
	Condition of communications systems (telephone, mail, FAX, e-mail services).
Marketing system	Availability of marketing options and alternative outlets.
	Direct sales vs. consignment (Who is responsible for losses? Who suffers financially?).
	Degree of cooperation and coordination among buyers and/or sellers.
	Establishment and viability of marketing cooperatives.
	Reliability of recordkeeping to help trace problems back to sources.
Market information	Availability of resources for collecting and analyzing data.
	Reliability of information.
	Timeliness of information.

about when and where the cause of losses occurred and who may be responsible. This task becomes even more difficult if the extension agent is unfamiliar with the location, crop, or cultural habits of the community.

Once the extension professional helps clients close these information gaps, it is often found that a simple change can solve a major problem. For example, quality problems at final destination could be traced to a packinghouse cooling system that is being overloaded, causing internal temperatures of produce to be higher than desired when produce is loaded for transport.

A variety of methods of postharvest loss assessment can be used to pinpoint the problem's source and to identify potential constraints to changing handling practices. Most involve direct observation of handling practices and the interviewing of key individuals regarding their standard postharvest practices. The United Nations Food and Agriculture Organization (FAO) has published loss assessment manuals for various commodities that focus on measuring physical losses (changes in weight or quantity of produce) and losses in value (quality changes or decrease in market price per unit). Any method used for loss assessment must attempt to understand postharvest losses within the context of the whole system of production, handling, and marketing of the commodity in question, since what are considered losses will vary by culture and economic situation.

Postharvest Systems Research (PSR) is a loss assessment method that follows a given lot of produce from harvest through packing, storage, and transport to the processing facility or distribution center, all the while measuring changes in quantity and quality attributes (Shewfelt and Prussia 1993). Implementation of PSR requires the long-term use of a vehicle equipped with scientific instruments for quality assessment as well as sleeping accommodations for the researcher(s), since the many steps of postharvest handling may take days or weeks. To date, this method has been used primarily for research studies in horticulture at the University of Georgia. Another systems approach, developed by LaGra (1990), is a practical team-based method used in fieldwork worldwide known as Commodity Systems Assessment Methodology (CSAM). Loss assessment using CSAM is within the capabilities of most extension services, with its rela-

tively low cost and focus on face-to-face information gathering. The framework and practice of CSAM is introduced and discussed in the section of this chapter on extension methods.

Postharvest loss assessment studies have provided information that suggests there are a variety of incentives and barriers to the adoption of new postharvest practices. Table 38.2 lists many factors that may affect the implementation of postharvest technology, some of which are discussed in detail in the following sections of this chapter.

SOCIOECONOMIC FACTORS
The potential for changing most improper handling practices will be affected by a wide range of socioeconomic factors. In general, socioeconomic factors have to do with costs and benefits (actual, perceived, short-term, and long-term). Some of the many examples found in the literature are listed in the first column of table 38.3.

In many countries, although curing is known to extend the postharvest life of root and tuber crops, curing can be viewed as too costly and unnecessary by growers intending to market their produce immediately after harvest. Yet if marketing is delayed for any reason, or uncured sweet potatoes are placed into storage by the buyer, postharvest losses due to water loss and decay can be severe.

Growers of horticultural crops in many countries are tempted to harvest earlier than their neighbors (before adequate maturity) in order to take advantage of high market prices that can be obtained early in the growing season. While profits may be higher for the grower, the postharvest life and quality of the produce is greatly reduced, leading to higher postharvest losses and making handling and marketing more difficult.

Sometimes existing circumstances contribute to the continuing use of a detrimental postharvest practice. In the Caribbean and in East Africa, freight charges for produce are set by the number of containers, rather than by the container's weight or size. This policy provides an incentive for handlers to use the largest possible container in order to reduce their shipping costs, regardless of the lack of protection offered by such packages to the produce within. A recent example comes from Tanzania, where cabbages are loaded into jute sacks that have been expanded beyond their intended size so traders can avoid paying for

Table 38.3. Factors affecting implementation of improved postharvest handling and food processing practices, leading to increased losses and quality deterioration

Socioeconomic factors	Cultural factors	Institutional factors
High cost of plastic sheeting forces handlers to sun-dry their vegetables directly on the soil (dust, insects, molds). (Morocco; Kitinoja 1996)	Onions are harvested very early, leading to high losses— but responding to consumer demand for the flavor of early onions. (Chad; Kitinoja 1992)	Government-established fees set for potato storage facilities was so low that storages sat empty (operators would have lost money by running the facility). (UP, India; Kitinoja 1995a)
Very early harvest (before maturity) to take advantage of higher market prices at beginning of season. (India; Reid et al. 1997) (Morocco; Kitinoja 1996)	Health and environmental concerns (irradiation, waxes, pesticides, non-recyclable packaging materials). (USA; Krimsky and Plough 1988)	Women cannot own property and therefore often have no collateral to offer in return for loans. (Africa; Madeley 1987)
Inter-island freight rates charged by unit rather than by weight or volume of produce. (Caribbean; Schurr 1988)	Washing produce is associated with the use of toxic chemicals, specifically pesticides (lowers perceived value). (Morocco; Kitinoja 1996)	Poor roads, lack of loading docks, cooling and storage facilities, telephone services, inadequate power supply. (Costa Rica; Breslin 1996)
Overloading of transport vehicles, packing containers because of limited availability. (LDCs; UN FAO 1989)	Canned goods are actively disliked by consumers (fear of contamination, preference for fresh, in-season foods). (India; Reid et al. 1997)	Government facilities for forced-air cooling and refrigerated transport remain unused for years due to bureaucratic delays. (Punjab, India; Reid et al. 1997)
Using less or lower quality packaging materials to cut costs. (Worldwide; LaGra 1990)	Late morning harvests due to womens' responsibilities in child care and water bearing. (Senegal; Kitinoja 1995a)	Market information is collected but not disseminated to potential users in a timely manner. (LDCs; UN FAO 1989b)
Very high interest rates for loans. (LDCs; UN FAO 1989b)	Women don't usually operate machines, so any change in practice may take women out of the postharvest system. (Africa; Chinsman and Fiagan 1987)	Lack of standardized grades and inspection services. (LDCs; LaGra 1990)
Lack of financial opportunity to make investments in onion storage facilities. (Chad; Yarnell 1990)	Male extension agents have difficulty working with women farmers and marketers in Islamic communities. (Senegal; Kitinoja 1995a) (Africa; IFPRI 1995)	Government horticultural research and extension services focus solely on production aspects. (India; Kitinoja 1995a)
Sprinkling leafy greens to reduce wilting is considered "cheating" since produce is sold by weight. (India; Kitinoja 1995a)	Farmers prefer to store their crops inside their own homes due to security concerns (leaving large-scale storages unused). (Peru; Rhoades 1984)	New technologies offered with little or no consultation with locals. (Worldwide; Wiggins 1994) (Ghana; Compton 1997)

freight on additional packages (fig. 38.2).

Even when the overloading of containers and transport vehicles is a well-known cause of losses due to compression damage and reduced ventilation, overloading is common in an effort to reduce immediate handling and transportation costs. While handlers may explain that they use these practices because the number of vehicles is limited or because the handling of a few very large packages is quicker than hand-stacking many small packages, these practices always compromise quality later on down the postharvest chain.

Large-scale growers or shippers may be using a particular practice because it is cost-effective, and losses of a certain magnitude may be deemed acceptable. Redesigning and producing a new package for specialty fruits, for example, might cost more than the new package saves in reduced losses. Studies have shown that profit-motivated decisions, such as choosing to use less packaging materials in order to maximize individual return on investment, can increase postharvest losses later in the commodity system. Yet any suggestion that packages be modified to better suit the commodity's needs will run into serious opposition, since the added cost may be viewed as excessive, especially if the produce is sold to an intermediary who then benefits from the investment.

Sometimes the reasons for the choice of a given postharvest practice are complex.

Figure 38.2

Packing cabbage in Tanzania for transport.

While field-packing is known to reduce losses by decreasing the number of times produce is handled, choosing to field-pack vegetables or to use a packinghouse will depend on many factors. The practical decision may be based on factors such as the amount of new investment required, what equipment and facilities are already owned or available, and the expertise of managers and packers. In the United States, another typical dilemma occurs when a choice must be made between performing a postharvest handling practice such as grading by using manual labor or with machinery designed to enhance efficiency. Which is a better choice for a given operation and its community depends on the philosophy of the owners, the cost of equipment, the expected returns, the availability of trained labor, and the level of local unemployment.

Extension educators must always consider many socioeconomic factors whenever assessing postharvest technology for its suitability for their clientele. Small-scale handlers may not be able to afford to implement well-known technical practices such as sorting, cooling, or improved packaging on their own without reducing profits. If the technology requires the use of credit, prevailing interest rates or access to collateral can affect the decision of whether or not to make a change in handling practices. The cost-effectiveness of postharvest technology will always be a primary factor in whether it will be considered appropriate.

CULTURAL FACTORS

In general, cultural factors affecting the implementation of changes in postharvest technology have to do with beliefs, preferences, traditions, or cultural norms. The second column of table 38.3 lists some examples of cultural factors that may interfere with making changes in postharvest technology.

Recommended postharvest technologies must fit into the existing cultural environment. For example, technical factors may suggest packaging a product in a certain way to reduce losses and protect quality, while consumers prefer to make their own selection from a bulk display. Some consumer groups may prefer processed or fresh-cut produce and be willing to pay more for the convenience offered by these products, while other consumers will avoid them due to perceived nutritional or health concerns.

Beliefs regarding the environment may influence food choices and can affect how buyers perceive produce wrapped in plastic or packaged in Styrofoam containers. Handlers may find that using recycled materials in packaging may be profitable from a monetary as well as a goodwill point of view. Even though concerns over the use of pesticides, food additives such as waxes or colorings, or postharvest treatments such as gamma irradiation are considered by scientists and the health care profession to be largely irrational, their use may make produce less valuable to specific groups of consumers. Market research studies undertaken by extension professionals can help identify key consumer preferences and the level of demand for fresh produce.

Sometimes local customs or consumer preferences lead directly to quality problems. For example, it is often recommended that harvests be performed early in the morning in order to reduce the heat load on produce and make precooling faster and less expensive. Vegetables in West Africa, however, are often harvested at midmorning and endure the heat of the day while awaiting transport from the field. In this case, the women who harvest the vegetables cannot come earlier to the fields, since family responsibilities such as water bearing and cooking must be given first priority. Another example of the effect of a cultural factor on postharvest handling occurs in New Delhi, India, where fresh produce is sold by weight. Leaving a few protective outer leaves on cole crops or sprinkling leafy greens to prevent wilting is viewed as "cheating" since the leaves or water adds weight to the produce.

Other cultural factors that can affect whether handlers will adopt changes in postharvest technology include religious traditions, gender barriers, the local definition of losses, and traditional secondary uses for low-quality produce (e.g., animal feed, food banks). If agroprocessing is to play a role in reducing food losses, recommended technologies must result in high-quality, healthful foods that consumers find to be good-tasting and easy to prepare. Extension professionals must be aware of these many cultural factors in order to best identify scale-appropriate postharvest technology and before attempting to develop educational programs targeting specific clientele.

INSTITUTIONAL FACTORS

The adoption of postharvest technology can be affected by existing laws or regulations providing incentives or disincentives, as well as by facilitating services provided (or not provided) by governments, universities, and the private sector. The third column of table 38.3 provides some examples of various institutional factors that may affect whether people make changes in postharvest technology.

Institutional factors include public sector policies such as commodity price supports or controls, land tenure and property rights, regulations regarding the use of pesticides and acceptable levels of residues, and types or sizes of packages. Secondary institutions can become involved when, for example, fees are charged as part of mandatory marketing orders. The status of these policies and regulations within a state or country may affect whether someone is willing or able to make individual changes in their postharvest practices. For example, the decision to provide price supports may increase production, which may lead to the requirement for more supplies for postharvest treatments, packaging, and storage facilities. If any of these are inadequate or lacking, postharvest losses may increase.

Also of concern are the availability and condition of facilitating services such as the physical infrastructure (e.g., roads, storage facilities), power supply, quality control and inspection services, availability of credit or loans, extension systems, and communica-

tions and market information systems. When services related to postharvest handling are nonexistent, skewed to serve certain groups over others, or poorly managed, adoption of postharvest technology can be negatively affected. For example, in many countries, market information is collected but not analyzed and hence goes unused. The FAO suggests limiting the scope of a market information system to a few commodities in major markets, then aiming to collect, analyze, and disseminate information (on prices, supply, and movements of produce) to users immediately (on the same day or at latest the next day). And while many of the people involved in horticultural production, handling, and marketing are women, most extension agents are men. A 1995 International Food Policy Research Institute (IFPRI) report notes that "if more extension agents and agricultural research scientists were women, extension services and agricultural technologies could be made more appropriate to female farmers. The representation of women in these fields is currently 'miniscule'" (IFPRI 1995). Several trainers have suggested that home economics agents (also known as family resource or home advisors) may be better suited to working directly with women than are most agricultural agents, since their contacts with rural women have already been established.

EXTENSION OF POSTHARVEST INFORMATION

Whenever extension efforts are undertaken in postharvest technology, it is important that the appropriate audience be targeted. For example, harvesters need to learn about maturity indices, while transport operators need information on temperature management during loading and shipping. California, with its excellent array of facilitating services, including the University of California Cooperative Extension system, serves as a model for successful cooperation between government and university research and extension and the horticultural industry. In many regions of the world, however, poorly developed infrastructure hinders the movement of produce and market information. The lack of linkages between agencies or a lack of resources for extension work and agricultural education leaves handlers without basic knowledge regarding causes of losses and with a shortage of skills related to improved postharvest handling practices. Postharvest educational needs assessments undertaken in many countries point to a wide range of training requirements before local handlers are ready to make improvements in postharvest technology within their operations.

Extension workers worldwide have an opportunity to extend appropriate postharvest technology to a variety of clientele who can then use recommended practices to reduce losses, maintain produce quality, and increase profits. The final decision of whether to adopt any given improved postharvest technology will be made on an individual basis, and it depends on many factors that affect its ultimate usefulness. The more information extension workers have on the postharvest problem, the client's operation, and the local situation, the better they will be able to identify potential solutions. Involvement of clientele in postharvest loss assessment can assist extension workers in pinpointing the sources and causes of losses and in identifying potential constraints to changing practices required to solve the problem. Commodity assessments can help identify who (i.e., men, women, growers, traders, retailers) needs what kind of postharvest information to solve the problem. Market research can help identify consumer preferences and demand characteristics that will affect the feasibility of certain solutions. Only then can effective postharvest extension programs be developed to meet the needs of local clientele. Specific postharvest handling recommendations, therefore, are best made by local development agents or extension workers who have had a chance to review the entire commodity system in question, gather information from those involved with postharvest handling, and become familiar with the local cultural, institutional, and socioeconomic environment.

REFERENCES

Abbott, J. C. 1986. Marketing improvement in the developing world: what happens and what we have learned. FAO Economic and Social Development Series no. 37. Rome: United Nations Food and Agriculture Organization (FAO). 237 pp.

Adams, M. E. 1982. Agricultural extension in developing countries. Intermediate Tropical Agricultural Series. London: Longman. 108 pp.

ASEAN-PHTRC. 1984. Village level handling of fruits and vegetables: traditional practices and technological innovations. Los Banos, Philippines: Postharv. Hort. Training and Res. Ctr. Ext. Bull. No. 1. 23 pp.

Baritelle, J. L., and P. D. Gardner. 1984. Economic losses in the food and fiber system: From the perspective of an economist. In H. E. Moline, ed., Postharvest pathology of fruits and vegetables: Postharvest losses in perishable crops. Oakland: Univ. Calif. Div. Ag. and Nat. Res. Bulletin 1914. 4–10.

Berardi, G. M., and C. C. Geisler, eds. 1984. The social consequences and challenges of new agricultural technologies. Boulder, CO: Westview Press. 376 pp.

Breslin, P. 1996. Costa Rican farmers find their mini-niche. Grassroots Dev. 20(2): 27–33.

Cernea, M. M. 1991. Putting people first: Sociological variables in rural development. 2nd ed. New York: Oxford University Press. 575 pp.

Chambers, R. 1991. Shortcut and participatory methods for gaining social information for projects. In M. M. Cernea, ed., Putting people first: Sociological variables in rural development. 2nd ed. New York: Oxford University Press. 515–537.

Chambers, R., A. Pacey., and L. A. Thrupp. 1989. Farmer first: Farmer innovation and agricultural research. London: Intermediate Technology Publications. 218 pp.

Chinsman, B., and Y. S. Fiagan. 1987. Postharvest technologies of root and tuber crops in Africa. In E. E. Terry et al., eds., Tropical root crops: Root crops and the African food crisis. Proc. 3rd Triennial Symp. of the International Society for Tropical Root Crops, Africa Branch. Owerri, Nigeria: ISTRC.

Compton, J. 1997. Problems, technologies and participation. Int'l. Ag. Dev. 17(4): 21.

Coursey, D. G. 1983. Postharvest losses in perishable foods of the developing world. In M. Lieberman, ed., Postharvest physiology and crop preservation. New York: Plenum. 485–514.

Eklund, P. 1983. Technology development and adoption rates: Systems for agricultural research and extension. Food Policy 8:141–153.

International Food Policy Research Institute (IFPRI). 1995. Report highlights barriers to women. Inter. Ag. Dev. 15(5): 4.

International Institute of Tropical Agriculture (IITA). 1994. Annual report. Ibadan, Nigeria: IITA. 54 pp.

Johnson, S. H., III, and E. D. Kellogg. 1984. Extension's role in adapting and evaluating new technology for farmers. In B. E. Swanson, ed., Agricultural extension: A reference manual. 2nd ed. Rome: FAO. 40–55.

Kitinoja, L. 1991. Research and extension programs for improved postharvest handling practices in West Africa. In: Proc. Symposium on Sustainable Agriculture for Africa, Center for African Studies, Ohio State University. Columbus: Ohio State University. 218–235.

———. 1992. U.S. AID consultancy report on food processing in the Ouaddhai, Chad, Central Africa. Washington, D.C.: AFRICARE. 76 pp.

———. 1995a. Improving postharvest technology for horticultural crops in Uttar Pradesh, India. UP-DASP Postharvest Consultancy Working Paper. Washington, D.C.: World Bank. 137 pp.

———. 1995b. Postharvest handling, storage and processing of vegetable crops in Senegal. Farmer-to-Farmer Program Consultancy Report. Dakar: Winrock International/OFPEP. 27 pp.

———. 1996. Small-scale postharvest handling and processing practices for horticultural crops in Morocco—Short course. Davis: USAID/ORM-VAT/Chemonics International/Univ. Calif. International Training and Education Ctr. 88 pp.

Krimsky, S., and A. Plough. 1988. Environmental hazards: Communicating risks as a social process. Dover, MA: Auburn House. 333 pp.

LaGra J. 1990. A commodity system assessment methodology for problem and project identification. Moscow, ID: Postharvest Institute for Perishables. 98 pp.

Lemoine, R. 1995. Faut-il avoir peur des aliments ionises (Should one be afraid of irradiated foods?) Revue Laitière Française No. 547:17–18.

Madeley, J. 1987. The African farmer...and her husband. Int'l. Agric. Dev. 7(2): 1.

Narayanan, A. 1991. Enhancing farmers' income through extension services for agricultural marketing. In W. M. Rivera and D. J. Gustafson, ed., Agricultural extension: Worldwide institutional evolution and forces for change. Amsterdam: Elsevier. 151–162.

National Academy of Sciences. 1978. Postharvest food losses in developing countries. Washington, D.C.: NAS. 206 pp.

Oakley, P. 1988. Extension and technology transfer: The need for an alternative. HortSci. 23(3): 482–485.

Oakley, P., and C. Garforth. 1985. Guide to extension training. Rome: FAO. 110 pp.

Odell, M. J. 1986. People, power and a new role for agricultural extension: Issues and options involving local participation and groups. In G. E. Jones, ed., Investing in rural extension: Strategies and goals. London: Elsevier. 169–178.

Quisumbing, A. R. 1995. Women: The key to food security. IFPRI Report. Washington, D.C.: International Food Policy Research Institute. 22 pp.

Reid, M., J. Gorny, L. Kitinoja, and D. Picha. 1997. Postharvest handling of horticultural produce training program. ACE/USAID/Punjab Agricultural University/UC Davis. 24 pp.

Rhoades, R. E. 1984. Breaking new ground: Agricultural anthropology. Lima, Peru: International Potato Center. 71 pp.

Rhoades, R. E., and R. H. Booth. 1982. Farmer-back-to-farmer: A model for generating acceptable agricultural technology. Ag. Admin. 11:127–137.

Schurr, C. C. M. 1988. Packaging for fruits, vegetables and root crops. Filed Document. Bridgetown, Barbados: FAO. 14 pp.

Sharma, R. K., S. K. Sharma, and R. K. Thakur. 1995. Marketing vegetables in Himachal Pradesh. Indian J. Ag. Mktng. 9(1): 44–50.

Shewfelt, R. L. 1987. Quality of minimally processed fruits and vegetables. J. of Food Quality 10 (1987): 143–156.

Shewfelt, R. L., and S. E. Prussia, eds. 1993. Postharvest handling: A systems approach. San Diego: Academic Press. 356 pp.

Sparks, A. L. 1995. Market values for the major characteristics of fresh pears. J. Food Prod. Mktg. 2(4): 45–55.

Torres, F., and G. Gallopin. 1994. Systems research methods and approaches at CIAT—Current and planned involvement. In P. Goldsworthy and F. P. de Vries, eds., Opportunities, use and transfer of systems research methods in agriculture to developing countries. London: Kluwer. 271–280.

United Nations Food and Agriculture Organization (FAO). 1985. Prevention of post-harvest food losses: A training manual. Rome: FAO Training Series 10. 120 pp

———. 1989a. Horticultural marketing: A resource and training manual for extension officers. Rome: FAO Agricultural Services Bulletin 76. 118 pp.

———. 1989b. Prevention of food losses: Fruit, vegetable and root crops. A training manual. Rome: FAO. 157 pp.

United Nations Food and Agriculture Organization/ United Nations Environment Program (FAO/UNEP). 1981. Food loss prevention in perishable crops. Rome: FAO Agricultural Services Bulletin 43. 72 pp.

Wiggins, S. 1994. When agricultural projects go badly wrong. In N. Maddock and F. A. Wilson, eds., Project design for agricultural development. Brookfield: Avebury. 5–25.

World Bank. 1985. Agricultural research and extension. Washington, D.C.: International Board on Research and Development (IBRD)/World Bank. 110 pp.

Yarnell, J. 1990. Ouaddhai onion storage and marketing study. Washington, D.C.: AFRICARE. 37 pp.

II. Extension Methods for Transferring Postharvest Technology

Lisa Kitinoja and Robert F. Kasmire

There is a great need for effective extension programming in postharvest technology to help solve the industry's problems and to disseminate research results and other information useful to fresh-market horticultural crops growers and those involved in transportation and marketing. This section of this chapter provides detailed information on a variety of methods of extension education. The objective of extension is to create a two-way link between information providers and information seekers. Extension professionals must ensure that developers of postharvest technology are made aware of priority problems of large-, medium-, and small-scale clientele, and that potential users of postharvest information find the technology accessible, easy to understand, and suited to their needs and constraints. To fulfill this objective, extension personnel must be highly motivated, well trained, and equipped with the latest resource materials, reliable transportation, and communications technology. They must be located where they can readily and effectively interact with researchers and those involved in the preparation, shipping, distribution, and marketing of horticultural crops.

THE ROLE OF EXTENSION

The objectives of an extension postharvest program are to improve the quality and value of horticultural crops available to consumers, reduce marketing losses of horticultural crops, and improve marketing efficiency. All of these relate to the profit motive. Most growers and handlers do not make changes in their postharvest operations just because they want to reduce losses—the changes they make must also lead to improved profits. Specific extension objectives will focus on solving a particular problem affecting one commodity or a group of related commodities in a specific location or commodity system. These could be focused on any aspect

Table 38.4. Examples of extension postharvest program objectives, extension messages, and related research topics for postharvest extension programs

Postharvest activity	Extension objective	Extension message	Related applied research topics
Crop selection	Ensure that farmers grow crops appropriate to market conditions and demand.	Introduction of "new" crops with market potential.	Prices, quality, yields, returns, and marketing strategies.
Harvesting	Ensure that produce is harvested at the time or quality required by the market.	Correct time and quality level to harvest crops for marketing.	Adaptation of technology to meet time or market requirements.
Preparation for market	Maximize value to growers.	On-farm cleaning, trimming, and selection of produce for market.	Adaptation of technology to meet time or market requirements.
Grading	Allow pooling of output and collective marketing. Improve output quality.	Grading methods for horticultural products, advantages of selling graded produce.	Formulate and monitor standards for marketable produce.
Packaging	Maximize returns to producers and handlers by minimizing damage.	Demonstrate packaging products for transport, storage, and sales.	Adapt traditional packaging materials to crop and market requirements.
Transport	Ensure that produce reaches buyers without delay or loss.	Encourage improved handling and packaging procedures in order to reduce losses during transport.	Cost-effective means of transport.
Storage	Reduce growers' dependence on intermediaries.	How to determine when, where, what, and how to store.	Adaptation of storage procedures to local conditions and crops.
Sales/marketing	Improve growers' bargaining position, reduce dependence.	Encourage collective marketing.	Market surveys, analyses of prices, and demand patterns and characteristics.

Source: Adapted from Narayanan 1991.

of postharvest extension, from ensuring quality and food safety at harvest to helping clients meet consumer preferences and the level of demand in new markets (table 38.4).

The clientele of the typical postharvest extension specialist consists of local extension workers (farm advisors, county agents, agricultural technicians, or village-level workers) and industry personnel involved in preparation, shipping, and distribution of fresh-market horticultural products. It may also include university or agricultural college students, via teaching, guidance, and counseling responsibilities. The local extension worker's clientele consists primarily of horticultural industry personnel within a specific county or region.

To plan any postharvest extension program, one must first identify a specific problem and then identify and understand the needs of the clientele being addressed. Sometimes the audience's composition, background, and distribution are major constraints. The fresh produce shipping, marketing, and distribution industry is very heterogeneous and widespread, often including handlers at distant points in the same country or in more than one country. Farm workers or market intermediaries may be less well educated than average, or extension programs might target clientele who speak a language different from the mainstream. Initially, clientele may know little, if anything, about extension's role and objectives and may be suspicious of any attempts to change their handling practices, techniques, or facilities. Outside of the United States, the reputation of extension has been tarnished by poorly designed extension systems that failed to serve their clientele. Indeed, extension work is now commonly termed "outreach" to avoid the negative connotations associated with the word "extension." To gain the attention and confidence of a reluctant clientele, a program must first demonstrate that it can benefit them, most often economically.

SIX STEPS TO DEVELOPING HIGH-QUALITY POSTHARVEST EXTENSION PROGRAMS

After identifying and learning about the intended audience for the postharvest extension program, the following steps can be helpful.

Step 1. Identify the postharvest problems to be targeted and work with stakeholders (planners, funding agencies, cooperators, clientele) to determine their priorities. A traditional educational needs assessment or commodity systems assessment, undertaken with the participation of representatives of the intended audience, can help identify key postharvest and marketing problems. It is not possible to develop extension educational programs on all problems at one time. Determine which are the most important for a given audience and which can be realistically resolved by providing educational information on postharvest principles and/or practices.

Step 2. Develop the long- and short-range (1-year) objectives of the extension program and the program's theory of action. The typical chain of program events is described by a model developed by Bennett (1979): INPUTS and resources must be utilized to get the program started; ACTIVITIES are then implemented to INVOLVE people in programs; participants REACT to what they experience; which leads to CHANGES in KNOWLEDGE, SKILLS, ATTITUDES, or ASPIRATIONS; PRACTICE CHANGES; and END RESULTS or overall impacts.

For example, a long-range objective might be to improve the nutritional quality of fresh fruits and vegetables sold to consumers; a short-range objective might be to improve cooling practices and facilities used before loading.

Step 3. Describe the specific postharvest principles, practices, and technologies to be offered during the extension program. Take into consideration any known or suspected constraints that may influence the adoption of postharvest technology. Paying attention to the many factors discussed in the first part of this chapter will help extension workers focus on those technologies that will be most appropriate for their clientele.

Step 4. Describe the extension methods that will be employed to reach the program's objectives. Indicate whether methods will be individual, group, mass media, or based on information technology (computers). Identify information sources and available teaching materials. It is best to involve cooperators from the produce industry or related organizations in extension programs whenever possible, both in order to benefit from their expertise and to gain their support for program objectives.

Step 5. Determine the resources needed to conduct the program. Compare the manpower, equipment, and facilities that will be needed with those that are currently available, and identify possible sources of funding, resource persons, tours, demonstration sites, and supplies.

Step 6. Develop a plan for evaluating the program during implementation in order to determine whether objectives are being met and how the program might be improved. Include the program's stakeholders in the evaluation to enhance the chances that the results of the evaluation will be utilized.

More information on the topics of educational needs assessment, commodity systems assessment, extension methods and program evaluation can be found later in this chapter. For more detailed information on the extension program planning process, see Blackburn (1994); Van der Ban and Hawkins (1996).

EXTENSION METHODS

Working within any extension system, there are many extension methods available for conducting postharvest programs. Successful extension work is an art as well as a science. The relative effectiveness of any method depends on the level of interest and voluntary involvement of postharvest researchers, public and private extension workers, and industry clients. Always, the work of extension is to build new links and strengthen existing links between information generators and information users. These linkages help make postharvest researchers aware of industry problems, perspectives, and constraints, and also help increase industry awareness of the problem-solving assistance available from scientists. In this section of the chapter is a description and discussion of some of the methods most commonly used in postharvest extension programs.

APPLIED OR ADAPTIVE RESEARCH

Many common handling practices and conditions found in the horticultural industry cause increased product deterioration and marketing losses (see part I of this chapter). Some problems are relatively simple and can be studied easily by comparing existing postharvest technologies under controlled conditions. Other postharvest problems will require long-term study to determine the underlying cause of the problem and to develop a suitable technology.

Applied research studies in postharvest technology seek to identify the causes or magnitude of deterioration or losses and to develop and evaluate possible corrective measures. Adaptive research attempts to modify an existing technology to better fit the exact conditions in which it will be used in practice. While some research must be conducted in laboratories, and other research can be conducted in industry facilities, all research must use scientifically sound methods and procedures.

Collaborative and participatory research and development methods ensure that the postharvest information developed by extension specialists is useful and appropriate for their clientele. The skills needed in applied research include patience and attention to detail, careful planning, and willingness to listen to cooperators and clientele.

Public extension services face a unique challenge whenever they work with clientele associated with large, technically sophisticated companies involved in produce handling and marketing. This clientele may be unwilling to cooperate fully with postharvest extension specialists since their competitive outlook makes it undesirable to share the results of the applied research studies with others.

CONSULTATIONS

Requests for consultations most often originate with industry leaders or groups who want to improve their postharvest operations. Consultations are usually done in person, but they can also be handled by telephone, through the mail, or even by FAX or e-mail and can involve individuals, companies, or other groups such as grower cooperatives. Consultations generally deal with specific subjects (e.g., decay problems or cooling methods) and individual problem diagnosis. Despite the high time requirement associated with individual consultations, these continue to be an important extension method. Extension workers can use the time they spend with individuals to obtain informal information on needs and feedback on prior or ongoing extension programs, and to build linkages with clientele.

APPLIED POSTHARVEST RESEARCH

Answering the following questions can guide researchers as they plan and design postharvest studies. Researchers must describe the exact conditions of their original studies so that others can replicate the research.

Research Hypothesis
- Is the hypothesis conceptually clear, specific, and subject to empirical testing?
- Is the hypothesis related to a body of theory and a specific postharvest problem?

Objectives
- Are the research objectives clear, specific, and measurable?
- Does the problem statement encompass and agree with all the relevant facts, explanatory concepts, and relationships among the variables under study?
- How will learning the answers help to solve a postharvest problem?

Research Design
- How will you set up the study to observe the hypothesized relationship between variables?
- Have you accounted for intervening variables that might also affect the dependent variable?
- How will you measure or control threats to the internal validity and external validity of the study?
- What experimental controls will you use?

Subject Selection
- How will you select your produce samples for experimental studies? (Describe sample size, characteristics, sampling site, sampling method).

- Are the samples are an adequate representation of the whole population?

Outcome Measures
- What methods and equipment or panel of experts will you use to measure outcomes (the dependent variables)?
- Are measurements objective (weight or size) or subjective (flavor or mouthfeel)?

Conditions of Testing
- When, where, and under what conditions will data be gathered?

Treatments
- How will treatments (independent variables) be administered?
- Have you described both the "traditional" and "new" postharvest methods in detail, explaining how they are different from one another?

Data Analysis
- Are the statistical tools selected for analysis suited to your research design and specific objectives?
- What level of significance is required to test your hypothesis?

For detailed information on applied and adaptive research design and analysis, see Andrew and Hildebrand (1993); Hildebrand and Russell (1996); Mead et al. (1993).

GROUP MEETINGS

Meetings with groups allow the extension worker to present information to larger audiences than through consultations. Regularly scheduled group meetings allow clientele the opportunity to feel connected and receive updates and progress reports on ongoing applied postharvest research and extension work. Group meetings also encourage direct audience participation and discussion. When postharvest researchers are invited to participate, clientele can offer direct feedback regarding the information presented, express their needs and concerns, and explain the constraints they face.

Reaching growers and large-scale shippers and retailers is relatively easy, but sometimes, postharvest handlers and small-scale marketing intermediaries fall through the cracks. Intermediaries are mobile by nature and may work in various regions at different times of the year and sell in several markets. In many developing countries, women do the majority of food marketing, while men handle cash crops. Market intermediaries there are often illiterate and learn their jobs by apprenticeship with an elder. Typically, government agencies, universities, and extension services focus on growers (through ministries, departments, and colleges of agriculture) and wholesale or retail vendors (via ministries or departments of commerce or marketing), leaving no one to specifically target market intermediaries.

As with most extension methods, the language and socioeconomic status of the extension worker should match that of his or her clientele for best results. The use of nonformal and hands-on teaching methods whenever clients are illiterate would help increase participation and comprehension. When women are part of the target audience for postharvest extension programs, meeting times must be scheduled with the needs of women in mind. Locations of meetings for educational purposes must be accessible and provide some form of child care facilities. The marketplace is perhaps the best location for many extension activities related to postharvest handling practices. As a natural gathering place, wholesale and retail markets are excellent locations for information gathering, tours, and demonstrations of recommended postharvest technologies. Holding regular group meetings in local wholesale and retail marketplaces would also help to reduce the transportation costs associated with extension work and participation.

DEMONSTRATIONS

Extension workers can use demonstrations to show how to use a new practice, procedure, or facility or to illustrate the results of recommended postharvest technology. Demonstrations are often used to extend the results of applied or adaptive research. A hands-on or experiential learning approach can enhance program results, since many people "learn by doing." Careful attention to the equipment, facilities, and visual aids used in demonstrations can increase their effectiveness.

The message of any given demonstration should be simple and clearly presented to participants. One example is the effect of various temperatures on the postharvest life of selected produce; another is the effect of chlorine on the vase life of roses. In one short course offered in India by UC postharvest specialists during the hot (dry) season, Punjabi flower producers were enthusiastic about their training when they witnessed a large vase of wilting roses seemingly brought back to life when a dash of chlorine bleach was added to the vase solution.

An important component of any demonstration is a cost-benefit analysis, in which a comparison is made between the current handling practice and the new postharvest technology under consideration. A simple chart for comparing any two practices can be constructed and used to calculate the comparative costs and the expected benefits. Costs include required equipment, labor, supplies, and power; benefits might include increased volume of produce, higher quality grade, or better market price.

A summary worksheet is included as figure 38.3. For more details, please refer to Kitinoja (1999) and Kitinoja and Gorny (1999).

Research has shown that there is a strong correlation between the characteristics of an innovation and its rate of adoption (Rogers 1995). Table 38.5 lists the characteristics that are known to be important. Postharvest technologies that have these characteristics can be demonstrated successfully in extension programs.

SHORT COURSES

Extension audiences can benefit from intensive, broad coverage of specific subjects in short courses through classroom lectures, laboratory demonstrations, or tours. The Postharvest Technology Short Course offered annually in June by the University of California, Davis, is an example, with 5 days of classroom instruction followed by 5 days of tours. The University of the West Indies conducts short courses throughout the Caribbean on postharvest topics via their Continuing Education Program in Agricultural Technology (CEPAT). Other short courses are conducted by individual marketing firms for their own personnel, or by trade association specialists. The UC Postharvest Technology and Research Information Center (PTRIC) website on the Internet (http://postharvest.ucdavis.edu) is an excellent source of information on current postharvest course offerings, and it also has links to other sites.

Short courses may be from 2 days to about 2 weeks long. They may be used to refresh the audience with previously learned information and to provide updated or new information on a subject. When the course meets for several days or more, the extension specialists can prepare demonstrations of the use of postharvest practices, and

Figure 38.3 Worksheet: Comparison of costs and benefits.

Does one practice cost more than the other for materials, power, equipment, labor, storage, transport, marketing, etc.? Calculate the difference based on expected yield, hourly labor costs, expected volumes to be handled.

	Current practice	New practice
1. Cost of equipment	$_____	$_____
2. Cost of supplies	$_____	$_____
3. Cost of labor	$_____	$_____
4. Cost of power	$_____	$_____
5. Other	$_____	$_____
Total direct costs	$_____	$_____

Benefits

Base these figures on expected yield, quality, amount of produce at various grades, and predicted market prices. Use either wholesale or retail prices or a combination if you will sell both ways.

	Current practice	New practice
1. Expected sales (wholesale)		
Highest grade	$_____	$_____
Second grade	$_____	$_____
Lowest grade	$_____	$_____
Subtotal: Sales (wholesale)	$_____	$_____
2. Expected sales (retail)		
Highest grade	$_____	$_____
Second grade	$_____	$_____
Lowest grade	$_____	$_____
Subtotal: Sales (retail)	$_____	$_____
3. Total expected sales	$_____	$_____
4. Comparative advantage		
(Total sales − Total direct costs)		
Which practice is most profitable?	$_____	$_____

Recovery of Invested Capital (ROIC)

How long will it take to pay for your investment in the new practice or technology?

1. Actual capital outlay for new practice = $_____
(Difference in total direct costs for new equipment, facilities, power costs, supplies, and labor requirements when compared to current practice)

2. Interest rate (if capital is borrowed)
= _____% per annum; or _____% per month
cost of capital at three months = $_____; or at six months = $_____

3. Change in sales using the new practice = $_____ per month
(Subtract total expected sales using the current practice from total expected sales using the new practice; divide the difference by number of months of sales)

4. Calculate ROIC in months to pay for investment:
(Actual capital outlay + any interest paid) ÷ Change in sales per month = Months to pay for investment.
($_____ + $_____) / $_____ per month = _____ months

Table 38.5. Characteristics of an innovation that enhance adoption

Characteristic	Comments
Relative advantage	Does the innovation enable the client to achieve his goals better or at lower cost than he could previously? Postharvest technology that is clearly cost-effective will be of most interest to potential users.
Compatibility	Is the innovation compatible with sociocultural values and beliefs, with previously introduced ideas, and/or with clients' felt needs? Any new postharvest technology must not cause more problems than it solves.
Complexity	Can the innovation be adopted without complex knowledge or skills? If the postharvest technology is difficult to understand or use, clients will be less likely to want to try it for themselves.
Trialability	Can the client try the innovation on a small scale on his own before making the decision on whether to make large-scale changes in practices? If a large investment is required before the user can see any results, the postharvest technology will remain a training exercise.
Observability	Can the client see the effects of changes made by others when they adopted the innovation?

Source: Rogers 1995; Van der Ban and Hawkins 1996.

clients can see for themselves the results of certain technologies.

Effective short courses require much planning, considerable professional involvement and input, proper facilities, and follow-up to evaluate their effectiveness. A good mix of extension methods should be used to maintain interest. The use of visual media (slides, videotapes, or CD-ROM) and group exercises or discussion helps keep participants involved. Printed materials, including a syllabus and a list of current references for further information, are generally provided to each participant in a short course.

WORKSHOPS

Workshops can improve the skills in postharvest technology of individuals or groups. For example, one might conduct a specific workshop on harvesting, cooling, or careful handling. Often, a workshop will focus on a single commodity and provide written materials, as well as use visual aids and include discussions. Workshops can last for 1 or more hours, and can meet only once or as a series over a period of time. Organized meetings and workshops in California have led to changes in practices within the apple industry, where growers have learned the importance of harvesting earlier and storage

operators have learned to monitor and maintain low levels of CO_2 in storage. These simple changes in postharvest practices have resulted in a decreased incidence of internal browning in Fuji apples.

The scheduling of workshops can be a sensitive issue, since clients may be too busy to attend during a time that is optimal for the extension worker in terms of demonstrating pertinent postharvest handling practices for a certain commodity.

TOURS

Postharvest facilities and operations are often easier to understand after a well-run tour. Tours can be a part of a demonstration, short course, or workshop. The University of Florida's Horticultural Sciences Department offers an annual 4-day-long Florida Postharvest Horticulture Institute and Industry Tour in the early spring, during which participants visit harvest, packing, shipping, and warehousing operations throughout south and central Florida.

A tour can be an effective way to introduce a new subject to an audience. Providing a brief outline of what participants might expect to see can help them to focus their observations. Figure 38.4 is an observation schedule used by short course participants to organize their notes on observations and interview results. This example was used during tours of organic produce growers, shippers, and marketers in northern California. Industry cooperation is essential for a successful tour, since participants will have many questions and will appreciate the opportunity for direct interaction with the owners or managers of the site.

PUBLICATIONS

Publishing in professional journals, trade publications, newsletters, and other media outlets helps the extension worker extend information to a greater, more distant audience than can be reached through personal contacts. Effective articles present information clearly, succinctly, and in an appealing manner; they are addressed to specific audiences and for a specific purpose, and they do not impose the writer's personal bias on the information presented. A publication can be written in a variety of formats, depending on the intended audience. Results from specific studies or other relevant information are

Figure 38.4 Interview and observation schedule for postharvest tours.

Small Scale Postharvest Handling of Organic Horticultural Crops
Interviews and Observations

Site _____

Date _____

Commodity _____

Use the following outline to take notes as you ask questions about the commodity system during field visits and market tours.

Production Management
Choice of cultivars _____
Cultural practices _____
Field sanitation _____

Harvest Practices
Maturity indices _____
Containers/tools _____
Handling methods _____

Preparation for Marketing
Sorting/grading _____
Sanitation _____
Quality control practices _____
Pest management _____

Packaging/Shipping Containers
Packing methods _____
Packaging materials _____

Cooling/ Temperature Management
Cooling methods _____
Temperature/RH control _____

Transport
Vehicle cooling _____
Loading methods _____

On-Farm Storage
Structures _____
Temperature/RH control _____
Pest control _____

Marketing
Pricing policies _____
Customer relations _____
Finding new markets _____

Uses for Surpluses
What happens to produce you can't sell? _____

often published in two or more types of publications to reach a broader audience and achieve the maximum effect. For example, the scientific details of a study might be reported in a professional society journal, the semitechnical aspects published in a university or government report, and a popular report of the study, showing its relevance to the postharvest industry, might be published in an extension newsletter and one or more trade publications. Following are examples of types of publications (see also the reference list in chapter 1):

Technical and semitechnical articles. These include articles on applied research published in professional societies' publications and in university and government technical reports. They are written in a scientific style, primarily for the benefit of professional researchers and extension workers; few industry representatives read these publications.

Brief, single-subject guides. This type of publication addresses a single specific subject or development in a direct, simple style. The UC Postharvest Technology Research and Information Center produce fact sheets are examples, where information is provided regarding recommendations for maintaining postharvest quality of an individual commodity (fig. 38.5).

Figure 38.5

An example of a "Produce Facts" publication.

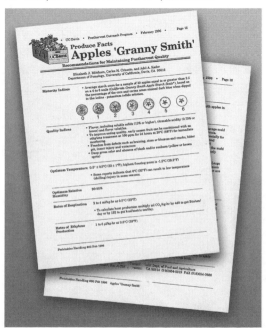

Progress reports. These publications extend current information on ongoing projects to cooperators, research sponsors, industry personnel, and others. An example might be results of a just-completed preliminary study on the effects of a questionable, presently used industry practice. They are usually brief reports, no more than a few pages long. Their main advantages are timeliness, brevity, and directness, and they keep interested persons informed about ongoing results.

Newsletters and quarterlies. These periodical publications extend information to broad audiences on a regular basis, typically four to six times a year. A good newsletter is an effective route for extension of brief, pertinent reports and articles to the postharvest industry and to fellow extension and research workers. The University of California *Perishables Handling Quarterly* has been an effective informational tool since its founding as a newsletter in 1962. Issues usually contain a review article on a specific subject related to postharvest handling, brief articles on recent research results, book reviews, and a list of recent postharvest publications and reports. The University of Florida postharvest newsletters for industry, *Packinghouse Newsletter* and *Handling Florida Vegetables*, is another effective newsletter. The horticultural industry also has its own postharvest newsletters, mostly intended for internal communication.

Trade publications. Articles in trade magazines and newspapers are most effective for extending information to industry handlers, who read them regularly. Magazines and bulletins produced by grower-shipper associations include *The Western Grower and Shipper*, published by the Western Growers Association; the monthly *PMA Bulletin*, published by the Produce Marketing Association; and *Fresh Outlook*, published quarterly by the United Fresh Fruit and Vegetable Association. The weekly newspapers of the fresh produce industry, *The Packer* and *Produce News*, are also included in this group. Articles by extension workers in trade publications extend information to a broad audience that may otherwise be hard to reach. These trade associations and their publications can be very effective media for extending postharvest information. Several postharvest extension specialists have regularly published columns in these types of publications.

Visual media. Videotapes and CD-ROM programs are becoming increasingly popular methods of extension as costs of production decline and computer technology makes production and editing feasible on a small scale. The United Nations Food and Agriculture Organization (FAO) is currently producing a series of postharvest publications on CD-ROM, which will be available worldwide in several languages. The University of California markets a wide selection of slide sets and videos related to recommended postharvest handling practices. Examples include the slide set *Postharvest Care and Handling of Cut Flowers* and the videotape *Harvesting Fruits and Vegetables*. The California Tomato Board has produced a video on the care and handling of tomatoes, and the University of California Agricultural Issues Center offers one entitled *Alternatives to Chemical Use: Postharvest Handling and Processing*. The Produce Marketing Association (PMA) markets several series of videotapes as training aids for the retail produce industry. Many of these teaching aids can be incorporated into an extension education program. When used as tutorials, audiovisual programs such as slide sets, videotapes, and CD-ROM programs can be valuable as part of short courses and workshops.

PARTICIPATION IN COMMITTEES AND ADVISORY BOARDS

Extension postharvest technology programs can be effectively enhanced through participation in professional society and trade association programs and committees. This means more work for extension personnel, but it can also mean greater program effectiveness. Important professional organizations include the American Society for Horticultural Science (ASHS); industry trade associations include the United Fresh Fruit and Vegetable Association (UFFVA), Produce Marketing Association (PMA), Western Growers' Association (WGA), International Fresh-Cut Produce Association (IFPA), and the Refrigerated Transportation Foundation. Extension personnel should also work closely with committees and advisory boards for fresh-market horticultural crops whenever these have been established by state marketing order programs. These groups can help extend information more effectively and widely than is possible by working alone.

DEVELOPING PROFESSIONAL EXTENSION SKILLS

Postharvest extension workers must maintain a high degree of professionalism if the educational programs they plan and implement are to be effective in transferring postharvest technology. It is important to stay current regarding principles and practices of loss assessment and postharvest biology and technology; upgrade skills in extension work, including program planning, extension methods, and evaluation practices; and be highly motivated to identify and assist clientele.

There are many tools and skills that can assist the extension worker in successfully transferring postharvest technology. In addition to a thorough understanding of the principles and practice of extension methods discussed earlier in this chapter, the methods of postharvest loss assessment, cost-benefit analysis, and program development are among skills of primary importance. These professional activities require direct and regular interaction with various clientele groups. Keeping in mind the typical difficulties (see Chambers 1991) can help extension workers avoid common problems (see below). For a new extension worker, finding a mentor among older, more experienced agents can be a key element in long-term success, since mentors support, challenge, and provide vision to their protégés.

DIFFICULTIES COMMONLY FACED BY EXTENSION WORKERS

The following difficulties can often be encountered when attempting to investigate local conditions and assess opportunities for program development (see Chambers 1991).

- Misleading replies. Respondents may give misleading replies (differential, prudent, designed to avoid penalties or gain benefits, evade sensitive topics, state ideals rather than actual practice, etc.) and tell extension workers what they think they want to hear.

- The know-it-all. Investigators may "know it all" and fail to listen and learn, reinforcing their own misperceptions and prejudices by projecting their own ideas and selecting their own meanings. It is important for extension workers to ask people why they are using the postharvest practices they are currently using.

- Overlooking the invisible. Investigators may overlook what they cannot see—they observe physical things and activities but not social and cultural settings and relationships (patron-client relations, factions, informal organizations, traditions, norms, indebtedness, interest rates, control of assets, decision-making practices). The more you understand about the local conditions for postharvest handling, the better chances will be for identifying the most appropriate changes.

- Overlooking cycles and trends. Investigators focus on only one moment in time—cyclical and periodic events such as crop rotations and weekly markets may be overlooked; trends and seasonal changes can be easily missed. Sometimes market price fluctuations can be the key to understanding when certain postharvest practices can be used to best effect (and profit).

COMMODITY SYSTEMS ASSESSMENT METHODOLOGY (CSAM)

This postharvest loss assessment method sets the stage for productive postharvest extension work by assessing the technical, socioeconomic, cultural, and institutional factors related to handling a given commodity in a specific locale. The end products of CSAM encompass both traditional loss assessment and cost-benefit analysis and lead to productive extension program and project development.

The commodity system is made up of 26 components that together account for all the steps associated with the production, postharvest handling, and marketing of any given commodity. The method was developed over the course of many years and was tested extensively in the Caribbean before being introduced worldwide to field personnel via an excellent training manual (LaGra 1990). The manual includes sample data collection instruments and detailed explanations of each of the components. Ideally, teams of people work together while investigating a commodity system—for example, a horticultural production researcher might be teamed up with a marketing specialist and an extension agent. CSAM can help build links between agencies and individuals, close information gaps, and help people solve problems while focusing on usable postharvest technology.

Sample CSAM questions

Table 38.6 is a list of system components and sample questions for investigating the commodity system. The team begins by considering these questions in relation to any commodity of interest, and then adds any other information that is pertinent to the situation. Some of the questions can be answered directly by extension personnel or others who are knowledgeable about the commodity, or information can be found in available literature. Other questions may require the data collection team to observe actual postharvest handling practices and ask questions of those people who harvest, handle, and market the product. Information on the costs and expected benefits of various postharvest technologies can be collected directly or estimated from applied or adaptive research studies.

Expected outcomes of CSAM

CSAM can assist a postharvest loss assessment team to determine the sources of postharvest losses (when, where, and who within the marketing chain is responsible); the causes of those losses (what handling or marketing practices are responsible); and the economic value of the losses compared to the costs of current and proposed postharvest practices. Once this kind of information has been collected, extension educators can target the responsible handlers with appropriate information on cost-effective, improved postharvest technical practices.

In the occasional situation where there is no existing appropriate technical solution for the handling or marketing problem uncovered using CSAM, the problem can be passed on to horticultural researchers in the universities or regional agricultural research centers. The more information provided regarding the commodity system, the better chance the researchers will have to identify solutions that are appropriate to the specific socioeconomic and cultural setting where the postharvest losses occur.

Table 38.6. Components of the commodity system, with sample questions

PREPRODUCTION

1. Importance of the crop	What is the relative importance of the crop (number of producers, amount produced, area of production, value)?
2. Governmental policies	Are there any laws, regulations, incentives, or disincentives related to producing or marketing the crop (e.g., existing price supports or controls, banned pesticides, or residue limits)?
3. Relevant institutions	Are there any organizations involved in projects related to production or marketing the crop? What are the goals of the projects? How many people are participating?
4. Facilitating services	What services are available to producers and marketers (e.g., credit, inputs, technical advice, subsidies)?
5. Producer/shipper organizations	Are there any producer or marketer organizations involved with the crop? What benefits or services do they provide to participants? At what cost?
6. Environmental conditions	Does the local climate, soil, or other factors limit the quality of production? Are the cultivars produced appropriate for the location?
7. Availability of planting materials	Are seeds or planting materials of adequate quality? Can growers obtain adequate supplies when needed?

PRODUCTION

8. Farmers' general cultural practices	Do any farming practices in use have an effect on produce quality (irrigation, weed control, fertilization practices, field sanitation)?
9. Pests and diseases	Are there any insects, fungi, bacteria, weeds, or other pests present that affect the quality of produce?
10. Preharvest treatments	What kinds of preharvest treatments might affect postharvest quality (use of pesticides, pruning practices, thinning)?
11. Production costs	Estimate the total cost of production (inputs, labor, rent, etc). What are the costs of any proposed alternative methods?

POSTHARVEST

12. Harvest	When and how is produce harvested? By whom? At what time of day? Why? What sort of containers are used? Is the produce harvested at the proper maturity for the intended market?
13. Grading and inspection	How is produce sorted? By whom? Does value (price) change as quality or size grades change? Do local, regional, or national standards (voluntary or mandatory) exist for inspection? What happens to culled produce?
14. Postharvest treatments	What kinds of postharvest treatments are used? (Describe any curing practices, cleaning, trimming, hot water dips, etc.) Are treatments appropriate for the product?
15. Packaging	How is produce packed for transport and storage? What kind of packages are used? Are packages appropriate for the product? Can they be reused or recycled?

EXTENSION PROGRAM DEVELOPMENT

Postharvest extension work requires knowledge and skill in planning, implementing, and evaluating educational programs. Extension workers need to use creativity and initiative in program approaches, teaching techniques, and extension methods. They also must be highly self-motivated and must seek to involve clientele throughout. Using informal and formal needs assessments, extension workers must seek out postharvest handling problems and the groups that face them. Including clientele in this first stage of program planning is essential, since industry handlers generally will not reveal their problems until extension agents gain their confidence. Encouraging active participation during programs helps people gain maximum value from educational opportunities and increases the likelihood that clientele will provide constructive feedback. For readers interested in more information on this topic, a variety of references related to extension program development are listed at the end of this chapter.

Staying current

Extension workers should continue to take courses and short courses in postharvest technology when possible and use sabbatical leaves to advance their capabilities. They must be familiar with modern research techniques, postharvest equipment, and instrumentation. New postharvest technologies such as smart films and ethylene-absorbing films for packaging are on the horizon, and

Table 38.6. Cont.

16. Cooling	When and how is produce cooled? To what temperature? Using which method(s)? Are methods appropriate for the product?
17. Storage	Where and for how long is produce stored? In what type of storage facility? Under what conditions (packaging, temperature, RH, physical setting, hygiene, inspections, etc.)?
18. Transport	How and for what distance is produce transported? In what type of vehicle? How many times is produce transported? How is produce loaded and unloaded?
19. Delays/waiting	Are there any delays during handling? How long and under what conditions (temperature, RH, physical setting) does produce wait between steps?
20. Other handling	What other types of handling does the produce undergo? Is there sufficient labor available? Is the labor force well trained for proper handling from harvest through transport? Would alternative handling methods reduce losses? Would these methods require new workers or displace current workers?
21. Agroprocessing	How is produce processed (methods, processing steps) and into what kinds of products? How much value is added? Are sufficient facilities, equipment, fuel, packaging materials, and labor available for processing? Is there consumer demand for processed products?

MARKETING

22. Market intermediaries	Who are the handlers of the crop between producers and consumers? How long do they have control of produce and how do they handle it? Who is responsible for losses (who suffers financially)? Is produce handled on consignment, marketed via direct sales, or moved through wholesalers?
23. Market information	Do handlers and marketers have access to current prices and volumes in order to plan their marketing strategies? Who does the recordkeeping? Is information accurate, reliable, timely, and useful to decision makers?
24. Consumer demand	Do consumers have specific preferences for produce sizes, flavors, colors, maturities, quality grades, packages types, package sizes, or other characteristics? Are there any signs of unmet demand and or oversupply? How do consumers react to the use of postharvest treatments (pesticides, irradiation, coatings, etc.) or certain packaging methods (plastic, Styrofoam, recyclables)?
25. Exports	Is this commodity produced for export? What are the specific requirements for export (regulations of importing country with respect to grades, packaging, pest control, etc.)?
26. Marketing costs	Estimate the total marketing costs for the crop (inputs and labor for harvest, packaging, grading, transport, storage, processing, etc.). What are the costs for any alternative handling or marketing methods proposed? Do handlers and marketers have access to credit? Are prevailing market interest rates at a level that allows the borrower to repay the loan and still make a profit? Is supporting infrastructure adequate (roads, marketing facilities, management skills of staff, communication systems such as telephone, FAX, e-mail services)?

it is expected that each year will bring new ideas and innovations.

In the United States, Cooperative Extension personnel are invited to attend the University of California Postharvest Technology Short Course. In Europe, the Natural Resources Institute (NRI) regularly offers postharvest training courses. If local extension capabilities in postharvest in-service training are limited, it is possible to work with a variety of development agencies to design and fund training programs. The FAO and the World Bank announce currently available programs on their websites. The U.S. Agency for International Development (AID) regularly funds postharvest training projects in developing countries through the Farmer-to-Farmer Program via Winrock International and through voluntary agencies such as Agricultural Cooperatives Development Institute/ Volunteers in Overseas Cooperative Assistance (ACDI/VOCA), in which postharvest extension specialists are invited to participate as speakers and resource persons.

Educational needs assessment

A common complaint raised against extension systems is that programs are related more to the interests of the extension worker than to the needs of the clientele. During the 1980s many evaluations pointed to ineffective extension programs, underutilized postharvest facilities, and disillusioned clientele. In order to avoid this situation, extension workers must regularly use educational needs assessment to find out what clients need to learn in order to solve the current postharvest problems they face. Needs assessment may be formal, involving surveys and literature reviews, or they may be informal, based on thorough dialog with representative clients. If mail or telephone surveys are undertaken, following the methods described by Dillman (1978) can greatly increase response rates.

Finding resources

Extension workers will no doubt be required to locate sources of funding for their applied research and extension educational programs. Grant writing skills can be developed by taking courses and seminars on the topic, which are offered at many colleges and universities worldwide. Most libraries have access to the *Foundation Directory* and the *Foundation Grants Index* in printed or electronic form. Both of these references contain thousands of entries describing foundations that offer grants to eligible organizations. In addition, The *National Data Book* lists over 30,000 foundations that have awarded a cash grant during the most recent fiscal year, and the *Catalog of Federal Domestic Assistance* lists grants given by U.S. government agencies.

It is important to write an effective letter proposal or concept paper when seeking funding for an extension program; much more detailed proposals will be required when applying for most government grants. It can be helpful to ask a trusted colleague to review grant proposals before they are submitted. A letter proposal should contain the following components:

- Summary of the proposal: Who are you, whom do you represent, what is your extension organizations' strong suit, what type of postharvest extension program do you want to offer, how much money are you requesting, and what will the major program outcomes be?

- Appeal to the sponsor: Focus on values the sponsor supports; mention key funding patterns that attracted you to the sponsor.

- The problem: Focus on the postharvest problem or need and how funding the extension program will help the sponsor reach its mission; document the need for the program.

- The solution: Summarize the program objectives you will meet with your approach and procedures; convey confidence that you can solve the postharvest problem.

- Capabilities: Establish the credibility of your organization, your idea, and the program coordinator. What can specifically demonstrate that you will be able to solve the problem?

- Budget: Ask for a specific amount, and express your request in terms of outcomes (number of people reached, postharvest handling skills learned, amount of increased income due to reduced postharvest losses, or better quality of produce offered for sale).

- Conclusion: Identify the desired action by the sponsor, identify a contact person should more details be requested, mention appendices if included (detailed procedures, time and task chart, organizational brochure highlighting past successes, etc).

Informal teaching and nontechnical writing

Various programs might offer topics specifically aimed, for example, at the problems faced by large-scale growers or shippers of cool-season crops, by small-scale direct marketers of a mixed lot of seasonal fruits and vegetables, or by the fresh-cut industry. Whatever clientele the extension worker intends to serve, the program must be well designed, interestingly presented, and full of useful information. Informal teaching methods and nontechnical writing are tools that can help an extension worker reach a wide range of clientele.

Extension workers who are interested in developing new informal teaching skills or in reading about the successful methods of others can find information in the *Journal of Extension* and the American Society for Horticultural Science's publication *Hort-Technology*. Recent articles have dealt with the topics of on-farm demonstration, small-group learning activities, task analysis, and using computers in extension programs. An enormous amount of published and unpublished information on informal teaching methods is available in most educational libraries in the form of microfiche (ERIC documents).

An entire field is dedicated to agricultural communication, including nontechnical writing for newsletters, extension manuals, fact sheets, and multimedia. Some of the most recent literature on agricultural communication is available via the University of Illinois at Champaign-Urbana and the Illinois Cooperative Extension Service. Back issues of this publication (*Agricultural Communication*) and a bibliography of literature on extension methods can be accessed free of charge via the Internet (http://www.ag.uiuc.edu). The Department of Agricultural Education at the Ohio State University publishes the quarterly newsletter *Agricultural Communication* that offers updates on extension methods, writing techniques, and multimedia productions. The editors also review books and short courses dealing with professional skill development.

Marketing extension programs

Even the best postharvest extension programs will be less effective than they can be if people don't know about them. Marketing extension programs requires planning ahead, developing high-quality promotional materials, and getting the word out via direct mail, posters or brochures, word-of-mouth, or the Internet. The four P's of marketing (product, place, promotion, and price) have been interpreted for postharvest extension programs in table 38.7. Similar to quality program planning, where the needs of the clientele determine the objectives of the extension effort, the more an extension worker knows about the target audience, the more likely it is that the program will be marketed to those who will benefit most from participation.

Program evaluation

Extension workers must be willing to do the work associated with evaluating their ongoing projects and extension programs in order to determine what is working well and what needs to be changed. Evaluation requires gathering information during implementation to determine whether the program has met its objectives. Figure 38.6 illustrates Bennett's model (1979) and the seven levels of evidence related to a program's theory of action, which can help an extension worker focus the evaluation on the key aspects of the program. A theory of action attempts to understand causal linkages between the program inputs and activities and the program's intended outcomes. It is important to focus attention and resources during evaluation on higher-level outcomes, such as actual or intended practice changes, rather than counting the number of participants and assuming they learned and used the postharvest technologies offered during the program. According to Patton (1986, p. 171):

> This model explicitly and deliberately places the highest value on attaining ultimate social or economic goals. Actual adoption of recommended practices and specific changes in client behaviors are necessary to achieve ultimate goals and are valued over knowledge, attitude and skill changes. People may learn about some new agricultural technique (knowledge change), believe it's a good idea (attitude change), and know how to apply it (skill change)—but the higher-level criterion is whether they actually begin using the new technique (i.e.,

Table 38.7. The marketing mix

Product	What will the postharvest extension program involve? What new information is being offered to clientele?
Place	How and where will the program be delivered to growers, shippers and/or marketers?
Promotion	How will the planned target audience be informed about the program and encouraged to participate?
Price	What costs or fees are part of the extension program?

Source: Adapted from Brown 1984.

Figure 38.6

Program theory of action/evaluation hierarchy. (Adapted from Bennett 1979; Patton 1986)

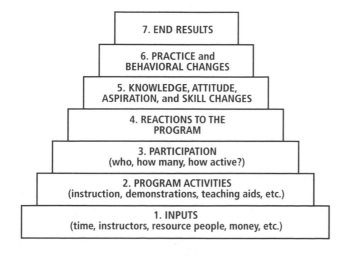

7. END RESULTS

6. PRACTICE and BEHAVIORAL CHANGES

5. KNOWLEDGE, ATTITUDE, ASPIRATION, and SKILL CHANGES

4. REACTIONS TO THE PROGRAM

3. PARTICIPATION
(who, how many, how active?)

2. PROGRAM ACTIVITIES
(instruction, demonstrations, teaching aids, etc.)

1. INPUTS
(time, instructors, resource people, money, etc.)

change their practices). Participant reactions (satisfaction, likes, and dislikes) are lower still on the hierarchy. All of these are outcomes, but they are not equally valued outcomes. The bottom part of the hierarchy identifies means necessary for accomplishing higher-level ends; namely, in descending order, getting people to participate, providing program activities and organizing the basic resources and inputs to get started.

Many resources are available to those interested in learning how to perform well-designed program evaluations. Several useful references are the books by Patton (1982; 1986) and Rossi and Freeman (1985) on program evaluation practices, and the journal *Evaluation and Program Planning*.

PLANNING A POSTHARVEST EXTENSION PROGRAM EVALUATION

An evaluation of the results of the postharvest extension program can help generate ideas for improving the program. When planning an evaluation, consider the following steps (see Kitinoja 1992).

Step 1. Identify and describe the extension program.

- What are the objectives of the program?

- Where and when will it be held, and who will be involved?

Step 2. Purpose of the evaluation.

- Is the evaluation for program improvement or accountability?

Step 3. Audience(s) for the evaluation.

- Who wants to know about the results of the evaluation?

- What specific evaluation questions do these stakeholders have?

    How do they plan to use the information?

Step 4. Levels of evidence needed.

- Consider Bennett's (1979) "levels of evidence" as they relate to the program's theory of action (inputs, activities, involvement, reactions, changes in knowledge or skills, practice changes, end results).

- What levels of evidence are required to answer evaluation questions for each audience?

Step 5. How will evidence be evaluated?

- What criteria will be used to judge the evaluation results (comparison to known standards, past experience with similar programs, wishful thinking)?

- How will you determine if program objectives were met?

Step 6. Tools for measurement.

- Describe the data collection methods and instruments and why they were selected (pre/post tests, surveys, interviews, observations, etc.)

- Describe sources of data (existing literature, program staff, participants, nonparticipants, etc.)

Step 7. Resources and time needed.

- Develop a timeline and estimate the time needed for each phase of the evaluation (planning, stakeholder input, design and testing of data collection instruments, data collection, data analyses, stakeholder feedback, reporting).

Step 8. Gather evidence (collect data).

- Who will collect data, and when will collection begin and conclude?

Step 9. Analyze, summarize, interpret, and draw implications.

- What analysis methods will you use?

- How will you seek assistance in determining what the results mean or their implications for the program?

Step 10. Report results.

- Describe the outcomes of steps 1 through 9.

- Who will receive the report(s)?

- When will the report be completed?

POSTHARVEST EXTENSION IN CALIFORNIA

The University of California Postharvest Technology Research and Information Center (PTRIC), formerly known as the Postharvest Outreach Program, is a research and extension center where fourteen UC Cooperative Extension specialists work collaboratively with industry partners to perform research studies and provide educational programs in the interdisciplinary field of postharvest technology of horticultural crops. Members of the program are housed throughout California, at the campuses of UC Davis and UC Riverside and at the Kearney Agricultural Center in Parlier. The specialists represent seven academic departments, including vegetable crops, pomology, food science, agricultural economics, environmental horticulture, plant science, and biological and agricultural engineering. A host of laboratories and controlled-temperature chambers are available for applied postharvest research and demonstration activities. The PTRIC cooperates closely with a variety of organizations representing the horticultural industry and seeks assistance, as required, from emeritus extension specialists and private, California-based postharvest consultants.

PURPOSE OF THE PROGRAM

The Postharvest Outreach Program focuses on particular problems the industry faces in handling fruits, vegetables, and ornamental crops. California growers supply nearly 50% of the fresh fruits and vegetables consumed in the United States. They also supply millions of cut flowers and other ornamentals that contribute to quality of life. The program's goals are to improve the quality and value of horticultural crops available to the consumer, reduce postharvest losses, and improve marketing efficiency.

PROGRAM DEVELOPMENT AND MANAGEMENT

Through the postharvest research and extension leadership efforts of the fourteen specialists the program strives to provide relevant information for all California growers, shippers, marketers, carriers, distributors, retailers, processors, and consumers of horticultural crops. Members meet monthly to discuss research results and program evaluations and to plan a variety of research and educational activities in response to industry needs.

These activities include

- problem-solving research related to maintaining the quality and safety of horticultural produce between harvest and consumption

- short courses, workshops, and seminars on topics related to postharvest technology

- publications such as textbooks, manuals, produce fact sheets, *Perishables Handling Quarterly*, and newsletters

- audiovisual materials such as slide sets and videotapes

- industry meetings where program members present research results and postharvest information

- consultations by telephone and in person with individual handlers

Several publications and videos have been translated into Spanish and other languages. Outreach activities include a website containing a calendar of events, information on the program, and links to a wide selection of related websites. Occasionally, members are invited to plan and implement short courses and workshops in foreign countries.

RESOURCES

The availability and sources of funding for the program differ widely from year to year. At any given time, funding for research and educational programs may come from the following:

- competitive grants from state and federal government agencies, industry sources, university funds, or private foundations

- participant fees from short courses, seminars, and workshops

- sales of postharvest manuals, educational materials, and quarterly and newsletter subscriptions and back issues

- grants from the UC administration funding sources for support of UC Research and Information Centers

- support for specific projects from organizations such as AID, the World Bank, and FAO

- interest income from the UC Postharvest Program Endowment fund, recently established to support teaching, research, and extension activities related to postharvest biology and technology of fruits, vegetables, and ornamental crops

CONSTRAINTS

When considering the UC Postharvest Technology Research and Information Center as a model for extension efforts, it is important to point out several weaknesses in the program. The relative autonomy of the fourteen specialists and their diverse locations sometimes makes it difficult to accomplish planned tasks. In the absence of a strong leader, many projects would stagnate. Much of the success of the PTRIC is due to the leader's dedication. He or she must spend much time and energy motivating members and overseeing projects to ensure their timely completion.

The lack of reliable funding is another weakness. The program coordinator (leader) and other members require strong fundraising skills in order to complete their applied research, outreach, and service objectives. Planning long-term activities is especially difficult due to the variation of funding sources and amounts each year.

SUCCESSES

The collaborative efforts of the PTRIC have resulted in the development of several postharvest handling practices that are now used extensively throughout the horticultural industry to reduce losses and maintain produce quality. Tight-fill packing was introduced in the 1960s, and a forced-air cooling system (originally designed by Rene Guillou in the 1950s) was adapted for use with cut flowers during the 1970s and is now the industry standard for many highly perishable crops. The elucidation of the ethylene biosynthesis pathway during the 1980s led to the development of methods for manipulating the ripening of fruits and to the increasing use of commercial ripening facilities. Recently, when the horticulture industry in California moved into the fresh-cut produce sector, the PTRIC added applied research and outreach activities aimed at maintaining quality and fresh-cut product safety. The educational activities of the PTRIC reach hundreds of people annually in short courses and workshops, and thousands of texts and manuals have been distributed. Each year several new titles are added to the collection of postharvest extension publications, and older publications, slide sets, and videos are regularly updated.

Industry partners as well as the entire horticultural community benefit from their collaboration with the PTRIC. Their mutual goals of reducing losses, ensuring safety and freshness of horticultural products, and improving marketing efficiency are critical to maintaining California's reputation for produce quality.

In summary, many extension methods have been used successfully to assist clientele to identify key problems and assess potential solutions for their technical and economic feasibility. The goal of extension is to make the link between those who are developing new postharvest technologies and those who are interested in adopting them. Postharvest extension specialists and extension workers worldwide have the opportunity to identify and transfer appropriate postharvest technologies that will assist their clientele to reduce produce losses, better maintain quality and value, and most importantly, to increase profits during handling and marketing.

REFERENCES

Adhikarya, R. 1994. Strategic extension campaign: A participatory-oriented method of agricultural extension (a case study of FAO's experiences). Rome: United Nations Food and Agriculture Organization (FAO). 209 pp.

Ameur, C. 1994. Agricultural extension: A step beyond the next step. Washington, D.C.: World Bank Technical Paper No. 247.

Andrew, C. O., and P. E. Hildebrand. 1993. Planning and conducting applied agricultural research. Boulder, CO: Westview Press. 223 pp.

Ayers, T. D. 1987. Stakeholders as partners in evaluation: A stakeholder-collaborative approach. Eval. and Prog. Plan. 10:263–271.

Barao, S. C. 1992. Behavioral aspects of technology adoption: The role of on-farm demonstration. J. of Extension 30(summer): 13–15.

Bennett, C. F. 1979. Analyzing impacts of extension programs. Washington, D.C.: USDA Extension Service.

———. 1993. Interdependence models: Overcoming barriers to collaboration with other agencies. J. of Extension 31(summer): 23–25.

Bhalla, A. S., and D. James. 1988. New technologies and development: Experiences in "technology blending." Boulder: Lynne Rienner. 336 pp.

Blackburn, D. J. 1984. Extension handbook. Guelph, Ontario: University of Guelph. 167 pp.

Blackburn, D. J., ed. 1994. Extension handbook: Processes and practices. Toronto: Thompson Educational Publishing. 407 pp.

Brown, S. A. 1984. Marketing extension programs. In D. J. Blackburn, ed., Extension handbook. Guelph, Ontario: University of Guelph. 141–149.

Cernea, M. M. 1991. Using knowledge from social science in development projects. Washington, D.C.: World Bank Discussion Papers No. 114. 54 pp.

Chambers, R. 1991. Shortcut and participatory methods for gaining social information for projects. In M. M. Cernea, ed., Putting people first:

Sociological variables in rural development. 2nd ed. New York: Oxford University Press. 515–537.

Dichter, D., R. Husbands, A. Arenson, and M. Frey. 1988. A guide to technology transfer for small and medium-sized enterprises. Aldershot, UK: Gower. 156 pp.

Dillman, D. A. 1978. Mail and telephone surveys: A total design method. New York: Wiley. 325 pp.

Ewell, P. 1990. Links between on-farm research and extension in nine countries. In D. Kaimowitz, ed., Making the link: Agricultural research and technology transfer in developing countries. Boulder: Westview Press. 151–196.

Hartley, R., and P. Hayman. 1992. Information without the transfer-A common problem? J. of Extension 30(spring): 28–30.

Hildebrand, P. E., and J. T. Russell. 1996. Adaptability analysis: A method for the design, analysis and interpretation of on-farm research-extension. Ames: Iowa State University Press. 189 pp.

Kaimowitz, D., ed. 1990. Making the link: Agricultural research and technology transfer in developing countries. Boulder: Westview Press. 278 pp.

Kitinoja, L. 1992. Short course on evaluating extension programs. Ames: University of Northern Iowa. 121 pp.

———. 1999. Costs and benefits of fresh handling practices. In L. Kitinoja, ed., Perishables Handling Quarterly, special issue: Costs and benefits of postharvest technologies. No. 97:7–13.

Kitinoja, L., and J. R. Gorny. 1999. Small-scale postharvest technology: Economic opportunities, quality and food safety. Davis: Univ. Calif. Postharv. Hort. Ser. 21.

LaGra J. 1990. A commodity system assessment methodology for problem and project identification. Moscow, ID: Postharvest Institute for Perishables. 98 pp.

Lionberger, H. F., and P. H. Gwin. 1991. Technology transfer: A textbook of successful research extension strategies used to develop agriculture. Publ. MX381. Columbus: University of Missouri. 189 pp.

Mead, R., R. N. Curnow, and A. M. Hasted. 1993. Statistical methods in agriculture and experimental biology. 2nd ed. London: Chapman and Hall. 415 pp.

Merrill-Sans, D., and D. Kaimowitz. 1989. The technology triangle: Linking farmers, technology transfer agents, and agricultural researchers. The Hague, Neth.: International Service for National Agricultural Research (ISNAR). 104 pp.

Narayanan, A. 1991. Enhancing farmers' income through extension services for agricultural marketing. In W. M. Rivera and D. J. Gustafson, eds., Agricultural extension: Worldwide institutional evolution and forces for change. Amsterdam: Elsevier. 151–162.

Odell, M. J., Jr., 1986. People, power and a new role for agricultural extension: Issues and options involving local participation and groups. In G. W. Jones, ed., Investing in rural extension: Strategies and goals. London: Elsevier. 169–184.

Patton, M. Q. 1982. Practical evaluation. Beverly Hills: Sage.

———. 1986. Utilization-focused evaluation. 2nd ed. Beverly Hills: Sage. 319 pp.

Rogers, E. M. 1995. Diffusion of innovations. 4th ed. New York: Free Press. 519 pp.

Roling, N. 1988. Extension science: Information systems in agricultural development. Cambridge: Cambridge University Press. 233 pp.

Rossi, P. H., and H. E. Freeman. 1985. Evaluation: A systematic approach. 3rd ed. Beverly Hills: Sage. 423 pp.

Scoones, I., and J. Thompson. 1994. Beyond farmer first: Rural peoples' knowledge, agricultural research and extension practice. London: Intermediate Technology Publications. 301 pp.

Sumberg, J., and C. Okali. 1997. Farmers' experiments: Creating local knowledge. Boulder: Lynne Rienner. 186 pp.

Swanson, B. E. 1984. Agricultural extension: A reference manual 2nd ed. Rome: FAO. 247 pp.

Swanson, B. E., ed. 1996. Improving agricultural extension. Rome: FAO. 220 pp.

Stringer, R., and H. Carey. 1992. Desktop publishing—A new tool for agricultural extension and training. Rome: FAO. 28 pp.

Van der Ban, A. W., and H. S. Hawkins. 1996. Agricultural extension. 2nd ed. Oxford: Blackwell Science. 294 pp.

Vossen, P. 1992. Starting a county agricultural marketing program: Cooperation helps sell "Sonoma County Select." J. of Extension 30(fall): 25–27.

Warner, P., and J. Christenson. 1984. The Cooperative Extension Service: A national assessment. Boulder: Westview Press. 195 pp.

Weidermann, C. J. 1987. Agricultural extension for women farmers. In: W. M. Rivera and S. Schram, eds., Agricultural extension worldwide. London: Croom Helm. 175–198.

Wentling, T. 1993. Planning for effective training—A guide to curriculum development. Rome: FAO. 271 pp.

Zimmer, B. P., and K. L. Smith. 1992. Successful mentoring for new agents: Dedicated mentors make the difference. J. of Extension 30(spring): 25–27.

Appendix: Summary Table of Optimal Handling Conditions for Fresh Produce *Compiled by Marita Cantwell*

Common name	Scientific name	Storage temperature		Relative humidity	Highest freezing temperature		Ethylene production*	Ethylene sensitivity†	Approximate storage life	Observations and beneficial CA conditions
		°C	°F	%	°C	°F				
Acerola, Barbados cherry	*Malpighia glabra*	0	32	85–90	−1.4	29.4			6–8 weeks	
African horned melon, kiwano	*Cucumis metuliferus*	13–15	55–60	90			L	M	6 months	
Amaranth, pigweed	*Amaranthus* spp.	0–2	32–36	95–100			VL	M	10–14 days	
Anise, fennel	*Foeniculum vulgare*	0–2	32–36	90–95	−1.1	30.0			2–3 weeks	
Apple	*Malus pumila*									2–3% O_2 + 1–2% CO_2
not chilling sensitive		−1.1	30.0	90–95	−1.5	29.3	VH	H	3–6 months	
chilling sensitive	cv. Yellow Newtown, Grimes Golden, McIntosh	4	40	90–95	−1.5	29.3	VH	H	1–2 months	
Apricot	*Prunus armeniaca*	−0.5–0	31–32	90–95	−1.1	30.0	M	M	1–3 weeks	2–3% O_2 + 2–3% CO_2
Artichoke										
Globe artichoke	*Cynara scolymus*	0	32	95–100	−1.2	29.9	VL	L	2–3 weeks	2–3% O_2 + 3–5% CO_2
Chinese artichoke	*Stachys affinis*	0	32	90–95			VL	VL	1–2 weeks	
Jerusalem artichoke	*Helianthus tuberosus*	−0.5–0	31–32	90–95	−2.5	27.5	VL	L	4 months	
Arugula	*Eruca vesicaria* var. *sativa*	0	32	95–100			VL	H	7–10 days	
Asian pear, nashi	*Pyrus serotina; P. pyrifolia*	1	34	90–95	−1.6	29.1	H	H	4–6 months	
Asparagus, green, white	*Asparagus officinalis*	2.5	36	95–100	−0.6	31.0	VL	M	2–3 weeks	5–12% CO_2 in air
Atemoya	*Annona squamosa* × *A. cherimola*	13	55	85–90			H	H	4–6 weeks	3–5% O_2 + 5–10% CO_2
Avocado	*Persea americana*									
cv. Fuchs, Pollock		13	55	85–90	−0.9	30.4	H	H	2 weeks	
cv. Fuerte, Hass		3–7	37–45	85–90	−1.6	29.1	H	H	2–4 weeks	2–5% O_2 + 3–10% CO_2
cv. Lula, Booth		4	40	90–95	−0.9	30.4	H	H	4–8 weeks	
Babaco, mountain papaya	*Carica candamarcensis*	7	45	85–90					1–3 weeks	
Banana	*Musa paradisiaca* var. *sapientum*	13–15	56–59	90–95	−0.8	30.5	M	H	1–4 weeks	2–5% O_2 + 2–5% CO_2
Barbados cherry	*see* Acerola									
Beans										
Fava, broad beans	*Vicia faba*	0	32	90–95					1–2 weeks	
Lima beans	*Phaseolus lunatus*	5–6	41–43	95	−0.6	31.0	L	M	5–7 days	
Long bean, yardlong bean	*Vigna sesquipedalis*	4–7	40–45	90–95			L	M	7–10 days	
Snap; wax; green	*Phaseolus vulgaris*	4–7	40–45	95	−0.7	30.7	L	M	7–10 days	2–3% O_2 + 4–7% CO_2
Winged bean	*Psophocarpus tetragonolobus*	10	50	90					4 weeks	

Common name	Scientific name	Storage temperature		Relative humidity %	Highest freezing temperature		Ethylene production*	Ethylene sensitivity†	Approximate storage life	Observations and beneficial CA conditions
		°C	°F		°C	°F				
Beet, bunched	*Beta vulgaris*	0	32	98–100	−0.4	31.3	VL	L	10–14 days	
beet, topped		0	32	98–100	−0.9	30.3	VL	L	4 months	
Berries										
Blackberry	*Rubus* spp.	−0.5–0	31–32	90–95	−0.8	30.6	L	L	3–6 days	5–10% O_2 + 15–20% CO_2
Blueberry	*Vaccinium corymbosum*	−0.5–0	31–32	90–95	−1.3	29.7	L	L	10–18 days	2–5% O_2 + 12–20% CO_2
Cranberry	*Vaccinium macrocarpon*	2–5	35–41	90–95	−0.9	30.4	L	L	8–16 weeks	1–2% O_2 + 0–5% CO_2
Dewberry	*Rubus* spp.	−0.5–0	31–32	90–95	−1.3	29.7	L	L	2–3 days	
Elderberry	*Rubus* spp.	−0.5–0	31–32	90–95	−1.1	30.0	L	L	5–14 days	
Loganberry	*Rubus* spp.	−0.5–0	31–32	90–95	−1.7	29.0	L	L	2–3 days	
Raspberry	*Rubus idaeus*	−0.5–0	31–32	90–95	−0.9	30.4	L	L	3–6 days	5–10% O_2 + 15–20% CO_2
Strawberry	*Fragaria* spp.	0	32	90–95	−0.8	30.6	L	L	7–10 days	5–10% O_2 + 15–20% CO_2
Bittermelon, bitter gourd	*Momordica charantia*	10–12	50–54	85–90			L	M	2–3 weeks	2–3% O_2 + 5% CO_2
Black salsify, scorzonera	*Scorzonera hispanica*	0–1	32–34	95–98			VL	L	6 months	
Bok choy	*Brassica chinensis*	0	32	95–100			VL	H	3 weeks	
Breadfruit	*Artocarpus altilis*	13–15	55–59	85–90					2–4 weeks	
Broccoli	*Brassica oleracea* var. *italica*	0	32	95–100	−0.6	31.0	VL	H	10–14 days	1–2% O_2 + 5–10% CO_2
Brussels sprouts	*Brassica oleracea* var. *gemnifera*	0	32	95–100	−0.8	30.5	VL	H	3–5 weeks	1–2% O_2 + 5–7% CO_2
Cabbage										
Chinese, Napa	*Brassica campestris* var. *pekinensis*	0	32	95–100	−0.9	30.4	VL	M–H	2–3 months	1–2% O_2 + 0–5% CO_2
common, early crop	*B. oleracea* var. *capitata*	0	32	98–100	−0.9	30.4	VL	H	3–6 weeks	
late crop	*B. oleracea* var. *capitata*	0	32	95–100	−0.9	30.4	VL	H	5–6 months	3–5% O_2 + 3–7% CO_2
Cactus leaves, nopalitos	*Opuntia* spp.	5–10	41–50	90–95			VL	M	2–3 weeks	
Cactus fruit, prickly pear fruit	*Opuntia* spp.	5	41	85–90	−1.8	28.7	VL	M	3 weeks	
Caimito	*see* Sapotes									
Calamondin	*see* Citrus									
Canistel	*see* Sapotes									
Carambola, starfruit	*Averrhoa carambola*	9–10	48–50	85–90	−1.2	29.8			3–4 weeks	
Carrots, topped	*Daucus carota*	0	32	98–100	−1.4	29.5	VL	H	6–8 months	no CA benefit
bunched, immature	*D. carota*	0	32	98–100	−1.4	29.5	VL	H	10–14 days	ethylene causes bitterness
Cashew apple	*Anacardium occidentale*	0–2	32–36	85–90					5 weeks	
Cassava, yucca, manioc	*Manihot esculenta*	0–5	32–41	85–90			VL	L	1–2 months	no CA benefit
Cauliflower	*Brassica oleracea* var. *botrytis*	0	32	95–98	−0.8	30.6	VL	H	3–4 weeks	2–5% O_2 + 2–5% CO_2
Celeriac	*Apium graveolens* var. *rapaceum*	0	32	98–100	−0.9	30.4	VL	L	6–8 months	2–4% O_2 + 2–3% CO_2

Common name	Scientific name	Storage temperature °C	Storage temperature °F	Relative humidity %	Highest freezing temperature °C	Highest freezing temperature °F	Ethylene production*	Ethylene sensitivity†	Approximate storage life	Observations and beneficial CA conditions
Celery	*Apium graveolens* var. *Dulce*	0	32	98–100	–0.5	31.1	VL	M	1–2 months	1–4% O_2 + 3–5% CO_2
Chard	*Beta vulgaris* var. *Cicla*	0	32	95–100			VL	H	10–14 days	
Chayote	*Sechium edule*	7	45	85–90					4–6 weeks	
Cherimoya, custard apple	*Annona cherimola*	13	55	90–95	–2.2	28.0	H	H	2–4 weeks	3–5% O_2 + 5–10% CO_2
Cherries, sour	*Prunus cerasus*	0	32	90–95	–1.7	29.0	VL	L	3–7 days	3–10% O_2 + 10–12% CO_2
Cherries, sweet	*Prunus avium*	–1–0	30–32	90–95	–2.1	28.2	VL	L	2–3 weeks	10–20% O_2 + 20–25% CO_2
Chicory	*see* Endive									
Chiles	*see* Pepper									
Chinese broccoli, gailan	*Brassica alboglabra*	0	32	95–100			VL	H	10–14 days	
Chives	*Allium schoenoprasum*	0	32	95–100			VL	H	2–3 weeks	
Cilantro, chinese parsley	*Coriandrum sativum*	0–1	32–34	95–100			VL	H	2 weeks	
Citrus										
Calamondin orange	*Citrus reticulata* × *Fortunella* spp.	9–10	48–50	90	–2.0	28.3			2 weeks	
Grapefruit	*C. paradisi*									3–10% O_2 + 5–10% CO_2
CA, AZ, dry areas		14–15	58–60	85–90	–1.1	30.0	VL	M	6–8 weeks	
FL, humid areas		10–15	50–60	85–90	–1.1	30.0	VL	M	6–8 weeks	
Kumquat	*Fortunella japonica*	4	40	90–95			VL	M	2–4 weeks	
Lemon	*C. limon*	10–13	50–55	85–90	–1.4	29.4			1–6 months	5–10% O_2 + 0–10% CO_2
Lime, Mexican, Tahitian, or Persian	*C. aurantifolia; C. latifolia*	9–10	48–50	85–90	–1.6	29.1			6–8 weeks	5–10% O_2 + 0–10% CO_2
Orange	*C. sinensis*									5–10% O_2 + 0–5% CO_2
CA, AZ, dry areas		3–9	38–48	85–90	–0.8	30.6	VL	M	3–8 weeks	
FL; humid regions		0–2	32–36	85–90	–0.8	30.6	VL	M	8–12 weeks	
Blood orange		4–7	40–44	90–95	–0.8	30.6			3–8 weeks	
Seville, Sour	*C. aurantium*	10	50	85–90	–0.8	30.6	L	N	12 weeks	
Pummelo	*C. grandis*	7–9	45–48	85–90	–1.6	29.1			12 weeks	
Tangelo, minneola	*C. reticulata* × *paradisi*	7–10	45–50	85–95	–0.9	30.3				
Tangerine, mandarin	*C. reticulata*	4–7	40–45	90–95	–1.1	30.1	VL	M	2–4 weeks	
Coconut	*Cocos nucifera*	0–2	32–36	80–85	–0.9	30.4			1–2 months	
Collards	*Brassica oleracea* var. *acephala*	0	32	95–100	–0.5	31.1	VL	H	10–14 days	
Corn, sweet and baby	*Zea mays*	0	32	95–98	–0.6	31.0	VL	L	5–8 days	2–4% O_2 + 5–10% CO_2
Cucumber, slicing	*Cucumis sativus*	10–12	50–54	85–90	–0.5	31.1	L	H	10–14 days	3–5% O_2 + 0–5% CO_2
pickling		4	40	95–100			L	H	7 days	3–5% O_2 + 3–5% CO_2

Common name	Scientific name	Storage temperature		Relative humidity %	Highest freezing temperature		Ethylene production*	Ethylene sensitivity†	Approximate storage life	Observations and beneficial CA conditions
		°C	°F		°C	°F				
Currants	*Ribes sativum; R. nigrum; R. rubrum*	−0.5–0	31–32	90–95	−1.0	30.2	L	L	1–4 weeks	CA can extend storage life to 3–6 months
Custard apple	*see* Cherimoya									
Daikon, Oriental radish	*Raphanus sativus*	0–1	32–34	95–100			VL	L	4 months	
Dasheen	*see* Taro									
Date	*Phoenix dactylifera*	−18–0	0–32	75	−15.7	3.7	VL	L	6–12 months	
Dill	*see* Herbs									
Durian	*Durio zibethinus*	4–6	39–42	85–90					6–8 weeks	3–5% O_2 + 5–15% CO_2
Eggplant	*Solanum melongena*	10–12	50–54	90–95	−0.8	30.6	L	M	1–2 weeks	3–5% O_2 + 0% CO_2
Endive, escarole	*Cichorium endivia*	0	32	95–100	−0.1	31.7	VL	M	2–4 weeks	
Belgian endive, Witloof chicory	*C. intybus*	2–3	36–38	95–98			VL	M	2–4 weeks	light causes greening; 3–4% O_2 + 4–5% CO_2
Feijoa, pineapple guava	*Feijoa sellowiana*	5–10	41–50	90			M	L	2–3 weeks	
Fennel, *see* anise										
Fig, fresh	*Ficus carica*	−0.5–0	31–32	85–90	−2.4	27.6	M	L	7–10 days	5–10% O_2 + 15–20% CO_2
Garlic	*Allium sativum*	0	32	65–70	−0.8	30.5	VL	L	6–7 months	0.5% O_2 + 5–10% CO_2
Ginger	*Zingiber officinale*	13	55	65			VL	L	6 months	no CA benefit
Gooseberry	*Ribes grossularia*	−0.5–0	31–32	90–95	−1.1	30.0	L	L	3–4 weeks	
Granadilla	*see* Passionfruit									
Grape	*Vitis vinifera*	−0.5–0	31–32	90–95	−2.7 fruit −2.0 stem	27.1 fruit 28.4 stem	VL	L	1–6 months	2–5% O_2 + 1–3% CO_2; to 4 wks, 5–10% O_2 + 10–15% CO_2
American grape	*V. labrusca*	−1 to −0.5	30–31	90–95	−1.4	29.4	VL	L	2–8 weeks	
Grapefruit	*see* Citrus									
Guava	*Psidium guajava*	5–10	41–50	90			L	M	2–3 weeks	
Herbs, fresh culinary	*See* specific herb									5–10% O_2 + 5–10% CO_2
Basil	*Ocimum basilicum*	10	50	90			VL	H	7 days	2% O_2 + 0 to<10% CO_2
Chives	*Allium schoenoprasum*	0	32	95–100	−0.9	30.4	L	M		
Cilantro, Chinese parsley	*Coriandrum sativum*	0–1	32–34	95–100			VL	H	2 weeks	3% O_2 +7–10% CO_2; air + 7–10% CO_2
Dill	*Anethum graveolens*	0	32	95–100	−0.7	30.7	VL	H	1–2 weeks	
Epazote	*Chenopodium ambrosioides*	0–5	32–41	90–95			VL	M	1–2 weeks	
Mint	*Mentha* spp.	0	32	95–100			VL	H	2–3 weeks	
Oregano	*Origanum vulgare*	0–5	32–41	90–95			VL	M	1–2 weeks	
Parsley	*Petroselinum crispum*	0	32	95–100	−1.1	30.0	VL	H	1–2 months	
Perilla, shiso	*Perilla frutescens*	10	50	95			VL	M	7 days	
Sage	*Salvia officinalis*	0	32	90–95					2–3 weeks	
Thyme	*Thymus vulgaris*	0	32	90–95					2–3 weeks	

Common name	Scientific name	Storage temperature		Relative humidity	Highest freezing temperature		Ethylene production*	Ethylene sensitivity†	Approximate storage life	Observations and beneficial CA conditions
		°C	°F	%	°C	°F				
Horseradish	*Armoracia rusticana*	−1–0	30–32	98–100	−1.8	28.7	VL	L	10–12 mo.	
Husk tomato	*see* Tomatillo									
Jaboticaba	*Myrciaria cauliflora = Eugenia cauliflora*	13–15	55–59	90–95					2–3 days	
Jackfruit	*Artocarpus heterophyllus*	13	55	85–90			M	M	2–4 weeks	
Jerusalem artichoke	*see* Artichoke									
Jicama, yambean	*Pachyrrhizus erosus*	13–18	55–65	85–90			VL	L	1–2 months	
Jujube; Chinese date	*Ziziphus jujuba*	2.5–10	36–50	85–90	−1.6	29.2	L	M	1 month	
Kaki	*see* Persimmon									
Kale	*Brassica oleracea var. acephala*	0	32	95–100	−0.5	31.1	VL	M		
Kiwano	*see* African horned melon									
Kiwifruit, Chinese gooseberry	*Actinidia chinensis*	0	32	90–95	−0.9	30.4	L	H	3–5 months	1–2% O_2 + 3–5% CO_2
Kohlrabi	*Brassica oleracea var. Gongylodes*	0	32	98–100	−1.0	30.2	VL	L	2–3 months	no CA benefit
LoBok	*see* Daikon									
Langsat, lanzone	*Aglaia* spp.; *Lansium* spp.	11–14	52–58	85–90					2 weeks	
Leafy greens										
Cool-season	*various*	0	32	95–100	−0.6	31.0	VL	H	10–14 days	
Warm-season	*various*	7–10	45–50	95–100	−0.6	31.0	VL	H	5–7 days	
Leek	*Allium porrum*	0	32	95–100	−0.7	30.7	VL	M	2 months	1–2% O_2 +2–5% CO_2
Lemon	*see* Citrus									
Lettuce	*Lactuca sativa*	0	32	98–100	−0.2	31.7	VL	H	2–3 weeks	2–5% O_2 + 0% CO_2
Lime	*see* Citrus									
Longan	*Dimocarpus longan = Euphoria longan*	4–7	39–45	90–95	−2.4	27.7			2–4 weeks	
Loquat	*Eriobotrya japonica*	0	32	90–95	−1.9	28.6			3 weeks	
Luffa, Chinese okra	*Luffa* spp.	10–12	50–54	90–95			L	M	1–2 weeks	
Lychee, litchi	*Litchi chinensis*	1–2	34–36	90–95			M	M	3–5 weeks	3–5% O_2 + 3–5% CO_2
Malanga, tania, new cocoyam	*Xanthosoma sagittifolium*	7	45	70–80			VL	L	3 months	
Mamey	*see* Sapotes									
Mandarin	*see* Citrus									
Mango	*Mangifera indica*	13	55	85–90	−1.4	29.5	M	M	2–3 weeks	3–5% O_2 + 5–10% CO_2
Mangosteen	*Garcinia mangostana*	13	55	85–90			M	H	2–4 weeks	

Common name	Scientific name	Storage temperature		Relative humidity	Highest freezing temperature		Ethylene production*	Ethylene sensitivity†	Approximate storage life	Observations and beneficial CA conditions
		°C	°F	%	°C	°F				
Melons										
Cantaloupes and other netted melons	*Cucumis melo var. reticulatus*	2–5	36–41	95	−1.2	29.9	H	M	2–3 weeks	3–5% O$_2$ + 10–15% CO$_2$
Casaba	*C. melo*	7–10	45–50	85–90	−1.0	30.3	L	L	3–4 weeks	3–5% O$_2$ + 5–10% CO$_2$
Crenshaw	*C. melo*	7–10	45–50	85–90	−1.1	30.1	M	H	2–3 weeks	3–5% O$_2$ + 5–10% CO$_2$
Honeydew, orange-flesh	*C. melo*	5–10	41–50	85–90	−1.1	30.1	M	H	3–4 weeks	3–5% O$_2$ + 5–10% CO$_2$
Persian	*C. melo*	7–10	45–50	85–90	−0.8	30.6	M	H	2–3 weeks	3–5% O$_2$ + 5–10% CO$_2$
Mint	*see* Herbs									
Mombin	*see* Spondias									
Mushrooms	*Agaricus*, other genera	0	32	90	−0.9	30.4	VL	M	7–14 days	3–21% O$_2$ + 5–15% CO$_2$
Mustard greens	*Brassica juncea*	0	32	90–95			VL	H	7–14 days	
Nashi	*see* Asian pear									
Nectarine	*Prunus persica*	−0.5–0	31–32	90–95	−0.9	30.3	M	M	2–4 weeks	1–2% O$_2$ + 3–5% CO$_2$; internal breakdown 3–10°C
Okra	*Abelmoschus esculentus*	7–10	45–50	90–95	−1.8	28.7	L	M	7–10 days	air + 4–10% CO$_2$
Olives, fresh	*Olea europea*	5–10	41–50	85–90	−1.4	29.4	L	M	4–6 weeks	2–3% O$_2$ + 0–1% CO$_2$
Onions	*Allium cepa*									
Mature bulbs, dry		0	32	65–70	−0.8	30.6	VL	L	1–8 months	1–3% O$_2$ + 5–10% CO$_2$
Green onions		0	32	95–100	−0.9	30.4	L	H	3 weeks	2–4% O$_2$ + 10–20% CO$_2$
Orange	*see* Citrus									
Papaya	*Carica papaya*	7–13	45–55	85–90	−0.9	30.4	M	M	1–3 weeks	2–5% O$_2$ + 5–8% CO$_2$
Parsley	*see* Herbs									
Parsnip	*Pastinaca sativa*	0	32	95–100	−0.9	30.4	VL	H	4–6 months	ethylene causes bitterness
Passionfruit	*Passiflora* spp.	10	50	85–90			VH	M	3–4 weeks	
Peach	*Prunus persica*	−0.5–0	31–32	90–95	−0.9	30.3	M	M	2–4 weeks	1–2% O$_2$ + 3–5% CO$_2$; internal breakdown 3–10°C
Pear, European	*Pyrus communis*	−1.5 to 0.5	29–31	90–95	−1.7	29.0	H	H	2–7 months	cultivar variations; 1–3% O$_2$ + 0–5% CO$_2$
Peas in pods; snow, snap & sugar peas	*Pisum sativum*	0–1	32–34	90–98	−0.6	30.9	VL	M	1–2 weeks	2–3% O$_2$ + 2–3% CO$_2$
Southern peas; cowpeas	*Vigna sinensis = V. unguiculata*	4–5	40–41	95					6–8 days	
Pepino, melon pear	*Solanum muricatum*	5–10	41–50	95			L	M	4 weeks	
Peppers										
Bell pepper, paprika	*Capsicum annuum*	7–10	45–50	95–98	−0.7	30.7	L	L	2–3 weeks	2–5% O$_2$ + 2–5% CO$_2$
Hot peppers, chiles	*Capsicum annuum* and *C. frutescens*	5–10	41–50	85–95	−0.7	30.7	L	M	2–3 weeks	3–5% O$_2$ + 5–10% CO$_2$
Persimmon, kaki	*Diospyros kaki*									3–5% O$_2$ + 5–8% CO$_2$
Fuyu		0	32	90–95	−2.2	28.0	L	H	1–3 months	
Hachiya		0	32	90–95	−2.2	28.0	L	H	2–3 months	
Pineapple	*Ananas comosus*	7–13	45–55	85–90	−1.1	30.0	L	L	2–4 weeks	2–5% O$_2$ + 5–10% CO$_2$

Common name	Scientific name	Storage temperature		Relative humidity %	Highest freezing temperature		Ethylene production*	Ethylene sensitivity†	Approximate storage life	Observations and beneficial CA conditions
		°C	°F		°C	°F				
Plantain	*Musa paradisiaca* var. *paradisiaca*	13–15	56–59	90–95	–0.8	30.6	L	H	1–5 weeks	
Plums and prunes	*Prunus domestica*	–0.5–0	31–32	90–95	–0.8	30.5	M	M	2–5 weeks	1–2% O_2 + 0–5% CO_2
Pomegranate	*Punica granatum*	5–7.2	41–45	90–95	–3.0	26.6	VL	L	2–3 months	3–5% O_2 + 5–10% CO_2
Potato	*Solanum tuberosum*									no CA benefit
early crop		10–15	50–59	90–95	–0.8	30.5	VL	M	10–14 days	
late crop		4–12	40–54	95–98	–0.8	30.5	VL	M	5–10 months	
Pumpkin	*Cucurbita maxima*	12–15	54–59	50–70	–0.8	30.5	L	M	2–3 months	
Quince	*Cydonia oblonga*	–0.5–0	31–32	90	–2.0	28.4	L	H	2–3 months	
Radicchio	*Cichorium intybus*	0–1	32–34	95–100					3–4 weeks	
Radish	*Raphanus sativus*	0	32	95–100	–0.7	30.7	VL	L	1–2 months	1–2% O_2 + 2–3% CO_2
Rambutan	*Nephelium lappaceum*	12	54	90–95			H	H	1–3 weeks	3–5% O_2 + 7–12% CO_2
Rhubarb	*Rheum rhaponticum*	0	32	95–100	–0.9	30.3	VL	L	2–4 weeks	
Rutabaga	*Brassica napus* var. *napobrassica*	0	32	98–100	–1.1	30.1	VL	L	4–6 months	
Sage	*see Herbs*									
Salsify, vegetable oyster	*Trapopogon porrifolius*	0	32	95–98	–1.1	30.1	VL	L	2–4 months	
Sapotes										
Caimito, star-apple	*Chrysophyllum cainito*	3	38	90	–1.2	29.9			3 weeks	
Canistel, eggfruit	*Pouteria campechiana*	13–15	55–60	85–90	–1.8	28.7			3 weeks	
Black sapote	*Diospyros ebenaster*	13–15	55–59	85–90	–2.3	27.8			2–3 weeks	
White sapote	*Casimiroa edulis*	20	68	85–90	–2.0	28.4			2–3 weeks	
Mamey sapote	*Calocarpum mammosum*	13–15	55–60	90–95			H	H	2–3 weeks	
Sapodilla, chicosapote	*Achras sapota*	15–20	59–68	85–90			H	H	2 weeks	
Scorzonera	*see Black salsify*									
Shallots	*Allium cepa* var. *ascalonicum*	0–2.5	32–36	65–70	–0.7	30.7	L	L		
Soursop	*Annona muricata*	13	55	85–90					1–2 weeks	
Spinach	*Spinacia oleracea*	0	32	95–100	–0.3	31.5	VL	H	10–14 days	5–10% O_2 + 5–10% CO_2
Spondias, mombin, wi apple, jobo, hogplum	*Spondias* spp.	13	55	85–90					1–2 weeks	
Sprouts from seeds	*various genera*	0	32	95–100					5–9 days	
Alfalfa sprouts	*Medicago sativa*	0	32	95–100					7 days	
Bean sprouts	*Phaseolus* spp.	0	32	95–100					7–9 days	
Radish sprouts	*Raphanus* spp.	0	32	95–100					5–7days	
Squash										
Summer (soft rind), courgette	*Cucurbita pepo*	7–10	45–50	95	–0.5	31.1	L	M	1–2 weeks	3–5% O_2 + 5–10% CO_2
Winter (hard rind), calabash	*Cucurbita moschata; C. maxima*	12–15	54–59	50–70	–0.8	30.5	L	M	2–3 months	large differences among varieties

Common name	Scientific name	Storage temperature °C	°F	Relative humidity %	Highest freezing temperature °C	°F	Ethylene production*	Ethylene sensitivity†	Approximate storage life	Observations and beneficial CA conditions
Star-apple	*see* Sapotes									
Starfruit	*see* Carambola									
Sweet potato, yam	*Ipomea batatas*	13–15	55–59	85–95	–1.3	29.7	VL	L	4–7 months	
Sweetsop, sugar apple, custard apple	*Annona squamosa; Annona* spp.	7	45	85–90			H	H	4 weeks	3–5% O_2 + 5–10% CO_2
Tamarillo, tree tomato	*Cyphomandra betacea*	3–4	37–40	85–95			L	M	10 weeks	
Tamarind	*Tamarindus indica*	2–7	36–45	90–95	–3.7	25.4	VL	VL	3–4 weeks	
Taro, cocoyam, eddoe, dasheen	*Colocasia esculenta*	7–10	45–50	85–90	–0.9	30.3			4 months	no CA benefit
Thyme	*see* Herbs									
Tomatillo, husk tomato	*Physalis ixocarpa*	7–13	45–55	85–90			VL	M	3 weeks	
Tomato	*Lycopersicon esculentum*									
Mature-green		10–13	50–55	90–95	–0.5	31.1	VL	H	2–5 weeks	3–5% O_2 + 2–3% CO_2
Firm-ripe		8–10	46–50	85–90	–0.5	31.1	H	L	1–3 weeks	3–5% O_2 + 3–5% CO_2
Turnip root	*Brassica campestris* var. *rapifera*	0	32	95	–1.0	30.1	VL	L	4–5 months	
Water chestnuts	*Eleocharis dulcis*	1–2	32–36	85–90					2–4 months	
Watercress, garden cress	*Lepidium sativum; Nasturtium officinalis*	0	32	95–100	–0.3	31.5	VL	H	2–3 weeks	
Watermelon	*Citrullus vulgaris*	10–15	50–59	90	–0.4	31.3	VL	H	2–3 weeks	no CA benefit
Yam	*Dioscorea* spp.	15	59	70–80	–1.1	30.0	VL	L	2–7 months	
Yucca	*see* Cassava									

Sources:

Facciola, S. 1990. Cornucopia. A source book of edible plants. Vista, CA: Kampong. 678 pp.

Hardenburg, R., A. E. Watada, C. Y. Wang. 1986. The commercial storage of fruits, vegetables, and florist and nursery stocks. Washington, D.C.: USDA Agric. Handb. 66. 130 pp.

Kader, A. A., ed. 1997. CA '97 Proceedings, Vol. 3: Fruits other than apples and pears. Davis: Univ. Calif. Postharv. Hort. Ser. 17. 263 pp.

Kays, S. J., and J. C. Silva Dias. 1996. Cultivated vegetables of the world. Athens, GA: Exon Press. 170 pp.

McGregor, B. M. 1987. Manual de transporte de productos tropicales. Washington, D.C.: USDA Agric. Handb. No. 668. 148 pp.

Mitcham, E. J., ed.1997. CA '97 Proceedings, Vol. 2: Apples and pears. Davis: Univ. Calif. Postharv. Hort. Ser.16. 308 pp.

Rubatzky, V. E., and M. Yamaguchi. 1997. World vegetables: Principles, production and nutritive values. 2nd ed. New York: Chapman and Hall. 843 pp.

Saltveit, M.E., ed. CA '97 Proceedings, Vol.4: Vegetables and ornamentals. Davis: Univ. Calif. Postharv. Hort. Ser. 18. 168 pp.

Whiteman, T. M. 1957. Freezing points of fruits, vegetables, and florist stocks. Washington, D.C.: USDA Mktng. Res. Rpt. No. 196. 32 pp.

Notes:

*Ethylene production rate:

VL = very low (<0.1 µl/kg-hr at 20°C)

L = low (0.1=1.0 µl/kg-hr)

M = moderate (1.0–10.0 µl/kg-hr)

H = high (10–100 µl/kg-hr)

VH = very high (>100 µl/kg-hr)

†Ethylene sensitivity (detrimental effects include yellowing, softening, increased decay, abcission, browning)

L = low sensitivity

M = moderately sensitive

H = highly sensitive

Index

Page numbers in *italics* refer to photographs and captions.